# Computer-Controlled Systems with Delay

Efim N. Rosenwasser · Bernhard P. Lampe ·
Torsten Jeinsch

# Computer-Controlled Systems with Delay

## A Transfer Function Approach

Springer

Efim N. Rosenwasser
Department of Shipping Automation Marine Technology
State Marine Technical University
Saint Petersburg, Russia

Bernhard P. Lampe (Deceased)
Rostock, Mecklenburg-Vorpommern
Germany

Torsten Jeinsch
Lehrstuhl Regelungstechnik
Universität Rostock
Rostock, Mecklenburg-Vorpommern
Germany

ISBN 978-3-030-15044-0 ISBN 978-3-030-15042-6 (eBook)
https://doi.org/10.1007/978-3-030-15042-6

Library of Congress Control Number: 2019933711

MATLAB® is a registered trademark of The MathWorks, Inc., 1 Apple Hill Drive, Natick, MA 01760-2098, USA, http://www.mathworks.com

Mathematics Subject Classification (2010): 93C05, 93C35, 93C57, 93C80, 93C83, 93D15, 93E20



This Springer imprint is published by the registered company Springer Nature Switzerland AG.
The registered company address is: Gewerbestrasse 11, 6330 Cham, Switzerland

# Preface

The traditional approach for the solution of analysis and synthesis problems for linear sampled-data (SD) systems is the transition to a discrete model [1, 2] which assigns discrete input values to their corresponding output values at the sampling points. In principle, such a transition is always possible if the external excitation acts on the system through the sampler. In cases when some external excitations act directly on the continuous parts, the rigorous transition to such a discrete model becomes impossible.

In such cases, approximated discrete models are usually used for the solution of practical problems. However, the practice shows that the solution of analysis and especially synthesis problems on the base of such approximate discrete models can be unsatisfactory. This fact is substantiated by examples given in Appendices B and C of this book.

In connection with the above, recently it was realized the need to develop exact methods for the study of SD systems that do not depend on the structure of the system and the points of attack of external excitations. It turned out that such methods should be based on mathematical models focused on the continuous process. The construction of such models requires taking into account the behavior of the system in continuous time, including the intervals between the sampling instants.

Currently, there are two types of methods for building direct models of SD systems. The first type includes the lifting method and the FR-operator method.

The lifting method is based on a system description in state space, and it is connected with the application of intermittent transformations utilizing matrices and determinants of infinite dimension. The description of this method and related literature are given in [3].

The frequency analog of the lifting method is the FR-operator (frequency response) method proposed in [4]. It also uses matrices and determinants of infinite dimension.

The solution of optimization tasks based on lifting methods and FR-operator is connected with the construction and solution of associated Riccati equations.

Methods of the second type are based on the concept of the parametric transfer function (PTF), which at all steps of investigation uses only finite-dimensional matrices and determinants.

The basic idea of this method applied to continuous linear nonstationary systems was already given in [5, 6], where the author introduced the transfer function $W(s,t)$, which differs from the ordinary transfer function $W(p)$ for continuous linear time-invariant (LTI) system by its dependence on the time t as a parameter. The paper [7] mentions the principal possibility to extend the PTF concept to SD systems.

The further realizations of this idea have been elaborated in [8]. Obviously, the first detailed presentation of the theory of SISO SD systems based on the PTF is provided in [9].

Further progress and generalization of the PTF method concerning SISO SD systems, including the solution of $\mathscr{H}_2, \mathscr{H}_\infty$ and $\mathscr{L}_2$ optimization problems, development of corresponding software tools, and solution of applied problems, are provided in [10, 11]. Hereby, the solution of the optimization problems is based on the application of the Wiener-Hopf method, connected with the use of separation and factorization operations. With the results of these works, we are able to state particular features of the PTF method, which make it useful for the solution of practical problems. In this regard, we note the following:

1. The PTF method has a universal character. Its applicability does not depend on the structure of the SD system and the entry points of external excitations.
2. The PTF method is based completely on frequency representations and does not require an intermediate transition to the state space.
3. The PTF method opens the possibility of taking into account the influence of the structure of SD system on its dynamical characteristics and the results of optimization.
4. The application of the PTF method allows to find the subset of fixed poles of H2 and L2 optimal systems independently of the type of the stochastic input signal. The fixed poles are determined only by the properties of the poles of the continuous elements of the SD system and its structure. It is proved that this set of fixed poles remains unchanged for SISO SD systems, which are optimal by the criterion $H_\infty$.
5. The PTF method allows the direct generalization to SD systems, in which the inputs and outputs of their elements have arbitrary constant delays. In particular, this property opens the possibility to approximately estimate the effect of computational delay on the quality of processes in SD systems and results of their optimization.
6. The existing relation between the PTF and the Laplace transformed output of the SD system allows to extend the Wiener-Hopf method to the L2 optimization problem.
7. On the basis of the PTF concept, we can formulate and solve an optimization problem that guarantees the quality of an SD system for a certain set of external disturbances.

The present book can be seen as a logical extension of the monographs [11, 12], where the PTM method is extended to the MIMO SD systems with delay. The basis for the content of the book is built upon papers and reports of the authors over the years 2007–2014 [13–20]. Also, we notice that an analog generalization of the lifting and FR-operator methods to systems with delay does not exist up to now.

The main content of the book is divided into three parts.

Part I spans Chaps. 1–3. Chaps. 1 and 2 treat some facts on theory of polynomial and rational matrices necessary for the understanding of further chapters. This material is mainly known. It is presented, however, in a form that simplifies the reading of the book for a person who is not familiar with the corresponding parts of algebra.

Chapter 3 considers from a general position a special class of polynomial equations, named determinant and establishes its tie with Diophantine polynomial equations, which are traditionally applied in applications. On this basis, the general modal control problem for discrete objects is investigated. The discrete objects are given by backward models described in the backward shift operator $\zeta$, or by forward models given in the forward shift operator $z$. It is ascertained that despite the principal equivalence of both methods for describing discrete models, the technique of the solution of the modal control problem based on backward and forward models is essentially different. In particular, it is claimed that all stabilizing controllers for strictly causal backward object are causal, whereas for strictly causal forward objects, an analog claim does not hold.

Part II of the book contains Chaps. 4–6. Chapter 4 considers the modal control problem and the backward stabilization of multidimensional SD systems with several delays and generalized holds of arbitrary order. By decomposing the hold equation and applying the discretization of functions of a continuous argument, the modal control problem leads to the problem of finding the solution of a determinant polynomial equation. The successive application of the results of Chap. 3 allows to construct the set of backward stabilizing controllers. Hereby, it is stated that for the considered class of SD systems with delay, all backward stabilizing controllers are causal. At the end of Chap. 4, the problem of building discrete forward model of continuous object in state space is studied. It is shown that the often used in literature standard procedure for the construction of those models, see, e.g., [1, 21, 22] may yield not completely controllable or observable models. A general method for construction of minimal discrete forward models is derived, and an example is given.

Chapter 5 considers from a general point of view the properties and construction methods for parametric (time-dependent) discrete models of continuous processes with delay. On this basis, we provide a closed solution for calculating problems of the stationary reaction to an exponential-periodic input signal of an LTI system with delay and of open SD systems with delay.

Chapter 6 introduces the concepts of standard structure and standard model for SD systems with delay. A standard model is always associated with a standard structure. The reverse relation, however, does not always hold. For the standard structure, in Chap. 6, the PTM $W_{zx}(p, t)$ from the input $x(t)$ to the output $z(t)$ is constructed, which is called standard. For the fundamental case, when the original is a standard model, the properties of the matrix $W_{zx}(p,t)$ as function of the arguments $t$ and $s$ are investigated. Hereby we claim the principally important fact that for any value of the parameter t the PTM $W_{zx}(s,t)$ establishes as a meromorphic function of the argument $s$, with its poles located in a certain set, independent of $t$. In this sense, we can say that the poles of the matrix $W_{zx}(p,t)$ do not depend on $t$. This chapter also provides the solution of the backward modal control and the stabilization problems as well as the construction of the parametrized set of stabilizing controllers, which all turn out to be causal. Chapter 6 introduces the concept of the system function, investigates its properties, and expresses the PTM $W_{zx}(p,t)$ through the system function. Finally, Chap. 6 also considers the question of standardization of SD systems with delay of arbitrary structure. Arbitrary SD system with delay is called structural standardizable, if its PTM $W(p,t)$ has standard form. If there exists a standard model with PTM $W_{zx}(p,t)$ that coincides with $W(p,t)$, the corresponding standard-structural standardizable SD system is called model standardizable. Necessary and sufficient conditions, under which a structural standardizable SD system with a standard PTM is model standartdizable are provided. By means of an example, Chap. 6 provides sufficient conditions for the model standardizability of a single-loop multivariable SD system with delay with the various choice of the output vector.

Part III of the book covers Chaps. 7–10, in which optimization problems for standard models of SD systems with delay are considered. Chapter 7 investigates the $\mathscr{H}_2$ optimization problem of a standard model when vectorial white noise acts on its input. When representing the PTM by its system function, the $\mathscr{H}_2$ optimization leads to the minimization of a quadratic functional defined over the set of stabilizing system functions. The optimal system function is detected using the Wiener-Hopf method by applying factorization and separation operations. On this basis, we are allowed to state an important for applications fact of existence of closed $\mathscr{H}_2$ optimal system a set of fixed poles, simply connected with poles of certain continuous elements of a system, which are called fixing poles. The existence of fixed poles imposes certain constraints to the achievable performance of H2 optimal system, which as a consequence leads to the necessity of a preliminary choice of the fixing elements in practical design of SD system on the basis of $\mathscr{H}_2$ optimization methods. Drawing on an example, in Chap. 7, the problem of determining the set of fixed poles for a single-loop multivariable SD system with delay for the various choice of output vector is considered.

Chapter 8 generalizes the results from the previous chapter for the case, in which the input of the standard model is excited by colored stochastic input signal. In the existing literature, this problem is solved by extending the state vector of the continuous object. In this way, however, the model can lose its controllability or observability. Chapter 8 presents a procedure for the solution of this problem, which allows to overcome this difficulty by some transformation of the PTM taking into account the coloration of the input signal. Hereby, all successive actions related to the design of the optimal controller follow in analogy to Chap. 7. The central moment of Chap. 8 consists in the proof of the independence of set of the fixed poles of the $\mathscr{H}_2$ optimal system from the form of the spectral density of the input signal. This fact makes it plausible that the fixed set of poles of a multivariable $\mathscr{H}_\infty$ optimal system retains all fixed poles of the $\mathscr{H}_2$ optimal system. For the SISO case, this property was proved in [23].

Chapter 10 of the book considers the solution of the $\mathscr{L}_2$ tracking problem for the standard model. A smart formula between the PTM and the Laplace transform of the transient process under zero initial energy establishes itself as a fundamental instrument for finding the solution of the problem. It allows to lead the $\mathscr{L}_2$ optimization problem to a minimization problem for a certain quadratic functional depending on the system function. Hereby, the obtained functional proves to be singular, and a direct application of the Wiener-Hopf method for its minimization is not possible. Chapter 9 provides a special polynomial transformation which generates a certain regular quadratic functional, allowing to use the Wiener-Hopf method. Hereby, the book shows that the optimal system function is not uniquely determined, which is a consequence of the singularity of the original quadratic functional. At the same time, the chapter proves that the $\mathscr{L}_2$ optimal transient process in the system does not depend on the choice of the optimal system function and is uniquely determined.

Chapter 10 extends the results of Chaps. 7–9 to generalized standard model with delay, which differs from standard model with delay in the fact that the individual components of the output vector can possess different time shifts in relation of the vector of states of continuous object and the control vector. Such systems most often arise in applications. It is shown that when fulfilling the conditions of modal controllability the solution of the $\mathscr{H}_2$ and $\mathscr{L}_2$ optimization problems for generalized standard system leads to the solution of analog problems for certain equivalent standard system, such that the results of Chaps. 7–9 can be applied. Hereby, all quality features of optimal standard systems are transferred on the optimal generalized standard system. In particular, this is related to the fact of existence of the set of fixed poles. By means of illustration of the described approach in Chap. 10, standardizibility problems and $\mathscr{H}_2$ and $\mathscr{L}_2$ optimization problems for MIMO single-loop SD system with three delays are described.

The book closes with three appendices. Appendix A, written by the authors of this monograph, illustrates the choice of sampling period for modal control of sampled-data systems with generalized hold and control delay. Appendix B, written by Dr. R. Cepeda-Gomez, presents the *DIRSD* Toolbox. This toolbox provides numerical solutions for the problem of optimal direct design of SD systems. Appendix C, composed by Dr. V. Rybinskii, describes Toolbox. This toolbox contains numerical algorithms for analysis and design of SD systems with guaranteed performance.

The authors hope that the book will be beneficial for scientists and engineers working in the field of investigation and design of modern control systems, because the solution of problems in connection with the development of technology and complex control systems requires the application of new and advanced mathematical instruments. In the authors' opinion, the methods exposed in the book can be effective tools for the solution of versatile problems, for which alternative methods are not known today. The book should also be useful for students of higher classes and aspirants, specializing in theory and application of control systems.

For understanding the book content, the reader needs a sound mathematical background normally conveyed by courses at higher technical education institution. Subsidial material from algebra, concerning polynomial and rational matrices, and also function of matrices is sufficiently exposed in Chaps. 1, 2 and 5.

The authors gratefully acknowledge the financial support from the German science foundation (Deutsche Forschungsgemeinschaft). We are also grateful for the additional funds made available by the German Academic Exchange Service (Deutscher Akademischer Austauschdienst). Wolfgang Drewelow and Renate Ziegler from the Rostock side have helped us very much with the completion of this book. On the side of St. Petersburg, the significant support was provided by Vladislav Rybinskii, Mikhail Karavainikov, and Nikolay Zhivotnev. The authors express their sincere gratitude to all of them.

Oliver Jackson from Springer helped us to overcome various editorial problems with great dedication. We are very grateful to him for this.

Saint Petersburg, Russia — Efim N. Rosenwasser
Rostock, Germany — Bernhard P. Lampe
June 2017 — Torsten Jeinsch

## References

1. K.J. Åström, B. Wittenmark. *Computer Controlled Systems: Theory and Design*, 3rd edn. (Prentice-Hall, Englewood Cliffs, NJ, (1997)
2. J.S. Tsypkin. *Sampling Systems Theory*. (Pergamon Press, New York, 1964)
3. T. Chen and B.A. Francis. *Optimal Sampled-data Control Systems*. (Springer, Berlin, Heidelberg, New York, 1995)
4. T. Hagiwara, M. Araki. FR-operator approach to the $\mathscr{H}_2$-analysis and synthesis of sampled-data systems. IEEE Trans. Autom. Contr, AC-**40**(8), 1411–1421 (1995)
5. L.A. Zadeh. Frequency analysis of variable networks. Proc. IRE, **39**(March):291–299 (1950)
6. L.A. Zadeh. Stability of linear varying-parameter systems. J. Appl. Phys. **22**(4):202–204 (1951)
7. J.R. Ragazzini, L.A. Zadeh. The analysis of sampled-data systems. AIEE Trans. **71**:225–234 (1952)
8. E.N. Rosenwasser. *Periodically Nonstationary Control Systems*. (Nauka, Moscow, 1973). (in Russian)
9. E.N. Rosenwasser. *Linear Theory of Digital Control in Continuous Time*. (Nauka, Moscow, 1994). (in Russian)
10. E.N. Rosenwasser, K.Y. Polyakov, and B.P. Lampe. Entwurf optimaler Kursregler mit Hilfe von Parametrischen Übertragungsfunktionen. Automatisierungstechnik, **44**(10), 487–495 (1996)
11. E.N. Rosenwasser, B.P. Lampe. *Computer Controlled Systems: Analysis and Design with Process-orientated Models*. (Springer, London Berlin Heidelberg, 2000)
12. E.N. Rosenwasser, B.P. Lampe. *Multivariable Computer-controlled Systems—A Transfer Function Approach*. (Springer, London, 2006)

13. B.P. Lampe, E.N. Rosenwasser. Polynomial modal control for sampled-data systems with delay. Autom. Remote Control **69**(6), 953–967 (Jun 2008)
14. B.P. Lampe, E.N. Rosenwasser. $\mathscr{H}_2$-optimisation of MIMO sampled-data systems with pure time-delays. Int. J. Control **82**(10), 1899–1916 (2009)
15. B.P. Lampe, E.N. Rosenwasser. $\mathscr{H}_2$-optimization of time-delayed sampled-data systems on basis of the parametric transfer matrix method. Autom. Remote Control **71**(1), 49–69 (2010)
16. B.P. Lampe, E.N. Rosenwasser. Parametric Transfer Matrix and Statistical Analysis of Sampled-data Systems with Delay. J. Vibr. Control. **16**(7/8), 1023–1048 (June/July 2010)
17. B.P. Lampe, E.N. Rosenwasser. $\mathscr{H}_2$-optimization and fixed poles for sampled-data systems with generalized hold. Automa. Remote Control **73**(1), 31–55 (2012)
18. B.P. Lampe, E.N. Rosenwasser. Singular $\mathscr{L}_2$-optimization problem and fixed poles for sampled-data systems with delay. In*7th IFAC Symp. on Robust Control Design*, 172–177, Aalborg, DK, (Jun 2012)
19. B.P. Lampe, E.N. Rosenwasser. $\mathscr{L}_2$-optimization and fixed poles for sampled-data systems with generalized digital to analog converters. Autom. Remote Control **75**(1), 34–56 (2014)
20. B.P. Lampe, E.N. Rosenwasser. $\mathscr{L}_2$-optimization and fixed poles of multivariable sampled-data systems with delay. Int. J. Control **88**(4), 815–831 (2015). (Published online Dec 2014)
21. J. Ackermann. *Sampled-Data Control Systems: Analysis and Synthesis, Robust System Design*. (Springer, Berlin, 1985)
22. H. Kwakernaak, R. Sivan. *Linear Optimal Control Systems*. (Wiley-Interscience, New York, 1972)
23. K.Y. Polyakov. *Polynomial methods for direct`design of optimal sampled-data control systems*. Thesis for Doctor degree of Technological Sciences, Saint Petersburg Maritime Technological University, (2006)

# Acknowledgement

To my deepest regret, the book comes out after the early death of my dear friend and colleague Prof. Bernhard Lampe. My collaboration with Berhard lasted about 30 years. The result of this work is over one hundred publications, including five monographs. May this book be a monument to this wonderful scientist and man.

Saint Petersburg
November 2018

Efim N. Rosenwasser

# Contents

**Part I Determinant Polynomial Equations and Modal Control of Discrete Processes**

**1 Polynomial Matrices** . . . . . 3
1.1 Scalar Polynomials . . . . . 3
1.2 Polynomial Matrices . . . . . 4
1.3 Equivalence and Pseudo-equivalence . . . . . 6
1.4 Properties of the Normal Rank . . . . . 8
1.5 Invariant Polynomials and Elementary Divisors . . . . . 10
1.6 Latent Values and Latent Equations . . . . . 13
1.7 Row and Column Reduced Matrices . . . . . 15
1.8 Pairs of Polynomial Matrices . . . . . 18
1.9 Polynomial Matrices of First Order . . . . . 20
1.10 Reciprocal Polynomials and Reciprocal Matrices . . . . . 23
References . . . . . 28

**2 Fractional-Rational Matrices** . . . . . 29
2.1 Scalar Rational Fractions . . . . . 29
2.2 Rational Matrices . . . . . 34
2.3 McMillan Canonical Form . . . . . 37
2.4 Matrix Fraction Description (MFD) . . . . . 41
2.5 General Properties of the MFD . . . . . 44
2.6 Properties of the McMillan Index . . . . . 47
2.7 Excess of Rational Matrices . . . . . 54
2.8 Strictly Proper Rational Matrices . . . . . 58
2.9 Division of Polynomial Matrices and Degree Reduction of Polynomial Pairs . . . . . 60
2.10 McMillan Index of Block Rational Matrices . . . . . 66
2.11 Associated Rational Matrices . . . . . 76
References . . . . . 83

**3 Determinant Polynomial Equations and Modal Control Problem for Discrete Processes** . . . 85
3.1 Determinant Polynomial Equations . . . 85
3.2 Basic Controllers . . . 86
3.3 Dual Processes and Dual Basic Controllers . . . 91
3.4 General Solution of the Determinant Polynomial Equation . . . 94
3.5 Solution of DPE Under Constraints on the Structure of the Set of Invariant Polynomials of the Characteristic Matrix . . . 99
3.6 Solution of DPE by Means of Transfer Matrices . . . 102
3.7 Causal Modal Control of Discrete Backward Processes . . . 106
3.8 Causal Stabilization of Backward Processes . . . 111
3.9 Causal Modal Control of Discrete Forward Processes . . . 117
3.10 Causal Stabilization of Discrete Forward Processes . . . 124
3.11 Causal Stabilization of the Double Integrator . . . 128
3.12 Causal Modal Control and Stabilization of Backward PMD Processes . . . 130
3.13 Causal Modal Control for SISO Discrete Forward Models . . . 139
References . . . 153

**Part II Parametric Discrete Models of Multivariable Continous Processes and Mathematical Description of Standard SD Systems with Delay**

**4 Modal Control and Stabilization of Multivariable Sampled-Data Systems with Delay** . . . 157
4.1 Linearized Model of Multivariable Digital Controllers (DC) . . . 157
4.2 Decomposition of the DAC Equation . . . 162
4.3 Mathematical Model of an SD System with Delay in Absence of External Disturbances . . . 165
4.4 Discretization of Functions of a Continuous Argument . . . 166
4.5 Backward Discrete Model for System $\mathscr{S}_\tau$ . . . 170
4.6 Modal Control of System $\mathscr{S}_\tau$ . . . 175
4.7 Stability and Backward Stabilization of System $\mathscr{S}_\tau$ . . . 179
4.8 Forward Discrete Model of a Continuous Process with Delay in State Space . . . 181
References . . . 193

**5 Parametric Discrete Models of Multivariable Continuous Processes with Delay** . . . 195
5.1 Functions of Matrices . . . 195
5.2 Index Function of a Matrix . . . 200
5.3 Discrete Laplace Transformation (DLT) of Functions of Continuous Arguments . . . 203
5.4 Discrete Laplace Transforms (DLT) of Images . . . 204

5.5 DPL and DPFR of Rational Matrices . . . . . . . . . . 206
5.6 Parametric Discrete Models of Continuous Processes . . . . . . . . 209
5.7 Parametric Discrete Models of Continuous Processes with Delay . . . . . . . . . . 212
5.8 DLT and DPFR of Modulated Processes . . . . . . . . . . 215
5.9 Parametric Discrete Models for Modulated Processes . . . . . . . . 221
5.10 Parametric Discrete Models of Modulated Processes with Delay . . . . . . . . . . 224
5.11 Response of LTI System with Delay to Exponential-Periodic Inputs . . . . . . . . . . 226
5.12 Response of Open Sampled-Data System with Delay to EP Input . . . . . . . . . . 229
References . . . . . . . . . . 232

**6 Mathematical Description of Standard Sampled-Data Systems with Delay** . . . . . . . . . . 235
6.1 Standard Model of SD Systems with Delay . . . . . . . . . . 235
6.2 Standard Structure of SD Systems with Delay . . . . . . . . . . 238
6.3 Parametric Transfer Matrices (PTM) for the Standard Structure of SD Systems with Delay . . . . . . . . . . 241
6.4 PTM of the Standard Model as a Function of Arguments $t$ and $\lambda$ . . . . . . . . . . 250
6.5 Modal Control and Stabilization of the Standard Model for SD System with Delay . . . . . . . . . . 258
6.6 Representing PTM $W_{zx}(\lambda, t)$ of the Standard Model of SD System with Delay by a System Function . . . . . . . . . . 265
6.7 Structural Standardizable SD Systems with Delay . . . . . . . . . . 270
6.8 Model-Standardizable SD Systems with Delay . . . . . . . . . . 279
References . . . . . . . . . . 287

**Part III Optimal SD Control with Delay**

**7 $\mathscr{H}_2$ Optimization and Fixed Poles of the Standard Model for SD Systems with Delay** . . . . . . . . . . 291
7.1 Advanced Statistical Analysis for the Standard Model of SD Systems with Delay . . . . . . . . . . 291
7.2 Mean Variance and $\mathscr{H}_2$ Norm . . . . . . . . . . 294
7.3 Representation of the $\mathscr{H}_2$ Norm by a System Function . . . . . . . 298
7.4 Properties of Matrix $A_M(\zeta)$ . . . . . . . . . . 304
7.5 Properties of Matrices $A_L(\zeta)$ and $C(\zeta)$ . . . . . . . . . . 313
7.6 Wiener–Hopf Method . . . . . . . . . . 323

7.7 Characteristic Polynomial and Fixed Poles of the $\mathscr{H}_2$ Optimal System . . . . . 330
7.8 Examples for Constructing the Set of Fixed Poles . . . . . 338
References . . . . . 342

**8 $\mathscr{H}_2$ Optimization and Fixed Poles for the Standard Model of SD Systems with Delay Under Colored Noise** . . . . . 345
8.1 Statement of the $\mathscr{H}_2$ Optimization Problem Under Colored Noise . . . . . 345
8.2 PTM and $\mathscr{H}_2$ Norm of the System $\mathscr{S}_d$ . . . . . 347
8.3 Properties of Matrices $A_{M\Phi}(\zeta)$ and $C_{\Phi}(\zeta)$ . . . . . 349
8.4 Wiener–Hopf Method . . . . . 355
8.5 Cancellation of External Poles . . . . . 357
8.6 General Properties of $\mathscr{H}_2$ Optimal Systems . . . . . 370
References . . . . . 376

**9 $\mathscr{L}_2$ Tracking Problem for the Standard Model of SD Systems with Delay** . . . . . 377
9.1 Operational Relations for the Standard Model of SD Systems with Delay . . . . . 377
9.2 Laplace Transforms of Processes with Zero Initial Energy . . . . . 382
9.3 Properties of the Image $Z(s)$ . . . . . 386
9.4 Statement of the $\mathscr{L}_2$-Optimal Tracking Problem for the Standard Model of SD Systems with Delay . . . . . 390
9.5 Construction of the Cost Functional . . . . . 391
9.6 Transformation and Minimization of the Cost Functional . . . . . 397
9.7 General Properties and Fixed Poles of $\mathscr{L}_2$ Optimal Systems . . . 402
References . . . . . 406

**10 Mathematical Description and Optimization for the Generalized Standard Model of SD Systems with Delay** . . . . . 407
10.1 Mathematical Description of the Generalized Standard Model . . . . . 407
10.2 $\mathscr{H}_2$ Optimization of System $\mathscr{S}_g$ . . . . . 411
10.3 $\mathscr{H}_2$ Optimization of the Generalized Standard Model for Colored Noise . . . . . 421
10.4 $\mathscr{L}_2$ Optimization of the Generalized Standard Model . . . . . 425
10.5 Mathematical Description of a Single-Loop Multivariable SD Systems with Delay as Generalized Standard Model . . . . . 431
10.6 $\mathscr{H}_2$ Optimization of Single-Loop SD Systems with Delay . . . . . 438
10.7 $\mathscr{H}_2$ Optimization of a Single-Loop SD System with Delay Under Colored Noise . . . . . 444
10.8 $\mathscr{L}_2$ Optimization of a Single-Loop SD System with Delay . . . . 446

**Appendix A: Choice of the Sampling Period for Modal Control of Sampled-Data Systems with Generalized Hold and Control Delay** . . . . . 449

**Appendix B: *DIRSD*–A Toolbox for Direct Design of SD Systems** . . . . . 469

**Appendix C: `GarSD`–A MATLAB Toolbox for Analysis and Design of Sampled-Data System with Guaranteed Performance** . . . . . 479

**Bibliography** . . . . . 503

**Index** . . . . . 513

# Part I
# Determinant Polynomial Equations and Modal Control of Discrete Processes

# Chapter 1
# Polynomial Matrices

**Abstract** This chapter summarizes the most important basics of the theory of polynomial matrices, which are necessary for understanding the essential content of the book. It basically contains well-known material, but is presented in a form adapted to the purposes of this monograph. Familiarity with the material in this chapter allows a less prepared reader to obtain the necessary information from mathematics without referring to special literature.

## 1.1 Scalar Polynomials

(1) In the following, considerations, the notation $\lambda \in \mathbb{C}$ ($\lambda \in \mathbb{R}$) means that the number $\lambda$ belongs to the set of complex (real) numbers. The notation $\lambda \in \mathbb{F}$ means that $\lambda \in \mathbb{C}$ or $\lambda \in \mathbb{R}$. By $\mathbb{C}[\lambda]$ or $\mathbb{R}[\lambda]$ we denote the set of complex (real) polynomials over $\mathbb{C}$ or $\mathbb{R}$, respectively. Consequently, $f(\lambda) \in \mathbb{F}[\lambda]$ means that $f(\lambda) \in \mathbb{C}[\lambda]$ or $f(\lambda) \in \mathbb{R}[\lambda]$. It is well known that any polynomial $f(\lambda) \in \mathbb{F}[\lambda]$ can be represented as product of the form

$$f(\lambda) = a_0(\lambda - \lambda_1) \cdots (\lambda - \lambda_n), \tag{1.1}$$

where $a_0$ is a constant, and among the numbers $\lambda_1, \ldots, \lambda_n$ might be equal ones. The integer $n$ in (1.1) is called the degree of the polynomial $f(\lambda)$, and it is denoted by $\deg f(\lambda)$. The numbers $\lambda_i$ are called the roots of the polynomial $f(\lambda)$. If the common factors in (1.1) are collected, we obtain

$$f(\lambda) = a_0(\lambda - \lambda_1)^{\mu_1} \cdots (\lambda - \lambda_q)^{\mu_q}, \quad \mu_1 + \cdots + \mu_q = n, \tag{1.2}$$

where all numbers $\lambda_1, \ldots, \lambda_q$ are distinct. The number $\mu_i$ in (1.2) is called the multiplicity of the root $\lambda_i$.

If $f(\lambda) \in \mathbb{R}[\lambda]$, and product (1.2) has a complex root $\lambda_i = a_i + b_i\mathrm{j}$ with multiplicity $\mu_i$, where $\mathrm{j} = \sqrt{-1}$, then the conjugate complex numbers $\overline{\lambda}_i = a_i - b_i\mathrm{j}$ with multiplicity $\mu_i$ also exist among the roots of the polynomial $f(\lambda)$. For $a_0 = 1$ polynomial (1.2) is called monic.



E. N. Rosenwasser et al., *Computer-Controlled Systems with Delay*,
https://doi.org/10.1007/978-3-030-15042-6_1

(2) For arbitrary polynomials $f(\lambda)$, $d(\lambda)$, there exists a pair of polynomials $q(\lambda)$, $r(\lambda)$ such that

$$f(\lambda) = q(\lambda)d(\lambda) + r(\lambda), \tag{1.3}$$

where

$$\deg r(\lambda) < \deg d(\lambda). \tag{1.4}$$

Here the polynomial $q(\lambda)$ is called the entire part, and the polynomial $r(\lambda)$ the remainder at the division of $f(\lambda)$ by $d(\lambda)$.

(3) If $r(\lambda) = 0$, i.e.,

$$f(\lambda) = q(\lambda)d(\lambda) \tag{1.5}$$

we say that the polynomial $q(\lambda)$ is a divisor of the polynomial $f(\lambda)$.

Let the polynomials $f_1(\lambda),\ f_2(\lambda) \in \mathbb{F}[\lambda]$ be given. Then the polynomial $p(\lambda)$ is called a common divisor (CD) of $f_1(\lambda)$ and $f_2(\lambda)$ if it is a divisor of each of them. A CD of greatest possible degree is called a greatest common divisor (GCD).

The GCD of the polynomials $f_1(\lambda)$ and $f_2(\lambda)$ can be represented in the form

$$p(\lambda) = f_1(\lambda)m_1(\lambda) + f_2(\lambda)m_2(\lambda), \tag{1.6}$$

where $m_1(\lambda),\ m_2(\lambda)$ are certain polynomials.

The monic GCD of the polynomials $f_1(\lambda)$ and $f_2(\lambda)$ is uniquely defined.

(4) The polynomials $f_1(\lambda)$ and $f_2(\lambda)$ are called coprime if they do not possess common roots. The monic GCD of coprime polynomials is equal to one. For the fact that the polynomials $f_1(\lambda)$ and $f_2(\lambda)$ are coprime, the following condition is necessary and sufficient: There exist polynomials $m_1(\lambda)$ and $m_2(\lambda)$ such that

$$f_1(\lambda)m_1(\lambda) + f_2(\lambda)m_2(\lambda) = 1. \tag{1.7}$$

If

$$f_1(\lambda) = p(\lambda)\tilde{f}_1(\lambda), \quad f_2(\lambda) = p(\lambda)\tilde{f}_2(\lambda), \tag{1.8}$$

where $p(\lambda)$ is the GCD, then the polynomials $\tilde{f}_1(\lambda)$ and $\tilde{f}_2(\lambda)$ are coprime.

## 1.2 Polynomial Matrices

(1) The set of constant $n \times m$ matrices with entries from $\mathbb{C},\ \mathbb{R},\ \mathbb{F}$ are denoted by $\mathbb{C}^{nm},\ \mathbb{R}^{nm},\ \mathbb{F}^{nm}$, respectively. The set of $n \times m$ matrices with entries from $\mathbb{C}[\lambda]$, $\mathbb{R}[\lambda]$, $\mathbb{F}[\lambda]$ are denoted by $\mathbb{C}^{nm}[\lambda]$, $\mathbb{R}^{nm}[\lambda]$, $\mathbb{F}^{nm}[\lambda]$, respectively. In the following, these matrices are called polynomial matrices. A polynomial matrix $A(\lambda) \in \mathbb{F}^{nm}[\lambda]$ can be written in the form

$$A(\lambda) = \begin{bmatrix} a_{11}(\lambda) & a_{12}(\lambda) & \dots & a_{1m}(\lambda) \\ \vdots & \vdots & \ddots & \vdots \\ a_{n1}(\lambda) & a_{n2}(\lambda) & \dots & a_{nm}(\lambda) \end{bmatrix}, \tag{1.9}$$

where the $a_{ik}$ are polynomials from the corresponding set.

For $n > m$ the matrix $A(\lambda)$ is called vertical, for $n < m$ horizontal, and for $n = m$ quadratic.

(2) Matrix (1.9) can be written in the form

$$A(\lambda) = A_0\lambda^q + A_1\lambda^{q-1} + \cdots + A_q\,, \tag{1.10}$$

where the $A_i$, $(i = 0, \dots, q)$ are constant $n \times m$ matrices. The matrix $A_0$ is called the highest coefficient of the polynomial matrix $A(\lambda)$.

For $A_0 \neq 0_{nm}$, where $0_{nm}$ is the $n \times m$ zero matrix, the number $q$ is called the degree of the polynomial matrix $A(\lambda)$, and we write

$$q = \deg A(\lambda). \tag{1.11}$$

If $n \neq m$ or $n = m$ and $\det A(\lambda) \equiv 0$, then matrix (1.10) is called singular. For $n = m$ and $\det A(\lambda) \not\equiv 0$, the matrix $A(\lambda)$ is called non-singular. When $\det A_0 \neq 0$, the non-singular matrix is called regular. Non-singular matrices with $\det A_0 = 0$ are called anomal.

(3) Let $A(\lambda) \in \mathbb{F}^{np}[\lambda]$, $B(\lambda) \in \mathbb{F}^{pm}[\lambda]$. Then the product $C(\lambda) = A(\lambda)B(\lambda) \in \mathbb{F}^{nm}$ is defined. Then, in the general case, we obtain

$$\deg C(\lambda) = \deg[A(\lambda)B(\lambda)] \leq \deg A(\lambda) + \deg B(\lambda). \tag{1.12}$$

However, if one of the factors in (1.12) is regular, we achieve the equality

$$\deg C(\lambda) = \deg[A(\lambda)B(\lambda)] = \deg A(\lambda) + \deg B(\lambda). \tag{1.13}$$

(4) Let the matrix $A(\lambda) \in \mathbb{F}^{nn}[\lambda]$ be non-singular. Then the determinant $\det A(\lambda)$ is a scalar polynomial, and the number

$$\operatorname{ord} A(\lambda) \stackrel{\triangle}{=} \deg \det A(\lambda) \tag{1.14}$$

is called the order of the matrix $A(\lambda)$. The degree and the order of any matrix $A(\lambda)$ are related by the inequality

$$\operatorname{ord} A(\lambda) \leq n \deg A(\lambda). \tag{1.15}$$

For regular matrices $A(\lambda)$ relation (1.15) becomes an equality. In general, the degree of a matrix $A(\lambda)$ for a given order can be rather high. A non-singular matrix $A(\lambda)$ with ord $A(\lambda) = 0$, i.e., $\det A(\lambda) = \text{const.} \neq 0$ is called unimodular.

(5) For a non-singular matrix $A(\lambda)$, the polynomial

$$\Delta_A(\lambda) \stackrel{\triangle}{=} \det A(\lambda) \tag{1.16}$$

is called its characteristic polynomial. The roots of the characteristic polynomial are called the eigenvalues of the matrix $A(\lambda)$. The totality of all eigenvalues of the matrix $A(\lambda)$ is called its spectrum. Two non-singular matrices $A_1(\lambda)$, $A_2(\lambda)$ (possibly of different size) are called spectral independent if their characteristic polynomials are coprime, i.e., their spectra are disjunct.

(6) Let $A$ be a constant $n \times n$ matrix. Then the regular polynomial matrix

$$A_\lambda = \lambda I_n - A, \tag{1.17}$$

where $I_n$ stands for the $n \times n$ identity matrix, is called the characteristic matrix of the matrix $A$. The polynomial

$$d_A(\lambda) = \det(\lambda I_n - A) \tag{1.18}$$

is called the characteristic polynomial of the matrix $A$. The set of all roots of the polynomial $d_A(\lambda)$ is called the spectrum of the matrix $A$.

## 1.3 Equivalence and Pseudo-equivalence

(1) As normal rank of a polynomial matrix $A(\lambda) \in \mathbb{F}^{nm}[\lambda]$ we understand the highest rank of the number matrix, which can be reached from (1.9) by inserting all possible values of $\lambda$. Further on, the normal rank of the matrix is denoted as Rank $A(\lambda)$. Beside, we denote the ordinary rank of a constant matrix $A$ by rank $A$. With the help of these concepts, the normal rank can be defined by the relation

$$\operatorname{Rank} A(\lambda) = \max_{\lambda \in \mathbb{C}} \operatorname{rank} A(\lambda). \tag{1.19}$$

It can be shown that Rank $A(\lambda)$ is equal to the highest dimension of the non-singular minors of the matrix $A(\lambda)$.

(2) Two matrices $A_1(\lambda)$, $A_2(\lambda)$ in $\mathbb{F}^{nm}[\lambda]$ are called equivalent, if they are connected by the formula

$$A_2(\lambda) = p(\lambda)A_1(\lambda)q(\lambda), \tag{1.20}$$

where $p(\lambda)$, $q(\lambda)$ are unimodular matrices of appropriate size. To denote the fact that the matrices $A_1(\lambda)$ and $A_2(\lambda)$ are equivalent, we write

$$A_1(\lambda) \sim A_2(\lambda). \tag{1.21}$$

In particular, if both $A_1(\lambda)$ and $A_2(\lambda)$ are scalar polynomials, then equivalence relation (1.21) means $A_2(\lambda) = kA_1(\lambda)$, $k = \text{const.} \neq 0$.

Classical results on equivalence are formulated in the next statements [1].

**Theorem 1.1** *Any matrix $A(\lambda) \in \mathbb{F}^{nm}[\lambda]$ with* Rank $A(\lambda) = \rho$ *is equivalent to the matrix*

$$S_A(\lambda) \triangleq \begin{bmatrix} S_\rho(\lambda) & 0_{\rho,m-\rho} \\ 0_{n-\rho,n} & 0_{n-\rho,m-\rho} \end{bmatrix}, \tag{1.22}$$

*where the matrix $S_\rho(\lambda) \in \mathbb{F}_{\rho\rho}[\lambda]$ has the form*

$$S_\rho(\lambda) = \operatorname{diag}\{a_1(\lambda), \ldots, a_\rho(\lambda)\}, \tag{1.23}$$

*where $a_i(\lambda)$, $(i = 1, \ldots, \rho)$ are monic polynomials such that $a_i(\lambda)$ is divisible by $a_{i-1}(\lambda)$. Moreover, in (1.23) and below,* diag *denotes a diagonal matrix. Moreover, if the matrix $A(\lambda)$ is real, i.e., $A(\lambda) \in \mathbb{R}[\lambda]$, then the matrices $p(\lambda), q(\lambda)$ corresponding to (1.20), and all polynomials $a_i(\lambda)$ are real. The matrix $S_\rho(\lambda)$ is uniquely defined by the matrix $A(\lambda)$.*

The matrix $S_A(\lambda)$ is called the Smith canonical form of the matrix $A(\lambda)$, and the polynomials $a_i(\lambda)$, $(i = 0, \ldots, \rho)$ are their invariant polynomials.

**Theorem 1.2** *For the equivalence of polynomial matrices of the same size, it is necessary and sufficient that their sets of invariant polynomials coincide.*

(3)

**Definition 1.1** Two quadratic matrices $A \in \mathbb{F}^{nn}[\lambda]$, $B \in \mathbb{F}^{mm}[\lambda]$ are called pseudo-equivalent, if the sets of their invariant polynomials coincide.

Further on, the pseudo-invariance of the two matrices $A(\lambda)$, $B(\lambda)$ is denoted by

$$A(\lambda) \overset{p}{\sim} B(\lambda). \tag{1.24}$$

*Example 1.1* Let $d(\lambda)$ be an arbitrary monic polynomial. Then the matrices

$$A(\lambda) = \operatorname{diag}\{1, 1, d(\lambda)\}, \quad B(\lambda) = \operatorname{diag}\{1, d(\lambda)\} \tag{1.25}$$

are pseudo-equivalent, and we can write (1.24).

*Remark 1.1* If in (1.24) the matrices $A(\lambda)$, $B(\lambda)$ possess equal size, then the relation of pseudo-equivalence becomes equivalence.

## 1.4 Properties of the Normal Rank

(1)

**Theorem 1.3** *The following statements hold:*

*(i) Let a matrix* $A \in \mathbb{F}^{nm}[\lambda]$ *be given with* $\operatorname{Rank} A(\lambda) = \rho$*. Then for all fixed values* $\lambda = \lambda_0$*, excluding possibly some finite set, we obtain*

$$\operatorname{rank} A(\lambda_0) = \rho = \operatorname{Rank} A(\lambda). \tag{1.26}$$

*(ii) Let a finite set of matrices* $A_1(\lambda), \ldots, A_q(\lambda)$ *be given, where* $\operatorname{Rank} A_i(\lambda) = \rho_i$, $(i = 1, \ldots, q)$*. Then for all fixed values* $\lambda = \lambda_0$*, possibly excluding a finite set, we obtain at the same time*

$$\operatorname{rank} A_i(\lambda_0) = \rho_i, \ (i = 1, \ldots, n). \tag{1.27}$$

*Proof* (i) Write the matrix $A(\lambda)$ in the form

$$A(\lambda) = p(\lambda) S_A(\lambda) q(\lambda), \tag{1.28}$$

where $p(\lambda)$, $q(\lambda)$ are unimodular matrices, and $S_A(\lambda)$ is the Smith canonical form. Then for any fixed value $\lambda = \lambda_0$, we obtain

$$\operatorname{rank} A(\lambda_0) = \operatorname{rank} S_A(\lambda_0), \tag{1.29}$$

that implies $\operatorname{Rank} A(\lambda) = \operatorname{Rank} S_A(\lambda)$. But from (1.22), it follows that $\operatorname{Rank} S_A(\lambda_0) = \rho$ for all $\lambda_0$, with the exception of the roots of the invariant polynomials. Thus (i) follows. Statement (ii) can be proven analogously. ■

(2)

**Theorem 1.4** *Let* $A(\lambda) \in \mathbb{F}^{nm}[\lambda]$*. Then*

$$\operatorname{Rank} A(\lambda) \leq \min\{n, m\}. \tag{1.30}$$

*Proof* Assume contradictorily that $\operatorname{Rank} A(\lambda) > \min\{n, m\}$. Then due to Theorem 1.3, there exists a $\lambda_0$, such that $\operatorname{rank} A(\lambda_0) > \min\{n, m\}$, which is impossible. This contradiction proves (1.38). ■

**Theorem 1.5** *Let* $A(\lambda) \in \mathbb{F}^{nl}[\lambda]$*,* $B(\lambda) \in \mathbb{F}^{lm}[\lambda]$*, and*

$$C(\lambda) = A(\lambda) B(\lambda). \tag{1.31}$$

*Then*

$$\operatorname{Rank} C(\lambda) \leq \min\{\operatorname{Rank} A(\lambda), \operatorname{Rank} B(\lambda)\} \tag{1.32}$$

*and*

$$\operatorname{Rank} C(\lambda) \geq \operatorname{Rank} A(\lambda) + \operatorname{Rank} B(\lambda) - l. \tag{1.33}$$

*Relation (1.33) is called Silvester inequality.*

*Proof* Let

$$\operatorname{Rank} A(\lambda) = \rho_A, \ \operatorname{Rank} B(\lambda) = \rho_B, \ \operatorname{Rank} C(\lambda) = \rho_C.$$

Suppose that for the matrix $C(\lambda)$ inequality (1.40) does not hold, i.e.,

$$\operatorname{Rank} C(\lambda) > \min\{\operatorname{Rank} A(\lambda), \operatorname{Rank} B(\lambda)\}. \tag{1.34}$$

Then follows from Theorem 1.3 the existence of a number $\lambda_0$, such that

$$\operatorname{rank} S_A(\lambda_0) = \rho_A, \ \operatorname{rank} S_B(\lambda_0) = \rho_B, \ \operatorname{rank} S_C(\lambda_0) = \rho_C. \tag{1.35}$$

Hence from (1.34)

$$\operatorname{rank} \left[ \, A(\lambda_0) \ B(\lambda_0) \, \right] > \min\{\operatorname{rank} A(\lambda_0), \operatorname{rank} B(\lambda_0)\}, \tag{1.36}$$

what contradicts the known rank property of constant matrices. This shows (1.32). Relation (1.33) is proven by analog reasoning. ■

From inequalities (1.32), (1.33), we find

$$\operatorname{Rank}[A(\lambda_0)B(\lambda_0)] = \operatorname{Rank} B(\lambda), \ \text{ for } \operatorname{Rank} A(\lambda) = l \tag{1.37}$$

and

$$\operatorname{Rank}[A(\lambda_0)B(\lambda_0)] = \operatorname{Rank} A(\lambda), \ \text{ for } \operatorname{Rank} B(\lambda) = l. \tag{1.38}$$

It is evident that the relation $\operatorname{Rank} A(\lambda) = l$ can only be true if $n \geq l$, and the relation $\operatorname{Rank} B(\lambda) = l$ if $m \geq l$. In particular, condition (1.37) holds if $n = l$ and the matrix $A(\lambda)$ is regular, and relation (1.38) is true if $m = l$ and the matrix $B(\lambda)$ is regular.

(3) Using the arguments in the proofs for Theorems 1.4 and 1.5, we can state the following facts [1, 2].

**Theorem 1.6** *Let the polynomial matrix $C(\lambda) \in \mathbb{F}^{n,l+m}[\lambda]$ possess the block form*

$$C(\lambda) = \left[ \, A(\lambda) \ B(\lambda) \, \right] \tag{1.39}$$

*where $A(\lambda) \in \mathbb{F}^{nl}[\lambda]$, $B(\lambda) \in \mathbb{F}^{nm}[\lambda]$. Then*

$$\operatorname{Rank} C(\lambda) \leq \operatorname{Rank} A(\lambda) + \operatorname{Rank} B(\lambda). \tag{1.40}$$

**Theorem 1.7** *For two polynomial matrices of the same size* $A(\lambda),\ B(\lambda) \in \mathbb{F}^{mn}[\lambda]$, *we have*

$$\mathrm{Rank}[A(\lambda) + B(\lambda)] \leq \mathrm{Rank}\ A(\lambda) + \mathrm{Rank}\ B(\lambda). \tag{1.41}$$

(4) Let $A(\lambda) \in \mathbb{F}^{nm}[\lambda]$. Introduce

$$\mathrm{def}\ A(\lambda) = \min\{n, m\} - \mathrm{Rank}\ A(\lambda). \tag{1.42}$$

**Definition 1.2** The number def $A(\lambda)$ is called the normal defect of the matrix $A(\lambda)$. It follows from Theorem 1.4 that def $A(\lambda) \geq 0$. Moreover, the normal defect of a non-singular matrix is equal to zero.

## 1.5 Invariant Polynomials and Elementary Divisors

(1) Let us consider the matrix $A(\lambda) \in \mathbb{F}^{nm}[\lambda]$ with Rank $A(\lambda) = \rho$. Let also $a_1(\lambda), \ldots, a_\rho(\lambda)$ be the sequence of invariant polynomials of the matrix $A(\lambda)$. Introduce the monic polynomials

$$h_1(\lambda) = a_1(\lambda),\ h_2(\lambda) = \frac{a_2(\lambda)}{a_1(\lambda)},\ \ldots,\ h_\rho(\lambda) = \frac{a_\rho(\lambda)}{a_{\rho-1}(\lambda)}. \tag{1.43}$$

Hence, we obtain

$$a_1(\lambda) = h_1(\lambda),\ a_2(\lambda) = h_1(\lambda)h_2(\lambda),\ \ldots,\ a_\rho(\lambda) = h_1(\lambda)\cdots h_\rho(\lambda), \tag{1.44}$$

and matrix $S_\rho(\lambda)$ in (1.23) can be represented in the form

$$S_\rho(\lambda) = \mathrm{diag}\{h_1(\lambda), h_1(\lambda)h_2(\lambda), \ldots, h_1(\lambda)\cdots h_\rho(\lambda)\}. \tag{1.45}$$

Since the invariant polynomials of the matrix $A(\lambda)$ do not change under elementary transformations, the polynomials $h_i(\lambda)$, $(i = 1, \ldots, \rho)$ dispose of analog properties.

(2) The monic GCDs of all $r \times r$ minors different from zero of the matrix $A(\lambda)$ are called determinant divisors of order $r$ and are denoted by $d_r(\lambda)$. If

$$\mathrm{Rank}\ A(\lambda) = \rho,$$

then we have $\rho$ determinant divisors

$$d_1(\lambda), \ldots, d_\rho(\lambda).$$

The polynomial $d_\rho(\lambda)$ is called the greatest determinant divisor of the matrix $A(\lambda)$. The determinant divisors $d_i(\lambda)$ and the invariant polynomials $a_i(\lambda)$ are connected by the well-known relation

$$d_1(\lambda) = a_1(\lambda),\ d_2(\lambda) = a_1(\lambda)a_2(\lambda),\ \ldots,\ d_\rho(\lambda) = a_1(\lambda)a_2(\lambda)\cdots a_\rho(\lambda), \quad (1.46)$$

from which we conclude that the sequence of determinant divisors does not change under elementary transformations. From (1.46), we derive the recursive relations

$$a_i(\lambda) = \frac{d_i(\lambda)}{d_{i-1}(\lambda)},\quad d_0(\lambda) = 1,\ (i = 1, \ldots, \rho). \quad (1.47)$$

(3) Consider the product

$$a_\rho(\lambda) = (\lambda - \lambda_1)^{\nu_{\rho 1}} \cdots (\lambda - \lambda_q)^{\nu_{\rho q}}, \quad (1.48)$$

where all numbers $\lambda_i$, $(i = 1, \cdots, q)$ are distinct. Then (1.47) yields

$$a_i(\lambda) = (\lambda - \lambda_1)^{\nu_{i1}} \cdots (\lambda - \lambda_q)^{\nu_{iq}},\quad (i = 1, \ldots, \rho), \quad (1.49)$$

where

$$0 \le \nu_{ik} \le \nu_{i+1,k} \le \nu_{\rho k},\quad (k = 1, \ldots, q). \quad (1.50)$$

The factors different from one in product (1.49) are called elementary divisors of the matrix $A(\lambda)$ (in the field of complex numbers). In general, one root $\lambda_k$ can be related to several elementary divisors. From the above follows that the set of invariant polynomials uniquely determines the set of elementary divisors. The reverse is also true, if only the normal rank and the size of the matrix $A(\lambda)$ are known.

*Example 1.2* Let the elementary divisors of the $5 \times 5$ matrix $A(\lambda)$ possess the form

$$(\lambda - 2)^2,\ (\lambda - 2)^2,\ \lambda - 2,\ \lambda - 3,\ \lambda - 3,\ \lambda - 4$$

and Rank $A(\lambda) = 4$. Due to Rank $A(\lambda) = 4$, the set of invariant polynomials becomes

$$a_4(\lambda) = (\lambda - 2)^2(\lambda - 3)(\lambda - 4),\ a_3(\lambda) = (\lambda - 2)^2(\lambda - 3),\ a_2(\lambda) = \lambda - 2,\ a_1(\lambda) = 1.$$

The Smith canonical form of the matrix $A(\lambda)$ takes the form

$$S_A(\lambda) = \begin{bmatrix} 1 & 0 & 0 & 0 & 0 \\ 0 & \lambda - 2 & 0 & 0 & 0 \\ 0 & 0 & (\lambda - 2)^2(\lambda - 3) & 0 & 0 \\ 0 & 0 & 0 & (\lambda - 2)^2(\lambda - 3)(\lambda - 4) & 0 \\ 0 & 0 & 0 & 0 & 0 \end{bmatrix}. \quad (1.51)$$

(4) For the construction of the Smith canonical form of diagonal and block diagonal matrices it is sufficient to know the elementary divisors of the blocks.

**Theorem 1.8** ([2]) *The system of elementary divisors of a diagonal matrix is the unification of the elementary divisors of its elements.*

*Example 1.3* Assume the diagonal matrix

$$A(\lambda) = \begin{bmatrix} \lambda^2 + a^2 & 0 & 0 & 0 \\ 0 & \lambda^2 - a^2 & 0 & 0 \\ 0 & 0 & (\lambda - a)^2 & 0 \\ 0 & 0 & 0 & \lambda - a \end{bmatrix} \tag{1.52}$$

where $a \neq 0$ is a real number. The set of elementary divisors in the complex plane consists of the polynomials

$$(\lambda + a\mathrm{j}),\ (\lambda - a\mathrm{j}),\ (\lambda - a),\ (\lambda + a),\ (\lambda - a)^2,\ (\lambda - a). \tag{1.53}$$

Therefore, the Smith canonical form of $A(\lambda)$ takes the form

$$S_A(\lambda) = \begin{bmatrix} 1 & 0 & 0 & 0 \\ 0 & \lambda - a & 0 & 0 \\ 0 & 0 & \lambda - a & 0 \\ 0 & 0 & 0 & (\lambda^2 + a^2)(\lambda - a)^2(\lambda + a) \end{bmatrix} \tag{1.54}$$

For $a = 0$ instead of (1.54), we obtain

$$S_A(\lambda) = \begin{bmatrix} \lambda & 0 & 0 & 0 \\ 0 & \lambda^2 & 0 & 0 \\ 0 & 0 & \lambda^2 & 0 \\ 0 & 0 & 0 & \lambda^2 \end{bmatrix}. \tag{1.55}$$

**Theorem 1.9** ([2]) *The system of elementary divisors of the block diagonal matrix*

$$A_d(\lambda) = \begin{bmatrix} A_1(\lambda) & 0 & \cdots & 0 \\ 0 & A_2(\lambda) & \cdots & 0 \\ \cdots & \cdots & \ddots & \cdots \\ 0 & 0 & \cdots & A_n(\lambda) \end{bmatrix} = \mathrm{diag}\{A_1(\lambda), \ldots, A_n(\lambda)\}, \tag{1.56}$$

*where $A_i(\lambda)$, $(i = 1, \ldots, n)$ are rectangular matrices, is the unification of the set of elementary divisors of the matrices $A_i(\lambda)$.*

*Example 1.4* Let $n = 2$ and

$$A_1(\lambda) = \begin{bmatrix} \lambda & 1 & 0 \\ 0 & \lambda & 1 \\ 0 & 0 & \lambda \end{bmatrix}, \quad A_2(\lambda) = \begin{bmatrix} \lambda & 0 & 0 \\ 0 & 1 & 0 \\ 0 & 0 & \lambda - a \end{bmatrix}. \tag{1.57}$$

It is immediately seen that the matrix $A_1(\lambda)$ as one elementary divisor $\lambda^3$. The matrix $A_2(\lambda)$ has for $a \neq 0$ the two elementary divisors $\lambda$ and $\lambda - a$, but for $a = 0$ two equal elementary divisors $\lambda$ and $\lambda$. Therefore, in the case, $a \neq 0$, the set of elementary divisors of the matrix $A_d(\lambda)$ is $\lambda^3,\ \lambda,\ \lambda - a$. Hence for $a \neq 0$, we obtain

$$S_{A_d}(\lambda) = \text{diag}\{1, 1, 1, 1, \lambda, \lambda^3(\lambda - a)\}. \tag{1.58}$$

However for $a = 0$, an analog reflection leads to

$$S_{A_d}(\lambda) = \text{diag}\{1, 1, 1, \lambda, \lambda, \lambda^3\}. \tag{1.59}$$

*Remark 1.2* Examples 1.3 and 1.4 show that the Smith canonical form $S_A(\lambda)$ and the structure of the set of elementary divisors are unstable for small changes in the entries of the matrix $A(\lambda)$.

## 1.6 Latent Values and Latent Equations

(1) Consider the matrix $A(\lambda) \in \mathbb{F}^{nm}[\lambda]$, Rank $A(\lambda) = \rho$. When there exists a number $\lambda_0$, for which

$$\text{rank}\, A(\lambda_0) < \rho,$$

then the matrix $A(\lambda)$ is called latent, and the number $\lambda_0$ is called its latent value. The latent values of a non-singular matrix coincide with their eigenvalues. A matrix $A(\lambda)$ that does not possess latent values, is called alatent. Alatent non-singular matrices are unimodular.

(2) Let the Smith canonical form of the matrix $A(\lambda)$ have the form (1.30), (1.31). Then the equation

$$a_1(\lambda) \cdots a_\rho(\lambda) = 0$$

is called latent equation of the matrix $A(\lambda)$.

**Theorem 1.10** *The set of latent values of the matrix $A(\lambda)$ coincides with the set of roots of its latent equation. Moreover, if $\lambda_i$ is a latent value, then we have*

$$\text{rank}\, A(\lambda_i) = \rho - d_i = \text{Rank}\, A(\lambda) - d_i, \tag{1.60}$$

*where $d_i$ is the number of invariant polynomials for which the number $\lambda_i$ is a root, and $\rho =$ Rank $A(\lambda)$.*

*Proof* From (1.22) and (1.23) we conclude, that if $\lambda_i$ is not a root of one of its invariant polynomials, then rank $A(\lambda) = \rho$. Thus Rank $A(\lambda_i) = \rho$. When, however, for $\lambda = \lambda_i$ $d_i$ polynomials are transmuted to zero, then rank $S_A(\lambda_i) = \rho - d_i$. Since

$$A(\lambda_i) = p(\lambda_i) S_A(\lambda_i) q(\lambda_i), \tag{1.61}$$

we obtain (1.60), because $\det p(\lambda_i) \neq 0,\ \det q(\lambda_i) \neq 0$. ∎

(3) We will derive some features of alatent matrices that we need later.

**Theorem 1.11** *For the matrix* $A(\lambda) \in \mathbb{F}^{nm}[\lambda]$ *with* Rank $A(\lambda) = \rho$ *to be alatent, it is necessary and sufficient, that the matrix* $S_\rho(\lambda)$ *in (1.22) takes the form*

$$S_\rho(\lambda) = I_\rho. \tag{1.62}$$

*Proof* Sufficiency: If (1.62) is true, then it follows from (1.22) and (1.61) that for all $\lambda$

$$\operatorname{rank} A(\lambda) = \rho.$$

The necessity can be seen by reverse reasoning.

The matrix $A(\lambda) \in \mathbb{F}^{nm}[\lambda]$ is called non-degenerated if

$$\operatorname{Rank} A(\lambda) = \min\{n, m\},$$

i.e., its normal rank has the highest possible value. Let the non-degenerated matrix $A(\lambda)$ be alatent and $n \le m$. Then from (1.20), (1.22), we obtain

$$A(\lambda_i) = p(\lambda_i) \left[\, S_\rho(\lambda_i)\ 0_{n,m-n} \,\right] q(\lambda) = a(\lambda) b(\lambda), \tag{1.63}$$

where $p(\lambda),\ q(\lambda)$ are unimodular matrices and

$$a(\lambda) = p(\lambda_i)\operatorname{diag}\{a_1(\lambda), \ldots, a_n(\lambda)\}, \quad b(\lambda) = \left[\, I_n\ 0_{n,m-n} \,\right] q(\lambda). \tag{1.64}$$

Obviously, the $n \times n$ matrix $a(\lambda)$ is latent and non-singular, but the $n \times m$ matrix $b(\lambda)$ is alatent. An analog representation is possible for $n > m$ too.

(4)

**Theorem 1.12** *Let the* $n \times m$ *matrix* $A(\lambda)$ *be alatent and* $n \le m$. *Then any submatrix* $A_p(\lambda)$, *consisting of* $p < n$ *rows is alatent too. The same claim is valid for the columns of matrix* $A(\lambda)$ *when* $n > m$.

*Proof* We show the claim for rows. Without loss of generality, we assume, that the matrix $A_p(\lambda)$ is built by the first $p$ rows of the matrix $A(\lambda)$ and $p < n,\ n \le m$. Then the matrix $A(\lambda)$ can be written in the form

$$A(\lambda) = \begin{bmatrix} a_{11}(\lambda) & \ldots & a_{1m}(\lambda) \\ \ldots & \ldots & \ldots \\ a_{p1}(\lambda) & \ldots & a_{pm}(\lambda) \\ a_{p+1,1}(\lambda) & \ldots & a_{p+1,m}(\lambda) \\ \ldots & \ldots & \ldots \\ a_{n1}(\lambda) & \ldots & a_{nm}(\lambda) \end{bmatrix} = \begin{bmatrix} A_p(\lambda) \\ A_1(\lambda) \end{bmatrix}. \tag{1.65}$$

We will show that the matrix $A_p(\lambda)$ is alatent. Suppose the contrary, that the matrix $A_p(\lambda)$ is latent. Then using (1.63), we obtain

$$A_p(\lambda) = a_p(\lambda)b_p(\lambda), \tag{1.66}$$

where the matrix $a_p(\lambda)$ is latent. With the help of this relation, matrix (1.65) can be written in the form

$$A(\lambda) = \begin{bmatrix} a_p(\lambda) & 0_{p,n-p} \\ 0_{n-p,p} & I_{n-p} \end{bmatrix} \begin{bmatrix} b_p(\lambda) \\ A_1(\lambda) \end{bmatrix}. \tag{1.67}$$

Let $\tilde{\lambda}$ be an eigenvalue of the matrix $a_p(\lambda)$, then

$$A(\tilde{\lambda}) = \begin{bmatrix} a_p(\tilde{\lambda}) & 0_{p,n-p} \\ 0_{n-p,p} & I_{n-p} \end{bmatrix} \begin{bmatrix} b_p(\tilde{\lambda}) \\ A_1(\tilde{\lambda}) \end{bmatrix}. \tag{1.68}$$

Since the rank of the first factor is smaller than $n$, also rank $A(\tilde{\lambda}) < n$. At the same time for all $\lambda$ $rank A(\lambda) = n$ is true, because the matrix $A(\lambda)$ is alatent. This contradiction proves the claim of the theorem for rows. The proof for columns runs analogously. ■

**Corollary 1.1** *Arbitrary submatrices consisting of rows or columns of unimodular matrices are alatent.*

## 1.7 Row and Column Reduced Matrices

(1) Consider the non-singular matrix $A(\lambda) \in \mathbb{R}^{nn}[\lambda]$, where $b_1(\lambda), \dots, b_n(\lambda)$ are its rows. Denote

$$\beta_i \overset{\triangle}{=} \deg b_i(\lambda), \quad (i = 1, \dots, n). \tag{1.69}$$

Then the matrix $A(\lambda)$ can be represented in the form

$$A(\lambda) = \operatorname{diag}\{\lambda^{\beta_1}, \dots, \lambda^{\beta_n}\}A_l + A_1(\lambda). \tag{1.70}$$

Here $A_1(\lambda)$ is a polynomial matrix, where the degree of its $i$th row is lower than $\beta_i$, but $A_l$ is a constant $n \times n$ matrix. Moreover, the number

$$\beta_l \overset{\triangle}{=} \beta_1 + \cdots + \beta_n \tag{1.71}$$

is called left order of the matrix $A(\lambda)$. With the notation

$$\beta_{\max} \overset{\triangle}{=} \max_{1 \le i \le n} \{\beta_i\}, \quad 1 \le i \le n, \tag{1.72}$$

we find

$$\deg A(\lambda) = \beta_{\max}. \tag{1.73}$$

(2) The matrix $A(\lambda)$ is called row reduced, if in (1.70)

$$det A_l \neq 0. \tag{1.74}$$

If this condition is fulfilled, then by taking the determinant on both sides of (1.70), we obtain

$$\det A(\lambda) = \lambda^{\beta_l} \det A_l + d_1(\lambda), \tag{1.75}$$

where $d_1(\lambda)$ is a polynomial and $\deg d_1(\lambda) < \beta_l$. Comparing (1.73) and (1.75), we find that for row reduced matrices

$$\beta_{\max} = \deg A(\lambda) \leq \deg \det A(\lambda) = \beta_l. \tag{1.76}$$

(3) Let $c_1(\lambda), \ldots, c_n(\lambda)$ be the columns of the matrix $A(\lambda) \in \mathbb{R}^{nn}[\lambda]$ and

$$\delta_i = \deg c_i(\lambda), \ (i = q, \ldots, n). \tag{1.77}$$

Then the matrix $A(\lambda)$ can be represented in the form

$$A(\lambda) = A_r \operatorname{diag}\{\lambda^{\delta_1}, \ldots, \lambda^{\delta_n}\} + A_2(\lambda), \tag{1.78}$$

where $A_r$ is a constant $n \times n$ matrix without zero columns, and $A_2(\lambda)$ is a polynomial matrix, where the degree of its $i$th column is lower than $\delta_i$. Moreover, the number

$$\delta_r = \delta_1 + \cdots + \delta_n \tag{1.79}$$

is called the right order of the matrix $A(\lambda)$. In general, the left and the right order of a polynomial matrix are not equal.

(4) Denote

$$\delta_{\max} \stackrel{\triangle}{=} \max_{1 \leq i \leq n} \{\delta_i\}, \tag{1.80}$$

then in analogy to (1.73) we obtain

$$\deg A(\lambda) = \delta_{\max}. \tag{1.81}$$

Since the degree of a polynomial matrix does not depend on the form of its notation, from (1.73) and (1.81) we find

$$\beta_{\max} = \delta_{\max}. \tag{1.82}$$

(5) The matrix $A(\lambda)$ is called column reduced, if in representation (1.78)

$$\det A_r \neq 0. \tag{1.83}$$

For column reduced matrices, we obtain in analogy to (1.76)

$$\delta_{\max} = \deg A(\lambda) \leq \deg \det A(\lambda) = \delta_r. \tag{1.84}$$

If the matrix $A(\lambda)$ is row reduced, then obviously the matrix $A'(\lambda)$ is column reduced, where the prime stands for the transposition operator. Therefore, in the following, we will mainly consider row reduced matrices, whereas the analog properties for column reduced matrices can be derived by transposition.

(6) Below we formulate some statements concerning row reduced (column reduced) matrices.

Two polynomial matrices $A(\lambda)$ and $B(\lambda)$ are called left equivalent (right equivalent), when they are related by

$$A(\lambda) = p_l(\lambda)B(\lambda) \quad (A(\lambda) = B(\lambda)p_r(\lambda))$$

where $p_l(\lambda)$, $p_r(\lambda)$ are unimodular matrices of appropriate size. The fact of left (right) equivalence is denoted by $A(\lambda) \overset{l}{\sim} B(\lambda)$, $A(\lambda) \overset{r}{\sim} B(\lambda)$.

**Theorem 1.13** ([3]) *For any non-singular polynomial matrix, there exists a left equivalent (right equivalent) row reduced (column reduced) matrix.*

**Theorem 1.14** ([1]) *Let the matrices*

$$\begin{aligned} A(\lambda) &= \operatorname{diag}\{\lambda^{\alpha_1}, \ldots, \lambda^{\alpha_n}\}A_0 + A_1(\lambda) \\ B(\lambda) &= \operatorname{diag}\{\lambda^{\beta_1}, \ldots, \lambda^{\beta_n}\}B_0 + B_1(\lambda) \end{aligned} \tag{1.85}$$

*be row reduced and left equivalent. Then the sets of numbers* $\{\alpha_1, \ldots, \alpha_n\}$ *and* $\{\beta_1, \ldots, \beta_n\}$ *coincide. Moreover,*

$$\deg A(\lambda) = \deg B(\lambda) = \alpha_{\max}. \tag{1.86}$$

**Theorem 1.15** ([1]) *If matrices (1.85) are left equivalent and the matrix* $A(\lambda)$ *is row reduced, then*

$$\sum_{i=1}^{n} \alpha_i \leq \sum_{i=1}^{n} \beta_i \tag{1.87}$$

*and*

$$\deg A(\lambda) \leq \deg B(\lambda). \tag{1.88}$$

*Moreover, the equality sign stands in (1.87) and (1.88) if and only if the matrix* $B(\lambda)$ *is also row reduced.*

## 1.8 Pairs of Polynomial Matrices

(1) Consider the matrices $a(\lambda) \in \mathbb{F}^{nn}[\lambda]$ and $b(\lambda) \in \mathbb{F}^{nm}[\lambda]$. The totality of these matrices is called a horizontal pair, and it is denoted by $(a(\lambda), b(\lambda))$. When we have $a(\lambda) \in \mathbb{F}^{nn}[\lambda]$ and $c(\lambda) \in \mathbb{F}^{mn}[\lambda]$, we speak about a vertical pair, which is denoted by $[a(\lambda), c(\lambda)]$. The pairs $(a(\lambda), b(\lambda))$ and $[a(\lambda), c(\lambda)]$ can be written as compound matrices

$$R_h(\lambda) = \begin{bmatrix} a(\lambda) & b(\lambda) \end{bmatrix}, \quad R_v(\lambda) = \begin{bmatrix} a(\lambda) \\ c(\lambda) \end{bmatrix}, \tag{1.89}$$

where the first is horizontal, and the second is vertical. Since a vertical pair can be generated from a horizontal one by transposition, the properties of vertical pairs can be held from corresponding properties of horizontal pairs by transposition. Therefore, below we mainly investigate horizontal pairs.

The pairs $(a(\lambda), b(\lambda))$, $[a(\lambda), c(\lambda)]$ are called non-degenerated, if corresponding compound matrices (1.89) are non-degenerated. In the following, when not accentuated, we will always consider non-degenerated pairs. The non-degenerated pairs $(a(\lambda), b(\lambda))$, $[a(\lambda), c(\lambda)]$ are called irreducible, if the corresponding compound matrices are alatent.

(2) Let the pair $(a(\lambda), b(\lambda))$ fulfill the relations

$$a(\lambda) = g(\lambda)a_1(\lambda), \quad b(\lambda) = g(\lambda)b_1(\lambda), \tag{1.90}$$

where the matrix $g(\lambda) \in \mathbb{F}^{nn}[\lambda]$ is non-singular. Then the matrix $g(\lambda)$ is called a left divisor of the pair $(a(\lambda), b(\lambda))$. Hereby, if the polynomial matrix

$$\begin{bmatrix} a_1(\lambda) & b_1(\lambda) \end{bmatrix} \stackrel{\triangle}{=} g^{-1}(\lambda) \begin{bmatrix} a(\lambda) & b(\lambda) \end{bmatrix} \tag{1.91}$$

is alatent, then the matrix $g(\lambda)$ is called a greatest left divisor (GLD) of the pair $(a(\lambda), b(\lambda))$.

As is known [1], all GLD of the pair $(a(\lambda), b(\lambda))$ are right equivalent. In analogy the concept of right and greatest right divisor is defined for a vertical pair $[a(\lambda), c(\lambda)]$. By transposition, we find that two GRD of the pair $[a(\lambda), c(\lambda)]$ are left equivalent. Let the pair $(a(\lambda), b(\lambda))$ be non-degenerated. Then, applying the Smith canonical form, we obtain

$$R_h(\lambda) = p(\lambda) \begin{bmatrix} \operatorname{diag}\{\alpha_1(\lambda), \ldots, \alpha_n(\lambda)\} & 0_{nm} \end{bmatrix} q(\lambda) = \begin{bmatrix} N(\lambda) & 0_{nm} \end{bmatrix} q(\lambda), \tag{1.92}$$

where $N(\lambda) \stackrel{\triangle}{=} p(\lambda) \operatorname{diag}\{\alpha_1(\lambda), \ldots, \alpha_n(\lambda)\}$, and $q(\lambda)$ is a unimodular matrix of the size $(n+m) \times (n+m)$. Moreover, the $n \times n$ matrix $N(\lambda)$ is the GLD of the pair $(a(\lambda), b(\lambda))$. Indeed, let

$$q(\lambda) = \begin{array}{c} \\ n \\ m \end{array} \begin{array}{c} \begin{array}{cc} n & m \end{array} \\ \begin{bmatrix} q_{11}(\lambda) & q_{12}(\lambda) \\ q_{21}(\lambda) & q_{22}(\lambda) \end{bmatrix} \end{array} \tag{1.93}$$

where the letters at the border of the matrix indicate the dimensions of the blocks. Thus

$$\begin{bmatrix} a(\lambda) \ b(\lambda) \end{bmatrix} = \begin{bmatrix} N(\lambda) \ 0_{nm} \end{bmatrix} \begin{bmatrix} q_{11}(\lambda) \ q_{12}(\lambda) \\ q_{21}(\lambda) \ q_{22}(\lambda) \end{bmatrix}, \tag{1.94}$$

i.e.,

$$a(\lambda) = N(\lambda)q_{11}(\lambda), \quad b(\lambda) = N(\lambda)q_{12}(\lambda). \tag{1.95}$$

Hence

$$\begin{bmatrix} a(\lambda) \ b(\lambda) \end{bmatrix} = N(\lambda)r(\lambda), \tag{1.96}$$

where

$$r(\lambda) \triangleq \begin{bmatrix} q_{11}(\lambda) \ q_{12}(\lambda) \end{bmatrix}. \tag{1.97}$$

From Corollary 1.1, it follows that the matrix $r(\lambda)$ is alatent, i.e., the pair$(q_{11}(\lambda), q_{12}(\lambda))$ is non-degenerated, and $N(\lambda)$ is the GLD of the matrices $a(\lambda)$, $b(\lambda)$.

(3) Below, we will formulate some conditions for the non-degeneration of the pair $(a(\lambda), b(\lambda))$.

**Theorem 1.16** *For the non-degeneration of the pair* $(a(\lambda), b(\lambda))$*, each of the following conditions (a)–(d) is necessary and sufficient:*

*(a) There exist unimodular matrices* $p(\lambda)$, $q(\lambda)$*, such that*

$$\begin{bmatrix} a(\lambda) \ b(\lambda) \end{bmatrix} = p(\lambda) \begin{bmatrix} I_n \ 0_{nm} \end{bmatrix} q(\lambda). \tag{1.98}$$

*(b) There exist polynomial matrices* $X(\lambda)$ *of size* $n \times n$ *and* $Y(\lambda)$ *of size* $m \times n$*, such that*

$$a(\lambda)X(\lambda) + b(\lambda)Y(\lambda) = I_n, \tag{1.99}$$

*where* $I_n$ *as above is the* $n \times n$ *identity matrix.*

*(c) The condition*

$$\operatorname{rank} \begin{bmatrix} a(\lambda_i) \ b(\lambda_i) \end{bmatrix} = n \tag{1.100}$$

*holds for all eigenvalues* $\lambda_i$, $(i = 1, \cdots, \rho)$ *of the matrix* $a(\lambda)$.

*(d) There exists a horizontal pair* $(\alpha(\lambda), \beta(\lambda))$ *such that the matrix*

$$\begin{array}{c} \\ n \\ m \end{array} \begin{array}{c} \begin{array}{cc} n & m \end{array} \\ \begin{bmatrix} a(\lambda) & -b(\lambda) \\ -\beta(\lambda) & \alpha(\lambda) \end{bmatrix} \end{array} \tag{1.101}$$

*becomes unimodular.*

(4) A necessary condition for the non-degeneration of a pair yields

**Theorem 1.17** *For the non-degeneration of the pair* $(a(\lambda), b(\lambda))$, *where* $a(\lambda) \in \mathbb{F}^{nn}[\lambda]$, $b(\lambda) \in \mathbb{F}^{nm}[\lambda]$ *and* $m < n$, *it is necessary that the matrix* $a(\lambda)$ *does not possess more than m invariant polynomials different from one.*

*Proof* Assume that the matrix $a(\lambda)$ possesses $k < n$ invariant polynomials $a_1(\lambda), \dots, a_k(\lambda)$ different from one. Then there exists a latent value $\lambda_0$, which is a root of all polynomials $a_1(\lambda), \dots, a_k(\lambda)$. Applying (1.40), we obtain

$$\operatorname{rank} R_h(\lambda_0) = \operatorname{rank} \left[ a(\lambda_0)\; b(\lambda_0) \right] \le \operatorname{rank} a(\lambda_0) + \operatorname{rank} b(\lambda_0). \tag{1.102}$$

Due to Theorem 1.10

$$\operatorname{rank} a(\lambda_0) = \operatorname{Rank} a(\lambda) - k. \tag{1.103}$$

Moreover,

$$\operatorname{rank} b(\lambda_0) \le m. \tag{1.104}$$

If $k > m$, then $\operatorname{rank} R_h(\lambda_0) \le \operatorname{Rank} a(\lambda) - k + m < \operatorname{Rank} a(\lambda)$, i.e., the matrix $R_h(\lambda)$ is not alatent, and so the pair $(a(\lambda), b(\lambda))$ is degenerated.

*Remark 1.3* By transposition, all the abovementioned can be claimed for vertical pairs.

## 1.9 Polynomial Matrices of First Order

(1) For $q = 1$ and $n = m$, the polynomial matrix (1.10) takes the form

$$A(\lambda) = A\lambda + B \tag{1.105}$$

with constant matrices $A$ and $B$. Such set of matrices is called a pencil [2]. The pencil $A(\lambda)$ is non-singular, when

$$\det(A\lambda + B) \not\equiv 0.$$

In accordance with the above definitions, non-singular matrix $A(\lambda)$ (1.105) is regular if $\det A \ne 0$ and anomal if $\det A = 0$.

(2) Starting from the general definitions, two pencils of the same dimension

$$A(\lambda) = A\lambda + B, \quad A_1(\lambda) = A_1\lambda + B_1 \tag{1.106}$$

are called left (right) equivalent if there exist unimodular matrices $p(\lambda)$, $(q(\lambda))$ such that

$$A(\lambda) = p(\lambda)A_1(\lambda), \quad (A(\lambda) = A_1(\lambda)q(\lambda)). \tag{1.107}$$

In this case, we write, as before

$$A(\lambda) \overset{l}{\sim} A_1(\lambda), \quad (A(\lambda) \overset{r}{\sim} A_1(\lambda)). \tag{1.108}$$

If, however,

$$A(\lambda) = p(\lambda)A_1(\lambda)q(\lambda), \tag{1.109}$$

where $p(\lambda)$ and $q(\lambda)$ are unimodular matrices, then the pencils $A(\lambda)$ and $A_1(\lambda)$ are equivalent, and we write

$$A(\lambda) \sim A_1(\lambda). \tag{1.110}$$

If relation (1.109) is fulfilled with $p(\lambda) = p,\ q(\lambda) = q$, where $p,\ q$ are constant matrices, then the pencils $A(\lambda)$, $A_1(\lambda)$ are called strictly equivalent. Conditions for strict equivalence of two regular pencils are suitably formulated using the concept of $n \times n$ (upper) Jordan blocks

$$J_n(a) = \begin{bmatrix} a & 1 & 0 & \cdots & 0 & 0 \\ 0 & a & 1 & \cdots & 0 & 0 \\ 0 & 0 & a & \cdots & 0 & 0 \\ \vdots & \vdots & \vdots & \ddots & \vdots & \vdots \\ 0 & 0 & 0 & \cdots & a & 1 \\ 0 & 0 & 0 & \cdots & 0 & a \end{bmatrix}, \tag{1.111}$$

where $a$ is a constant, in general complex.

A classical result is given by the next theorem.

**Theorem 1.18** ([1, 2]) *Let the regular pencil $A(\lambda) = A\lambda + B$ have the elementary divisors in the complex plane*

$$(\lambda - \lambda_1)^{\eta_1}, \ldots, (\lambda - \lambda_q)^{\eta_q}. \tag{1.112}$$

*Then the pencil $A(\lambda)$ is strictly equivalent to the block diagonal matrix*

$$A_J(\lambda) = \operatorname{diag}\{\lambda I_{\eta_1} - J_{\eta_1}(\lambda_1), \ldots, \lambda I_{\eta_q} - J_{\eta_q}(\lambda_q)\}. \tag{1.113}$$

(3) Let $A$ be a constant $n \times n$ matrix. As shown above, matrix $A$ can be assigned the regular pencil (characteristic matrix)

$$A_\lambda = \lambda I_n - A. \tag{1.114}$$

For the characteristic matrix $A_\lambda$, we can apply all definitions and statements, which are formulated above for polynomial matrices of general form. In particular, the characteristic polynomial of the matrix $A_\lambda$

$$\det A_\lambda = \det(\lambda I_n - A) = d_A(\lambda), \tag{1.115}$$

as above, we will call the characteristic polynomial of the matrix $A$. An analog terminology is used for the eigenvalues, minimal polynomial, invariant polynomials, elementary divisors, etc. Obviously,

$$\deg det A_\lambda = n. \tag{1.116}$$

For regular pencils of form (1.114), the concepts of equivalence and strict equivalence coincide. Moreover,

**Theorem 1.19** *The two characteristic matrices $A_\lambda = \lambda I_n - A$ and $A_{\lambda_1} = \lambda I_n - A_1$ are (strictly) equivalent, if and only if $A$ and $A_1$ are similar, i.e.,*

$$A_1 = LAL^{-1}, \tag{1.117}$$

*where $L$ is a non-singular matrix, i.e.,* $\det L \neq 0$.

**Corollary 1.2** *Let the $n \times n$ matrix $\lambda I_n - A$ have the sequence of elementary divisors*

$$(\lambda - \lambda_1)^{\nu_1}, \ldots, (\lambda - \lambda_q)^{\nu_q}, \; \nu_1 + \cdots + \nu_q = n. \tag{1.118}$$

*Then the matrix $A$ is similar to the block diagonal matrix $J_A$ of the form*

$$J_A = \text{diag}\{J_{\nu_1}(\lambda_1), \ldots, J_{\nu_q}(\lambda_q)\}. \tag{1.119}$$

*The matrix $J_A$ is called the Jordan canonical form of the associated matrix $A$. For any matrix $A$, its Jordan canonical form is uniquely determined but up to the sequence of the blocks.*

(4) Consider the constant matrices $A \in \mathbb{R}^{nn}$, $B \in \mathbb{R}^{nm}$, building a horizontal pair $(A, B)$. The pair $(A, B)$ is called controllable, if the polynomial pair $(\lambda I_n - A, B)$ is irreducible. This condition means that the pair $(A, B)$ is controllable if and only if the extended horizontal polynomial matrix

$$R_c(\lambda) = \begin{bmatrix} \lambda I_n - A & B \end{bmatrix} \tag{1.120}$$

is alatent. As is known [1], the pair $(A, B)$ is controllable if and only if

$$\text{rank } Q_c(A, B) = n, \tag{1.121}$$

where $Q_c(A, B)$ is a constant matrix of the form

$$Q_c(A, B) \triangleq \begin{bmatrix} B & AB & \cdots & A^{n-1}B \end{bmatrix}. \tag{1.122}$$

The matrix $Q_c(A, B)$ is called controllability matrix. Below, we formulate some important properties of controllable pairs.

(a) If the pair $(A, B)$ is controllable, and $L$ is a non-singular matrix, then the pair $(A_1, B_1)$, where $A_1 = LAL^{-1},\ B_1 = LB$, is also controllable.

(b) **Theorem 1.20** ([4]) *Consider the matrices $A \in R^{nn}$, $B \in R^{nm}$ and the matrix $L$, which is commutative with $A$, i.e., $AL = LA$. Then the following statements hold:*

*(i) If the pair $(A, B)$ is not controllable, then the pair $(A, LB)$ is also not controllable.*

*(ii) If the matrix $L$ is singular, i.e., $\det L = 0$, then the pair $(A, LB)$ is not controllable.*

*(iii) If the pair $(A, B)$ is controllable and the matrix $L$ is not singular, then the pair $(A, LB)$ is also controllable.*

(5) The vertical pair $[A, C]$, where $A \in \mathbb{R}^{mm}$, $C \in \mathbb{R}^{nm}$ is called observable, if the horizontal pair $(\lambda I_m - A', C')$ is irreducible. Obviously, the pair $[A, C]$ is observable, if and only if the horizontal pair $(A', C')$ is controllable. Therefore, about vertical pairs we can formulate adequate properties.

## 1.10 Reciprocal Polynomials and Reciprocal Matrices

(1) Consider the scalar polynomial $\tilde{f}(z) \in \mathbb{R}[z]$, $\deg \tilde{f}(z) = p$. Then, according to [5, 6], the polynomial

$$f(\zeta) = \zeta^p \tilde{f}(\zeta^{-1}) \tag{1.123}$$

is called reciprocal polynomial with respect to $\tilde{f}(z)$. Further on, when (1.123) is true, we will write

$$f(\zeta) = \operatorname{rec}\left[\tilde{f}(z)\right]. \tag{1.124}$$

In relation to that, the polynomial $\tilde{f}(z)$ is called the original one. Obviously, to a given original polynomial $\tilde{f}(z)$, the reciprocal polynomial $f(\zeta)$ is uniquely determined. The reverse claim does not hold. If $\deg \tilde{f}(z) = p$, then any polynomial

$$\tilde{f}_1(z) = z^{l+p} \tilde{f}(z^{-1}), \tag{1.125}$$

where $l$ is any nonnegative integer, corresponds to the same reciprocal polynomial $f(\zeta)$.

(2) Let the polynomial $\tilde{f}(z)$ be monic, and we have the product

$$\tilde{f}(z) = z^{\nu_0}(z - z_1)^{\nu_1} \ldots (z - z_\rho)^{\nu_\rho}, \tag{1.126}$$

where $\nu_i > 0,\ (i = 0, \dots, \rho)$ and all numbers $z_i$. $(i = 1, \dots, \rho)$ are distinct. Then from (1.123), we obtain

$$\operatorname{rec}\left[\tilde{f}(z)\right] = f(\zeta) = (1 - z_1\zeta)^{\nu_1} \dots (1 - z_\rho\zeta)^{\nu_\rho}, \tag{1.127}$$

from which we see that the roots of the polynomial $f(\zeta)$ are the numbers $z_1^{-1}, \dots, z_\rho^{-1}$ with the multiplicities $\nu_1, \dots, \nu_\rho$, respectively. Moreover, for any original polynomial $\tilde{f}(z)$, we find

$$f(0) \neq 0. \tag{1.128}$$

In general, reciprocal polynomial (1.127) is not monic. However, after multiplying (1.127) by a certain constant factor, we generate the equivalent monic polynomial

$$f_m(\zeta) = (\zeta - z_1^{-1})^{\nu_1} \dots (\zeta - z_\rho^{-1})^{\nu_\rho},$$

which will be called monic reciprocal polynomial. For this operation, we write

$$f_m(\zeta) = \operatorname{Rec}\left[\tilde{f}(z)\right]. \tag{1.129}$$

(3) Consider the non-singular matrix $\tilde{a}(z) \in \mathbb{R}^{nn}[z]$, and the sequence of its invariant polynomials $\tilde{a}_1(z), \dots, \tilde{a}_n(z)$ should have the form

$$\tilde{a}_i(z) = z^{\nu_i}\tilde{b}_i(z),\ (i = 1, \dots, n), \tag{1.130}$$

where $\nu_i$ are nonnegative numbers, and $\nu_{i+1} \geq \nu_i$. Moreover, $\tilde{b}_i(z)$ are monic polynomials, such that $\tilde{b}_{i+1}(z)$ is divisible by $\tilde{b}_i(z)$, and

$$\tilde{b}_i(0) \neq 0,\ (i = 1, \dots, n). \tag{1.131}$$

Further on, the polynomials $\tilde{b}_i(z)$ are called basic invariant polynomials of the matrix $\tilde{a}(z)$.. Consider the sequence of monic reciprocal polynomials

$$a_i(\zeta) = \operatorname{Rec}\left[\tilde{b}_i(z)\right],\ (i = 1, \dots, n). \tag{1.132}$$

By construction, the polynomials $a_{i+1}(\zeta)$ are divisible by $a_i(\zeta)$. Thus, because the polynomials $a_i(\zeta)$ are monic, sequence (1.132) can be interpreted as the sequence of invariant polynomials of a certain polynomial $n \times n$ matrix $a(\zeta)$.

**Definition 1.3** Any $n \times n$ matrix $a(\zeta)$, for which the sequence of invariant polynomials possesses form (1.132), is called a reciprocal matrix to the original matrix $\tilde{a}(z)$, and we denote

$$a(\zeta) = \operatorname{rec}\left[\tilde{a}(z)\right]. \tag{1.133}$$

The configured matrix $\tilde{a}(z)$ in (1.133) is called original.

*Remark 1.4* According to Definition 1.3, in contrast to the scalar case, the reciprocal matrix $a(\zeta)$ is not uniquely determined by the original matrix $\tilde{a}(z)$. Indeed, let the matrix $a_0(\zeta)$ satisfy relation (1.133). Then any equivalent matrix

$$a(\zeta) = p(\zeta)a_0(\zeta)q(\zeta), \tag{1.134}$$

where $p(\zeta),\ q(\zeta)$ are unimodular matrices, also satisfies (1.133).

If $a(\zeta)$ is a reciprocal matrix, then it follows by construction from (1.134) that

$$\det a(\zeta) \sim a_1(\zeta)\cdots a_n(\zeta). \tag{1.135}$$

With the help of the general property (1.128) of reciprocal matrices we conclude $a_i(0) \neq 0$. Therefore, an arbitrary reciprocal matrix $a(\zeta)$ satisfies

$$\det a(0) \neq 0. \tag{1.136}$$

(4) For a given matrix $a(\zeta)$ we can associate the totality of all original matrices. Let $a_1(\zeta), \ldots, a_n(\zeta)$ be the sequence of invariant polynomials of the reciprocal matrix $a(\zeta)$, and $\deg a_i(\zeta) = p_i$. By definition $p_{i+1} \geq p_i$. Consider the matrix

$$\tilde{a}(z) = \tilde{p}(z)\tilde{S_A}(z)\tilde{q}(z), \tag{1.137}$$

where $\tilde{p}(z),\ \tilde{q}(z)$ are unimodular matrices and

$$\tilde{S_A}(z) = \operatorname{diag}\{k_1 z^{\rho_1+p_1} a_1(z^{-1}), \ldots, k_n z^{\rho_n+p_n} a_n(z^{-1})\}, \tag{1.138}$$

where $\rho_i \leq \rho_{i+1}$ are nonnegative integers, and the constants $k_1, \ldots, k_n$ are chosen such that matrices (1.133) become monic polynomials.

Obviously, $\tilde{a}(z)$ is original for $a(\zeta)$. It is easy to show, that the set of matrices (1.137) for all possible unimodular matrices $\tilde{p}(z),\ \tilde{q}(z)$ determines the complete set of matrices, which are original in relation to $a(\zeta)$.

(5) The next theorem yields a method for building the reciprocal matrix.

**Theorem 1.21** *For any matrix $A \in \mathbb{R}^{pp}$*

$$I_p - \zeta A = \operatorname{rec}\left[zI_p - A\right]. \tag{1.139}$$

*Proof* (a) At first consider the case, when the matrix $A$ is a Jordan block of size $n \times n$, i.e.,

$$A = J_n(a) = \begin{bmatrix} a & 1 & 0 & \cdots & 0 & 0 \\ 0 & a & 1 & \cdots & 0 & 0 \\ \vdots & \vdots & \vdots & \ddots & \vdots & \vdots \\ 0 & 0 & 0 & \cdots & a & 1 \\ 0 & 0 & 0 & \cdots & 0 & a \end{bmatrix}. \tag{1.140}$$

Hence

$$zI_n - A = \begin{bmatrix} z-a & -1 & 0 & \cdots & 0 & 0 \\ 0 & z-a & -1 & \cdots & 0 & 0 \\ \vdots & \vdots & \vdots & \ddots & \vdots & \vdots \\ 0 & 0 & 0 & \cdots & z-a & -1 \\ 0 & 0 & 0 & \cdots & 0 & z-a \end{bmatrix}. \tag{1.141}$$

Obviously

$$\det[zI_n - J_n(a)] = (z-a)^n.$$

Since

$$\operatorname{rank}[aI_n - J_n(a)] = n-1,$$

applying Theorem 1.10, the matrix $zI_n - A$ possesses only one elementary divisor different from one, namely, $(z-a)^n$, and the sequence of its invariant polynomials has the form

$$\tilde{a}_1(z) = 1, \ \ldots, \ \tilde{a}_{n-1}(z) = 1, \ \tilde{a}_n(z) = (z-a)^n. \tag{1.142}$$

In the present case

$$I_n - \zeta A = I_n - \zeta J_n(a) = \begin{bmatrix} 1-\zeta a & -\zeta & 0 & \cdots & 0 & 0 \\ 0 & 1-\zeta a & -\zeta & \cdots & 0 & 0 \\ \vdots & \vdots & \vdots & \ddots & \vdots & \vdots \\ 0 & 0 & 0 & \cdots & 1-\zeta a & -\zeta \\ 0 & 0 & 0 & \cdots & 0 & 1-\zeta a \end{bmatrix}. \tag{1.143}$$

Thus

$$\det[I_n - \zeta J_n(a)] = (1-\zeta a)^n. \tag{1.144}$$

Assume $a \neq 0$, then matrix (1.143) has the single eigenvalue $\zeta = a^{-1}$. Moreover, due to

$$\operatorname{rank}\left[I_n - a^{-1} J_n(a^{-1})\right] = n-1, \tag{1.145}$$

matrix (1.143) has the single elementary divisor $(\zeta - a^{-1})^n$, and therefore, its sequence of invariant polynomials takes the form

$$\tilde{a}_1(\zeta) = 1, \ \ldots, \ \tilde{a}_{n-1}(\zeta) = 1, \ \tilde{a}_n(\zeta) = (\zeta - a^{-1})^n. \tag{1.146}$$

If, however, $a = 0$, then

$$\det[I_n - \zeta J_n(0)] = 1, \tag{1.147}$$

and matrix (1.143) becomes unimodular. Therefore, in the case $a = 0$, the sequence of invariant polynomials of the matrix $I_n - \zeta J_n(a)$ takes the form

$$\tilde{a}_i(\zeta) = 1, \ (i = 1, \dots, n). \tag{1.148}$$

Comparing (1.142) with (1.146) yields

$$I_n - \zeta J_n(a) = \operatorname{rec}\left[zI_n - J_n(a)\right]. \tag{1.149}$$

(b) Consider now the general case. Assume the sequence of invariant polynomials of the matrix $A \in \mathbb{R}^{pp}$ as

$$\begin{aligned} a_1(z) &= z^{\mu_{10}}(z - z_1)^{\mu_{11}} \cdots (z - z_\rho)^{\mu_{1\rho}}, \\ &\vdots \;\vdots\; \vdots \\ a_p(z) &= z^{\mu_{p0}}(z - z_1)^{\mu_{p1}} \cdots (z - z_\rho)^{\mu_{p\rho}}, \end{aligned} \tag{1.150}$$

where

$$0 \leq \mu_{ik} \leq \mu_{i+1,k}, \ (k = 0, 1, \dots, \rho). \tag{1.151}$$

Therefore, $A_z \stackrel{\triangle}{=} zI_p - A$ allows the representation

$$A_z = V \operatorname{diag}\left\{zI_{\mu_{10}} - J_{\mu_{10}}(0), \dots, zI_{\mu_{p\rho}} - J_{\mu_{p\rho}}(z_\rho)\right\} V^{-1}, \tag{1.152}$$

where $V$ is a non-singular matrix. In harmony with the above shown, each block on the right side of (1.152) corresponds for the eigenvalue $z = 0$ to an elementary divisor one, and each block of the form $I_{\mu_{ik}} - \zeta J_{\mu_{ik}}(z_i)$ corresponds for $\mu_{ik} \neq 0$ to an elementary divisor $(\zeta - z_i^{-1})^{\mu_{ik}}$. Since the totality of elementary divisors of a block diagonal matrix is the unification of the sets of elementary divisors of the individual blocks, the totality of the individual elementary divisors different from one of the matrix $I_p - \zeta A$ takes the form

$$\begin{aligned} &(\zeta - z_1^{-1})^{\mu_{11}} \cdots (\zeta - z_1^{-1})^{\mu_{1\rho}}, \\ &(\zeta - z_2^{-1})^{\mu_{21}} \cdots (\zeta - z_2^{-1})^{\mu_{2\rho}}, \\ &\cdots \cdots \cdots \cdots \cdots \cdots \cdots \cdots \\ &(\zeta - z_\rho^{-1})^{\mu_{p1}} \cdots (\zeta - z_\rho^{-1})^{\mu_{p\rho}}. \end{aligned} \tag{1.153}$$

Moreover, since

$$\operatorname{Rank}(I_p - \zeta A) = p,$$

the sequence of invariant polynomials of the matrix $I_p - \zeta A$ takes the form

$$\begin{aligned} a_1(\zeta) &= (\zeta - z_1^{-1})^{\mu_{11}} \cdots (\zeta - z_\rho^{-1})^{\mu_{1\rho}}, \\ &\cdots \cdots \cdots \cdots \cdots \cdots \cdots \cdots \cdots \\ a_p(\zeta) &= (\zeta - z_1^{-1})^{\mu_{p1}} \dots (\zeta - z_\rho^{-1})^{\mu_{p\rho}}. \end{aligned} \tag{1.154}$$

Opposing (1.150) and (1.154), we find

$$a_i(\zeta) = \text{Rec}\left[\tilde{a}_i(z)\right]$$

that completes the proof. ■

## References


1. T. Kailath, *Linear Systems* (Prentice Hall, Englewood Cliffs, NJ, 1980)
2. F.R. Gantmacher, *The Theory of Matrices* (Chelsea, New York, 1959)
3. V.M. Popov, *Hyperstability of Control Systems* (Springer, Berlin, 1973)
4. E.N. Rosenwasser, B.P. Lampe, *Multivariable Computer-controlled Systems—A Transfer Function Approach* (Springer, London, 2006)
5. K.J. Åström, B. Wittenmark, *Computer Controlled Systems: Theory and Design*, 3rd edn. (Prentice-Hall, Englewood Cliffs, 1997)
6. E.N. Rosenwasser, B.P. Lampe, *Computer Controlled Systems: Analysis and Design with Process-orientated Models* (Springer, Berlin, 2000)

# Chapter 2
# Fractional-Rational Matrices

**Abstract** This chapter has an introductory character and provides necessary information from the theory of rational matrices in a form adapted to the main content of the book. Some questions are presented unconventionally in the chapter. This applies in particular to the concept of independence from rational matrices and to the question of calculating the McMillan index for block rational matrices.

## 2.1 Scalar Rational Fractions

(1) Relations of the form

$$l(\lambda) = \frac{m(\lambda)}{d(\lambda)} \tag{2.1}$$

in which $d(\lambda)$, $m(\lambda)$ are polynomials are called fractional-rational functions or, in short, rational functions. According to the particular set from which the polynomials $d(\lambda)$ and $m(\lambda)$ are selected, the set of rational fractions is denoted by $\mathbb{C}(\lambda)$, $\mathbb{R}(\lambda)$, or $\mathbb{F}(\lambda)$. As is well known, $\mathbb{C}(\lambda)$ and $\mathbb{R}(\lambda)$ are algebraic fields, in which the multiplication with a number, the addition and subtraction, as well as multiplication and division are all defined. The zero element of this field is the fraction $\frac{0}{1}$, and the inverse element $l^{-1}(\lambda)$ is defined by

$$l^{-1}(\lambda) = \frac{d(\lambda)}{m(\lambda)}. \tag{2.2}$$

(2) Two rational fractions

$$l_1(\lambda) = \frac{m_1(\lambda)}{d_1(\lambda)}, \quad l_2(\lambda) = \frac{m_2(\lambda)}{d_2(\lambda)} \tag{2.3}$$

are called equal, and we write

$$l_1(\lambda) = l_2(\lambda),$$

if they satisfy the equation



E. N. Rosenwasser et al., *Computer-Controlled Systems with Delay*,
https://doi.org/10.1007/978-3-030-15042-6_2

$$m_1(\lambda)d_2(\lambda) - m_2(\lambda)d_1(\lambda) = 0. \tag{2.4}$$

If, in particular

$$m_2(\lambda) = a(\lambda)m_1(\lambda),\ d_2(\lambda) = a(\lambda)d_1(\lambda), \tag{2.5}$$

then condition (2.4) is true, and the expressions

$$l_1(\lambda) = \frac{m_1(\lambda)}{d_1(\lambda)},\quad l_2(\lambda) = \frac{m_2(\lambda)}{d_2(\lambda)} = \frac{a(\lambda)m_1(\lambda)}{a(\lambda)d_1(\lambda)}, \tag{2.6}$$

define one and the same rational function. This fact means that reducing the numerator and denominator of a rational fraction by one and the same polynomial does not change this fraction.

Since any scalar polynomial $f(\lambda)$ can be written in the form

$$f(\lambda) = \frac{f(\lambda)}{1}, \tag{2.7}$$

the sets of polynomials $\mathbb{C}[\lambda]$ and $\mathbb{R}[\lambda]$ are subsets of the fields $\mathbb{C}(\lambda)$ and $\mathbb{R}(\lambda)$, respectively.

(3) Consider fraction (2.1), when

$$m(\lambda) = g(\lambda)m_1(\lambda),\ d(\lambda) = g(\lambda)d_1(\lambda), \tag{2.8}$$

where $g(\lambda)$ is the GCD, such that the polynomials $m_1(\lambda)$ and $d_1(\lambda)$ are coprime. Then

$$l(\lambda) = \frac{m_1(\lambda)}{d_1(\lambda)}. \tag{2.9}$$

Representation (2.9) is called irreducible form of the rational fraction. Let

$$d_1(\lambda) = d_0\lambda^n + d_1\lambda^{n-1} + \cdots + d_n,\ d_0 \neq 0. \tag{2.10}$$

Then the numerator and denominator of fraction (2.8) can be divided by $d_0$. As result, we obtain an irreducible representation of form (2.1)

$$l(\lambda) = \frac{m(\lambda)}{d(\lambda)}, \tag{2.11}$$

where the polynomial $d(\lambda)$ is monic. That representation of a rational fraction is called its standard form. The standard form of a rational fraction is uniquely determined.

(4) The sum of rational fractions (2.3) is defined by the formula

$$l_1(\lambda) + l_2(\lambda) = \frac{m_1(\lambda)}{d_1(\lambda)} + \frac{m_2(\lambda)}{d_2(\lambda)} = \frac{m_1(\lambda)d_2(\lambda) + m_2(\lambda)d_1(\lambda)}{d_1(\lambda)d_2(\lambda)}. \tag{2.12}$$

If the fractions $l_1(\lambda)$ and $l_2(\lambda)$ are irreducible and the polynomials $d_1(\lambda)$ and $d_2(\lambda)$ are coprime, then the fraction on the right side of (2.12) is irreducible.

The product of rational fractions (2.3) is defined by the relation

$$l_1(\lambda)l_2(\lambda) = \frac{m_1(\lambda)m_2(\lambda)}{d_1(\lambda)d_2(\lambda)}. \tag{2.13}$$

(5) The integer $\operatorname{exc} l(\lambda)$, for which the limit

$$\lim_{\lambda\to\infty} l(\lambda)\lambda^{\operatorname{exc} l(\lambda)} = l_0 \neq 0 \tag{2.14}$$

exists, is called the (pole) excess of rational fraction (2.1). It is easy to see that

$$\operatorname{exc} l(\lambda) = \deg d(\lambda) - \deg m(\lambda). \tag{2.15}$$

In the case, when $\operatorname{exc} l(\lambda) = 0$, or $\operatorname{exc} l(\lambda) > 0$, fraction (2.1) is called proper or strictly proper. For $\operatorname{exc} l(\lambda) \geq 0$ fraction (2.1) is called at least proper. It is easy to recognize, that fraction $l(\lambda)$ (2.1) is proper, when $\deg m(\lambda) = \deg d(\lambda)$, and strictly proper, when $\deg m(\lambda) < \deg d(\lambda)$. For $\deg m(\lambda) > \deg d(\lambda)$, fraction (2.1) is called improper. The zero fraction is defined as strictly proper.

(6) Any fraction (2.1) can be represented in the form

$$l(\lambda) = \frac{r(\lambda)}{d(\lambda)} + q(\lambda), \tag{2.16}$$

where $r(\lambda)$ and $q(\lambda)$ are polynomials, but

$$\deg r(\lambda) < \deg d(\lambda),$$

i.e., the fraction on the right side is strictly proper. From (2.1) and (2.16) we find

$$m(\lambda) = d(\lambda)q(\lambda) + r(\lambda), \tag{2.17}$$

from which we recognize, that in relation (2.16), $q(\lambda)$ is the entire part, and $r(\lambda)$ is the remainder, when the numerator of fraction (2.1) is divided by the denominator. Further on, the rational fraction on the right side of (2.16) is called broken part of the function $l(\lambda)$, and the polynomial $q(\lambda)$ is called its integral part.

(7) If an irreducible strictly proper rational fraction has the form

$$l(\lambda) = \frac{m(\lambda)}{d_1(\lambda)d_2(\lambda)}, \tag{2.18}$$

where $d_1(\lambda)$ and $d_2(\lambda)$ are coprime polynomials, then there exists the unique representation

$$l(\lambda) = \frac{m_1(\lambda)}{d_1(\lambda)} + \frac{m_2(\lambda)}{d_2(\lambda)}, \tag{2.19}$$

where $m_1(\lambda)$ and $m_2(\lambda)$ are polynomials and the fractions on the right side are strictly proper. Decomposition of form (2.19) can be extended for a more general case. Assume an irreducible strictly proper rational fraction of the form

$$l(\lambda) = \frac{m(\lambda)}{d_1(\lambda)\cdots d_n(\lambda)}, \tag{2.20}$$

where the polynomials $d_i(\lambda)$, $(i = 1, \ldots, n))$ are pairwise coprime, then there exists the unique representation, analogously to (2.19)

$$l(\lambda) = \frac{m_1(\lambda)}{d_1(\lambda)} + \cdots + \frac{m_n(\lambda)}{d_n(\lambda)}, \tag{2.21}$$

where all fractions on the right side are irreducible and strictly proper. In particular, if the fraction $l(\lambda)$ has form (2.11), but

$$d(\lambda) = (\lambda - \lambda_1)^{\nu_1} \ldots (\lambda - \lambda_\rho)^{\nu_\rho}, \tag{2.22}$$

where all $\lambda_i$ are distinct, then applying decomposition (2.21), we obtain

$$l(\lambda) = \sum_{i=1}^{\rho} \frac{m_i(\lambda)}{(\lambda - \lambda_i)^{\nu_i}}, \quad \deg m_i(\lambda) < \nu_i. \tag{2.23}$$

In the following, the numbers $\lambda_i$, $(i = 1, \ldots, \rho)$ are called poles of the fraction $l(\lambda)$, and the integers $\nu_i$ their multiplicities.

(8) Moreover, notice that the strictly proper irreducible fraction

$$l_i(\lambda) = \frac{m_i(\lambda)}{(\lambda - \lambda_i)^{\nu_i}} \tag{2.24}$$

can be uniquely represented in the form

$$l_i(\lambda) = \frac{m_{i1}}{(\lambda - \lambda_i)^{\nu_i}} + \frac{m_{i2}}{(\lambda - \lambda_i)^{\nu_i - 1}} + \cdots + \frac{m_{i\nu_i}}{\lambda - \lambda_i}, \tag{2.25}$$

where $m_{ik}$ are certain constants. Comparing (2.25) and (2.23), we find

$$l(\lambda) = \sum_{i=1}^{\rho} \left[ \frac{m_{i1}}{(\lambda - \lambda_i)^{\nu_i}} + \frac{m_{i2}}{(\lambda - \lambda_i)^{\nu_i - 1}} + \cdots + \frac{m_{i\nu_i}}{\lambda - \lambda_i} \right], \tag{2.26}$$

which is called expansion of the function $l(\lambda)$ into simple fractions. For the calculation of the coefficients $m_{ik}$, we can apply the formulae [1, 2]

$$m_{ik} = \frac{1}{(k-1)!}\left[\frac{\partial^{k-1}}{\partial\lambda^{k-1}}\frac{m(\lambda)(\lambda-\lambda_i)^{\nu_i}}{d(\lambda)}\right]_{\lambda=\lambda_i}. \tag{2.27}$$

(9) Consider the irreducible fraction (2.18), where the polynomials $d_1(\lambda)$ and $d_2(\lambda)$ are coprime. Then using (2.16) and (2.19) we obtain the unique representation

$$\frac{m(\lambda)}{d(\lambda)} = \frac{m_1(\lambda)}{d_1(\lambda)} + \frac{m_2(\lambda)}{d_2(\lambda)} + q(\lambda), \tag{2.28}$$

where the fractions on the right side are strictly proper and irreducible. Let $g(\lambda)$ be an arbitrary polynomial. Then (2.28) can be written in the form

$$l(\lambda) = \left[\frac{m_1(\lambda)}{d_1(\lambda)} + g(\lambda)\right] + \left[\frac{m_2(\lambda)}{d_2(\lambda)} + q(\lambda) - g(\lambda)\right], \tag{2.29}$$

or, what is equivalent

$$l(\lambda) = \frac{n_1(\lambda)}{d_1(\lambda)} + \frac{n_2(\lambda)}{d_2(\lambda)}, \tag{2.30}$$

where

$$\begin{aligned} n_1(\lambda) &= m_1(\lambda) + g(\lambda)d_1(\lambda), \\ n_2(\lambda) &= m_2(\lambda) + [q(\lambda) - g(\lambda)]d_2(\lambda). \end{aligned} \tag{2.31}$$

Any representation of form (2.30) is called separation of the fraction $l(\lambda)$ according to the polynomials $d_1(\lambda)$ and $d_2(\lambda)$. From (2.30) and (2.31) it can be seen that the separation according to coprime polynomials $d_1(\lambda)$ and $d_2(\lambda)$ is always possible, and is not uniquely determined. At the same time [3] it can be shown that formulae (2.30) and (2.31) define all possible separations according to the polynomials $d_1(\lambda)$ and $d_2(\lambda)$.

If we assume $g(\lambda) = 0$ in (2.29), we obtain the separation

$$l_1(\lambda) = \frac{m_1(\lambda)}{d_1(\lambda)}, \quad l_2(\lambda) = \frac{m_2(\lambda)}{d_2(\lambda)} + q(\lambda), \tag{2.32}$$

where $l_1(\lambda)$ is strictly proper. For $g(\lambda) = q(\lambda)$ we obtain

$$l_1(\lambda) = \frac{m_1(\lambda)}{d_1(\lambda)} + q(\lambda), \quad l_2(\lambda) = \frac{m_2(\lambda)}{d_2(\lambda)}, \tag{2.33}$$

where the fraction $l_2(\lambda)$ is strictly proper. Further on, separation (2.32) (resp. (2.33)) is called minimal with respect to $d_1(\lambda)$ (resp. $d_2(\lambda)$). From the above explanations, it follows that the minimal separation is uniquely determined. In the important special case, when fraction (2.18) is strictly proper, we have $q(\lambda) = 0$ and the minimal separations (2.32) and (2.33) coincide.

## 2.2 Rational Matrices

(1) The $n \times n$ matrix $L(\lambda)$

$$L(\lambda) = \begin{bmatrix} l_{11}(\lambda) & \dots & l_{1m}(\lambda) \\ \vdots & \ddots & \vdots \\ l_{n1}(\lambda) & \dots & l_{nm}(\lambda) \end{bmatrix}, \quad l_{ik}(\lambda) = \frac{\tilde{m}_{ik}(\lambda)}{d_{ik}(\lambda)}, \tag{2.34}$$

where $l_{ik}(\lambda) \in \mathbb{F}(\lambda)$, is called fractional rational or in short rational matrix. If all $l_{ik}(\lambda) \in \mathbb{C}(\lambda)$, $l_{ik}(\lambda) \in \mathbb{R}(\lambda)$, or $l_{ik}(\lambda) \in \mathbb{F}(\lambda)$, then the set of rational matrices is denoted by $\mathbb{C}^{nm}(\lambda)$, $\mathbb{R}^{nm}(\lambda)$, or $\mathbb{F}^{nm}(\lambda)$, respectively.

Let the polynomial $d(\lambda)$ be the least common multiple of the denominators $d_{ik}(\lambda)$. Then the representations

$$l_{ik}(\lambda) = \frac{m_{ik}(\lambda)}{d(\lambda)}, \tag{2.35}$$

are valid, where $m_{ik}(\lambda)$ are polynomials.

Inserting (2.35) into (2.34) yields

$$L(\lambda) = \frac{M(\lambda)}{d(\lambda)}, \tag{2.36}$$

where

$$M(\lambda) = \begin{bmatrix} m_{11}(\lambda) & \dots & m_{1m}(\lambda) \\ \vdots & \ddots & \vdots \\ m_{n1}(\lambda) & \dots & m_{nm}(\lambda) \end{bmatrix} \tag{2.37}$$

is a polynomial matrix in $\mathbb{C}^{nm}[\lambda]$, $\mathbb{R}^{nm}[\lambda]$, or $\mathbb{F}^{nm}[\lambda]$, respectively. Below, the matrix $M(\lambda)$ is called the numerator of the matrix $L(\lambda)$ and the polynomial $d(\lambda)$ its denominator.

The matrix $L(\lambda)$ is called proper, strictly proper, or at least proper, if $\deg M(\lambda) = \deg d(\lambda)$, $\deg M(\lambda) < \deg d(\lambda)$, or $\deg M(\lambda) \le \deg d(\lambda)$, respectively.

(2) Assume

$$d(\lambda) = (\lambda - \lambda_1)^{\nu_1} \cdots (\lambda - \lambda_\rho)^{\nu_\rho}, \tag{2.38}$$

where all $\lambda_i$ are distinct.

**Definition 2.1** Fraction (2.36) is called irreducible if

$$M(\lambda_i) \neq 0_{nm}, \quad (i = 1, \dots, \rho). \tag{2.39}$$

If, however for one $i$

$$M(\lambda_i) = 0_{nm}, \tag{2.40}$$

then fraction (2.36) is called reducible.

*Remark 2.1* From the above definition, it follows that fraction (2.36) is reducible if all fractions (2.35) are reducible by one and the same factor. If, however, there is one scalar fraction (2.35) irreducible, then fraction (2.36) is irreducible.

After all possible reductions, matrix (2.34) can be presented in form (2.36)

$$L(\lambda) = \frac{M_0(\lambda)}{d_0(\lambda)}, \tag{2.41}$$

where the fraction on the right side is irreducible, and the polynomial $d_0(\lambda)$ is monic. This representation of a rational matrix is called its standard representation. The standard form of a rational matrix is unique.

If standard form (2.41) contains the product

$$d_0(\lambda) = (\lambda - \lambda_1)^{\nu_1} \cdots (\lambda - \lambda_\rho)^{\nu_\rho}, \tag{2.42}$$

where all numbers $\lambda_i$ are distinct, then the roots $\lambda_1, \dots, \lambda_\rho$ are called the poles of the matrix $L(\lambda)$ and the numbers $\nu_1, \dots, \nu_\rho$ their multiplicity. In the following, we use the notation

$$\nu_i = \operatorname{mult} \lambda_i . \tag{2.43}$$

(3) For rational matrices represented in form (2.36) and possessing appropriate sizes, in analogy to scalar rational fractions, various operations can be defined. Some of those operations and their properties will be explained below.

(a) The sum of two rational $n \times m$ matrices

$$L_1(\lambda) = \frac{M_1(\lambda)}{d_1(\lambda)}, \quad L_2(\lambda) = \frac{M_2(\lambda)}{d_2(\lambda)} \tag{2.44}$$

is defined by the formula

$$L_1(\lambda) + L_2(\lambda) = \frac{M_1(\lambda)d_2(\lambda) + M_2(\lambda)d_1(\lambda)}{d_1(\lambda)d_2(\lambda)} . \tag{2.45}$$

Moreover, when matrices (2.44) are irreducible and the polynomials $d_1(\lambda)$ and $d_2(\lambda)$ are coprime, then the fraction on the right side is irreducible.

(b) If every element of $M_0(\lambda)$ in (2.41) is divided by $d_0(\lambda)$, an arbitrary matrix of form (2.41) can be written as

$$L(\lambda) = \frac{N(\lambda)}{d_0(\lambda)} + Q(\lambda), \tag{2.46}$$

where $N(\lambda)$ and $Q(\lambda)$ are polynomial matrices and

$$\deg N(\lambda) < \deg d_0(\lambda).$$

Below the strictly proper fraction on the right side of (2.46) is called the broken part of the matrix $L(\lambda)$, and the polynomial matrix $Q(\lambda)$ is called its integer part.

(c) If matrix (2.41) is strictly proper, and

$$d_0(\lambda) = d_1(\lambda) \cdots d_n(\lambda), \tag{2.47}$$

where the $d_i(\lambda)$ are pairwise coprime monic polynomials, then there exists the unique expansion

$$L(\lambda) = \sum_{i=1}^{n} \frac{M_i(\lambda)}{d_i(\lambda)}, \tag{2.48}$$

where all terms of the sum are strictly proper. Moreover, if the polynomial $d_0(\lambda)$ in (2.41) satisfies (2.42), then the partial fraction expansion

$$L(\lambda) = \sum_{i=1}^{n} \left[ \frac{M_{i1}}{(\lambda - \lambda_i)^{\nu_i}} + \frac{M_{i2}}{(\lambda - \lambda_i)^{\nu_{i-1}}} + \cdots + \frac{M_{i\nu_i}}{\lambda - \lambda_i} \right] \tag{2.49}$$

can take place, where the $M_{ik}$ are constant matrices, which are determined by the formula

$$M_{ik} = \frac{1}{(k-1)!} \left[ \frac{\partial^{k-1}}{\partial \lambda^{k-1}} \frac{M(\lambda)(\lambda - \lambda_i)^{\nu_i}}{d(\lambda)} \right]_{|\lambda = \lambda_i}. \tag{2.50}$$

The partial fraction expansion (2.49) is uniquely determined.

(d) Assume standard form (2.41), where

$$L(\lambda) = \frac{M(\lambda)}{d_1(\lambda) d_2(\lambda)}, \tag{2.51}$$

and $d_1(\lambda)$, $d_2(\lambda)$ are coprime. Then there exists a set of separations

$$L(\lambda) = \frac{N_1(\lambda)}{d_1(\lambda)} + \frac{N_2(\lambda)}{d_2(\lambda)}, \tag{2.52}$$

where $N_1(\lambda)$ and $N_2(\lambda)$ are polynomial matrices. The set of matrices $N_1(\lambda)$ and $N_2(\lambda)$, configured in (2.52), is defined by the relations

$$\begin{aligned} N_1(\lambda) &= M_1(\lambda) + G(\lambda)d_1(\lambda) \\ N_2(\lambda) &= M_2(\lambda) + [Q(\lambda) - G(\lambda)]d_2(\lambda). \end{aligned} \tag{2.53}$$

Here the polynomial matrices $M_1(\lambda)$, $M_2(\lambda)$ and $Q(\lambda)$ are determined by the expansion

$$L(\lambda) = \frac{M_1(\lambda)}{d_1(\lambda)} + \frac{M_2(\lambda)}{d_2(\lambda)} + Q(\lambda), \tag{2.54}$$

where the fractions on the right side are strictly proper, and $Q(\lambda)$ is a polynomial matrix. Moreover, $G(\lambda)$ in (2.53) stands for an arbitrary polynomial matrix of appropriate size.

The special separations

$$\begin{aligned} L(\lambda) &= \frac{M_1(\lambda)}{d_1(\lambda)} + \frac{M_2(\lambda) + Q(\lambda)d_2(\lambda)}{d_2(\lambda)}, \\ L(\lambda) &= \frac{M_1(\lambda) + Q(\lambda)d_1(\lambda)}{d_1(\lambda)} + \frac{M_2(\lambda)}{d_2(\lambda)} \end{aligned} \tag{2.55}$$

are called minimal according to $d_1(\lambda)$ or $d_2(\lambda)$, respectively. If the primary fraction $L(\lambda)$ is strictly proper, then both minimal separations (2.55) coincide.

## 2.3 McMillan Canonical Form

(1) Let the matrix $L(\lambda) \in \mathbb{R}^{nm}(\lambda)$ be given in standard form (2.41)

$$L(\lambda) = \frac{M_0(\lambda)}{d_0(\lambda)}. \tag{2.56}$$

Further on, we understand as the normal rank of the matrix $L(\lambda)$ the normal rank of the polynomial matrix $M(\lambda)$. Thus, when Rank $M(\lambda) = \rho$, then we write Rank $L(\lambda) = \rho$.

(2) Let $L(\lambda) \in \mathbb{R}^{nm}(\lambda)$ and Rank $L(\lambda) = \rho$. Then, using the Smith canonical form, we obtain

$$M_0(\lambda) = p(\lambda) \begin{bmatrix} \operatorname{diag}\{a_1(\lambda), \ldots, a_\rho(\lambda)\} & 0_{\rho,m-\rho} \\ 0_{n-\rho,\rho} & 0_{n-\rho,m-\rho} \end{bmatrix} q(\lambda), \tag{2.57}$$

where $p(\lambda)$, $q(\lambda)$ are real unimodular matrices of size $n \times n$ and $m \times m$, respectively, and $a_1(\lambda), \ldots, a_\rho(\lambda)$ is the corresponding sequence of real invariant polynomials. From (2.56) and (2.57), we find

$$L(\lambda) = p(\lambda)N_L(\lambda)q(\lambda), \tag{2.58}$$

where

$$N_L(\lambda) = \begin{bmatrix} N_\rho(\lambda) & 0_{\rho,m-\rho} \\ 0_{n-\rho,\rho} & 0_{n-\rho,m-\rho} \end{bmatrix} \tag{2.59}$$

and

$$N_\rho(\lambda) = \operatorname{diag}\left\{\frac{a_1(\lambda)}{d(\lambda)}, \ldots, \frac{a_\rho(\lambda)}{d(\lambda)}\right\}. \tag{2.60}$$

Performing herein all reductions, we obtain

$$N_\rho(\lambda) = \operatorname{diag}\left\{\frac{\xi_1(\lambda)}{\psi_1(\lambda)}, \ldots, \frac{\xi_\rho(\lambda)}{\psi_\rho(\lambda)}\right\}, \tag{2.61}$$

where the $\xi_i(\lambda)$, $(i = 1, \ldots, \rho)$ and $\psi_i(\lambda)$, $(i = 1, \ldots, \rho)$ are real monic polynomials. Hereby, the polynomials $\xi_i(\lambda)$ and $\psi_i(\lambda)$ are pairwise coprime, where $\xi_{i+1}(\lambda)$ is divisible by $\xi_i(\lambda)$, and $\psi_i(\lambda)$ divisible by $\psi_{i+1}(\lambda)$. Further on, the matrix $N_L(\lambda)$ is called the McMillan canonical form of the matrix $L(\lambda)$.

(3)

**Definition 2.2** The monic polynomial

$$\psi_L(\lambda) \stackrel{\triangle}{=} \psi_1(\lambda) \cdots \psi_\rho(\lambda) \tag{2.62}$$

is called the McMillan denominator of the matrix $L(\lambda)$, and the monic polynomial

$$\xi_L(\lambda) \stackrel{\triangle}{=} \xi_1(\lambda) \cdots \xi_\rho(\lambda) \tag{2.63}$$

its McMillan numerator.

**Lemma 2.1** *The relation*

$$\Delta_L(\lambda) \stackrel{\triangle}{=} \frac{\psi_L(\lambda)}{d_0(\lambda)} \tag{2.64}$$

*turns out to be a polynomial.*

*Proof* Since matrix (2.56) is irreducible, we find from (2.58), (2.59) that the polynomials $a_1(\lambda)$ and $d_0(\lambda)$ are coprime. Therefore

$$\xi_1(\lambda) = a_1(\lambda), \quad \psi_1(\lambda) = d_0(\lambda).$$

Hence

$$\psi_L(\lambda) = d_0(\lambda)\psi_2(\lambda)\cdots\psi_\rho(\lambda), \tag{2.65}$$

and expression $\Delta_L(\lambda)$ is a polynomial. ■

By construction, the polynomials $\psi_2(\lambda), \ldots, \psi_\rho(\lambda)$ are factors of the polynomial $d_0(\lambda)$. Therefore, when expansion (2.22) holds, then

$$\psi_L(\lambda) = (\lambda - \lambda_1)^{\bar{\mu}_1} \cdots (\lambda - \lambda_\rho)^{\bar{\mu}_\rho}, \tag{2.66}$$

where $\bar{\mu}_i \geq \nu_i$.

**Definition 2.3** The number $\bar{\mu}_i$ configured in (2.66) is called McMillan multiplicity of the pole $\lambda_i$, and we use the notation

$$\text{Mult}\,\lambda_i \stackrel{\triangle}{=} \bar{\mu}_i. \tag{2.67}$$

It can be shown, [4], that the number $\bar{\mu}_i$ is equal to the maximal multiplicity of the pole $\lambda_i$ among all nonzero minors of the matrix $L(\lambda)$.

By construction

$$\text{Mult}\,\lambda_i \geq \text{mult}\,\lambda_i. \tag{2.68}$$

(4)

**Lemma 2.2** *Assume in (2.58)* Rank $L(\lambda) = \rho$, *and expansion (2.42) is valid. Then for any pole* $\lambda_i$

$$\nu_i = \text{mult}\,\lambda_i \leq \text{Mult}\,\lambda_i \leq \rho\,\text{mult}\,\lambda_i = \rho\nu_i. \tag{2.69}$$

*Proof* The left side of relations (2.69) emerge from (2.65) for

$$\psi_2 = \cdots = \psi_\rho = 1,$$

and the right side for

$$\psi_2(\lambda) = \cdots = \psi_\rho(\lambda) = d_0(\lambda).$$

■

Further on, in case of equality

$$\text{Mult}\,\lambda_i = \rho\,\text{mult}\,\lambda_i, \tag{2.70}$$

we will say, that the pole $\lambda_i$ of matrix (2.46) has maximal McMillan multiplicity.

**Lemma 2.3** *For the fact that the pole $\lambda_i$ of matrix (2.56) has maximal McMillan multiplicity, it is necessary and sufficient that the number $\lambda_i$ is not a latent number of the matrix $M(\lambda)$.*

*Proof* Sufficiency: Let $\lambda_i$ not be a latent number of the matrix $M(\lambda)$. Then the number $\lambda_i$ is not a root of any of the invariant polynomials $a_k(\lambda)$ and hence all fractions $\dfrac{a_k(\lambda)}{d(\lambda)}$ in (2.60) are not reducible by $\lambda - \lambda_i$, and each of these fractions has at the point $\lambda_i$ the pole multiplicity $\nu_i$. Therefore, it follows from (2.65) that the McMillan denominator $\psi_L(\lambda)$ possesses in the point $\lambda = \lambda_i$ a root with multiplicity $\operatorname{Mult}\lambda_i = \rho\nu_i$.

Necessity: Assume (2.70). This means that in all fractions $\dfrac{a_k(\lambda)}{d(\lambda)}$ the pole $\lambda_i$ is not reducible. Hence, $\lambda_i$ is not a latent number of the numerator $M(\lambda)$.

(5) Let $\lambda_1, \dots, \lambda_\rho$ be the set of distinct poles of matrix (2.56) and $\bar{\mu}_i$, $(i = 1, \dots, q)$ their McMillan multiplicity.

**Definition 2.4** The number Mind $L(\lambda)$, defined by

$$\operatorname{Mind} L(\lambda) \stackrel{\triangle}{=} \sum_{i=1}^{\rho} \operatorname{Mult}\lambda_i = \sum_{i=1}^{q} \bar{\mu}_i \tag{2.71}$$

is called McMillan index of the matrix $L(\lambda)$.

From the above considerations, we find

$$\operatorname{Mind} L(\lambda) = \deg \psi_L(\lambda), \tag{2.72}$$

i.e., the McMillan index of a rational matrix is equal to the degree of its McMillan denominator.

**Lemma 2.4** *Let the $n \times n$ matrix $L(\lambda)$ in (2.56) possess the maximal normal rank equal to $n$. Then*

$$\det L(\lambda) = k\frac{\xi_L(\lambda)}{\psi_L(\lambda)}, \tag{2.73}$$

*where $\xi_L(\lambda)$ and $\psi_L(\lambda)$ are the McMillan numerator resp. denominator of the matrix $L(\lambda)$, and $k \neq 0$ is a constant.*

*Proof* Under the taken suppositions, the matrix $L(\lambda)$ allows the representation

$$L(\lambda) = p(\lambda)\operatorname{diag}\left\{\frac{\xi_1(\lambda)}{\psi_1(\lambda)}, \dots, \frac{\xi_n(\lambda)}{\psi_n(\lambda)}\right\} q(\lambda), \tag{2.74}$$

where $p(\lambda)$, $q(\lambda)$ are certain unimodular matrices. Calculating the determinant on both sides of the equation yields formula (2.73). ∎

*Remark 2.2* In general, the fraction on the right side of (2.73) can be reduced.

## 2.4 Matrix Fraction Description (MFD)

(1) Let us have the matrix $L(\lambda) \in \mathbb{R}^{nm}(\lambda)$ in standard form (2.56). Assume a non-singular matrix $a_l(\lambda) \in \mathbb{R}^{nn}[\lambda]$ such that the product

$$b_l(\lambda) \triangleq a_l(\lambda)L(\lambda) = a_l(\lambda)\frac{M(\lambda)}{d(\lambda)} \tag{2.75}$$

becomes a polynomial matrix. Then the matrix $a_l(\lambda)$ is called a left denominator of the matrix $L(\lambda)$, and $b_l(\lambda)$ is called its left numerator. Moreover, the representation

$$L(\lambda) = a_l^{-1}(\lambda)b_l(\lambda), \tag{2.76}$$

following [2] is called left matrix fraction description (LMFD).

In analogy, if the non-singular matrix $a_r(\lambda) \in \mathbb{R}^{mm}[\lambda]$ is found such that

$$b_r(\lambda) \triangleq L(\lambda)a_r(\lambda) = \frac{M(\lambda)}{d(\lambda)}a_r(\lambda) \tag{2.77}$$

is a polynomial matrix, then the matrix $a_r(\lambda)$ is called right denominator of the matrix $L(\lambda)$, and the matrix $b_r(\lambda)$ is its right numerator. Moreover, the representation

$$L(\lambda) = b_r(\lambda)a_r^{-1}(\lambda) \tag{2.78}$$

is called right matrix fraction description (RMFD).

(2) For any matrix $L(\lambda) \in \mathbb{R}^{nm}[\lambda]$, there exists a set of LMFD and RMFD. For instance, we can always choose

$$\begin{aligned} a_l(\lambda) &= I_n d(\lambda), \quad b_l(\lambda) = M(\lambda), \\ a_r(\lambda) &= I_m d(\lambda), \quad b_r(\lambda) = M(\lambda). \end{aligned} \tag{2.79}$$

In that case

$$\deg\det a_l(\lambda)) = n \deg d(\lambda)$$

and

$$\deg\det a_r(\lambda) = m \deg d(\lambda).$$

However, as examples show, in general, it is possible to find LMFD and RMFD with lower values of $\deg\det a_l(\lambda)$ and $\deg\det a_r(\lambda)$.

For the aims of this book, very important are the properties of LMFD and RMFD, where $\deg\det a_l(\lambda)$ and $\deg\det a_r(\lambda)$ take their minimal possible values. Below, those MFD are called irreducible (IMFD). Thus, we distinguish left IMFD (ILMFD) from right IMFD (IRMFD).

(3) Below, we will state some properties of ILMFD and IRMFD [2, 3].

(a) For the fact that LMFD (2.76) is irreducible, it is necessary and sufficient that the pair $(a_l(\lambda), b_l(\lambda))$ is irreducible, i.e., the matrix $\left[\, a_l(\lambda) \;\, b_l(\lambda) \,\right]$ is alatent. In analogy, RMFD (2.78) is irreducible, if and only if the pair $[a_r(\lambda), b_r(\lambda)]$ is irreducible, i.e., the matrix $\left[\, a'_r(\lambda) \;\, b'_r(\lambda) \,\right]$ is alatent.

(b) If we have for a matrix $L(\lambda)$ two LMFD

$$L(\lambda) = a_l^{-1}(\lambda) b_l(\lambda) = a_{l_1}^{-1}(\lambda) b_{l_1}(\lambda), \tag{2.80}$$

where the pair $(a_l(\lambda), b_l(\lambda))$ is irreducible, then there exists a non-singular polynomial $n \times n$ matrix $g_l(\lambda)$ such that

$$a_{l_1}(\lambda) = g_l(\lambda) a_l(\lambda), \quad b_{l_1}(\lambda) = g_l(\lambda) b_l(\lambda). \tag{2.81}$$

In particular, if the pair $(a_{l_1}(\lambda), b_{l_1}(\lambda))$ is irreducible too, then the matrix $g_l(\lambda)$ is unimodular, and the matrices $a_l(\lambda)$ and $a_{l_1}(\lambda)$ are left equivalent, i.e.,

$$a_{l_1}(\lambda) \overset{l}{\sim} a_l(\lambda). \tag{2.82}$$

Analogously, if we have two RMFD

$$L(\lambda) = b_r(\lambda) a_r^{-1}(\lambda) = b_{r_1}(\lambda) a_{r_1}^{-1}(\lambda), \tag{2.83}$$

where the pair $[a_r(\lambda), b_r(\lambda)]$ is irreducible, then there exists a non-singular matrix $g_r(\lambda) \in \mathbb{R}^{mm}[\lambda]$ such that

$$a_{r_1}(\lambda) = a_r(\lambda) g_r(\lambda), \quad b_{r_1}(\lambda) = b_r(\lambda) g_r(\lambda). \tag{2.84}$$

If, however, the pair $[a_{r_1}(\lambda), b_{r_1}(\lambda)]$ is also irreducible, then the matrix $g_r(\lambda)$ is unimodular, such that the matrices $a_r(\lambda)$ and $a_{r_1}(\lambda)$ are right equivalent

$$a_{r_1}(\lambda) \overset{r}{\sim} a_r(\lambda). \tag{2.85}$$

(4) A theoretical approach for the construction of ILMFD and IRMFD opens the application of McMillan canonical form (2.58)–(2.61), [2]. Indeed, from (2.59)–(2.61), we find the LMFD and RMFD

$$N_L(\lambda) = \tilde{a}_l^{-1}(\lambda) \tilde{b}(\lambda) = \tilde{b}(\lambda) \tilde{a}_r^{-1}(\lambda), \tag{2.86}$$

where

$$
\begin{aligned}
\tilde{a}_l(\lambda) &= \operatorname{diag}\{d(\lambda), \psi_2(\lambda), \ldots, \psi_\rho(\lambda), 1, \ldots, 1\}, \\
\tilde{a}_r(\lambda) &= \operatorname{diag}\{d(\lambda), \psi_2(\lambda), \ldots, \psi_\rho(\lambda), 1, \ldots, 1\}
\end{aligned}
\tag{2.87}
$$

and

$$
\tilde{b}(\lambda) = \begin{bmatrix} \operatorname{diag}\{\xi_1(\lambda), \ldots, \xi_\rho(\lambda)\} & 0_{\rho,m-\rho} \\ 0_{n-\rho,\rho} & 0_{n-\rho,m-\rho} \end{bmatrix}. \tag{2.88}
$$

Hence from (2.58), we obtain LMFD and RMFD for the matrix $L(\lambda)$

$$
L(\lambda) = a_l^{-1}(\lambda)b_l(\lambda) = b_r(\lambda)a_r^{-1}(\lambda), \tag{2.89}
$$

where

$$
\begin{aligned}
a_l(\lambda) &= \tilde{a}_l(\lambda)p^{-1}(\lambda), \quad b_l(\lambda) = \tilde{b}(\lambda)q(\lambda), \\
a_r(\lambda) &= q^{-1}(\lambda)\tilde{a}_r(\lambda), \quad b_r(\lambda) = p(\lambda)\tilde{b}(\lambda).
\end{aligned}
\tag{2.90}
$$

In [2], it is shown that pairs (2.90) are irreducible, and thus expressions (2.89) are ILMFD resp. IRMFD of the matrix $L(\lambda)$.

(5) From the above-realized way of construction for the IMFD, a number of important consequences follow and they are stated below.

**Lemma 2.5** *Assume that expressions (2.89) define ILMFD and IRMFD of matrix (2.56). Then the following statements hold:*

*(i) The matrices $a_l(\lambda)$ and $a_r(\lambda)$ are pseudo-equivalent, i.e.,*

$$
a_l(\lambda) \overset{p}{\sim} a_r(\lambda). \tag{2.91}
$$

*(ii) The matrices $b_l(\lambda)$ and $b_r(\lambda)$ are equivalent, i.e.,*

$$
b_l(\lambda) \sim b_r(\lambda). \tag{2.92}
$$

*(iii) The following equivalence relations are valid:*

$$
\det a_l(\lambda) \sim \det a_r(\lambda) \sim \psi_L(\lambda). \tag{2.93}
$$

*(iv) The following relations hold:*

$$
\deg \det a_l(\lambda) = \deg \det a_r(\lambda) = \operatorname{Mind} L(\lambda). \tag{2.94}
$$

*Proof* (i) Consider the sequence of polynomials

$$
\psi_\rho(\lambda), \psi_{\rho-1}(\lambda), \ldots, \psi_2(\lambda), d(\lambda). \tag{2.95}
$$

These polynomials are monic, where each follower is divisible by the predecessor. Therefore, sequence (2.95) uniquely defines the set of invariant polynomials of the matrices $a_l(\lambda)$ and $a_r(\lambda)$, which are different from one. This fact proves (2.91).

(ii) Relation (2.92) is a consequence from (2.88), (2.90).

(iii) Formula (2.93) follows from (2.87), (2.90) and (2.62).

(iv) Relation (2.94) is a direct consequence from (2.93) and (2.82).

## 2.5 General Properties of the MFD

(1)

**Lemma 2.6** *Assume the LMFD*

$$L(\lambda) = a_{l1}^{-1}(\lambda)b_{l1}(\lambda). \tag{2.96}$$

*Then, there exists an RMFD*

$$L(\lambda) = b_{r1}(\lambda)a_{r1}^{-1}(\lambda), \tag{2.97}$$

*such that* $\det a_{l1}(\lambda) \sim \det a_{r1}(\lambda)$. *The reverse statement is also true.*

*Proof* Assume $L(\lambda) \in \mathbb{R}^{nm}(\lambda)$, and let us have the IMFD

$$L(\lambda) = a_l^{-1}(\lambda)b_l(\lambda) = b_r(\lambda)a_r^{-1}(\lambda). \tag{2.98}$$

Then it follows from (2.81) that

$$a_{l1}(\lambda) = g_l(\lambda)a_l(\lambda),$$

where the matrix $g_l(\lambda)$ is non-singular. Denote $\det g_l(\lambda) = h(\lambda)$, and let $g_r(\lambda)$ be any non-singular $m \times m$ matrix, such that $\det g_r(\lambda) \sim h(\lambda)$. Then the pair $[a_{r1}(\lambda), b_{r1}(\lambda)]$, where

$$a_{r1}(\lambda) = a_r(\lambda)g_r(\lambda), \; b_{r1}(\lambda) = b_r(\lambda)g_r(\lambda), \tag{2.99}$$

possesses the required properties.

(2)

**Lemma 2.7** *The following statements are true:*

*(i) Let the rational matrix* $L(\lambda) \in \mathbb{R}^{nm}(\lambda)$ *have the form*

$$L(\lambda) = c(\lambda)a^{-1}(\lambda)b(\lambda),$$

*where* $a(\lambda), b(\lambda), c(\lambda)$ *are real polynomial matrices of sizes* $l \times l,\ l \times m,\ n \times l$, *respectively. Let the pair* $(a(\lambda), b(\lambda))$ *be irreducible. Assume the ILMFD*

$$c(\lambda)a^{-1}(\lambda) = a_1^{-1}(\lambda)c_1(\lambda). \tag{2.100}$$

*Then the pair* $(a_1(\lambda), c_1(\lambda)b(\lambda))$ *is irreducible, and hence the representation*

$$L(\lambda) = c(\lambda)a^{-1}(\lambda)b(\lambda) = a_1^{-1}(\lambda)[c_1(\lambda)b(\lambda)] \tag{2.101}$$

*is an ILMFD.*

*(ii) If, however, the pair* $[a(\lambda), c(\lambda)]$ *is irreducible, and we have the IRMFD*

$$a^{-1}(\lambda)b(\lambda) = b_1(\lambda)a_2^{-1}(\lambda), \tag{2.102}$$

*then the pair* $[a_2(\lambda), c(\lambda)b_1(\lambda)]$ *is irreducible, and the representation*

$$L(\lambda) = c(\lambda)a^{-1}(\lambda)b(\lambda) = [c(\lambda)b_1(\lambda)]a_2^{-1}(\lambda)] \tag{2.103}$$

*is an IRMFD.*

*Proof* (i) Since the pair $(a(\lambda), b(\lambda))$ is irreducible, from (1.107), it follows that there exist matrices $X(\lambda) \in \mathbb{R}^{ll}[\lambda]$ and $Y(\lambda) \in \mathbb{R}^{ml}[\lambda]$, such that

$$a(\lambda)X(\lambda) + b(\lambda)Y(\lambda) = I_l. \tag{2.104}$$

Moreover, from the irreducibility of the pair $(a_1(\lambda), c_1(\lambda))$ we conclude the existence of matrices $U(\lambda) \in \mathbb{R}^{nn}[\lambda]$ and $V(\lambda) \in \mathbb{R}^{ln}[\lambda]$ satisfying

$$a_1(\lambda)U(\lambda) + c_1(\lambda)V(\lambda) = I_n. \tag{2.105}$$

Using the last two relations, we find

$$\begin{aligned} a_1(\lambda)U(\lambda) + c_1(\lambda)V(\lambda) &= a_1(\lambda)U(\lambda) + c_1(\lambda)I_l V(\lambda) \\ &= a_1(\lambda)U(\lambda) + c_1(\lambda)[a(\lambda)X(\lambda) + b(\lambda)Y(\lambda)]V(\lambda) \\ &= a_1(\lambda)U(\lambda) + c_1(\lambda)a(\lambda)X(\lambda)V(\lambda) + c_1(\lambda)b(\lambda)Y(\lambda)V(\lambda) = I_n, \end{aligned}$$

which, using (2.100), can be represented in the form

$$a_1(\lambda)[U(\lambda) + c(\lambda)X(\lambda)V(\lambda)] + c_1(\lambda)b(\lambda)[Y(\lambda)V(\lambda)] = I_n. \tag{2.106}$$

From this together with (1.107), it follows that the pair $(a_1(\lambda), c_1(\lambda)b(\lambda))$ is irreducible. Statement (i) is proven.

(ii) The proof is analog to that for (i). ■

(3)

**Lemma 2.8** *For the matrix* $L(\lambda) \in \mathbb{R}^{nm}(\lambda)$ *let us have the IMFD*

$$L(\lambda) = a_l^{-1}(\lambda)b_l(\lambda) = b_r(\lambda)a_r^{-1}(\lambda), \tag{2.107}$$

*and a matrix* $d(\lambda) \in \mathbb{R}^{nm}[\lambda]$*. Then the right sides of the expressions*

$$L_1(\lambda) \triangleq L(\lambda) + d(\lambda) = a_l^{-1}(\lambda)[b_l(\lambda) + a_l(\lambda)d(\lambda)] \tag{2.108}$$

*and*

$$L_1(\lambda) \triangleq [b_r(\lambda) + d(\lambda)a_r(\lambda)]a_r^{-1}(\lambda) \tag{2.109}$$

*are ILMFD or IRMFD, respectively.*

*Proof* Consider the extended matrix

$$R_{h_1}(\lambda) = \left[\, a_l(\lambda)\ b_l(\lambda) + a_l(\lambda)d(\lambda) \,\right]. \tag{2.110}$$

It is easily seen that

$$R_{h_1}(\lambda) = R_h(\lambda)d_1(\lambda), \tag{2.111}$$

where

$$R_h(\lambda) = \left[\, a_l(\lambda)\ b_l(\lambda) \,\right], \quad d_1(\lambda) = \begin{bmatrix} I_n & d(\lambda) \\ 0_{mn} & I_m \end{bmatrix}. \tag{2.112}$$

Since the matrix $d_1(\lambda)$ is unimodular, from (2.111), we find

$$R_{h_1}(\lambda) \overset{r}{\sim} R_h(\lambda). \tag{2.113}$$

Under the taken assumptions, the matrix $R_h(\lambda)$ is alatent, and so $R_{h1}(\lambda)$ is alatent too, and the right part of (2.108) is an ILMFD. The property of (2.109) to be an IRMFD can be seen analogously. ∎

**Lemma 2.9** *Let us have IMFD (2.98) and the spectra of the matrices* $d_l(\lambda) \in \mathbb{R}^{nn}[\lambda]$ *and* $d_r(\lambda) \in \mathbb{R}^{mm}[\lambda]$ *have no intersection with the spectra of the matrices* $a_l(\lambda)$ *and* $a_r(\lambda)$*, respectively. Then the expressions*

$$\begin{aligned} L_1(\lambda) &= a_l^{-1}(\lambda)[b_l(\lambda)d_l(\lambda)], \\ L_2(\lambda) &= [d_r(\lambda)b_r(\lambda)]a_r^{-1}(\lambda) \end{aligned} \tag{2.114}$$

*are ILMFD or IRMFD, respectively.*

*Proof* Let $\lambda_1, \ldots, \lambda_q$ be the totality of different eigenvalues of the matrix $a_l(\lambda)$, forming its spectrum. Since the matrix

$$R_h(\lambda) = \left[ a_l(\lambda) \; b_l(\lambda) \right]$$

is alatent, using (1.100) we obtain

$$\operatorname{rank} \left[ a_l(\lambda_i) \; b_l(\lambda_i) \right] = n, \quad (i = 1, \ldots, q). \tag{2.115}$$

Consider the extended matrix

$$R_{h1}(\lambda) = \left[ a_l(\lambda) \; b_l(\lambda) d_1(\lambda) \right]. \tag{2.116}$$

The latent numbers of the matrix $R_{h1}(\lambda)$ are located in the spectrum of $a_l(\lambda)$. But, for $1 \leq i \leq q$, we have

$$R_{h1}(\lambda_i) = R_h(\lambda_i) F(\lambda_i), \quad (i = 1, \ldots, q), \tag{2.117}$$

where

$$F(\lambda) = \operatorname{diag}\{I_n, d_1(\lambda)\}. \tag{2.118}$$

Under the actual assumptions

$$\operatorname{rank} F(\lambda_i) = n + m.$$

Thus

$$\operatorname{rank} R_{h1}(\lambda_i) = n, \quad (i = 1, \ldots, q), \tag{2.119}$$

i.e., the matrix $R_{h1}(\lambda_i)$ fulfills condition (1.108). Hence, the first relation in (2.114) defines an ILMFD of the matrix $L_1(\lambda)$. In the same way, it can be seen that the second relation in (2.114) is an IRMFD. ■

## 2.6 Properties of the McMillan Index

(1) Assume a matrix $L(\lambda) \in \mathbb{R}^{nm}(\lambda)$, and the non-singular polynomial matrices $\tilde{a}_l(\lambda) \in \mathbb{R}^{nn}[\lambda]$, $\tilde{a}_r[\lambda] \in \mathbb{R}^{mm}[\lambda]$.

**Definition 2.5** The non-singular matrix $\tilde{a}_l(\lambda) \in \mathbb{R}^{nn}[\lambda]$ $(\tilde{a}_r(\lambda) \in \mathbb{R}^{mm}[\lambda])$ is called left reducing (right reducing) for the matrix $L(\lambda)$ if the matrix $\tilde{a}_l(\lambda) L(\lambda)$ $(L(\lambda)\tilde{a}_r(\lambda))$ becomes a polynomial matrix.

**Lemma 2.10** *Let the matrix $\tilde{a}_l(\lambda)$ $(\tilde{a}_r(\lambda))$ be left (right) reducing, then*

$$\operatorname{Mind} L(\lambda) \leq \deg \det \tilde{a}_l(\lambda) \quad (\operatorname{Mind} L(\lambda) \leq \deg \det \tilde{a}_r(\lambda)). \tag{2.120}$$

*Proof* Let the product

$$\tilde{b}_l(\lambda) = \tilde{a}_l(\lambda)L(\lambda) \tag{2.121}$$

be a polynomial matrix. Assume the ILMFD

$$L(\lambda) = a_l^{-1}(\lambda)b(\lambda). \tag{2.122}$$

Then due to the properties of ILMFD, we find

$$\tilde{a}_l(\lambda) = p(\lambda)a_l(\lambda), \tag{2.123}$$

where the matrix $p(\lambda) \in \mathbb{R}^{nn}(\lambda)$ is non-singular. Hence

$$\deg\det\tilde{a}(\lambda) \geq \deg\det a_l(\lambda) = \text{Mind}\, L(\lambda). \tag{2.124}$$

The dual inequality in (2.120) is proven analogously.

**Lemma 2.11** *The following statements hold:*

*(i) Let $L(\lambda) \in \mathbb{R}^{nm}(\lambda)$ and $d(\lambda) \in \mathbb{R}^{nm}[\lambda]$. Then*

$$\text{Mind}\, L(\lambda) = \text{Mind}[L(\lambda) + d(\lambda)]. \tag{2.125}$$

*(ii) Assume ILMFD (2.122), and the spectra of the matrices $a_l(\lambda)$ and $d(\lambda) \in \mathbb{R}^{mm}[\lambda]$ are disjunct. Then*

$$\text{Mind}\, L(\lambda) = \text{Mind}[L(\lambda)d(\lambda)]. \tag{2.126}$$

*Proof* The proof follows immediately from Lemmata 2.8 and 2.9.

(2)

**Lemma 2.12** *Let the standard form of the matrix $L(\lambda) \in \mathbb{R}^{nm}(\lambda)$ have the shape*

$$L(\lambda) = \frac{M(\lambda)}{d_1(\lambda)d_2(\lambda)}, \tag{2.127}$$

*where the polynomials $d_1(\lambda)$ and $d_2(\lambda)$ are coprime. Denote*

$$L_i(\lambda) \stackrel{\triangle}{=} \frac{M(\lambda)}{d_i(\lambda)}, \quad (i = 1, 2). \tag{2.128}$$

*Then*

$$\text{Mind}\, L(\lambda) = \text{Mind}\, L_1(\lambda) + \text{Mind}\, L_2(\lambda). \tag{2.129}$$

*Proof* Using relations (2.58)–(2.61), the matrix $L(\lambda)$ can be written in the form

$$L(\lambda) = p(\lambda)N_L(\lambda)q(\lambda), \tag{2.130}$$

where

$$N_L(\lambda) = \operatorname{diag}\left\{\operatorname{diag}\left\{\frac{\xi_1(\lambda)}{d_1(\lambda)d_2(\lambda)}, \frac{\xi_2(\lambda)}{\psi_{21}(\lambda)\psi_{22}(\lambda)}, \ldots, \frac{\xi_\rho(\lambda)}{\psi_{\rho 1}(\lambda)\psi_{\rho 2}(\lambda)}\right\}, 0_{n-\rho,n-\rho}\right\}, \tag{2.131}$$

where the monic polynomials $\psi_{21}(\lambda), \ldots, \psi_{\rho 1}(\lambda)$ are divisors of the polynomial $d_1(\lambda)$, and the monic polynomials $\psi_{22}(\lambda), \ldots, \psi_{\rho 2}(\lambda)$ are divisors of the polynomial $d_2(\lambda)$. Hereby, the McMillan denominator of the matrix $L(\lambda)$ is equal to

$$\psi_L(\lambda) = d_1(\lambda)d_2(\lambda)\psi_{21}(\lambda)\psi_{22}(\lambda)\cdots\psi_{\rho 1}(\lambda)\psi_{\rho 2}(\lambda). \tag{2.132}$$

Drawing analog conclusions, we can show that the McMillan denominator of the matrices $L_1(\lambda)$ and $L_2(\lambda)$ can be represented by

$$\psi_{L_i}(\lambda) = d_i(\lambda)\psi_{2i}(\lambda)\cdots\psi_{\rho i}(\lambda), \quad (i = 1, 2). \tag{2.133}$$

These two relations yield

$$\psi_L(\lambda) = \psi_{L_1}(\lambda)\psi_{L_2}(\lambda). \tag{2.134}$$

Hence

$$\deg\psi_L(\lambda) = \deg\psi_{L_1}(\lambda) + \deg\psi_{L_2}(\lambda), \tag{2.135}$$

which is equivalent to (2.129). ∎

**Lemma 2.13** *Assume the two matrices in $\mathbb{R}^{nm}(\lambda)$ in standard form*

$$L_i(\lambda) = \frac{M_i(\lambda)}{d_i(\lambda)}, \quad (i = 1, 2), \tag{2.136}$$

*where the polynomials $d_1(\lambda)$ and $d_2(\lambda)$ are coprime, and*

$$L(\lambda) = L_1(\lambda) + L_2(\lambda) = \frac{M_1(\lambda)}{d_1(\lambda)} + \frac{M_2(\lambda)}{d_2(\lambda)}. \tag{2.137}$$

*Then*

$$\operatorname{Mind} L(\lambda) = \operatorname{Mind} L_1(\lambda) + \operatorname{Mind} L_2(\lambda). \tag{2.138}$$

*Proof* From (2.136) and (2.137), we obtain

$$L(\lambda) = \frac{N(\lambda)}{d_1(\lambda)d_2(\lambda)}, \tag{2.139}$$

where

$$N(\lambda) \stackrel{\triangle}{=} M_1(\lambda)d_2(\lambda) + M_2(\lambda)d_1(\lambda). \tag{2.140}$$

Fraction (2.139) is irreducible, because, if we propose the opposite, then we find that at least one of fractions (2.136) contrary to the supposition is reducible. Hence from Lemma 2.12

$$\text{Mind}\, L(\lambda) = \text{Mind}\, N_1(\lambda) + \text{Mind}\, N_2(\lambda), \tag{2.141}$$

where

$$N_i(\lambda) \stackrel{\triangle}{=} \frac{N(\lambda)}{d_i(\lambda)}, \quad (i = 1, 2). \tag{2.142}$$

Applying (2.140) yields

$$\begin{aligned} N_1(\lambda) &= L_1(\lambda) d_2(\lambda) + M_2(\lambda), \\ N_2(\lambda) &= L_2(\lambda) d_1(\lambda) + M_1(\lambda). \end{aligned} \tag{2.143}$$

Since the polynomials $d_1(\lambda)$ and $d_2(\lambda)$ are coprime, using Lemmata 2.8 and 2.9, we obtain

$$\text{Mind}\, N_i(\lambda) = \text{Mind}\, L_i(\lambda), \quad (i = 1, 2). \tag{2.144}$$

From this and (2.141), we find (2.138). ∎

**Lemma 2.14** *Under the condition of Lemma 2.13 assume the ILMFD*

$$L(\lambda) = a_l^{-1}(\lambda) b_l(\lambda), \quad L_i(\lambda) = a_i^{-1}(\lambda) b_i(\lambda), \ (i = 1, 2). \tag{2.145}$$

*Then the representations*

$$a_l(\lambda) = \tilde{a}_2(\lambda) a_1(\lambda), \quad a_l(\lambda) = \tilde{a}_1(\lambda) a_2(\lambda) \tag{2.146}$$

*are possible, where* $\tilde{a}_1(\lambda), \tilde{a}_2(\lambda) \in \mathbb{R}^{nn}[\lambda]$ *satisfy*

$$\tilde{a}_1(\lambda) \sim a_1(\lambda), \quad \tilde{a}_2(\lambda) \sim a_2(\lambda). \tag{2.147}$$

*Proof* Using (2.145), we obtain

$$L(\lambda) = a_1^{-1}(\lambda) b_1(\lambda) + a_2^{-1}(\lambda) b_2(\lambda) = a_1^{-1}(\lambda)[b_1(\lambda) + a_1(\lambda) a_2^{-1}(\lambda) b_2(\lambda)]. \tag{2.148}$$

Since the polynomials $d_1(\lambda)$ and $d_2(\lambda)$ are coprime, the spectra of the matrices $a_1(\lambda)$ and $a_2(\lambda)$ are disjunct. Hence, the representation

$$L_a(\lambda) \stackrel{\triangle}{=} a_1(\lambda) a_2^{-1}(\lambda) \tag{2.149}$$

is an IRMFD, i.e., for all $\lambda$

$$\text{rank} \left[ a_1(\lambda) \ a_2(\lambda) \right] = n. \tag{2.150}$$

Build the ILMFD

$$L_a(\lambda) = \tilde{a}_2^{-1}(\lambda)\tilde{a}_1(\lambda). \tag{2.151}$$

Then, regarding that the matrices $a_2(\lambda)$ and $\tilde{a}_2(\lambda)$ are of the same size and possessing the same spectrum, we obtain

$$a_2(\lambda) \sim \tilde{a}_2(\lambda). \tag{2.152}$$

With the help of (2.151) and (2.148), we find the MFD

$$L(\lambda) = [\tilde{a}_2(\lambda)a_1(\lambda)]^{-1}[\tilde{a}_2(\lambda)b_1(\lambda) + \tilde{a}_1(\lambda)b_2(\lambda)]. \tag{2.153}$$

Under the taken propositions, we obtain

$$\begin{aligned} \deg\det[\tilde{a}_2(\lambda)a_1(\lambda)] &= \deg\det\tilde{a}_2(\lambda) + \deg\det a_1(\lambda) \\ &= \deg\det a_2(\lambda) + \deg\det a_1(\lambda) \\ &= \text{Mind}\, L_2(\lambda) + \text{Mind}\, L_1(\lambda) = \text{Mind}\, L(\lambda). \end{aligned} \tag{2.154}$$

From this, we conclude that the left side of (2.153) is an ILMFD, and the first relations in (2.146), (2.147) are proven. The remaining relations can be proven analogously. ■

(3) Let the matrices $L_1(\lambda) \in \mathbb{R}^{np}(\lambda)$ and $L_2(\lambda) \in \mathbb{R}^{pm}(\lambda)$ be given, and

$$L(\lambda) = L_1(\lambda)L_2(\lambda). \tag{2.155}$$

**Lemma 2.15** *The following claims hold:*

*(i) The inequality*

$$\text{Mind}\, L(\lambda) \le \text{Mind}\, L_1(\lambda) + \text{Mind}\, L_2(\lambda) \tag{2.156}$$

*is true.*

*(ii) If $\psi_L(\lambda)$, $\psi_{L_1}(\lambda)$, $\psi_{L_2}(\lambda)$ are the McMillan denominators of the matrices $L(\lambda)$, $L_1(\lambda)$, $L_2(\lambda)$, respectively, then the relation*

$$\chi(\lambda) \stackrel{\triangle}{=} \frac{\psi_{L_1}(\lambda)\psi_{L_2}(\lambda)}{\psi_L(\lambda)} \tag{2.157}$$

*turns out to be a polynomial.*

*Proof* (i) Assume the ILMFD

$$L(\lambda) = a_l^{-1}(\lambda)b_l(\lambda), \tag{2.158}$$

and also the ILMFD

$$L_i(\lambda) = a_i^{-1}(\lambda)b_i(\lambda), \quad (i = 1, 2). \tag{2.159}$$

Then for the corresponding McMillan denominators, we obtain

$$\text{Mind}\, L(\lambda) = \deg \det a_l(\lambda), \quad \text{Mind}\, L_i(\lambda) = \deg \det a_i(\lambda). \tag{2.160}$$

From (2.155) and (2.159), it follows

$$L(\lambda) = a_1^{-1}(\lambda) b_1(\lambda) a_2^{-1}(\lambda) b_2(\lambda). \tag{2.161}$$

Now, Lemma 2.7 yields the existence of the LMFD

$$a_3^{-1}(\lambda) b_3(\lambda) = b_1(\lambda) a_2^{-1}(\lambda), \tag{2.162}$$

where

$$\det a_3(\lambda) \sim \det a_2(\lambda) \sim \psi_{L_2}(\lambda)$$

and

$$\deg \det a_3(\lambda) = \deg \det a_2(\lambda) = \text{Mind}\, L_2(\lambda).$$

From (2.161) and (2.162), we find the LMFD of the matrix $L(\lambda)$

$$L(\lambda) = [a_3(\lambda) a_1(\lambda)]^{-1} b_3(\lambda) b_2(\lambda). \tag{2.163}$$

Hence, due to properties of LMFD (2.120), we obtain

$$\text{Mind}\, L(\lambda) \le \deg \det[a_3(\lambda) a_1(\lambda)] = \text{Mind}\, L_1(\lambda) + \text{Mind}\, L_2(\lambda), \tag{2.164}$$

and inequality (2.156) is shown.

(ii) Since (2.158) is an ILMFD, it follows from (2.163) that the expressions

$$\frac{\det a_3(\lambda) \det a_1(\lambda)}{\det a_l(\lambda)} \sim \frac{\psi_{L_1}(\lambda) \psi_{L_2}(\lambda)}{\psi_L(\lambda)} \tag{2.165}$$

are polynomials. ∎

Notice that if the products $L_1(\lambda) L_2(\lambda)$ and $L_2(\lambda) L_1(\lambda)$ exist at the same time, then in general

$$\text{Mind}\, [L_1(\lambda) L_2(\lambda)] \ne \text{Mind}\, [L_2(\lambda) L_1(\lambda)]. \tag{2.166}$$

Let us demonstrate the claim at hand of an example.

*Example 2.1* Assume

$$L_1(\lambda) = \begin{bmatrix} V_1(\lambda) & 0 \\ V_2(\lambda) & 0 \end{bmatrix}, \quad L_2(\lambda) = \begin{bmatrix} W(\lambda) & W(\lambda) \\ 0 & 0 \end{bmatrix}, \tag{2.167}$$

where

$$V_1(\lambda) = \frac{1}{\lambda - b} + \frac{1}{\lambda - c}, \quad V_2(\lambda) = \frac{1}{\lambda - b} - \frac{1}{\lambda - c}, \quad W(\lambda) = \frac{1}{\lambda - a} \tag{2.168}$$

and all numbers $a, b, c$ are different. Analyzing the multiplicity of the poles of the nonzero minors of the matrices $L_1(\lambda)$ and $L_2(\lambda)$, we find

$$\operatorname{Mind} L_1(\lambda) = 2, \quad \operatorname{Mind} L_2(\lambda) = 1. \tag{2.169}$$

After multiplication, we obtain

$$\begin{aligned} L_1(\lambda)L_2(\lambda) &= \begin{bmatrix} V_1(\lambda)W(\lambda) & V_1(\lambda)W(\lambda) \\ V_2(\lambda)W(\lambda) & V_2(\lambda)W(\lambda) \end{bmatrix}, \\ L_2(\lambda)L_1(\lambda) &= \begin{bmatrix} W(\lambda)\,[V_1(\lambda) + V_2(\lambda)]\,0 & \\ 0 & 0 \end{bmatrix}. \end{aligned} \tag{2.170}$$

Again, we consider the multiplicity of the roots of the nonzero minors of these products, to achieve

$$\operatorname{Mind}\,[L_1(\lambda)L_2(\lambda)] = 3, \quad \operatorname{Mind}\,[L_2(\lambda)L_1(\lambda)] = 2. \tag{2.171}$$

(4)

**Definition 2.6** The matrices $L_1(\lambda) \in \mathbb{R}^{np}(\lambda)$ and $L_2(\lambda) \in \mathbb{R}^{pm}(\lambda)$ are called independent, when

$$\operatorname{Mind}[L_1(\lambda)L_2(\lambda)] = \operatorname{Mind} L_1(\lambda) + \operatorname{Mind} L_2(\lambda). \tag{2.172}$$

In general, the property of independence of two rational matrices $L_1(\lambda)$ and $L_2(\lambda)$ depends on their particular order. So under the conditions of Example 2.1, the matrices $L_1(\lambda)$ and $L_2(\lambda)$ are independent, but the matrices $L_2(\lambda)$ and $L_1(\lambda)$ do not possess this property.

**Lemma 2.16** *Assume ILMFD (2.158) and (2.159). Then the following claims hold:*

*(i) For the independence of the matrices $L_1(\lambda)$ and $L_2(\lambda)$, it is necessary that the pair $[a_2(\lambda), b_1(\lambda)]$ is irreducible.*
*(ii) Let condition (i) be fulfilled and assume the ILMFD*

$$a_3^{-1}(\lambda)b_3(\lambda) = b_1(\lambda)a_2^{-1}(\lambda). \tag{2.173}$$

*Then for the independence of the matrices $L_1(\lambda)$ and $L_2(\lambda)$, it is necessary and sufficient that the pair $(a_3(\lambda)a_1(\lambda), b_3(\lambda)b_2(\lambda))$ is irreducible.*

*Proof* (i) Assume the contrary that the pair $[a_2(\lambda), b_1(\lambda)]$ is reducible, although the matrices $L_1(\lambda)$ and $L_2(\lambda)$ are independent. Since the pair $[a_2(\lambda), b_1(\lambda)]$ is reducible, there exists a ILMFD

$$b_1(\lambda)a_2^{-1}(\lambda) = a_4^{-1}(\lambda)b_4(\lambda), \tag{2.174}$$

where

$$\deg\det a_4(\lambda) < \deg\det a_2(\lambda) = \mathrm{Mind}\, L_2(\lambda). \tag{2.175}$$

Applying (2.174), we find the LMFD of the matrix $L(\lambda)$

$$L(\lambda) = [a_4(\lambda)a_1(\lambda)]^{-1}b_4(\lambda)b_2(\lambda). \tag{2.176}$$

Here, by construction, we obtain

$$\deg\det[a_4(\lambda)a_1(\lambda)] = \deg\det a_1(\lambda) + \deg\det a_4(\lambda) < \mathrm{Mind}\, L_1(\lambda) + \mathrm{Mind}\, L_2(\lambda), \tag{2.177}$$

which contradicts the proposition on the independence of the matrices $L_1(\lambda)$ and $L_2(\lambda)$.

(ii) Sufficiency: Let condition (i) be fulfilled, then for ILMFD (2.173), we obtain

$$\deg\det a_3(\lambda) = \deg\det a_2(\lambda) = \mathrm{Mind}\, L_2(\lambda).$$

Moreover, for LMFD (2.176), we find

$$\deg\det[a_3(\lambda)a_1(\lambda)] = \mathrm{Mind}\, L_1(\lambda) + \mathrm{Mind}\, L_2(\lambda).$$

This relation implies that LMFD (2.176) is an ILMFD, and the matrices $L_1(\lambda)$ and $L_2(\lambda)$ are independent.
The necessity is proven by reverse reasoning. ■

## 2.7 Excess of Rational Matrices

(1)

**Definition 2.7** Assume $L(\lambda) \in \mathbb{R}^{nm}(\lambda)$. Then the integer $\mathrm{exc}\, L(\lambda)$ with the property, that there exists the finite limit

$$\lim_{\lambda\to\infty} L(\lambda)\lambda^{\mathrm{exc}\, L(\lambda)} = L_\infty \neq 0_{nm}, \tag{2.178}$$

is called the excess of the matrix $L(\lambda)$.

If the matrix $L(\lambda)$ is given in form (2.41), then

$$\mathrm{exc}\, L(\lambda) = \deg d(\lambda) - \deg M(\lambda). \tag{2.179}$$

(2)

**Definition 2.8** For exc $L(\lambda) = 0$ or exc $L(\lambda) > 0$ the matrix $L(\lambda)$ is called proper or strictly proper, respectively. For exc $L(\lambda) \geq 0$, the matrix $L(\lambda)$ is called at least proper, and for exc $L(\lambda) < 0$ improper.

Helpful estimates for the number exc $L(\lambda)$ can be gained from the properties of LMFD and RMFD.

**Lemma 2.17** *Assume for the matrix* $L(\lambda) \in \mathbb{R}^{nm}(\lambda)$ *the MFD*

$$L(\lambda) = a_l^{-1}(\lambda)b_l(\lambda) = b_r(\lambda)a_r^{-1}(\lambda). \tag{2.180}$$

*Then for the number* exc $L(\lambda)$*, the following inequalities are true:*

$$\begin{aligned} \operatorname{exc} L(\lambda) &\leq \deg a_l(\lambda) - \deg b_l(\lambda), \\ \operatorname{exc} L(\lambda) &\leq \deg a_r(\lambda) - \deg b_r(\lambda). \end{aligned} \tag{2.181}$$

*Proof* Assume standard form (2.41) for the matrix $L(\lambda)$. Then, due to (2.180), we can write

$$d(\lambda)b_l(\lambda) = a_l(\lambda)M(\lambda), \tag{2.182}$$

with the consequence

$$\deg[d(\lambda)b_l(\lambda)] = \deg[a_l(\lambda)M(\lambda)]. \tag{2.183}$$

Since

$$\begin{aligned} \deg[d(\lambda)b_l(\lambda)] &= \deg d(\lambda) + \deg b_l(\lambda), \\ \deg[a_l(\lambda)M(\lambda)] &\leq \deg a_l(\lambda) + \deg M(\lambda), \end{aligned} \tag{2.184}$$

from (2.183), we find

$$\deg d(\lambda) + \deg b_l(\lambda) \leq \deg a_l(\lambda) + \deg M_l(\lambda), \tag{2.185}$$

which is equivalent to the first inequality in (2.181). The second inequality can be shown analogously. ■

**Corollary 2.1** *If the matrix* $L(\lambda)$ *is proper, i.e.,* exc $L(\lambda) = 0$*, then (2.181) yields*

$$\deg b_l(\lambda) \leq \deg a_l(\lambda), \quad \deg b_r(\lambda) \leq \deg a_r(\lambda). \tag{2.186}$$

*If, however, the matrix* $L(\lambda)$ *is strictly proper, i.e.,* exc $L(\lambda) > 0$*, then we obtain the strict inequalities*

$$\deg b_l(\lambda) < \deg a_l(\lambda), \quad \deg b_r(\lambda) < \deg a_r(\lambda). \tag{2.187}$$

(3) A complete information about the value of exc $L(\lambda)$ can be obtained in cases, when in MFD (2.180) the matrix $a_l(\lambda)$ is row reduced, and the matrix $a_r(\lambda)$ is column reduced.

**Theorem 2.1** ([3]) *The following statements hold:*

*(i) Let in MFD (2.180) the $n \times n$ matrix $a_l(\lambda)$ be row reduced. Denote by $\alpha_i(\lambda)$, $\beta_i(\lambda)$, $(i = 1, \dots, n)$ the rows of the matrices $a_l(\lambda)$ and $b_l(\lambda)$, respectively, and moreover,*

$$\alpha_i \triangleq \deg \alpha_i(\lambda), \quad \beta_i \triangleq \deg \beta_i(\lambda), \tag{2.188}$$

$$\gamma_i \triangleq \alpha_i - \beta_i, \quad (i = 1, \dots, n) \tag{2.189}$$

$$\delta_l = \min_{1 \le i \le n} \{\gamma_i\}. \tag{2.190}$$

*Then,*

$$\operatorname{exc} L(\lambda) = \delta_l. \tag{2.191}$$

*(ii) Let in MFD (2.180) the $m \times m$ matrix $a_r(\lambda)$ be columns reduced. Denote by $\psi_i(\lambda)$, $\xi_i(\lambda)$, $(i = 1, \dots, m)$ the columns of the matrix $a_r(\lambda)$ and $b_r(\lambda)$, respectively, and moreover,*

$$\psi_i \triangleq \deg \psi_i(\lambda), \quad \xi_i \triangleq \deg \xi_i(\lambda), \tag{2.192}$$

$$\eta_i \triangleq \psi_i - \xi_i, \quad (i = 1, \dots, m), \tag{2.193}$$

$$\delta_r = \min_{1 \le i \le m} \{\eta_i\}. \tag{2.194}$$

*Then,*

$$\operatorname{exc} L(\lambda) = \delta_r. \tag{2.195}$$

*Proof* (i) Using (1.78), we represent the row reduced matrix $a_l(\lambda)$ in the form

$$a_l(\lambda) = \operatorname{diag}\{\lambda^{\alpha_1}, \dots, \lambda^{\alpha_n}\}(A_0 + A_1\lambda^{-1} + A_2\lambda^{-2} + \cdots), \tag{2.196}$$

where $A_i$, $(i = 0, 1, 2, \dots)$ are constant $n \times n$ matrices and $\det A_0 \ne 0$. Assume

$$b_l(\lambda) = B_0\lambda^{\mu} + B_1\lambda^{\mu-1} + \cdots + B_{\mu}, \tag{2.197}$$

where $B_i$ are constant $n \times m$ and $B_0 \ne 0_{nm}$. Excluding from the rows of $b_l(\lambda)$ certain factors, we obtain

$$\begin{aligned} b_l(\lambda) &= \operatorname{diag}\{\lambda^{\alpha_1}, \dots, \lambda^{\alpha_n}\}(\tilde{B}_0\lambda^{-\delta_l} + \tilde{B}_1\lambda^{-\delta_l-1} + \tilde{B}_2\lambda^{-\delta_l-2} + \cdots) \\ &= \operatorname{diag}\{\lambda^{\alpha_1}, \dots, \lambda^{\alpha_n}\}\lambda^{-\delta_l}\left(\tilde{B}_0 + \tilde{B}_1\lambda^{-1} + \tilde{B}_2\lambda^{-2} + \cdots\right), \end{aligned} \tag{2.198}$$

where $\tilde{B}_i$ are constant $n \times m$ matrices, and $\tilde{B}_0 \neq 0_{nm}$. With the help of (2.196), (2.198), from (2.180), we find

$$\lambda^{\delta_l} L(\lambda) = (A_0 + A_1\lambda^{-1} + A_2\lambda^{-2} + \cdots)^{-1}(\tilde{B}_0 + \tilde{B}_1\lambda^{-1} + \tilde{B}_2\lambda^{-2} + \cdots). \tag{2.199}$$

Since $det A_0 \neq 0$, therefrom

$$\lim_{\lambda\to\infty} \lambda^{\delta_l} L(\lambda) = A_0^{-1}\tilde{B}_0 \neq 0_{nm}, \tag{2.200}$$

which, observing (2.178), proves (i).

(ii) Since the excess of a rational matrix does not change under transposition, so statement (ii) directly can be held by transposing the matrix $L(\lambda)$. ■

**Corollary 2.2** *Under the conditions of Theorem 2.1, for a strictly proper matrix $L(\lambda)$, the following conditions are necessary and sufficient*

$$\alpha_i > \beta_i, \ (i = 1, \ldots, n), \ \ or \ \psi_i > \xi_i, \ \ (i = 1, \ldots, m).$$

*For the properness of the matrix $L(\lambda)$, the following conditions are necessary and sufficient*

$$\alpha_i \geq \beta_i, \ (i = 1, \ldots, n), \ \ or \ \psi_i \geq \xi_i, \ (i = 1, \ldots, m), \tag{2.201}$$

*where in each case at least for one i equality takes place.*

*Example 2.2* Consider the matrices

$$a_l(\lambda) = \begin{bmatrix} 2\lambda^2+1 & \lambda+2 \\ 1 & \lambda+1 \end{bmatrix}, \quad b_l(\lambda) = \begin{bmatrix} 2 & 3\lambda^2+1 \\ 5 & 7 \end{bmatrix}.$$

Since

$$a_l(\lambda) = \begin{bmatrix} \lambda^2 & 0 \\ 0 & \lambda \end{bmatrix}\begin{bmatrix} 2 & 0 \\ 0 & 1 \end{bmatrix} + \begin{bmatrix} 1 & \lambda+2 \\ 1 & 1 \end{bmatrix},$$

the matrix $a_l(\lambda)$ is row reduced with $\alpha_1 = 2, \ \alpha_2 = 1$. Moreover, $\beta_1 = 2, \ \beta_2 = 0$. Thus, $\gamma_1 = 0, \ \gamma_2 = 1, \ \delta_l = 0$. This means that the matrix $a_l^{-1}(\lambda)b_l(\lambda)$ is proper. Since in the given case

$$A_0 = \begin{bmatrix} 2 & 0 \\ 0 & 1 \end{bmatrix}, \quad \tilde{B}_0 = \begin{bmatrix} 0 & 3 \\ 0 & 0 \end{bmatrix},$$

relation (2.200) yields

$$\lim_{\lambda\to\infty} a_l^{-1}(\lambda)b_l(\lambda) = A_0^{-1}\tilde{B}_0 = \begin{bmatrix} 0 & 1.5 \\ 0 & 0 \end{bmatrix}.$$

## 2.8 Strictly Proper Rational Matrices

(1) In harmony with the above definition, a matrix $L(\lambda) \in \mathbb{R}^{nm}(\lambda)$ is strictly proper, if $\operatorname{exc} L(\lambda) > 0$. Strictly proper matrices have a number of important special properties, which we will use later. At first, it should be mentioned, that the sum, difference, and product of strictly proper matrices are also strictly proper matrices.

As is known, [2], for any strictly proper matrix $L(\lambda) \in \mathbb{R}^{nm}(\lambda)$, there exists a set of real polynomial matrix descriptions (PMD)

$$\Pi(\lambda) = (\lambda I_n - A, B, C), \tag{2.202}$$

where $A \in \mathbb{R}^{pp}$, $B \in \mathbb{R}^{pm}$, $C \in \mathbb{R}^{np}$ are constant matrices, such that

$$L(\lambda) = C(\lambda I_n - A)^{-1}B. \tag{2.203}$$

When (2.203) is fulfilled, the triplet of constant matrices $(A, B, C)$ is called a realization of the matrix $L(\lambda)$, and the number $p$ is the dimension of this realization.

(2) The realization $(A, B, C)$, where the dimension $p$ takes its minimal possible value $p_{\min}$ is called a minimal realization. Below, we formulate two criteria for the minimality of a realization.

(a) The realization $(A, B, C)$ is minimal, if and only if the pair $(A, B)$ is completely controllable, and the pair $[A, C]$ is completely observable [2].
(b)

**Theorem 2.2** *The realization $(A, B, C)$ is minimal, if and only if*

$$\operatorname{Mind} L(\lambda) = p. \tag{2.204}$$

*Proof* Necessity: Let the realization $(A, B, C)$ be minimal. Then, in the ILMFD

$$a_l^{-1}(\lambda)b_l(\lambda) = C(\lambda I_p - A)^{-1} \tag{2.205}$$

we obtain

$$a_l(\lambda) \overset{p}{\sim} (\lambda I_p - A). \tag{2.206}$$

Moreover, due to Lemma 2.7, the representation

$$L(\lambda) = a_l^{-1}(\lambda)[b_l(\lambda)B] \tag{2.207}$$

is an ILMFD. Thanks to the properties of ILMFD, observing (2.206), we find

$$\deg \det a_l(\lambda) = \operatorname{Mind} L(\lambda) = p. \tag{2.208}$$

The necessity is shown.

Sufficiency: Let Eq. (2.204) be fulfilled. In addition, the pair $(A, B)$ is controllable, and the pair $[A, C]$ is observable. Then, if we would assume the contrary, that there exists ILMFD (2.207), wherein $\deg\det a_l(\lambda) < p$, then this would contradict (2.204).

Further on, the strictly proper matrix $L(\lambda)$ in (2.203) is called transfer matrix of the realization $(A, B, C)$. Two realizations $(A, B, C)$ and $(A_1, B_1, C_1)$ are called related, if their transfer matrices coincide. As is known, two related minimal realizations $(A, B, C)$ and $(A_1, B_1, C_1)$ are similar, i.e., there exists a non-singular matrix $R$, such that

$$A_1 = RAR^{-1}, \ B_1 = RB, \ C_1 = CR^{-1}. \tag{2.209}$$

Moreover,

$$\lambda I_p - A \sim \lambda I_p - A_1.$$

(3) For strictly proper matrices, a sufficient condition for their independence can be formulated.

**Theorem 2.3** *Assume that the strictly proper matrices $L_1(\lambda) \in \mathbb{R}^{nl}(\lambda)$, $L_2(\lambda) \in \mathbb{R}^{lm}(\lambda)$ are given by their minimal realizations $(A_1, B_1, C_1)$ and $(A_2, B_2, C_2)$, i.e.,*

$$L_1(\lambda) = C_1(\lambda I_p - A_1)^{-1}B_1, \quad L_2(\lambda) = C_2(\lambda I_q - A_2)^{-1}B_2. \tag{2.210}$$

*Introduce the realization $(\bar{A}, \bar{B}, \bar{C})$, defined by*

$$\bar{A} = \begin{array}{c} \\ p \\ q \end{array}\!\!\begin{array}{c} \begin{array}{cc} p & q \end{array} \\ \begin{bmatrix} A_1 & B_1C_2 \\ 0_{qp} & A_2 \end{bmatrix} \end{array}, \quad \bar{B} = \begin{array}{c} \\ p \\ q \end{array}\!\!\begin{array}{c} m \\ \begin{bmatrix} 0_{pm} \\ B_2 \end{bmatrix} \end{array}, \quad \bar{C} = n \begin{array}{c} \begin{array}{cc} p & q \end{array} \\ \begin{bmatrix} C_1 & 0_{nq} \end{bmatrix} \end{array}. \tag{2.211}$$

*Then for the independence of the matrices $L_1(\lambda)$ and $L_2(\lambda)$, it is necessary and sufficient that the realization (2.211) is minimal.*

*Proof* Sufficiency: Since the realizations $(A_1, B_1, C_1)$ and $(A_2, B_2, C_2)$ are minimal, we obtain

$$\operatorname{Mind} L_1(\lambda) = p, \quad \operatorname{Mind} L_2(\lambda) = q. \tag{2.212}$$

We easily verify

$$L(\lambda) = L_1(\lambda)L_2(\lambda) = \bar{C}(\lambda I_{p+q} - \bar{A})^{-1}\bar{B}. \tag{2.213}$$

If now realization (2.211) is minimal, we recognize

$$\operatorname{Mind} L(\lambda) = p + q = \operatorname{Mind} L_1(\lambda) + \operatorname{Mind} L_2(\lambda), \tag{2.214}$$

i.e., the matrices $L_1(\lambda)$ and $L_2(\lambda)$ are independent.

Necessity: Let the matrices $L_1(\lambda)$ and $L_2(\lambda)$ be independent, but realization (2.211) not minimal. This means that for the matrix $L(\lambda)$, there exists a representation

$$L(\lambda) = C(\lambda I_\alpha - A)^{-1}B, \tag{2.215}$$

where $\alpha < p + q$. Hence

$$\alpha = \text{Mind}\, L(\lambda) < \text{Mind}\, L_1(\lambda) + \text{Mind}\, L_2(\lambda) = p + q,$$

which contradicts the supposition on the independence of the matrices $L_1(\lambda)$ and $L_2(\lambda)$. Therefore, the independence of the matrices $L_1(\lambda)$ and $L_2(\lambda)$ needs the minimality of realization (2.211).

*Remark 2.3* Let the matrix $L(\lambda) \in \mathbb{R}^{nm}(\lambda)$ not be strictly proper. Then according to (2.46), it can be represented in the form

$$L(\lambda) = L_p(\lambda) + Q(\lambda), \tag{2.216}$$

where $L_p(\lambda) \in \mathbb{R}^{nm}(\lambda)$ is the broken part, and $Q(\lambda) \in \mathbb{R}^{nm}[\lambda]$ is the integral part. Hereby, because the matrix $L_p(\lambda)$ is strictly proper and

$$\text{Mind}\, L(\lambda) = \text{Mind}\, L_p(\lambda), \tag{2.217}$$

we find out that the McMillan index of the matrix $L(\lambda)$ is equal to the dimension of the minimal realization of its broken part.

## 2.9 Division of Polynomial Matrices and Degree Reduction of Polynomial Pairs

(1) The pairs $(a_l(\lambda), b_l(\lambda))$ and $[a_r(\lambda), b_r(\lambda)]$ are called non-singular, if the matrices $a_l(\lambda)$ and $a_r(\lambda)$ are non-singular, respectively. For the non-singular horizontal pair $(a_l(\lambda), b_l(\lambda))$, there exists the rational matrix

$$W_l(\lambda) = a_l^{-1}(\lambda) b_l(\lambda), \tag{2.218}$$

and for the non-singular vertical pair $[a_r(\lambda), b_r(\lambda)]$, there exists the rational matrix

$$W_r(\lambda) = b_r(\lambda) a_r^{-1}(\lambda). \tag{2.219}$$

Below, the matrices $W_l(\lambda)$ and $W_r(\lambda)$ are called transfer matrices of the corresponding pair. The right side of formulae (2.218), (2.219) can be seen as LMFD resp. RMFD of the matrices $W_l(\lambda)$ resp. $W_r(\lambda)$. From the above-stated properties of MFD, it follows that there exists a set of pairs $(a_l(\lambda), b_l(\lambda))$ and $[a_r(\lambda), b_r(\lambda)]$, having the same transfer matrix. Further on, those pairs are called related. Let the pairs $(a_l(\lambda), b_l(\lambda))$ and $(a_{l1}(\lambda), b_{l1}(\lambda))$ be related, where the first one is irreducible. Then, due to the properties of MFD

$$a_{l1}(\lambda) = g_l(\lambda)a_l(\lambda), \quad b_{l1}(\lambda) = g_l(\lambda)b_l(\lambda), \tag{2.220}$$

where the polynomial matrix $g_l(\lambda)$ is non-singular. Formula (2.220) defines the set of all related horizontal pairs.

By analogy, if the pair $[a_r(\lambda), b_r(\lambda)]$ is irreducible, then the set of all related vertical pairs is determined by the relations

$$a_{r1}(\lambda) = a_r(\lambda)g_r(\lambda), \quad b_{r1}(\lambda) = b_r(\lambda)g_r(\lambda), \tag{2.221}$$

where $g_r(\lambda)$ is any non-singular polynomial matrix.

(2) During the solution of practical problems, often the requirement appears for the construction of an equivalent pair with some additional properties. In particular, in many cases the question arises, to construct a related pair of minimal degree.

**Definition 2.9** As degree of the pair $(a_l(\lambda), b_l(\lambda))$ resp. $[a_r(\lambda), b_r(\lambda)]$, we understand the degree of the extended matrix

$$R_h(\lambda) = \begin{bmatrix} a_l(\lambda) & b_l(\lambda) \end{bmatrix}, \text{ resp. } R_v(\lambda) = \begin{bmatrix} a_r(\lambda) \\ b_r(\lambda) \end{bmatrix}. \tag{2.222}$$

**Lemma 2.18** *Let be the transfer matrix $W(\lambda)$ of the irreducible pair $(a_{l1}(\lambda), b_{l1}(\lambda))$ at least proper. Consider the irreducible pair $(a_l(\lambda), b_l(\lambda))$, where*

$$a_l(\lambda) = \chi(\lambda)a_{l1}(\lambda), \quad b_l(\lambda) = \chi(\lambda)b_{l1}(\lambda) \tag{2.223}$$

*and $\chi(\lambda)$ is a unimodular matrix, such that the matrix $a_l(\lambda)$ becomes row reduced. Then pair (2.223) has a minimal degree among all irreducible pairs that are related to $(a_l(\lambda), b_l(\lambda))$.*

*Proof* Since the pair $(a_{l1}(\lambda), b_{l1}(\lambda))$ is irreducible, the set of all related irreducible pairs to the pair $(a_l(\lambda), b_l(\lambda))$ is defined by the relations

$$a_l(\lambda) = \chi(\lambda)a_{l1}(\lambda), \quad b_l(\lambda) = \chi(\lambda)b_{l1}(\lambda), \tag{2.224}$$

where $\chi(\lambda)$ is any unimodular matrix. Let the matrix $\chi(\lambda)$ of such a kind, that the matrix $a_l(\lambda)$ becomes row reduced. Then, applying Corollary 2.1, we obtain

$$\deg b_l(\lambda) \le \deg a_{l1}(\lambda).$$

Moreover, due to Theorem 1.15, the matrix $a_l(\lambda)$ has the lowest degree among all matrices that are left equivalent to $a_{l1}(\lambda)$. From this, the claim of Lemma 2.18 is shown. ■

**Corollary 2.3** *Let be the transfer matrix of the irreducible $[a_r(\lambda), b_r(\lambda)]$ at least proper, and $\chi_1(\lambda)$ be a unimodular matrix, such that the matrix $a_r(\lambda)\chi_1(\lambda)$ becomes* *column reduced. Then the pair* $[a_r(\lambda)\chi_1(\lambda), b_r(\lambda)\chi_1(\lambda)]$ *has the lowest degree among all irreducible pairs, which are related to the pair* $[a_r(\lambda), b_r(\lambda)]$.

(3) In a number of cases, the related pairs of minimal degree can be obtained, without applying equivalence transformations. In particular, for this purpose, divisibility properties of polynomial matrices have to be investigated. The description of this approach needs some general concepts connected with this question, [3].

Consider the matrices $a(\lambda) \in \mathbb{R}^{nn}[\lambda]$, $b(\lambda) \in \mathbb{R}^{nm}[\lambda]$, where the matrix $a(\lambda)$ is non-singular. If the condition

$$b(\lambda) = a(\lambda)q_l(\lambda) + p_l(\lambda) \tag{2.225}$$

is fulfilled, where $q_l(\lambda),\ p_l(\lambda) \in \mathbb{R}^{nm}[\lambda]$, then we say that $q_l(\lambda)$ is the left quotient, and $p_l(\lambda)$ is the left remainder of the matrix $b(\lambda)$, when it is divided from left by the matrix $a(\lambda)$. In analogy, when we have $a(\lambda) \in \mathbb{R}^{mm}[\lambda]$, $b(\lambda) \in \mathbb{R}^{nm}[\lambda]$, and

$$b(\lambda) = q_r(\lambda)a(\lambda) + p_r(\lambda) \tag{2.226}$$

is fulfilled, where $q_r(\lambda),\ p_r(\lambda) \in \mathbb{R}^{nm}[\lambda]$, then we say that $q_r(\lambda)$ is the right quotient, and $p_r(\lambda)$ is the right remainder of the matrix $b(\lambda)$, when it is divided from the right by the matrix $a(\lambda)$.

Since right division can be obtained from left division by transposition, below we will mainly consider left division.

For the given polynomials $a(\lambda)$ and $b(\lambda)$, relation (2.225) could be understood as polynomial equation depending on the matrices $p_l(\lambda), q_l(\lambda)$. In general, this equation has infinitely many solutions. Therefore, in applications, we look for various variants of division, in which the matrices $p_l(\lambda)$ and $q_l(\lambda)$ are subject to constraints.

If the solution of Eq. (2.225) is being subject to the constraint

$$\deg p_l(\lambda) < \deg a(\lambda),$$

then this division is called Euclidean. If however, it is supposed that the matrix $a(\lambda)$ is non-singular, and the matrix $a^{-1}(\lambda)p_l(\lambda)$ is strictly proper, then such a division is called basic division.

**Theorem 2.4** *The following statements hold:*

*(i) The basic division problem is always solvable and has a unique solution.*
*(ii) The solution of the basic division, at the same time, is the solution of the Euclidean division.*
*(iii) If the matrix* $a(\lambda)$ *is regular, then the solutions of the basic and the Euclidean division coincide.*

*Proof* (i) Build a composition of form (2.46)

$$a^{-1}(\lambda)b(\lambda) = A(\lambda) + q_l(\lambda), \tag{2.227}$$

where the matrix $A(\lambda)$ is rational and strictly proper, $q_l(\lambda)$ is a polynomial matrix. Then the matrix

$$p_l(\lambda) \stackrel{\triangle}{=} a(\lambda)A(\lambda) \tag{2.228}$$

is a polynomial, i.e.,

$$p_l(\lambda) = a(\lambda)A(\lambda) = b(\lambda) - a(\lambda)q_l(\lambda). \tag{2.229}$$

Moreover, from (2.227) we find

$$b(\lambda) = a(\lambda)q_l(\lambda) + p_l(\lambda), \tag{2.230}$$

i.e., the matrices $q_l(\lambda)$ and $p_l(\lambda)$ fulfill (2.225). Here, by construction, the matrix $a^{-1}(\lambda)p_l(\lambda) = A(\lambda)$ is strictly proper. Hence, matrices $q_l(\lambda)$ and $p_l(\lambda)$ build a solution of the basic division problem.

We will show that the obtained solution is unique. For that, assume that the basic division problem has another solution $p_1(\lambda), q_1(\lambda)$, such that

$$b(\lambda) = a(\lambda)q_1(\lambda) + p_1(\lambda), \tag{2.231}$$

where the matrix $a^{-1}(\lambda)p_1(\lambda)$ is strictly proper. Then from (2.230) and (2.231), we derive the equation

$$q_l(\lambda) - q_1(\lambda) = a^{-1}(\lambda)p_l(\lambda) - a^{-1}(\lambda)p_1(\lambda).$$

The right side of this equation is a strictly proper rational matrix, and the left side is a polynomial matrix. This is possible if and only if $q_l(\lambda) = q_1(\lambda)$, $p_l(\lambda) = p_1(\lambda)$.

(ii) The claim follows from Corollary 2.3.

(iii) As shown in [3], if the matrix $a(\lambda)$ is regular, then the Euclidean division is solvable, and the corresponding solution is unique. Therefore (iii) follows from (i) and (ii). ∎

*Remark 2.4* Obviously, by transposition Theorem 2.4 can be formulated for the right division case.

(4) Consider the practically important case of division, when the matrix $a(\lambda)$ is regular and has the form

$$a(\lambda) = \lambda I_n - A, \tag{2.232}$$

where $A$ is a constant matrix. Let the $n \times m$ matrix $b(\lambda)$ possess the form

$$b(\lambda) = B_0\lambda^d + B_1\lambda^{d-1} + \cdots + B_d, \tag{2.233}$$

where $B_i$, $(i = 0, \ldots, d)$ are constant $n \times m$ matrices.

**Theorem 2.5** ([1]) *In case of (2.232), (2.233), the left remainder* $p_l(\lambda) = p_l$ *turns out to be a constant matrix with the value*

$$p_l = b(A), \tag{2.234}$$

*wherein we denoted*

$$b(A) \triangleq A^d B_0 + A^{d-1} B_1 + \cdots + B_d. \tag{2.235}$$

*Proof* In the considered case $\deg a(\lambda) = 1$. Therefore, due to Theorem 2.4, the left remainder becomes a constant matrix $p_l(\lambda) = p_l$. Moreover, from the equation

$$b(\lambda) - p_l = (\lambda I_n - A) q_l(\lambda) \tag{2.236}$$

and the regularity of the matrix $\lambda I_n - A$, we find out that $\deg q_l(\lambda) = d - 1$, i.e.,

$$q_l(\lambda) = q_1 \lambda^{d-1} + \cdots + q_d, \tag{2.237}$$

where $q_i$, $(i = 1, \ldots, d)$ are constant $n \times m$ matrices.

Inserting (2.233) and (2.237) into (2.236), we obtain

$$B_0 \lambda^d + B_1 \lambda^{d-1} + \cdots + B_d = (\lambda I_n - A_l)(q_1 \lambda^{d-1} + q_2 \lambda^{d-2} + \cdots + q_d) + p_l. \tag{2.238}$$

Comparing here the coefficients for equal powers of $\lambda$ on both sides, we find

$$\begin{aligned} B_0 &= q_1, \\ B_1 &= q_2 - A q_1, \\ \cdots &\quad \cdots \\ B_{d-1} &= q_d - A q_{d-1}, \\ B_d &= p_l - A q_d. \end{aligned} \tag{2.239}$$

Hence

$$\begin{aligned} q_1 &= B_0, \\ q_2 &= A q_1 + B_1 = A B_0 + B_1, \\ \cdots &\quad \cdots \\ q_d &= A^{d-1} B_0 + A^{d-2} B_1 + \cdots + B_{d-1}. \end{aligned} \tag{2.240}$$

Substituting here for $q_d$ the last expression in (2.239), formula (2.234) is achieved. ■

**Corollary 2.4** *Assume*

$$b(\lambda) = B_0 \lambda^d + \cdots + B_d, \quad a(\lambda) = \lambda I_m - A \tag{2.241}$$

*and the result of the right division*

$$b(\lambda) = q_r(\lambda)(\lambda I_m - A) + p_r(\lambda). \tag{2.242}$$

*Then, the right remainder* $p_r(\lambda) = b_r(A)$ *is a constant matrix with the value*

$$b_r(A) = B_0 A^d + \cdots + B_d. \tag{2.243}$$

(5)

**Theorem 2.6** *Assume the horizontal matrix of the form*

$$Q(\lambda) = \left[ a(\lambda)\ b(\lambda) \right], \tag{2.244}$$

*where* $a(\lambda) \in \mathbb{R}^{nn}[\lambda]$, $b(\lambda) \in \mathbb{R}^{nm}[\lambda]$. *Moreover, the matrix* $a(\lambda)$ *is non-singular,* $\deg\det a(\lambda) = \gamma$, *and the pair* $(a(\lambda), b(\lambda))$ *is irreducible. Then the matrix* $Q(\lambda)$ *is equivalent to a matrix*

$$\tilde{Q}(\lambda) \triangleq \left[ \tilde{a}(\lambda)\ \tilde{b}(\lambda) \right], \tag{2.245}$$

*where the matrix* $\tilde{a}(\lambda)$ *is left equivalent to the matrix* $a(\lambda)$, $\deg \tilde{a}(\lambda) \leq \gamma$, *and the matrix* $\tilde{a}(\lambda)^{-1}\tilde{b}(\lambda)$ *is strictly proper.*

*Proof* Let the result of the left division of the matrix $b(\lambda)$ by the matrix $a(\lambda)$ possess form (2.230), where the matrix $a^{-1}(\lambda)p_l(\lambda)$ is strictly proper. Introduce the unimodular matrix

$$R(\lambda) = \begin{bmatrix} I_n & -q_l(\lambda) \\ 0_{mn} & I_m \end{bmatrix}. \tag{2.246}$$

Thus

$$\begin{aligned} \left[ a(\lambda)\ b(\lambda) \right] R(\lambda) &= \left[ a(\lambda)\ b(\lambda) \right] \begin{bmatrix} I_n & -q_l(\lambda) \\ 0_{mn} & I_m \end{bmatrix} \\ &= \left[ a(\lambda)\ b(\lambda) - a(\lambda)q_l(\lambda) \right] = \left[ a(\lambda)\ p_l(\lambda) \right]. \end{aligned} \tag{2.247}$$

Multiplying the last equation from left by $\chi(\lambda)$, where $\chi(\lambda)$ is the unimodular matrix configured in Lemma 2.18, we obtain

$$\tilde{Q}(\lambda) = \chi(\lambda) \left[ a(\lambda)\ b(\lambda) \right] R(\lambda) = \left[ \chi(\lambda)a(\lambda)\ \chi(\lambda)p(\lambda) \right] = \left[ \tilde{a}(\lambda)\ \tilde{b}(\lambda) \right]. \tag{2.248}$$

The matrix $\tilde{Q}(\lambda)$ possesses all the above claimed properties. ■

*Remark 2.5* Under the condition of Theorem 2.6, the degree of each row of the matrix $\tilde{b}(\lambda)$ is lower than the degree of the corresponding row of the matrix $\tilde{a}(\lambda)$.

**Corollary 2.5** *The matrix of the form*

$$Q(\lambda) = \left[\, I_n - \lambda A \; b(\lambda) \,\right], \tag{2.249}$$

*where $A$ is a non-singular matrix and $b(\lambda)$ has form (2.233), is equivalent to the matrix*

$$\tilde{Q}(\lambda) = \left[\, I_n - \lambda A \; b(A^{-1}) \,\right], \tag{2.250}$$

*where*

$$b(A^{-1}) = A^{-d} B_0 + A^{-d+1} B_1 + \cdots + B_d. \tag{2.251}$$

*Proof* The proof emerges from the chain of relations

$$\begin{aligned} Q(\lambda) &= -A\left[\, I_n\lambda - A^{-1} \; -A^{-1}b(\lambda) \,\right] \\ &\sim -A\left[\, I_n\lambda - A^{-1} \; -A^{-1}b(A^{-1}) \,\right] = \left[\, I_n - \lambda A \; b(A^{-1}) \,\right]. \end{aligned} \tag{2.252}$$

■

## 2.10 McMillan Index of Block Rational Matrices

(1) Consider the block rational matrix

$$W(\lambda) = \begin{bmatrix} W_{11}(\lambda) & \cdots & W_{1m}(\lambda) \\ \cdots & \cdots & \cdots \\ W_{n1}(\lambda) & \cdots & W_{nm}(\lambda) \end{bmatrix}, \tag{2.253}$$

where

$$W_{ik}(\lambda), \;\; (i = 1, \ldots, n; k = 1, \ldots, m)$$

are rational matrices of appropriate size. Let $\psi_{ik}(\lambda)$ be the McMillan denominator of the matrix $W_{ik}(\lambda)$, and $\psi(\lambda)$ the McMillan denominator of the matrix $W(\lambda)$.

**Lemma 2.19** *All expressions*

$$d_{ik}(\lambda) \stackrel{\triangle}{=} \frac{\psi(\lambda)}{\psi_{ik}(\lambda)} \tag{2.254}$$

*turn out to be polynomials.*

*Proof* At first consider the block row

$$W(\lambda) = \left[\, W_1(\lambda) \ldots W_m(\lambda) \,\right]. \tag{2.255}$$

Build the ILMFD

$$W(\lambda) = a^{-1}(\lambda)b(\lambda), \tag{2.256}$$

where according to the properties of ILMFD

$$\det a(\lambda) \sim \psi(\lambda). \tag{2.257}$$

Thus

$$\text{Mind}\, W(\lambda) = \deg \det a(\lambda). \tag{2.258}$$

Since the matrix

$$b(\lambda) = a(\lambda)W(\lambda) = \left[ a(\lambda)W_1(\lambda) \ldots a(\lambda)W_m(\lambda) \right]$$

is a polynomial matrix, the polynomial matrix $a(\lambda)$ is a divisor for all matrices $W_i(\lambda)$. Let $\psi_i(\lambda)$ be the McMillan denominator of the matrix $W_i(\lambda)$. Then, due to (2.81), relation $\psi(\lambda)/\psi_i(\lambda)$ is a polynomial. So the claim of the lemma is shown for block rows. Obviously, it is also true for block columns. In the general case denote $\tilde{W}_i(\lambda)$, $(i = 1, \ldots, m)$ the block columns of matrix $W(\lambda)$ according to (2.253). Then

$$W(\lambda) = \left[ \tilde{W}_1(\lambda) \ldots \tilde{W}_m(\lambda) \right]. \tag{2.259}$$

Denote by $\tilde{\psi}_i(\lambda)$ the McMillan denominator of the matrix $\tilde{W}_i(\lambda)$. Thanks to the above-shown relations, $\psi(\lambda)/\tilde{\psi}_i(\lambda)$ are polynomials. At the same time, for the block column

$$\tilde{W}_i(\lambda) = \begin{bmatrix} W_{i1}(\lambda) \\ \vdots \\ W_{in}(\lambda) \end{bmatrix}, \tag{2.260}$$

the expressions $\tilde{\psi}_i(\lambda)/\psi_{ik}(\lambda)$ are also polynomials. Therefore, the relations

$$\frac{\psi(\lambda)}{\psi_{ik}(\lambda)} = \frac{\psi(\lambda)}{\tilde{\psi}_i(\lambda)} \frac{\tilde{\psi}_i(\lambda)}{\psi_{ik}(\lambda)}$$

are polynomials. ■

(2)

**Corollary 2.6** *For arbitrary entries of the block matrix $W(\lambda)$ (2.253)*

$$\text{Mind}\, W_{ik}(\lambda) \leq \text{Mind}\, W(\lambda). \tag{2.261}$$

For many problems, the situation is of interest, where some entries of block matrix (2.253) determine the McMillan index of that matrix.

**Definition 2.10** If there exists an entry $W_{ik}(\lambda)$ for the block matrix (2.253), such that

$$\psi_{ik}(\lambda) = \psi(\lambda), \tag{2.262}$$

then we say that the matrix $W_{ik}(\lambda)$ dominates in this matrix.

**Lemma 2.20** *The matrix $W_{ik}(\lambda)$ dominates in matrix (2.253), if and only if*

$$\text{Mind } W_{ik}(\lambda) = \text{Mind } W(\lambda). \tag{2.263}$$

*Proof* The necessity of Eq. (2.263) follows directly from (2.258). Assume (2.263), then

$$\deg \psi_{ik}(\lambda) = \deg \psi(\lambda). \tag{2.264}$$

Since the relation $\psi(\lambda)/\psi_{ik}(\lambda)$ is a polynomial, and this polynomial is monic, this expression is one, i.e., (2.262) is true.

(3)

**Definition 2.11** Let the matrices $W_1(\lambda) \in \mathbb{R}^{nm}(\lambda)$, $W_2(\lambda) \in \mathbb{R}^{nl}(\lambda)$ be given, and the ILMFD $W_1(\lambda) = a_l^{-1}(\lambda)b_l(\lambda)$ takes place. Then, we say that the matrix $W_2(\lambda)$ is subordinated from left to the matrix $W_1(\lambda)$, and we write

$$W_2(\lambda) \overset{l}{\prec} W_1(\lambda), \quad (W_1(\lambda) \overset{l}{\succ} W_2(\lambda)), \tag{2.265}$$

if the matrix

$$\tilde{b}(\lambda) \overset{\triangle}{=} a_l(\lambda) W_2(\lambda) \tag{2.266}$$

is a polynomial. In analogy, if we have the IRMFD $W_1(\lambda) = b_r(\lambda)a_r^{-1}(\lambda)$, then we say that the matrix $W_2(\lambda)$ is subordinated from right to the matrix $W_1(\lambda)$, and we write

$$W_2(\lambda) \overset{r}{\prec} W_1(\lambda), \quad (W_1(\lambda) \overset{r}{\succ} W_2(\lambda)), \tag{2.267}$$

if the matrix

$$\tilde{b}_r(\lambda) \overset{\triangle}{=} W_2(\lambda) a_r(\lambda) \tag{2.268}$$

is a polynomial.

**Theorem 2.7** *Assume the block row*

$$W(\lambda) = \left[\, W_1(\lambda) \;\; W_2(\lambda) \,\right] \tag{2.269}$$

*and the ILMFD*

$$W_1(\lambda) = a_1^{-1}(\lambda) b_1(\lambda). \tag{2.270}$$

*Then, for the fact that the matrix $W_1(\lambda)$ dominates in the matrix $W(\lambda)$, it is necessary and sufficient, that the condition*

$$W_2(\lambda) \overset{l}{\prec} W_1(\lambda) \tag{2.271}$$

*is true, i.e.,* $\tilde{b}_l(\lambda) = a_1(\lambda)W_2(\lambda)$ *is a polynomial.*

*Proof* Necessity: Assume

$$\text{Mind}\, W(\lambda) = \text{Mind}\, W_1(\lambda) \tag{2.272}$$

and let be given the ILMFD

$$W(\lambda) = a^{-1}(\lambda)b(\lambda). \tag{2.273}$$

Hence, from (2.272), it follows that $a(\lambda) = \chi(\lambda)a_1(\lambda)$, where $\chi(\lambda)$ is a unimodular matrix. Moreover, because the matrix $a(\lambda)W_2(\lambda)$ is a polynomial, so matrix $\tilde{b}_l(\lambda)$ (2.266) is also a polynomial.

Sufficiency: Let us have ILMFD (2.270), and $\tilde{b}_l(\lambda)$ be a polynomial matrix. Then, we obtain

$$W(\lambda) = \left[\, a_1^{-1}(\lambda)b_1(\lambda)\; a_1^{-1}(\lambda)\tilde{b}_l(\lambda)\,\right] = a_1^{-1}(\lambda)\left[\, b_1(\lambda)\; \tilde{b}_l(\lambda)\,\right]. \tag{2.274}$$

The right part of the last relation is an ILMFD. Indeed, if we assume the contrary, so that an ILMFD has the form

$$W(\lambda) = \bar{a}^{-1}(\lambda)\left[\, \bar{b}_1(\lambda)\; \bar{b}_2(\lambda)\,\right], \tag{2.275}$$

where $\deg\det \bar{a}(\lambda) = \alpha < \deg\det a(\lambda)$. Then we would obtain the LMFD

$$W_1(\lambda) = \bar{a}^{-1}(\lambda)\bar{b}_1(\lambda), \tag{2.276}$$

for which $\deg\det \bar{a}(\lambda) < \text{Mind}\, a(\lambda)$ that contradicts the assumption that (2.270) is an ILMFD. Thus, the right side of (2.274) is an ILMFD, and we obtain

$$\text{Mind}\, W(\lambda) = \deg\det a(\lambda) = \text{Mind}\, W_1(\lambda). \tag{2.277}$$

■

**Lemma 2.21** *Assume (2.265), and* $d_1(\lambda), d_2(\lambda)$ *are any polynomial matrices of appropriate size. Then,*

$$W_2(\lambda) + d_2(\lambda) \overset{l}{\prec} W_1(\lambda) + d_1(\lambda). \tag{2.278}$$

*Proof* Assume ILMFD (2.270). Then, due to Lemma 2.8, the representation

$$W_1(\lambda) + d_1(\lambda) = a_1^{-1}(\lambda)\left[b_1(\lambda) + a_1(\lambda)d_1(\lambda)\right] \tag{2.279}$$

is an ILMFD. Hence, from (2.265), it follows that the product

$$a_1(\lambda)\,[W_2(\lambda) + d_2(\lambda)] \tag{2.280}$$

is a polynomial matrix, which proves (2.278). ■

*Remark 2.6* An analogous claim could be formulated for subordination from right. Moreover, if the matrices $W_1(\lambda)$ and $W_2(\lambda)$ satisfy relations (2.267) or (2.271), then its fractional part possesses an analog property.

(4)

**Theorem 2.8** *Let the strictly proper matrix $W_1(\lambda) \in \mathbb{R}^{nm}(\lambda)$ be represented in the minimal realization $(A, B, C)$:*

$$W_1(\lambda) = C(\lambda I_p - A)^{-1}B. \tag{2.281}$$

*Then for the validity of the relation*

$$W_2(\lambda) \overset{l}{\prec} W_1(\lambda), \tag{2.282}$$

*where $W_2(\lambda) \in \mathbb{R}^{nq}(\lambda)$ is strictly proper, it is necessary and sufficient, that the matrix $W_2(\lambda)$ allows the representation*

$$W_2(\lambda) = C(\lambda I_p - A)^{-1}B_1, \tag{2.283}$$

*where $B_1$ is a constant $p \times q$ matrix.*

*Proof* Sufficiency: Build the ILMFD

$$C(\lambda I_p - A)^{-1} = a^{-1}(\lambda)b(\lambda). \tag{2.284}$$

Then, due to Lemma 2.7, the representation

$$W_1(\lambda) = a^{-1}(\lambda)\,[b(\lambda)B] \tag{2.285}$$

is an ILMFD. Moreover, from (2.283) we obtain that

$$a(\lambda)W_2(\lambda) = b(\lambda)B_1 \tag{2.286}$$

is a polynomial matrix, thus (2.282) follows.

Necessity: Assume (2.282). Build the strictly proper matrix

$$W(\lambda) = \left[\, W_1(\lambda) \;\; W_2(\lambda) \,\right] \in \mathbb{R}^{n,m+q}. \tag{2.287}$$

From the minimality of realization (2.281) together with relations (2.282), it follows that

$$\operatorname{Mind} W(\lambda) = \operatorname{Mind} W_1(\lambda) = p. \tag{2.288}$$

From this, because the matrix $W(\lambda)$ is strictly proper, it follows the existence of a minimal realization

$$W(\lambda) = \tilde{C}(\lambda I_p - \tilde{A})^{-1}\tilde{B}, \tag{2.289}$$

where $\tilde{C} \in \mathbb{R}^{np}$, $\tilde{B} \in \mathbb{R}^{p,m+q}$, and the pair $(\tilde{A}, \tilde{B})$ is controllable, and the pair $[\tilde{A}, \tilde{C}]$ is observable. Represent the matrix $\tilde{B}$ in the block form

$$\tilde{B} = p \begin{array}{c} \begin{matrix} m & q \end{matrix} \\ \begin{bmatrix} \tilde{B}_1 & \tilde{B}_2 \end{bmatrix} \end{array} \tag{2.290}$$

where $\tilde{B}_1 \in \mathbb{R}^{pm}$, $\tilde{B}_2 \in \mathbb{R}^{pq}$. Then from (2.289), we find

$$W(\lambda) = \begin{bmatrix} \tilde{C}(\lambda I_p - \tilde{A})^{-1}\tilde{B}_1 & \tilde{C}(\lambda I_p - \tilde{A})^{-1}\tilde{B}_2 \end{bmatrix}, \tag{2.291}$$

from which, we obtain

$$W_1(\lambda) = \tilde{C}(\lambda I_p - \tilde{A})^{-1}\tilde{B}_1. \tag{2.292}$$

Since $\operatorname{Mind} W_1(\lambda) = p$, realization (2.292) is minimal. Using that realization (2.281) is minimal too, we obtain

$$\tilde{A} = RAR^{-1}, \quad \tilde{B}_1 = RB, \quad \tilde{C} = CR^{-1}, \tag{2.293}$$

where $R \in \mathbb{R}^{pp}$ is a non-singular matrix. Moreover, from (2.291), we find

$$W_2(\lambda) = \tilde{C}(\lambda I_p - \tilde{A})^{-1}\tilde{B}_2 = C(\lambda I_p - A)^{-1}B_2, \quad B_2 = R^{-1}\tilde{B}_2, \tag{2.294}$$

which completes the proof. ■

An analogous claim as in Theorem 2.8 is true for right subordination.

**Theorem 2.9** *For the compliance with relation (2.267), when the matrices $W_1(\lambda)$ and $W_2(\lambda)$ are strictly proper, and the minimal representation (2.281) takes place, it is necessary and sufficient, that the matrix $W_2(\lambda)$ allows a representation*

$$W_2(\lambda) = C_1(\lambda I_p - A)^{-1}B, \tag{2.295}$$

*where $C_1$ is a constant matrix of appropriate size.* ■

(5)

**Theorem 2.10** *Assume the rational matrices of appropriate size*

$$F(\lambda), \; G(\lambda), \quad H(\lambda) = F(\lambda)G(\lambda) \tag{2.296}$$

*and we have the ILMFD*

$$F(\lambda) = a_1^{-1}(\lambda)b_1(\lambda), \quad G(\lambda) = a_2^{-1}(\lambda)b_2(\lambda), \tag{2.297}$$

*and in addition*

$$b_1(\lambda)a_2^{-1}(\lambda) = a_3^{-1}(\lambda)b_3(\lambda). \tag{2.298}$$

*Then the relation*

$$F(\lambda) \overset{l}{\prec} H(\lambda), \tag{2.299}$$

*is true, if the matrix*

$$R_h(\lambda) = \left[\, a_3(\lambda)a_1(\lambda) \;\; b_3(\lambda)b_2(\lambda) \,\right] \tag{2.300}$$

*is alatent, i.e., the pair* $(a_3(\lambda)a_1(\lambda), b_3(\lambda)b_2(\lambda))$ *is irreducible.*

*Proof* Assume the ILMFD

$$H(\lambda) = a^{-1}(\lambda)b(\lambda). \tag{2.301}$$

At the same time, using (2.297) and (2.298), we obtain

$$\begin{aligned} H(\lambda) &= a_1^{-1}(\lambda)b_1(\lambda)a_2^{-1}(\lambda)b_2(\lambda) = a_1^{-1}(\lambda)a_3^{-1}(\lambda)b_3(\lambda)b_2(\lambda) \\ &= [a_3(\lambda)a_1(\lambda)]^{-1}\, b_3(\lambda)b_2(\lambda). \end{aligned} \tag{2.302}$$

Let matrix (2.300) be alatent. Then the right side of (2.302) is an ILMFD. Since (2.301) is an ILMFD too, we obtain

$$a(\lambda) = \xi(\lambda)a_3(\lambda)a_1(\lambda), \tag{2.303}$$

where $\xi(\lambda)$ is a unimodular matrix. Applying (2.303), we find out that the product

$$a(\lambda)F(\lambda) = \xi(\lambda)a_3(\lambda)b_1(\lambda) \tag{2.304}$$

is a polynomial matrix. ∎

A corresponding claim holds for right MFD.

**Theorem 2.11** *Assume (2.296), and let us have the IRMFD*

$$F(\lambda) = \tilde{b}_1(\lambda)\tilde{a}_1^{-1}(\lambda), \; G(\lambda) = \tilde{b}_2(\lambda)\tilde{a}_2^{-1}(\lambda), \quad H(\lambda) = \tilde{b}(\lambda)\tilde{a}^{-1}(\lambda). \tag{2.305}$$

*Moreover, the IRMFD*

$$\tilde{a}_1^{-1}(\lambda)\tilde{b}_2(\lambda) = \tilde{b}_3(\lambda)\tilde{a}_3^{-1}(\lambda) \tag{2.306}$$

*should be given. Then for the compliance with the relation*

$$G(\lambda) \overset{r}{\prec} H(\lambda), \tag{2.307}$$

*it is sufficient, that the pair*

$$[\tilde{a}_2(\lambda)\tilde{a}_3(\lambda), \tilde{b}_1(\lambda)\tilde{b}_3(\lambda)] \tag{2.308}$$

*is irreducible.* ∎

**Corollary 2.7** *Let the matrices $L_1(\lambda) \in \mathbb{R}^{nl}(\lambda)$, $L_2(\lambda) \in \mathbb{R}^{lm}(\lambda)$ be independent, and $L(\lambda) = L_1(\lambda)L_2(\lambda)$. Then*

$$L_1(\lambda) \overset{l}{\prec} L(\lambda), \quad L_2(\lambda) \overset{r}{\prec} L(\lambda). \tag{2.309}$$

*Proof* The proof immediately follows from Lemma 2.16, because the condition on irreducibility of the pairs $(a_3(\lambda)a_1(\lambda), b_3(\lambda)b_2(\lambda))$ and $[\tilde{a}_2(\lambda)\tilde{a}_3(\lambda), \tilde{b}_1(\lambda)\tilde{b}_2(\lambda)]$ is a necessary condition for the independence of the matrices $L_1(\lambda)$ and $L_2(\lambda)$. ∎

*Remark 2.7* Under the condition of Theorem 2.10, applying (2.303), we obtain

$$a(\lambda)F(\lambda) = \zeta(\lambda)a_3(\lambda)b_3(\lambda), \tag{2.310}$$

which means that the matrix $\zeta(\lambda)a_3(\lambda)$ is a left divisor of the polynomial matrix $a(\lambda)F(\lambda)$. In the same way, it follows from Corollary 2.7 that

$$G(\lambda)\tilde{a}(\lambda) = \tilde{b}_2(\lambda)\tilde{a}_3(\lambda)\psi(\lambda) \tag{2.311}$$

with a unimodular matrix $\psi(\lambda)$, which means that the matrix $\tilde{a}_3(\lambda)\psi(\lambda)$ is a right divisor of the polynomial matrix $G(\lambda)\tilde{a}(\lambda)$.

(6) The next theorem is a generalization of Theorem 2.8.

**Theorem 2.12** *Assume the strictly proper block matrix*

$$\bar{W}(\lambda) \overset{\triangle}{=} \begin{matrix} \\ i \\ n \end{matrix} \begin{matrix} \begin{matrix} l & \quad m \end{matrix} \\ \begin{bmatrix} K(\lambda) & L(\lambda) \\ M(\lambda) & N(\lambda) \end{bmatrix} \end{matrix} \tag{2.312}$$

*and the minimal realization*

$$N(\lambda) = C(\lambda I_p - A)^{-1}B. \tag{2.313}$$

*Then for the fact that the matrix $N(\lambda)$ dominates in the matrix $\bar{W}(\lambda)$, i.e., for the validity of*

$$\text{Mind}\, \bar{W}(\lambda) = \text{Mind}\, N(\lambda) = p, \tag{2.314}$$

*it is necessary and sufficient that there exist a $i \times p$ matrix $C_1$ and a $p \times l$ matrix $B_1$ such that the following set of relations holds:*

$$\begin{aligned} K(\lambda) &= C_1(\lambda I_p - A)^{-1}B_1, \\ L(\lambda) &= C_1(\lambda I_p - A)^{-1}B, \\ M(\lambda) &= C(\lambda I_p - A)^{-1}B_1. \end{aligned} \tag{2.315}$$

*Proof* Necessity: Assume (2.314). Then, because the matrix $\bar{W}(\lambda)$ is strictly proper, there exists a minimal realization

$$\bar{W}(\lambda) = \tilde{C}(\lambda I_p - \tilde{A})^{-1}\tilde{B}, \tag{2.316}$$

where $\tilde{A}, \tilde{B}, \tilde{C}$ are constant matrices of dimensions $(i+n) \times p, p \times p, p \times (l+m)$, respectively. Take the partitioning into blocks

$$\tilde{C} = \begin{matrix} \\ i \\ n \end{matrix} \overset{p}{\begin{bmatrix} \tilde{C}_1 \\ \tilde{C}_2 \end{bmatrix}}, \quad \tilde{B} = p \; \overset{\begin{matrix} l & m \end{matrix}}{\begin{bmatrix} \tilde{B}_1 & \tilde{B}_2 \end{bmatrix}}. \tag{2.317}$$

Substituting these relations in (2.316), we obtain

$$\bar{W}(\lambda) = \begin{bmatrix} \tilde{C}_1(\lambda I_p - \tilde{A})^{-1}\tilde{B}_1 & \tilde{C}_1(\lambda I_p - \tilde{A})^{-1}\tilde{B}_2 \\ \tilde{C}_2(\lambda I_p - \tilde{A})^{-1}\tilde{B}_1 & \tilde{C}_2(\lambda I_p - \tilde{A})^{-1}\tilde{B}_2 \end{bmatrix}. \tag{2.318}$$

Comparing this expression with (2.312), we find the realization

$$N(\lambda) = \tilde{C}_2(\lambda I_p - \tilde{A})^{-1}\tilde{B}_2. \tag{2.319}$$

Since realizations (2.313) and (2.319) have the same dimension, and realization (2.313) is minimal, also realization (2.319) must be minimal. Hence, we obtain

$$\tilde{A} = RAR^{-1}, \; \tilde{B}_2 = RB, \; \tilde{C}_2 = CR^{-1} \tag{2.320}$$

with a certain non-singular matrix $R$. Substituting this relation in (2.318), we arrive at formula (2.315), where

$$C_1 = \tilde{C}_1 R, \quad B_1 = R^{-1}\tilde{B}_1, \tag{2.321}$$

which proves the necessity.

Sufficiency: Let relations (2.313), (2.315) be fulfilled. Then, we obtain

$$\bar{W}(\lambda) = \begin{bmatrix} C_1(\lambda I_p - A)^{-1}B_1 & C_1(\lambda I_p - A)^{-1}B \\ C(\lambda I_p - A)^{-1}B_1 & C(\lambda I_p - A)^{-1}B \end{bmatrix}, \tag{2.322}$$

which can be represented in the form

$$\bar{W}(\lambda) = \begin{bmatrix} C_1 \\ C \end{bmatrix} (\lambda I_p - A)^{-1} \begin{bmatrix} B_1 & B \end{bmatrix}. \tag{2.323}$$

From (2.323), it follows that Mind $\bar{W}(\lambda) \leq p$. On the other side, because realization (2.313) is minimal, due to Corollary 2.6, Mind $\bar{W}(\lambda) \geq p$. Thus, Mind $\bar{W}(\lambda) = p$. ■

**Corollary 2.8** *When conditions (2.315) are fulfilled, then the following subordinations are true:*

$$\begin{aligned} K(\lambda) &\overset{l}{\prec} L(\lambda), \quad K(\lambda) \overset{r}{\prec} M(\lambda), \\ L(\lambda) &\overset{r}{\prec} N(\lambda), \quad M(\lambda) \overset{l}{\prec} N(\lambda). \end{aligned} \tag{2.324}$$

*Proof* Consider the ILMFD

$$C_1(\lambda I_p - A)^{-1} = a^{-1}(\lambda)b(\lambda). \tag{2.325}$$

Since the pair $(A, B)$ is controllable, the right side of the relation

$$L(\lambda) = a^{-1}(\lambda)[b(\lambda)B], \tag{2.326}$$

due to Lemma 2.7, is an ILMFD. Hence, the product

$$a(\lambda)K(\lambda) = b(\lambda)B_1 \tag{2.327}$$

is a polynomial matrix. This fact proves the first claim of relations (2.324). The remaining claims can be seen analogously. ■

**Theorem 2.13** *Let the matrix $\bar{W}(\lambda)$ in (2.312) not be strictly proper. Let $K_1(\lambda)$, $L_1(\lambda)$, $M_1(\lambda)$, $N_1(\lambda)$ be the fractional parts of the matrices $K(\lambda)$, $L(\lambda)$, $M(\lambda)$, $N(\lambda)$, respectively. Then Eq. (2.314) becomes true, if and only if the matrix*

$$\bar{W}_1(\lambda) = \begin{bmatrix} K_1(\lambda) & L_1(\lambda) \\ M_1(\lambda) & N_1(\lambda) \end{bmatrix} \tag{2.328}$$

*satisfies all conditions of Theorem 2.12.*

*Proof* Obviously, we can dissect

$$\bar{W}(\lambda) = \bar{W}_1(\lambda) + D_1(\lambda), \tag{2.329}$$

where $\bar{W}_1(\lambda)$ is the fractional part of $\bar{W}(\lambda)$, and $D_1(\lambda)$ is its integer part. Hence, the claim follows from Lemma 2.8. ■

## 2.11 Associated Rational Matrices

(1) In this section, we will use the two complex variables $\lambda \triangleq z$ and $\lambda \triangleq \zeta = z^{-1}$.

**Definition 2.12** The rational matrix $\tilde{W}(z) \in \mathbb{R}^{nm}(z)$ is called causal, if it is at least proper, and strictly causal, if it is strictly proper.

**Definition 2.13** The rational matrix $W(\zeta) \in \mathbb{R}^{nm}(\zeta)$ is called causal, it does not possess a pole at $\zeta = 0$, and strictly causal, if

$$W(0) = 0_{nm}. \tag{2.330}$$

**Definition 2.14** The matrices $\tilde{W}(z) \in \mathbb{R}^{nm}(z)$ and $W(\zeta) \in \mathbb{R}^{nm}(\zeta)$ are called associated, if they are related by

$$\tilde{W}(z) = W(z^{-1}), \quad W(\zeta) = \tilde{W}(\zeta^{-1}). \tag{2.331}$$

**Lemma 2.22** *Associated matrices are at the same time causal or strictly causal.*

*Proof* Let the matrix $\tilde{W}(z)$ be causal. Then, there exists the finite limit

$$\lim_{z\to\infty} \tilde{W}(z) = \tilde{W}_{\infty}. \tag{2.332}$$

Hence from (2.331) and (2.332)

$$\lim_{\zeta\to 0} W(\zeta) = W(0) = \tilde{W}_{\infty}, \tag{2.333}$$

i.e., the matrix $W(\zeta)$ is causal. If the matrix $\tilde{W}(z)$ is strictly causal, then $\tilde{W}_{\infty} = 0_{nm}$, thus

$$W(0) = 0_{nm}, \tag{2.334}$$

i.e., the matrix $W(\zeta)$ is strictly causal. The further run of the proof is obvious. ∎

(2)

**Theorem 2.14** *Let the matrix $\tilde{W}(z) \in \mathbb{R}^{nm}(z)$ be causal, and assume the ILMFD*

$$\tilde{W}(z) = \tilde{a}^{-1}(z)\tilde{b}(z). \tag{2.335}$$

*Moreover, let the associated matrix $W(\zeta)$ and its ILMFD*

$$W(\zeta) = a^{-1}(\zeta)b(\zeta) \tag{2.336}$$

*be given. Then the relation*

$$a(\zeta) = \operatorname{rec}[\tilde{a}(z)] \tag{2.337}$$

*holds, i.e., $a(\zeta)$ is the reciprocal matrix according to $\tilde{a}(z)$.*

*Proof* The proof runs over several steps.

(a) Since the matrix $\tilde{W}(z)$ is at least causal, it allows a realization

$$\tilde{W}(z) = C(zI_p - A)^{-1}B + D, \tag{2.338}$$

where $(A, B, C)$ is minimal. Build the ILMFD

$$C(zI_p - A)^{-1} = \tilde{a}_1^{-1}(z)\tilde{b}_1(z). \tag{2.339}$$

Then,

$$\tilde{W}(z) = \tilde{a}_1^{-1}(z)[\tilde{b}_1(z)B + \tilde{a}_1(z)D], \tag{2.340}$$

where the right side is an ILMFD. Due to the fact that the right side of (2.335) is an ILMFD too, we obtain

$$\tilde{a}(z) \sim \tilde{a}_1(z). \tag{2.341}$$

At the same time, since the left and right side of (2.339) are IMFD, we realize that the matrices $\tilde{a}_1(z)$ and $zI_p - A$ are pseudo-equivalent, i.e.,

$$\tilde{a}_1(z) \overset{p}{\sim} zI_p - A. \tag{2.342}$$

Thus

$$\tilde{a}(z) \overset{p}{\sim} zI_p - A. \tag{2.343}$$

The last relation means that the sequences of invariant polynomials different from one for the matrices $a(z)$ and $zI_p - A$ coincide.

(b) Substituting in (2.338) $\zeta^{-1}$ for $z$, we obtain the associated matrix

$$W(\zeta) = \zeta C(I_p - \zeta A)^{-1}B + D. \tag{2.344}$$

We will show that under the taken propositions, the pairs $(I_p - \zeta A, \zeta B)$ and $[I_p - \zeta A, C]$ are irreducible. Indeed, let $z_0 = 0,\ z_i \neq 0,\ (i = 1, \dots, q)$ are all different eigenvalues of the matrix $A$. Then, as follows from Theorem 1.21, the eigenvalues of the polynomial matrix $I_p - \zeta A$ consist of the numbers $\zeta_i = z_i^{-1}$, $(i = 1, \dots, q)$. Moreover, for all $1 \leq i \leq q$, we obtain

$$\begin{aligned} \operatorname{rank}\left[\, I_p - \zeta_i A \ \ \zeta_i B \,\right] &= \operatorname{rank}\left[\, I_p - z_i^{-1} A \ \ z_i^{-1} B \,\right] \\ &= \operatorname{rank}\left[\, z_i I_p - A \ \ B \,\right] = p, \\ \operatorname{rank}\begin{bmatrix} I_p - \zeta_i A \\ C \end{bmatrix} &= \operatorname{rank}\begin{bmatrix} z_i I_p - A \\ C \end{bmatrix} = p. \end{aligned} \tag{2.345}$$

The irreducibility of the pairs $(I_p - \zeta A, \zeta B)$, $[I_p - \zeta A, C]$ follows from claim (c) of Theorem 1.16.

(c) Consider the ILMFD

$$C(I_p - \zeta A)^{-1} = a_1^{-1}(\zeta)b(\zeta). \tag{2.346}$$

Then, in analogy to (2.343), it can be shown that in (2.336)

$$a(\zeta) \sim a_1(\zeta) \overset{p}{\sim} I_p - \zeta A. \tag{2.347}$$

Therefore, thanks to Theorem 1.21

$$I_p - \zeta A = \text{rec}[zI_p - A], \tag{2.348}$$

so that (2.341), (2.347) yield (2.337). ■

*Remark 2.8* If the matrix $\tilde{W}(z)$ is noncausal, then the claim of Theorem 2.14 is no longer true. Indeed, let the matrix $\tilde{W}(z)$ be noncausal. Then instead of (2.338), we obtain the representation

$$\tilde{W}(z) = C(zI_p - A)^{-1}B + \tilde{D}(z), \tag{2.349}$$

where $\tilde{D}(z)$ is a polynomial matrix and $\deg \tilde{D}(z) > 0$. Hence, the corresponding associated matrix

$$W(\zeta) = \zeta C(I_p - \zeta A)^{-1}B + \tilde{D}(\zeta^{-1}) \tag{2.350}$$

possesses a pole at the point $\zeta = 0$. In this case in ILMFD (2.336) the matrix $a(\zeta)$ has the eigenvalue $\zeta = 0$. Now, Eq. (2.337) cannot be satisfied, because the reciprocal matrix cannot possess eigenvalues equal to zero.

(3) The next statement, on basis of Theorem 2.14, yields a procedure for the construction of the reciprocal polynomial matrix.

**Theorem 2.15** *Assume the non-singular matrix $\tilde{a}(z) \in \mathbb{R}^{nn}[z]$. Let $\tilde{\chi}(z)$ be a unimodular matrix, such that the matrix*

$$\tilde{a}_\rho(z) \overset{\triangle}{=} \tilde{\chi}(z)\tilde{a}(z) \tag{2.351}$$

*is row reduced, i.e., it can be represented in the form*

$$\tilde{a}_\rho(z) = \text{diag}\,\{z^{\alpha_1}, \ldots, z^{\alpha_n}\}\, A_0 + \tilde{a}_1(z), \quad \det A_0 \neq 0, \tag{2.352}$$

*where $\alpha_i$ are nonnegative integers, and moreover,*

$$\alpha_i > \beta_i, \tag{2.353}$$

*where $\beta_i$ is the degree of the ith row of the matrix $\tilde{a}_1(z)$. Define the matrix*

$$a_\rho(\zeta) \triangleq \operatorname{diag}\{\zeta^{\alpha_1}, \dots, \zeta^{\alpha_n}\}\, \tilde{a}_\rho(\zeta^{-1}) = A_0 + \operatorname{diag}\{\zeta^{\alpha_1}, \dots, \zeta^{\alpha_n}\}\tilde{a}_1(\zeta^{-1}). \quad (2.354)$$

*Then, $a_\rho(\zeta)$ is a polynomial matrix, and*

$$a_\rho(\zeta) = \operatorname{rec}[\tilde{a}_\rho(z)]. \quad (2.355)$$

*Proof* The proof runs over several steps.

(a) From (2.352) and (2.353), it follows that $a_\rho(\zeta)$ is a polynomial matrix. Denote

$$\tilde{R}_h(z) \triangleq \left[\tilde{a}_\rho(z)\ I_n\right]. \quad (2.356)$$

Obviously, the matrix $\tilde{R}_h(z)$ is alatent. Consider the polynomial matrix

$$R_h(\zeta) = \operatorname{diag}\{\zeta^{\alpha_1}, \dots, \zeta^{\alpha_n}\}\left[\tilde{a}_\rho(\zeta^{-1})\ I_n\right] = \left[a_\rho(\zeta)\ \operatorname{diag}\{\zeta^{\alpha_1}, \dots, \zeta^{\alpha_n}\}\right]. \quad (2.357)$$

We will show that the matrix $R_h(\zeta)$ is alatent. Indeed, from (2.354), it follows that

$$\operatorname{rank} a_\rho(0) = \operatorname{rank} A_0 = n. \quad (2.358)$$

Moreover, for all $\zeta \neq 0$, $\operatorname{rank} R_h(\zeta) = n$. Together, we realize that for all $\zeta$, the relation $\operatorname{rank} R_h(\zeta) = n$ is true, i.e., $R_h(\zeta)$, is alatent.

(b) Consider the rational matrices

$$\tilde{W}(z) \triangleq \tilde{a}_\rho^{-1}(z), \quad W(\zeta) \triangleq a_\rho^{-1}(\zeta)\operatorname{diag}\{\zeta^{\alpha_1}, \dots, \zeta^{\alpha_n}\}. \quad (2.359)$$

We will show that the matrices $\tilde{W}(z)$ and $W(\zeta)$ are associated. Indeed, from (2.354), after substituting $z^{-1}$ for $\zeta$, we obtain

$$\tilde{a}_\rho(z) = \operatorname{diag}\{z^{\alpha_1}, \dots, z^{\alpha_n}\}a_\rho(z^{-1}).$$

Hence

$$\begin{aligned}\tilde{W}(z) &= \tilde{a}_\rho^{-1}(z) = a_\rho^{-1}(z^{-1})\operatorname{diag}\{z^{-\alpha_1}, \dots, z^{-\alpha_n}\}\\ &= [a_\rho^{-1}(\zeta)\operatorname{diag}\{\zeta^{\alpha_1}, \dots, \zeta^{\alpha_n}\}]|_{\zeta=z^{-1}} = W(z^{-1}).\end{aligned}$$

Moreover, the matrix $\tilde{W}(z)$ is causal. Indeed, the matrix $\tilde{W}(z)$ can be written in form of the ILMFD

$$\tilde{W}(z) = \tilde{a}_\rho^{-1}(z)I \quad (2.360)$$

and, because the matrix $\tilde{a}_\rho(z)$ is row reduced, it follows from Theorem 2.1 that the matrix $\tilde{W}(z)$ is at least proper, i.e., causal.

(c) Since matrix (2.357) is alatent, the representation

$$W(\zeta) = a_\rho^{-1}(\zeta)\,\mathrm{diag}\,\{\zeta^{\alpha_1}, \ldots, \zeta^{\alpha_n}\} \tag{2.361}$$

is an ILMFD. Then, comparing ILMFD (2.360) and (2.361), on basis of Theorem 2.14, we win

$$a_\rho(\zeta) = \mathrm{rec}[\tilde{a}_\rho(z)]. \tag{2.362}$$

From this together with (2.351), the wanted relation (2.355) is achieved. ■

(4)

**Theorem 2.16** *Let the matrix $\tilde{W}(z) \in \mathbb{R}^{nm}(z)$ be strictly causal, and the matrix $\tilde{W}_d(z) \in \mathbb{R}^{mn}(z)$ be causal, and let us have the ILMFD*

$$\tilde{W}(z) = \tilde{a}^{-1}(z)\tilde{b}(z), \quad \tilde{W}_d(z) = \tilde{\alpha}^{-1}(z)\tilde{\beta}(z). \tag{2.363}$$

*Build the associated matrices*

$$W(\zeta) = \tilde{W}(\zeta^{-1}), \quad W_d(\zeta) = \tilde{W}_d(\zeta^{-1}) \tag{2.364}$$

*and the related ILMFD*

$$W(\zeta) = a^{-1}(\zeta)\zeta b(\zeta), \quad W_d(\zeta) = \alpha^{-1}(\zeta)\beta(\zeta). \tag{2.365}$$

*Introduce the polynomial matrices*

$$\tilde{Q}(z, \tilde{\alpha}, \tilde{\beta}) \stackrel{\triangle}{=} \begin{matrix} \\ n \\ m \end{matrix} \overset{\begin{matrix} n & \quad m \end{matrix}}{\begin{bmatrix} \tilde{a}(z) & -\tilde{b}(z) \\ -\tilde{\beta}(z) & \tilde{\alpha}(z) \end{bmatrix}} \tag{2.366}$$

$$Q(\zeta, \alpha, \beta) \stackrel{\triangle}{=} \begin{matrix} \\ n \\ m \end{matrix} \overset{\begin{matrix} n & \quad m \end{matrix}}{\begin{bmatrix} a(\zeta) & -b(\zeta) \\ -\beta(\zeta) & \alpha(\zeta) \end{bmatrix}}. \tag{2.367}$$

*Then the following relation holds:*

$$Q(\zeta, \alpha, \beta) = \mathrm{rec}[\tilde{Q}(z, \tilde{\alpha}, \tilde{\beta})]. \tag{2.368}$$

*Proof* The proof runs again over several steps.

(a) Let $\tilde{\chi}(z)$ be the unimodular matrix configured in (2.351), and $\tilde{k}(z)$ be such a unimodular matrix that $\tilde{k}(z)\tilde{\alpha}(z)$ becomes row reduced, and we obtain

$$\tilde{k}(z)\tilde{\alpha}(z) \stackrel{\triangle}{=} \tilde{\alpha}_\rho(z) = \operatorname{diag}\left\{z^{l_1}, \ldots, z^{l_m}\right\} B_0 + \tilde{\alpha}_1(z), \tag{2.369}$$

where $l_1, \ldots, l_m$ are nonnegative integers, $\det B_0 \neq 0$ and

$$\deg \tilde{\alpha}_i(z) < l_i, \tag{2.370}$$

where $\tilde{\alpha}_i(z),\ (i = 1, \ldots, m)$ are the rows of the matrix $\tilde{\alpha}_1(z)$.

(b) Denote

$$\tilde{L}(z) \stackrel{\triangle}{=} \operatorname{diag}\left\{\tilde{\chi}(z), \tilde{k}(z)\right\} \tag{2.371}$$

and consider the matrix

$$\tilde{Q}_\rho(z, \tilde{\alpha}_\rho, \tilde{\beta}_\rho) \stackrel{\triangle}{=} \tilde{L}(z)\tilde{Q}(z, \tilde{\alpha}, \tilde{\beta}), \tag{2.372}$$

which can be written in the form

$$\tilde{Q}_\rho(z, \tilde{\alpha}_\rho, \tilde{\beta}_\rho) = \begin{matrix} \\ n \\ m \end{matrix} \begin{matrix} \quad n \qquad\quad m \\ \begin{bmatrix} \tilde{a}_\rho(z) & -\tilde{b}_\rho(z) \\ -\tilde{\beta}_\rho(z) & \tilde{\alpha}_\rho(z) \end{bmatrix} \end{matrix} \tag{2.373}$$

where

$$\begin{aligned} \tilde{a}_\rho(z) &= \tilde{\chi}(z)\tilde{a}(z), \quad \tilde{\alpha}_\rho(z) = \tilde{k}(z)\tilde{\alpha}(z). \\ \tilde{b}_\rho(z) &= \tilde{\chi}(z)\tilde{b}(z), \quad \tilde{\beta}_\rho(z) = \tilde{k}(z)\tilde{\beta}(z). \end{aligned} \tag{2.374}$$

We will show that the matrix $\tilde{Q}_\rho(z, \tilde{\alpha}_\rho, \tilde{\beta}_\rho)$ is row reduced. By construction, the representations

$$\tilde{W}(z) = \tilde{a}_\rho^{-1}(z)\tilde{b}_\rho(z), \quad \tilde{W}_d(z) = \tilde{\alpha}_\rho^{-1}(z)\tilde{\beta}_\rho(z) \tag{2.375}$$

are ILMFD. Moreover, because the matrices $\tilde{a}_\rho(z)$ and $\tilde{\alpha}_\rho(z)$ are row reduced, due to Theorem 2.1, we obtain

$$\begin{aligned} &\deg \tilde{b}_{\rho_i}(z) < \alpha_i,\ (i = 1, \ldots, n), \\ &\deg \tilde{\beta}_{\rho_i}(z) \le l_i,\ (i = 1, \ldots, m), \end{aligned} \tag{2.376}$$

where $\tilde{b}_{\rho_i}(z)$ are the rows of the matrix $\tilde{b}_\rho(z)$, $\tilde{\beta}_{\rho_i}(z)$ are the rows of the matrix $\tilde{\beta}_\rho(z)$, and the numbers $\alpha_i, l_i$ are determined by (2.352) and (2.369), respectively. Therefore, using (2.352) and (2.369), the matrix $\tilde{Q}_\rho(z, \tilde{\alpha}_\rho, \tilde{\beta}_\rho)$ can be written in the form

$$\tilde{Q}_\rho(z, \tilde{\alpha}_\rho, \tilde{\beta}_\rho) = \operatorname{diag}\left\{z^{\lambda_1}, \ldots, z^{\lambda_{n+m}}\right\} Q_0 + \tilde{Q}_1(z), \tag{2.377}$$

where

$$\begin{aligned}\lambda_i &= \alpha_i, \ (i = 1, \ldots, n), \\ \lambda_{n+i} &= l_i, \ (i = 1, \ldots, m),\end{aligned} \tag{2.378}$$

and $Q_0$ is a constant $(n + m) \times (n + m)$ matrix of the form

$$Q_0 = \begin{bmatrix} A_0 & 0_{nm} \\ C_0 & B_0 \end{bmatrix}. \tag{2.379}$$

Moreover, the degree of each row $\tilde{Q}_i(z)$ of the matrix $\tilde{Q}_1(z)$ is lower than $\lambda_i$. Since

$$\det Q_0 = \det A_0 \det B_0 \neq 0, \tag{2.380}$$

matrix (2.373) is row reduced.

(c) Since the matrix $\tilde{Q}_\rho(z, \tilde{\alpha}_\rho, \tilde{\beta}_\rho)$ is row reduced, the matrix

$$\begin{aligned}Q_\rho(\zeta, \alpha, \beta) &\triangleq \operatorname{diag}\left\{\zeta^{\alpha_1}, \ldots, \zeta^{\alpha_n}, \zeta^{l_1}, \ldots, \zeta^{l_m}\right\} \tilde{Q}_\rho(z^{-1}, \tilde{\alpha}, \tilde{\beta})|_{z=\zeta^{-1}} \\ &= \begin{bmatrix} \operatorname{diag}\{\zeta^{\alpha_1}, \ldots, \zeta^{\alpha_n}\} \tilde{a}_\rho(\zeta^{-1}) & -\operatorname{diag}\{\zeta^{\alpha_1}, \ldots, \zeta^{\alpha_n}\} \tilde{b}_\rho(\zeta^{-1}) \\ -\operatorname{diag}\left\{\zeta^{l_1}, \ldots, \zeta^{l_m}\right\} \tilde{\beta}_\rho(\zeta^{-1}) & \operatorname{diag}\left\{\zeta^{l_1}, \ldots, \zeta^{l_m}\right\} \tilde{\alpha}_\rho(\zeta^{-1}) \end{bmatrix} \\ &\triangleq \begin{bmatrix} a_\rho(\zeta) & -\zeta b_\rho(\zeta) \\ -\beta_\rho(\zeta) & \alpha_\rho(\zeta) \end{bmatrix}\end{aligned} \tag{2.381}$$

is a polynomial matrix. Hence, applying Theorem 2.15 yields

$$Q_\rho(\zeta) = \operatorname{rec}[\tilde{Q}_\rho(z)]. \tag{2.382}$$

(d) Since the pairs $(\tilde{a}_\rho(z), \tilde{b}_\rho(z))$ and $(\tilde{\alpha}_\rho(z), \tilde{\beta}_\rho(z))$ are irreducible, the matrices

$$\tilde{R}_h(z) = \begin{bmatrix} \tilde{a}_\rho(z) & -\tilde{b}_\rho(z) \end{bmatrix}, \quad \tilde{R}_{h_d}(z) = \begin{bmatrix} -\tilde{\beta}_\rho(z) & \tilde{\alpha}_\rho(z) \end{bmatrix} \tag{2.383}$$

are alatent. We claim that the matrices

$$R_h(\zeta) = \begin{bmatrix} a_\rho(\zeta) & -b_\rho(\zeta) \end{bmatrix}, \quad R_{h_d}(\zeta) = \begin{bmatrix} -\beta_\rho(\zeta) & \alpha_\rho(\zeta) \end{bmatrix} \tag{2.384}$$

are alatent too, and will show that for the matrix $R_h(\zeta)$. Let $z_0 = 0$, $z_i \neq 0$, $(i = 1, \ldots, q)$ be the totality of eigenvalues of the matrix $\tilde{a}(z)$. Then, due to $a_\rho(\zeta) = \operatorname{rec}[\tilde{a}(z)]$, the eigenvalues of the matrix $a_\rho(\zeta)$ are the numbers $\zeta_i = z_i^{-1}$, $(i = 1, \ldots, q)$. For $i = 1, \ldots, q$, we obtain

$$\begin{aligned}\operatorname{rank} R_h(\zeta_i) &= \operatorname{rank}\begin{bmatrix} a_\rho(\zeta_i) & -b_\rho(\zeta_i) \end{bmatrix} \\ &= \operatorname{rank}\begin{bmatrix} \operatorname{diag}\left\{\zeta_i^{\alpha_1}, \ldots, \zeta_i^{\alpha_n}\right\} \tilde{a}_\rho(z_i) & -\tilde{b}_\rho(z_i) \end{bmatrix} = n,\end{aligned} \tag{2.385}$$

and applying Theorem 1.16 we win that $R_h(\zeta)$ is alatent. The alatence of the matrix $R_{h_d}(\zeta)$ could be shown analogously.

(e) The alatence of matrix (2.384) implies, that the representations

$$\begin{aligned} W(\zeta) &= \tilde{a}_\rho^{-1}(\zeta^{-1})\tilde{b}_\rho(\zeta^{-1}) = a_\rho^{-1}(\zeta)b_\rho(\zeta), \\ W_d(\zeta) &= \tilde{\alpha}_\rho^{-1}(\zeta^{-1})\tilde{\beta}_\rho(\zeta^{-1}) = \alpha_\rho^{-1}(\zeta)\beta_\rho(\zeta) \end{aligned} \tag{2.386}$$

are ILMFD. At the same time, we have ILMFD (2.365), so that we obtain

$$\begin{aligned} a_\rho(\zeta) &= \mu(\zeta)a(\zeta), \quad b_\rho(\zeta) = \mu(\zeta)b(\zeta), \\ \alpha_\rho(\zeta) &= \nu(\zeta)\alpha(\zeta), \quad \beta_\rho(\zeta) = \nu(\zeta)\beta(\zeta), \end{aligned} \tag{2.387}$$

where $\mu(\zeta)$ and $\nu(\zeta)$ are unimodular matrices. Inserting (2.387) into (2.381), we find

$$Q_\rho(\zeta) = \operatorname{diag}\{\mu(\zeta), \nu(\zeta)\}\, Q(\zeta), \tag{2.388}$$

i.e., the matrices $Q_\rho(\zeta)$ and $Q(\zeta)$ are equivalent. Therefore, (2.382) and (2.388) yields (2.362).

*Remark 2.9* The proposition in Theorem 2.16 on strict causality of the matrix $\tilde{W}(z)$ is essential. If this proposition is violated, then generally speaking, the claim of the theorem is wrong. Indeed, if the matrix $\tilde{W}(z)$ is causal (but not strictly causal), then in representation (2.377) instead of (2.379), we would obtain

$$Q_0 = \begin{bmatrix} A_0 & D_0 \\ C_0 & B_0 \end{bmatrix}, \tag{2.389}$$

where $D_0 \neq 0_{nm}$. Herein, the condition $\det Q_0 \neq 0$ could, generally speaking, not be fulfilled, and the further proof runs into trouble.

## References


1. F.R. Gantmacher, *The Theory of Matrices* (Chelsea, New York, 1959)
2. T. Kailath, *Linear Systems* (Prentice Hall, Englewood Cliffs, NJ, 1980)
3. E.N. Rosenwasser, B.P. Lampe, *Multivariable Computer-controlled Systems–A Transfer Function Approach* (Springer, London, 2006)
4. A.E. Barabanov, *Synthesis of minimax controllers* (State University Press, St.-Peterburg, 1996). (in Russian)

# Chapter 3
# Determinant Polynomial Equations and Modal Control Problem for Discrete Processes

**Abstract** The chapter presents the theory of determinant polynomial equations (DPEs) and based on this the general methodology for solving the problems of modal control and stabilization of linear discrete processes, described by backward and forward models. Essential consequences of using these model types for solving modal control problems are highlighted. In particular, it is shown that all stabilizing regulators are causal for a strictly causal backward process. The same does not apply to strictly causal processes specified by discrete forward models.

## 3.1 Determinant Polynomial Equations

(1) Assume the polynomial pair $(a_l(\lambda), b_l(\lambda))$, where $a_l(\lambda) \in \mathbb{R}^{nn}[\lambda]$, $b_l(\lambda) \in \mathbb{R}^{nm}[\lambda]$. Further on, this pair is called left process. Another arbitrary pair $(\alpha_l(\lambda), \beta_l(\lambda))$, where $\alpha_l(\lambda) \in \mathbb{R}^{mm}[\lambda]$, $\beta_l(\lambda) \in \mathbb{R}^{mn}[\lambda]$, is called left controller. The polynomial $(n+m) \times (n+m)$ matrix

$$Q_l(\lambda, \alpha_l, \beta_l) = \begin{matrix} \\ n \\ m \end{matrix} \begin{matrix} \quad n \qquad\quad m \\ \begin{bmatrix} a_l(\lambda) & -b_l(\lambda) \\ -\beta_l(\lambda) & \alpha_l(\lambda) \end{bmatrix} \end{matrix} \tag{3.1}$$

is called the left characteristic matrix of the closed loop, and the polynomial

$$\Delta(\lambda, \alpha_l, \beta_l) = \det Q_l(\lambda, \alpha_l, \beta_l) \tag{3.2}$$

the characteristic polynomial of the closed loop. The eigenvalues of the matrix $Q_l(\lambda, \alpha_l, \beta_l)$ are called modes of the closed system. The totality of modes forms the spectrum of this matrix. For the control theory and its applications, the modal control problem is fundamental, and it can be formulated as follows: For a given process $(a_l(\lambda), b_l(\lambda))$ find the set of controllers $(\alpha_l(\lambda), \beta_l(\lambda))$, such that the spectrum of the matrix $Q_l(\lambda, \alpha_l, \beta_l)$ takes a given form.



E. N. Rosenwasser et al., *Computer-Controlled Systems with Delay*,
https://doi.org/10.1007/978-3-030-15042-6_3

(2) Let the desired spectrum of matrix (3.1) consist of the numbers $\lambda_1, \ldots, \lambda_\rho$ with multiplicities $\nu_1, \ldots, \nu_\rho$, respectively. Then the modal control problem is formulated as follows:

Let be given the left process $(a_l(\lambda), b_l(\lambda))$ and the monic polynomial

$$d(\lambda) = (\lambda - \lambda_1)^{\nu_1} \cdots (\lambda - \lambda_\rho)^{\nu_\rho}. \tag{3.3}$$

Find the set of all left controllers $(\alpha_l(\lambda), \beta_l(\lambda))$, such that

$$\det Q_l(\lambda, \alpha_l, \beta_l) = \det \begin{bmatrix} a_l(\lambda) & -b_l(\lambda) \\ -\beta_l(\lambda) & \alpha_l(\lambda) \end{bmatrix} \sim d(\lambda). \tag{3.4}$$

For given left process $(a_l(\lambda), b_l(\lambda))$ and monic polynomial $d(\lambda)$, Eq. (3.4) can be considered as an equation depending on the polynomial matrices $\alpha_l(\lambda)$, $\beta_l(\lambda)$. Further on, this type is called a determinant polynomial equation (DPE).

(3) In analogy to the above said, the vertical pair $[a_r(\lambda), b_r(\lambda)]$, where $a_r(\lambda) \in \mathbb{R}^{mm}[\lambda]$, $b_l(\lambda) \in \mathbb{R}^{nm}[\lambda]$, is called a right process, and the pair $[\alpha_r(\lambda), \beta_r(\lambda)]$, where $\alpha_r(\lambda) \in \mathbb{R}^{nn}[\lambda]$, $\beta_r(\lambda) \in \mathbb{R}^{nm}[\lambda]$, a right controller. The matrix

$$Q_r(\lambda, \alpha_r, \beta_r) = \begin{bmatrix} \alpha_r(\lambda) \; b_r(\lambda) \\ \beta_r(\lambda) \; a_r(\lambda) \end{bmatrix}, \tag{3.5}$$

is called the right characteristic matrix of the closed loop. Moreover, as above, we can formulate the right modal control problem:

Let be given the right process $[a_r(\lambda), b_r(\lambda)]$ and monic polynomial (3.3). Find the set of all right controllers $[\alpha_r(\lambda), \beta_r(\lambda)]$, such that

$$\det Q_r(\lambda, \alpha_r, \beta_r) \sim d(\lambda). \tag{3.6}$$

It is easy to see that the matrix $Q_r(\lambda, \alpha_r, \beta_r)$ can be represented in form (3.1) by transposition and applying elementary transforms. Since the determinant does not change under those transforms, the solution of DPE (3.6) can always be related to a solution of Eq. (3.4). Therefore, in the following, we will mainly consider equations of form (3.4).

## 3.2 Basic Controllers

(1) If the assumed characteristic matrix of the closed loop (3.1) does not have eigenvalues, i.e., it is unimodular, then DPE (3.4) takes the form

$$det \begin{bmatrix} a_l(\lambda) & -b_l(\lambda) \\ -\beta_l(\lambda) & \alpha_l(\lambda) \end{bmatrix} \sim 1. \tag{3.7}$$

Further on, any controller which satisfies relation (3.7) is called a left basic controller, or shortly, a left basis.

(2) As follows from Theorem 1.16, for the existence of a basic controller, it is necessary and sufficient, that the pair $(a_l(\lambda), b_l(\lambda))$ is irreducible. In that case, the process $(a_l(\lambda), b_l(\lambda))$ is also called irreducible. The following claim yields a general representation of the set of all left basic controllers for a given irreducible process $(a_l(\lambda), b_l(\lambda))$.

**Theorem 3.1** *Let $(\alpha_{l0}^B(\lambda), \beta_{l0}^B(\lambda))$ be any left basic controller for the irreducible process $(a_l(\lambda), b_l(\lambda))$. Then the set of all left basic controllers $(\alpha_l^B(\lambda), \beta_l^B(\lambda))$ is defined by the relations*

$$\begin{aligned} \alpha_l^B(\lambda) &= D_l(\lambda)\alpha_{l0}^B(\lambda) - M_l(\lambda)b_l(\lambda), \\ \beta_l^B(\lambda) &= D_l(\lambda)\beta_{l0}^B(\lambda) - M_l(\lambda)a_l(\lambda), \end{aligned} \tag{3.8}$$

*where $M_l(\lambda) \in \mathbb{R}^{mn}[\lambda]$, $D_l(\lambda) \in \mathbb{R}^{mm}[\lambda]$, and the matrix $M_l(\lambda)$ is arbitrary, but $D_l(\lambda)$ is any unimodular matrix.*

*Proof* Denote the set of all left basic controllers by $\mathscr{R}^B$ and the set of all controllers of form (3.8) by $\mathscr{R}^0$. At first, we will show $\mathscr{R}^B \subset \mathscr{R}^0$. Let $(\alpha_{l0}^B(\lambda), \beta_{l0}^B(\lambda))$ a certain basic controller. Then

$$Q_l(\lambda, \alpha_{l0}^B, \beta_{l0}^B) = \begin{matrix} \\ n \\ m \end{matrix} \overset{\begin{matrix} n & \quad m \end{matrix}}{\begin{bmatrix} a_l(\lambda) & -b_l(\lambda) \\ -\beta_{l0}^B(\lambda) & \alpha_{l0}^B(\lambda) \end{bmatrix}}, \tag{3.9}$$

is an unimodular matrix. Denote

$$Q_r(\lambda, \alpha_{r0}^B, \beta_{r0}^B) \triangleq Q_l^{-1}(\lambda, \alpha_{l0}^B, \beta_{l0}^B) = \begin{matrix} \\ n \\ m \end{matrix} \overset{\begin{matrix} n & \quad m \end{matrix}}{\begin{bmatrix} \alpha_{r0}^B(\lambda) & b_r(\lambda) \\ \beta_{r0}^B(\lambda) & a_r(\lambda) \end{bmatrix}}. \tag{3.10}$$

Obviously, the matrix $Q_r(\lambda, \alpha_{r0}^B, \beta_{r0}^B)$ is unimodular. By construction

$$Q_l(\lambda, \alpha_{l0}^B, \beta_{l0}^B) Q_r(\lambda, \alpha_{r0}^B, \beta_{r0}^B) = I_{n+m}. \tag{3.11}$$

Comparing now (3.9) and (3.10), we find

$$\begin{aligned} a_l(\lambda)\alpha_{r0}^B(\lambda) - b_l(\lambda)\beta_{r0}^B(\lambda) &= I_n, \\ a_l(\lambda)b_r(\lambda) - b_l(\lambda)a_r(\lambda) &= 0_{nm}. \end{aligned} \tag{3.12}$$

Let $(\alpha_l^B(\lambda), \beta_l^B(\lambda))$ be any left basic controller, and

$$Q_l(\lambda, \alpha_l^B, \beta_l^B) = \begin{bmatrix} a_l(\lambda) & -b_l(\lambda) \\ -\beta_l^B(\lambda) & \alpha_l^B(\lambda) \end{bmatrix}, \tag{3.13}$$

the associated left characteristic matrix. Then, using (3.12), we obtain

$$Q_l(\lambda, \alpha_l^B, \beta_l^B) Q_r(\lambda, \alpha_{r0}^B, \beta_{r0}^B) = \begin{bmatrix} I_n & 0_{nm} \\ M_l(\lambda) & D_l(\lambda) \end{bmatrix}, \tag{3.14}$$

where

$$\begin{aligned} D_l(\lambda) &= -\beta_l^B(\lambda) b_r(\lambda) + \alpha_l^B(\lambda) a_r(\lambda), \\ M_l(\lambda) &= -\beta_l^B(\lambda) \alpha_{r0}^B(\lambda) + \alpha_l^B(\lambda) \beta_{r0}^B(\lambda), \end{aligned} \tag{3.15}$$

are polynomial matrices. Moreover, applying (3.11), yields

$$\begin{aligned} Q_l(\lambda, \alpha_l^B, \beta_l^B) &= \begin{bmatrix} I_n & 0_{nm} \\ M_l(\lambda) & D_l(\lambda) \end{bmatrix} Q_r^{-1}(\lambda, \alpha_{r0}^B, \beta_{r0}^B) \\ &= \begin{bmatrix} I_n & 0_{nm} \\ M_l(\lambda) & D_l(\lambda) \end{bmatrix} Q_l(\lambda, \alpha_{l0}^B, \beta_{l0}^B). \end{aligned} \tag{3.16}$$

Comparing now (3.13) and (3.9), relations (3.8) are achieved. Therefore, from (3.14) by calculating the determinant of the left and right side, we find

$$\det D(\lambda) = \det Q_l(\lambda, \alpha_l^B, \beta_l^B) \det Q_r(\lambda, \alpha_{r0}^B, \beta_{r0}^B) = \text{const.} \neq 0, \tag{3.17}$$

i.e., the matrix $D(\lambda)$ is unimodular. Thus $\mathscr{R}^B \subset \mathscr{R}^0$.

On the other side, let the controller $(\alpha_l^B(\lambda), \beta_l^B(\lambda))$ be determined by relations (3.8) with an unimodular matrix $D_l(\lambda)$. Then from (3.16), we obtain

$$\begin{aligned} Q_l(\lambda, \alpha_l^B, \beta_l^B) &= \begin{bmatrix} a_l(\lambda) & -b_l(\lambda) \\ -D_l(\lambda)\beta_{l0}^B(\lambda) + M_l(\lambda) a_l(\lambda) & D_l(\lambda)\alpha_{l0}^B(\lambda) - M_l(\lambda) b_l(\lambda) \end{bmatrix} \\ &= \begin{bmatrix} I_n & 0_{nm} \\ M_l(\lambda) & D_l(\lambda) \end{bmatrix} Q_l(\lambda, \alpha_{l0}^B, \beta_{l0}^B). \end{aligned} \tag{3.18}$$

Since the matrix $D_l(\lambda)$ is unimodular, the right side of Eq. (3.18) is the product of two unimodular matrices. Therefore, the matrix $Q_l(\lambda, \alpha_l^B, \beta_l^B)$ is unimodular, and the controller $(\alpha_l^B(\lambda), \beta_l^B(\lambda))$ is a basic controller. Thus $\mathscr{R}^0 \subset \mathscr{R}^B$. Because each set contains the other one, we conclude $\mathscr{R}^B = \mathscr{R}^0$, and formulae (3.8) define the complete set of left basic controllers. ■

(3) For the practical application of formulae (3.8), it is necessary to find some left basic controller. In [1], a recursive algorithm for the construction of such a basic controller is described. Another method, based on the Diophantine polynomial equations, yields the following theorem.

**Theorem 3.2** *Assume the irreducible left process $(a_l(\lambda), b_l(\lambda))$. Then the following statements hold:*

*(i) There exists an irreducible right process $[a_r(\lambda), b_r(\lambda)]$, such that*

$$a_l(\lambda)b_r(\lambda) = b_l(\lambda)a_r(\lambda). \tag{3.19}$$

*(ii) The set of all left basic controllers $(\alpha_l^B(\lambda), \beta_l^B(\lambda))$ coincides with the unification of the sets sof solutions of the polynomial equations of the form*

$$-\beta_l^B(\lambda)b_r(\lambda) + \alpha_l^B(\lambda)a_r(\lambda) = D_l(\lambda), \tag{3.20}$$

*for any possible unimodular matrices $D_l(\lambda)$.*

*Proof* (i) The claim follows directly from the second formula in (3.12).
(ii) Denote as above, by $\mathscr{R}^B$ the set of all left basic controllers, and by $\mathscr{R}^d$ the unification of the solution sets of all Eq. (3.20) for any possible unimodular matrices $D_l(\lambda)$. At first, we will show $\mathscr{R}^B \subset \mathscr{R}^d$. For this notice that from (3.8), we obtain

$$\begin{aligned}
&\alpha_l^B(\lambda)a_r(\lambda) - \beta_l^B(\lambda)b_r(\lambda) \\
&\quad = [D_l(\lambda)\alpha_{l0}^B(\lambda) - M_l(\lambda)b_l(\lambda)]a_r(\lambda) - [D_l(\lambda)\beta_{l0}^B(\lambda) + M_l(\lambda)a_l(\lambda)]b_r(\lambda) \\
&\quad = D_l(\lambda)[\alpha_{l0}^B(\lambda)a_r(\lambda) - \beta_{l0}^B(\lambda)b_r(\lambda)] + M_l(\lambda)[b_l(\lambda)a_r(\lambda) - a_l(\lambda)b_r(\lambda)].
\end{aligned} \tag{3.21}$$

Owing to (3.10)

$$\begin{bmatrix} a_l(\lambda) & -b_l(\lambda) \\ -\beta_{l0}^B(\lambda) & \alpha_{l0}^B(\lambda) \end{bmatrix} \begin{bmatrix} \alpha_{r0}^B(\lambda) & b_r(\lambda) \\ \beta_{r0}^B(\lambda) & a_r(\lambda) \end{bmatrix} = \begin{bmatrix} I_n & 0_{nm} \\ 0_{mn} & I_m \end{bmatrix}, \tag{3.22}$$

such that

$$-\beta_{l_0}^B(\lambda)b_r(\lambda) + \alpha_{l_0}^B(\lambda)a_r(\lambda) = I_m. \tag{3.23}$$

Using this relation and (3.19), from (3.21), we find that the controller $(\alpha_l^B(\lambda), \beta_l^B(\lambda))$ satisfies Eq. (3.20). Thus $\mathscr{R}^B \subset \mathscr{R}^d$.
Conversely, assume the irreducible process $(a_l(\lambda), b_l(\lambda))$. Moreover, let $[\alpha_{r0}^B(\lambda), \beta_{r0}^B(\lambda)]$ be a vertical pair configured in (3.22). Then we obtain

$$a_l(\lambda)\alpha_{r0}^B(\lambda) - b_l(\lambda)\beta_{r0}^B(\lambda) = I_n. \tag{3.24}$$

Let also the pair $(\alpha_l^B(\lambda), \beta_l^B(\lambda))$ be a solution of Diophantine Eq. (3.20) with an unimodular matrix $D_l(\lambda)$. Achieve the product

$$
\begin{aligned}
&\begin{bmatrix} a_l(\lambda) & -b_l(\lambda) \\ -\beta_l^B(\lambda) & \alpha_l^B(\lambda) \end{bmatrix} \begin{bmatrix} \alpha_{r0}^B(\lambda) & b_r(\lambda) \\ \beta_{r0}^B(\lambda) & a_r(\lambda) \end{bmatrix} \\
&\quad = \begin{bmatrix} a_l(\lambda)\alpha_{r0}^B(\lambda) - b_l(\lambda)\beta_r^B(\lambda) & a_l(\lambda)b_r(\lambda) - b_l(\lambda)a_r(\lambda) \\ -\beta_l^B(\lambda)\alpha_{r0}^B(\lambda) + \alpha_l^B(\lambda)\beta_{r0}^B(\lambda) & -\beta_l^B(\lambda)b_r(\lambda) + \alpha_l^B(\lambda)a_r(\lambda) \end{bmatrix},
\end{aligned}
\tag{3.25}
$$

which with the help of (3.19), (3.20), and (3.24) takes the form

$$
\begin{bmatrix} a_l(\lambda) & -b_l(\lambda) \\ -\beta_l^B(\lambda) & \alpha_l^B(\lambda) \end{bmatrix} \begin{bmatrix} \alpha_{r0}^B(\lambda) & b_r(\lambda) \\ \beta_{r0}^B(\lambda) & a_r(\lambda) \end{bmatrix} = \begin{bmatrix} I_n & 0_{nm} \\ N(\lambda) & D_l(\lambda) \end{bmatrix},
$$

where $N(\lambda)$ is a polynomial matrix. Since the matrix $D_l(\lambda)$ is unimodular, the matrix on the right side is unimodular. Hence the first factor on the left side is an unimodular matrix and the pair $(\alpha_{l0}^B(\lambda), \beta_l^B(\lambda))$ is a basic controller. Thus $\mathscr{R}^d \subset \mathscr{R}^0$. Since both sets contain each other, $\mathscr{R}^B$ and $\mathscr{R}^d$ coincide. ■

(4) In analogy, the pair $[\alpha_r^B(\lambda), \beta_r^B(\lambda)]$ is called a right basic controller for the right process $[a_r(\lambda), b_r(\lambda)]$, when the matrix $Q_r(\lambda, \alpha_r^B, \beta_r^B)$ is unimodular, i.e.,

$$
\det \begin{bmatrix} \alpha_r^B(\lambda) & b_r(\lambda) \\ \beta_r^B(\lambda) & a_r(\lambda) \end{bmatrix} = \det Q_r(\lambda, \alpha_r^B, \beta_r^B) \sim 1,
$$

where $Q_r(\lambda, \alpha_r^B, \beta_r^B)$ is the right characteristic equation.

**Theorem 3.3** *Let $[\alpha_{r0}^B(\lambda), \beta_{r0}^B(\lambda)]$ be any right basic controller for the irreducible process $[a_r(\lambda), b_r(\lambda)]$. Then the set of all right basic controllers $[\alpha_r^B(\lambda), \beta_r^B(\lambda)]$ is defined by the formulae*

$$
\begin{aligned}
\alpha_r^B(\lambda) &= \alpha_{r0}^B(\lambda)D_r(\lambda) - b_r(\lambda)M_r(\lambda), \\
\beta_r^B(\lambda) &= \beta_{r0}^B(\lambda)D_r(\lambda) - a_r(\lambda)M_r(\lambda),
\end{aligned}
\tag{3.26}
$$

*where the $n \times m$ matrix $M_r(\lambda) \in \mathbb{R}^{nm}[\lambda]$ is arbitrary, and the matrix $D_r(\lambda) \in \mathbb{R}^{nn}[\lambda]$ is any unimodular matrix.*

The proof runs analogously to that of Theorem 3.1. ■

**Theorem 3.4** *The set of all basic controllers $[\alpha_r^B(\lambda), \beta_r^B(\lambda)]$ for the irreducible process $[a_r(\lambda), b_r(\lambda)]$ coincides with the unification of the solution sets of the Diophantine equation*

$$
-b_l(\lambda)\beta_r^B(\lambda) + a_l(\lambda)\alpha_r^B(\lambda) = D_r(\lambda),
$$

*where $D_r(\lambda)$ is any unimodular $n \times n$ matrix, and $(a_l(\lambda), b_l(\lambda))$ is any irreducible pair satisfying condition (3.19).*

The proof runs analogously to that of Theorem 3.2. ■

## 3.3 Dual Processes and Dual Basic Controllers

(1)

**Definition 3.1** The irreducible left and right processes $(a_l(\lambda), b_l(\lambda))$ and $[a_r(\lambda), b_r(\lambda)]$ are called dual, when they are related by condition (3.19).

Assume $(\alpha_l^B(\lambda), \beta_l^B(\lambda))$ and $[\alpha_r^B(\lambda), \beta_r^B(\lambda)]$ are basic controllers for dual processes, then $Q_l(\lambda, \alpha_l^B, \beta_l^B)$ and $Q_r(\lambda, \alpha_r^B, \beta_r^B)$ are unimodular.

**Definition 3.2** The basic controllers $(\alpha_l^B(\lambda), \beta_l^B(\lambda))$ and $[\alpha_r^B(\lambda), \beta_r^B(\lambda)]$ are called dual, if equation

$$Q_l(\lambda, \alpha_l^B, \beta_l^B)Q_r(\lambda, \alpha_r^B, \beta_r^B) = Q_r(\lambda, \alpha_r^B, \beta_r^B)Q_l(\lambda, \alpha_l^B, \beta_l^B) = I_{n+m}, \tag{3.27}$$

is true, which is equivalently written as

$$\begin{aligned} &\begin{bmatrix} a_l(\lambda) & -b_l(\lambda) \\ -\beta_l^B(\lambda) & \alpha_l^B(\lambda) \end{bmatrix} \begin{bmatrix} \alpha_r^B(\lambda) & b_r(\lambda) \\ \beta_r^B(\lambda) & a_r(\lambda) \end{bmatrix} \\ &\qquad = \begin{bmatrix} \alpha_r^B(\lambda) & b_r(\lambda) \\ \beta_r^B(\lambda) & a_r(\lambda) \end{bmatrix} \begin{bmatrix} a_l(\lambda) & -b_l(\lambda) \\ -\beta_l^B(\lambda) & \alpha_l^B(\lambda) \end{bmatrix} = I_{n+m}. \end{aligned} \tag{3.28}$$

Performing the matrix multiplications, we obtain two groups of relations, which are satisfied from left and right dual processes and dual basic controllers

$$\begin{aligned} a_l(\lambda)\alpha_r^B(\lambda) - b_l(\lambda)\beta_r^B(\lambda) &= I_n, \\ a_l(\lambda)b_r(\lambda) - b_l(\lambda)a_r(\lambda) &= 0_{nm}, \\ -\beta_l^B(\lambda)\alpha_r^B(\lambda) + \alpha_l^B(\lambda)\beta_r^B(\lambda) &= 0_{mn}, \\ -\beta_l^B(\lambda)b_r(\lambda) + \alpha_l^B(\lambda)a_r(\lambda) &= I_m, \end{aligned} \tag{3.29}$$

and also

$$\begin{aligned} \alpha_r^B(\lambda)a_l(\lambda) - b_r(\lambda)\beta_l^B(\lambda) &= I_n, \\ -\alpha_r^B(\lambda)b_l(\lambda) + b_r(\lambda)\alpha_l^B(\lambda) &= 0_{nm}, \\ \beta_r^B(\lambda)a_l(\lambda) - a_r(\lambda)\beta_l^B(\lambda) &= 0_{mn}, \\ -\beta_r^B(\lambda)b_l(\lambda) + a_r(\lambda)\alpha_l^B(\lambda) &= I_m. \end{aligned} \tag{3.30}$$

Relations (3.29) and (3.30) are called direct and inverse Bézout's identities, respectively. The validity of the direct or inverse Bézout's identity is a necessary and sufficient condition for the duality of the processes $(a_l(\lambda), b_l(\lambda))$, $[a_r(\lambda), b_r(\lambda)]$ and their associated basic controllers $(\alpha_l^B(\lambda), \beta_l^B(\lambda))$, $[\alpha_r^B(\lambda), \beta_r^B(\lambda)]$.

(2)

**Theorem 3.5** *Assume the dual processes* $(a_l(\lambda), b_l(\lambda))$ *and* $[a_r(\lambda), b_r(\lambda)]$. *Then the following statements hold:*

*(i) For any pair of dual processes, there exists a nonempty set of dual controllers.*
*(ii) For the fact, that for a given pair of dual processes the left basic controller* $(\alpha_l^B(\lambda), \beta_l^B(\lambda))$ *possesses a dual basic controller, it is necessary and sufficient, that it fulfills the Diophantine equation*

$$\alpha_l^B(\lambda)a_r(\lambda) - \beta_l^B(\lambda)b_r(\lambda) = I_m. \tag{3.31}$$

*(iii) If the left basic controller* $(\alpha_l^B(\lambda), \beta_l^B(\lambda))$ *satisfies Eq. (3.31), then the associated dual pair* $[\alpha_r^B(\lambda), \beta_r^B(\lambda)]$ *is unique.*

*Proof*

(i) Since the pair $[a_r(\lambda), b_r(\lambda)]$ is irreducible, the Diophantine Eq. (3.31) is solvable. Therefore, claim (i) is a consequence of claim (ii).
(ii) Sufficiency: Let the basic controller $(\alpha_l^B(\lambda), \beta_l^B(\lambda))$ be a solution of Eq. (3.31). By construction $(\alpha_l^B(\lambda), \beta_l^B(\lambda))$ is irreducible. From the irreducibility of the pair $(\alpha_l^B(\lambda), \beta_l^B(\lambda))$ and (3.28) it follows the existence of a right basic controller $[\tilde{\alpha}_r(\lambda), \tilde{\beta}_r(\lambda)]$, satisfying

$$a_l(\lambda)\tilde{\alpha}_r(\lambda) - b_l(\lambda)\tilde{\beta}_r(\lambda) = I_n. \tag{3.32}$$

Using (3.19), (3.31), and (3.32), we obtain

$$\begin{bmatrix} a_l(\lambda) & -b_l(\lambda) \\ -\beta_l^B(\lambda) & \alpha_l^B(\lambda) \end{bmatrix} \begin{bmatrix} \tilde{\alpha}_r(\lambda) & b_r(\lambda) \\ \tilde{\beta}_r(\lambda) & a_r(\lambda) \end{bmatrix} = \begin{bmatrix} I_n & 0_{nm} \\ -\beta_l^B(\lambda)\tilde{\alpha}_r(\lambda) + \alpha_l^B(\lambda)\tilde{\beta}_r(\lambda) & I_m \end{bmatrix}.$$

Hence,

$$\begin{bmatrix} a_l(\lambda) & -b_l(\lambda) \\ -\beta_l^B(\lambda) & \alpha_l^B(\lambda) \end{bmatrix} \begin{bmatrix} \alpha_r^B(\lambda) & b_r(\lambda) \\ \beta_r^B(\lambda) & a_r(\lambda) \end{bmatrix} = \begin{bmatrix} I_n & 0_{nm} \\ 0_{mn} & I_m \end{bmatrix}, \tag{3.33}$$

where

$$\begin{aligned} \alpha_r^B(\lambda) &= \tilde{\alpha}_r(\lambda) + b_r(\lambda)[\beta_l^B(\lambda)\tilde{\alpha}_r(\lambda) - \alpha_l^B(\lambda)\tilde{\beta}_r(\lambda)], \\ \beta_r^B(\lambda) &= \tilde{\beta}_r(\lambda) + a_r(\lambda)[\beta_l^B(\lambda)\tilde{\alpha}_r(\lambda) - \alpha_l^B(\lambda)\tilde{\beta}_r(\lambda)]. \end{aligned} \tag{3.34}$$

Obviously, $[\alpha_r^B(\lambda), \beta_r^B(\lambda)]$ is a right basic controller, which is dual to the basic controller $(\alpha_l^B(\lambda), \beta_l^B(\lambda))$. The sufficiency is shown.
The necessity directly follows from the direct Bézout identity (3.29).
(iii) The uniqueness of the dual basic controller follows from the uniqueness of the inverse matrix. ■

*Remark 3.1* In analogy to Theorem 3.5, it can be shown that for the right basic controller $[\alpha_r^B(\lambda), \beta_r^B(\lambda)]$ there exists a dual left basic controller if and only if the condition

$$a_l(\lambda)\alpha_r^B(\lambda) - b_l(\lambda)\beta_r^B(\lambda) = I_n, \tag{3.35}$$

is true. Moreover, if the pair $(\tilde{\alpha}_l(\lambda), \tilde{\beta}_l(\lambda))$ is a solution of

$$\tilde{\alpha}_l(\lambda)a_r(\lambda) - \tilde{\beta}_l(\lambda)b_r(\lambda) = I_m, \tag{3.36}$$

then the dual left basic controller $(\alpha_l^B(\lambda), \beta_l^B(\lambda))$ can be determined by the formulae

$$\begin{aligned}\alpha_l^B(\lambda) &= \tilde{\alpha}_l(\lambda) - [\tilde{\beta}_l(\lambda)\alpha_r^B(\lambda) - \tilde{\alpha}_l(\lambda)\beta_r^B(\lambda)]b_l(\lambda),\\ \beta_l^B(\lambda) &= \tilde{\beta}_l(\lambda) - [\tilde{\beta}_l(\lambda)\alpha_r^B(\lambda) - \tilde{\alpha}_l(\lambda)\beta_r^B(\lambda)]a_l(\lambda).\end{aligned} \tag{3.37}$$

(3) The next claim defines a parametrization of the set of all pairs of dual basic controllers.

**Theorem 3.6** *For a given pair of dual processes $(a_l(\lambda), b_l(\lambda))$ and $[a_r(\lambda), b_r(\lambda)]$ assume a pair of dual basic controllers $(\alpha_{l0}^B(\lambda), \beta_{l0}^B(\lambda))$ and $(\alpha_{r0}^B(\lambda), \beta_{r0}^B(\lambda))$. Then the set of all pairs of dual basic controllers $(\alpha_l^B(\lambda), \beta_l^B(\lambda))$ and $[\alpha_r^B(\lambda), \beta_r^B(\lambda)]$ is determined by the relations*

$$\begin{aligned}\alpha_l^B(\lambda) &= \alpha_{l0}^B(\lambda) - M(\lambda)b_l(\lambda), \quad & \beta_l^B(\lambda) &= \beta_{l0}^B(\lambda) - M(\lambda)a_l(\lambda),\\ \alpha_r^B(\lambda) &= \alpha_{r0}^B(\lambda) - b_r(\lambda)M(\lambda), \quad & \beta_r^B(\lambda) &= \beta_{r0}^B(\lambda) - a_r(\lambda)M(\lambda),\end{aligned} \tag{3.38}$$

*where $M(\lambda)$ is an arbitrary polynomial matrix of appropriate size.*

*Proof* From (3.8) and (3.26), we obtain

$$\begin{aligned}\alpha_l^B(\lambda) &= D_l(\lambda)\alpha_{l0}^B(\lambda) - M_l(\lambda)b_l(\lambda), \quad & \beta_l^B(\lambda) &= D_l(\lambda)\beta_{l0}^B(\lambda) - M_l(\lambda)a_l(\lambda),\\ \alpha_r^B(\lambda) &= \alpha_{r0}^B(\lambda)D_r(\lambda) - b_r(\lambda)M_r(\lambda), \quad & \beta_r^B(\lambda) &= \beta_{r0}^B(\lambda)D_r(\lambda) - a_r(\lambda)M_r(\lambda),\end{aligned} \tag{3.39}$$

that can be written in the form

$$\begin{aligned}Q_l(\lambda, \alpha_l^B, \beta_l^B) &= \begin{bmatrix} I_n & 0_{nm} \\ M_l(\lambda) & D_l(\lambda) \end{bmatrix} Q_l(\lambda, \alpha_{l0}^B, \beta_{l0}^B),\\ Q_r(\lambda, \alpha_r^B, \beta_r^B) &= Q_r(\lambda, \alpha_{r0}^B, \beta_{r0}^B) \begin{bmatrix} D_r(\lambda) & 0_{nm} \\ -M_r(\lambda) & I_m \end{bmatrix}.\end{aligned} \tag{3.40}$$

Since the basic controllers $[\alpha_r^B(\lambda), \beta_r^B(\lambda)]$ and $(\alpha_l^B(\lambda), \beta_l^B(\lambda))$ are dual, they satisfy

$$Q_l(\lambda, \alpha_l^B, \beta_l^B)Q_r(\lambda, \alpha_r^B, \beta_r^B) = \begin{bmatrix} I_n & 0_{nm} \\ 0_{mn} & I_m \end{bmatrix}. \tag{3.41}$$

On the other side, using the duality of the basic controllers in $Q_l(\lambda, \alpha_{l0}^B, \beta_{l0}^B)$ and $Q_r(\lambda, \alpha_{r0}^B, \beta_{r0}^B)$, from (3.40) we find

$$Q_l(\lambda, \alpha_l^B, \beta_l^B)Q_r(\lambda, \alpha_r^B, \beta_r^B) = \begin{bmatrix} D_r(\lambda) & 0_{nm} \\ M_l(\lambda)D_r(\lambda) - D_l(\lambda)M_r(\lambda) & D_l(\lambda) \end{bmatrix}. \tag{3.42}$$

Comparing the right sides of the last two equations, then it is clear that (3.41) is true if and only if

$$D_r(\lambda) = I_n, \quad D_l(\lambda) = I_m, \quad M_l(\lambda) = M_r(\lambda) = M(\lambda), \tag{3.43}$$

which leads to (3.38). ∎

(4)

*Remark 3.2* Consider the arbitrary dual processes $(a_l(\lambda), b_l(\lambda))$, $[a_r(\lambda), b_r(\lambda)]$, and the unimodular matrices $d_l(\lambda)$, $d_r(\lambda)$ of size $n \times n$, $m \times m$, respectively. Assume

$$\begin{aligned} a_{l1}(\lambda) &= d_l(\lambda)a_l(\lambda), \quad b_{l1}(\lambda) = d_l(\lambda)b_l(\lambda), \\ a_{r1}(\lambda) &= a_r(\lambda)d_r(\lambda), \quad b_{r1}(\lambda) = b_r(\lambda)d_r(\lambda). \end{aligned} \tag{3.44}$$

Due to

$$a_{l1}(\lambda)b_{r1}(\lambda) = b_{l1}(\lambda)a_{r1}(\lambda), \tag{3.45}$$

pairs (3.44) are dual. In particular, the matrices $d_l(\lambda)$ and $d_r(\lambda)$ can be chosen in such a way that the matrix $a_{l1}(\lambda)$ becomes row reduced, and the matrix $a_{r1}(\lambda)$ column reduced.

## 3.4 General Solution of the Determinant Polynomial Equation

(1) In this section, we consider the general solution of the determinant polynomial equation (DPE)

$$\det Q_l(\lambda, \alpha_l, \beta_l) = \det \begin{bmatrix} a_l(\lambda) & -b_l(\lambda) \\ -\beta_l(\lambda) & \alpha_l(\lambda) \end{bmatrix} \sim d(\lambda), \tag{3.46}$$

where $(a_l(\lambda), b_l(\lambda))$ is a given left process, and $d(\lambda)$ is a given monic polynomial. The next claim yields the general solution of DPE (3.46) for an irreducible process $(a_l(\lambda), b_l(\lambda))$.

**Theorem 3.7** *Let the process $(a_l(\lambda), b_l(\lambda))$ be irreducible. Then the following statements hold:*

(i) *DPE (3.46) is solvable for any polynomial* $d(\lambda)$.

(ii) *Let* $(\alpha_l^B(\lambda), \beta_l^B(\lambda))$ *be any basic controller for the process* $(a_l(\lambda), b_l(\lambda))$. *Then the set of all controllers satisfying Eq. (3.46), is defined by the formulae*

$$\begin{aligned} \alpha_l(\lambda) &= D_l(\lambda)\alpha_l^B(\lambda) - M_l(\lambda)b_l(\lambda), \\ \beta_l(\lambda) &= D_l(\lambda)\beta_l^B(\lambda) - M_l(\lambda)a_l(\lambda), \end{aligned} \tag{3.47}$$

*where* $D_l(\lambda)$ *and* $M_l(\lambda)$ *are polynomial matrices of size* $m \times m$ *and* $m \times n$, *respectively. Moreover, the matrix* $M_l(\lambda)$ *is arbitrary, but the matrix* $D_l(\lambda)$ *is subject to the only condition*

$$\det D_l(\lambda) \sim d(\lambda). \tag{3.48}$$

(iii) *The controller* $(\alpha_l(\lambda), \beta_l(\lambda))$ *is irreducible, if and only if the pair* $(D_l(\lambda), M_l(\lambda))$ *is irreducible.*

*Proof*

(i) Inserting (3.47) into (3.46), we find

$$\begin{aligned} \begin{bmatrix} a_l(\lambda) & -b_l(\lambda) \\ -\beta_l(\lambda) & \alpha_l(\lambda) \end{bmatrix} &= \begin{bmatrix} a_l(\lambda) & -b_l(\lambda) \\ -D_l(\lambda)\beta_l^B(\lambda) + M_l(\lambda)a_l(\lambda) & D_l(\lambda)\alpha_l^B(\lambda) - M_l(\lambda)b_l(\lambda) \end{bmatrix} \\ &= \begin{bmatrix} I_n & 0_{nm} \\ M_l(\lambda) & D_l(\lambda) \end{bmatrix} \begin{bmatrix} a_l(\lambda) & -b_l(\lambda) \\ -\beta_l^B(\lambda) & \alpha_l^B(\lambda) \end{bmatrix} \end{aligned} \tag{3.49}$$

Calculating the determinants on both sides of this equation, we obtain

$$\det Q_l(\lambda, \alpha_l, \beta_l) = \det D_l(\lambda) \det Q_l(\lambda, \alpha_l^B, \beta_l^B). \tag{3.50}$$

With the help of (3.48), the last equation is equivalent to (3.46), because the matrix $Q_l(\lambda, \alpha_l^B, \beta_l^B)$ is unimodular. Thus, relation (3.47) defines a solution of DPE (3.46).

(ii) We show that formulae (3.47) generate the set of all solutions of DPE (3.46). Denote by $\mathscr{N}$ the set of all solutions of (3.46), and by $\mathscr{N}^p$ the set of all controllers (3.47). Let $[a_r(\lambda), b_r(\lambda)]$ be the dual process, and $(\alpha_{l0}^B(\lambda), \beta_{l0}^B(\lambda))$ and $(\alpha_{r0}^B(\lambda), \beta_{r0}^B(\lambda))$ be a pair of dual basic controllers. Then, by construction the matrices $Q_l(\lambda, \alpha_{l0}^B, \beta_{l0}^B)$ and $Q_r(\lambda, \alpha_{r0}^B, \beta_{r0}^B) = Q_l^{-1}(\lambda, \alpha_{l0}^B, \beta_{l0}^B)$ are unimodular and satisfy

$$Q_l(\lambda, \alpha_{l0}^B, \beta_{l0}^B) Q_r(\lambda, \alpha_{r0}^B, \beta_{r0}^B) = I_{n+m}. \tag{3.51}$$

Assume that controller $(\alpha_l(\lambda), \beta_l(\lambda))$ satisfies Eq. (3.46). Using direct Bézout identity (3.29), we obtain

$$Q_l(\lambda, \alpha_l, \beta_l) Q_r(\lambda, \alpha_{r0}^B, \beta_{r0}^B) = \begin{bmatrix} I_n & 0_{nm} \\ M_l(\lambda) & D_l(\lambda) \end{bmatrix} \triangleq N_l(\lambda), \tag{3.52}$$

where

$$D_l(\lambda) = -\beta_l(\lambda)b_r(\lambda) + \alpha_l(\lambda)a_r(\lambda),$$
$$M_l(\lambda) = -\beta_l(\lambda)\alpha_{r0}^B(\lambda) + \alpha_l(\lambda)\beta_{r0}^B(\lambda). \tag{3.53}$$

From (3.52) and (3.51), we find

$$Q_l(\lambda, \alpha_l, \beta_l) = N_l(\lambda)Q_l(\lambda, \alpha_{l0}^B, \beta_{l0}^B). \tag{3.54}$$

Hence, as above, we realize that the controller $(\alpha_l(\lambda), \beta_l(\lambda))$ has form (3.47). Moreover, from (3.54), we obtain

$$\det Q_l(\lambda, \alpha_l, \beta_l) = \det N_l(\lambda) \det Q(\lambda, \alpha_{l0}^B, \beta_{l0}^B) \sim \det D_l(\lambda) \sim d(\lambda).$$

From these explanations, $\mathscr{N} \subset \mathscr{N}^p$ is achieved. On the other side, from the proof of (i), we conclude $\mathscr{N}^p \subset \mathscr{N}$, that finishes the proof of (ii).

(iii) Formulae (3.47) can be written in the form

$$\begin{bmatrix} -\beta_l(\lambda) & \alpha_l(\lambda) \end{bmatrix} = \begin{bmatrix} M_l(\lambda) & D_l(\lambda) \end{bmatrix} Q_l(\lambda, \alpha_l^B, \beta_l^B). \tag{3.55}$$

Since the matrix $Q_l(\lambda, \alpha_l^B, \beta_l^B)$ is unimodular, the matrices $\begin{bmatrix} M_l(\lambda) & D_l(\lambda) \end{bmatrix}$ and $\begin{bmatrix} -\beta_l(\lambda) & \alpha_l(\lambda) \end{bmatrix}$ are equivalent. Therefore, the pairs $(\alpha_l(\lambda), \beta_l(\lambda))$ and $(D_l(\lambda), M_l(\lambda))$ are at the same time irreducible or reducible. ∎

(3) The following theorem states a connection between DPE (3.46) and certain sets of Diophantine polynomial equations.

**Theorem 3.8** *Let $(a_l(\lambda), b_l(\lambda))$ and $[a_r(\lambda), b_r(\lambda)]$ be dual processes. Then the set of solutions of DPE (3.46) coincides with the unification of the solution sets for the Diophantine equations*

$$\alpha_l(\lambda)a_r(\lambda) - \beta_l(\lambda)b_r(\lambda) = D_l(\lambda), \tag{3.56}$$

*where the $m \times m$ matrix $D_l(\lambda)$ satisfies*

$$\det D_l(\lambda) \sim d(\lambda). \tag{3.57}$$

*Proof* Denote the set of all solutions of Eq. (3.46) by $\mathscr{N}$, and the unification of all solution sets of (3.56) by $\mathscr{N}^d$. Assume a dual pair of basic controller $(\alpha_l^B(\lambda), \beta_l^B(\lambda))$, $[\alpha_r^B(\lambda), \beta_r^B(\lambda)]$. Let the controllers $(\alpha_l(\lambda), \beta_l(\lambda))$ be determined by relations (3.47). Then, using the direct Bézout identity (3.29), we find

$$\alpha_l(\lambda)a_r(\lambda) - \beta_l(\lambda)b_r(\lambda) = D_l(\lambda)[\alpha_l^B(\lambda)a_r(\lambda) - \beta_l^B(\lambda)b_r(\lambda)]$$
$$-M_l(\lambda)[b_l(\lambda)a_r(\lambda) - a_l(\lambda)b_r(\lambda)] = D_l(\lambda). \tag{3.58}$$

Thus, $\mathscr{N} \subset \mathscr{N}^d$. On the other side, let the controller $(\alpha_l(\lambda), \beta_l(\lambda))$ be a solution of Eq. (3.56), for which condition (3.57) is fulfilled. Then, applying (3.56) and (3.29), we obtain

$$\begin{aligned}
Q_l(\lambda, \alpha_l, \beta_l) Q_r(\lambda, \alpha_r^B, \beta_r^B) &= \begin{bmatrix} a_l(\lambda) & -b_l(\lambda) \\ -\beta_l(\lambda) & \alpha_l(\lambda) \end{bmatrix} \begin{bmatrix} \alpha_r^B(\lambda) & b_r(\lambda) \\ \beta_r^B(\lambda) & a_r(\lambda) \end{bmatrix} \\
&= \begin{bmatrix} a_l(\lambda)\alpha_r^B(\lambda) - b_l(\lambda)\beta_r^B(\lambda) & a_l(\lambda)b_r(\lambda) - b_l(\lambda)a_r(\lambda) \\ -\beta_l(\lambda)\alpha_r^B(\lambda) + \alpha_l(\lambda)\beta_r^B(\lambda) & -\beta_l(\lambda)b_r(\lambda) + \alpha_l(\lambda)a_r(\lambda) \end{bmatrix} \\
&= \begin{bmatrix} I_n & 0_{nm} \\ M_l(\lambda) & D_l(\lambda) \end{bmatrix}.
\end{aligned} \tag{3.59}$$

Hence

$$\det Q_l(\lambda, \alpha_l, \beta_l) \sim \det D_l \sim d(\lambda). \tag{3.60}$$

Thus $\mathscr{N}^d \subset \mathscr{N}$. Since both sets contain each other, the theorem is proven. ■

(4) Let us consider now the question about the solutions of DPE (3.4) in the case the process $(a_l(\lambda), b_l(\lambda))$ is reducible. As explained in Sect. 1.8, in that case there exists the representation

$$a_l(\lambda) = q(\lambda) a_{l1}(\lambda), \quad b_l(\lambda) = q(\lambda) b_{l1}(\lambda), \tag{3.61}$$

where $q(\lambda)$ is the greatest common divisor (GCD) of the matrices $a_l(\lambda)$ and $b_l(\lambda)$, and $(a_{l1}(\lambda), b_{l1}(\lambda))$ is irreducible.

The condition for the solvability of DPE (3.4) for that case is stated in the next theorem.

**Theorem 3.9** *The following statements are true:*

*(i) For the solvability of Eq. (3.4), it is necessary and sufficient that the expression*

$$\tilde{d}(\lambda) \triangleq \frac{d(\lambda)}{\det q(\lambda)}, \tag{3.62}$$

*is a polynomial.*

*(ii) Let $(\alpha_l^B(\lambda), \beta_l^B(\lambda))$ be a basic controller for the irreducible process $(a_{l1}(\lambda), b_{l1}(\lambda))$. Then under condition (i) the set of solutions of Eq. (3.4) is defined by the formulae*

$$\begin{aligned}
\alpha_l(\lambda) &= \tilde{D}_l(\lambda)\alpha_l^B(\lambda) - \tilde{M}_l(\lambda) b_{l1}(\lambda), \\
\beta_l(\lambda) &= \tilde{D}_l(\lambda)\beta_l^B(\lambda) - \tilde{M}_l(\lambda) a_{l1}(\lambda),
\end{aligned} \tag{3.63}$$

*where the $m \times n$ matrix $\tilde{M}_l(\lambda)$ is arbitrary, but the matrix $\tilde{D}_l(\lambda)$ satisfies the single condition*

$$\det \tilde{D}_l(\lambda) \sim \tilde{d}(\lambda).$$

*Proof* (i) Necessity: Assume (3.61), and $(\alpha_l^B(\lambda), \beta_l^B(\lambda))$ as a solution of (3.4). Then, using (3.61) Eq. (3.46) can be written in the form

$$\det\left\{\begin{bmatrix} q(\lambda) & 0_{nm} \\ 0_{mn} & I_m \end{bmatrix} \tilde{Q}_l(\lambda, \alpha_l^B, \beta_l^B)\right\} \sim d(\lambda), \tag{3.64}$$

where

$$\tilde{Q}_l(\lambda, \alpha_l^B, \beta_l^B) = \begin{bmatrix} a_{l1}(\lambda) & -b_{l1}(\lambda) \\ -\beta_l(\lambda) & \alpha_l(\lambda) \end{bmatrix}. \tag{3.65}$$

Calculating in (3.64) the determinant on the left side, we find

$$\det q(\lambda) \det \tilde{Q}_l(\lambda, \alpha_l^B, \beta_l^B) \sim d(\lambda). \tag{3.66}$$

Hence, the polynomial $\det q(\lambda)$ is a divisor of the polynomial $d(\lambda)$, and expression (3.62) is a polynomial.

Sufficiency: Assume, that expression (3.62) is a polynomial. Then, from (3.66) after reduction, the DPE

$$\det \tilde{Q}_l(\lambda, \alpha_l, \beta_l) \sim \tilde{d}(\lambda). \tag{3.67}$$

is achieved. Since the pair $(a_{l1}(\lambda), b_{l1}(\lambda))$ is irreducible, Eq. (3.67) is solvable.

(ii) Claim (ii) is a consequence of Theorem 3.7. ■

(5) Let $(a_l(\lambda), b_l(\lambda))$ be an irreducible left process, and $(\alpha_l(\lambda), \beta_l(\lambda))$ a controller, such that $\det Q_l(\lambda, \alpha_l, \beta_l) \sim d(\lambda)$, where $d(\lambda) \neq 0$ is a monic polynomial. Moreover, assume a certain basic controller $(\alpha_l^B(\lambda), \beta_l^B(\lambda))$. Then, in harmony with Theorem 3.7 there exist polynomial matrices $D_l(\lambda)$ and $M_l(\lambda)$ of sizes $m \times m$ and $m \times n$, respectively, such that $\det D_l(\lambda) \sim d(\lambda)$ and

$$\begin{aligned} \alpha_l(\lambda) &= D_l(\lambda)\alpha_l^B(\lambda) - M_l(\lambda)b_l(\lambda), \\ \beta_l(\lambda) &= D_l(\lambda)\beta_l^B(\lambda) - M_l(\lambda)a_l(\lambda). \end{aligned} \tag{3.68}$$

**Definition 3.3** Relations (3.68) are called basic representation of the controller $(\alpha_l(\lambda), \beta_l(\lambda))$ by the basis $(\alpha_l^B(\lambda), \beta_l^B(\lambda))$.

**Theorem 3.10** *For a fixed basis $(\alpha_l^B(\lambda), \beta_l^B(\lambda))$, the basic representation (3.68) is unique in the sense, that if we would have in addition to (3.68) another representation*

$$\begin{aligned} \alpha_l(\lambda) &= D_{l1}(\lambda)\alpha_l^B(\lambda) - M_{l1}(\lambda)b_l(\lambda), \\ \beta_l(\lambda) &= D_{l1}(\lambda)\beta_l^B(\lambda) - M_{l1}(\lambda)a_l(\lambda), \end{aligned} \tag{3.69}$$

*then*

$$D_{l1}(\lambda) = D_l(\lambda), \quad M_{l1}(\lambda) = M_l(\lambda). \tag{3.70}$$

*Proof* Let us have at the same time both representations (3.68) and (3.69). Then we obtain

$$\begin{aligned} &[D_l(\lambda) - D_{l1}(\lambda)]\,\alpha_l^B(\lambda) - [M_l(\lambda) - M_{l1}(\lambda)]b_l(\lambda) = 0_{mm}, \\ &[D_l(\lambda) - D_{l1}(\lambda)]\beta_l^B(\lambda) - [M_l(\lambda) - M_{l1}(\lambda)]a_l(\lambda) = 0_{mn}, \end{aligned} \tag{3.71}$$

that can be written in the form

$$\left[\, M_l(\lambda) - M_{l1}(\lambda)\ D_l(\lambda) - D_{l1}(\lambda)\,\right] Q_l(\lambda, \alpha_l^B, \beta_l^B) = 0_{m,m+n}. \tag{3.72}$$

Hence, (3.70) becomes true, because $Q_l(\lambda, \alpha_l^B, \beta_l^B)$ is unimodular. ∎

## 3.5 Solution of DPE Under Constraints on the Structure of the Set of Invariant Polynomials of the Characteristic Matrix

(1) Let the determinant polynomial equation

$$\det Q_l(\lambda, \alpha_l, \beta_l) = \det \begin{bmatrix} a_l(\lambda) & -b_l(\lambda) \\ -\beta_l(\lambda) & \alpha_l(\lambda) \end{bmatrix} \sim d(\lambda), \tag{3.73}$$

where $d(\lambda)$ is monic polynomial, have a set of solutions, which is denoted by $\mathscr{N}$. Then, in principle for different controllers $(\alpha_l(\lambda), \beta_l(\lambda)) \in \mathscr{N}$, the matrix $Q_l(\lambda, \alpha_l, \beta_l)$ can have different Smith canonical forms, i.e., different sequences of invariant polynomials.

**Definition 3.4** An arbitrary sequence of polynomials $a_1(\lambda), \ldots, a_q(\lambda)$, where $q \le n + m$ is called admissible for Eq. (3.73), if it fulfills the following conditions:

(a) The polynomials $a_1(\lambda) \ldots a_q(\lambda)$ are monic.
(b) The polynomial $a_i(\lambda)$ is a divisor of the polynomial $a_{i+1}(\lambda)$.
(c) $a_1(\lambda) \cdots a_q(\lambda) = d(\lambda)$.

On basis of the above concepts, the following problem for the solution of DPE (3.73) under constraints on the structure of the set of invariant polynomials can be formulated:

**Advanced pole placement:** Let DPE (3.73) be solvable for the given polynomial $d(\lambda)$. Find the set of controllers $\mathscr{N}_q \subset \mathscr{N}$, such that the sequence of invariant polynomials of the matrix $Q_l(\lambda, \alpha_l, \beta_l)$, coincides with a given admissible sequence $a_1(\lambda), \ldots, a_q(\lambda)$.

The actual subsection provides the general solution of the formulated problem for an irreducible process.

(2)

**Theorem 3.11** *Assume the irreducible process* $(a_l(\lambda), b_l(\lambda))$, *and formulae (3.47) define the set of solutions of DPE (3.73). Then for fixed matrix* $D_l(\lambda)$, *the relation*

$$Q_l(\lambda, \alpha_l, \beta_l) \overset{p}{\sim} D_l(\lambda). \tag{3.74}$$

*is true. Here, as earlier, the symbol* $\overset{p}{\sim}$ *means that the matrices* $Q_l(\lambda, \alpha_l, \beta_l)$ *and* $D_l(\lambda)$ *are pseudo-equivalent, i.e., the invariant polynomials different from one of* $Q_l(\lambda, \alpha_l, \beta_l)$ *and* $D_l(\lambda)$ *coincide.*

*Proof* We will show that the matrices $Q_l(\lambda, \alpha_l, \beta_l)$ and

$$S(\lambda) \stackrel{\triangle}{=} \operatorname{diag}\{I_n, D_l(\lambda)\}, \tag{3.75}$$

are equivalent. Indeed, since

$$N_l(\lambda) = \begin{bmatrix} I_n & 0_{nm} \\ M_l(\lambda) & D_l(\lambda) \end{bmatrix} = \begin{bmatrix} I_n & 0_{nm} \\ M_l(\lambda) & I_m \end{bmatrix} \begin{bmatrix} I_n & 0_{nm} \\ 0_{mn} & D_l(\lambda) \end{bmatrix}, \tag{3.76}$$

we find from (3.54)

$$Q_l(\lambda, \alpha_l, \beta_l) = \begin{bmatrix} I_n & 0_{nm} \\ M_l(\lambda) & I_m \end{bmatrix} S(\lambda) Q_l(\lambda, \alpha_{l0}^B, \beta_{l0}^B). \tag{3.77}$$

Notice that the first and the last factor on the right side are unimodular matrices. Hence

$$Q_l(\lambda, \alpha_l, \beta_l) \sim S(\lambda). \tag{3.78}$$

In addition, from (3.75), it follows

$$S(\lambda) \overset{p}{\sim} D_l(\lambda). \tag{3.79}$$

The last two formulae yield (3.74). ■

The last insight allows to formulate the following statements.

**Theorem 3.12** *Under the conditions of Theorem 3.11 the following statements are true:*

*(i) Independently on the choice of the matrices $D_l(\lambda)$ and $M_l(\lambda)$, in formulae (3.47) the first n invariant polynomials of the matrix $Q_l(\lambda, \alpha_l, \beta_l)$ are equal to one, i.e.,*

$$a_1(\lambda) = \cdots = a_n(\lambda) = 1. \tag{3.80}$$

*(ii) The sequence of invariant polynomials $a_{n+1}(\lambda), \ldots, a_{n+m}(\lambda)$ of the matrix $Q_l(\lambda, \alpha_l, \beta_l)$ coincides with the sequence of invariant polynomials of the matrix $D_l(\lambda)$.*

*Proof* Claim (i) follows immediately from (3.78), and (ii) is a consequence of (3.74).

(3) The next theorem connects the investigated problem with the properties of Diophantine polynomial equations.

**Theorem 3.13** *Assume the dual models $(a_l(\lambda), b_l(\lambda))$ and $[a_r(\lambda), b_r(\lambda)]$ of a process. Then for the fact that the sequence of invariant polynomials of the matrix $Q_l(\lambda, \alpha_l, \beta_l)$ fulfills*

$$a_{n+k}(\lambda) = d_k(\lambda), \quad (k = 1, \ldots, m), \tag{3.81}$$

*where $d_1(\lambda), \ldots, d_m(\lambda)$ is an admissible sequence of invariant polynomials, it is necessary and sufficient, that the controller $(\alpha_l(\lambda), \beta_l(\lambda))$ is a solution of the equation*

$$-\beta_l(\lambda)b_r(\lambda) + \alpha_l(\lambda)a_r(\lambda) = D_l(\lambda), \tag{3.82}$$

*where $D_l(\lambda)$ is an arbitrary polynomial matrix, whose sequence of invariant polynomials coincides with the sequence $d_1(\lambda), \ldots, d_m(\lambda)$.*

*Proof* Sufficiency: Let $(\alpha_l^B(\lambda), \beta_l^B(\lambda))$, $[\alpha_r^B(\lambda), \beta_r^B(\lambda)]$ be a pair of dual basic controllers. Then, from (3.59), we obtain

$$Q_l(\lambda, \alpha_l, \beta_l)Q_r(\lambda, \alpha_r^B, \beta_r^B) = \begin{bmatrix} I_n & 0_{nm} \\ M_l(\lambda) & D_l(\lambda) \end{bmatrix} = N_l(\lambda), \tag{3.83}$$

where

$$D_l(\lambda) = -\beta_l(\lambda)b_r(\lambda) + \alpha_l(\lambda)a_r(\lambda). \tag{3.84}$$

Since the matrix $Q_r(\lambda, \alpha_r^B, \beta_r^B)$ is unimodular, we find

$$Q_l(\lambda, \alpha_l, \beta_l) \sim N_l(\lambda). \tag{3.85}$$

Moreover,

$$D_l(\lambda) \overset{p}{\sim} N_l(\lambda). \tag{3.86}$$

Thus

$$Q_l(\lambda, \alpha_l, \beta_l) \overset{p}{\sim} D_l(\lambda), \tag{3.87}$$

which is equivalent to (3.81).
The necessity of the conditions is achieved by reverse reasoning. ■

## 3.6 Solution of DPE by Means of Transfer Matrices

(1) In those cases, when $\det \alpha_l(\lambda) \not\equiv 0$, i.e., the controller is non-singular, there exists the transfer function (matrix) of the left controller

$$W_{dl}(\lambda) = \alpha_l^{-1}(\lambda)\beta_l(\lambda). \tag{3.88}$$

Assume the transfer matrix $W_{dl}(\lambda)$ in standard form

$$W_{dl}(\lambda) = \frac{M_d(\lambda)}{\Delta_d(\lambda)}. \tag{3.89}$$

Then the right side of (3.88) can be seen as LMFD of matrix (3.89). Moreover, if the transfer matrix of the controller is given in standard form (3.89), each LMFD (3.88) can be related to the matrix

$$Q_l(\lambda, \alpha_l, \beta_l) = \begin{bmatrix} a_l(\lambda) & -b_l(\lambda) \\ -\beta_l(\lambda) & \alpha_l(\lambda) \end{bmatrix}. \tag{3.90}$$

Hence, each LMFD (3.88) corresponds to a characteristic polynomial

$$\Delta_d(\lambda) = \det Q_l(\lambda, \alpha_l, \beta_l). \tag{3.91}$$

**Definition 3.5** We will say that the rational matrix $W_{dl}(\lambda)$ defines a solution of DPE (3.4), if it allows LMFD (3.88), in which the pair $(\alpha_l(\lambda), \beta_l(\lambda))$ is a solution of Eq. (3.4).

(2) The next theorem provides a general description of the set of transfer matrices of the controllers as solution of DPE (3.4).

**Theorem 3.14** *Let $(a_l(\lambda), b_l(\lambda))$ be irreducible, and $(\alpha_{l0}(\lambda), \beta_{l0}(\lambda))$ be some left basic controller. Then the matrix $W_{dl}(\lambda)$ defines a solution of DPE (3.4), if and only if it allows the representation*

$$W_{dl}(\lambda) = [\alpha_{l0}(\lambda) - \theta_l(\lambda)b_l(\lambda)]^{-1}[\beta_{l0}(\lambda) - \theta_l(\lambda)a_l(\lambda)], \tag{3.92}$$

*where $\theta_l(\lambda)$ is either the zero matrix or any rational matrix permitting the LMFD*

$$\theta_l(\lambda) = D_l(\lambda)^{-1}M_l(\lambda), \tag{3.93}$$

*in which* $\det D_l(\lambda) \sim d(\lambda)$.

*Proof* Sufficiency: Assume (3.92) and (3.93). Then, we obtain

$$W_{dl}(\lambda) = [D_l(\lambda)\alpha_{l0}(\lambda) - M_l(\lambda)b_l(\lambda)]^{-1}[D_l(\lambda)\beta_{l0}(\lambda) - M_l(\lambda)a_l(\lambda)]. \tag{3.94}$$

The right side of this equation is an LMFD, in which the pair

$$\begin{aligned}\alpha_l(\lambda) &= D_l(\lambda)\alpha_{l0}(\lambda) - M_l(\lambda)b_l(\lambda),\\ \beta_l(\lambda) &= D_l(\lambda)\beta_{l0}(\lambda) - M_l(\lambda)a_l(\lambda)\end{aligned} \tag{3.95}$$

is a solution of DPE (3.4).
Necessity: Assume for LMFD (3.88) the relation $\det Q_l(\lambda, \alpha_l, \beta_l) \sim d(\lambda)$. Since the process $(a_l(\lambda), b_l(\lambda))$ is irreducible, the controller $(\alpha_l(\lambda), \beta_l(\lambda))$ is defined by formulae (3.95), where $\det D_l(\lambda) \sim d(\lambda)$. Hence, from (3.95) we find

$$W_{dl}(\lambda) = [\alpha_{0l}(\lambda) - D_l(\lambda)^{-1}M_l(\lambda)b_l(\lambda)]^{-1}[\beta_{0l}(\lambda) - D_l(\lambda)^{-1}M_l(\lambda)a_l(\lambda)], \tag{3.96}$$

which is a special case of (3.92) for $\theta_l(\lambda) = D_l(\lambda)^{-1}M_l(\lambda)$. ■

**Definition 3.6** The in (3.92) configured matrix $\theta_l(\lambda)$ is called left system parameter.

(3) Assume in addition to LMFD (3.88) the RMFD

$$W_{dr}(\lambda) = \beta_r(\lambda)\alpha_r^{-1}(\lambda), \tag{3.97}$$

and the dual irreducible right process $[a_r(\lambda), b_r(\lambda)]$. Then, in analogy to (3.4), we can consider the right DPE

$$\det Q_r(\lambda, \alpha_r, \beta_r) \sim d(\lambda), \tag{3.98}$$

where

$$Q_r(\lambda, \alpha_r, \beta_r) = \begin{bmatrix} \alpha_r(\lambda) & b_r(\lambda) \\ \beta_r(\lambda) & a_r(\lambda) \end{bmatrix}, \tag{3.99}$$

and Eq. (3.98) takes the form

$$\det \begin{bmatrix} \alpha_r(\lambda) & b_r(\lambda) \\ \beta_r(\lambda) & a_r(\lambda) \end{bmatrix} \sim d(\lambda). \tag{3.100}$$

In analogy to the above explained, we will say that the matrix $W_{dr}(\lambda)$ defines a solution of right DPE (3.98), if it allows RMFD (3.97), satisfying relation (3.98). In analogy to above, we can derive that the set of those matrices is defined by the formula

$$W_{dr}(\lambda) = [\beta_{r0}(\lambda) - a_r(\lambda)\theta_r(\lambda)][\alpha_{r0}(\lambda) - b_r(\lambda)\theta_r(\lambda)]^{-1}, \tag{3.101}$$

where $[\alpha_{r0}(\lambda), \beta_{r0}(\lambda)]$ is a right basic controller, and $\theta_r(\lambda)$ is either the zero matrix or any rational matrix permitting a RMFD

$$\theta_r(\lambda) = M_r(\lambda)D_r^{-1}(\lambda), \tag{3.102}$$

where $\det D_r(\lambda) \sim d(\lambda)$. Further on, the matrix $\theta_r(\lambda)$ is called right system parameter.

**Theorem 3.15** *The sets of matrices (3.92) and (3.101) coincide.*

*Proof* For the proof it is sufficient to show that for $\theta_l(\lambda) = \theta_r(\lambda) = \theta(\lambda)$ sets of transfer matrices (3.94) and (3.101) coincide. Without loss of generality, we suppose that in formulae (3.94) and (3.101) the basic controllers $(\alpha_{l0}(\lambda), \beta_{l0}(\lambda))$ and $[\alpha_{r0}(\lambda), \beta_{r0}(\lambda)]$ are dual. Then, using the direct Bézout identity, we obtain

$$\begin{aligned}
&[\beta_{l0}(\lambda) - \theta(\lambda)a_l(\lambda)]\,[\alpha_{r0}(\lambda) - b_r(\lambda)\theta(\lambda)] - [-\alpha_l(\lambda) - \theta(\lambda)b_l(\lambda)]\cdot \\
&\qquad\qquad [\beta_{r0}(\lambda) - a_r(\lambda)\theta(\lambda)] \\
&= \beta_{l0}(\lambda)\alpha_{r0}(\lambda) - \alpha_{l0}(\lambda)\beta_{r0}(\lambda) - \theta(\lambda)[a_l(\lambda)\alpha_{r0}(\lambda) - b_l(\lambda)\beta_{r0}(\lambda)] + \\
&+ [\beta_{l0}(\lambda)b_r(\lambda) - b_l(\lambda)a_r(\lambda)]\theta(\lambda) + \theta(\lambda)[a_l(\lambda)b_r(\lambda) - b_l(\lambda)a_r(\lambda)] = 0.
\end{aligned} \tag{3.103}$$

The last equation can be written as

$$\begin{aligned}
&[\beta_{l0}(\lambda) - \theta(\lambda)a_l(\lambda)][\alpha_{r0}(\lambda) - b_r(\lambda)\theta(\lambda)] \\
&\qquad = [\alpha_{l0}(\lambda) - \theta(\lambda)b_l(\lambda)][\beta_{r0}(\lambda) - a_r(\lambda)\theta(\lambda)].
\end{aligned} \tag{3.104}$$

Thus

$$W_{dl}(\lambda) = W_{dr}(\lambda). \tag{3.105}$$

**Corollary 3.1** *When studying left and right dual DPE, we can speak about one set of transfer matrices of controllers $W_d(\lambda)$, and about one system parameter $\theta(\lambda)$.*

(4) Let the transfer matrix of the controller $W_d(\lambda)$ be represented in form of MFD (3.92) or (3.101). Moreover, assume the MFD

$$\theta_l(\lambda) = \theta_r(\lambda) = \theta(\lambda) = D_l^{-1}(\lambda)M_l(\lambda) = M_r(\lambda)D_r^{-1}(\lambda). \tag{3.106}$$

Then, any matrix $W_d(\lambda)$ can be represented in the form of MFD

$$W_d(\lambda) = \alpha_l^{-1}(\lambda)\beta_l(\lambda) = \beta_r(\lambda)\alpha_r^{-1}(\lambda), \tag{3.107}$$

where the pairs $(\alpha_l(\lambda), \beta_l(\lambda))$ and $[\alpha_r(\lambda), \beta_r(\lambda)]$ are defined by

$$\begin{aligned}
\alpha_l(\lambda) &= D_l(\lambda)\alpha_{l0}(\lambda) - M_l(\lambda)b_l(\lambda), \\
\beta_l(\lambda) &= D_l(\lambda)\beta_{l0}(\lambda) - M_l(\lambda)a_l(\lambda)
\end{aligned} \tag{3.108}$$

and

$$\begin{aligned}\alpha_r(\lambda) &= \alpha_{r0}(\lambda)D_r(\lambda) - b_r(\lambda)M_r(\lambda),\\ \beta_r(\lambda) &= \beta_{r0}(\lambda)D_r(\lambda) - a_r(\lambda)M_r(\lambda).\end{aligned} \tag{3.109}$$

Therefore, relations (3.94), (3.101) are ILMFD, if and only if relation (3.106) defines an IMFD of the matrix $\theta(\lambda)$.

(5) From the above explanations, it follows that any matrix $W_d(\lambda)$ can be represented in the equivalent forms

$$\begin{aligned}W_d(\lambda) &= [\alpha_{l0}(\lambda) - \theta(\lambda)a_l(\lambda)]^{-1}[\beta_{l0}(\lambda) - \theta(\lambda)b_l(\lambda)],\\ W_d(\lambda) &= [\beta_{r0}(\lambda) - b_r(\lambda)\theta(\lambda)][\alpha_{r0}(\lambda) - a_r(\lambda)\theta(\lambda)]^{-1},\end{aligned} \tag{3.110}$$

which are called basis representations of the matrix $W_d(\lambda)$ in the dual bases $(\alpha_{l0}(\lambda), \beta_{l0}(\lambda))$ and $[\alpha_{r0}(\lambda), \beta_{r0}(\lambda)]$.

**Theorem 3.16** *For fixed bases, representations (3.110) are unique.*

*Proof* Assume that beside of the first formula in (3.110), we have

$$W_d(\lambda) = [\alpha_{l0}(\lambda) - \theta_1(\lambda)a_l(\lambda)]^{-1}[\beta_{l0}(\lambda) - \theta_1(\lambda)b_l(\lambda)]. \tag{3.111}$$

Suppose, that the matrix $\theta(\lambda)$ is given by an ILMFD of form (3.106) and, in addition, assume the LMFD

$$\theta_1(\lambda) = D_{l1}^{-1}(\lambda)M_{l1}(\lambda). \tag{3.112}$$

Then, at the same time, we obtain

$$\begin{aligned}W_d(\lambda) &= [D_l(\lambda)\alpha_{l0}(\lambda) - M_l(\lambda)b_l(\lambda)]^{-1}[D_l(\lambda)\beta_{l0}(\lambda) - M_l(\lambda)a_l(\lambda)],\\ W_d(\lambda) &= [D_{l1}(\lambda)\alpha_{l0}(\lambda) - M_{l1}(\lambda)b_l(\lambda)]^{-1}[D_{l1}(\lambda)\beta_{l0}(\lambda) - M_{l1}(\lambda)a_l(\lambda)].\end{aligned} \tag{3.113}$$

These relations define two LMFD of the matrix $W_d(\lambda)$. Moreover, by construction the first relation in (3.113) is an ILMFD. Therefore, by virtue of the results in Sect. 2.4, we obtain

$$\begin{aligned}D_{l1}(\lambda)\alpha_{l0}(\lambda) - M_{l1}(\lambda)b_l(\lambda) &= U(\lambda)[D_l(\lambda)\alpha_{l0}(\lambda) - M_l(\lambda)b_l(\lambda)],\\ D_{l1}(\lambda)\beta_{l0}(\lambda) - M_{l1}(\lambda)a_l(\lambda) &= U(\lambda)[D_l(\lambda)\beta_{l0}(\lambda) - M_l(\lambda)a_l(\lambda)],\end{aligned} \tag{3.114}$$

where $U(\lambda)$ is a non-singular $m \times m$ polynomial matrix. The last relation can be written in the form

$$\left[\, M_{l1}(\lambda) - U(\lambda)M_l(\lambda) \;\; D_{l1}(\lambda) - U(\lambda)D_l(\lambda) \,\right] Q_l(\lambda, \alpha_{l0}, \beta_{l0}) = 0_{m,n+m}, \tag{3.115}$$

where, as before

$$Q_l(\lambda, \alpha_{l0}, \beta_{l0}) = \begin{bmatrix} a_l(\lambda) & -b_l(\lambda) \\ -\beta_{l0}(\lambda) & \alpha_{l0}(\lambda) \end{bmatrix}.$$

Since the matrix $Q_l(\lambda, \alpha_{l0}, \beta_{l0})$ by construction is unimodular, from (3.115) we arrive at

$$M_{l1}(\lambda) = U(\lambda)M_l(\lambda), \quad D_{l1}(\lambda) = U(\lambda)D_l(\lambda). \tag{3.116}$$

Hence

$$\theta_1(\lambda) = D_{l1}^{-1}(\lambda)M_{l1}(\lambda) = D_l^{-1}(\lambda)M_l(\lambda) = \theta(\lambda). \tag{3.117}$$

## 3.7 Causal Modal Control of Discrete Backward Processes

(1) Discrete models for many technological processes are described by difference equations of the form [2]

$$a_0 y_k + a_1 y_{k-1} + \cdots + a_\rho y_{k-\rho} = b_0 x_k + a_1 x_{k-1} + \cdots + b_\rho x_{k-\rho}. \tag{3.118}$$

Here the quantities $y_i, x_i$ are $n \times 1, m \times 1$ vectors; $a_i, b_i$ are constant $n \times n,\ n \times m$ matrices, respectively. Introduce the shift operator for one step backward $g$, such that

$$g y_k = y_{k-1}, \quad g x_k = x_{k-1}, \tag{3.119}$$

then Eq. (3.118) can be written in operator form

$$a_l(g) y_k = b_l(g) x_k, \tag{3.120}$$

where

$$\begin{aligned} a_l(g) &= a_0 + a_1 g + \cdots + a_\rho g^\rho, \\ b_l(g) &= b_0 + b_1 g + \cdots + b_\rho g^\rho \end{aligned} \tag{3.121}$$

are polynomial matrices.

In the following investigations, the operator $g$ is identified with the complex variable $\zeta$ according to the $\zeta$ transformation. Then Eq. (3.120) can be associated with the process $(a_l(\zeta), b_l(\zeta))$, where

$$a_l(\zeta) = a_l(g)|_{g=\zeta}, \quad b_l(\zeta) = b_l(g)|_{g=\zeta}. \tag{3.122}$$

**Definition 3.7** The polynomial pair $(a_l(\zeta), b_l(\zeta))$ is called backward process, or shortly B-process.

(2) A B-process is called non-singular, if $\det a_l(\zeta) \not\equiv 0$. A non-singular process can be associated to an $n \times m$ transfer matrix

$$W(\zeta) = a_l^{-1}(\zeta) b_l(\zeta). \tag{3.123}$$

**Definition 3.8** A non-singular process $(a_l(\zeta), b_l(\zeta))$ is called causal, if its transfer matrix has the analog property. It means, that process (3.122) is causal, if matrix (3.123) does not possess poles at $\zeta = 0$. A causal process is strictly causal, when

$$W(0) = 0. \tag{3.124}$$

**Theorem 3.17** *For a process $(a_l(\zeta), b_l(\zeta))$ the following claims hold:*

*(i) If the B-process is causal and irreducible, then*

$$\det a_l(0) = \det a_0 \neq 0. \tag{3.125}$$

*(ii) If the B-process is irreducible and strictly causal, then in addition to (3.125)*

$$b_l(\zeta) = \zeta b_{l1}(\zeta), \tag{3.126}$$

*where $b_{l1}(\zeta)$ is a polynomial matrix.*

*Proof* (i) From the irreducibility of the B-process, it follows that the right side of formula (3.123) is an ILMFD. Hence, from (2.93)

$$\det a_l(\zeta) \sim \psi_W(\zeta), \tag{3.127}$$

where $\psi_W(\zeta)$ is the McMillan denominator of $W(\zeta)$. If we now assume, that the matrix $a_l(\zeta)$ would have an eigenvalue $\zeta = 0$, then this number would be a root of the polynomial $\psi_W(\zeta)$. But then from the above stated properties of the McMillan denominator, the matrix $W(\zeta)$ would possess a pole at $\zeta = 0$, and this property would contradict the supposition and the causality of the process. Thus (i) is shown.

(ii) From (3.123) and (3.125), it follows that condition (3.124) is fulfilled if $b_l(0) = 0_{nm}$, which is equivalent to relation (3.126).

(3)

**Definition 3.9** An arbitrary object $(\alpha_l(\zeta), \beta_l(\zeta))$, where $\alpha_l(\zeta), \beta_l(\zeta)$ are $m \times m$, $m \times n$ polynomial matrices, is called a left B-controller. For given B-process and B-controller, we can build the matrix

$$Q_l(\zeta, \alpha_l, \beta_l) = \begin{bmatrix} a_l(\zeta) & -b_l(\zeta) \\ -\beta_l(\zeta) & \alpha_l(\zeta) \end{bmatrix}, \tag{3.128}$$

which is called left B-characteristic matrix of the closed loop.

For a non-singular left B-controller, the transfer matrix

$$W_d(\zeta) = \alpha_l^{-1}(\zeta)\beta_l(\zeta), \tag{3.129}$$

is defined. Let the B-process $(a_l(\zeta), b_l(\zeta))$ and B-controller $(\alpha_l(\zeta), \beta_l(\zeta))$ be irreducible. Then the right sides of formulae (3.123) and (3.129) are ILMFD.

Consider the IRMFD

$$W(\zeta) = b_r(\zeta)a_r^{-1}(\zeta), \quad W_d(\zeta) = \beta_r(\zeta)\alpha_r^{-1}(\zeta). \tag{3.130}$$

Further on, without loss of generality, we assume that the controllers $(\alpha_l(\zeta), \beta_l(\zeta))$ and $[\alpha_r(\zeta), \beta_r(\zeta)]$ are dual.

The pairs $[a_r(\zeta), b_r(\zeta)]$ and $[\alpha_r(\zeta), \beta_r(\zeta)]$ correspond to the matrix

$$Q_r(\zeta, \alpha_r, \beta_r) = \begin{bmatrix} \alpha_r(\zeta) & b_r(\zeta) \\ \beta_r(\zeta) & a_r(\zeta) \end{bmatrix}, \tag{3.131}$$

which is called right B-characteristic matrix of the closed loop.

**Theorem 3.18** *Assume IMFD (3.123), (3.129), (3.130). Then*

$$\det Q_l(\zeta, \alpha_l, \beta_l) \sim \det Q_r(\zeta, \alpha_r, \beta_r). \tag{3.132}$$

*Proof* Applying Schur's formula [3], we obtain

$$\begin{aligned} \det Q_l(\zeta, \alpha_l, \beta_l) &= \det a_l(\zeta) \det \alpha_l(\zeta) \det[I_n - a_l^{-1}(\zeta)b_l(\zeta)\alpha_l^{-1}(\zeta)\beta_l(\zeta)] \\ &= \det a_l(\zeta) \det \alpha_l(\zeta) \det[I_n - W(\zeta)W_d(\zeta)]. \end{aligned} \tag{3.133}$$

An analog transform of the matrix $Q_r(\zeta, \alpha_r, \beta_r)$ yields

$$\begin{aligned} \det Q_r(\zeta, \alpha_r, \beta_r) &= \det a_r(\zeta) \det \alpha_r(\zeta) \det[I_n - b_r(\zeta)a_r^{-1}(\zeta)\beta_r(\zeta)\alpha_r^{-1}(\zeta)] \\ &= \det a_r(\zeta) \det \alpha_r(\zeta) \det[I_n - W(\zeta)W_d(\zeta)]. \end{aligned} \tag{3.134}$$

Under the taken suppositions, we find

$$\det a_l(\zeta) \sim \det a_r(\zeta), \quad \det \alpha_l(\zeta) \sim \det \alpha_r(\zeta). \tag{3.135}$$

Hence, allocating (3.133) and (3.134) leads to relation (3.132).

(4) It is well known that due to the time regime in digital control systems, discrete mathematical models of real continuous-time processes are strictly causal, and digital controllers have to be causal, when they should be realized in real time. With regard to these facts, the following causal backward modal control problem is important.

**B-modal control problem:** Let the strictly causal B-process $(a_l(\zeta), \zeta b(\zeta))$ and the monic polynomial $\Delta(\zeta)$ be given. Find the set of causal controllers $(\alpha_l(\zeta), \beta_l(\zeta))$, such that

$$\det Q_l(\zeta, \alpha_l, \beta_l) = \det \begin{bmatrix} a_l(\zeta) & -\zeta b(\zeta) \\ -\beta_l(\zeta) & \alpha_l(\zeta) \end{bmatrix} \sim \Delta(\zeta). \tag{3.136}$$

An analog problem can be formulated for right processes and right controllers. According to Theorem 3.18, left and right problems are equivalent. Therefore, in the following, we consider only left problems, and the corresponding results for right problems are provided without proofs.

Relation (3.136) is a DPE for the unknown polynomial matrices $\alpha_l(\zeta)$ and $\beta_l(\zeta)$. However, in contrast to the DPE considered in the sections above, in the actual case, the demanded solution has to fulfill the additional condition on causality. Below it will be shown that for irreducible strictly causal B-processes, this difficulty is easy to overcome.

(5)

**Definition 3.10** An arbitrary scalar polynomial $\Delta(\zeta)$ is called causal, when

$$\Delta(0) \neq 0. \tag{3.137}$$

**Theorem 3.19** *Let in DPE (3.136) the process $(a_l(\zeta), \zeta b(\zeta))$ be irreducible and strictly causal, and the polynomial $\Delta(\zeta)$ be causal. Then all solutions of DPE (3.136) are causal.*

*Proof* When (3.137) is fulfilled, then from (3.136) for $\zeta = 0$ we obtain

$$\det \begin{bmatrix} a_l(0) & 0_{nm} \\ -\beta_l(0) & \alpha_l(0) \end{bmatrix} \sim \Delta(0) \neq 0, \tag{3.138}$$

i.e.,

$$\det a_l(0) \det \alpha_l(0) \neq 0. \tag{3.139}$$

Since the process $(a_l(\zeta), \zeta b(\zeta))$ is irreducible, we know that $\det a_l(0) \neq 0$. Thus, $\det \alpha_l(0) \neq 0$, i.e., the controller $(\alpha_l(\zeta), \beta_l(\zeta))$ is causal. ∎

**Corollary 3.2** *All basic controllers for irreducible strictly causal B-processes are causal.*

*Proof* An arbitrary basis $(\alpha_{l0}(\zeta), \beta_{l0}(\zeta))$ is a solution of the DPE

$$\det Q_l(\zeta, \alpha_{l0}, \beta_{l0}) \sim 1. \tag{3.140}$$

Since the right side of this DPE is a causal polynomial, the claim of the corollary follows from Theorem 3.19. ∎

(6) Notice that the postulate on strictly causality of the B-process in Theorem 3.19 is essential. If the process is only causal, and not strictly causal, this theorem expires.

*Example 3.1* For the causal B-process $(\zeta + 5,\ \zeta^2 + 5\zeta - 1)$ Eq. (3.140) takes the form

$$\det \begin{bmatrix} \zeta + 5 & -\zeta^2 - 5\zeta + 1 \\ -\beta(\zeta) & \alpha(\zeta) \end{bmatrix} \sim 1, \tag{3.141}$$

i.e.,

$$(\zeta + 5)\alpha(\zeta) - (\zeta^2 + 5\zeta - 1)\beta(\zeta) \sim 1. \tag{3.142}$$

This equation possesses the non-causal solution $\alpha(\zeta) = \zeta$, $\beta(\zeta) = 1$.

(7) On basis of the above investigations, the general result is formulated regarding the solution of the modal causal control problem for irreducible strictly causal B-processes.

**Theorem 3.20** *Let $(a_l(\zeta), \zeta b(\zeta))$ be an irreducible strictly causal process, and $(\alpha_{l0}(\zeta), \beta_{l0}(\zeta))$ be some basic controller. Moreover, let $\Delta(\zeta)$ be a causal scalar polynomial. Then the following statements hold:*

*(i) The set of solutions of B-modal control problem (3.136) exists and is determined by the relations*

$$\begin{aligned} \alpha_l(\zeta) &= D_l(\zeta)\alpha_{l0}(\zeta) - \zeta M_l(\zeta) b(\zeta), \\ \beta_l(\zeta) &= D_l(\zeta)\beta_{l0}(\zeta) - M_l(\zeta) a_l(\zeta), \end{aligned} \tag{3.143}$$

*where $D_l(\zeta)$ and $M_l(\zeta)$ are polynomial $m \times m$ resp. $m \times n$ matrices, where the matrix $M_l(\zeta)$ is arbitrary, but the matrix $D_l(\zeta)$ satisfies*

$$\det D_l(\zeta) \sim \Delta(\zeta). \tag{3.144}$$

*All controllers (3.143) are causal. Moreover, the controller $(\alpha_l(\zeta), \beta_l(\zeta))$ is irreducible, if and only if the pair $(D_l(\zeta), M_l(\zeta))$ is irreducible.*

*(ii) Let $a_1(\zeta), \ldots, a_{n+m}(\zeta)$ be the sequence of invariant polynomials of matrix $Q_l(\zeta, \alpha_l, \beta_l)$ (3.128). Then we obtain*

$$a_1(\zeta) = a_2(\zeta) = \cdots = a_n(\zeta) = 1, \tag{3.145}$$

*and, moreover,*

$$a_{n+i}(\zeta) = d_i(\zeta), \quad (i = 1, \ldots, m), \tag{3.146}$$

*where $d_1(\zeta), \ldots, d_m(\zeta)$ are the invariant polynomials of the matrix $D_l(\zeta)$.*

*(iii) The set of transfer matrices $W_d(\zeta)$ of controllers (3.143) is defined by the relations*

$$
\begin{aligned}
W_d(\zeta) &= \alpha_l^{-1}(\zeta)\beta_l(\zeta) \\
&= [\alpha_{l0}(\zeta) - \zeta\theta(\zeta)b(\zeta)]^{-1}[\beta_{l0}(\zeta) - \theta(\zeta)a_l(\zeta)],
\end{aligned}
\tag{3.147}
$$

*where $\theta(\zeta)$ is the zero matrix or any rational matrix allowing an LMFD of the form*

$$
\theta(\zeta) = D_l^{-1}(\zeta)M_l(\zeta), \tag{3.148}
$$

*where the matrix $M_l(\zeta)$ is arbitrary, but the matrix $D_l(\zeta)$ satisfies condition (3.144). All matrices (3.147) are causal.* ■

**Corollary 3.3** *If the process $(a_l(\zeta), \zeta b(\zeta))$ is reducible, then the question about the solution of the B-modal control can be considered on basis of Theorem 3.9.*

## 3.8 Causal Stabilization of Backward Processes

(1)

**Definition 3.11** The quadratic polynomial matrix $a(\zeta)$ is called stable, if it is free of eigenvalues on the disc $|\zeta| \leq 1$. A rational matrix $W(\zeta)$ of any size is called stable, if it does not possess poles on the disc $|\zeta| \leq 1$. From the viewpoint of this definition, any polynomial matrix $a(\zeta)$ is a stable rational matrix. Moreover, scalar stable polynomials and stable rational matrices $W(\zeta)$ are causal. Polynomial and rational matrices $a(\zeta)$ and $W(\zeta)$ are called unstable, if they are not stable.

In control theory and in numerous applications, the problem of causal stabilization has fundamental importance. This problem is exemplarily formulated for B-processes as follows:

**B-stabilization problem 1:** For a given B-process $(a_l(\zeta), b_l(\zeta))$ find the set of all causal controllers $(\alpha_l(\zeta), \beta_l(\zeta))$, for which the characteristic matrix of the closed loop $Q_l(\zeta, \alpha_l, \beta_l)$ (3.128) becomes stable.

In the following, when the formulated problem is solvable, the process $(a_l(\zeta), b_l(\zeta))$ is called stabilizable, and the causal controller yielding the solution of the stabilization problem, is called stabilizing controller.

(2) The causal stabilization problem can also be formulated in terms of DPE.

**B-stabilization problem 2:** For a given B-process $(a_l(\zeta), b_l(\zeta))$ find the set of all causal solutions of polynomial Eq. (3.136) for arbitrary stable polynomials $\Delta(\zeta)$.

Since any stable polynomial is causal, the general solution of the formulated problem for strictly causal irreducible B-processes can be immediately derived from Theorem 3.20.

**Theorem 3.21** *Assume the strictly causal irreducible B-process* $(a_l(\zeta), \zeta b_l(\zeta))$. *Then, the following claims hold:*

*(i) The process* $(a_l(\zeta), \zeta b_l(\zeta))$ *is stabilizable. All stabilizing controllers* $(\alpha_l(\zeta), \beta_l(\zeta))$ *are causal. The set of all B-stabilizable controllers* $(\alpha_l(\zeta), \beta_l(\zeta))$ *is determined by the relations*

$$\begin{aligned} \alpha_l(\zeta) &= D_l(\zeta)\alpha_{l0}(\zeta) - \zeta M_l(\zeta) b_l(\zeta), \\ \beta_l(\zeta) &= D_l(\zeta)\beta_{l0}(\zeta) - M_l(\zeta) a_l(\zeta), \end{aligned} \tag{3.149}$$

*where* $D_l(\zeta)$ *and* $M_l(\zeta)$ *are polynomial matrices of size* $m \times m$ *resp.* $m \times n$, *where the matrix* $M_l(\zeta)$ *is arbitrary, but the matrix* $D_l(\zeta)$ *is stable.*

*(ii) For the sequence of invariant polynomials of the characteristic matrix* $Q_l(\zeta, \alpha_l, \beta_l)$ *relations (3.145), (3.146) are valid.*

*(iii) The set of all transfer matrices of stabilizing controllers is defined by formula (3.147), where* $\theta(\zeta)$ *is the zero matrix or any stable rational* $m \times n$ *matrix.*

(3) The solution of the stabilization problem in terms of Diophantine polynomial equations is given next.

**Theorem 3.22** *Let* $(a_l(\zeta), \zeta b_l(\zeta))$ *and* $[a_r(\zeta), \zeta b_r(\zeta)]$ *be strictly causal dual processes. Then the following claims hold:*

*(i) For the controller* $(\alpha_l(\zeta), \beta_l(\zeta))$ *to be stabilizing for the irreducible left process* $(a_l(\zeta), \zeta b_l(\zeta))$, *it is necessary and sufficient that the pair* $(\alpha_l(\zeta), \beta_l(\zeta))$ *is a solution of the Diophantine polynomial equation*

$$\alpha_l(\zeta) a_r(\zeta) - \zeta \beta_l(\zeta) b_r(\zeta) = D_l(\zeta), \tag{3.150}$$

*where* $D_l(\zeta)$ *is any stable* $m \times m$ *matrix.*

*(ii) For the controller* $[\alpha_r(\zeta), \beta_r(\zeta)]$ *to be stabilizing for the irreducible right process* $[a_r(\zeta), \zeta b_r(\zeta)]$, *it is necessary and sufficient that the pair* $[\alpha_r(\zeta), \beta_r(\zeta)]$ *is a solution of the Diophantine polynomial equation*

$$a_l(\zeta)\alpha_r(\zeta) - \zeta b_l(\zeta)\beta_r(\zeta) = D_r(\zeta), \tag{3.151}$$

*where* $D_r(\zeta)$ *is any stable* $n \times n$ *matrix.*

The proof is a direct consequence of Theorem 3.8.

(4) The next two theorems provide subsidiary criteria concerning the definition of the set of transfer matrices of causal stabilizing controllers.

**Theorem 3.23** *For the rational* $m \times n$ *matrix* $W_d(\zeta)$ *to be the transfer matrix of a causal stabilizing controller for the strictly causal irreducible left process* $(a_l(\zeta), \zeta b_l(\zeta))$, *it is necessary and sufficient that there exists a representation*

$$W_d(\zeta) = F_1^{-1}(\zeta)F_2(\zeta), \tag{3.152}$$

*where $F_1(\zeta)$ and $F_2(\zeta)$ are stable rational matrices of size $m \times m$ resp. $m \times n$, and the right dual process $[a_r(\zeta), \zeta b_r(\zeta)]$ fulfills the condition*

$$F_1(\zeta)a_r(\zeta) - \zeta F_2(\zeta)b_r(\zeta) = I_m. \tag{3.153}$$

*Proof* Necessity: Let $(\alpha_l(\zeta), \beta_l(\zeta))$ be a causal stabilizing controller. Then the matrices $\alpha_l(\zeta), \beta_l(\zeta)$ build a solution of Eq. (3.150) with stable matrix $D_l(\zeta)$. Hereby, the rational matrices

$$F_1(\zeta) = D_l^{-1}(\zeta)\alpha_l(\zeta), \quad F_2(\zeta) = D_l^{-1}(\zeta)\beta_l(\zeta), \tag{3.154}$$

are stable, fulfill Eq. (3.153), and the matrix $F_1(\zeta)$ is invertible. Hence

$$F_1^{-1}(\zeta)F_2(\zeta) = \alpha_l^{-1}(\zeta)\beta_l(\zeta) = W_d(\zeta). \tag{3.155}$$

Sufficiency: Let the rational $m \times m$ resp. $m \times n$ matrices $F_1(\zeta)$ and $F_2(\zeta)$ be stable and satisfy relation (3.153). Then the rational matrix

$$F(\zeta) = \begin{bmatrix} F_1(\zeta) & F_2(\zeta) \end{bmatrix}, \tag{3.156}$$

is stable. Consider the ILMFD

$$F(\zeta) = a_F^{-1}(\zeta)b_F(\zeta) = a_F^{-1}(\zeta)\begin{bmatrix} d_1(\zeta) & d_2(\zeta) \end{bmatrix}, \tag{3.157}$$

where by construction the polynomial matrix $a_F(\zeta)$ is stable. From (3.157), we find that the matrices

$$d_1(\zeta) = a_F(\zeta)F_1(\zeta), \quad d_2(\zeta) = a_F(\zeta)F_2(\zeta), \tag{3.158}$$

are polynomial matrices. Since

$$d_1(\zeta)a_r(\zeta) - \zeta d_2(\zeta)b_r(\zeta) = a_F(\zeta)[F_1(\zeta)a_r(\zeta) - \zeta F_2(\zeta)b_r(\zeta)] = a_F(\zeta), \tag{3.159}$$

by virtue of Theorem 3.22, the pair $(d_1(\zeta), d_2(\zeta))$ is a stabilizing controller with the transfer function

$$W_d(\zeta) = d_1(\zeta)^{-1}d_2(\zeta) = F_1(\zeta)^{-1}F_2(\zeta). \tag{3.160}$$

*Remark 3.3* In analogy to Theorem 3.23, it can be shown that the matrix $W_d(\zeta)$ is stabilizing for the irreducible dual right process $[a_r(\zeta), \zeta b_r(\zeta)]$ if and only if it allows a presentation

$$W_d(\zeta) = G_2(\zeta)G_1(\zeta)^{-1}, \tag{3.161}$$

where the rational matrices $G_1(\zeta)$ and $G_2(\zeta)$ are stable and fulfill the condition

$$a_l(\zeta)G_1(\zeta) - \zeta b_l(\zeta)G_2(\zeta) = I_n. \tag{3.162}$$

**Theorem 3.24** *Assume the irreducible process $(a_l(\zeta), \zeta b_l(\zeta))$ and the irreducible controller $(\alpha_l(\zeta), \beta_l(\zeta))$, which define the characteristic matrix of the closed loop*

$$Q_l(\zeta, \alpha_l, \beta_l) = \begin{matrix} & n & m \\ n & a_l(\zeta) & -\zeta b_l(\zeta) \\ m & -\beta_l(\zeta) & \alpha_l(\zeta) \end{matrix} \quad \left[\begin{matrix} a_l(\zeta) & -\zeta b_l(\zeta) \\ -\beta_l(\zeta) & \alpha_l(\zeta) \end{matrix}\right]. \tag{3.163}$$

*Then the following statements are true:*

*(i) For the controller $(\alpha_l(\zeta), \beta_l(\zeta))$ to be stabilizing, it is necessary that matrix (3.163) is invertible.*

*(ii) Let (i) be fulfilled, and*

$$Q_l^{-1}(\zeta, \alpha_l, \beta_l) = \begin{matrix} & n & m \\ n & & \\ m & & \end{matrix} \left[\begin{matrix} v_1(\zeta) & q_1(\zeta) \\ v_2(\zeta) & q_2(\zeta) \end{matrix}\right]. \tag{3.164}$$

*Then for the controller $(\alpha_l(\zeta), \beta_l(\zeta))$ to be stabilizing, it is necessary and sufficient that the rational matrices $v_1(\zeta)$ and $v_2(\zeta)$ are stable.*

*Proof* (i) Let the controller $(\alpha_l(\zeta), \beta_l(\zeta))$ be stabilizing. Then the polynomial matrix $Q_l(\zeta, \alpha_l, \beta_l)$ is stable, i.e., it does not possess eigenvalues on the disc $|\zeta| \leq 1$. Moreover, for $\zeta = 0$ we have $\det \alpha_l(0) \neq 0$ and the matrix $Q_l(\zeta, \alpha_l, \beta_l)|_{\zeta=0}$ is non-singular. Hence, the matrix $Q_l(\zeta, \alpha_l, \beta_l)$ is invertible.

(ii) Necessity: Let condition (i) be fulfilled. Then, since the controller $(\alpha_l(\zeta), \beta_l(\zeta))$ is stabilizing, the matrix $Q_l^{-1}(\zeta, \alpha_l, \beta_l)$ does not possess poles on the disc $|\zeta| \leq 1$, i.e., it is stable. The stability of the matrix $Q_l^{-1}(\zeta, \alpha_l, \beta_l)$ implies the stability of its block entries. Therefore, the matrices $v_1(\zeta)$ and $v_2(\zeta)$ are stable. So the necessity of (ii) is proven.

Sufficiency: Since by construction

$$Q_l(\zeta, \alpha_l, \beta_l)Q_l^{-1}(\zeta, \alpha_l, \beta_l) = \left[\begin{matrix} I_n & 0_{nm} \\ 0_{mn} & I_m \end{matrix}\right], \tag{3.165}$$

so using (3.163) and (3.164), we obtain

$$a_l(\zeta)v_1(\zeta) - \zeta b_l(\zeta)v_2(\zeta) = I_n. \tag{3.166}$$

From Remark 3.3, we find that the matrix

$$W_d(\zeta) = v_2(\zeta)v_1^{-1}(\zeta), \tag{3.167}$$

is the transfer matrix of a stabilizing controller for the irreducible dual right process. Since the sets of transfer matrices of stabilizing controllers for irreducible dual processes coincide, matrix (3.167) is also a stabilizing for the process $(a_l(\zeta), \zeta b_l(\zeta))$. Hereby, it follows from (3.165)

$$-\beta_l(\zeta)v_1(\zeta) + \alpha_l(\zeta)v_2(\zeta) = 0_{mn}. \tag{3.168}$$

Herein, matrix $\alpha_l(\zeta)$ is invertible by construction. Moreover, the matrix $v_1(\zeta)$ is also invertible, which follows from (3.166) for $\zeta = 0$. Therefore, from (3.168) we find

$$W_d(\zeta) = v_2(\zeta)v_1^{-1}(\zeta) = \alpha_l^{-1}(\zeta)\beta_l(\zeta). \tag{3.169}$$

Hence, the controller $(\alpha_l(\zeta), \beta_l(\zeta))$ is stabilizing.

(5) Considering relations (3.166) and (3.168) as a system of equations depending on the matrices $v_1(\zeta)$ and $v_2(\zeta)$, obvious expressions for them can be derived. From (3.169), we obtain

$$v_2(\zeta) = W_d(\zeta)v_1(\zeta). \tag{3.170}$$

Substituting this expression in (3.166), with the help of (3.170), we find

$$\begin{aligned} v_1(\zeta) &= [a_l(\zeta) - \zeta b_l(\zeta)W_d(\zeta)]^{-1}, \\ v_2(\zeta) &= W_d(\zeta)[a_l(\zeta) - \zeta b_l(\zeta)W_d(\zeta)]^{-1}. \end{aligned} \tag{3.171}$$

Moreover, under the conditions of Theorem 3.24, the matrix $W_d(\zeta)$ is stabilizing, if and only if matrices (3.171) are stable.

(6) On basis of Theorem 3.8, solutions for the stabilization problem can be obtained for a process $(a_l(\zeta), \zeta b_l(\zeta))$, which is not irreducible. In that case, we build

$$a_l(\zeta) = l(\zeta)a_{l1}(\zeta), \quad \zeta b_l(\zeta) = l(\zeta)b_{l1}(\zeta), \tag{3.172}$$

where $l(\zeta)$ is the GCLD of the matrices $a_l(\zeta)$ and $\zeta b_l(\zeta)$, and the pair $(a_{l1}(\zeta), b_{l1}(\zeta))$ are irreducible.

**Theorem 3.25** *The following statements are true:*

*(i) For the stabilizability of the process $(a_l(\zeta), b_l(\zeta))$, it is necessary and sufficient that the polynomial matrix $l(\zeta)$ is stable.*
*(ii) When (3.172) is fulfilled, we obtain*

$$b_{l1}(\zeta) = \zeta b_{l2}(\zeta), \tag{3.173}$$

*where $b_{l2}(\zeta)$ is a polynomial matrix. Moreover, the set of stabilizing controllers coincides with the set of stabilizing controllers for the strictly causal irreducible process $(a_{l1}(\zeta), \zeta b_{l2}(\zeta))$.*

*Proof* (i) Necessity: When (3.172) is fulfilled, we obtain

$$Q_l(\zeta, \alpha_l, \beta_l) = \begin{bmatrix} l(\zeta) & 0_{nm} \\ 0_{mn} & I_m \end{bmatrix} \begin{bmatrix} a_{l1}(\zeta) & -b_{l1}(\zeta) \\ -\beta_l(\zeta) & \alpha_l(\zeta) \end{bmatrix}. \tag{3.174}$$

Thus

$$\det Q_l(\zeta, \alpha_l, \beta_l) = \det l(\zeta) \det \begin{bmatrix} a_{l1}(\zeta) & -b_{l1}(\zeta) \\ -\beta_l(\zeta) & \alpha_l(\zeta) \end{bmatrix}, \tag{3.175}$$

i.e., the characteristic polynomial of the closed loop, independently on the selection of the controller, contains the factor $\det l(\zeta)$. If the polynomial $l(\zeta)$ is unstable, then the characteristic polynomial of the closed loop cannot become stable for any controller, and consequently the process cannot be stabilized.

Sufficiency: When (3.175) is fulfilled, the DPE for the determination of the set of stabilizing controllers takes the form

$$\det l(\zeta) \det \begin{bmatrix} a_{l1}(\zeta) & -b_{l1}(\zeta) \\ -\beta_l(\zeta) & \alpha_l(\zeta) \end{bmatrix} \sim \Delta_+(\zeta), \tag{3.176}$$

where $\Delta_+(\zeta)$ is a stable polynomial. Assume

$$\Delta_+(\zeta) = \det l(\zeta) \Delta_{+1}(\zeta), \tag{3.177}$$

where $\Delta_{+1}(\zeta)$ is a stable polynomial. Substituting (3.177) in (3.176), after reduction, we find the DPE

$$\det \begin{bmatrix} a_{l1}(\zeta) & -b_{l1}(\zeta) \\ -\beta_l(\zeta) & \alpha_l(\zeta) \end{bmatrix} \sim \Delta_{+1}(\zeta). \tag{3.178}$$

Since the pair $(a_{l1}(\zeta), b_{l1}(\zeta))$ is irreducible, DPE (3.178) is solvable. If $(\alpha_l(\zeta), \beta_l(\zeta))$ is one of the solutions of this equation, then from (3.176), (3.177) follows that

$$\det \begin{bmatrix} a_l(\zeta) & -b_l(\zeta) \\ -\beta_l(\zeta) & \alpha_l(\zeta) \end{bmatrix} \sim \Delta_+(\zeta), \tag{3.179}$$

and the right side of this relation is a stable polynomial. Therefore, the process $(a_l(\zeta), b_l(\zeta))$ is stabilizable.

(ii) Since $\det l(0) \neq 0$, so the second Eq. (3.172) implies (3.173), and DPE (3.178) can be written in the form

$$\det \begin{bmatrix} a_{l1}(\zeta) & -\zeta b_{l2}(\zeta) \\ -\beta_l(\zeta) & \alpha_l(\zeta) \end{bmatrix} \sim \Delta_{+1}(\zeta). \tag{3.180}$$

From this, the claim arises immediately. ∎

## 3.9 Causal Modal Control of Discrete Forward Processes

(1) In many cases the control of discrete processes is investigated, which are given different from (3.118), but in the form

$$a_0 y_{k+\rho} + a_1 y_{k+\rho-1} + \ldots + a_\rho y_k = b_0 x_{k+\rho} + \cdots + b_\rho x_k. \tag{3.181}$$

Here $a_i$, $b_i$ are constant $n \times n$ resp. $n \times m$ matrices configured in (3.118). Moreover, $y_k$, $x_k$ are $n \times 1$ resp. $m \times 1$ vectors. Introduce the forward shift operator $q$ for one step ahead by the condition

$$q y_k = y_{k+1}, \quad q x_k = x_{k+1}, \tag{3.182}$$

then Eq. (3.181) can be written in operator form as

$$\tilde{a}_l(q) y_k = \tilde{b}_l(q) x_k, \tag{3.183}$$

where

$$\begin{aligned} \tilde{a}_l(q) &= a_0 q^\rho + a_1 q^{\rho-1} + \cdots + a_\rho, \\ \tilde{b}_l(q) &= b_0 q^\rho + b_1 q^{\rho-1} + \cdots + b_\rho \end{aligned} \tag{3.184}$$

are polynomial matrices.

In the following investigations the operator $q$ will be identified with the complex variable $z$ according to the $z$ transformation [2]. Then Eq. (3.183) corresponds to the process $(\tilde{a}_l(z), \tilde{b}_l(z))$, where

$$\tilde{a}_l(z) = \tilde{a}_l(q)|_{q=z}, \quad \tilde{b}_l(z) = \tilde{b}_l(q)|_{q=z}. \tag{3.185}$$

**Definition 3.12** The pair $(\tilde{a}(z), \tilde{b}(z))$ is called forward process, or shortly, F-process.

(2) Starting from the general definitions, the above introduced F-process is called non-singular, if $\det \tilde{a}(z) \neq 0$, and irreducible, if the pair $(\tilde{a}(z), \tilde{b}(z))$ is irreducible. For a non-singular process, the rational matrix

$$\tilde{W}(z) = \tilde{a}_l^{-1}(z) \tilde{b}_l(z), \tag{3.186}$$

is defined, which is called its transfer matrix. A non-singular F-process is called causal, if its transfer matrix $\tilde{W}(z)$ disposes of the analog property.

In the following investigations, when not explicitly explained different, we always suppose that in (3.184)

$$\det a_0 \neq 0. \tag{3.187}$$

**Theorem 3.26** *Under condition (3.187) the following claims hold:*

*(i) F-process* $(\tilde{a}(z), \tilde{b}(z))$ *(3.181) is causal.*
*(ii) For strict causality of the F-process* $(\tilde{a}(z), \tilde{b}(z))$ *it is necessary and sufficient that*

$$\deg \tilde{a}_l(z) > \deg \tilde{b}_l(z). \tag{3.188}$$

*Hereby, inequality (3.188) is equivalent to*

$$b_0 = 0_{nm}. \tag{3.189}$$

*Proof* When (3.187) is fulfilled, then the matrix $\tilde{a}(z)$ is regular and row reduced. Therefore, from (3.181) and Corollary 2.2 it follows, that the matrix $\tilde{W}(z)$ is causal, when $\deg \tilde{a}(z) = \deg \tilde{b}(z)$ and strictly causal, when $\deg \tilde{a}(z) > \deg \tilde{b}(z)$, which is equivalent to (3.188), (3.189). ■

(3) The B-process $(a_l(\zeta), b_l(\zeta))$, defined by (3.118), and F-object $(\tilde{a}_l(z), \tilde{b}_l(z))$ (3.183) are called associated.

**Theorem 3.27** *The transfer matrices* $W(\zeta)$ *and* $\tilde{W}(z)$ *of associated processes (3.123) and (3.186) are associated rational matrices, i.e., the relations*

$$W(\zeta) = \tilde{W}(z)|_{z=\zeta^{-1}}, \quad \tilde{W}(z) = W(\zeta)|_{\zeta=z^{-1}} \tag{3.190}$$

*are true.*

*Proof* By definition, we obtain

$$\tilde{W}(z) = \tilde{a}_l^{-1}(z)\tilde{b}_l(z) = (a_0 z^\rho + \cdots + a_\rho)^{-1}(b_0 z^\rho + \cdots + b_\rho). \tag{3.191}$$

Substituting $z$ by $\zeta^{-1}$, first equation in (3.190) is achieved. The second equation is proven analogously.

**Corollary 3.4** *Associated B- and F-processes are at the same time causal or strictly causal.*

(4) An arbitrary pair $(\tilde{\alpha}_l(z), \tilde{\beta}_l(z))$, where $\tilde{\alpha}_l(z), \tilde{\beta}_l(z)$ are polynomial $m \times m$ resp. $m \times n$ matrices, is called a left forward (F) controller. Moreover, the matrix

$$\tilde{Q}_l(z, \tilde{\alpha}_l, \tilde{\beta}_l) = \begin{bmatrix} \tilde{a}_l(z) & -\tilde{b}_l(z) \\ -\tilde{\beta}_l(z) & \tilde{\alpha}_l(z) \end{bmatrix} \tag{3.192}$$

is called left forward (F) characteristic matrix of the closed loop. For F-processes, in analogy to above, a causal modal control problem can be formulated.

**F-modal control problem:** Let be given the strictly causal $(\tilde{a}(z), \tilde{b}(z))$. Find the set of causal F-controllers $(\tilde{\alpha}_l(z), \tilde{\beta}_l(z))$, such that

$$\det \tilde{Q}_l(z, \tilde{\alpha}_l, \tilde{\beta}_l) = \det \begin{bmatrix} \tilde{a}_l(z) & -\tilde{b}_l(z) \\ -\tilde{\beta}_l(z) & \tilde{\alpha}_l(z) \end{bmatrix} \sim \tilde{\Delta}(z), \tag{3.193}$$

where $\tilde{\Delta}(z)$ is a given monic polynomial.

A formal solution of DPE (3.193) can be obtained on basis of the above achieved insight. In particular, let the process $(\tilde{a}_l(z), \tilde{b}_l(z))$ be irreducible. Then there exists the set of basic controllers $(\tilde{\alpha}_{l0}(z), \tilde{\beta}_{l0}(z))$, as the solution of the DPE

$$\det \tilde{Q}_l(z, \tilde{\alpha}_{l0}, \tilde{\beta}_{l0}) = \det \begin{bmatrix} \tilde{a}_l(z) & -\tilde{b}_l(z) \\ -\tilde{\beta}_{l0}(z) & \tilde{\alpha}_{l0}(z) \end{bmatrix} \sim 1. \tag{3.194}$$

Moreover, the set of solutions of DPE (3.194) is determined by the relations

$$\begin{aligned} \tilde{\alpha}_l(z) &= \tilde{D}_l(z)\tilde{\alpha}_{l0}(z) - \tilde{M}_l(z)\tilde{b}_l(z), \\ \tilde{\beta}_l(z) &= \tilde{D}_l(z)\tilde{\beta}_{l0}(z) - \tilde{M}_l(z)\tilde{a}_l(z), \end{aligned} \tag{3.195}$$

where the matrix $\tilde{M}_l(z)$ is arbitrary, but the matrix $\tilde{D}_l(z)$ satisfies $\det \tilde{D}_l(z) \sim \tilde{\Delta}(z)$.

(5) Considered outwardly, solution (3.195) does not differ from the solution of the B-modal control problem for a strictly causal irreducible process

$$Q_l(\zeta, \alpha_l, \beta_l) = \det \begin{bmatrix} a_l(\zeta) & -\zeta b_l(\zeta) \\ -\beta_l(\zeta) & \alpha_l(\zeta) \end{bmatrix} \sim \Delta(\zeta), \tag{3.196}$$

possessing the form

$$\begin{aligned} \alpha(\zeta) &= D_l(\zeta)\alpha_{l0}(\zeta) - \zeta M_l(\zeta) b_l(\zeta), \\ \beta(\zeta) &= D_l(\zeta)\beta_{l0}(\zeta) - M_l(\zeta) a_l(\zeta), \end{aligned} \tag{3.197}$$

where $(\alpha_{l0}(\zeta), \beta_{l0}(\zeta))$ is a basic controller satisfying the DPE

$$\det \begin{bmatrix} a_l(\zeta) & -\zeta b_l(\zeta) \\ -\beta_{l0}(\zeta) & \alpha_{l0}(\zeta) \end{bmatrix} \sim 1. \tag{3.198}$$

Nevertheless, between B-solution (3.197) and F-solution (3.195) there exists a principal difference. As was shown above, for a strictly causal irreducible B-process, all basic controllers are causal. For a strictly causal irreducible F-process, the situation turns out to be the contrary.

**Theorem 3.28** *Let the F-process $(\tilde{a}_l(z), \tilde{b}_l(z))$ be strictly causal and irreducible. Then all basic controllers $(\tilde{\alpha}_{l0}(z), \tilde{\beta}_{l0}(z))$ are non-causal.*

*Proof* Assume controller $(\tilde{\alpha}_l(z), \tilde{\beta}_l(z))$ in (3.192) is causal. Then

$$\deg\det \tilde{Q}_l(z, \tilde{\alpha}_l, \tilde{\beta}_l) = \deg\det \tilde{a}_l(z) + \deg\det \tilde{\alpha}_l(z). \tag{3.199}$$

Indeed, using Schur's formula [3], we obtain

$$\begin{aligned}\det\begin{bmatrix} \tilde{a}_l(z) & -\tilde{b}_l(z) \\ -\tilde{\beta}_l(z) & \tilde{\alpha}_l(z) \end{bmatrix} &= \det\tilde{a}_l(z)\det\tilde{\alpha}_l(z)\det[I_n - \tilde{a}_l^{-1}(z)\tilde{b}_l(z)\tilde{\alpha}_l^{-1}(z)\tilde{\beta}_l(z)] \\ &= \det\tilde{a}_l(z)\det\tilde{\alpha}_l(z)\det[I_n - \tilde{W}(z)\tilde{W}_d(z)],\end{aligned} \tag{3.200}$$

where

$$\tilde{W}_d(z) = \tilde{\alpha}_l^{-1}(z)\tilde{\beta}_l(z) \tag{3.201}$$

is the transfer matrix of the controller. Under the taken propositions, the matrix $\tilde{W}(z)$ is strictly proper and the matrix $\tilde{W}_d(z)$ at least proper. Hence,

$$\lim_{z\to\infty} \det[I_n - \tilde{W}(z)\tilde{W}_d(z)] = 1. \tag{3.202}$$

Thus, there exists a representation of the form

$$\det[I_n - \tilde{W}(z)\tilde{W}_d(z)] = \frac{z^\mu + g_1 z^{\mu-1} + \cdots + g_\mu}{z^\mu + f_1 z^{\mu-1} + \cdots + f_\mu}, \tag{3.203}$$

where $\mu$ is a nonnegative integer. From (3.200) and (3.203), we find

$$\det\tilde{Q}_l(z, \tilde{\alpha}_l, \tilde{\beta}_l) = \det\tilde{a}_l(z)\det\tilde{\alpha}_l(z)\frac{z^\mu + g_1 z^{\mu-1} + \cdots + g_\mu}{z^\mu + f_1 z^{\mu-1} + \cdots + f_\mu}. \tag{3.204}$$

Since both sides of this equation are polynomials, so from (3.204) formula (3.199) arises. Let $(\tilde{\alpha}_{l0}(z), \tilde{\beta}_{l0}(z))$ be any basic F-controller as solution of DPE (3.194). From (3.187) it follows that $\deg\tilde{a}(z) > 0$ and, due to (3.199), $\deg\det\tilde{Q}_l(z, \tilde{\alpha}_{l0}, \tilde{\beta}_{l0}) > 0$. Therefore, for causal basic controllers $(\tilde{\alpha}_{l0}(z), \tilde{\beta}_{l0}(z))$ condition (3.194) cannot be fulfilled. ∎

**Corollary 3.5** *Under the condition of Theorem 3.28, for any polynomial $\tilde{\Delta}(z)$, the set of solutions of DPE (3.195) contains a nonempty subset of non-causal solutions.*

*Proof* Let $(\tilde{\alpha}_{l0}(z), \tilde{\beta}_{l0}(z))$ be a non-singular basic F-controller. Because this controller is non-causal, the matrix

$$\tilde{W}_{d0}(z) = \tilde{\alpha}_{l0}^{-1}(z)\tilde{\beta}_{l0}(z) \tag{3.205}$$

is improper. For $M_l(z) = 0_{nm}$ from (3.195), we find the family of solutions

$$\tilde{\alpha}(z) = \tilde{D}_l(z)\tilde{\alpha}_{l0}(z), \quad \tilde{\beta}(z) = \tilde{D}_l(z)\tilde{\beta}_{l0}(z). \tag{3.206}$$

The transfer matrix for any controller of family (3.206) is equal to

$$W_d(z) = \alpha^{-1}(z)\beta(z) = \tilde{\alpha}_{l0}^{-1}(z)\tilde{\beta}_{l0}(z). \tag{3.207}$$

Hence, all solutions (3.206) are non-causal. ∎

(6) As follows from Corollary 3.5, for practical applications the general solution of (3.195) is unsuitable, because not all controllers (3.195) are causal. The approach described below will provide the general solution of the causal F-modal control problem. It is based on the properties of associated rational matrices.

For a given strictly causal irreducible F-process $(\tilde{a}_l(z), \tilde{b}_l(z))$ consider the family of DPE depending on the parameter $\lambda$

$$\det\begin{bmatrix} \tilde{a}_l(z) & -\tilde{b}_l(z)) \\ -\tilde{\beta}_l(z) & \tilde{\alpha}_l(z) \end{bmatrix} \sim z^{\lambda}\tilde{\Delta}_1(z) \triangleq \tilde{\Delta}_{\lambda}(z), \tag{3.208}$$

where $\lambda$ is a nonnegative integer, and $\tilde{\Delta}_1(z)$ is a polynomial such that $\tilde{\Delta}_1(0) \neq 0$. Let the matrix

$$W(\zeta) = [\tilde{a}_l^{-1}(z)\tilde{b}_l(z)]|_{z=\zeta^{-1}}, \tag{3.209}$$

be associated with the transfer matrix of the F-process $\tilde{W}(z)$. Since by proposition the matrix $\tilde{W}(z)$ is strictly proper, due to Lemma 2.22, the matrix $W(\zeta)$ is strictly causal and has an ILMFD of the form

$$W(\zeta) = \zeta a_l^{-1}(\zeta)b_l(\zeta). \tag{3.210}$$

Besides of DPE (3.208) consider the DPE

$$\det\begin{bmatrix} a_l(\zeta) & -\zeta b_l(\zeta) \\ -\beta_l(\zeta) & \alpha_l(\zeta) \end{bmatrix} \sim \Delta_1(\zeta), \tag{3.211}$$

where

$$\Delta_1(\zeta) \triangleq \operatorname{Rec} \Delta_1(z) \sim \zeta^{\deg \tilde{\Delta}_1(z)}\tilde{\Delta}_1(\zeta^{-1}) \tag{3.212}$$

is the monic reciprocal polynomial of the polynomial $\tilde{\Delta}_1(z)$.

**Definition 3.13** The above explained processes $(\tilde{a}_l(z), \tilde{b}_l(z))$ and $(a_l(\zeta), \zeta b_l(\zeta))$, and also DPE (3.208) and (3.211) are called associated.

Under the taken propositions, DPE (3.211) is solvable for any polynomial $\Delta_1(\zeta)$. Moreover, since by construction $\Delta_1(0) \neq 0$, i.e., the polynomial $\Delta_1(\zeta)$ is causal, so all solutions of DPE (3.211) are causal. As was shown earlier, the set of all transfer matrices $W_d(\zeta)$ as solutions of DPE (3.211) are determined by relation (3.147)

$$W_d(\zeta) = [\alpha_{l0}(\zeta) - \zeta\theta(\zeta)b_l(\zeta)]^{-1}[\beta_{l0}(\zeta) - \theta(\zeta)a_l(\zeta)], \tag{3.213}$$

where $\theta(\zeta)$ is a rational matrix satisfying condition (3.144), (3.148), where $\Delta(\zeta) = \Delta_1(\zeta)$ and $(\alpha_{l0}(\zeta), \beta_{l0}(\zeta))$ is any basic B-controller.

**Theorem 3.29** *The set of all causal irreducible solutions $(\tilde{\alpha}_l(z), \tilde{\beta}_l(z))$ of the family of DPE (3.208) for the variety of $\lambda \geq 0$ is determined by the ILMFD*

$$\tilde{W}_d(z) = \tilde{\alpha}_l^{-1}(z)\tilde{\beta}_l(z), \tag{3.214}$$

*where the matrix*

$$\tilde{W}_d(z) = W_d(z^{-1}) = [\alpha_{l0}(z^{-1}) - z^{-1}\theta(z^{-1})b_l(z^{-1})]^{-1}[\beta_{l0}(z^{-1}) - \theta(z^{-1})a_l(z^{-1})] \tag{3.215}$$

*is associated with matrix $W_d(\zeta)$ (3.213).*

*Proof* Assume (3.214) and (3.215). From Theorem 2.16 (formula (2.366)–(2.368)), under the taken propositions, we obtain

$$\begin{bmatrix} a_l(\zeta) & -\zeta b_l(\zeta) \\ -\beta_l(\zeta) & \alpha_l(\zeta) \end{bmatrix} = \operatorname{rec} \begin{bmatrix} \tilde{a}_l(z) & -\tilde{b}_l(z) \\ -\tilde{\beta}_l(z) & \tilde{\alpha}_l(z) \end{bmatrix}. \tag{3.216}$$

Hence,

$$\det \begin{bmatrix} a_l(\zeta) & -\zeta b_l(\zeta) \\ -\beta_l(\zeta) & \alpha_l(\zeta) \end{bmatrix} \sim \operatorname{rec} \det \begin{bmatrix} \tilde{a}_l(z) & -\tilde{b}_l(z) \\ -\tilde{\beta}_l(z) & \tilde{\alpha}_l(z) \end{bmatrix} \sim \operatorname{rec} \left[ z^{\lambda} \tilde{\Delta}_1(z) \right], \tag{3.217}$$

for some $\lambda \geq 0$. Due to $\tilde{\Delta}_1(0) \neq 0$, we find

$$\operatorname{rec}[z^{\lambda}\tilde{\Delta}_1(z)] = \operatorname{rec}[\tilde{\Delta}_{\lambda}(z)] = \zeta^{\deg \tilde{\Delta}_1(z)} \tilde{\Delta}_1(\zeta^{-1}) \sim \Delta_1(\zeta). \tag{3.218}$$

From (3.212) it follows that $\Delta_1(0) \neq 0$, i.e., the polynomial $\Delta_1(\zeta)$ is causal. Therefore, all solutions of DPE (3.211) are causal, and their transfer matrices allow representation (3.213). Thus, relations (3.214), (3.215) are true. ∎

**Corollary 3.6** *Let in the conditions of Theorem 3.29 $(\alpha_l(\zeta), \beta_l(\zeta))$ be an irreducible causal B-controller, and $W_d(\zeta)$ its transfer matrix. Assume the ILMFD*

$$W_d(z^{-1}) = \tilde{\alpha}_l^{-1}(z)\tilde{\beta}_l(z). \tag{3.219}$$

*Then, if*

$$\det \begin{bmatrix} a_l(\zeta) & -\zeta b_l(\zeta) \\ -\beta_l(\zeta) & \alpha_l(\zeta) \end{bmatrix} \sim \Delta(\zeta), \quad \Delta(0) \neq 0, \tag{3.220}$$

*then*

$$\det \begin{bmatrix} \tilde{a}_l(z) & -\tilde{b}_l(z) \\ -\tilde{\beta}_l(z) & \tilde{\alpha}_l(z) \end{bmatrix} \sim z^{\lambda} z^{\deg \Delta(\zeta)} \Delta(z^{-1}), \tag{3.221}$$

*where $\lambda$ is a nonnegative integer.*

(7) At first glance, the above considered solution of the F-modal control problem has an indeterminacy, because the undetermined integer parameter $\lambda$ is configured therein. We will explain that primarily the introduction of this parameter allows a complete understanding of the real reasons. Let $(\tilde{\alpha}_l(z), \tilde{\beta}_l(z))$ be an irreducible causal solution of Eq. (3.208), then the closed loop corresponds to the system of difference equations

$$\begin{aligned} \tilde{a}_l(q)y_k - \tilde{b}_l(q)x_k &= 0_{n1}, \\ -\tilde{\beta}_l(q)y_k + \tilde{\alpha}_l(q)x_k &= 0_{m1}, \end{aligned} \tag{3.222}$$

where $q$ is F-operator of shift (3.182). Under the taken propositions, for the solution of Eq. (3.222) the methods of $z$ transformation can be applied. Hereby, the corresponding $z$ transforms $y^*(z), x^*(z)$ for arbitrary initial conditions can be represented as expressions of the form

$$\begin{aligned} y^*(z) &= \frac{k_1(z)}{z^\lambda} + \frac{k_2(z)}{\tilde{\Delta}_1(z)}, \\ x^*(z) &= \frac{l_1(z)}{z^\lambda} + \frac{l_2(z)}{\tilde{\Delta}_1(z)}, \end{aligned} \tag{3.223}$$

where $k_i(z),\ l_i(z),\ (i = 1, 2)$ are polynomial vectors depending on the initial conditions. Applying the inverse $z$ transformation on $z$ transforms (3.223), we obtain [2], that the terms with zero poles correspond to deadbeat processes of finite length $\{y_k\}, \{x_k\}$, vanishing for $k > \lambda$. Hence, for $k > \lambda$ the character of the motions in the closed loop is completely determined by the roots of the polynomial $\tilde{\Delta}_1(z)$, i.e., the nonzero modes of the closed loop. Below, those modes are called fundamental. Thus, closed systems constructed with the help of the solution of DPE (3.208) for different $\lambda$ differ only in the character and length of their deadbeat processes.

(8) Consider the question about the structure of the characteristic matrix of the closed loop $\tilde{Q}_l(z, \tilde{\alpha}_l, \tilde{\beta}_l)$. Let $(\alpha_l(\zeta), \beta_l(\zeta))$ be a causal irreducible B-controller, as a solution of DPE (3.211) for the irreducible process $(a_l(\zeta), \zeta b_l(\zeta))$. Then for the chosen basis $(\alpha_{l0}(\zeta), \beta_{l0}(\zeta))$, we obtain the basic representation

$$\begin{aligned} \alpha_l(\zeta) &= D_l(\zeta)\alpha_{l0}(\zeta) - \zeta M_l(\zeta)b_l(\zeta), \\ \beta_l(\zeta) &= D_l(\zeta)\beta_{l0}(\zeta) - M_l(\zeta)a_l(\zeta). \end{aligned} \tag{3.224}$$

Moreover, if $d_1(\zeta), \ldots, d_m(\zeta)$ is the sequence of invariant polynomials of the matrix $D_l(\zeta)$, and $q_1(\zeta), \ldots, q_{n+m}(\zeta)$ is the sequence of invariant polynomials of the characteristic matrix $Q_l(\zeta, \alpha_l, \beta_l)$, then we obtain, as was shown earlier

$$q_1(\zeta) = q_2(\zeta) = \cdots = q_n(\zeta) = 1, \quad q_{n+i}(\zeta) = d_i(\zeta), \ (i = 1, \ldots, m). \quad (3.225)$$

Besides, for the controller $(\tilde{\alpha}_l(z), \tilde{\beta}_l(z))$ relations (3.216) are fulfilled, which means, that the sequence of invariant polynomials of the matrix $\tilde{Q}_l(z, \tilde{\alpha}_l, \tilde{\beta}_l)$ named $\tilde{q}_1(z), \ldots, \tilde{q}_{n+m}(z)$ is connected with the sequence $q_1(\zeta), \ldots, q_{n+m}(\zeta)$ by the relations

$$q_i(\zeta) = \text{Rec}\,\tilde{q}_i(z). \quad (3.226)$$

Utilizing that the process $(\tilde{a}_l(z), \tilde{b}_l(z))$ is irreducible and that in (3.225) $d_i(0) \neq 0, \ (i = 1, \ldots, m)$, from (3.226) we can derive

$$\tilde{q}_1(z) = \cdots = \tilde{q}_n, \quad \tilde{q}_{n+i}(z) = z^{\lambda_i}\tilde{q}_{n+i,1}(z), \ (i = 1, \ldots, m), \quad (3.227)$$

where $\tilde{q}_{n+i,1}(z)$ is a monic polynomial determined by

$$\tilde{q}_{n+i,1}(z) \sim z^{\deg d_i(\zeta)} d_i(z^{-1}). \quad (3.228)$$

In these formulae the quantities $\lambda_i$ are nonnegative integers satisfying

$$\lambda_{i-1} \leq \lambda_i, \quad \lambda_1 + \lambda_2 + \cdots + \lambda_m = \lambda.$$

Further on, polynomials $\tilde{q}_{n+i,1}(z)$ (3.228) are called fundamental (basic?) invariant polynomials of the matrix $\tilde{Q}_l(z, \tilde{\alpha}_l, \tilde{\beta}_l)$.

## 3.10 Causal Stabilization of Discrete Forward Processes

(1) The polynomial $\tilde{f}(z)$ is called stable, if it does not possess roots $z_i$ outside or on the unit circle, i.e., with $|z_i| \geq 1$. The polynomial $n \times n$ matrix $\tilde{f}(z)$ is called stable, if the polynomial $\det \tilde{f}(z)$ is stable. Polynomials, which are not stable, are called unstable.

**Theorem 3.30** *The polynomials $\tilde{f}(z)$ and $f(\zeta) = rec\, \tilde{f}(z)$ are at the same time stable or unstable.*

*Proof* Let the roots of the polynomial $\tilde{f}(z)$ be the numbers $z_0 = 0$ and $z_i \neq 0, \ (i = 1, \ldots, \rho)$. Then the roots of the polynomial $f(\zeta)$ are the numbers $\zeta_i = z_i^{-1}, \ (i = 1, \ldots, \rho)$. Therefore, when the polynomial $\tilde{f}(z)$ is stable, i.e., $|z_i| < 1$, then we obtain $|\zeta_i| > 1$ and the polynomial $f(\zeta)$ is stable too. However, if for some $\alpha$ we have $|z_\alpha| \geq 1$, then $|\zeta_\alpha| \leq 1$, and the polynomial $f(\zeta)$ is also unstable. The inverse claim can be proven analogously noticing that $|z_0| < 1$. ■

An important special case of causal F-modal control is the causal stabilization problem for F-processes, which can be formulated as follows.

**F-stabilization problem:** For a given strictly causal irreducible forward process $(\tilde{a}_l(z), \tilde{b}_l(z))$ find the set of irreducible causal controllers $(\tilde{\alpha}_l(z), \tilde{\beta}_l(z))$, for which the characteristic matrix of the closed loop $\tilde{Q}_l(z, \tilde{\alpha}_l, \tilde{\beta}_l)$ is stable.

(2) The formulated F-stabilization problem leads to the construction of the set of causal solutions of the DPE

$$\det \tilde{Q}_l(z, \tilde{\alpha}_l, \tilde{\beta}_l) = \det \begin{bmatrix} \tilde{a}_l(z) & -\tilde{b}_l(z) \\ -\tilde{\beta}_l(z) & \tilde{\alpha}_l(z) \end{bmatrix} \sim \tilde{\Delta}_+(z), \tag{3.229}$$

where $\tilde{\Delta}_+(z)$ is any stable monic polynomial, under the condition that the transfer matrix of the controller

$$\tilde{W}_d(z) = \tilde{\alpha}^{-1}(z)\tilde{\beta}(z) \tag{3.230}$$

is causal (at least proper). The solution of the formulated problem can be achieved on basis of the results in Sect. 3.9. Hereby, the circumstance plays an essential role, that the polynomials $\tilde{f}(z)$ and rec $\tilde{f}(z) = f(\zeta)$ are at the same time stable or unstable.

(3) Assume the irreducible strictly causal F-process $(\tilde{a}_l(z), \tilde{b}_l(z))$ with the transfer function

$$\tilde{W}(z) = \tilde{a}_l^{-1}(z)\tilde{b}_l(z), \tag{3.231}$$

which is strictly proper. Let

$$W(\zeta) = \tilde{W}(\zeta^{-1}), \tag{3.232}$$

be the associated rational matrix, and its ILMFD

$$W(\zeta) = \zeta a_l^{-1}(\zeta)b_l(\zeta), \tag{3.233}$$

defines the irreducible strictly causal B-process $(a_l(\zeta), \zeta b_l(\zeta))$. As was shown above, the process $(a_l(\zeta), \zeta b_l(\zeta))$ is stabilizable, all B-stabilizing controllers are causal, and the set of transfer matrices of all stabilizing B-controllers $W_d(\zeta)$ is determined by the relations

$$W_d(\zeta) = [\alpha_{l0}(\zeta) - \zeta\theta(\zeta)b_l(\zeta)]^{-1}[\beta_{l0}(\zeta) - \theta(\zeta)a_l(\zeta)], \tag{3.234}$$

where $(\alpha_{l0}(\zeta), \beta_{l0}(\zeta))$ is a certain basic B-controller, and $\theta(\zeta)$ is an arbitrary stable rational matrix, i.e., it does not possess poles on the disc $|\zeta| \leq 1$. All matrices (3.234) are causal. Moreover, the set of all irreducible stabilizing B-controllers is determined from the ILMFD

$$W_d(\zeta) = \alpha_l^{-1}(\zeta)\beta_l(\zeta). \tag{3.235}$$

**Theorem 3.31** *The irreducible strictly causal F-process* $(\tilde{a}_l(z), \tilde{b}_l(z))$ *is stabilizable, and for it there exists a set of causal stabilizing controllers. The set of all irreducible causal stabilizing F-controllers* $(\tilde{\alpha}_l(z), \tilde{\beta}_l(z))$ *is determined by the ILMFD*

$$\tilde{W}_d(z) = W_d(z^{-1}) = \tilde{\alpha}_l^{-1}(z)\tilde{\beta}_l(z), \tag{3.236}$$

*where $W_d(\zeta)$ is any matrix of set of matrices (3.234).*

*Proof* Let $(\tilde{\alpha}_l(z), \tilde{\beta}_l(z))$ be a causal irreducible stabilizing F-controller. Then the transfer matrix of the controller

$$\tilde{W}_d(z) = \tilde{\alpha}_l^{-1}(z)\tilde{\beta}_l(z), \tag{3.237}$$

is causal. Hence, the associated matrix

$$W_d(\zeta) = \tilde{W}_d(\zeta^{-1}) \tag{3.238}$$

is also causal. Build ILMFD (3.235). Then, using (2.368), we obtain

$$\begin{bmatrix} a_l(\zeta) & -\zeta b_l(\zeta) \\ -\beta_l(\zeta) & \alpha_l(\zeta) \end{bmatrix} = \operatorname{rec} \begin{bmatrix} \tilde{a}_l(z) & -\tilde{b}_l(z) \\ -\tilde{\beta}_l(z) & \tilde{\alpha}_l(z) \end{bmatrix}. \tag{3.239}$$

Since the matrix on the right side is stable, so the matrix on the left side is stable too. Thus, $(\alpha_l(\zeta), \beta_l(\zeta))$ is an irreducible stabilizing B-controller. The transfer matrix of this controller has form (3.234) for a certain matrix $\theta(\zeta)$. Therefore, controller $(\tilde{\alpha}_l(z), \tilde{\beta}_l(z))$ satisfies condition (3.236). By reverse reasoning, it can be shown that any matrix of form (3.234) with the help of (3.236) defines a irreducible causal stabilizing F-controller $(\tilde{\alpha}_l(z), \tilde{\beta}_l(z))$. ■

(4) Next we consider the question about the stabilization of the strictly causal F-process in the case, when the pair $(\tilde{a}_l(z), \tilde{b}_l(z))$ is reducible.

**Theorem 3.32** *Assume the non-irreducible strictly causal F-process $(\tilde{a}_l(z), \tilde{b}_l(z))$. Let $\tilde{\varphi}(z)$ be the GCD of the matrices $\tilde{a}_l(z)$ and $\tilde{b}_l(z)$, so that we can write*

$$\begin{bmatrix} \tilde{a}(z) & \tilde{b}(z) \end{bmatrix} = \tilde{\varphi}(z) \begin{bmatrix} \tilde{a}_{l1}(z) & \tilde{b}_{l1}(z) \end{bmatrix}, \tag{3.240}$$

*where the pair $(\tilde{a}_{l1}(z), \tilde{b}_{l1}(z))$is irreducible. Then for the stabilizability of the process $(\tilde{a}_l(z), \tilde{b}_l(z))$, it is necessary and sufficient that the matrix $\tilde{\varphi}(z)$ is stable, i.e., the polynomial* $\det \tilde{\varphi}(z)$ *has all its roots $z_i$ inside the unit circle, i.e., $|z_i| \leq 1$.*

*Proof* Necessity: When (3.240) is fulfilled, then the set of stabilizing controllers is determined by the DPE

$$\det \tilde{\varphi}(z) \det \begin{bmatrix} \tilde{a}_{l1}(z) & -\tilde{b}_{l1}(z) \\ -\tilde{\beta}_l(z) & \tilde{\alpha}_l(z) \end{bmatrix} \sim \tilde{\Delta}_+(z), \tag{3.241}$$

where $\tilde{\Delta}_+(z)$ is any stable polynomial. Obviously, this equation can only have a solution, when the polynomial $\det \tilde{\varphi}(z)$ is stable.

Sufficiency: Let the polynomial $\det \tilde{\varphi}(z)$ be stable. Then for

$$\tilde{\Delta}_{+}(z) = \tilde{\Delta}_{+1}(z) \det \tilde{\varphi}(z),$$

where $\tilde{\Delta}_{+1}(z)$ is a stable polynomial, from (3.241) we obtain a DPE, which has a causal solution according to Theorem 3.31. ■

(6) From Theorem 3.31 and (3.234), we achieve that the set of causal irreducible stabilizing F-controllers for the strictly causal irreducible process $(\tilde{a}_l(z), \tilde{b}_l(z))$ is determined by formulae (3.236), where

$$\tilde{W}_d(z) = [\alpha_{l0}(z^{-1}) - z^{-1}\theta(z^{-1})b_l(z^{-1})]^{-1}[\beta_{l0}(z^{-1}) - \theta(z^{-1})a_l(z^{-1})], \quad (3.242)$$

and $\theta(\zeta)$ is any stable rational matrix. The structure of the characteristic matrix of the closed loop for a concrete selection of the matrix $\theta(\zeta)$ is determined by the next claim.

**Theorem 3.33** *Assume in (3.242) for the matrix $\theta(\zeta)$ the ILMFD*

$$\theta(\zeta) = D_l^{-1}(\zeta)M_l(\zeta), \quad (3.243)$$

*where the matrix $D_l(\zeta) \in \mathbb{R}^{mm}(\zeta)$ is stable.*

*Let $d_1(\zeta), \ldots, d_m(\zeta)$ be the sequence of invariant polynomials of the matrix $D_l(\zeta)$, and $\tilde{q}_1(z), \ldots, \tilde{q}_m(z)$ be monic polynomials such that*

$$d_i(\zeta) = Rec\, \tilde{q}_i(z). \quad (3.244)$$

*Then the sequence of invariant polynomials of the characteristic matrix of the closed loop (3.192) is defined by formulae (3.225)–(3.228) for certain sequence of nonnegative numbers $\lambda_i$.*

*Proof* When (3.243) is fulfilled, for ILMFD (3.235), we obtain

$$\begin{aligned} \alpha_l(\zeta) &= D_l(\zeta)\alpha_{l0}(\zeta) - \zeta M_l(\zeta)b_l(\zeta), \\ \beta_l(\zeta) &= D_l(\zeta)\beta_{l0}(\zeta) - M_l(\zeta)a_l(\zeta). \end{aligned} \quad (3.245)$$

Moreover,

$$\det Q_l(\zeta, \alpha_l, \beta_l) = \det \begin{bmatrix} a_l(\zeta) & -\zeta b_l(\zeta) \\ -\beta_l(\zeta) & \alpha_l(\zeta) \end{bmatrix} \sim \Delta_{+}(\zeta), \quad (3.246)$$

where $\Delta_{+}(\zeta) \sim det\, D_l(\zeta)$ is a monic stable polynomial. Besides, for the set of invariant polynomials of matrix $Q_l(\zeta, \alpha_l, \beta_l)$, relations (3.145), (3.146) hold. Therefore from relation (3.239), we achieve relations (3.225)–(3.228) for some nonnegative numbers $\lambda_i$. ■

*Remark 3.4* The matrix $\tilde{\theta}(z) = \theta(z^{-1})$ configured in (3.242) is stable and causal. Indeed, since the matrix $\theta(\zeta)$ is stable, it is causal. Therefore, the associated matrix $\tilde{\theta}(z)$ is also causal. Moreover, when $\zeta_1, \ldots, \zeta_\rho$ are the poles of the matrix $\theta(\zeta)$, then by definition $|\zeta_i| > 1,\ (i = 1, \ldots, \rho)$. Then the poles of the matrix $\tilde{\theta}(z)$ are the numbers $z_1 = \zeta^{-1},\ (i = 1, \ldots, \rho)$ and, possibly, the number $z_0 = 0$. All these poles are located inside the circle $|z| = 1$, therefore, the matrix $\tilde{\theta}(z)$ is stable and causal.

## 3.11 Causal Stabilization of the Double Integrator

(1) In this section, we consider the stabilization problem for the discrete F-model of the double integrator [2], which is described by the differential equation

$$\ddot{y}(t) = u(t). \tag{3.247}$$

As was shown in [2], the transfer function of the discrete F-model of the continuous process (3.247), when using a zero-order hold, has the form

$$\tilde{W}(z) = K\frac{z+1}{(z-1)^2}, \tag{3.248}$$

where

$$K = \frac{T^2}{2}, \tag{3.249}$$

and $T$ is the sampling period.

Since the fraction on the right side of (3.248) is irreducible, we can configure

$$\tilde{a}_l(z) = (z-1)^2, \quad \tilde{b}_l(z) = K(z+1). \tag{3.250}$$

Now we have to find the set of all irreducible causal controllers $(\tilde{\alpha}_l(z), \tilde{\beta}_l(z))$, which is determined by

$$\det \tilde{Q}_l(z, \tilde{\alpha}_l, \tilde{\beta}_l) = \det \begin{bmatrix} \tilde{a}_l(z) & -\tilde{b}_l(z) \\ -\tilde{\beta}_l(z) & \tilde{\alpha}_l(z) \end{bmatrix} \sim z^\lambda \tilde{\Delta}_+(z), \tag{3.251}$$

where $\tilde{\Delta}_+(z)$ is any stable monic polynomial, but $\tilde{\Delta}_+(0) \neq 0$, and $\lambda \geq 0$ is any nonnegative integer.

(2) Substituting in (3.248) $z$ by $\zeta^{-1}$, we obtain the transfer function of the discrete B-model

$$W_d(\zeta) = K\frac{\zeta(\zeta+1)}{(1-\zeta)^2}. \tag{3.252}$$

Since here the fraction on the right side is irreducible too, we find

$$a_l(\zeta) = (1-\zeta)^2, \quad b_l(\zeta) = K(\zeta+1). \tag{3.253}$$

We construct a basic controller $(\alpha_{l0}(\zeta), \beta_{l0}(\zeta))$ for B-process (3.253) as solution of the DPE

$$\det \begin{bmatrix} (1-\zeta)^2 & -K\zeta(\zeta+1) \\ -\beta_{l0}(\zeta) & \alpha_{l0}(\zeta) \end{bmatrix} = 1. \tag{3.254}$$

DPE (3.254) is equivalent to the Diophantine equation

$$(1-\zeta)^2\alpha_{l0}(\zeta) - K\zeta(\zeta+1)\beta_{l0}(\zeta) = 1. \tag{3.255}$$

As a concrete solution of Eq. (3.255), we choose the solution with minimal degree

$$\alpha_{l0}(\zeta) = \frac{3}{4}\zeta + 1, \quad \beta_{l0}(\zeta) = \frac{1}{4K}(3\zeta - 5). \tag{3.256}$$

Therefore, according to (3.234) the set of transfer functions of stabilizing B-controllers for process (3.253) can be represented in the form

$$W_d(\zeta) = \frac{\frac{1}{4K}(3\zeta-5) - \theta(\zeta)(1-\zeta)^2}{\frac{3}{4}\zeta + 1 - K\zeta\theta(\zeta)(\zeta+1)}, \tag{3.257}$$

where $\theta(\zeta)$ is zero or any stable rational function, i.e., is free of poles at $\zeta \le 1$. Substituting in (3.257) $\zeta$ by $z^{-1}$, we obtain

$$\tilde{W}_d(z) = W_d(z^{-1}) = \frac{\frac{1}{4K}(3z^{-1}-5) - \theta(z^{-1})(1-z^{-1})^2}{\frac{3}{4}z^{-1} + 1 - Kz^{-1}\theta(z^{-1})(z^{-1}+1)}, \tag{3.258}$$

which can be written as

$$\tilde{W}_d(z) = \frac{\frac{1}{4K}z(3-5z) - \tilde{\theta}(z)(z-1)^2}{z(\frac{3}{4}+z) - K\tilde{\theta}(z)(z+1)}, \tag{3.259}$$

where $\tilde{\theta}(z) = \theta(z^{-1})$ is zero or any at least proper rational function, where its poles are located inside the circle $|z| = 1$. When applying the irreducible representation

$$\tilde{\theta}(z) = \frac{\tilde{m}(z)}{\tilde{\Delta}_+(z)}, \tag{3.260}$$

where $\tilde{m}(z)$ is an arbitrary polynomial, but $\tilde{\Delta}_+(z)$ is any stable polynomial, such that $\deg \tilde{\Delta}_+(z) \ge \deg \tilde{m}(z)$, we obtain

$$\tilde{W}_d(z) = \frac{\tilde{\beta}_l(z)}{\tilde{\alpha}_l(z)}, \tag{3.261}$$

where

$$
\begin{aligned}
\tilde{\alpha}_l(z) &= \left(\frac{3}{4}z + z^2\right)\tilde{\Delta}_+(z) - K(z+1)\tilde{m}(z),\\
\tilde{\beta}_l(z) &= \frac{1}{4K}(3z - 5z^2)\tilde{\Delta}_+(z) - (z-1)^2\tilde{m}(z).
\end{aligned}
\tag{3.262}
$$

Hereby, any pair (3.262) for which the fraction on the right side of (3.261) is irreducible, defines an irreducible causal stabilizing controller for the forward discrete model of double integrator (3.247).

(3) Investigate the question about the structure of the characteristic matrix of the closed loop

$$
\tilde{Q}_l(z, \tilde{\alpha}_l, \tilde{\beta}_l) = \begin{bmatrix} (z-1)^2 & -K(z+1) \\ -\tilde{\beta}_l(z) & \tilde{\alpha}_l(z) \end{bmatrix}. \tag{3.263}
$$

Since process (3.248) is irreducible, sequence of invariant polynomials of matrix (3.263) takes the form

$$
\tilde{q}_1(z) = 1, \quad \tilde{q}_2(z) = z^{\lambda}\tilde{\Delta}_+(z), \tag{3.264}
$$

where $\lambda$ is any nonnegative integer.

## 3.12 Causal Modal Control and Stabilization of Backward PMD Processes

(1) This section considers the modal control and stabilization problem for backward (B) processes described by the equations

$$
\begin{aligned}
a(\zeta)x_k &= b(\zeta)u_k,\\
y_k &= c(\zeta)x_k + d(\zeta)u_k,
\end{aligned}
\tag{3.265}
$$

where $x_k, y_k, u_k$ are discrete vector sequences of dimensions $p \times 1, n \times 1, m \times 1$, respectively. Moreover, $a(\zeta), b(\zeta), c(\zeta), d(\zeta)$ are polynomial matrices of size $p \times p, p \times m, n \times p, n \times m$, respectively. As a whole these matrices build the polynomial matrix description (PMD) [4]

$$
\Pi_b(\zeta) = (a(\zeta), b(\zeta), c(\zeta), d(\zeta)). \tag{3.266}
$$

For $d(\zeta) = 0_{nm}$, PMD $\Pi_b(\zeta)$ is called homogenous.

(2) As above, any pair $(\alpha(\zeta), \beta(\zeta))$, where $\alpha(\zeta) \in \mathbb{R}^{mm}[\zeta]$, $\beta(\zeta) \in \mathbb{R}^{mn}[\zeta]$ is called B-controller, and the equation

$$\alpha(\zeta)u_k = \beta(\zeta)y_k \tag{3.267}$$

is called the B-controller equation. All together, process equations and controller equation build the equations of the closed discrete system

$$\begin{aligned} a(\zeta)x_k - b(\zeta)u_k &= 0, \\ -c(\zeta)x_k + y_k - d(\zeta)u_k &= 0, \\ -\beta(\zeta)y_k + \alpha(\zeta)u_k &= 0, \end{aligned} \tag{3.268}$$

where the right sides are zero vectors of appropriate sizes.

**Definition 3.14** The $(p+n+m) \times (p+n+m)$ polynomial matrix

$$Q_\Pi(\zeta, \alpha, \beta) \stackrel{\triangle}{=} \begin{bmatrix} a(\zeta) & 0_{pn} & -b(\zeta) \\ -c(\zeta) & I_n & -d(\zeta) \\ 0_{mp} & -\beta(\zeta) & \alpha(\zeta) \end{bmatrix} \tag{3.269}$$

is called the characteristic matrix of the closed loop, and the polynomial

$$\Delta_\Pi(\zeta) \stackrel{\triangle}{=} \det Q_\Pi(\zeta, \alpha, \beta) \tag{3.270}$$

its characteristic polynomial.

(3) One of the fundamental problems concerning the control of PMD processes is the causal modal control.

**Modal control of PMD processes:** For given PMD (3.266) and given monic polynomial $\Delta_\Pi(\zeta)$, find the set of causal controllers $(\alpha(\zeta), \beta(\zeta))$ satisfying

$$\det Q_\Pi(\zeta, \alpha, \beta) = \det \begin{bmatrix} a(\zeta) & 0_{pn} & -b(\zeta) \\ -c(\zeta) & I_n & -d(\zeta) \\ 0_{mp} & -\beta(\zeta) & \alpha(\zeta) \end{bmatrix} \sim \Delta_\Pi(\zeta). \tag{3.271}$$

Relation (3.271) can be considered as DPE depending on the matrices $\alpha(\zeta)$ and $\beta(\zeta)$. Introduce the polynomial matrices

$$\begin{aligned} \bar{a}(\zeta) &\stackrel{\triangle}{=} \begin{bmatrix} a(\zeta) & 0_{pn} \\ -c(\zeta) & I_n \end{bmatrix}, \quad \bar{b}(\zeta) \stackrel{\triangle}{=} \begin{bmatrix} b(\zeta) \\ d(\zeta) \end{bmatrix}, \\ \bar{\beta}(\zeta) &\stackrel{\triangle}{=} \begin{bmatrix} 0_{mp} & \beta(\zeta) \end{bmatrix}, \quad \bar{\alpha}(\zeta) \stackrel{\triangle}{=} \alpha(\zeta). \end{aligned} \tag{3.272}$$

Then DPE (3.271) can be written in form (3.4)

$$\det \begin{bmatrix} \bar{a}(\zeta) & -\bar{b}(\zeta) \\ -\bar{\beta}(\zeta) & \bar{\alpha}(\zeta) \end{bmatrix} \sim \Delta_\Pi(\zeta). \tag{3.273}$$

However, the general approach for the solution of DPE (3.4), as considered above, cannot be directly applied in the given case, because due to allocation (3.272), the matrix $\bar{\beta}(\zeta)$ arising in DPE (3.273) has a special structure. Therefore, the causal modal control problem for PMD (3.266) requires a separate investigation.

(4)

**Definition 3.15** PMD (3.266) is called non-singular, if the matrix $a(\zeta)$ is non-singular, i.e.,

$$\det a(\zeta) \neq 0, \tag{3.274}$$

and minimal, if the pairs $[a(\zeta), b(\zeta)]$ and $[a(\zeta), c(\zeta)]$ are irreducible, i.e., for all $\zeta$

$$\operatorname{rank}\begin{bmatrix} a\zeta) & b(\zeta) \end{bmatrix} = p, \quad \operatorname{rank}\begin{bmatrix} a(\zeta) \\ c(\zeta) \end{bmatrix} = p. \tag{3.275}$$

For non-singular PMD (3.266), the rational $n \times m$ matrix

$$W(\zeta) = c(\zeta)a^{-1}(\zeta)b(\zeta) + d(\zeta) \tag{3.276}$$

is defined.

**Definition 3.16** Matrix $W(\zeta)$ (2.76) is called transfer matrix of PMD (3.266), which is called causal or strictly causal, if the transfer matrix $W(\zeta)$ disposes of the analog property.

**Theorem 3.34** *Let the PMD $\Pi(\zeta)$ be minimal and causal. Then*

$$\det a(0) \neq 0. \tag{3.277}$$

*Proof* Consider the ILMFD

$$a_1^{-1}(\zeta)c_1(\zeta) = c(\zeta)a^{-1}(\zeta). \tag{3.278}$$

Since the right side of this relation is an IRMFD, we obtain

$$a_1(\zeta) \overset{p}{\sim} a(\zeta), \tag{3.279}$$

and it follows from Lemmas 2.8 to 2.14 that the representation

$$W(\zeta) = a_1^{-1}(\zeta)[c_1(\zeta)b_1(\zeta) + a_1(\zeta)d(\zeta)] \tag{3.280}$$

is an ILMFD. Therefore

$$\det a_1(\zeta) \sim \psi_W(\zeta), \tag{3.281}$$

where $\psi_W(\zeta)$ is the McMillan denominator of the transfer matrix $W(\zeta)$. If we would assume that the matrix $a_1(\zeta)$ has an eigenvalue $\zeta = 0$, then this number would become a pole of the matrix $W(\zeta)$, which is not allowed, because the matrix $W(\zeta)$ is causal. Thus, the matrix $a_1(\zeta)$ does not possess an eigenvalue equal to zero, and due to (3.279), the matrix $a(\zeta)$ disposes of the same property. ■

**Theorem 3.35** *Let PMD (3.266) be minimal and non-singular. Then the following claims are valid:*

*(i) The set of pairs $(\alpha(\zeta), \beta(\zeta))$ as solution of solutions of DPE (3.271) coincides with the set of solutions of DPE (3.4)*

$$\det Q_1(\zeta, \alpha, \beta) \sim \Delta_\Pi(\zeta). \tag{3.282}$$

*Here $Q_1(\zeta, \alpha, \beta)$ is the polynomial matrix*

$$Q_1(\zeta, \alpha, \beta) \triangleq \begin{bmatrix} a_1(\zeta) & -c_1(\zeta)\zeta b(\zeta) - a_1(\zeta)\zeta d(\zeta) \\ -\beta(\zeta) & \alpha(\zeta) \end{bmatrix}, \tag{3.283}$$

*where the matrices $a_1(\zeta)$, $c_1(\zeta)$ are determined by ILMFD (3.278).*

*(ii) DPE (3.271) is solvable for any polynomial $\Delta\Pi(\zeta)$.*

The proof is subdivided into several steps.

**Lemma 3.1** *The following formula is true:*

$$\det Q_\Pi(\zeta, \alpha, \beta) \sim \det a(\zeta) \det\left[\alpha(\zeta) - \beta(\zeta)\ W(\zeta)\right], \tag{3.284}$$

*where*

$$W(\zeta) = c(\zeta)a^{-1}(\zeta)\zeta b(\zeta) + \zeta d(\zeta) \tag{3.285}$$

*is the transfer matrix of PMD (3.266).*

*Proof* Notice that in (3.273) the matrix $\bar{a}(\zeta)$ is invertible, such that due to (3.272)

$$\det \bar{a}(\zeta) = \det a(\zeta) \neq 0. \tag{3.286}$$

Therefore, applying to matrix (3.273) Schur's formula, we find

$$\det Q_\Pi(\zeta, \alpha, \beta) = \det \bar{a}(\zeta) \det\left[\bar{\alpha}(\zeta) - \bar{\beta}(\zeta)\bar{a}^{-1}(\zeta)\zeta\bar{b}(\zeta)\right]. \tag{3.287}$$

Since

$$\tilde{a}^{-1}(\zeta) = \begin{bmatrix} a^{-1}(\zeta) & 0_{pn} \\ c(\zeta)a^{-1}(\zeta) & I_n \end{bmatrix}, \tag{3.288}$$

using (3.272), we obtain

$$\begin{aligned} \tilde{\beta}(\zeta)a^{-1}(\zeta)\tilde{b}(\zeta) &= \left[0_{mp}\ \beta(\zeta)\right]\begin{bmatrix} a^{-1}(\zeta) & 0_{pn} \\ c(\zeta)a^{-1}(\zeta) & I_n \end{bmatrix}\begin{bmatrix} \zeta b(\zeta) \\ \zeta d(\zeta) \end{bmatrix} \\ &= \beta(\zeta)\left[c(\zeta)a^{-1}(\zeta)\zeta b(\zeta) + \zeta d(\zeta)\right] = \beta(\zeta)W(\zeta), \end{aligned} \tag{3.289}$$

so that with the help of (3.286), (3.287), formula (3.284) is achieved. ■

*Proof* (*Theorem* 3.35)

(i) Since representation (3.278) is an ILMFD, the pair $(a_1(\zeta),\ c_1(\zeta)\zeta b(\zeta) + a_1(\zeta)\zeta d(\zeta))$ is irreducible. Applying to matrix $Q_1(\zeta, \alpha, \beta)$ Schur's formula, we obtain

$$\begin{aligned}\det Q_1(\zeta,\alpha,\beta) &= \\ &= \det a_1(\zeta)\det\left[\alpha(\zeta) - \beta(\zeta)a^{-1}(\zeta)(c_1(\zeta)\zeta b(\zeta) + a_1(\zeta)\zeta d(\zeta))\right] \\ &= \det a_1(\zeta)\det\left[\alpha(\zeta) - \beta(\zeta)W_d(\zeta)\right].\end{aligned} \tag{3.290}$$

Since $\det a_1(\zeta) \sim \det a(\zeta)$, comparing (3.288) with (3.295), we find that the sets of solutions of DPE (3.271) and (3.286) coincide, which proves claim (i).

(ii) The irreducibility of the pair $(a(\zeta), c_1(\zeta)\zeta b(\zeta) + a_1(\zeta)\zeta d(\zeta))$ implies that Eq. (3.286) is solvable for any polynomial $\Delta(\zeta)$. Therefore from (i) claim (ii) is achieved. ■

(7) Introduce the notation

$$b_1(\zeta) \stackrel{\triangle}{=} c_1(\zeta)b(\zeta) + a_1(\zeta)d(\zeta). \tag{3.291}$$

Then DPE (3.286) can be written in the form

$$\det\begin{bmatrix} a_1(\zeta) & -b_1(\zeta) \\ -\beta(\zeta) & \alpha(\zeta)\end{bmatrix} \sim \Delta_\Pi(\zeta). \tag{3.292}$$

Since by construction the pair $(a_1(\zeta), \zeta b_1(\zeta))$ is irreducible, the set of solutions of Eq. (3.298) is determined by the relations

$$\begin{aligned}\alpha(\zeta) &= D(\zeta)\alpha_{l0}(\zeta) - \zeta M(\zeta)b_1(\zeta), \\ \beta(\zeta) &= D(\zeta)\beta_{l0}(\zeta) - M(\zeta)a_1(\zeta),\end{aligned} \tag{3.293}$$

where the matrix $M(\zeta)$ is arbitrary, $\det D(\zeta) \sim \Delta(\zeta)$, and the pair $(\alpha_{l0}(\zeta), \beta_{l0}(\zeta))$ is any basic controller for the pair $(a_1(\zeta), \zeta b_1(\zeta))$. It follows from the proof of Theorem 3.35, that for a fixed controller $(\alpha(\zeta), \beta(\zeta))$ the equivalence

$$\det Q_\Pi(\zeta, \bar{\alpha}, \bar{\beta}) \sim \det Q_1(\zeta, \alpha, \beta). \tag{3.294}$$

Therefore, when controller $(\alpha(\zeta), \beta(\zeta))$ is a solution of (3.298), then $(\bar{\alpha}(\zeta), \bar{\beta}(\zeta))$ with

$$\bar{\alpha}(\zeta) = \alpha(\zeta), \quad \bar{\beta}(\zeta) = \left[\, 0_{mp}\ \beta_(\zeta)\,\right] \tag{3.295}$$

is a solution of the DPE (3.273). The reverse claim also holds. Therefore, the set of pairs $(\alpha(\zeta), \beta(\zeta))$ in solutions of the DPE (3.271) is determined by formulae (3.293).

Inserting (3.293) into (3.295) yields the general solution of DPE (3.271) in the form

$$\begin{aligned}\bar{\alpha}(\zeta) &= D(\zeta)\alpha_{l0}(\zeta) - M(\zeta)b_1(\zeta),\\ \bar{\beta}(\zeta) &= D(\zeta)\left[0_{mp}\ \beta_{l0}(\zeta)\right] - M(\zeta)\left[0_{mp}\ a_1(\zeta)\right].\end{aligned} \tag{3.296}$$

(5) The question about the structure of the set of invariant polynomials of the matrix $Q_\Pi(\zeta, \bar{\alpha}, \bar{\beta})$ (3.273) by using a controller in form (3.296) is answered below.

**Theorem 3.36** *Under the condition of Theorem 3.35, when using the controller (3.300), the first* $p+n$ *invariant polynomials* $q_1(\zeta), \ldots, q_{p+n}(\zeta)$ *of the matrix* $Q_\Pi(\zeta, \alpha, \beta)$ *are equal to one. Moreover, the remaining sequence of invariant polynomials satisfies*

$$q_{p+n+i}(\zeta) = d_i(\zeta), \quad (i = 1, \ldots, m), \tag{3.297}$$

*where* $d_1(\zeta), \ldots, d_m(\zeta)$ *is the sequence of invariant polynomials of the matrix* $D(\zeta)$.

*Proof* Under the conditions of Theorem 3.35, the pair $(\bar{a}(\zeta), \bar{b}(\zeta))$,where the matrices $\bar{a}(\zeta)$, $\bar{b}(\zeta)$ are determined from (3.272), is irreducible. This fact follows from the circumstance, that there exists a basic controller $(\bar{\alpha}_{l0}(\zeta), \bar{\beta}_{l0}(\zeta))$, for which

$$\bar{\alpha}_{l0}(\zeta) = \alpha_{l0}(\zeta), \quad \bar{\beta}_{l0}(\zeta) = \left[0_{mp}\ \beta_{l0}(\zeta)\right]. \tag{3.298}$$

Therefore, the set of solutions of the DPE (3.273) in addition to (3.300) can be represented by the formulae

$$\begin{aligned}\tilde{\alpha}(\zeta) &= D_1(\zeta)\bar{\alpha}_{l0}(\zeta) - M_1(\zeta)\bar{b}(\zeta),\\ \tilde{\beta}(\zeta) &= D_1(\zeta)\bar{\beta}_{l0}(\zeta) - M_1(\zeta)\bar{a}(\zeta),\end{aligned} \tag{3.299}$$

where $M_1(\zeta)$ is any $m \times (p+n)$ polynomial matrix, but $D_1(\zeta)$ is an arbitrary $m \times m$ polynomial matrix satisfying $\det D_1(\zeta) \sim \Delta_\Pi(\zeta)$.

Since the structure of the set of invariant polynomials of the matrix $Q_\Pi(\zeta, \alpha, \beta)$ does not depend on the choice of the matrix $M_1(\zeta)$, for the determination of this structure we can assume that in (2.293) $M_1(\zeta) = 0_{m,p+n}$. Thus

$$\begin{aligned}\tilde{\alpha}(\zeta) &= D_1(\zeta)\bar{\alpha}_{l0}(\zeta) = D_1(\zeta)\alpha_{l0}(\zeta),\\ \tilde{\beta}(\zeta) &= D_1(\zeta)\bar{\beta}_{l0}(\zeta) = D_1(\zeta)\left[0_{mp}\ \beta_{l0}(\zeta)\right].\end{aligned} \tag{3.300}$$

Moreover, due to Theorem 3.11, it follows from the irreducibility of the pair $(\bar{a}(\zeta), \bar{b}(\zeta))$ that the $(p+n)$ first invariant polynomials of the matrix $Q_\Pi(\zeta, \alpha, \beta)$ are equal to one, and the remaining invariant polynomials satisfy Eq. (3.301), when the matrix $D(\zeta)$ is substituted by the matrix $D_1(\zeta)$. At the same time, from (3.300), for $M(\zeta) = 0_{mn}$, we obtain

$$\begin{aligned}\bar{\alpha}(\zeta) &= D(\zeta)\alpha_{l0}(\zeta),\\ \bar{\beta}(\zeta) &= D(\zeta)\left[0_{mp}\ \beta_{l0}(\zeta)\right].\end{aligned} \tag{3.301}$$

Comparing relations (3.300) and (3.301), we find that the polynomials $D(\zeta)$ and $D_1(\zeta)$ coincide, which completes the proof. ■

**Corollary 3.7** *When applying the controllers (3.293) and (3.296), the matrices $Q_{\Pi}(\zeta, \alpha, \beta)$ and $Q_1(\zeta, \alpha, \beta)$ are pseudo-equivalent.*

(6) Necessary and sufficient conditions for the solvability of the DPE (3.271) in the case, when PMD (3.266) is not minimal are given next.

**Theorem 3.37** *Let non-singular PMD (3.266) be not minimal. Assume $\psi_1(\zeta)$ to be the GCD of $a(\zeta)$, $b(\zeta)$, such that*

$$\left[a(\zeta)\ b(\zeta)\right] = \psi_1(\zeta)\left[a_1(\zeta)\ b_1(\zeta)\right], \tag{3.302}$$

*where the pair $(a_1(\zeta), b_1(\zeta))$ is irreducible. Let also $\psi_2(\zeta)$ be the GCD of $a_1(\zeta)$ and $c(\zeta)$, such that*

$$\begin{bmatrix} a_1(\zeta) \\ c(\zeta) \end{bmatrix} = \begin{bmatrix} a_2(\zeta) \\ c_1(\zeta) \end{bmatrix} \psi_2(\zeta), \tag{3.303}$$

*where the pair $[a_2(\zeta), c_1(\zeta)]$ is irreducible. Denote*

$$\det \psi_1(\zeta) \stackrel{\triangle}{=} \gamma_1(\zeta), \quad \det \psi_2(\zeta) \stackrel{\triangle}{=} \gamma_2(\zeta). \tag{3.304}$$

*Then the following statements hold:*

*(i) The PMD*

$$\Pi_1(\zeta) = (a_3(\zeta), b_2(\zeta), c_2(\zeta), d(\zeta)) \tag{3.305}$$

*is minimal.*

*(ii) For the solvability of the DPE (3.271), it is necessary and sufficient that the relation*

$$\Delta_{12}(\zeta) \stackrel{\triangle}{=} \frac{\Delta_{\Pi}(\zeta)}{\gamma_1(\zeta)\gamma_2(\zeta)} \tag{3.306}$$

*is a polynomial.*

*(iii) When (ii) is fulfilled, the set of solutions of Eq. (3.271) coincides with the set of solutions of the DPE*

$$\det \begin{bmatrix} a_3(\zeta) & 0_{pm} & -b_2(\zeta) \\ -c_2(\zeta) & I_n & d(\zeta) \\ 0_{mp} & -\beta(\zeta) & \alpha(\zeta) \end{bmatrix} \sim \Delta_{12}(\zeta). \tag{3.307}$$

*The solution of (3.307) can be found on the basis of Theorem 3.35.*

*Proof*

(i) Under the taken suppositions, we obtain

$$\begin{aligned} a(\zeta) &= \psi_1(\zeta)a_3(\zeta)\psi_2(\zeta), \quad b(\zeta) = \psi_1(\zeta)b_2(\zeta), \\ c(\zeta) &= c_2(\zeta)\psi_2(\zeta), \quad a_2(\zeta) = a_3(\zeta)\psi_2(\zeta). \end{aligned} \tag{3.308}$$

Hence, we can write

$$W(\zeta) = c(\zeta)a^{-1}(\zeta)b(\zeta) + d(\zeta) = c_2(\zeta)a_3^{-1}b_2(\zeta) + d(\zeta) \stackrel{\triangle}{=} W_1(\zeta), \tag{3.309}$$

where $W_1(\zeta)$ is the transfer matrix of the PMD (3.306). We show that the PMD $\Pi_1(\zeta)$ is minimal. Indeed, the pair $(a_3(\zeta), c_2(\zeta))$ is irreducible by construction, because $\psi_2(\zeta)$ is the GCD. Moreover, since the pair $(a_2(\zeta), c_2(\zeta))$ is irreducible, so from (1.99), we obtain

$$a_2(\zeta)X(\zeta) + b_2(\zeta)Y(\zeta) = I, \tag{3.310}$$

where $X(\zeta)$ and $Y(\zeta)$ are polynomial matrices of appropriate size. Using (3.308), the last relation can be written in the form

$$a_3(\zeta)\,[\psi_2(\zeta)X(\zeta)] + b_2(\zeta)Y(\zeta) = I. \tag{3.311}$$

This means that the pair $(a_3(\zeta), b_2(\zeta))$ is irreducible. This fact proves (i).

(ii) With the help of (3.302) , (3.303), the matrix $Q_\Pi(\zeta, \alpha, \beta)$ (3.269) can be represented in the form

$$Q_\Pi(\zeta, \alpha, \beta) = G_1(\zeta)Q_2(\zeta, \alpha, \beta)G_2(\zeta), \tag{3.312}$$

where we denote

$$G_1(\zeta) \stackrel{\triangle}{=} \operatorname{diag}\{\psi_1(\zeta), I, I\}, \quad G_2(\zeta) \stackrel{\triangle}{=} \operatorname{diag}\{\psi_2(\zeta), I, I\} \tag{3.313}$$

and

$$Q_2(\zeta, \alpha, \beta) = \begin{bmatrix} a_3(\zeta) & 0_{pn} & -b_2(\zeta) \\ -c_2(\zeta) & I_n & -d(\zeta) \\ 0_{mp} & -\beta(\zeta) & \alpha(\zeta) \end{bmatrix}. \tag{3.314}$$

Using (3.312) and (3.313) together with (3.305), DPE (3.271) can be written in the form

$$\gamma_1(\zeta)\gamma_2(\zeta) \det Q_2(\zeta, \alpha, \beta) \sim \Delta_\Pi(\zeta). \tag{3.315}$$

Since the PMD (3.306) is irreducible, we immediately obtain that this equation is solvable, if and only if relation (3.307) is a polynomial.

(iii) When (ii) is valid, Eq. (3.314) after reduction takes the form

$$\det Q_2(\zeta, \alpha, \beta) \sim \Delta_{12}(\zeta). \tag{3.316}$$

Since the PMD (3.306) is minimal, Eq. (3.308) is solvable for any polynomial $\Delta_{12}(\zeta)$, and the corresponding set of solutions can be constructed on basis of Theorem 3.35. ■

(7) Let the PMD (3.266) be non-singular, minimal and strictly causal, in harmony with the above said, we obtain

$$W(0) = 0_{nm}. \tag{3.317}$$

This relation is fulfilled if

$$b(\zeta) = \zeta b_3(\zeta), \quad d(\zeta) = \zeta d_1(\zeta), \tag{3.318}$$

where $b_3(\zeta)$ and $d_1(\zeta)$ are polynomial matrices. In the following investigations, when not explicitly noticed different, we always consider strictly causal PMD, for which relations (3.318) hold.

Under condition (3.318), DPE (3.271) can be represented in the form

$$\det \begin{bmatrix} a(\zeta) & 0_{pn} & -\zeta b_3(\zeta) \\ -c(\zeta) & I_n & -\zeta d_1(\zeta) \\ 0_{mp} & -\beta(\zeta) & \alpha(\zeta) \end{bmatrix} \sim \Delta_\Pi(\zeta). \tag{3.319}$$

In addition, the next statement holds.

**Theorem 3.38** *If in Eq. (3.319) the polynomial $\Delta_\Pi(\zeta)$ is causal, i.e., $\Delta(0) \neq 0$, then all solutions of this equation are causal.*

*Proof* For $\zeta = 0$, from (3.319) we obtain

$$\det a(0) \det \alpha(0) = \Delta_\Pi(0) \neq 0. \tag{3.320}$$

Since PMD (3.266) is non-singular, we have $\det a(0) \neq 0$. Thus $\det \alpha(0) \neq 0$. ■

(8) On basis of the above investigations, we are able to answer the question about the stabilizability of PMD B-processes and the construction of the set of dedicated stabilizing B-controllers. In the most important case for applications, when condition (3.318) is true, the corresponding result is as follows.

**Theorem 3.39** *Let the strictly causal PMD*

$$\Pi_2(\zeta) = (a(\zeta), \zeta b_3(\zeta), c(\zeta), \zeta d_1(\zeta)), \tag{3.321}$$

*be non-singular and minimal, and, in analogy to (3.308), we have*

$$\begin{aligned} a(\zeta) &= \psi_1(\zeta) a_2(\zeta), \quad \zeta b_3(\zeta) = \psi_1(\zeta) \zeta b_4(\zeta), \\ a_2(\zeta) &= a_3(\zeta) \psi_2(\zeta), \quad c(\zeta) = c_2(\zeta) \psi_2(\zeta), \end{aligned} \tag{3.322}$$

*where $\psi_1(\zeta)$ and $\psi_2(\zeta)$ are GCLD and GCRD, respectively, such that the pairs $(a_3(\zeta), b_4(\zeta))$ and $[a_3(\zeta), c_2(\zeta)]$ are irreducible. Then the following statements hold:*

*(i) For the stabilizability of the PMD (3.321), it is necessary and sufficient that the matrices $\psi_1(\zeta)$ and $\psi_2(\zeta)$ are stable.*

*(ii) When (i) is true, then for the polynomial matrix $b_4(\zeta)$ configured in (3.322), the representation*

$$b_4(\zeta) = \zeta b_5(\zeta) \tag{3.323}$$

*is possible, where $b_5(\zeta)$ is a polynomial matrix. Then, the set of stabilizing controllers coincides with the set of solutions of the DPE*

$$\det \begin{bmatrix} a_3(\zeta) & 0_{pn} & -\zeta b_5(\zeta) \\ -c_2(\zeta) & I_n & \zeta d_1(\zeta) \\ 0_{mp} & -\beta(\zeta) & \alpha(\zeta) \end{bmatrix} \sim \Delta_+(\zeta), \tag{3.324}$$

*where $\Delta_+(\zeta)$ is any stable polynomial.*

*(iii) All stabilizing controllers are causal.*

*(iv) Independent on the concrete selection of the polynomial $\Delta_+(\zeta)$, the characteristic polynomial of the closed loop $\Delta_\Pi(\zeta)$ (3.270) satisfies*

$$\Delta_\Pi(\zeta) \stackrel{\triangle}{=} \det \begin{bmatrix} a_3(\zeta) & 0_{pn} & -\zeta b_3(\zeta) \\ -c(\zeta) & I_n & -\zeta d_1(\zeta) \\ 0_{mp} & -\beta(\zeta) & \alpha(\zeta) \end{bmatrix} \sim \det \psi_1(\zeta) \det \psi_2(\zeta) \Delta_+(\zeta). \tag{3.325}$$

*Proof*

(i) This statement is a consequence of Theorem 3.37.

(ii) For $\zeta = 0$, from (3.322), we obtain

$$\psi_1(0) b_4(0) = 0_{pm}. \tag{3.326}$$

When (i) is true, then $\det \psi_1(0) \neq 0$, because the matrix $\psi_1(\zeta)$ is stable. Thus $b_4(0) = 0_{pm}$, which is equivalent to (3.323).

(iii) This claim is a consequence from Theorem 3.38.

(iv) Formula (3.325) follows from formula (3.307). ■

## 3.13 Causal Modal Control for SISO Discrete Forward Models

(1) This section provides a direct solution of the modal control problem for SISO discrete forward models, which does not utilize the transfer to the associated discrete backward model.

Let the dynamics of a one dimensional linear process be described by the scalar operator equation of form (3.183)

$$\tilde{a}(z)y_k = \tilde{b}(z)u_k, \tag{3.327}$$

where $z$ is the shift operator for one step ahead. Moreover, in (3.327) the quantities $\tilde{a}(z)$, $\tilde{b}(z)$ are coprime polynomials. As earlier, the forward process $(\tilde{a}(z), \tilde{b}(z)$ is called strictly causal, if

$$\deg \tilde{a}(z) > \deg \tilde{b}(z). \tag{3.328}$$

Beside process (3.327), we consider an object described by the equation

$$\tilde{\alpha}(z)u_k = \tilde{\beta}(z)y_k, \tag{3.329}$$

where $\tilde{\alpha}(z)$, $\tilde{\beta}(z)$ are polynomials. Below, (3.329) is called controller equation, and the pair $(\tilde{\alpha}(z), \tilde{\beta}(z))$ is called controller. In accordance with the above consideration, the controller $(\tilde{\alpha}(z), \tilde{\beta}(z))$ is called causal, if

$$\deg \tilde{\alpha}(z) \geq \deg \tilde{\beta}(z). \tag{3.330}$$

Together, Eqs. (3.327) and (3.329) define the operational equations of the closed loop

$$\begin{aligned} \tilde{a}(z)y_k - \tilde{b}(z)u_k &= 0, \\ -\tilde{\beta}(z)y_k + \tilde{\alpha}(z)u_k &= 0. \end{aligned} \tag{3.331}$$

The dynamical properties of the closed loop are mainly determined by the properties of the characteristic polynomial

$$\tilde{\Delta}_0(z) \triangleq \det \begin{bmatrix} \tilde{a}(z) & -\tilde{b}(z) \\ -\tilde{\beta}(z) & \tilde{\alpha}(z) \end{bmatrix}. \tag{3.332}$$

The roots of the polynomial $\tilde{\Delta}_0(z)$ are called the poles of the closed system.

Let $\tilde{\Delta}(z)$ be a certain desired characteristic polynomial of the closed system. Then, the modal control problem for process (3.327) consists in the construction of the set of causal controllers $(\tilde{\alpha}(z), \tilde{\beta}(z))$, which satisfy the equality

$$\det \begin{bmatrix} \tilde{a}(z) & -\tilde{b}(z) \\ -\tilde{\beta}(z) & \tilde{\alpha}(z) \end{bmatrix} = \tilde{\Delta}(z). \tag{3.333}$$

Relation (3.333), for a given process $(\tilde{a}(z), \tilde{b}(z))$ can be considered an equation depending on the polynomials $\tilde{\alpha}(z)$, $\tilde{\beta}(z)$ that fulfills (3.330). Evaluating the determinant in (3.333), we find the Diophantine polynomial equation [2]

$$\tilde{a}(z)\tilde{\alpha}(z) - \tilde{b}(z)\tilde{\beta}(z) = \tilde{\Delta}(z), \tag{3.334}$$

and the considered causal pole placement problem can be formulated as follows:

**Modal control problem for SISO forward process:** Let be given a strictly causal process $(\tilde{a}(z), \tilde{b}(z)$ and a polynomial $\tilde{\Delta}(z)$. Find the set of solutions of Eq. (3.334) satisfying condition (3.330).

Let $\tilde{\alpha}_0(z)$, $\tilde{\beta}_0(z)$ be any solution of Eq. (3.334). Then the set of all solutions of this equation is determined by the relations [2]

$$\begin{aligned}\tilde{\alpha}(z) &= \tilde{\alpha}_0(z) + \tilde{b}(z)\tilde{\xi}(z),\\ \tilde{\beta}(z) &= \tilde{\beta}_0(z) + \tilde{a}(z)\tilde{\xi}(z),\end{aligned} \tag{3.335}$$

where $\tilde{\xi}(z) = 0$ or any polynomial.

Hence, the considered causal modal control problem leads to selecting from set of controllers (3.335) the subset of controllers satisfying the causality condition (3.330).

We will show that set of controllers (3.335) always contains non-causal controllers. Indeed, from (3.335) we find

$$\begin{aligned}\deg \tilde{\alpha}(z) &= \deg \left[\tilde{\alpha}_0(z) + \tilde{b}(z)\tilde{\xi}(z)\right],\\ \deg \tilde{\beta}(z) &= \deg \left[\tilde{\beta}_0(z) + \tilde{a}(z)\tilde{\xi}(z)\right].\end{aligned} \tag{3.336}$$

For sufficiently large values of $\deg \xi(z)$, we obtain

$$\begin{aligned}\deg \tilde{\alpha}_0(z) &< \deg \left[\tilde{b}(z)\tilde{\xi}(z)\right],\\ \deg \tilde{\beta}_0(z) &< \deg \left[\tilde{a}(z)\tilde{\xi}(z)\right].\end{aligned} \tag{3.337}$$

Thus

$$\begin{aligned}\deg \tilde{\alpha}(z) &= \deg \left[\tilde{b}(z)\tilde{\xi}(z)\right] = \deg \tilde{b}(z) + \deg \tilde{\xi}(z),\\ \deg \tilde{\beta}(z) &= \deg \left[\tilde{a}(z)\tilde{\xi}(z)\right] = \deg \tilde{a}(z) + \deg \tilde{\xi}(z).\end{aligned} \tag{3.338}$$

Due to the strict properness of the process, from (3.338) we obtain $\deg \tilde{\alpha}(z) < \deg \tilde{\beta}(z)$, i.e., for sufficiently large values of $\deg \tilde{\xi}(z)$ all controllers are non-causal.

(2) Introduce the rational function

$$\tilde{W}(z) = \frac{\tilde{\Delta}(z)}{\tilde{a}(z)\tilde{b}(z)}. \tag{3.339}$$

Since the polynomials $\tilde{a}(z)$, $\tilde{b}(z)$ are coprime, there exists the separation

$$\tilde{W}(z) = \frac{\tilde{\Delta}(z)}{\tilde{a}(z)\tilde{b}(z)} = \frac{\tilde{\alpha}(z)}{\tilde{b}(z)} - \frac{\tilde{\beta}(z)}{\tilde{a}(z)}, \tag{3.340}$$

where $\tilde{\alpha}(z)$, $\tilde{\beta}(z)$ are polynomials. We immediately verify that any pair of polynomials $\tilde{\alpha}(z)$, $\tilde{\beta}(z)$ configured in (3.340) defines a solution of Eq. (3.334). Also the reverse is true: any pair of polynomials as solution of Eq. (3.334), corresponds to a separation of form (3.340). Among all possible separations (3.340), there exists a unique one, in which the polynomials $\tilde{\alpha}(z) = \tilde{\alpha}^*(z)$, $\tilde{\beta}(z) = \tilde{\beta}^*(z)$ fulfill the condition

$$\tilde{W}(z) = \frac{\tilde{\alpha}^*(z)}{\tilde{b}(z)} - \frac{\tilde{\beta}^*(z)}{\tilde{a}(z)} \tag{3.341}$$

and, moreover,

$$\deg \tilde{\beta}^*(z) < \deg \tilde{a}(z). \tag{3.342}$$

The so constructed solution $\tilde{\alpha}^*(z)$, $\tilde{\beta}^*(z)$ of Eq. (3.334) is called $\beta$-minimal controller, and it is denoted by $\left(\tilde{\alpha}^*(z), \tilde{\beta}^*(z)\right)$.

Introduce the constants

$$\begin{aligned} D_1 &= \deg \tilde{\Delta}(z) - \deg \tilde{a}(z), \\ D_2 &= \deg \tilde{\Delta}(z) - \deg \tilde{a}(z) - \deg \tilde{b}(z), \\ D_3 &= \deg \tilde{\Delta}(z) - 2 \deg \tilde{a}(z). \end{aligned} \tag{3.343}$$

Under the taken assumptions, we have $D_1 \geq D_2 > D_3$.

Using notations (3.343), we can formulate claims, characterizing the properties of the $\beta$-minimal controller.

**Theorem 3.40** *The following claims hold:*

*(i) For*

$$D_1 < 0 \tag{3.344}$$

*the $\beta$-minimal controller is non-causal.*

*(ii) For*

$$D_1 \geq 0, \quad D_3 < 0 \tag{3.345}$$

*the $\beta$-minimal controller can be causal or non-causal depending on the selection of the polynomial $\tilde{\Delta}(z)$.*

*(iii) For*

$$D_3 \geq 0 \tag{3.346}$$

*the $\beta$-minimal controller is causal.*

*Proof*

(i) Let $(\tilde{\alpha}(z), \tilde{\beta}(z))$ be any causal solution of Eq. (3.334). Since $\deg \tilde{a}(z) > \deg \tilde{b}(z)$ and $\deg \tilde{\alpha}(z) \geq \deg \tilde{\beta}(z)$, we obtain

$$\deg \left[\tilde{a}(z)\tilde{\alpha}(z)\right] > \deg \left[\tilde{b}(z)\tilde{\beta}(z)\right], \tag{3.347}$$

and from (3.334), we find

$$\deg \left[\tilde{a}(z)\tilde{\alpha}(z)\right] = \deg \tilde{a}(z) + \deg \tilde{\alpha}(z) = \deg \tilde{\Delta}(z). \tag{3.348}$$

Hence, any causal controller $(\tilde{\alpha}(z), \tilde{\beta}(z))$ satisfies $D_1 = \deg \tilde{\Delta}(z) - \deg \tilde{a}(z) \geq 0$, because $\deg \tilde{\alpha}(z) \geq 0$. Therefore, when (3.344) is true, then Eq. (3.334) in general does not possess causal solutions. Thus, the $\beta$-minimal controller as a special solution of Eq. (3.334) is also non-causal.

(ii) For the proof of this claim notice that under condition (3.345) two cases are possible:

$$D_1 \geq 0, \quad D_2 < 0, \tag{3.349}$$

and also

$$D_2 \geq 0, \quad D_3 < 0. \tag{3.350}$$

In order to avoid cumbersome constructions, we verify claim ii) at hand of an example.

For case (3.349) consider the equation

$$(z+1)^5\tilde{\alpha}(z) - (z-2)^2\tilde{\beta}(z) = (z+1)^5 z - (z-2)^2(z+2)^k, \tag{3.351}$$

where $k \geq 0$ is an integer. For $0 \leq k \leq 3$, we obtain $\deg \tilde{\Delta}(z) = 6$, $\deg \tilde{a}(z) = 5$, $\deg \tilde{b}(z) = 2$, i.e., $D_1 = 1$, $D_2 = -1$. Under this condition, the $\beta$-minimal controller of Eq. (3.351) takes the form

$$\tilde{\alpha}^*(z) = z, \quad \tilde{\beta}^*(z) = (z+2)^k. \tag{3.352}$$

Hence, for $0 \leq k \leq 1$ the constructed $\beta$-minimal controller is causal, but for $2 \leq k \leq 3$ it is non-causal.

Analogously, for the illustration of case (3.350), consider the equation

$$(z-1)^7\tilde{\alpha}(z) - (z-2)^2\tilde{\beta}(z) = (z-1)^7 z^3 - (z-2)^2(z+1)^k, \tag{3.353}$$

where $k$ is again a nonnegative integer. For $0 \leq k \leq 6$, we obtain $\deg \tilde{\Delta}(z) = 10$, $\deg \tilde{a}(z) = 7$, $\deg \tilde{b}(z) = 2$, and therefore $D_2 = 1$, $D_3 = -4$.

Then the $\beta$-minimal controller becomes

$$\tilde{\alpha}^*(z) = z^3, \quad \tilde{\beta}^*(z) = (z+1)^k.$$

This controller is causal for $0 \leq k \leq 3$, but non-causal for $4 \leq k \leq 6$.

(iii) For the proof of this claim, at first notice that for $D_3 \geq 0$, we have

$$D_2 > 0, \tag{3.354}$$

because of

$$D_2 - D_3 = \deg \tilde{a}(z) - \deg \tilde{b}(z) > 0.$$

When condition (3.354) holds, then

$$\deg \tilde{\alpha}^*(z) = \deg \tilde{\Delta}(z) - \deg \tilde{a}(z) = D_1. \tag{3.355}$$

Indeed, since the second term on the right side of (3.341) is a strictly proper fraction, so owing to (3.339), we obtain

$$\deg \tilde{\Delta}(z) - \deg \tilde{a}(z) - \deg \tilde{b}(z) = \deg \tilde{\alpha}^*(z) - \deg \tilde{b}(z),$$

which is equivalent to (3.355). At the same time, by definition of the $\beta$-minimal controller, we find

$$\deg \tilde{\beta}^*(z) = \deg \tilde{a}(z) - \rho, \tag{3.356}$$

where $\rho$ is a nonnegative integer. From (3.355) and (3.356), we find

$$\deg \tilde{\alpha}^*(z) - \deg \tilde{\beta}^*(z) = D_3 + \rho > 0,$$

which was to be shown. ■

(3) The next statements provide the complete solution for the construction of the set of causal solutions of Eq. (3.344).

**Theorem 3.41** *The following statements are true:*

*(i) Under condition (3.344), Eq. (3.334) does not possess causal solutions.*

*(ii) When condition (3.345) holds, then the set causal solutions of Eq. (3.334) consists of the $\beta$-minimal controller, if this controller is causal, and it is empty, if this controller is non-causal.*

*(iii) When condition (3.346) is true, then there exists a set of causal solutions of Eq. (3.334), which is determined by the relations*

$$\begin{aligned} \tilde{\alpha}(z) &= \tilde{\alpha}^*(z) + \tilde{b}(z)\tilde{\xi}(z), \\ \tilde{\beta}(z) &= \tilde{\beta}^*(z) + \tilde{a}(z)\tilde{\xi}(z), \end{aligned} \tag{3.357}$$

*where $(\tilde{\alpha}^*(z), \tilde{\beta}^*(z))$ is the $\beta$-minimal controller, and $\tilde{\xi}(z)$ is zero or any polynomial, which satisfies*

$$0 \leq \deg \tilde{\xi}(z) \leq D_3. \tag{3.358}$$

*Proof* (i) This fact follows from the proof of claim (i) in Theorem 3.40.
(ii) Assume (3.349). Then fraction (3.339) is strictly proper. In this case both fractions on the right side of (3.341) are strictly proper, so that we obtain

$$\deg \tilde{\alpha}^*(z) \le \deg \tilde{b}(z) - 1, \quad \deg \tilde{\beta}^*(z) \le \deg \tilde{a}(z) - 1. \tag{3.359}$$

Choose $\tilde{\alpha}_0(z) = \tilde{\alpha}^*(z), \tilde{\beta}_0(z) = \tilde{\beta}^*(z)$, and write general solution (3.335) in the form

$$\begin{aligned} \tilde{\alpha}(z) &= \tilde{\alpha}^*(z) + \tilde{b}(z)\tilde{\xi}(z), \\ \tilde{\beta}(z) &= \tilde{\beta}^*(z) + \tilde{a}(z)\tilde{\xi}(z). \end{aligned} \tag{3.360}$$

Then, owing to (3.359), we obtain for $\tilde{\xi}(z) \ne 0$

$$\begin{aligned} \deg \tilde{\alpha}(z) &= \deg \tilde{b}(z) + \deg \tilde{\xi}(z), \\ \deg \tilde{\beta}(z) &= \deg \tilde{a}(z) + \deg \tilde{\xi}(z). \end{aligned} \tag{3.361}$$

Since $\deg \tilde{a}(z) > \deg \tilde{b}(z)$, we find that for $\tilde{\xi}(z) \ne 0$ all controllers (3.360) are non-causal. For $\tilde{\xi}(z) = 0$, formulae (3.360) yield the $\beta$-minimal controller, which, on basis of Theorem 3.40, in the given situation can be causal or non-causal.

In order to consider case (3.350), generate the dependence of the quantity $\deg \tilde{\alpha}(z)$ on the value $\deg \tilde{\xi}(z)$. Notice that the first equation in (3.360) implies

$$\deg \tilde{\alpha}(z) = \deg[\tilde{\alpha}^*(z) + \tilde{b}(z)\tilde{\xi}(z)]. \tag{3.362}$$

Since $D_2 \ge 0$, in the given case relation (3.360) becomes

$$\deg \tilde{\alpha}^*(z) = \deg \tilde{\Delta}(z) - \deg \tilde{a}(z). \tag{3.363}$$

Applying this formula, we find

$$\begin{aligned} \deg \tilde{\alpha}^*(z) - \deg[\tilde{b}(z)\tilde{\xi}(z)] &= \deg \tilde{\Delta}(z) - \deg \tilde{a}(z) - \deg \tilde{b}(z) - \deg \tilde{\xi}(z) \\ &= D_2 - \deg \tilde{\xi}(z). \end{aligned} \tag{3.364}$$

Hence for

$$0 < \deg \tilde{\xi}(z) < \deg \tilde{\Delta}(z) - \deg \tilde{a}(z) - \deg \tilde{b}(z) = D_2, \tag{3.365}$$

we obtain

$$\deg \tilde{\alpha}^*(z) > \deg[\tilde{b}(z)\tilde{\xi}(z)], \tag{3.366}$$

and finally, due to (3.362)

$$\deg \tilde{\alpha}(z) = \deg \tilde{\alpha}^*(z) = \deg \tilde{\Delta}(z) - \deg \tilde{a}(z) = D_1. \tag{3.367}$$

For $\deg \tilde{\xi}(z) = D_2$, from (3.364), we find

$$\deg \tilde{\alpha}^*(z) = \deg[\tilde{b}(z)\tilde{\xi}(z)] = D_1.$$

Thus, owing to (3.362)

$$\deg \tilde{\alpha}(z) = \deg[\tilde{\alpha}^*(z) + \tilde{b}(z)\tilde{\xi}(z)] \leq D_1. \tag{3.368}$$

Finally, we arrive at the relations

$$\begin{aligned} &\deg \tilde{\alpha}(z) = D_1, \quad 0 \leq \deg \tilde{\xi}(z) < D_2, \\ &\deg \tilde{\alpha}(z) \leq D_1, \quad \deg \tilde{\xi}(z) = D_2. \end{aligned} \tag{3.369}$$

For

$$\deg \tilde{\xi}(z) > D_2, \tag{3.370}$$

we calculate

$$\deg(\tilde{b}(z)\tilde{\xi}(z)) > \deg \tilde{\Delta}(z) - \deg \tilde{a}(z) = \deg \tilde{\alpha}^*(z), \tag{3.371}$$

and from (3.362), we obtain

$$\deg \tilde{\alpha}(z) = \deg \tilde{b}(z) + \deg \tilde{\xi}(z). \tag{3.372}$$

Applying (3.367), (3.369) and (3.370), (3.372), we are able to construct the dependence of the quantity $\deg \tilde{\alpha}(z)$ on the quantity $\deg \tilde{\xi}(z)$ for any solution of Eq. (3.357) for $\tilde{\xi}(z) \neq 0$

$$\begin{aligned} \deg \tilde{\alpha}(z) = D_1, &\quad 0 \leq \deg \tilde{\xi}(z) < D_2, \\ \deg \tilde{\alpha}(z) \leq D_1, &\quad \deg \tilde{\xi}(z) = D_2, \\ \deg \tilde{\alpha}(z) = \deg \tilde{b}(z) + \deg \tilde{\xi}(z), &\quad \deg \tilde{\xi}(z) > D_2. \end{aligned} \tag{3.373}$$

This means that the integers $\deg \tilde{\alpha}(z)$ are located for $0 \leq \deg \tilde{\xi}(z) < D_1$, and $\deg \tilde{\xi}(z) > D_1$, on the folded line in Fig. 3.1. Hereby, the integers $\deg \tilde{\beta}(z)$ are located on the straight line

$$\deg \tilde{\beta}(z) = \deg \tilde{a}(z) + \deg \tilde{\xi}(z). \tag{3.374}$$

in Fig. 3.1.

Under condition (3.350), we obtain

$$D_3 = \deg \tilde{\Delta}(z) - 2\deg \tilde{a}(z) < 0, \tag{3.375}$$

i.e.,

$$D_1 = \deg \tilde{\Delta}(z) - \deg \tilde{a}(z) < \deg \tilde{a}(z). \tag{3.376}$$

The last inequality says that for $\tilde{\xi}(z) \neq 0$, the straight line, corresponding to Eq. (3.374), is located above the folded line related to (3.373). This statement means that in case (3.350) for $\tilde{\xi}(z) \neq 0$, there are no causal solutions in set controllers (3.357). This result proves claim (ii).

Figure 3.1 shows the folded line according to (3.373) for the case $\deg \tilde{\Delta}(z) = 7$, $\deg \tilde{a}(z) = 4$, $\deg \tilde{b}(z) = 2$. Hence $D_1 = 3$, $D_2 = 1$, $D_3 = -1$. Moreover, Fig. 3.1 shows the straight line (3.374) $\deg \tilde{\beta}(z) = 4 + \deg \tilde{\xi}(z)$.

(iii) Under condition (3.346) straight line (3.374) intersects folded line (3.373) at

$$\deg \tilde{a}(z) + \deg \tilde{\xi}(z) = D_1 = \deg \tilde{\Delta}(z) - \deg \tilde{a}(z). \tag{3.377}$$

This means that in the point of intersection

$$\deg \tilde{\xi}(z) = \deg \tilde{\Delta}(z) - 2\deg \tilde{a}(z) = D_3. \tag{3.378}$$

Then for $0 \leq \deg \tilde{\xi}(z) \leq D_3$, we obtain

$$\deg \tilde{\beta}(z) \leq \deg \tilde{\alpha}(z), \tag{3.379}$$

i.e., solution (3.357) is causal. At the same time, for $\deg \tilde{\xi}(z) > D_3$

$$\deg \tilde{\beta}(z) > \deg \tilde{\alpha}(z), \tag{3.380}$$

and the corresponding controller (3.357) becomes non-causal. Finally, notice that due to Theorem 3.40, when (3.346) is true and for $\tilde{\xi}(z) = 0$ the related $\beta$-minimal controller $(\tilde{\alpha}^*(z), \tilde{\beta}^*(z))$ is causal, so claim (iii) is proven.

In Fig. 3.2 the situation of case (iii) is shown for the values $\deg \tilde{\Delta}(z) = 12$, $\deg \tilde{a}(z) = 4$, $\deg \tilde{b}(z) = 2$, i.e., $D_1 = 8$, $D_2 = 6$, $D_3 = 4$. ∎

As a consequence of the fundamental theorem, we formulate the following corollary, which is convenient for the solution of application problems.

**Corollary 3.8** *Under condition (3.328), Eq. (3.334) possesses causal solutions, if and only if the $\beta$-minimal controller is causal.*

(4) As an example, let us consider the causal modal control problem for the discrete model of a double integrator with input delay, which is described by the differential-difference equation

$$\frac{\mathrm{d}^2 y(t)}{\mathrm{d}t^2} = u(t - \tau). \tag{3.381}$$

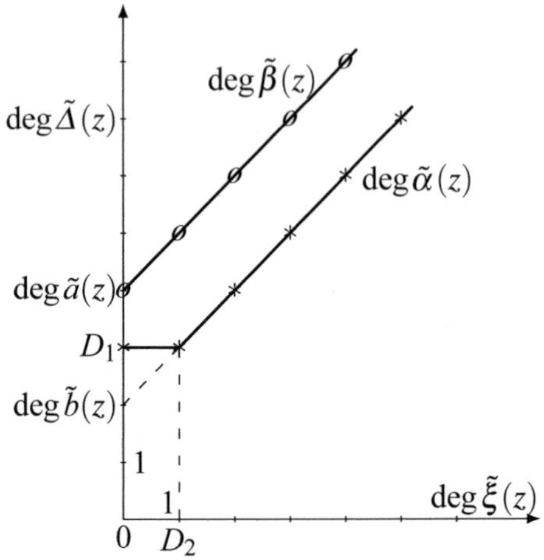


**Fig. 3.1** Controller numerator and denominator degrees over degree of parameter $\tilde{\xi}(z)$, case (ii)

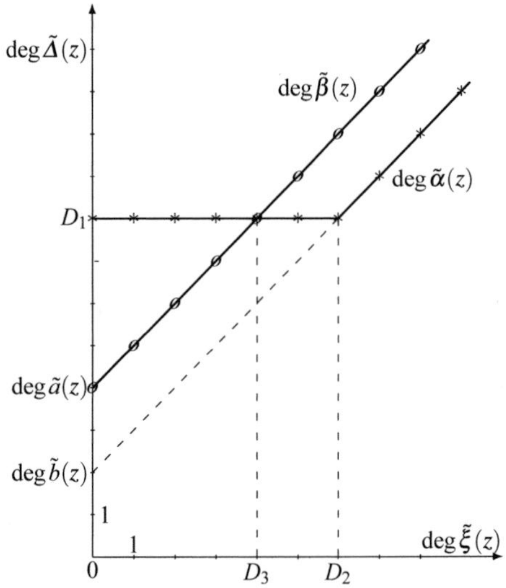


**Fig. 3.2** Controller numerator and denominator degrees over degree of parameter $\tilde{\xi}(z)$, case (iii)

As was shown in [2], when using a zero-order hold, sampling period $T = 1$ and delay $\tau = 0, 5$, the transfer function of the forward discrete process model becomes

$$\tilde{W}(z) = \frac{\tilde{b}(z)}{\tilde{a}(z)}, \tag{3.382}$$

where

$$\tilde{a}(z) = z(z-1)^2, \quad \tilde{b}(z) = 0.125(z^2 + 6z + 1). \tag{3.383}$$

The causal modal control problem consists in seeking out the set of causal solutions of the Diophantine polynomial equation

$$z(z-1)^2\tilde{\alpha}(z) - 0.125(z^2+6z+1)\tilde{\beta}(z) = \tilde{\Delta}(z), \tag{3.384}$$

for a given polynomial $\tilde{\Delta}(z)$.

Since in the given case $\deg \tilde{a}(z) = 3$, $\deg \tilde{b}(z) = 2$, so according to (3.343), we obtain

$$\begin{aligned} D_1 &= \deg \tilde{\Delta}(z) - 3, \\ D_2 &= \deg \tilde{\Delta}(z) - 5, \\ D_3 &= \deg \tilde{\Delta}(z) - 6. \end{aligned} \tag{3.385}$$

It follows from Theorem 3.40, that for $\deg \tilde{\Delta}(z) < 3$, Eq. (3.384) does not possess causal solutions. For

$$3 \le \deg \tilde{\Delta}(z) < 6, \tag{3.386}$$

according to Theorem 3.41, the solution of the causal modal control problem is determined by the properties of the separation

$$\tilde{W}(z) = \frac{\tilde{\alpha}^*(z)}{0,125(z^2+6z+1)} - \frac{\tilde{\beta}^*(z)}{z(z-1)^2}, \tag{3.387}$$

where $\deg \tilde{\beta}^*(z) \le 2$. When we obtain $\deg \tilde{\alpha}^*(z) \ge \deg \tilde{\beta}^*(z)$, then the controller $(\tilde{\alpha}^*(z), \tilde{\beta}^*(z))$ defines the unique causal solution of Eq. (3.384). If, however $\deg \alpha^*(z) < \deg \tilde{\beta}^*(z)$, then a causal solution of Eq. (3.384) does not exist. For $\deg \tilde{\Delta}(z) \ge 6$, the $\beta$-minimal controller $(\tilde{\alpha}^*(z), \tilde{\beta}^*(z))$ is causal owing to Theorems 3.40, and 3.41 yields the existence of a set of causal solutions of Eq. (3.384), which is determined by the relations

$$\begin{aligned} \alpha(z) &= \alpha^*(z) + 0.125(z^2+6z+1)\tilde{\xi}(z), \\ \beta(z) &= \beta^*(z) + z(z-1)^2\tilde{\xi}(z), \end{aligned} \tag{3.388}$$

where $\tilde{\xi}(z)$ is zero or any polynomial satisfying $0 \le \deg \tilde{\xi}(z) \le \deg \tilde{\Delta}(z) - 6$.

(5) Using Theorems 3.40 and 3.41, we can obtain additional results with respect to the causal modal control and stabilization of the strictly causal forward process $(\tilde{a}(z), \tilde{b}(z))$.

**Definition 3.17** The polynomial $\tilde{\Delta}(z)$ is called admissible for the strictly causal process $(\tilde{a}(z), \tilde{b}(z))$, if the Diophantine polynomial equation

$$\tilde{a}(z)\tilde{\alpha}(z) - \tilde{b}(z)\tilde{\beta}(z) = \tilde{\Delta}(z), \tag{3.389}$$

possesses at least one causal solution, i.e., a solution satisfying $\deg \tilde{\alpha}(z) \geq \deg \tilde{\beta}(z)$.

**Theorem 3.42** *The following statements hold:*

(i) *There do not exist admissible polynomials for the process* $(\tilde{a}(z), \tilde{b}(z))$, *when*

$$\deg \tilde{\Delta}(z) < \deg \tilde{a}(z). \tag{3.390}$$

(ii) *For*

$$\deg \tilde{a}(z) \leq \deg \tilde{\Delta}(z) < 2 \deg \tilde{a}(z), \tag{3.391}$$

*there exists a set of admissible polynomials* $\mathcal{M}_d$. *Hereby, for* $\tilde{\Delta}(z) \in \mathcal{M}_d$ *it is necessary and sufficient that the polynomial* $\tilde{\Delta}(z)$ *allows a representation of the form*

$$\tilde{\Delta}(z) = \tilde{a}(z)\tilde{\alpha}^*(z) - \tilde{b}(z)\tilde{\beta}^*(z), \tag{3.392}$$

*where* $\tilde{\alpha}^*(z)$, $\tilde{\beta}^*(z)$ *are any polynomials satisfying*

$$\begin{aligned} &\deg \tilde{\alpha}^*(z) = \deg \tilde{\Delta}(z) - \deg \tilde{a}(z), \\ &\deg \tilde{\beta}^*(z) \leq \deg \tilde{\alpha}^*(z), \quad \deg \tilde{\beta}^*(z) < \deg \tilde{a}(z). \end{aligned} \tag{3.393}$$

(iii) *For*

$$\deg \tilde{\Delta}(z) \geq 2 \deg \tilde{a}(z), \tag{3.394}$$

*any polynomial* $\tilde{\Delta}(z)$ *is admissible and allows a representation of form (3.392).*

(iv) *For any polynomial satisfying conditions (3.392), (3.393), or (3.394), representation (3.392) is unique.*

*Proof*

(i) This claim is an implication of Theorem 3.41, because under condition (3.390), Eq. (3.389) does not possess causal solutions.

(ii) Necessity. Let condition (3.391) be fulfilled, and polynomial $\tilde{\Delta}(z)$ be admissible. Then, owing to Theorem 3.41, there exists a unique causal solution of Eq. (3.389), which coincides with the related $\beta$-minimal controller. Hereby, since the causal $\beta$-minimal controller satisfies condition (3.393), the polynomial $\tilde{\Delta}(z)$ can be written in form (3.392).
Sufficiency. Let the polynomial $\tilde{\Delta}(z)$ be representable in form (3.392), where the polynomials $\alpha^*(z)$, $\beta^*(z)$ satisfy (3.393). Then, we can build the separation

$$\frac{\tilde{\Delta}(z)}{\tilde{a}(z)\tilde{b}(z)} = \frac{\tilde{\alpha}^*(z)}{\tilde{b}(z)} - \frac{\tilde{\beta}^*(z)}{\tilde{a}(z)}, \tag{3.395}$$

from which due to (3.393) it follows that $(\tilde{\alpha}^*(z), \tilde{\beta}^*(z))$ is a causal $\beta$-minimal solution of Eq. (3.389). Thus, polynomial (3.392) under condition (3.393) is admissible.

(iii) This claim can be proven with the help of Theorem 3.40 in analogy to (ii).
(iv) The uniqueness of representations (3.392), (3.393) follows form the uniqueness of separation (3.395), when $\deg \tilde{\beta}^*(z) < \deg \tilde{a}(z)$.

*Example 3.2* As an example for the application of Theorem 4.2 consider the construction problem for the set of admissible polynomials for the double integrator, when

$$\tilde{a}(z) = (z-1)^2, \quad \tilde{b}(z) = \frac{T^2}{2}(z+1). \tag{3.396}$$

Without loss of generality, we assume that the polynomial $\tilde{\Delta}(z)$ is monic. Assume $\deg \tilde{\Delta}(z) = 2$. Then $\deg \tilde{\Delta}(z) - \deg \tilde{a}(z) = 0$ and condition (3.393) can only be fulfilled for $\tilde{\alpha}^*(z) = \alpha_0 = \text{const.}$, $\tilde{\beta}^*(z) = \beta_0 = \text{const.}$ Moreover, from (3.393) and (3.396) we obtain

$$\Delta(z) = (z^2 - 2z + 1)\alpha_0 - \frac{T^2}{2}(z+1)\beta_0. \tag{3.397}$$

Since by proposition the polynomial $\tilde{\Delta}(z)$ is monic, so $\alpha_0 = 1$. Consequently, the set of admissible polynomials of second degree for the double integrator is determined by

$$\tilde{\Delta}(z) = z^2 - \left(\frac{T^2}{2}\beta_0 + 2\right)z + 1 - \frac{T^2}{2}\beta_0, \tag{3.398}$$

where $\beta_0$ is any real parameter.

For $\deg \tilde{\Delta}(z) = 3$, by analog reasoning as above, we find that representation (3.392) has to be in harmony with (3.393)

$$\tilde{\alpha}^*(z) = z + \alpha_1, \quad \tilde{\beta}^*(z) = \beta_0 z + \beta_1, \tag{3.399}$$

where $\alpha_1$, $\beta_0$, $\beta_1$ are real parameters. Moreover, formula (3.392) permits a general representation of the set of admissible monic polynomials

$$\begin{aligned} \tilde{\Delta}(z) &= (z-1)^2(z+\alpha_1) - \frac{T^2}{2}(z+1)(\beta_0 z + \beta_1) \\ &= z^3 + z^2\left(-2 + \alpha_1 - \frac{T^2}{2}\beta_0\right) + z\left(1 - 2\alpha_1 - \frac{T^2}{2}\beta_0 - \frac{T^2}{2}\beta_1\right) + \alpha_1 - \frac{T^2}{2}\beta_1. \end{aligned} \tag{3.400}$$

Assume that the chosen polynomial $\tilde{\Delta}(z)$ has the form

$$\tilde{\Delta}(z) = z^3 + d_1 z^2 + d_2 z + d_3. \tag{3.401}$$

Comparing expressions (3.401) and (3.400), we achieve the system of linear equations

$$
\begin{aligned}
\alpha_1 - \frac{T^2}{2}\beta_0 &= d_1 + 2,\\
-2\alpha_1 - \frac{T^2}{2}\beta_0 - \frac{T^2}{2}\beta_1 &= d_2 - 1,\\
\alpha_1 - \frac{T^2}{2}\beta_1 &= d_3.
\end{aligned}
\tag{3.402}
$$

The determinant of the system of Eq. (3.402) is different from zero for $T^2 \neq 0$. Therefore, for any values of the polynomial coefficients (3.401), Eq. (3.402) possess the unique solution

$$
\begin{aligned}
\alpha_1 &= \frac{3 + d_1 - d_2 + d_3}{4},\\
\beta_0 &= \frac{-5 - 3d_1 - d_2 + d_3}{2T^2},\\
\beta_1 &= \frac{3 + d_1 - d_2 - 3d_3}{2T^2}.
\end{aligned}
\tag{3.403}
$$

Thus, any polynomial (3.401) can be written in form (3.392), where

$$
\begin{aligned}
\alpha^*(z) &= z + \frac{3 + d_1 - d_2 + d_3}{4},\\
\beta^*(z) &= -\frac{5 + 3d_1 + d_2 - d_3}{2T^2}z + \frac{3 + d_1 - d_2 - 3d_3}{2T^2}.
\end{aligned}
\tag{3.404}
$$

The obtained result means that any polynomial (3.401) can be presented in form (3.392), and therefore, it is admissible for the double integrator (3.396). Moreover, it follows from Theorem 3.42, that for any polynomial (3.401) the related polynomial Eq. (3.389) possesses a unique causal solution (3.404), which is $\beta$-minimal.

For $\deg \tilde{\Delta}(z) > 3$, with respect to Theorem 3.42, any polynomial $\tilde{\Delta}(z)$ is admissible. Hereby, there exists a set of causal solutions of the equation

$$
(z-1)^2\alpha(z) - \frac{T^2}{2}(z+1)\beta(z) = \tilde{\Delta}(z).
$$

This set is determined by formulae (3.357), (3.358).

(6) Applying Theorem 3.42, we derive additional results in connection with the stabilization problem for a strictly causal forward process.

**Definition 3.18** A stable admissible polynomial $\tilde{\Delta}(z)$ is called stabilizing for the process $(\tilde{a}(z), \tilde{b}(z))$.

The question about the existence of stabilizing polynomials can be solved by applying available stability criteria [2, 5] to the polynomial (3.392). Obviously, the process $(\tilde{a}(z), \tilde{b}(z))$ is stabilizable for a fixed value of $\deg \tilde{\Delta}(z)$, if and only if the corresponding set of polynomials (3.392) contains stable ones.

*Example 3.3* We will use the introduced concepts for the solution of the causal stabilization problem for the discrete forward model of the double integrator for a given degree of the characteristic polynomial $\tilde{\Delta}(z)$. Let $\deg \tilde{\Delta}(z) = 2$, then the set of admissible polynomials has the form (3.398) and the question about the existence of a stabilizing polynomial $\tilde{\Delta}(z)$ with $\deg \Delta(z) = 2$ leads to the question whether among the polynomials of form (3.398) there are stable ones.

Denote

$$d_1 = -\frac{T^2}{2}\beta_0 - 2, \quad d_2 = -\frac{T^2}{2}\beta_0 + 1. \tag{3.405}$$

Then the necessary and sufficient condition for the stability of polynomial (3.397) can be written in the form

$$d_2 < 1, \quad d_2 > -1 + d_1, \quad d_2 > -1 - d_1. \tag{3.406}$$

The first condition in (3.406), according to (3.405), is equivalent to $T^2\beta_0 > 0$. The last inequality in (3.406) with the help of (3.405) yields

$$-\frac{T^2}{2}\beta_0 + 1 > -1 + 2 + \frac{T^2}{2}\beta_0, \tag{3.407}$$

which is equivalent to $T^2\beta_0 < 0$. Since the first and the last condition in (3.406) contradict, we conclude that the set of polynomials (3.398) does not contain stable ones. This means that a double integrator cannot be stabilized by the use of a zero-order hold alone in the feedback. This assertion does not depend on the gain coefficient of the feedback or on the chosen sampling period $T$.

For $\deg \tilde{\Delta}(z) \geq 3$ the situation mutates. Indeed, as follows from the results of Example 3.2, in this case any stable polynomial $\tilde{\Delta}(z)$ becomes stabilizing. Hereby, in case $\deg \Delta(z) = 3$, where the characteristic polynomial of the closed loop has form (3.401), any stable polynomial is stabilizing. Moreover, there exists only one unique stabilizing controller (3.404), which is $\beta$-minimal. For $\deg \tilde{\Delta}(z) > 3$, there exists a set of stabilizing controllers determined by formulae (3.357), (3.358).

## References

1. E.N. Rosenwasser, B.P. Lampe, *Multivariable Computer-controlled Systems—A Transfer Function Approach* (Springer, London, 2006)
2. K.J. Åström, B. Wittenmark, *Computer Controlled Systems: Theory and Design*, 3rd edn. (Prentice-Hall, Englewood Cliffs, 1997)
3. F.R. Gantmacher, *The Theory of Matrices* (Chelsea, New York, 1959)
4. T. Kailath, *Linear Systems* (Prentice Hall, Englewood Cliffs, 1980)
5. J.S. Tsypkin, *Sampling Systems Theory* (Pergamon Press, New York, 1964)

# Part II
# Parametric Discrete Models of Multivariable Continous Processes and Mathematical Description of Standard SD Systems with Delay

# Chapter 4
# Modal Control and Stabilization of Multivariable Sampled-Data Systems with Delay

**Abstract** In this chapter, the results of Chap. 3 are applied to the discrete model of a multidimensional sampled-data (SD) system with delays at the input and output of a continuous process. The solution is based on the discrete backward model of the system, expressed by the inverse shift operator $\zeta$. Necessary and sufficient conditions for modal controllability and stabilization are given. At the end of the chapter, the construction of a minimal realization of the considered system in the state space is considered with the forward shift operator $z$, since the method traditionally used to solve this problem does not always lead to the desired results.

## 4.1 Linearized Model of Multivariable Digital Controllers (DC)

(1) The characteristic feature of a DC as an element of automatic control systems consists of the fact that it transforms continuous to discrete signals, and reversely, discrete to continuous signals. If the quantization of the magnitude is neglected during sampling and hold, then the simplified linear model of the DC can be represented by the structure shown in Fig. 4.1.

In Fig. 4.1, the following notations are used:

ADC is the analog/digital converter, which is a sampler, i.e., it transforms the continuous input $n \times 1$ vector signal $y(t)$ into the input $n \times 1$ vector sequence $\{\xi_k\}$.
CP is the control program, which generates the $q \times 1$ control vector sequence $\{\psi_k\}$ from the discrete-time input signal $\{\xi_k\}$.
DAC is the digital/analog converter, which is a hold element, i.e., it transforms the sequence $\{\psi_k\}$ into the continuous $m \times 1$ control vector signal $u(t)$.

Below, it is always supposed that the sequence $\{\xi_k\}$ is generated by the law

$$\xi_k = Ny(kT), \quad (k = 0, \pm 1, \ldots), \tag{4.1}$$



E. N. Rosenwasser et al., *Computer-Controlled Systems with Delay*,
https://doi.org/10.1007/978-3-030-15042-6_4

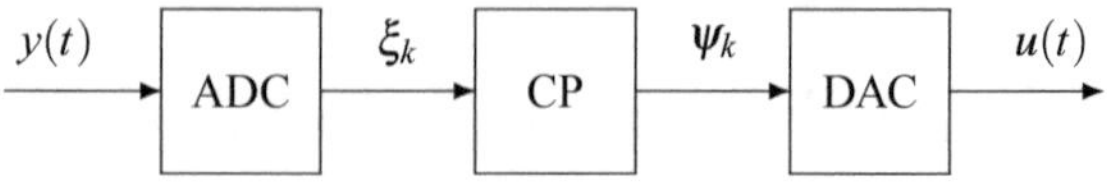


**Fig. 4.1** Linearized model of a digital controller (DC)

where $N$ is a constant $n \times n$ matrix and $T$ is the sampling period. Moreover, we assume that the input signal $y(t)$ is continuous at the sampling instants $t_k = kT$. This assumption imposes limitations on the structure of sampled-data systems and on the properties of their continuous elements.

(2) The control program (CP) shown in Fig. 4.1 is a finite dimensional discrete process generating the control sequence $\{\psi_k\}$ as solution of the linear difference equation

$$\alpha_0\psi_k + \alpha_1\psi_{k-1} + \cdots + \alpha_\rho\psi_{k-\rho} = \beta_0\xi_k + \beta_1\xi_{k-1} + \cdots + \beta_\rho\xi_{k-\rho}, \tag{4.2}$$

where $\alpha_i$, $\beta_i$ are constant matrices of size $q \times q$ and $q \times p$, respectively. When applying the backward shift operator $\zeta$, which has been introduced in (3.119), Eq. (4.2) can be written in the form

$$\alpha(\zeta)\psi_k = \beta(\zeta)\xi_k, \tag{4.3}$$

where according to (3.121), the quantities $\alpha(\zeta)$, $\beta(\zeta)$ are polynomial matrices of sizes $q \times q$ and $q \times p$, respectively, taking the form

$$\begin{aligned} \alpha(\zeta) &= \alpha_0 + \alpha_1\zeta + \cdots + \alpha_\rho\zeta^\rho, \\ \beta(\zeta) &= \beta_0 + \beta_1\zeta + \cdots + \beta_\rho\zeta^\rho . \end{aligned} \tag{4.4}$$

Using the terminology of Chap. 3, (4.2) is called backward (B) equation of the discrete controller, and the polynomial pair $(\alpha(\zeta), \beta(\zeta))$ is the discrete backward controller, or shortly, B-controller. Hereby, in the following investigations all concepts and results obtained in Chap. 3 in connection with B-models hold.

In many applications, and also in scientific literature, the equation of the discrete controller (4.2) is written in the form

$$\alpha_0\psi_{k+\rho} + \alpha_1\psi_{k+\rho-1} + \cdots + \alpha_\rho\psi_k = \beta_0\xi_{k+\rho} + \beta_1\xi_{k+\rho-1} + \cdots + \beta_\rho\xi_k. \tag{4.5}$$

When applying the forward shift operator $z$, which has been introduced in (3.182), then Eq. (4.5) can be written in the form

$$\tilde{\alpha}(z)\psi_k = \tilde{\beta}(z)\xi_k, \tag{4.6}$$

where $\tilde{\alpha}(z)$ and $\tilde{\beta}(z)$ are polynomial matrices of sizes $q \times q$ and $q \times p$, respectively, taking the form

$$\begin{aligned} \tilde{\alpha}(z) &= \alpha_0 z^\rho + \alpha_1 z^{\rho-1} + \cdots + \alpha_\rho, \\ \tilde{\beta}(z) &= \beta_0 z^\rho + \beta_1 z^{\rho-1} + \cdots + \beta_\rho . \end{aligned} \tag{4.7}$$

Equation (4.6) is called forward (F) equation of the discrete controller, and the polynomial pair $(\tilde{\alpha}(z), \tilde{\beta}(z))$ is a discrete forward controller (F-controller).

(3) Represent B-controller (4.2) in the form

$$\alpha_0\psi_k = \beta_0\xi_k + \cdots + \beta_\rho\xi_{k-\rho} - \alpha_1\psi_{k-1} - \cdots - \alpha_\rho\psi_{k-\rho}. \tag{4.8}$$

Assume

$$\det\alpha_0 \neq 0. \tag{4.9}$$

Then, the initial values $\psi_{k-1}, \ldots, \psi_{k-\rho}, \xi_{k-1}, \ldots, \xi_{k-\rho}$, together with the input sequence $\xi_k, \xi_{k+1}, \ldots$ uniquely determine the control sequence $\psi_k, \psi_{k+1}, \ldots$. However, if in (4.8) $\det\alpha_0 = 0$, then the construction of the sequence $\psi_k, \psi_{k+1}, \ldots$ is only possible for special combinations of the initial conditions of the input sequence $\xi_k, \xi_{k-1}, \ldots, \xi_{k-\rho}$. This case leads to algebra-difference equations, which are sometimes called descriptor systems. Obviously, for controllers that have to be realized in real time, this situation is inconvenient. Therefore, in our future investigations, we will always assume that in controller Eqs. (4.2), (4.5) condition (4.9) is fulfilled. In view of (4.4), this equation can be formulated as

$$\det\alpha(0) \neq 0. \tag{4.10}$$

**Definition 4.1** When the conditions (4.9), (4.10) hold, controllers (4.2) or (4.5) are called regular.

(4) When condition on regularity (4.10) is fulfilled, the matrices $\alpha(\zeta)$ and $\tilde{\alpha}(z)$ are invertible. Therefore, for regular B- and F-controllers the rational matrices

$$\begin{aligned} W_d(\zeta) &= \alpha^{-1}(\zeta)\beta(\zeta) \text{ and} \\ \tilde{W}_d(z) &= \tilde{\alpha}^{-1}(z)\tilde{\beta}(z) \end{aligned} \tag{4.11}$$

are defined, which are called B- and F-transfer matrices of the controller, respectively.

B- and F-transfer matrices (4.11) are associated rational matrices in the sense of Definition 2.14, i.e., they are connected by the relations

$$\begin{aligned} W_d(\zeta) &= \tilde{W}_d(z) \mid_{z=\zeta^{-1}} \text{ and} \\ \tilde{W}_d(z) &= W_d(\zeta) \mid_{\zeta=z^{-1}} \end{aligned} \tag{4.12}$$

that have been shown in Chap. 3.

**Definition 4.2** B- and F-controllers are called causal, if their transfer matrices are causal in the sense of Definitions 2.12, 2.13.

**Lemma 4.1** *Regular B- and F-controllers are causal.*

*Proof* Assume (4.10), then the polynomial matrix $a(\zeta)$ does not possess an eigenvalue at $\zeta = 0$. Therefore, due to (4.11), the matrix $W_d(\zeta)$ has no pole at $\zeta = 0$, i.e., it is causal. The causality of the F-controller then follows from Lemma 2.22 that associated rational matrices at the same time are causal or noncausal.

B- and F-controllers are called irreducible, if their polynomial pairs $(a(\zeta), b(\zeta))$ and $(\tilde{a}(\zeta), \tilde{b}(\zeta))$, respectively, are irreducible.

Notice that for irreducible B-controllers, condition (4.10) is not only sufficient, but also necessary for causality. Indeed, let the pair $(a(\zeta), b(\zeta))$ be irreducible. Then, representation (4.11) for $W_d(\zeta)$ is an ILMFD. Now, if the matrix $W_d(\zeta)$ does not possess a pole at $\zeta = 0$, then the matrix $a(\zeta)$ has no eigenvalue $\zeta = 0$, i.e., condition (4.10) is fulfilled. For F-controllers, regularity condition (4.10) is not necessary for causality.

*Example 4.1* Let an F-controller be described by the equations

$$\begin{aligned} \psi_{1,k+2} + \psi_{1,k} + \psi_{2,k} &= \xi_{1,k+2}, \\ \psi_{1,k} + 2\psi_{2,k+1} &= \xi_{1,k} + \xi_{2,k+1}. \end{aligned} \tag{4.13}$$

Then, we obtain

$$\begin{aligned} \tilde{\alpha}(z) &= \alpha_0 z^2 + \alpha_1 z + \alpha_2, \\ \tilde{\beta}(z) &= \beta_0 z^2 + \beta_1 z + \beta_2, \end{aligned} \tag{4.14}$$

where

$$\begin{aligned} \alpha_0 &= \begin{bmatrix} 1 & 0 \\ 0 & 0 \end{bmatrix}, \quad \alpha_1 = \begin{bmatrix} 0 & 0 \\ 0 & 2 \end{bmatrix}, \quad \alpha_2 = \begin{bmatrix} 1 & 1 \\ 1 & 0 \end{bmatrix}, \\ \beta_0 &= \begin{bmatrix} 1 & 0 \\ 0 & 0 \end{bmatrix}, \quad \beta_1 = \begin{bmatrix} 0 & 0 \\ 0 & 1 \end{bmatrix}, \quad \beta_2 = \begin{bmatrix} 0 & 0 \\ 1 & 0 \end{bmatrix}, \end{aligned} \tag{4.15}$$

and the matrices $\tilde{\alpha}(z)$, $\tilde{\beta}(z)$ have the form

$$\tilde{\alpha}(z) = \begin{bmatrix} z^2+1 & 1 \\ 1 & 2z \end{bmatrix}, \quad \tilde{\beta}(z) = \begin{bmatrix} z^2 & 0 \\ 1 & z \end{bmatrix}. \tag{4.16}$$

In the actual case $\det \alpha_0 = 0$, however a direct calculation shows that the transfer matrix of controller (4.13) takes the form

$$\tilde{W}_d(z) = \frac{\begin{bmatrix} 2z^3 - 1 & -z \\ 1 & z^3 + z \end{bmatrix}}{2z^3 + 2z - 1}, \tag{4.17}$$

i.e., it is causal. The result of this example meets Corollary 2.3, i.e., the matrix $\tilde{\alpha}(z)$ is row reduced, and the degree of each row of the matrix $\tilde{\alpha}(z)$ is at least so high as the degree of the corresponding row of $\tilde{\beta}(z)$.

In the actual case $\det \alpha_0 = 0$, but the matrix $\tilde{\alpha}(z)$ is row reduced. Moreover, the degree of each row of the matrix $\tilde{\alpha}(z)$ is equal to the degree of the corresponding row of the matrix $\tilde{\beta}(z)$.

(4) Below, it is always supposed that the equation of the DAC has the form

$$u(t) = \sum_{i=0}^{P} h_i(t - kT)\psi_{k-i}, \quad kT < t < (k+1)T. \tag{4.18}$$

Here, $P$ is a nonnegative integer, which is called the order of the DAC. Moreover, in (4.18), the quantities $h_i(t)$ are $m \times q$ matrices, where their entries are functions of finite variation in the interval $+0 \leq t \leq T - 0$. In summary, the DC in Fig. 4.1 can be described by the set of equations

$$\begin{aligned} \xi_k &= Ny(kT), \\ \alpha(\zeta)\psi_k &= \beta(\zeta)\xi_k, \\ u(t) &= \sum_{i=0}^{P} h_i(t - kT)\psi_{k-i}, \quad kT < t < (k+1)T \end{aligned} \tag{4.19}$$

when using a B-controller, and by

$$\begin{aligned} \xi_k &= Ny(kT), \\ \tilde{\alpha}(z)\psi_k &= \tilde{\beta}(z)\xi_k, \\ u(t) &= \sum_{i=0}^{P} h_i(t - kT)\psi_{k-i}, \quad kT < t < (k+1)T \end{aligned} \tag{4.20}$$

when using an F-controller. In the future considerations, without loss of generality, we assume in (4.19), (4.20) $N = I_n$, so that we can write

$$\begin{aligned} \xi_k &= y(kT), \\ \alpha(\zeta)\psi_k &= \beta(\zeta)\xi_k \quad (\tilde{\alpha}(z)\psi_k = \tilde{\beta}(z)\xi_k), \\ u(t) &= \sum_{i=0}^{P} h_i(t - kT)\psi_{k-i}, \quad kT < t < (k+1)T. \end{aligned} \tag{4.21}$$

Further on, the equations of the digital controller will always be applied in form (4.21).

## 4.2 Decomposition of the DAC Equation

(1) In the following investigations, the digital/analog converter (DAC) is called hold element, or shortly hold. In the literature and in technological applications [1], normally a zero-order hold is used, working according to the formula

$$u(t) = h(t - kT)\psi_k, \quad kT < t < (k+1)T, \tag{4.22}$$

where $h(t)$ is a matrix, where its entries define the shape of the control impulses. In most applications, the special case of (4.22) is applied, when $q = m$ and $h(t) = I_m$

$$u(t) = \psi_k, \quad kT < t < (k+1)T. \tag{4.23}$$

When not otherwise explicitly mentioned, we understand by zero-order hold this special case. For practical applications, this hold has serious theoretical and practical importance [1].

(2) Formula (4.22) can be written in operational form

$$u(t) = \mathscr{L}_0\left[\{\psi_k\}\right], \tag{4.24}$$

where $\mathscr{L}_0$ maps the set of input sequences $\{\psi_k\}$ into the set of continuous functions $u(t)$. Obviously, this operator is linear. Moreover, this operator disposes of the important feature, that it is invariant according to a shift of the input signal by a multiple of the sampling period $T$. This property will be explained in more detail. Assume the sequence $\{\psi_k\}$ as input of a zero-order hold. Moreover, let $\ell$ be any integer, and $\{\psi_{k-\ell}\}$ be the shifted sequence by $\ell$ steps. Then,

$$\mathscr{L}_0[\{\psi_{k-\ell}\}] = u(t - \ell T), \tag{4.25}$$

because this follows directly from (4.22). Formula (4.25) means that the operator $\mathscr{L}_0$ is $T$-periodic in the sense of [2, 3]. For $\ell > 0$, formula (4.25) yields the equivalence of the structures shown in Fig. 4.2. In Fig. 4.2, the symbol $\zeta$ means backward shift operator (3.119), $\mathscr{L}_0$ is operator (4.22), and $e^{-s\ell T}$ is the pure delay operator in continuous time by $\ell T$.

By using the equivalence of the structures in Fig. 4.2, we will show that a general hold of order $r$ described by (4.18) can be represented as superposition of zero-order holds and pure delays in continuous time. Indeed, denote by $\mathscr{L}_{0i}$ the ZOH (4.22) with matrix $h_i(t)$. Then, formula (4.18) can be written as the decomposition

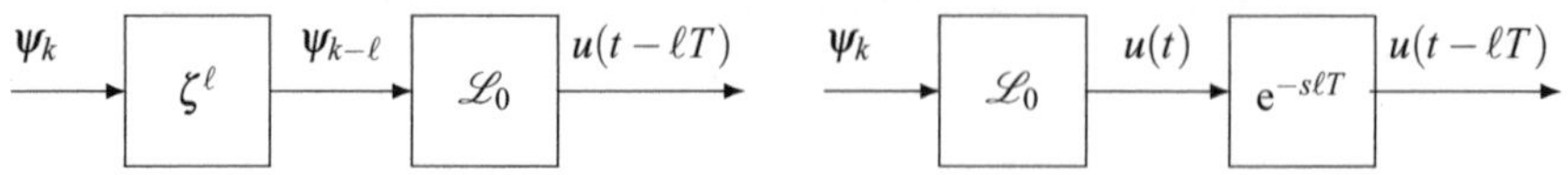


**Fig. 4.2** Equivalent structures of hold with shifting of input sequence

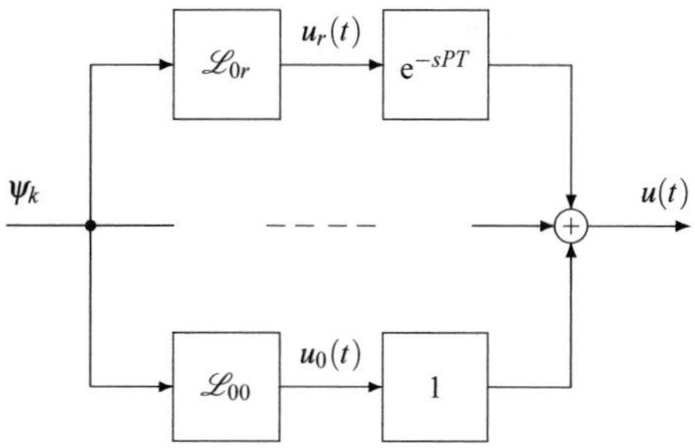


**Fig. 4.3** Decomposition of general hold of order $P$

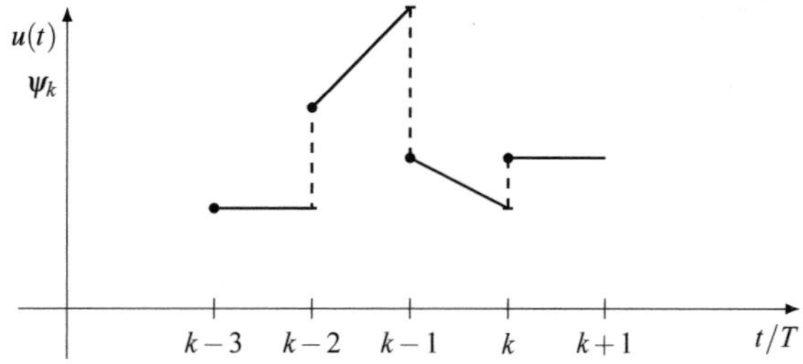


**Fig. 4.4** Operating scheme of first-order hold (extrapolator) (4.27)

$$u(t) = \sum_{i=0}^{P} \mathscr{L}_{0i}[\{\psi_{k-i}\}]. \tag{4.26}$$

With regard to Fig. 4.2, formula (4.26) corresponds to the structure presented in Fig. 4.3.

(3)

*Example 4.2* Consider the standard scalar first-order hold (extrapolator), operating according to the scheme in Fig. 4.4, where the relation between the input $\{\psi_k\}$ and the output $u(t)$ is determined by formula [1]

$$u(t) = \psi_k + \frac{\psi_k - \psi_{k-1}}{T} \cdot (t - kT), \quad kT < t < (k+1)T. \tag{4.27}$$

This formula can also be written in form (4.18). Hence, formula (4.27) can be associated with a structure shown in Fig. 4.5, where $\mathscr{L}_{0i}$ are zero-order holds, for which the functions $h_i(t)$ are determined by

$$u(t) = h_0(t - kT)\psi_k + h_1(t - kT)\psi_{k-1}, \quad kT < t < (k+1)T, \tag{4.28}$$

where

$$h_0(t) = 1 + \frac{t}{T}, \quad h_1(t) = -\frac{t}{T}. \tag{4.29}$$

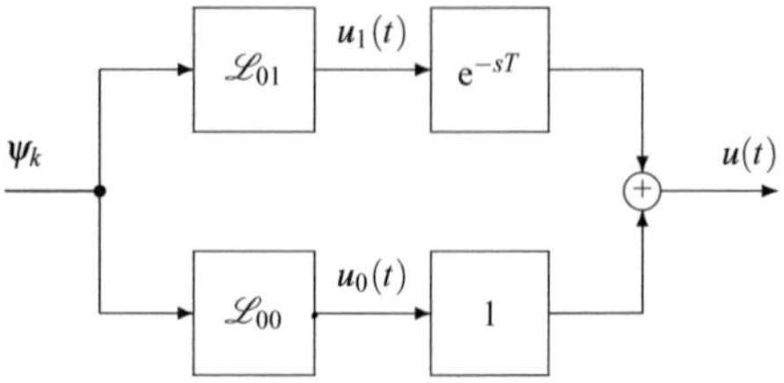


**Fig. 4.5** Structure of first-order hold

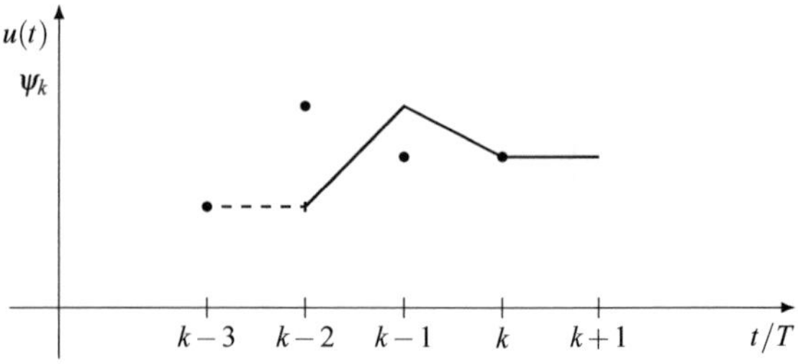


**Fig. 4.6** Operating scheme of first-order hold (interpolator)

(4)

*Example 4.3* Consider the first-order hold (interpolator) described in [4] that operates according to the scheme in Fig. 4.6. Its input and output are related by the formula

$$u(t) = \frac{\psi_k - \psi_{k-1}}{T}(t - kT) + \psi_{k-1}, \quad kT < t < (k+1)T, \tag{4.30}$$

which can be presented in form (4.28), where

$$h_0(t) = \frac{t}{T}, \quad h_1(t) = 1 - \frac{t}{T}. \tag{4.31}$$

In contrast to extrapolator (4.27), the output $u(t)$ of the interpolator is a continuous signal, independent of the input sequence $\{\psi_k\}$. Indeed, from (4.30), we obtain on the interval $(k-1)T < t < kT$

$$u(t) = \frac{\psi_{k-1} - \psi_{k-2}}{T}(t - kT + T) + \psi_{k-2}, \quad (k-1)T < t < kT.$$

From this and (4.30) for $t = kT$, we obtain

$$u(kT - 0) = \psi_{k-1}, \quad u(kT + 0) = \psi_{k-1}. \tag{4.32}$$

Applying decomposition (4.26), the equations of the DC (4.21) can be represented in the form

$$\xi_k = y(kT),$$

$$\alpha(\zeta)\psi_k = \beta(\zeta)\xi_k, \qquad (\tilde{\alpha}(z)\psi_k = \tilde{\beta}(z)\xi_k),$$

$$u_i(t) = h_i(t - kT)\psi_k, \quad kT < t < (k+1)T,$$   (4.33)

$$u(t) = \sum_{i=0}^{P} u_i(t - iT).$$

## 4.3   Mathematical Model of an SD System with Delay in Absence of External Disturbances

(1) Below, the continuous process is assumed to be described in state space

$$\frac{d\upsilon(t)}{dt} = A\upsilon(t) + Bu(t),$$

$$y_1(t) = C\upsilon(t).$$   (4.34)

Here, $\upsilon(t)$ is the $\chi \times 1$ state vector of the process, $y_1(t)$ is the $n \times 1$ output vector, and $u(t)$ is the $m \times 1$ control vector. Moreover, in (4.34) $A, B, and C$ are constant matrices of appropriate sizes. When pure delay acts onto the output, the actual output takes the form

$$y(t) = \sum_{\ell=0}^{\Lambda} C_\ell \upsilon(t - \tau_\ell),$$   (4.35)

where the $C_\ell$ are constant $n \times \chi$ matrices and $\tau_\ell$ are nonnegative constants. Below, without loss of generality, we suppose

$$\tau_{\ell-1} < \tau_\ell, \quad (\ell = 1, ...\Lambda).$$   (4.36)

(2) Further on, we always assume that the control vector $u(t)$ is generated by a digital controller, described by Eq. (4.33). When pure delay acts on various input channels of the process, the control vector $u(t)$ can then be given with the help of the ideas in Sect. 4.2 by the formula

$$u(t) = \sum_{i=0}^{\Omega} B_i u_i(t - \mu_i).$$   (4.37)

Herein, $\mu_i$ $(i = 0, \ldots, \Omega)$ are nonnegative real numbers, such that

$$\mu_{i-1} \leq \mu_i, \quad (i = 1, \ldots, \Omega).$$   (4.38)

Moreover, in (4.37), the $B_i$ are constant matrices, and

$$u_i(t) = h_i(t - kT)\psi_{k-i}, \quad kT < t < (k+1)T \tag{4.39}$$

are zero-order holds. From Sect. 4.2, we realize that representation (4.37) is a special case of a generalized hold of arbitrary order.

(3) In total, Eqs. (4.33)–(4.39) build a system of differential–difference equations, which in case of using a B-controller takes the form

$$\begin{aligned}
\frac{dv(t)}{dt} &= Av(t) + \sum_{i=0}^{\Omega} B_i u_i(t - \mu_i),\\
y(t) &= \sum_{\ell=0}^{\Lambda} C_\ell v(t - \tau_\ell),\\
\xi_k &= y(kT),\\
\alpha(\zeta)\psi_k &= \beta(\zeta)\xi_k,\\
u_i(t) &= h_i(t - kT)\psi_{k-i}, \quad kT < t < (k+1)T.
\end{aligned} \tag{4.40}$$

In the following, Eq. (4.40) is named system $\mathscr{S}_\tau$.

(4) Further on, we always assume the segmentations

$$\tau_\ell = m_\ell T + \theta_\ell, \quad \mu_i = n_i T + \sigma_i, \tag{4.41}$$

where $T$ is the sampling period, $m_\ell, n_i$ are nonnegative integers, and

$$0 \le \theta_\ell < T, \quad 0 \le \sigma_i < T. \tag{4.42}$$

Moreover, we introduce the notations

$$\begin{aligned}
T - \theta_\ell &\stackrel{\triangle}{=} \gamma_\ell, \quad 0 < \gamma_\ell \le T,\\
T - \sigma_i &\stackrel{\triangle}{=} \eta_i, \quad 0 < \eta_i \le T.
\end{aligned} \tag{4.43}$$

## 4.4 Discretization of Functions of a Continuous Argument

(1) Let $f(t)$ be a piecewise continuous function of the continuous argument $t$, defined on the unbounded interval $-\infty < t < \infty$. Build the set of functions $f_k(\varepsilon)$, which is defined for all integers $k$ and for $+0 \le \varepsilon \le T - 0$ by the equation

$$f_k(\varepsilon) = f(kT + \varepsilon), \quad +0 \leq \varepsilon \leq T - 0. \tag{4.44}$$

**Definition 4.3** The sequence $\{f_k(\varepsilon)\}$ depending on the parameter $\varepsilon$ is called discretization of the function $f(t)$. Hereby, the function $f_k(\varepsilon)$ is called an element of the discretization $\{f_k(\varepsilon)\}$.

Obviously, a given function $f(t)$ uniquely determines its discretization $\{f_k(\varepsilon)\}$. The reverse also is true. Indeed, for a given discretization $\{f_k(\varepsilon)\}$, we could find an original function $f(t)$ by the relation

$$f(t) = f_k(t - kT), \quad kT + 0 \leq t \leq (k+1)T - 0. \tag{4.45}$$

If the function $f(t)$ is continuous at $t = kT$, then we obtain

$$f_{k-1}(T) = f_k(0) = f(kT). \tag{4.46}$$

Reversely, from (4.46) it follows the continuity of the function $f(t)$ at $t = kT$.

(2)

*Example 4.4* Assume

$$f(t) = t, \quad -\infty < t < \infty. \tag{4.47}$$

Then

$$f_k(\varepsilon) = kT + \varepsilon, \quad +0 \leq \varepsilon \leq T - 0. \tag{4.48}$$

If, however

$$f(t) = \begin{cases} t, & t \geq T + 0 \\ 0, & t \leq T - 0 \end{cases} \tag{4.49}$$

is the function shown in Fig. 4.7, then we obtain

$$f_k(\varepsilon) = \begin{cases} 0, & k \leq 0, \\ kT + \varepsilon, & k \geq 1. \end{cases} \tag{4.50}$$

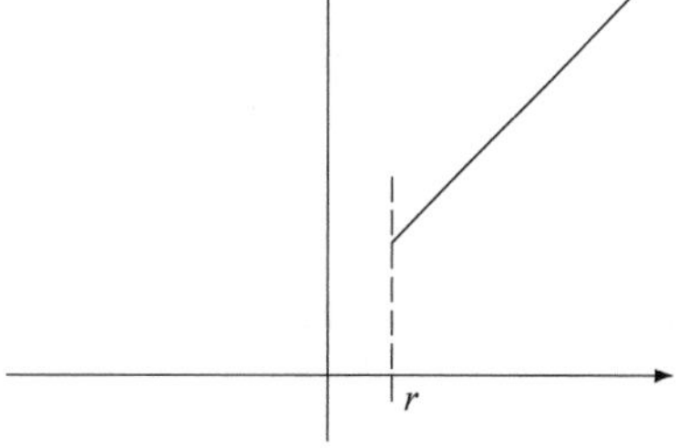


**Fig. 4.7** Discontinuous function in Example 4.4

Thus, relation (4.46) is not fulfilled for $k = 1$, because at $t = T$ function (4.49) has a jump discontinuity.

(3) Let the function $f(t)$ and its discretization $\{f_k(\varepsilon)\}$ be given, and in addition the nonnegative number $\tau$. We construct the discretization $\{f_{\tau k}(\varepsilon)\}$ for the function

$$f_\tau(t) \stackrel{\triangle}{=} f(t-\tau). \tag{4.51}$$

The next statement yields the required result.

**Lemma 4.2** *Assume the representation*

$$\tau = mT + \theta = (m+1)T - \gamma, \quad 0 \le \theta < T, \quad 0 < \gamma = T - \theta \le T, \tag{4.52}$$

*where m is a nonnegative integer. Then, the elements of the discretization* $\{f_{\tau k}(\varepsilon)\}$ *are determined by the relations*

$$f_{\tau k}(\varepsilon) = \begin{cases} f_{k-m-1}(\varepsilon+\gamma), & +0 \le \varepsilon \le \theta - 0, \\ f_{k-m}(\varepsilon-\theta), & \theta + 0 \le \varepsilon \le T - 0. \end{cases} \tag{4.53}$$

*Proof* Assume $\theta = 0$, i.e., $\tau = mT$. Then

$$f_{\tau k}(\varepsilon) = f_\tau(kT+\varepsilon) = f(kT - mT + \varepsilon) = f_{k-m}(\varepsilon). \tag{4.54}$$

Assume now $m = 0$, i.e., $\tau = \theta > 0$. Then, we have the situation shown in Fig. 4.8.
We realize from Fig. 4.8 that for $m = 0,\ 0 < \theta < T,\ 0 < \gamma < T$

$$f_{\tau k}(\varepsilon) = \begin{cases} f_{k-1}(\varepsilon+\gamma), & +0 \le \varepsilon \le \theta - 0, \\ f_{k}(\varepsilon-\theta), & \theta + 0 \le \varepsilon \le T - 0. \end{cases} \tag{4.55}$$

Applying relations (4.54), (4.55) and Fig. 4.8, formula (4.53) is achieved.

(4)

*Example 4.5* Let the function $f(t)$ and its discretization $\{f_k(\varepsilon)\}$ be determined by formulae (4.49) and (4.50). Then, under condition (4.52), we obtain from (4.50)

$$\begin{aligned} f_{k-m-1}(\varepsilon+\gamma) &= \begin{cases} 0, & k \le m+1, \\ (k-m-1)T + \varepsilon + \gamma, & k \ge m+2, \end{cases} \\ f_{k-m}(\varepsilon-\theta) &= \begin{cases} 0, & k \le m, \\ (k-m)T + \varepsilon - \theta, & k \ge m+1. \end{cases} \end{aligned} \tag{4.56}$$

Hence, applying (4.52), (4.53), we find that the discretization of the function

$$f_\tau(t) = \begin{cases} 0, & t \le T + \tau - 0 \\ t - \tau, & t \ge T + \tau + 0 \end{cases} \tag{4.57}$$

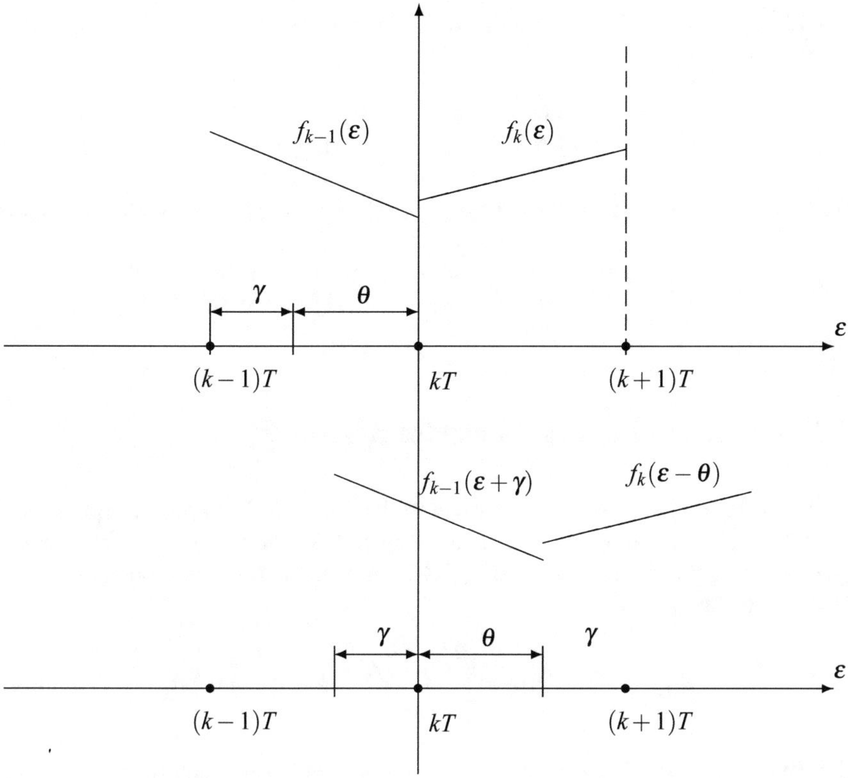


**Fig. 4.8** Discretization of function $f_\tau(t)$ at $m = 0$

is determined by the formulae

$$f_{\tau k}(\varepsilon) = \begin{cases} 0, & +0 \le \varepsilon \le T - 0, \quad k \le m, \\ 0, & +0 \le \varepsilon \le \theta - 0, \quad k = m + 1, \\ T + \varepsilon - \theta, & \theta + 0 \le \varepsilon \le T - 0, \quad k = m + 1, \\ (k - m - 1)T + \varepsilon + \gamma, & +0 \le \varepsilon \le \theta - 0, \quad k > m + 1, \\ (k - m)T + \varepsilon - \theta, & \theta + 0 \le \varepsilon \le T - 0, \quad k > m + 1. \end{cases} \tag{4.58}$$

(5) Build the discretization of the output of zero-order hold (4.22) in the presence of delay. By definition, we find

$$u(t) = h(t - kT)\psi_k, \quad kT < t < (k + 1)T. \tag{4.59}$$

Assuming $t = kT + \varepsilon,\ 0 \le \varepsilon \le T$, from (4.44), (4.49), we obtain the discretizations of the functions $u_i(t)$

$$u_k(\varepsilon) = h(\varepsilon)\psi_k. \tag{4.60}$$

Therefore, when (4.52) is fulfilled, we obtain from (4.53) for the discretization of the function $u_\tau(t) = u(t-\tau)$

$$u_{\tau k}(\varepsilon) = \begin{cases} h(\varepsilon+\gamma)\psi_{k-m-1}, & +0 \le \varepsilon \le \theta - 0, \\ h(\varepsilon-\theta)\psi_{k-m}, & \theta + 0 \le \varepsilon \le T - 0. \end{cases} \tag{4.61}$$

Applying herein (4.45), (4.52) and substituting $\varepsilon$ by $t - kT$, we find the expression

$$u_\tau(t) = u(t-\tau) = \begin{cases} h(t-kT+T+\gamma)\psi_{k-m-1}, & kT+0 \le t \le kT+\theta-0, \\ h(t-kT-\theta)\psi_{k-m}, & kT+\theta+0 \le t \le (k+1)T-0. \end{cases} \tag{4.62}$$

## 4.5 Backward Discrete Model for System $\mathscr{S}_\tau$

(1) In this section, on basis of the system of the differential–difference equations of the system $\mathscr{S}_\tau$, we construct an equivalent system of difference equations. For this goal, we integrate the first equation in (4.40) under the initial condition $v(kT) = v_k$. Hence, we obtain

$$v(t) = \mathrm{e}^{A(t-kT)}v_k + \sum_{i=0}^{\Omega}\int_{kT}^{t}\mathrm{e}^{A(t-\nu)}B_i u_i(\nu-\mu_i)\,\mathrm{d}\nu. \tag{4.63}$$

Substituting here $t - kT = \varepsilon,\ 0 \le \varepsilon \le T$, we find for the elements of the discretization of the vector $v(t)$

$$v_k(\varepsilon) = v(kT+\varepsilon) = \mathrm{e}^{A\varepsilon}v_k + \sum_{i=0}^{\Omega}\int_{kT}^{kT+\varepsilon}\mathrm{e}^{A(kT+\varepsilon-\nu)}B_i u_i(\nu-\mu_i)\,\mathrm{d}\nu. \tag{4.64}$$

From (4.62) with notation (4.41), we obtain

$$u_i(\nu-\mu_i) = \begin{cases} h_i(\nu-kT+\eta_i)\psi_{k-n_i-1}, & kT < \nu < kT+\sigma_i, \\ h_i(\nu-kT-\sigma_i)\psi_{k-n_i}, & kT+\sigma_i < \nu < (k+1)T. \end{cases} \tag{4.65}$$

Therefore,

$$\begin{aligned} &\int_{kT}^{kT+\varepsilon}\mathrm{e}^{A(kT+\varepsilon-\nu)}B_i u_i(\nu-\mu_i)\,\mathrm{d}\nu \\ &= \begin{cases} \int_{kT}^{kT+\varepsilon}\mathrm{e}^{A(kT+\varepsilon-\nu)}B_i h_i(\nu-kT+\eta_i)\,\mathrm{d}\nu\,\psi_{k-n_i-1}, & 0 \le \varepsilon \le \sigma_i, \\ \int_{kT}^{kT+\sigma_i}\mathrm{e}^{A(kT+\varepsilon-\nu)}B_i h_i(\nu-kT+\eta_i)\,\mathrm{d}\nu\,\psi_{k-n_i-1} & \\ \quad + \int_{kT+\sigma_i}^{kT+\varepsilon}\mathrm{e}^{A(kT+\varepsilon-\nu)}B_i h_i(\nu-kT-\sigma_i)\,\mathrm{d}\nu\,\psi_{k-n_i}, & \sigma_i \le \varepsilon \le T. \end{cases} \end{aligned} \tag{4.66}$$

After renaming the integration variable, this expression can be written in the form

$$\int_{kT}^{kT+\varepsilon} \mathrm{e}^{A(kT+\varepsilon-\nu)} B_i u_i(\nu-\mu_i)\,\mathrm{d}\nu$$
$$= \begin{cases} \mathrm{e}^{A(\varepsilon+\eta_i)} \displaystyle\int_{\eta_i}^{\eta_i+\varepsilon} \mathrm{e}^{-Au} B_i h_i(u)\,\mathrm{d}u\,\psi_{k-n_i-1}, & 0 \le \varepsilon \le \sigma_i, \\ \mathrm{e}^{A(\varepsilon+\eta_i)} \displaystyle\int_{\eta_i}^{T} \mathrm{e}^{-Au} B_i h_i(u)\,\mathrm{d}u\,\psi_{k-n_i-1} & \\ \quad + \mathrm{e}^{A(\varepsilon-\lambda_i)} \displaystyle\int_{0}^{\varepsilon-\sigma_i} \mathrm{e}^{-Au} B_i h_i(u)\,\mathrm{d}u\,\psi_{k-n_i}, & \sigma_i \le \varepsilon \le T. \end{cases} \tag{4.67}$$

Inserting (4.67) into (4.64), we find

$$v_k(\varepsilon) = \mathrm{e}^{A\varepsilon} v_k + \sum_{i=0}^{\Omega} \Gamma_{2i}(\varepsilon)\psi_{k-n_i-1} + \sum_{i=0}^{\Omega} \Gamma_{1i}(\varepsilon)\psi_{k-n_i}, \quad 0 \le \varepsilon \le T, \tag{4.68}$$

where

$$\Gamma_{2i}(\varepsilon) = \begin{cases} \mathrm{e}^{A(\varepsilon+\eta_i)} \displaystyle\int_{\eta_i}^{\eta_i+\varepsilon} \mathrm{e}^{-Au} B_i h_i(u)\,\mathrm{d}u, & 0 \le \varepsilon \le \sigma_i, \\ \mathrm{e}^{A(\varepsilon+\eta_i)} \displaystyle\int_{\eta_i}^{T} \mathrm{e}^{-Au} B_i h_i(u)\,\mathrm{d}u, & \sigma_i \le \varepsilon \le T, \end{cases} \tag{4.69}$$

and

$$\Gamma_{1i}(\varepsilon) = \begin{cases} 0, & 0 \le \varepsilon \le \sigma_i, \\ \mathrm{e}^{A(\varepsilon-\sigma_i)} \displaystyle\int_{0}^{\varepsilon-\sigma_i} \mathrm{e}^{-Au} B_i h_i(u)\,\mathrm{d}u, & \sigma_i \le \varepsilon \le T. \end{cases} \tag{4.70}$$

For $\varepsilon = T$, from (4.68)–(4.70), after renaming $k$ by $k-1$, we obtain

$$v_k = \mathrm{e}^{AT} v_{k-1} + \sum_{i=0}^{\Omega} \Gamma_{2i}(T)\psi_{k-n_i-2} + \sum_{i=0}^{\Omega} \Gamma_{1i}(T)\psi_{k-n_i-1}, \tag{4.71}$$

where

$$\Gamma_{2i}(T) = \mathrm{e}^{A(T+\eta_i)} \int_{\eta_i}^{T} \mathrm{e}^{-Au} B_i h_i(u)\,\mathrm{d}u,$$
$$\Gamma_{1i}(T) = \mathrm{e}^{A\eta_i} \int_{0}^{\eta_i} \mathrm{e}^{-Au} B_i h_i(u)\,\mathrm{d}u. \tag{4.72}$$

**Definition 4.4** Equation (4.71) is called backward discrete model of the state vector $v(t)$, and Eq. (4.68) its modified backward discrete model.

(2) In order to construct a backward discrete model of the output vector $y(t)$, we introduce the notation

$$y_\tau^{(\ell)}(t) \triangleq C_\ell v(t-\tau_\ell) = C_\ell v(t-m_\ell T-\theta_\ell). \tag{4.73}$$

Applying the general formula (4.53) and notations (4.41)–(4.43), we immediately obtain that the elements of the discretization $\{y_{\tau k}^{(\ell)}(\varepsilon)\}$ of the vector (4.73) are determined by the relations

$$y_{\tau k}^{(\ell)}(\varepsilon) = \begin{cases} C_\ell v_{k-m_\ell-1}(\varepsilon+\gamma_\ell), & 0 \le \varepsilon \le \theta_\ell, \\ C_\ell v_{k-m_\ell}(\varepsilon-\theta_\ell), & \theta_\ell \le \varepsilon \le T. \end{cases} \tag{4.74}$$

From (4.68), we find

$$\begin{aligned} v_{k-m_\ell-1}(\varepsilon+\gamma_\ell) &= \mathrm{e}^{A(\varepsilon+\gamma_\ell)} v_{k-m_\ell-1} \\ &+ \sum_{i=0}^{\Omega} \Gamma_{2i}(\varepsilon+\gamma_\ell)\psi_{k-n_i-m_\ell-2} + \sum_{i=0}^{\Omega} \Gamma_{1i}(\varepsilon+\gamma_\ell)\psi_{k-n_i-m_\ell-1}, \\ v_{k-m_\ell}(\varepsilon-\theta_\ell) &= \mathrm{e}^{A(\varepsilon-\theta_\ell)} v_{k-m_\ell} \\ &+ \sum_{i=0}^{\Omega} \Gamma_{2i}(\varepsilon-\theta_\ell)\psi_{k-n_i-m_\ell-1} + \sum_{i=0}^{\Omega} \Gamma_{1i}(\varepsilon-\theta_\ell)\psi_{k-n_i-m_\ell}. \end{aligned} \tag{4.75}$$

Using that, due to (4.40), (4.73),

$$y(t) = \sum_{\ell=0}^{\Lambda} y_\tau^{(\ell)}(t) = \sum_{\ell=0}^{\Lambda} C_\ell v(t-m_\ell T-\theta_\ell), \tag{4.76}$$

with the help of (4.74), (4.75), we find the elements of the discretization $\{y_k(\varepsilon)\}$ of the vector $y(t)$. Obviously, we have

$$y_k(\varepsilon) = \sum_{\ell=0}^{\Lambda} C_\ell y_{\tau k}^{(\ell)}(\varepsilon), \tag{4.77}$$

Substituting herein (4.74) and (4.75), an expression of the form

$$\begin{aligned} y_k(\varepsilon) = &\sum_{\ell=0}^{\Lambda} P_\ell^{(1)}(\varepsilon) v_{k-m_\ell-1} + \sum_{\ell=0}^{\Lambda} P_\ell^{(0)}(\varepsilon) v_{k-m_\ell} + \sum_{\ell=0}^{\Lambda}\sum_{i=0}^{\Omega} Q_{\ell i}^{(2)}(\varepsilon)\psi_{k-n_i-m_\ell-2} \\ &+ \sum_{\ell=0}^{\Lambda}\sum_{i=0}^{\Omega} Q_{\ell i}^{(1)}(\varepsilon)\psi_{k-n_i-m_\ell-1} + \sum_{\ell=0}^{\Lambda}\sum_{i=0}^{\Omega} Q_{\ell i}^{(0)}(\varepsilon)\psi_{k-n_i-m_\ell}, \quad 0 \le \varepsilon \le T, \end{aligned} \tag{4.78}$$

is achieved, where $P_\ell^{(\alpha)}(\varepsilon)$, $(\alpha = 1, 0)$, $Q_{\ell i}^{(\beta)}$, $(\beta = 2, 1, 0)$ are matrices that are known functions of the argument $\varepsilon$. For $\varepsilon = 0$, from (4.74), we obtain

$$y_{\tau k}^{(\ell)} = y_{\tau k}^{(\ell)}(\varepsilon)_{|\varepsilon=0} = C_\ell v_{k-m_\ell-1}(\gamma_\ell). \tag{4.79}$$

Moreover, (4.75) for $\varepsilon = 0$ yields

$$y_{\tau k}^{(\ell)} = C_\ell \left[ \mathrm{e}^{A\gamma_\ell} v_{k-m_\ell-1} + \sum_{i=0}^{\Omega} \Gamma_{2i}(\gamma_\ell)\psi_{k-n_i-m_\ell-2} + \sum_{i=0}^{\Omega} \Gamma_{1i}(\gamma_\ell)\psi_{k-n_i-m_\ell-1} \right]. \tag{4.80}$$

Since from (4.77) for $\varepsilon = 0$, we obtain

$$y_k(0) = \sum_{\ell=0}^{\Lambda} y_{\tau k}^{(\ell)}(0), \tag{4.81}$$

so, using (4.80), we find

$$\begin{aligned} y_k = y_k(0) &= \sum_{\ell=0}^{\Lambda} C_\ell \mathrm{e}^{A\gamma_\ell} v_{k-m_\ell-1} + \sum_{\ell=0}^{\Lambda}\sum_{i=0}^{\Omega} C_\ell \Gamma_{2i}(\gamma_\ell)\psi_{k-n_i-m_\ell-2} \\ &+ \sum_{\ell=0}^{\Lambda}\sum_{i=0}^{\Omega} C_\ell \Gamma_{1i}(\gamma_\ell)\psi_{k-n_i-m_\ell-1}. \end{aligned} \tag{4.82}$$

(3)

**Definition 4.5** Equation (4.78) is called modified backward (B) discrete model of the process output, and Eq. (4.82) its backward discrete model.

Assembling the B-discrete models of the state vector and the output vector with the discrete controller, the B-discrete model of the system $\mathscr{S}_\tau$ is achieved:

$$\begin{aligned} v_k &= \mathrm{e}^{AT} v_{k-1} + \sum_{i=1}^{\Omega} \Gamma_{2i}(T)\psi_{k-n_i-2} + \sum_{i=1}^{\Omega} \Gamma_{1i}(T)\psi_{k-n_i-1}, \\ \xi_k &= \sum_{\ell=0}^{\Lambda} C_\ell \mathrm{e}^{A\gamma_\ell} v_{k-m_\ell-1} + \sum_{\ell=0}^{\Lambda}\sum_{i=0}^{\Omega} C_\ell \Gamma_{2i}(\gamma_\ell)\psi_{k-n_i-m_\ell-2} + \sum_{\ell=0}^{\Lambda}\sum_{i=0}^{\Omega} C_\ell \Gamma_{1i}(\gamma_\ell)\psi_{k-n_i-m_\ell-1}, \\ & \alpha_0\psi_k + \alpha_1\psi_{k-1} + \cdots + \alpha_\rho\psi_{k-\rho} = \beta_0\xi_k + \beta_1\xi_{k-1} + \cdots + \beta_\rho\xi_{k-\rho}, \end{aligned} \tag{4.83}$$

where we remember that $y_k = \xi_k$.

**Lemma 4.3** *Let the controller* $(\alpha(\zeta), \beta(\zeta))$ *be regular, i.e.,*

$$\det \alpha_0 \neq 0. \tag{4.84}$$

*Then for any initial conditions*

$$\begin{aligned} &\upsilon_0, \upsilon_{-1}, \ldots, \upsilon_{-m_\Lambda - 1}, \\ &\psi_0, \psi_{-1}, \ldots, \psi_{-\varkappa}, \quad \varkappa = \max\{\rho, n_\Omega + m_\Lambda + 2\}, \\ &\xi_0, \ldots, \xi_{-\rho} \end{aligned} \tag{4.85}$$

*there exists a unique solution of difference Eq. (4.83), defined for all $k \geq 1$.*

*Proof* Inserting the second equation of (4.83) into the third and using (4.84), an equivalent system of difference equations is achieved

$$\begin{aligned} \upsilon_k &= \mathrm{e}^{AT}\upsilon_{k-1} + \sum_{i=0}^{\Omega} \Gamma_{2i}(T)\psi_{k-n_i-2} + \sum_{i=0}^{\Omega} \Gamma_{1i}(T)\psi_{k-n_i-1} \\ \xi_k &= \sum_{\ell=0}^{\Lambda} C_\ell \mathrm{e}^{A\gamma_\ell}\upsilon_{k-m_\ell-2} + \sum_{\ell=0}^{\Lambda}\sum_{i=0}^{\Omega} C_\ell \Gamma_{2i}(\gamma_\ell)\psi_{k-n_i-m_\ell-2} + \sum_{\ell=0}^{\Lambda}\sum_{i=0}^{\Omega} C_\ell \Gamma_{1i}(\gamma_\ell)\psi_{k-n_i-m_\ell-1}, \end{aligned} \tag{4.86}$$

and, moreover

$$\begin{aligned} \psi_k = &-\alpha_0^{-1}(\alpha_1\psi_{k-1} + \cdots + \alpha_\rho\psi_{k-\rho}) + \alpha_0^{-1}(\beta_1\xi_{k-1} + \cdots + \beta_\rho\xi_{k-\rho}) + \\ &+\alpha_0^{-1}\beta_0 \left[ \sum_{\ell=0}^{\Lambda} C_\ell \mathrm{e}^{A\gamma_\rho}\upsilon_{k-m_\ell-1} + \sum_{\ell=0}^{\Lambda}\sum_{i=0}^{\Omega} C_\ell \Gamma_{2i}(\gamma_\rho)\psi_{k-n_i-m_\ell-2} \right. \\ &\left. + \sum_{\ell=0}^{\Lambda}\sum_{i=0}^{\Omega} C_\ell \Gamma_{1i}(\gamma)\psi_{k-n_i-m_\ell-1} \right]. \end{aligned} \tag{4.87}$$

The system of difference Eqs. (4.86), (4.87) establishes a recursive algorithm. Beginning with any initial values (4.85), Eqs. (4.86), (4.87) yield uniquely step by step the sequences $\{\upsilon_k\}$, $\{\psi_k\}$, $\{\xi_k\}$ for all $k \geq 1$.

(4) Assembling Eqs. (4.68), (4.78), we obtain

$$\begin{aligned} \upsilon_k(\varepsilon) &= \mathrm{e}^{A\varepsilon}\upsilon_k + \sum_{i=0}^{\Omega} \Gamma_{2i}(\varepsilon)\psi_{k-n_i-1} + \sum_{i=0}^{\Omega} \Gamma_{1i}(\varepsilon)\psi_{k-n_i} \\ y_k(\varepsilon) &= \sum_{\ell=0}^{\Lambda} P_\ell^{(1)}(\varepsilon)\upsilon_{k-m_\ell-1} + \sum_{\ell=0}^{\Lambda} P_\ell^{(0)}(\varepsilon)\upsilon_{k-m_\ell} + \sum_{\ell=0}^{\Lambda}\sum_{i=0}^{\Omega} Q_{\ell i}^{(2)}(\varepsilon)\psi_{k-n_i-m_\ell-2} \\ &+ \sum_{\ell=0}^{\Lambda}\sum_{i=0}^{\Omega} Q_{\ell i}^{(1)}(\varepsilon)\psi_{k-n_i-m_\ell-1} + \sum_{\ell=0}^{\Lambda}\sum_{i=0}^{\Omega} Q_{\ell i}^{(0)}(\varepsilon)\psi_{k-n_i-m_\ell}. \end{aligned} \tag{4.88}$$

Using the solution $\{\upsilon_k\}$, $\{\psi_k\}$ of the B-discrete model, on basis of formulae (4.86), (4.87), we construct with the help of (4.88) the discretizations $\{\upsilon_k(\varepsilon)\}$, $\{y_k(\varepsilon)\}$ of the vectors $\upsilon(t)$, $y(t)$, which on the other hand can be constructed on the basis of relations (4.35). It can be shown [5], that the so generated vector signals $v(t)$, $y(t)$ are continuous.

(5) By construction between the continuous solutions of system of differential–difference Eq. (4.40) and system of difference Eq. (4.83) there exists a one-to-one connection, which consists of the following. Let $v(t)$, $y(t)$, $\{\psi_k\}$ be a solution of Eq. (4.40), where the signals $v(t)$, $y(t)$ by definition are continuous. Then, the sequences $\{v_k\} = \{v(kT)\}$, $\{\xi_k\} = \{y(kT)\}$, $\{\psi_k\}$ are a solution of Eq. (4.83). Reversely, due to the above considerations, every solution of the B-discrete model corresponds to a solution of the system $\mathscr{S}_\tau$. From the above said it follows that, when condition (4.84) is fulfilled, any set of initial conditions (4.85) corresponds to a unique solution of the system $\mathscr{S}_\tau$.

## 4.6 Modal Control of System $\mathscr{S}_\tau$

(1) Introduce the polynomial matrices

$$
\begin{aligned}
a(\zeta) &= I_\chi - \zeta \mathrm{e}^{AT}, \\
b(\zeta) &= \sum_{i=0}^{\Omega} \Gamma_{2i}(T)\zeta^{n_i+1} + \sum_{i=0}^{\Omega} \Gamma_{1i}(T)\zeta^{n_i}, \\
c(\zeta) &= \zeta \sum_{\ell=0}^{\Lambda} C_\ell \mathrm{e}^{A\gamma_\ell}\zeta^{m_\ell}, \\
d(\zeta) &= \sum_{\ell=0}^{\Lambda}\sum_{i=0}^{\Omega} C_\ell \Gamma_{2i}(\gamma_\ell)\zeta^{n_i+m_\ell+1} + \sum_{\ell=0}^{\Lambda}\sum_{i=0}^{\Omega} C_\ell \Gamma_{1i}(\gamma_\ell)\zeta^{n_i+m_\ell}.
\end{aligned}
\tag{4.89}
$$

Then, the first two equations of (4.83), when using the backward shift operator $\zeta$, can be written as

$$
\begin{aligned}
a(\zeta)v_k &= \zeta b(\zeta)\psi_k, \\
\xi_k &= c(\zeta)v_k + \zeta d(\zeta)\psi_k.
\end{aligned}
\tag{4.90}
$$

These equations can be interpreted as the PMD

$$
\Pi_\tau(\zeta) = (a(\zeta), b_1(\zeta), c(\zeta), d_1(\zeta)) = (a(\zeta), \zeta b(\zeta), c(\zeta), \zeta d(\zeta)). \tag{4.91}
$$

Since all numbers $n_i$, $m_\ell$ are nonnegative, it follows directly from (4.89), (4.91)

$$
\det a(0) \neq 0, \quad b_1(0) = 0_{\chi q}, \quad d_1(0) = 0_{nq}. \tag{4.92}
$$

These relations say that the PMD $\Pi_\tau(\zeta)$ is non-singular and strictly causal.

(2) Pooling Eq. (4.90) with the operator equation of the controller (4.3), a B-discrete model of the system $\mathscr{S}_\tau$ in polynomial form is achieved as follows:

$$\begin{aligned} a(\zeta)\upsilon_k &= \zeta b(\zeta)\psi_k, \\ \xi_k &= c(\zeta)\upsilon_k + \zeta d(\zeta)\psi_k, \\ \alpha(\zeta)\psi_k &= \beta(\zeta)\xi_k. \end{aligned} \tag{4.93}$$

**Definition 4.6** The characteristic matrix of equation system (4.93)

$$Q_\tau(\zeta, \alpha, \beta) = \begin{bmatrix} a(\zeta) & O_{\chi n} & -\zeta b(\zeta) \\ -c(\zeta) & I_n & -\zeta d(\zeta) \\ O_{q\chi} & -\beta(\zeta) & \alpha(\zeta) \end{bmatrix} \tag{4.94}$$

is called the characteristic matrix of the system $\mathscr{S}_\tau$. The eigenvalues of the characteristic matrix are called modes of the system $\mathscr{S}_\tau$.

It is well known that the modes play for the system $\mathscr{S}_\tau$ a comparable role as the roots of the characteristic equation for continuous LTI systems.

(3) One of the fundamental control problems for the system $\mathscr{S}_\tau$ is the modal control problem, which can be formulated as follows:

**Modal control problem:** For a given PMD (4.91) and monic polynomial $f(\zeta)$, find the set of all causal controllers $(\alpha(\zeta), \beta(\zeta))$, fulfilling the condition

$$\det Q_\tau(\zeta, \alpha, \beta) = \det \begin{bmatrix} a(\zeta) & O_{\chi n} & -\zeta b(\zeta) \\ -c(\zeta) & I_n & -\zeta d(\zeta) \\ O_{q\chi} & -\beta(\zeta) & \alpha(\zeta) \end{bmatrix} \sim f(\zeta). \tag{4.95}$$

Obviously, the formulated problem leads to the general causal control problem for PMD processes considered in Sect. 3.12. However, in the present case, we can benefit from the concrete structure of PMD (4.91). It allows to find further relations, which are helpful in applications.

(a) At first notice that for the solvability of the formulated problem, it is necessary that the polynomial $f(\zeta)$ is causal, i.e., $f(0) \neq 0$ is true. Indeed, assume that the formulated problem is solvable for a certain polynomial $f(\zeta)$. Then, from (4.95), we obtain

$$\det Q(\zeta, \alpha, \beta)|_{\zeta=0} = \det a(0) \det \alpha(0) = f(0). \tag{4.96}$$

Here, $\det a(0) \neq 0$ and by proposition $\det \alpha(0) \neq 0$. Therefore, $f(0) \neq 0$, i.e., the polynomial $f(\zeta)$ is causal.

(b) If the polynomial $f(\zeta)$ is causal, then any irreducible controller satisfies relation (4.95), and hence it is causal. Indeed, under the taken propositions, in (4.96) we have $\det a(0) \neq 0$, $f(0) \neq 0$, and thus $\det \alpha(0) \neq 0$.

**Definition 4.7** The system $\mathscr{S}_\tau$ is called completely modal controllable, if PMD (4.95) is solvable for any polynomial $f(\zeta)$.

Necessary and sufficient conditions for the complete modal controllability of system $\mathscr{S}_\tau$ are provided by the next theorem.

**Theorem 4.1** *Introduce the matrices*

$$\begin{aligned} B &\triangleq \sum_{i=0}^{\Omega} \mathrm{e}^{-A\mu_i} \int_0^T \mathrm{e}^{-A\nu} B_i h_i(\nu)\,\mathrm{d}\nu, \\ C &\triangleq \sum_{\ell=0}^{\Lambda} C_\ell \mathrm{e}^{-A\tau_\ell}. \end{aligned} \tag{4.97}$$

*Then the system $\mathscr{S}_\tau$ is completely modal controllable, if and only if the pairs $(\mathrm{e}^{-AT}, B)$ and $[\mathrm{e}^{-AT}, C]$ are minimal, i.e., the pair $(\mathrm{e}^{-AT}, B)$ is controllable, and the pair $[\mathrm{e}^{-AT}, C]$ is observable.*

*Proof* As it follows from Theorem 3.35, for the solvability of the DPE (4.95) for any polynomial $f(\zeta)$, it is necessary and sufficient that the polynomial pairs $(a(\zeta), \zeta b(\zeta))$, and $[a(\zeta), c(\zeta)]$ are irreducible, i.e., for all $\zeta$ the relations

$$\begin{aligned} \operatorname{rank}\left[\, a(\zeta)\ \ \zeta b(\zeta) \,\right] &= \operatorname{rank}\left[\, I_\chi - \zeta \mathrm{e}^{AT}\ \ \zeta b(\zeta) \,\right] = \chi, \\ \operatorname{rank}\begin{bmatrix} a(\zeta) \\ c(\zeta) \end{bmatrix} &= \operatorname{rank}\begin{bmatrix} I_\chi - \zeta \mathrm{e}^{AT} \\ c(\zeta) \end{bmatrix} = \chi \end{aligned} \tag{4.98}$$

are true. Since the matrix $\mathrm{e}^{AT}$ is non-singular, it follows from (4.89) and Corollary 2.5 that the equivalence relation,

$$\left[\, I_\chi - \zeta \mathrm{e}^{AT}\ \ \zeta b(\zeta) \,\right] \sim \left[\, I_\chi - \zeta \mathrm{e}^{AT}\ \ \mathrm{e}^{-AT} b_\ell(\mathrm{e}^{-AT}) \,\right], \tag{4.99}$$

is valid, where

$$\mathrm{e}^{-AT} b_\ell(\mathrm{e}^{-AT}) = \sum_{i=0}^{\Lambda} \mathrm{e}^{-(n_i+1)AT} \Gamma_{1i}(T) + \sum_{i=0}^{\Lambda} \mathrm{e}^{-(n_i+2)AT} \Gamma_{2i}(T). \tag{4.100}$$

Using (4.72), from (4.100), we find

$$
\begin{aligned}
\mathrm{e}^{-AT} b_\ell(\mathrm{e}^{-AT}) &= \sum_{i=0}^{\Omega} \mathrm{e}^{-(n_i+1)AT} \mathrm{e}^{A\eta_i} \int_0^{\eta_i} \mathrm{e}^{-Au} B_i h_i(u)\,\mathrm{d}u \\
&\quad + \sum_{i=0}^{\Omega} \mathrm{e}^{-(n_i+2)AT} \mathrm{e}^{A(T+\eta_i)} \int_{\eta_i}^{T} \mathrm{e}^{-Au} B_i h_i(u)\,\mathrm{d}u \qquad (4.101) \\
&= \sum_{i=0}^{\Omega} \mathrm{e}^{A(\eta_i - n_i T - T)} \int_0^T \mathrm{e}^{-Au} B_i h_i(u)\,\mathrm{d}u.
\end{aligned}
$$

Since, owing to (4.41)–(4.43)

$$
\eta_i - n_i T - T = -\mu_i, \tag{4.102}
$$

so from (4.101), it follows

$$
\mathrm{e}^{-AT} b_\ell(\mathrm{e}^{-AT}) = \sum_{i=0}^{\Omega} \mathrm{e}^{-A\mu_i} \int_0^T \mathrm{e}^{-Au} B_i h_i(u)\,\mathrm{d}u = B. \tag{4.103}
$$

Therefore, the pair $(a(\zeta), \zeta b(\zeta))$ is irreducible, if and only if the pair $(I_\chi - \zeta \mathrm{e}^{AT}, B)$ is irreducible. Since the matrix $\mathrm{e}^{AT}$ is non-singular, the pair $(I_\chi - \zeta \mathrm{e}^{AT}, B)$ is irreducible in and only in the case, when the pair $(\mathrm{e}^{-AT}, B)$ is completely controllable. In analogy, with the help of (4.89), it can be shown that

$$
\begin{bmatrix} I_\chi - \zeta \mathrm{e}^{AT} \\ c(\zeta) \end{bmatrix} \sim \begin{bmatrix} I_\chi - \zeta \mathrm{e}^{AT} \\ c_r(\mathrm{e}^{-AT}) \end{bmatrix}, \tag{4.104}
$$

where

$$
c_r(\mathrm{e}^{-AT}) = \sum_{\ell=0}^{\Lambda} C_\ell \mathrm{e}^{A(\gamma_\ell - m_\ell T - T)}. \tag{4.105}
$$

From (4.41)–(4.43), we obtain

$$
\gamma_\ell - m_\ell T - T = -\tau_\ell. \tag{4.106}
$$

Hence, from (4.105)

$$
c_r(\mathrm{e}^{-AT}) = \sum_{\ell=0}^{\Lambda} C_\ell \mathrm{e}^{-\tau_\ell}. \tag{4.107}
$$

The remaining part of the proof repeats analog thoughts. ■

(4) If the PMD (4.91) is not minimal, then the system $\mathscr{S}_\tau$ is not completely modal controllable and the question about causal modal control can be solved on basis of the results of Theorem 3.37.

## 4.7 Stability and Backward Stabilization of System $\mathscr{S}_\tau$

(1)

**Definition 4.8** The system $\mathscr{S}_\tau$ is called stable, if for any initial conditions the corresponding solutions $v(t),\ y(t),\ \{\psi_k\}$ of Eq. (4.40) for $t > 0,\ k > 0$ satisfy the estimates

$$||v(t)|| \le c_v \mathrm{e}^{-\delta t}, \quad ||y(t)|| < c_y \mathrm{e}^{-\delta t}, \quad ||\psi_k|| < c_\psi \mathrm{e}^{-\delta kT}, \tag{4.108}$$

where $||\cdot||$ is any norm for finite dimensional number vectors, $c_v, c_y, c_\psi, \delta$ are positive constants, where $\delta$ does not depend on the initial conditions.

**Theorem 4.2** *For the stability of system $\mathscr{S}_\tau$, it is necessary and sufficient that discrete model (4.93) is stable, i.e., that the polynomial matrix $Q_\tau(\zeta, \alpha, \beta)$ (4.94) does not possess eigenvalues located on the disc $|\zeta| \le 1$ .*

*Proof* Necessity: Let estimates (4.108) hold. Then at $t = kT,\ k > 0$ for any solution $v(t),\ y(t),\ \{\psi_k\}$, we obtain

$$\begin{aligned} ||v_k|| &= ||v(kT)|| < c_v \mathrm{e}^{-\delta kT}, \\ ||y_k|| &= ||y(kT)|| < c_y \mathrm{e}^{-\delta kT}, \\ ||\psi_k|| &< c_\psi e^{-\delta kT}. \end{aligned} \tag{4.109}$$

Hence, because all the sequences $\{v_k\}, \{\xi_k\}, \{\psi_k\}$ coincide with the set of solutions of the B-discrete model (4.83), from (4.109) it follows that the B-discrete model is asymptotically stable.
Sufficiency: Let the matrix $Q_\tau(\zeta, \alpha, \beta)$ be stable. Then, as follows from [5], for any solution of B-discrete model (4.81) estimates of form (4.109) are valid. Due to (4.68)

$$v_k(\varepsilon) = v(kT + \varepsilon) = \mathrm{e}^{A\varepsilon} v_k + \sum_{i=0}^{\Omega} \Gamma_{2i}(\varepsilon)\psi_{k-n_i-1} + \sum_{i=0}^{\Omega} \Gamma_{1i}(\varepsilon)\psi_{k-n_i}, \quad 0 \le \varepsilon \le T. \tag{4.110}$$

Estimating the left and right sides of this relation according to the norm, we find that for $0 \le \varepsilon \le T$

$$||v(kT + \varepsilon)|| \le L_1 ||v_k|| + \sum_{i=0}^{\Omega} L_{2i} ||\psi_{k-n_i-1}|| + \sum_{i=0}^{\Omega} L_{1i} ||\psi_{k-n_i}||, \tag{4.111}$$

where we used the notations

$$L_1 = \max_{0 \le \varepsilon \le T} ||e^{A\varepsilon}||, \quad L_{\alpha i} = \max_{0 \le \varepsilon \le T} ||\Gamma_{\alpha i}(\varepsilon)||, \quad (\alpha = 1, 2). \tag{4.112}$$

When (4.109) is true, we obtain for nonnegative integers $p$ and $k - p > 0$

$$||\psi_{k-p}|| \le c_\psi \mathrm{e}^{\delta pT} \mathrm{e}^{-k\delta T}. \tag{4.113}$$

Therefore, for sufficiently large $k$, from (4.111)–(4.113), we obtain

$$||v(t)|| \leq L\mathrm{e}^{-k\delta T}, \quad kT \leq t \leq (k+1)T, \tag{4.114}$$

where

$$L = L_1 c_v + c_\psi \left[ \sum_{i=0}^{\Omega} L_{2i} e^{(n_i+1)\delta T} + \sum_{i=0}^{\Omega} L_{1i} \mathrm{e}^{n_i \delta T} \right] \tag{4.115}$$

is a positive constant, not depending on $k$. From (4.114), we conclude that for $t > 0$

$$||v(t)|| < L_v \mathrm{e}^{-\delta t} \tag{4.116}$$

with a positive constant $L_v$. From relation

$$y(t) = \sum_{\ell=0}^{\Lambda} c_\ell v(t - \tau_\ell) \tag{4.117}$$

and (4.116), we find an estimation of the form

$$||y(t)|| \leq L_y \mathrm{e}^{-\delta t}, \quad t > 0, \tag{4.118}$$

which completes the proof. ■

(2) Assume that the polynomial pairs

$$(a^*(\zeta), b^*(\zeta)), \quad [a^*(\zeta), c^*(\zeta)] \tag{4.119}$$

are not irreducible.

**Definition 4.9** Let $\phi_l(\zeta)$ be the GCD of the matrices $a^*(\zeta), b^*(\zeta)$ and $\phi_r(\zeta)$ the GCD of the matrices $a^*(\zeta), c^*(\zeta)$. Then pairs (4.119) are called stabilizable, if the matrices $\phi_l(\zeta)$ and $\phi_r(\zeta)$ are stable, respectively.

**Definition 4.10** The pairs of constant matrices $(A, B)$, $[A, C]$ are called stabilizable, if the polynomial pairs $(I - \zeta A, B)$, $(I - \zeta A', C')$ are stabilizable.

(3) The causal stabilization problem for system $\mathscr{S}_\tau$ is a special case of the modal control problem for PMD processes considered in Chap. 3, and can be formulated in the following way:

> For given PMD (4.91), find the set of causal controllers $(\alpha(\zeta), \beta(\zeta))$, such that relations (4.95) are fulfilled for some stable polynomial $f(\zeta)$.

As earlier, the system $\mathscr{S}_\tau$ is called stabilizable, if there exists a controller $(\alpha(\zeta), \beta(\zeta))$, such that the system $\mathscr{S}_\tau$ becomes stable. The adequate controller is called stabilizing. Applying Theorem 3.39 yields the following result.

**Theorem 4.3** *The system $\mathscr{S}_\tau$ is stabilizable, if and only if the pairs* $(\mathrm{e}^{-AT}, B)$, $[\mathrm{e}^{-AT}, C]$ *are stabilizable, where $B, C$ are matrices (4.97).* ∎

## 4.8 Forward Discrete Model of a Continuous Process with Delay in State Space

(1) For the solution of a number of control problems for sampled-data systems with delay, the use of forward discrete models of the continuous process in state space is convenient. The method for the construction of those models is described in [6] and is applied in [7]. Furthermore, the explained method is called standard. In this section, the standard method is analyzed from a general point of view, and it is established that its application not always leads to realizations of minimal dimension. Necessary and sufficient conditions are derived, under which the standard method yields minimal realizations. Moreover, a general approach for the construction of minimal realizations is provided, when these conditions are violated.

(2) Consider system $\mathscr{S}_\tau$ (4.40) in absence of delay at the output of the continuous process, which is described by

$$\begin{aligned} \frac{\mathrm{d}v(t)}{\mathrm{d}t} &= Av(t) + \sum_{i=0}^{\Omega} B_i u_i(t-\mu_i), \\ y(t) &= Cv(t). \end{aligned} \tag{4.120}$$

Moreover,

$$u_i(t) = h_i(t-kT)\psi_k, \quad kT < t < (k+1)T. \tag{4.121}$$

In the considered case, the discrete model of the continuous process, using (4.68) for $\varepsilon = T$, can be represented in the form

$$\begin{aligned} v_{k+1} &= \mathrm{e}^{AT} v_k + \sum_{i=0}^{\Omega} \Gamma_{1i}(T)\psi_{k-n_i} + \sum_{i=0}^{\Omega} \Gamma_{2i}(T)\psi_{k-n_i-1}, \\ \xi_k &= y(kT) = Cv_k. \end{aligned} \tag{4.122}$$

The standard method for the construction of the forward model in state space on basis of Eq. (4.122) consists of the following [1]. Collecting the terms with equal indices, the first equation in (4.122) can be written in the form

$$v_{k+1} = \mathrm{e}^{AT} v_k + \sum_{\lambda=0}^{R} D_\lambda \psi_{k-\lambda}, \quad R = n_\Omega + 1, \tag{4.123}$$

where $D_\lambda$ are constant $\chi \times q$ matrices, some of which could be zero. Comparing (4.122) and (4.123), we obtain

$$\sum_{\lambda=0}^{R} D_\lambda \psi_{k-\lambda} = \sum_{i=0}^{\Omega} \Gamma_{1i}(T)\psi_{k-n_i} + \sum_{i=0}^{\Omega} \Gamma_{2i}(T)\psi_{k-n_i-1}. \tag{4.124}$$

With the help of the operator $\zeta$, the last equation can be written in the form

$$\left(\sum_{\lambda=0}^{R} D_\lambda \zeta^\lambda\right)\psi_k = \left(\sum_{i=0}^{\Omega} \Gamma_{1i}(T)\zeta^{n_i} + \sum_{i=0}^{\Omega} \Gamma_{2i}(T)\zeta^{n_i+1}\right)\psi_k\,, \tag{4.125}$$

which leads to the equation

$$\sum_{\lambda=0}^{R} D_\lambda \zeta^\lambda = \sum_{i=0}^{\Omega} \Gamma_{1i}(T)\zeta^{n_i} + \sum_{i=0}^{\Omega} \Gamma_{2i}(T)\zeta^{n_i+1}. \tag{4.126}$$

Using (4.124), Eq. (4.122) can be written in the form

$$\begin{aligned} v_{k+1} &= \mathrm{e}^{AT}\psi_k + \sum_{\lambda=0}^{R} D_\lambda \psi_{k-\lambda} \\ \xi_k &= C v_k\,. \end{aligned} \tag{4.127}$$

The standard approach for building a forward model of system (4.127) in state space consists of the following [1].

Introduce the $q \times 1$ vectors

$$\eta_{1,k} \stackrel{\triangle}{=} \psi_{k-R}, \quad \eta_{2,k} \stackrel{\triangle}{=} \psi_{k-R+1} \quad, \ldots, \quad \eta_{R,k} \stackrel{\triangle}{=} \psi_{k-1}. \tag{4.128}$$

Then by construction

$$\eta_{i,k+1} = \eta_{i+1,k}, \ (i = 1, 2, \ldots, R-1), \quad \eta_{R,k+1} = \psi_k\,. \tag{4.129}$$

Introduce the extended state vector $x_k$ of size $(\chi + Rq) \times 1$

$$x_k \stackrel{\triangle}{=} \begin{bmatrix} v_k \\ \psi_{k-R} \\ \cdots \\ \psi_{k-2} \\ \psi_{k-1} \end{bmatrix} = \begin{bmatrix} v_k \\ \eta_{1,k} \\ \cdots \\ \eta_{R-1,k} \\ \eta_{R,k} \end{bmatrix}. \tag{4.130}$$

Then, using (4.127) and (4.129), we obtain

$$x_{k+1} = \begin{bmatrix} \upsilon_{k+1} \\ \eta_{1,k+1} \\ \cdots \\ \eta_{R-1,k+1} \\ \eta_{R,k+1} \end{bmatrix} = \begin{bmatrix} e^{AT}\upsilon_k + D_R\eta_{1,k} + D_{R-1}\eta_{2,k} + \cdots + D_1\eta_{R,k} + D_0\psi_k \\ \eta_{2,k} \\ \cdots \\ \eta_{R,k} \\ \psi_k \end{bmatrix}. \tag{4.131}$$

The last relation can be written in state space form

$$x_{k+1} = \bar{A}x_k + \bar{B}\psi_k, \tag{4.132}$$

where $\bar{A}$ and $\bar{B}$ are constant matrices of sizes $(\chi + Rq) \times (\chi + Rq)$ and $(\chi + Rq) \times q$, respectively. The matrix $\bar{A}$ can be represented in block form

$$\bar{A} = \begin{bmatrix} e^{AT} & D \\ 0_{Rq,\chi} & J_{Rq} \end{bmatrix}, \tag{4.133}$$

where

$$D \triangleq \begin{bmatrix} D_R \; D_{R-1} \; \ldots \; D_1 \end{bmatrix} \tag{4.134}$$

is a $\chi \times Rq$ matrix. Moreover

$$J_{Rq} = \begin{bmatrix} 0_{qq} & I_q & 0_{qq} & \ldots & 0_{qq} & 0_{qq} \\ 0_{qq} & 0_{qq} & I_q & \ldots & 0_{qq} & 0_{qq} \\ \cdots & \cdots & \cdots & \cdots & \cdots & \cdots \\ 0_{qq} & 0_{qq} & 0_{qq} & \ldots & 0_{qq} & I_q \\ 0_{qq} & 0_{qq} & 0_{qq} & \ldots & 0_{qq} & 0_{qq} \end{bmatrix} \tag{4.135}$$

is a block Jordan chain of size $Rq \times Rq$. In addition, the matrix $\bar{B}$ in (4.132) has the form

$$\bar{B} \triangleq \begin{matrix} \\ \chi \\ (R-1)q \\ q \end{matrix} \begin{matrix} q \\ \begin{bmatrix} D_0 \\ 0 \\ I_q \end{bmatrix} \end{matrix}. \tag{4.136}$$

The right side of (4.127) can also be expressed by the extended state vector $x_k$. Obviously,

$$\xi_k = \bar{C}x_k, \tag{4.137}$$

where the $n \times (\chi + Rq)$ matrix $\bar{C}$ proves to be

$$\bar{C} = \begin{bmatrix} C \; 0_{n,Rq} \end{bmatrix}. \tag{4.138}$$

All together, Eqs. (4.132) and (4.138) form the state equations

$$\begin{aligned} x_{k+1} &= \bar{A}x_k + \bar{B}\psi_k, \\ \xi_k &= \bar{C}x_k, \end{aligned} \tag{4.139}$$

which are called system $\mathscr{S}_R$. The totality of matrices $(\bar{A}, \bar{B}, \bar{C})$ is called a realization of the system $\mathscr{S}_R$.

The technique for the conversion from the discrete model (4.122) to the state space description (4.139) is relatively simple. Apparently, this fact explains the wide application of this approach for the description and investigation of sampled-data systems with delay [1, 8]. But in this connection, the question arises, in which case the system $\mathscr{S}_R$ becomes minimal, i.e., the pair $(\bar{A}, \bar{B})$ should be completely controllable, and the pair $[\bar{A}, \bar{C}]$ completely observable. Below, a complete answer to this question will be derived. Moreover, in this section, a method for the construction of a minimal F-discrete model in state space for the general system $\mathscr{S}_\tau$ is presented. Here, the case is included, in which delay acts at the output of the continuous process.

(3)

**Theorem 4.4** *Introduce the rational $n \times q$ matrix*

$$\tilde{W}_R(z) \triangleq C\left(zI_\chi - \mathrm{e}^{AT}\right)^{-1} \sum_{i=0}^{R} D_i z^{-i}. \tag{4.140}$$

*Then for the minimality of the realization $(\bar{A}, \bar{B}, \bar{C})$, it is necessary and sufficient that*

$$\operatorname{Mind} \tilde{W}_R(z) = \chi + Rq. \tag{4.141}$$

*Proof* The proof is given in several steps.

(a)

**Lemma 4.4** *The following relation is true:*

$$\tilde{W}(z) \triangleq \bar{C}(zI_{\chi+Rq} - A)^{-1}\bar{B} = \tilde{W}_R(z). \tag{4.142}$$

*Proof* From (4.133), we obtain

$$zI_{\chi+Rq} - \bar{A} = \begin{bmatrix} zI_\chi - \mathrm{e}^{AT} & -\bar{D} \\ 0_{Rq,\chi} & zI_{Rq} - J_{Rq} \end{bmatrix}. \tag{4.143}$$

Applying the formula for the inverse of a triangular block matrix

$$\begin{bmatrix} X & Y \\ 0 & Z \end{bmatrix} = \begin{bmatrix} X^{-1} & -X^{-1}YZ^{-1} \\ 0 & Z^{-1} \end{bmatrix}, \tag{4.144}$$

from (4.143), we achieve

$$(zI_{\chi+Rq} - \bar{A})^{-1} = \begin{bmatrix} (zI_\chi - e^{AT})^{-1} & Q(z) \\ 0_{Rq,\chi} & (zI_{Rq} - J_{Rq})^{-1} \end{bmatrix}, \tag{4.145}$$

where

$$Q(z) \triangleq (zI_\chi - A)^{-1} \bar{D} (zI_{Rq} - J_{Rq})^{-1}. \tag{4.146}$$

Since from (4.135), we find

$$zI_{Rq} - J_{Rq} = \begin{bmatrix} zI_q & -I_q & 0_{qq} & \dots & 0_{qq} & 0_{qq} \\ 0_{qq} & zI_q & -I_q & \dots & 0_{qq} & 0_{qq} \\ \dots & \dots & \dots & \dots & \dots & \dots \\ 0_{qq} & 0_{qq} & 0_{qq} & \dots & zI_q & -I_q \\ 0_{qq} & 0_{qq} & 0_{qq} & \dots & 0_{qq} & zI_q \end{bmatrix}, \tag{4.147}$$

we immediately conclude

$$(zI_{Rq} - J_{Rq})^{-1} = \begin{bmatrix} z^{-1}I_q & z^{-2}I_q & \dots & z^{-R+1}I_q & z^{-R}I_q \\ 0_{qq} & z^{-1}I_q & \dots & z^{-R+2}I_q & z^{-R+1}I_q \\ \dots & \dots & \dots & \dots & \dots \\ 0_{qq} & 0_{qq} & \dots & z^{-1}I_q & z^{-2}I_q \\ 0_{qq} & 0_{qq} & \dots & 0_{qq} & z^{-1}I_q \end{bmatrix}. \tag{4.148}$$

Inserting this expression into (4.146), with regard to (4.143), we achieve

$$Q(z) = (zI_\chi - e^{AT})^{-1} Q_1(z), \tag{4.149}$$

where

$$Q_1(z) \triangleq \begin{bmatrix} z^{-1}D_R & z^{-2}D_R + z^{-1}D_{R-1} & \dots & z^{-R}D_R + z^{-R+1}D_{R-1} + \dots + z^{-1}D_1 \end{bmatrix}. \tag{4.150}$$

With the help of (4.145)–(4.150), we find

$$\tilde{W}_c(z) \triangleq \mathrm{diag}\{(zI_{\chi+Rq} - \bar{A})^{-1}\bar{B} = \mathrm{diag}\{(zI_\chi - e^{AT})^{-1}, I_{Rq}\} \times \begin{bmatrix} I_\chi & z^{-1}D_R & z^{-2}D_R + z^{-1}D_{R-1} & \dots & z^{-R}D_R + \dots + z^{-1}D_1 \\ 0_{q\chi} & z^{-1}I_q & z^{-2}I_q & \dots & z^{-R}I_q \\ 0_{q\chi} & 0_{qq} & z^{-1}I_q & \dots & z^{-R+1}I_q \\ \dots & \dots & \dots & \dots & \dots \\ 0_{q\chi} & 0_{qq} & 0_{qq} & \dots & z^{-1}I_q \end{bmatrix} \begin{bmatrix} D_0 \\ 0_{R-1,q} \\ I_q \end{bmatrix} \tag{4.151}$$

which after multiplying the block matrices results in

$$\tilde{W}_c(z) = \begin{bmatrix} (zI_\chi - e^{AT})^{-1}(D_0 + z^{-1}D_1 + \cdots + z^{-R}D_R) \\ z^{-R}I_q \\ z^{-R+1}I_q \\ \cdots \\ z^{-1}I_q \end{bmatrix}. \tag{4.152}$$

On basis of this relation, from (4.142), (4.152), and (4.138), we find

$$\tilde{W}(z) = \bar{C}\tilde{W}_c(z) = C(zI_\chi - \mathrm{e}^{AT})^{-1}(D_0 + z^{-1}D_1 + \cdots + z^{-R}D_R). \tag{4.153}$$

■

(b) *Proof of Theorem* 4.4: Since due to (4.147)

$$\det(zI_{Rq} - J_{Rq}) = z^{Rq}, \tag{4.154}$$

so from (4.143), we find

$$\det(zI_{\chi+Rq} - \bar{A}) = \det(zI_\chi - \mathrm{e}^{AT})z^{Rq}. \tag{4.155}$$

Relation (4.142) yields that a necessary and sufficient condition for the minimality of the realization $(\bar{A}, \bar{B}, \bar{C})$ consists of

$$\operatorname{Mind}\tilde{W}(z) = \chi + Rq. \tag{4.156}$$

Due to $\tilde{W}(z) = \tilde{W}_R(z)$, the proof is complete. ■

(4) Substituting in (4.153) $z$ by $\zeta^{-1}$, we obtain the associated matrix

$$W_R(\zeta) = \tilde{W}_R(z)\Big|_{z=\zeta^{-1}} = C\left(I_\chi - \zeta\mathrm{e}^{AT}\right)^{-1}\zeta D_1(\zeta), \tag{4.157}$$

where

$$D_1(\zeta) \triangleq D_0 + \zeta D_1 + \cdots + \zeta^R D_R.$$

**Theorem 4.5** *The following statements are true:*

*(i) Condition (4.141) can only be fulfilled, when*

$$\operatorname{Mind}W_R(\zeta) = \chi. \tag{4.158}$$

*(ii) For the fulfillment of (4.158) it is necessary and sufficient that the conditions of Theorem 4.1 are valid, i.e., the pairs $(\mathrm{e}^{-AT}, B)$ and $[\mathrm{e}^{-AT}, C]$ are minimal, where the matrices $B$ and $C$ are determined by formulae (4.97).*

*Proof* (i) Let $z_i \neq 0$, $i = 1, 2, \ldots, \ell$, be the totality of different eigenvalues of the matrix $zI_\chi - \mathrm{e}^{AT}$. Then the set of poles of the matrix $\tilde{W}_R(z)$ consists of the numbers $z_i$ and the number $z_0 = 0$. Hereby, we obtain

$$\mathrm{Mind}\,\tilde{W}_R(z) = \mathrm{Mult}(\chi) + \mathrm{Mult}(0), \tag{4.159}$$

where $\mathrm{Mult}(\chi)$ is the sum of the McMillan multiplicities of the poles $z_i$, $(i = 1, \ldots, \ell)$, and $\mathrm{Mult}(0)$ is the McMillan multiplicity of the pole $z_0 = 0$. From the properties of the associated matrices, we find out that, when (4.159) is true, then

$$\mathrm{Mind}\,W_R(\zeta) = \mathrm{Mult}(\chi). \tag{4.160}$$

Therefore, Eq. (4.141) is true, if and only if

$$\mathrm{Mind}\,W_R(\zeta) = \chi, \tag{4.161}$$

i.e., claim (i) is shown.

(ii) Notice that due to (4.157), the matrix $W_R(\zeta)$ can be considered as transfer matrix of the PMD

$$\Pi_R(\zeta) \stackrel{\triangle}{=} (I_\chi - \zeta \mathrm{e}^{AT}, \zeta D_1(\zeta), c). \tag{4.162}$$

Therefore, the results from Chap. 3 yield, that condition (4.161) is satisfied if and only if the PMD $\Pi_R(\zeta)$ is minimal, i.e., the polynomial pairs $[i - \zeta \mathrm{e}^{AT}, c]$ and $(I - \mathrm{e}\mathrm{e}^{AT}, \zeta D_1(\zeta))$ are irreducible. But the irreducibility of the pair $[I - \zeta \mathrm{e}^{AT}, C]$ is equivalent to the observability of $[\mathrm{e}^{-AT}, C]$. Moreover, from Corollary 2.5, we find that the irreducibility of the pair $(I_\chi - \zeta \mathrm{e}^{AT}, D_1(\zeta))$ is equivalent to the controllability of the pair $(\mathrm{e}^{-AT}, \mathrm{e}^{-AT} D_{1\ell}(\mathrm{e}^{-AT}))$, where $D_{1\ell}(\mathrm{e}^{-AT})$ is the result, when the matrix $\mathrm{e}^{-AT}$ is substituted from left for the argument $\zeta$. Finally, with the help of (4.126) and (4.100), (4.103), we obtain

$$\mathrm{e}^{-AT} D_{1\ell}(\mathrm{e}^{-AT}) = B, \tag{4.163}$$

so that claim (ii) is proven. ∎

The eventual outcome is formulated next.

**Theorem 4.6** *For the minimality of the realization $(\tilde{A}, \tilde{B}, \tilde{C})$, it is necessary and sufficient that the following two conditions (a) and (b) are fulfilled:*

*(a) The pairs $(\mathrm{e}^{-AT}, B)$ and $[\mathrm{e}^{-AT}, C]$ are minimal.*

*(b)*

$$\mathrm{rank}\, C \mathrm{e}^{AT} D_R = q. \tag{4.164}$$

*Proof Necessity:* As follows from Theorem 4.5, for the validity of (4.141), condition (a) is necessary. It remains to show that condition (b) is necessary. Thereto, notice that when (4.158) is valid, then

$$\mathrm{Mind}\,\tilde{W}_R(z) = \chi + \mathrm{Mult}(0), \tag{4.165}$$

where, as above, $\mathrm{Mult}(0)$ is the McMillan multiplicity of the zero pole of the matrix $\tilde{W}_R(z)$. Therefore, (4.141) becomes true, only if

$$\mathrm{Mult}(0) = Rq. \tag{4.166}$$

Using (4.140), we write the matrix $\tilde{W}_R(z)$ in the form

$$\tilde{W}_R(z) = \frac{\tilde{L}_R(z)}{\det(zI_\chi - \mathrm{e}^{AT})z^R}, \tag{4.167}$$

where $\tilde{L}_R(z)$ is a polynomial $n \times q$ matrix of the form

$$\tilde{L}_R(z) \triangleq C\mathrm{adj}(zI_\chi - \mathrm{e}^{AT})(z^R D_0 + z^{R-1} D_1 + \cdots + D_R), \tag{4.168}$$

where adj means the adjoint of the matrix. From (4.167) and (2.69), we find

$$\mathrm{Mult}(0) \le R \min\{n, q\}.$$

Since for $n < q$ Eq. (4.166) cannot be satisfied. For $n \ge q$, Eq. (4.166) means that the McMillan multiplicity of the zero pole of the matrix $\tilde{W}_R(z)$ should have the maximal possible value equal to $Rq$. With the help of Lemma 2.3, this fact takes place if and only if the number $z_0 = 0$ is a latent number of the matrix $\tilde{L}_R(z)$, which is equivalent to

$$\mathrm{rank}\tilde{L}_R(z) = q. \tag{4.169}$$

From (4.168) for $z = 0$, we find

$$\tilde{L}_R(0) = -C(\mathrm{adj}\mathrm{e}^{AT})D_R$$

and due to $\det \mathrm{e}^{AT} \ne 0$, relation (4.169) is equivalent to

$$\mathrm{rank}C(\mathrm{adj}\mathrm{e}^{AT})D_R = \mathrm{rank}A\mathrm{e}^{AT}D_R = q. \tag{4.170}$$

The necessity of the condition in Theorem 4.6 is shown.
*Sufficiency:* When condition (b) in (4.168) is fulfilled, we obtain $\mathrm{Mult}(0) = Rq$. Under condition (a) The PMD (4.162) is irreducible, and from (4.160), (4.161), we obtain $\mathrm{Mult}(\chi) = \chi$. Thus, Eq. (4.141) holds and the realization $(\tilde{A}, \tilde{B}, \tilde{C})$ is irreducible. ■

*Remark 4.1* From the proof of Theorem 4.6, the validity of Theorem 4.4 follows directly.

Relation (4.164) can be derived by the coefficients of the first equation of (4.122). For that reason notice that from (4.125) for $R = n_\Omega + 1$, we find

$$D_R = \Gamma_{2,n_\Omega}(T),$$

where the matrix $\Gamma_{2,n_\Omega}(T)$ is determined from (4.72). With the help of this formula, relation (4.170) takes the form

$$\operatorname{rank}\left[Ce^{2AT}e^{A\eta_\Omega}\int_{\eta_\Omega}^{T}e^{-Au}B_\Omega h_\Omega(u)\,du\right]=q. \tag{4.171}$$

(5) The following claim yields a general method for constructing an F-discrete model of process (4.120) in the general case, when at least one of the conditions of Theorem 4.6 is violated.

**Theorem 4.7** *Let at least one of the conditions of Theorem 4.6 be violated. Then the realization* $(\bar{A}, \bar{B}, \bar{C})$ *in (4.139) is not minimal. In that case the minimal realization of F-discrete model (4.139) in state space can be constructed as a minimal realization of the strictly proper rational matrix*

$$\tilde{W}_R(z)=C(zI_\chi-e^{AT})^{-1}\sum_{j=0}^{R}D_jz^{-j}. \tag{4.172}$$

*Proof* Renaming in (4.123) $k$ by $k-1$ and using the second equation in (4.122), we obtain the backward discrete model of the continuous process

$$\begin{aligned}
v_k &= e^{AT}v_{k-1}+\sum_{j=0}^{R}D_j\psi_{k-j-1},\\
\xi_k &= Cv_k,
\end{aligned} \tag{4.173}$$

which can be written in operator form

$$\begin{aligned}
v_k &= \zeta e^{AT}v_k+\zeta\sum_{j=0}^{R}D_j\zeta^j\psi_k,\\
\xi_k &= Cv_k.
\end{aligned} \tag{4.174}$$

Then the transfer matrix of this model satisfies

$$W_R(\zeta)=\zeta C(I_\chi-\zeta e^{AT})^{-1}\sum_{j=0}^{R}D_j\zeta^j. \tag{4.175}$$

Since $W_R(0)=0_{nq}$, the matrix $W_R(\zeta)$ is strictly causal. It is easy to see that the strictly causal matrix (4.172) is associated to the matrix $W_R(\zeta)$. Hereby, the realization $(\bar{A}, \bar{B}, \bar{C})$ in (4.139) is a realization of the matrix $\tilde{W}_R(z)$, which has the order $\chi + Rq$.

If the conditions of Theorems 4.6 are violated, then this realization is not minimal. In this case, a fair minimal realization of the forward discrete model can be held as minimal realization of matrix (4.172). ∎

(7)

*Example 4.6* We consider an example connecting with the digital control of a mixing tank with delay of one sampling period [8]. The continuous process is described by

$$\frac{\mathrm{d}\upsilon(t)}{\mathrm{d}t} = A\upsilon(t) + B_0 u_0(t) + B_1 u(t-T), \quad y(t) = C\upsilon(t), \tag{4.176}$$

where $\upsilon(t), u(t), y(t)$ are $2 \times 1$ vectors. Moreover,

$$A = \begin{bmatrix} \alpha_1 & 0 \\ 0 & \alpha_2 \end{bmatrix}, \quad B_0 = \begin{bmatrix} a & a \\ 0 & 0 \end{bmatrix}, \quad B_1 = \begin{bmatrix} 0 & 0 \\ c & b \end{bmatrix}, \quad C = I_2 = \begin{bmatrix} 1 & 0 \\ 0 & 1 \end{bmatrix}, \tag{4.177}$$

where $\alpha_1, \alpha_2, a, b, c$ are real constants, and $\alpha_1 \neq \alpha_2,\ c \neq b,\ a \neq 0$. Moreover, assume that the control system applies a zero-order hold, such that

$$\begin{aligned} u(t) &= \psi_k, \quad kT < t < (k+1)T, \\ u(t-T) &= \psi_{k-1}, \quad kT < t < (k+1)T, \end{aligned} \tag{4.178}$$

where $\psi_k$ is a $2 \times 1$ vector. Equation (4.176) is a special case of Eq. (4.40) under the conditions

$$\begin{aligned} &R = 1, \quad \chi = 2, \quad \Omega = 1, \quad \mu_0 = 0, \quad \mu_1 = T, \quad n_0 = 0, \quad \lambda_0 = 0, \\ &n_1 = 1, \quad \lambda_1 = 0, \quad \Lambda = 0, \quad \tau_0 = 0, \quad m_0 = 0, \quad \theta_0 = 0. \end{aligned} \tag{4.179}$$

By not so hard calculations, we observe that the discrete process model in the present case (4.176) is obtained from (4.71) when substituting $k$ by $k+1$

$$\upsilon_{k+1} = \mathrm{e}^{AT}\upsilon_k + \Gamma_{20}(T)\psi_{k-1} + \Gamma_{10}(T)\psi_k + \Gamma_{21}(T)\psi_{k-2} + \Gamma_{11}(T)\psi_{k-1}, \tag{4.180}$$

where the matrices $\Gamma_{ik}(T)$ are defined by formulae (4.72). Using that in our case $\eta_0 = \eta_1 = T, h_i(T) = I_2$, from (4.72), we derive

$$\begin{aligned} \Gamma_{10}(T) &= \mathrm{e}^{AT}\int_0^T \mathrm{e}^{-Ax}\,\mathrm{d}x\, B_0 \\ \Gamma_{11}(T) &= \mathrm{e}^{AT}\int_0^T \mathrm{e}^{-Ax}\,\mathrm{d}x\, B_1 \\ \Gamma_{20}(T) &= \Gamma_{21}(T) = 0_{22}. \end{aligned} \tag{4.181}$$

Then from (4.180), we obtain

$$\upsilon_{k+1} = \mathrm{e}^{AT}\upsilon_k + \Gamma_{11}(T)\psi_{k-1} + \Gamma_{10}(T)\psi_k, \tag{4.182}$$

which is equivalent to representation (4.123) with

$$D_0 = \Gamma_{10}(T), \quad D_R = D_1 = \Gamma_{11}(T). \tag{4.183}$$

Since from (4.177), we find

$$\mathrm{e}^{At} = \begin{bmatrix} \mathrm{e}^{\alpha_1 t} & 0 \\ 0 & \mathrm{e}^{\alpha_2 t} \end{bmatrix}, \quad \mathrm{e}^{-At} = \begin{bmatrix} \mathrm{e}^{-\alpha_1 t} & 0 \\ 0 & \mathrm{e}^{-\alpha_2 t} \end{bmatrix}, \tag{4.184}$$

so

$$\int_0^T \mathrm{e}^{-Ax}\,\mathrm{d}x = A^{-1}(I_2 - \mathrm{e}^{-AT}) = \begin{bmatrix} \frac{1-\mathrm{e}^{-\alpha_1 T})}{\alpha_1} & 0 \\ 0 & \frac{1-\mathrm{e}^{-\alpha_2 T}}{\alpha_2} \end{bmatrix}. \tag{4.185}$$

Introduce

$$q_1 \stackrel{\triangle}{=} \frac{\mathrm{e}^{\alpha_1 T - 1}}{\alpha_1}, \quad q_2 \stackrel{\triangle}{=} \frac{\mathrm{e}^{\alpha_2 T - 1}}{\alpha_2}. \tag{4.186}$$

Thus

$$D_0 = \begin{bmatrix} aq_1 & aq_1 \\ 0 & 0 \end{bmatrix}, \quad D_1 = \begin{bmatrix} 0 & 0 \\ cq_2 & bq_2 \end{bmatrix}. \tag{4.187}$$

Formulae (4.133), (4.136), and (4.13) with the help of (4.170) lead to a realization $(\bar{A}, \bar{B}, \bar{C})$ of fourth order, where the matrices are determined by the relations

$$\bar{A} = \begin{bmatrix} \mathrm{e}^{\alpha_1 T} & 0 & 0 & 0 \\ 0 & \mathrm{e}^{\alpha_2 T} & cq_2 & bq_2 \\ 0 & 0 & 0 & 1 \\ 0 & 0 & 0 & 0 \end{bmatrix}, \quad \bar{B} = \begin{bmatrix} aq_1 & aq_1 \\ 0 & 0 \\ 1 & 0 \\ 0 & 1 \end{bmatrix}, \quad \bar{C} = \begin{bmatrix} 1 & 0 & 0 & 0 \\ 0 & 1 & 0 & 0 \end{bmatrix}. \tag{4.188}$$

In the given case, we find from (4.187) that $\operatorname{rank} D_R = 1$. Therefore, the necessary condition for the minimality of realization (4.188) is not fulfilled, and hence this realization is not minimal. In order to construct a minimal realization, we apply Theorem 4.7. Since in the present example

$$(zI_2 - A)^{-1} = \begin{bmatrix} \frac{1}{z-\mathrm{e}^{\alpha_1 T}} & 0 \\ 0 & \frac{1}{z-\mathrm{e}^{\alpha_2 T}} \end{bmatrix} \tag{4.189}$$

so from (4.167), (4.168), we obtain

$$\tilde{W}_R(z) = \frac{\tilde{L}_R(z)}{\tilde{\Delta}(z)}, \tag{4.190}$$

where

$$\tilde{L}_R(z) \triangleq \begin{bmatrix} (z-\mathrm{e}^{\alpha_2 T})zaq_1 & (z-\mathrm{e}^{\alpha_2 T})zaq_1 \\ (z-\mathrm{e}^{\alpha_1 T})cq_2 & (z-\mathrm{e}^{\alpha_1 T})bq_2 \end{bmatrix},$$
$$\tilde{\Delta}(z) \triangleq z(z-\mathrm{e}^{\alpha_1 T})(z-\mathrm{e}^{\alpha_2 T}). \tag{4.191}$$

In order to construct a relevant minimal realization of the matrix $\tilde{W}_R(z)$, we pick up on a method presented in [9]. Hence, we build the partial fraction decomposition of matrix (4.190)

$$\tilde{W}_R(z) = \frac{P_1}{z-\mathrm{e}^{\alpha_1 T}} + \frac{P_2}{z-\mathrm{e}^{\alpha_2 T}} + \frac{P_0}{z}, \tag{4.192}$$

where

$$P_1 = \begin{bmatrix} q_1 a & q_1 a \\ 0 & 0 \end{bmatrix}, \quad P_2 = \begin{bmatrix} 0 & 0 \\ \mathrm{e}^{-\alpha_2 T} q_2 c & \mathrm{e}^{-\alpha_2 T} q_2 b \end{bmatrix},$$
$$P_0 = \begin{bmatrix} 0 & 0 \\ -\mathrm{e}^{-\alpha_2 T} q_2 c & -\mathrm{e}^{-\alpha_2 T} q_2 b \end{bmatrix} = -p_2. \tag{4.193}$$

Obviously,

$$P_i = Q_i R_i', \quad (i = 0, 1, 2), \tag{4.194}$$

where

$$Q_1 \triangleq \begin{bmatrix} 1 \\ 0 \end{bmatrix}, \quad Q_2 \triangleq \begin{bmatrix} 0 \\ 1 \end{bmatrix}, \quad Q_0 \triangleq \begin{bmatrix} 0 \\ 1 \end{bmatrix},$$
$$R_1 \triangleq \begin{bmatrix} aq_1 \\ aq_1 \end{bmatrix}, \quad R_2 \triangleq \begin{bmatrix} \mathrm{e}^{-\alpha_2 T} cq_2 \\ \mathrm{e}^{-\alpha_2 T} bq_2 \end{bmatrix}, \quad R_0 = -R_2. \tag{4.195}$$

From (4.192)–(4.195) and [6], it becomes obvious that the transfer matrix $\tilde{W}_R(z)$ must be represented by a minimal realization of third order $(\bar{A}_1, \bar{B}_1, \bar{C}_1)$, where

$$\bar{A}_1 = \begin{bmatrix} \mathrm{e}^{\alpha_1 T} & 0 & 0 \\ 0 & \mathrm{e}^{\alpha_2 T} & 0 \\ 0 & 0 & 0 \end{bmatrix}, \quad \bar{B}_1 = \begin{bmatrix} R_1' \\ R_2' \\ R_0' \end{bmatrix} = \begin{bmatrix} aq_1 & aq_1 \\ \mathrm{e}^{-\alpha_2 T} cq_2 & \mathrm{e}^{-\alpha_2 T} bq_2 \\ -\mathrm{e}^{-\alpha_2 T} cq_2 & -\mathrm{e}^{-\alpha_2 T} bq_2 \end{bmatrix},$$
$$\bar{C}_1 = \begin{bmatrix} Q_1 & Q_2 & Q_0 \end{bmatrix} = \begin{bmatrix} 1 & 0 & 0 \\ 0 & 1 & 1 \end{bmatrix} \tag{4.196}$$

which corresponds to a minimal realization of dimension three

$$\tilde{W}_R(z) = \bar{C}_1 (zI_3 - \bar{A}_1)^{-1} \bar{B}_1 \tag{4.197}$$

and a minimal discrete forward model in state space

$$\begin{aligned} \upsilon_{k+1} &= \bar{A}_1 \upsilon_k + \bar{B}_1 \psi_k, \\ \xi_k &= \bar{C}_1 \upsilon_k. \end{aligned} \tag{4.198}$$

## References


1. K.J. Åström, B. Wittenmark, *Computer Controlled Systems: Theory and Design*, 3rd edn. (Prentice-Hall, Englewood Cliffs, 1997)
2. E.N. Rosenwasser, B.P. Lampe, *Digitale Regelung in kontinuierlicher Zeit - Analyse und Entwurf im Frequenzbereich* (B.G. Teubner, Stuttgart, 1997)
3. E.N. Rosenwasser, B.P. Lampe, *Computer Controlled Systems: Analysis and Design with Process-orientated Models* (Springer, Berlin, 2000)
4. W.B. Cleveland, First-order-hold interpolation dogital-to-analog converter with application to aircraft simulation, NASA technical note, NASA TN D-8331 (1982) Washington, D.C
5. E.N. Rosenwasser, B.P. Lampe, *Multivariable Computer-controlled Systems–A Transfer Function Approach* (Springer, London, 2006)
6. B.P. Lampe, E.N. Rosenwasser, Controllability and observability of discrete-time models for sampled continuous-time processes with higher order holds and delay. Autom. Remote Control **68**(4), 593–609 (2007)
7. B.P. Lampe, E.N. Rosenwasser, Polynomial modal control for sampled-data systems with delay. Autom. Remote Control **69**(6), 953–967 (2008)
8. H. Kwakernaak, R. Sivan, *Linear Optimal Control Systems* (Wiley-Interscience, New York, 1972)
9. E.G. Gilbert, Controllability and observability in multivariable control systems. SIAM J. Control **A**(1), 128–151 (1963)

# Chapter 5
# Parametric Discrete Models of Multivariable Continuous Processes with Delay

**Abstract** The chapter is devoted to the construction of discrete parametric (i.e., depending on the continuous time as a parameter) models of multidimensional continuous processes with delay. The method used to construct such models is based on the concept of the discrete Laplace transform (DLT) of continuous argument functions and the extension of the concept of the displaced pulse frequency response (DPFR) introduced in Rosenwasser and Lampe (Computer Controlled Systems: Analysis and Design with Process-orientated Models. Springer, Berlin, 2000.) for processes with delay. On this basis, a general solution of the problem of the transfer of an exponentially periodic (EP) signal through a linear stationary system with delay and through an open SD system with delay and a generalized holding element is obtained.

## 5.1 Functions of Matrices

(1) This section on basis of [2] presents some results of the theory of matrix functions, which are necessary for the further investigations. Let $A$ be a constant $\chi \times \chi$ matrix, then

$$A_\lambda \stackrel{\triangle}{=} \lambda I_\chi - A \tag{5.1}$$

is its characteristic matrix,

$$d_A(\lambda) = \det(\lambda I_\chi - A), \tag{5.2}$$

its characteristic polynomial, and $d_{\min}(\lambda)$ its minimal polynomial. As already known [2], matrix $A_\lambda^{-1} = (\lambda I_\chi - A)^{-1}$ is strictly proper and it allows the irreducible representation

$$(\lambda I_\chi - A)^{-1} = \frac{\tilde{\text{adj}}(\lambda I_\chi - A)}{d_{\min}(\lambda)}, \tag{5.3}$$

where the adjoint matrix arises in the numerator. Moreover, we use the representation



E. N. Rosenwasser et al., *Computer-Controlled Systems with Delay*,
https://doi.org/10.1007/978-3-030-15042-6_5

$$d_{\min}(\lambda) = (\lambda - \lambda_1)^{\nu_1} \cdots (\lambda - \lambda_q)^{\nu_q}, \tag{5.4}$$

where $\lambda_1, \ldots, \lambda_q$ are the different eigenvalues of the matrix $A$. Since the matrix $A_\lambda^{-1}$ is strictly proper, we may decompose it into partial fractions

$$(\lambda I_\chi - A)^{-1} = \frac{M_{11}}{\lambda - \lambda_1} + \cdots + \frac{M_{1,\nu_1}}{(\lambda - \lambda_1)^{\nu_1}} + \cdots + \frac{M_{q1}}{\lambda - \lambda_q} + \cdots + \frac{M_{q,\nu_q}}{(\lambda - \lambda_q)^{\nu_q}}, \tag{5.5}$$

where $M_{ij}$ are constant matrices, which can be calculated with the help of (2.50). The irreducibility of representation (5.3) implies

$$M_{i,\nu_i} \neq 0_{\chi\chi}, \quad (i = 1, \ldots, q). \tag{5.6}$$

(2) Introduce

$$Z_{ij} \triangleq \frac{M_{ij}}{(j-1)}, \quad (i = 1, \ldots, q; \quad j = 1, \ldots, \nu_q). \tag{5.7}$$

The constant matrices $Z_{ij}$ are called generating matrices of the matrix $A$. Each eigenvalue $\lambda_i$ of the matrix $A$ is related to $\nu_i$ generating matrices

$$Z_{i1}, \ldots, Z_{i,\nu_i}, \quad (i = 1, \ldots, q). \tag{5.8}$$

The totality of matrices (5.8) is called the set of components of the matrix $A$ according to the eigenvalue $\lambda_i$. The total number of all components of the matrix $A$ is equal to the degree of its minimal polynomial.

Following [2, 3], we will list a number of important properties of components (5.8).

(a) If the matrix $A$ is real, and $\lambda_1, \lambda_2$ are two of its conjugate complex eigenvalues, then the components according to this eigenvalues are conjugate complex too. The components related to real eigenvalues are real.
(b) The components of the matrix $A$ are commutative to each other, i.e.,

$$Z_{in} Z_{\ell m} = Z_{\ell m} Z_{in}. \tag{5.9}$$

(c) None of components (5.8) is ever equal to the zero matrix.
(d) The components of the matrix $A$ are linearly independent, i.e., the equality

$$\sum_{i=1}^{q} \sum_{k=1}^{\nu_i} c_{ik} Z_{ik} = 0_{\chi\chi}, \tag{5.10}$$

where $c_{in}$ are scalar constants, is only possible if all numbers $c_{in}$ vanish.
(e) Introduce the notation

$$Z_{i1} \stackrel{\Delta}{=} Q_i, \quad (i = 1, \ldots, q). \tag{5.11}$$

The matrices $Q_i$ are called projectors of the matrix $A$. The projectors $Q_i$ possess a number of additional properties listed below.

(i) The equations

$$Q_i Z_{ik} = Z_{ik} Q_i = Z_{ik}, \quad (k = 1, \ldots, \nu_i), \tag{5.12}$$

hold, and also

$$Q_i Z_{\ell k} = Z_{\ell k} Q_i = 0_{\chi\chi}, \quad i \neq \ell. \tag{5.13}$$

(ii) From (5.12), we obtain for $k = 1$

$$Q_i^2 = Q_i, \tag{5.14}$$

and from (5.13) it follows

$$Q_i Q_\ell = Q_\ell Q_i = 0_{\chi\chi}, \quad (i \neq \ell). \tag{5.15}$$

Moreover,

$$\sum_{i=1}^{q} Q_i = I_\chi. \tag{5.16}$$

(3) Let the matrix $A$ have minimal polynomial (5.4), and $f(\lambda)$ be some scalar function. We will say that the function $f(\lambda)$ is defined on the spectrum of the matrix $A$, if there exist the numbers

$$\begin{array}{l} f(\lambda_1), f'(\lambda_1), \ldots, f^{(\nu_1-1)}(\lambda_1), \\ f(\lambda_2), f'(\lambda_2), \ldots, f^{(\nu_2-1)}(\lambda_2), \\ \qquad \cdots\cdots\cdots\cdots\cdots\cdots \\ f(\lambda_q), f'(\lambda_q), \ldots, f^{(\nu_q-1)}(\lambda_q). \end{array} \tag{5.17}$$

The totality of numbers (5.17) is called the values of the function $f(\lambda)$ on the spectrum of matrix $A$.

Assume that the function $f(\lambda)$ takes on the spectrum of matrix $A$ values (5.17). Then there exists a nonempty set of polynomials, such that each polynomial takes values (5.17) on the spectrum of $A$. Let $h(\lambda)$ be one of those polynomials. Then by construction

$$h(\lambda_1) = f(\lambda_1), \ \ldots, \ h^{(\nu_1-1)}(\lambda_1) = f^{(\nu_1-1)}(\lambda_1),$$
$$\cdots\cdots\cdots\cdots\cdots\cdots \tag{5.18}$$
$$h(\lambda_q) = f(\lambda_q), \ \ldots, h^{(\nu_q-1)}(\lambda_q) = f^{(\nu_q-1)}(\lambda_q).$$

The entirety of Eq. (5.18) is symbolically written as

$$h(\Lambda_A) = f(\Lambda_A). \tag{5.19}$$

When (5.19) is fulfilled, then the function $f(A)$ of the matrix $A$ is defined by

$$f(A) = h(A), \tag{5.20}$$

and moreover, if

$$h(\lambda) = h_0\lambda^n + h_{n-1}\lambda^{n-1} + \cdots + h_n,$$

then

$$h(A) = h_0 A^n + h_{n-1}A^{n-1} + \cdots + h_n I_\chi.$$

It is known that the matrix $f(A)$, defined by formula (5.20), does not depend on the special choice of the polynomial $h(\lambda)$, when condition (5.19) holds.

Aside from the above said, the matrix $f(A)$ can be build with the help of the components of the matrix $A$ by the formula

$$f(A) = \sum_{i=1}^{q}\left[f(\lambda_i)Z_{i1} + f'(\lambda_i)Z_{i2} + \cdots + f^{(\nu_i-1)}(\lambda_i)Z_{i,\nu_i}\right], \tag{5.21}$$

or, equivalently with the help of the coefficients of partial fraction decomposition (5.5)

$$f(A) = \sum_{i=1}^{q}\left[f(\lambda_i)M_{i1} + f'(\lambda_i)\frac{M_{i2}}{1!} + \cdots + f^{(\nu_i-1)}(\lambda_i)\frac{M_{i,\nu_i}}{(\nu-1)!}\right]. \tag{5.22}$$

Representations (5.21) and (5.22) can be applied, even in the case, when the function $f(\lambda)$ is given by a convergent infinite series. If we have the convergent series

$$f(\lambda) = \sum_{k=-\infty}^{\infty} f_k(\lambda), \tag{5.23}$$

where all functions $f_k(\lambda)$ and the sum $f(\lambda)$ are defined on the spectrum of the matrix $A$, then

$$f(A) = \sum_{k=-\infty}^{\infty} f_k(A), \tag{5.24}$$

(4) Because a function of the matrix $A$ is a polynomial of this matrix, so any functions $f_1(A)$ and $f_2(A)$ are commutative:

$$f_1(A)f_2(A) = f_2(A)f_1(A). \tag{5.25}$$

In many cases, nested functions of matrices are relevant. Let us have, for instance,

$$f(\lambda) = F(f_1(\lambda), \ldots, f_n(\lambda)) \tag{5.26}$$

where all $f_i(\lambda)$ and $F(f_1, \ldots, f_n)$ are known functions. Then, if the function $f(\lambda)$ is defined on the spectrum of the matrix $A$, then the function $f(A)$ makes sense even in the case, when some of the functions $f_i(\lambda)$ are not defined on the spectrum of the matrix $A$.

*Example 5.1* Consider the scalar function

$$f(\lambda) = \frac{\sin\lambda}{\lambda}, \tag{5.27}$$

which can be written in the form

$$f(\lambda) = f_1(\lambda)f_2(\lambda), \quad f_1(\lambda) = \sin\lambda, \;\; f_2(\lambda) = \lambda^{-1}. \tag{5.28}$$

The function $f_2(\lambda)$ is only defined for non-singular matrices, which are free of the eigenvalue zero. However, (5.27) is an integral function and defined on the spectra of arbitrary matrices $A$. Therefore, the function,

$$f(A) = (\sin A)A^{-1} = A^{-1}(\sin A), \tag{5.29}$$

makes sense and can be calculated with the help of formulae (5.21), (5.22).

(5) Assume the characteristic polynomial of the matrix $A$ in the form

$$d_A(\lambda) = \det(\lambda I_\chi - A) = (\lambda - \lambda_1)^{\mu_1} \cdots (\lambda - \lambda_q)^{\mu_q}, \tag{5.30}$$

where $\mu_i \geq \nu_i$. Then the characteristic polynomial $d_f(\lambda)$ of the matrix $f(A)$ takes the form

$$d_f(\lambda) \triangleq \det(\lambda I_\chi - f(A)) = (\lambda - f(\lambda_1))^{\mu_1} \cdots (\lambda - f(\lambda_q))^{\mu_q}. \tag{5.31}$$

It is important that this formula does not depend on the structure of the set of invariant polynomials of the matrix $\lambda I_\chi - A$. It follows from (5.31) that in the case, when $f(\lambda_i) \neq 0$, $(i = 1, \ldots, q)$, the matrix $f(A)$ is non-singular. If, however, for some $i$ we have $f(\lambda_i) = 0$, then the matrix $f(A)$ is singular.

(6) Consider in future mattering structure of the set of elementary divisors of the matrix $f(A)$ (matrix $\lambda I_\chi - f(A)$). Let the set of elementary divisors of the matrix $A_\lambda$ $(\lambda I_\chi - A)$ consists of the polynomials

$$(\lambda - \lambda_1)^{\theta_1}, \ldots, (\lambda - \lambda_\ell)^{\theta_\ell}, \tag{5.32}$$

where some of the numbers $\lambda_1, \ldots, \lambda_\ell$ could be equal. Then in those cases, when for all $\theta_i > 1$, we have $f^{'}(\lambda_i) \neq 0$, then each elementary divisor from the set (5.32) corresponds to an elementary divisor of the matrix $f(A)$ of the form $(\lambda - f(\lambda_i))^{\theta_i}$.

When this condition is fulfilled, we will say that the matrices $A$ and $f(A)$ possess a similar set of elementary divisors. When, however, for some $i$, we have $\theta_i > 1$ but $f^{'}(\lambda_i) = 0$, then the elementary divisor $(\lambda - \lambda_i)^{\theta_i}$ of the matrix $A$ corresponds to more than one elementary divisor of the matrix $f(A)$.

(7) Assume the sequence of invariant polynomials $a_1(\lambda), \ldots, a_\chi(\lambda)$ of the matrix $A$ in the form

$$a_i(\lambda) = (\lambda - \lambda_1)^{\mu_{i1}} \cdots (\lambda - \lambda_q)^{\mu_{iq}}, \quad (i = 1, \ldots, \chi), \tag{5.33}$$

where all factors are different. We will say that the matrices $A$ and $f(A)$ possess the same structure of the set of invariant polynomials, if the sequence $a_{f1}(\lambda), \ldots, a_{f\chi}(\lambda)$ of invariant polynomials of the matrix $f(A)$ has the form

$$a_{fi}(\lambda) = (\lambda - f(\lambda_1))^{\mu_{i1}} \cdots (\lambda - f(\lambda_q))^{\mu_{iq}}, \quad (i = 1, \ldots, \chi), \tag{5.34}$$

where all factors are different.

**Theorem 5.1** *For the fact that the matrices A and $f(A)$ possess the same structure of the set of invariant polynomials, it is necessary and sufficient that the following two conditions hold:*

*(i) The matrices A and $f(A)$ have a similar set of elementary divisors.*
*(ii) The eigenvalues of the matrix A must fulfill*

$$f(\lambda_i) \neq f(\lambda_k), \quad (i \neq k; \quad i, k = 1, \ldots, q). \tag{5.35}$$

The proof is obvious.

**Corollary 5.1** *Under the conditions of Theorem 5.1, from Eq. (5.4) it follows that the minimal polynomial $d_{f_{\min}}(\lambda)$ of the matrix $f(A)$ has the form*

$$d_{f_{\min}}(\lambda) = (\lambda - f(\lambda_1))^{\nu_1} \cdots (\lambda - f(\lambda_q))^{\nu_q} \tag{5.36}$$

*where all factors are different.*

## 5.2 Index Function of a Matrix

(1) Consider the scalar function

$$f(\lambda) = e^{\lambda t}, \tag{5.37}$$

where $t$ is a real parameter. The function $f(\lambda)$ is defined on the spectrum of any matrix $A$. Therefore, due to (5.22)

$$f(A) = \sum_{i=1}^{q} e^{\lambda_i t} \left[ M_{i1} + \frac{t}{1!} M_{i2} + \cdots + \frac{t^{(\nu_i - 1)}}{(\nu_i - 1)!} M_{i,\nu_i} \right], \tag{5.38}$$

where we used that in the given case $f(\lambda_i) = e^{\lambda_i t}$, $f'(\lambda_i) = t e^{\lambda_i t}, \ldots, f^{(\nu_i - 1)}(\lambda_i) = t^{(\nu_i - 1)} e^{\lambda_i t}$. Matrix (5.38) is called index function of the matrix $A$, and we denote

$$f(A) \stackrel{\triangle}{=} e^{At}. \tag{5.39}$$

(2) The function $e^{\lambda t}$ can be developed in the power series

$$e^{\lambda t} = 1 + \lambda t + \frac{\lambda^2 t^2}{2!} + \cdots$$

that converges for all $\lambda$ and $t$. The left and right side of this equation are defined on the spectrum of any matrix $A$. Therefore, beside of (5.38), we have for all $A$ and $t$ convergent representation

$$e^{At} = I_\chi + At + \frac{A^2 t^2}{2} + \cdots \tag{5.40}$$

(3) For $t = 0$ from (5.40) we obtain

$$e^{At}|_{t=0} = I_\chi. \tag{5.41}$$

Substituting in (5.40) $t$ by $-\tau$, we find

$$e^{-A\tau} = I_\chi - A\tau + \frac{A^2 \tau^2}{2!} + - \cdots. \tag{5.42}$$

Multiplying (5.40) and (5.42), we achieve

$$e^{At} e^{-A\tau} = I_\chi - A(t - \tau) + \frac{A^2 (t - \tau)^2}{2!} + \cdots,$$

i.e.,

$$e^{At} e^{-A\tau} = e^{A(t - \tau)}. \tag{5.43}$$

For $t = \tau$, the last equation yields

$$e^{At} e^{-At} = I_\chi. \tag{5.44}$$

This means that the matrix $\mathrm{e}^{At}$ is invertible for all $t$, whereas

$$(\mathrm{e}^{At})^{-1} = \mathrm{e}^{-At}. \tag{5.45}$$

(4) Differentiating expansion (5.40), we easily obtain

$$\frac{\mathrm{d}}{\mathrm{d}t}\mathrm{e}^{At} = A\mathrm{e}^{At} = \mathrm{e}^{At}A. \tag{5.46}$$

(5) Obviously

$$\int_a^b \mathrm{e}^{\lambda t}\,\mathrm{d}t = \lambda^{-1}(\mathrm{e}^{\lambda b} - \mathrm{e}^{\lambda a}), \tag{5.47}$$

where the left and the right side are also defined for $\lambda = 0$. Therefore,

$$\int_a^b \mathrm{e}^{At}\,\mathrm{d}t = A^{-1}(\mathrm{e}^{Ab} - \mathrm{e}^{Aa}) = (\mathrm{e}^{Ab} - \mathrm{e}^{Aa})A^{-1}. \tag{5.48}$$

where all expressions are also defined when $\det A = 0$.

(6)

**Theorem 5.2** *Let $T > 0$ be a constant, which is called period. Then for the fact that the matrices A and $\mathrm{e}^{AT}$ possess the same structure of the set of invariant polynomials, it is necessary and sufficient that the eigenvalues $\lambda_1, \ldots, \lambda_q$ of the matrix A satisfy the conditions*

$$\mathrm{e}^{\lambda_i T} \neq \mathrm{e}^{\lambda_k T}, \quad (i \neq k; \quad i, k = 1, \ldots, q), \tag{5.49}$$

*or, what is equivalent*

$$\lambda_i - \lambda_k \neq \frac{2n\pi \mathrm{j}}{T} = n\mathrm{j}\omega, \quad (i \neq k; \quad i, k = 1, \ldots, q), \tag{5.50}$$

*where n is any integer, and $\omega = 2\pi/T$.*

*Proof* Since $\frac{\mathrm{d}(\mathrm{e}^{\lambda T})}{\mathrm{d}\lambda} = T\mathrm{e}^{\lambda T} \neq 0$, condition (i) of Theorem 5.1 is true for all $\lambda$. Therefore, $A$ and $\mathrm{e}^{AT}$ possess the same structure of the set of invariant polynomials, if and only if condition (5.49) or (5.50) holds.

**Definition 5.1** Following [4, 5], a period $T$ is called non-pathologic, when conditions (5.49), (5.50) hold.

**Theorem 5.3** ([6]) *Let the pair $(A, B)$ be controllable and the pair $[A, C]$ be observable. Assume that the period $T$ is non-pathological. Then the pair $(\mathrm{e}^{AT}, B)$ is controllable and the pair $[\mathrm{e}^{AT}, C]$ is observable.*

## 5.3 Discrete Laplace Transformation (DLT) of Functions of Continuous Arguments

(1) Denote by $\Omega_{\alpha,\beta}$ the set of functions $f(t)$, disposing of the following properties:

(a) The function $f(t)$ is defined for all $-\infty < t < \infty$.
(b) The estimate

$$|f(t)| < \begin{cases} M\mathrm{e}^{\alpha t}, \; t \geq +0, \\ M\mathrm{e}^{\beta t}, \; t \leq t - 0 \end{cases} \tag{5.51}$$

is valid for constants $M, \alpha, \beta$, where $\alpha < \beta$.
(c) The function $f(t)$ is of bounded variation on any finite interval.

When $f(t)$ is a matrix, then we will write $f(t) \in \Omega_{\alpha,\beta}$, if all elements of this matrix belong to the set $\Omega_{\alpha,\beta}$.

(2) For $f(t) \in \Omega_{\alpha,\beta}$ there exists the two-sided Laplace transform [7]

$$F(s) = \int_{-\infty}^{\infty} f(t)\mathrm{e}^{-st}\,\mathrm{d}t, \quad \alpha < \mathrm{Re}s < \beta, \tag{5.52}$$

where the integral on the right side converges absolutely, and the inversion formula holds:

$$\lim_{\nu\to\infty} \frac{1}{2\pi \mathrm{j}} \int_{c-\mathrm{j}\nu}^{c+\mathrm{j}\nu} F(s)\mathrm{e}^{st}\,\mathrm{d}s = \frac{f(t+0) + f(t-0)}{2}. \tag{5.53}$$

Moreover, in any vertical stripe $\alpha < \alpha_1 \leq \mathrm{Re}s \leq \beta_1 < \beta$ for $s \to \infty$

$$|F(s)| < \frac{K}{|s|}, \quad K = \text{const}. \tag{5.54}$$

Further on, the function (matrix) $f(t) \in \Omega_{\alpha,\beta}$ is called the original, and the corresponding function $F(s)$ its image.

(3) As it was shown in [1, 8], for $f(t) \in \Omega_{\alpha,\beta}$ and $\alpha < \mathrm{Re}s < \beta$ on any finite interval $[t_1, t_2]$ the series

$$\tilde{D}_f(T, s, t) = \sum_{k=-\infty}^{\infty} f(t + kT)\mathrm{e}^{-ksT}, \quad -\infty < t < \infty \tag{5.55}$$

converges absolutely and uniformly.

**Definition 5.2** The function $\tilde{D}_f(T, s, t)$ is called the discrete Laplace transform (DLT) of the function of a continuous argument $f(t)$.

Notice, according to [1, 8], the fundamental properties of the DLT $\tilde{D}_f(T, s, t)$, which directly follow from (5.55):

(a)

$$\tilde{D}_f(T,s,t) = \tilde{D}_f(T,s+\mathrm{j}\omega,t), \quad \omega = \frac{2\pi}{T}. \tag{5.56}$$

(b) For any integer $m$

$$\tilde{D}_f(T,s,t+mT) = \tilde{D}_f(T,s,t)\mathrm{e}^{msT}. \tag{5.57}$$

Indeed, from (5.55) we find

$$\tilde{D}_f(T,s,t+mT) = \frac{1}{T}\sum_{k=-\infty}^{\infty} f(t+kT+mT)\mathrm{e}^{-ksT}. \tag{5.58}$$

Substituting here $k = k_1 - m$, formula (5.57) is achieved.

## 5.4 Discrete Laplace Transforms (DLT) of Images

(1) Consider the function

$$\varphi_f(T,s,t) \triangleq \tilde{D}_f(T,s,t)\mathrm{e}^{-st}. \tag{5.59}$$

It follows from (5.55) that function (5.59) is the sum of the series

$$\varphi_f(T,s,t) = \frac{1}{T}\sum_{k=-\infty}^{\infty} f(t+kT)\mathrm{e}^{-s(t+kT)}, \quad -\infty < t < \infty. \tag{5.60}$$

**Definition 5.3** The function $\varphi_f(T,s,t)$ is called the displaced pulse frequency response (DPFR) of the original $f(t)$.

From (5.60), we find

$$\varphi_f(T,s,t) = \varphi_f(T,s,t+T). \tag{5.61}$$

Denote by $\Omega^*_{\alpha,\beta}$ the subset of the set $\Omega_{\alpha,\beta}$ of functions $f(t)$, for which the periodic function $\varphi_f(T,s,t)$ has finite variation with respect to $t$. As it was shown in [1, 8], for $f(t) \in \Omega^*_{\alpha,\beta}$ and $\alpha < \mathrm{Re}s < \beta$ the function $\varphi_f(T,s,t)$ can be represented by the Fourier series

$$\varphi_F(T,s,t) = \frac{1}{T}\sum_{k=-\infty}^{\infty} F(s+k\mathrm{j}\omega)\mathrm{e}^{k\mathrm{j}\omega t}, \quad -\infty < t < \infty, \tag{5.62}$$

which converges for all $t$, where $F(s)$ is the image of the original $f(t)$. From the properties of Fourier series [9], it emerges that in points of continuity of $\varphi_f(T,s,t)$ according to $t$

$$\varphi_F(T,s,t) = \varphi_f(T,s,t). \tag{5.63}$$

However, if the function $\varphi_f(T,s,t)$ at $t = t_0$ has a jump discontinuity, then

$$\varphi_F(T,s,t_0) = \frac{\varphi_f(T,s,t_0+0) + \varphi_f(T,s,t_0-0)}{2}. \tag{5.64}$$

(2) Assume $f(t) \in \Omega^*_{\alpha,\beta}$. The we can define as follows.

**Definition 5.4** The function,

$$\tilde{D}_F(T,s,t) \triangleq \varphi_F(T,s,t)\mathrm{e}^{st} = \frac{1}{T}\sum_{k=-\infty}^{\infty} F(s+k\mathrm{j}\omega)\mathrm{e}^{(s+k\mathrm{j}\omega)t}, \tag{5.65}$$

is called the discrete Laplace transform (DLT) of the image $F(s)$.+, and function (5.62) is its displaced pulse frequency response (DPFR).

From (5.63), (5.64) we achieve, that in points of continuity of $\tilde{D}_f(t,s,t)$ according to $t$, we obtain

$$\tilde{D}_F(T,s,t) = \tilde{D}_f(T,s,t), \tag{5.66}$$

and at any jump point $t_0$

$$\tilde{D}_F(T,s,t_0) = \frac{\tilde{D}_f(T,s,t_0+0) + \tilde{D}_f(T,s,t_0-0)}{2}. \tag{5.67}$$

Since a function of finite variation does not posses more than a countable set of jump points, so for $f(t) \in \Omega^*_{\alpha,\beta}$ the functions $\tilde{D}_f(T,s,t)$ and $\tilde{D}_F(T,s,t)$ are equivalent, i.e., they differ only on a set of measure zero. Further on, we write this fact by

$$\tilde{D}_f(T,s,t) \overset{t}{\sim} \tilde{D}_F(T,s,t). \tag{5.68}$$

In the further investigations, functions satisfying condition (5.68) will be seen as equal.

(3) Notice, from (5.65), we can achieve the following properties of the function (matrix) $\tilde{D}_F(T,s,t)$:

$$\tilde{D}_F(T,s,t) = \tilde{D}_F(T,s+\mathrm{j}\omega,t), \quad \omega = \frac{2\pi}{T}, \tag{5.69}$$

$$\tilde{D}_F(T,s,t+mT) = \tilde{D}_F(T,s,t)\mathrm{e}^{msT}, \tag{5.70}$$

$$\tilde{D}_F(T,s,t) = \tilde{D}_F(T,s,t-mT)\mathrm{e}^{msT}, \quad mT < t < (m+1)T. \tag{5.71}$$

In (5.70), (5.71) the variable $m$ is any integer.

## 5.5 DPL and DPFR of Rational Matrices

(1) Let $W(s) \in \mathbb{R}^{nm}(s)$ be a strictly proper rational matrix, and $r_1, \ldots, r_q$ the real parts of the poles of the matrix $W(s)$. Denote $r_0 = -\infty$, $r_{q+1} = \infty$, and consider the open intervals $L_i : r_i < \mathrm{Re}s < r_{i+1}$, $(i = 0, \ldots, q)$. As it follows from [1, 8], in each of the intervals $L_i$ the matrix $W(s)$ is the image of some matrix $f_i(t) \in \Omega^*_{r_i, r_{i+1}}$. Therefore, the sum of the series,

$$\tilde{D}_W(T,s,t) = \frac{1}{T}\sum_{k=-\infty}^{\infty} W(s+k\mathrm{j}\omega)\mathrm{e}^{(s+k\mathrm{j}\omega)t}, \tag{5.72}$$

is for $\mathrm{Re}s \in L_i$ the DPL of the image of the function $f_i(t)$. At the same time, the sum of series (5.72) does not depend on $i$, so for it there must exist a general expression not depending on $i$. This fact is the claim of the next theorem.

**Theorem 5.4** *Assume for the matrix $W(s)$ the partial fraction decomposition (2.49)*

$$W(s) = \sum_{i=1}^{q}\sum_{k=1}^{\nu_i} \frac{W_{i\kappa}}{(s-\lambda_i)^\kappa} \tag{5.73}$$

*with constant matrices $W_{i\kappa}$. Then the sum of series (5.72) is determined by*

$$\tilde{D}_W(T,s,t) = \tilde{\mathscr{D}}_W(T,s,t), \quad 0 < t < T, \tag{5.74}$$

*where*

$$\tilde{\mathscr{D}}_W(T,s,t) \triangleq \sum_{i=1}^{q}\sum_{\kappa=1}^{\nu_i} \frac{W_{i\kappa}}{(\kappa-1)!}\left[\frac{\partial^{\kappa-1}}{\partial\lambda^{\kappa-1}}\frac{\mathrm{e}^{\lambda t}}{1-\mathrm{e}^{\lambda T}\mathrm{e}^{-sT}}\right]_{\lambda=\lambda_i}. \tag{5.75}$$

*Moreover, for $kT < t < (k+1)T$, where $m$ is an integer,*

$$\tilde{D}_W(T,s,t) = \tilde{\mathscr{D}}_W(T,s,t-kT)\mathrm{e}^{ksT}, \quad kT < t < (k+1)T. \tag{5.76}$$

*Proof* Substituting (5.73) in (5.72), we find

$$\tilde{D}_W(T,s,t) = \sum_{i=1}^{q}\sum_{\kappa=1}^{\nu_i} W_{i\kappa}\left[\frac{1}{T}\sum_{n=-\infty}^{\infty}\frac{\mathrm{e}^{(s+n\mathrm{j}\omega)t}}{(s+n\mathrm{j}\omega-\lambda_i)^\kappa}\right]. \tag{5.77}$$

As it was shown in [1, 8], for $0 < t < T$

$$\frac{1}{T}\sum_{n=-\infty}^{\infty}\frac{\mathrm{e}^{(s+nj\omega)t}}{(s+nj\omega-\lambda_i)^{\kappa}}=\frac{1}{(\kappa-1)!}\left[\frac{\partial^{\kappa-1}}{\partial\lambda^{\kappa-1}}\frac{\mathrm{e}^{\lambda t}}{1-\mathrm{e}^{(\lambda-s)T}}\right]_{\lambda=\lambda_i}. \tag{5.78}$$

Inserting this expression into (5.77), formula (5.75) is achieved. Relation (5.76) is a consequence of formula (5.71). ■

(2) The next statement considers an important special case of the achieved relations.

**Theorem 5.5** *Assume the matrix $W(s)$ in the form*

$$W(s)\stackrel{\triangle}{=}W_A(s)=(sI_\chi-A)^{-1}, \tag{5.79}$$

*where $A$ is a constant $\chi\times\chi$ matrix. Then for $0 < t < T$*

$$\tilde{D}_A(T,s,t)\stackrel{\triangle}{=}\tilde{\mathscr{D}}_A(T,s,t)=(I_\chi-\mathrm{e}^{-sT}\mathrm{e}^{AT})^{-1}\mathrm{e}^{At}=\mathrm{e}^{At}(I_\chi-\mathrm{e}^{-sT}\mathrm{e}^{AT})^{-1}. \tag{5.80}$$

*Moreover, for $kT < t < (k+1)T$*

$$\tilde{D}_A(T,s,t)=\tilde{\mathscr{D}}_A(T,s,t-kT)\mathrm{e}^{ksT}, \tag{5.81}$$

*that in detail has the form*

$$\tilde{D}_A(T,s,t)=\mathrm{e}^{A(t-kT)}\mathrm{e}^{ksT}(I_\chi-\mathrm{e}^{-sT}\mathrm{e}^{AT})^{-1}=(I_\chi-\mathrm{e}^{-sT}\mathrm{e}^{AT})\mathrm{e}^{A(t-kT)}\mathrm{e}^{ksT}. \tag{5.82}$$

*Proof* It follows from (5.5) that in the actual case, partial fraction decomposition (5.73) has the form

$$W_A(s)=(sI_\chi-A)^{-1}=\sum_{i=1}^{q}\sum_{\kappa=1}^{\nu_i}\frac{M_{i\kappa}}{(s-\lambda_i)^{\kappa}}, \tag{5.83}$$

where

$$M_{i\kappa}=(\kappa-1)!\,Z_{i\kappa}, \tag{5.84}$$

and $Z_{i\kappa}$ are the components of the matrix $A$. Therefore, in decomposition (5.73), we actually obtain

$$W_{i\kappa}=(\kappa-1)!\,Z_{i\kappa}. \tag{5.85}$$

Substituting this expression in (5.75), we find

$$\tilde{\mathscr{D}}_A(T,s,t)=\sum_{i=1}^{q}\sum_{\kappa=1}^{\nu_i}Z_{i\kappa}\left[\frac{\partial^{\kappa-1}}{\partial\lambda^{\kappa-1}}\frac{\mathrm{e}^{\lambda t}}{1-\mathrm{e}^{(\lambda-s)T}}\right]_{\lambda=\lambda_i}. \tag{5.86}$$

For the further calculations introduce the function

$$f(\lambda, s, t) \triangleq e^{\lambda t}(1 - e^{-sT}e^{\lambda T})^{-1}. \tag{5.87}$$

Then formula (5.86) can be written in the form

$$\tilde{\mathscr{D}}_A(T, s, t) = \sum_{i=1}^{q}\sum_{\kappa=1}^{\lambda_i} Z_{i\kappa}\left[\frac{\partial^{\kappa-1}}{\partial\lambda^{\kappa-1}} f(\lambda, s, t)\right]_{\lambda=\lambda_i}. \tag{5.88}$$

Inserting this expression into (5.21), we obtain

$$\tilde{\mathscr{D}}_A(T, s, t) = f(A, s, t), \tag{5.89}$$

which is equivalent to (5.80). Relation (5.82) is a direct consequence of formulae (5.81) and (5.80). ■

**Corollary 5.2** *If the matrix* $W(s) \in \mathbb{R}^{nm}(s)$ *is given by the realization* $(A, B, C)$

$$W(s) = C(sI_\chi - A)^{-1}B, \tag{5.90}$$

*then for* $0 < t < T$

$$\begin{aligned}\tilde{D}_W(T, s, t) &= \tilde{\mathscr{D}}_W(T, s, t)\\ &= C(I_\chi - e^{-sT}e^{AT})^{-1}e^{At}B = Ce^{At}(I_\chi - e^{-sT}e^{AT})^{-1}B.\end{aligned} \tag{5.91}$$

*Moreover, for* $kT < t < (k+1)T$

$$\begin{aligned}\tilde{D}_W(T, s, t) &= \tilde{\mathscr{D}}_W(T, s, t - kT)e^{ksT}\\ &= C(I_\chi - e^{-sT}e^{AT})^{-1}e^{A(t-kT)}e^{ksT}B\\ &= Ce^{A(t-kT)}e^{ksT}(I_\chi - e^{-sT}e^{AT})^{-1}B.\end{aligned} \tag{5.92}$$

(3) The answer to the question on the continuity of the matrix $\tilde{D}_W(T, s, t)$ with respect to $t$ gives the next claim.

**Theorem 5.6** *Assume the matrix* $W(s)$ *in form (5.90). Then for* $\operatorname{exc} W(s) > 1$ *the matrix* $\tilde{D}_W(T, s, t)$ *is continuous according to* $t$. *When* $\operatorname{exc} W(s) = 1$, *then the matrix* $\tilde{D}_W(T, s, t)$ *has finite jumps at the points* $t_k = kT,\ (k = 0, \pm 1, \ldots)$. *Hereby,*

$$\tilde{D}_W(T, s, kT + 0) - \tilde{D}_W(T, s, kT - 0) = e^{ksT}CB \neq 0_{nm}. \tag{5.93}$$

*Proof* From (5.90), we obtain

$$\lim_{s\to\infty} s^{-1}W(s) = CB. \tag{5.94}$$

Therefore, for matrix (5.90) we achieve exc $W(s) = 1$ for $CB \neq 0_{nm}$ and exc $W(s) > 1$ for $CB = 0_{nm}$. From (5.91), (5.92) for $t = kT + 0$ and $t = kT - 0$, we find

$$\tilde{D}_W(T, s, kT + 0) = C(I\chi - e^{-sT}e^{AT})^{-1}Be^{ksT},$$
$$\tilde{D}_W(T, s, kT - 0) = C(I\chi - e^{-sT}e^{AT})^{-1}e^{-sT}e^{AT}Be^{ksT}.$$

Hence

$$\tilde{D}_W(T, s, kT + 0) - \tilde{D}_W(T, s, kT - 0) = CBe^{-ksT},$$

that proves (5.93) for $CB \neq 0_{nm}$. At the same time for $CB = 0_{nm}$, the matrix $\tilde{D}_W(T, s, t)$ is continuous according to $t$.

*Remark 5.1* Because the sum of series (5.72) uniquely determines the matrix $W(s)$, the sum of this series does not depend on the choice of the concrete realization $(A, B, C)$, configured in formula (5.90). Therefore, when two realizations $(A, B, C)$ and $(A_1, B_1, C_1)$ define the same matrix (5.90), then

$$C(I - e^{-sT}e^{AT})^{-1}e^{At}B = C_1(I - e^{-sT}e^{A_1T})^{-1}e^{A_1t}B_1. \tag{5.95}$$

## 5.6 Parametric Discrete Models of Continuous Processes

(1) Let $F(\zeta)$ be a rational matrix. Then the matrices

$$\breve{F}(s) = F(\zeta)|_{\zeta=e^{-sT}}, \quad F(\zeta) = \breve{F}(s)|_{e^{-sT}=\zeta} \tag{5.96}$$

are called associated to each other. Hereby, following [1], the matrix $\breve{F}(s)$ is called rational periodic (RP). In this book, RP functions are equipped with ˘ sign. As it emerges from the above results, the DLT of a strictly proper rational matrix is a RP function of the argument $s$.

**Definition 5.5** Using the terminology in [1, 8], the connected rational matrix to the DLP $\breve{D}_W(T, s, t)$

$$D_W(T, \zeta, t) = \tilde{D}_W(T, s, t)|_{e^{-sT}=\zeta} \tag{5.97}$$

is called parametric (depending on the continuous time $t$ as parameter) discrete model of the process $W(s)$.

Applying formulae (5.91), (5.92) by substituting $e^{-sT}$ by $\zeta$, we obtain closed formulae for the matrix $D_W(T, \zeta, t)$. Obviously, for $0 < t < T$

$$D_W(T, \zeta, t) = \mathscr{D}_W(T, \zeta, t), \tag{5.98}$$

where

$$\begin{aligned}\mathscr{D}_W(T,\zeta,t) &= \tilde{\mathscr{D}}_W(T,s,t)_{|e^{-sT}=\zeta} \\ &= Ce^{At}(I_\chi - \zeta e^{AT})^{-1}B = C(I_\chi - \zeta e^{AT})^{-1}e^{At}B.\end{aligned} \tag{5.99}$$

Moreover, for $kT < t < (k+1)T$

$$\mathscr{D}_W(T,\zeta,t) = \mathscr{D}_W(T,\zeta,t-kT)\zeta^{-k}, \tag{5.100}$$

or more detailed

$$\begin{aligned}\mathscr{D}_W(T,\zeta,t) &= Ce^{A(t-kT)}(I_\chi - \zeta e^{AT})^{-1}B\zeta^{-k} \\ &= C(I_\chi - \zeta e^{AT})^{-1}e^{A(t-kT)}B\zeta^{-k}, \quad kT < t < (k+1)T.\end{aligned} \tag{5.101}$$

From (5.101), we can see that for $m \le 0$, the matrix $D_W(T,\zeta,t)$ can be considered in the interval $kT < t < (k+1)T$ as transfer matrix of the PMD

$$\Pi(\zeta) \triangleq (C, I_\chi - \zeta e^{AT}, \zeta^{-k}e^{A(t-kT)}B). \tag{5.102}$$

**Theorem 5.7** *Let the pair $(A,B)$ be controllable and the pair $[A,C]$ be observable, and the period $T$ should be non-pathological. Then for $k \le 0$ PMD (5.102) is minimal, i.e., the polynomial pairs*

$$(I_\chi - \zeta e^{AT}, \zeta^{-k}e^{A(t-kT)}B), \quad [I_\chi - \zeta e^{AT}, C] \tag{5.103}$$

*are irreducible.*

*Proof* We will show the irreducibility of the pair $(I_\chi - \zeta e^{AT}, \zeta^{-k}e^{A(t-kT)}B)$. At first, notice that it suffices to show the irreducibility of the pair $(I_\chi - \zeta e^{AT}, e^{A(t-kT)}B)$, because the matrix $I_\chi - \zeta e^{AT}$ does not possess zero eigenvalues. Since the period $T$ is non-pathological, Theorem 5.2 yields that the pair $(e^{AT}, B)$ is controllable, and hence the polynomial pair $(zI_\chi - e^{AT}, B)$ is irreducible. Therefore, for all eigenvalues $z_1, \ldots, z_q$ of the matrix $e^{AT}$

$$\operatorname{rank}\left[\, z_i I_\chi - e^{AT} \;\; B \,\right] = \chi, \quad (i = 1, \ldots, q). \tag{5.104}$$

Since among the numbers $z_1, \ldots, z_q$ are no zeros, so from (5.104)

$$\operatorname{rank}\left[\, I_\chi - z_i^{-1}e^{AT} \;\; B \,\right] = \chi, \quad (i = 1, .., q). \tag{5.105}$$

Hence, the matrix $\left[\, I_\chi - \zeta e^{AT} \;\; B \,\right]$ is alatent and the pair $(I_\chi - \zeta e^{AT}, B)$ is irreducible. Since the matrix $e^{A(t-kT)}$ is invertible for all $t$ and commutative with the matrix $e^{AT}$, we realize

$$\left[\, I_\chi - \zeta e^{AT} \;\; e^{A(t-kT)}B \,\right] = e^{A(t-kT)}\left[\, I_\chi - \zeta e^{AT} \;\; B \,\right]\begin{bmatrix} e^{-A(t-kT)} & 0_{\chi m} \\ 0_{m\chi} & I_m \end{bmatrix}. \tag{5.106}$$

Because the first and last factor on the right side are non-singular matrices, so

$$\left[\, I_\chi - \zeta \mathrm{e}^{AT}\ \mathrm{e}^{A(t-kT)} B \,\right] \sim \left[\, I_\chi - \zeta \mathrm{e}^{AT}\ B \,\right], \tag{5.107}$$

and the irreducibility of the pair $(I_\chi - \zeta \mathrm{e}^{AT}, B)$ implies the irreducibility of the pair $(I_\chi - \zeta \mathrm{e}^{AT}, \mathrm{e}^{A(t-kT)} B)$, and of the pair $(I_\chi - \zeta \mathrm{e}^{AT}, \zeta^{-k} \mathrm{e}^{A(t-kT)} B)$ for $k \leq 0$ too.

The irreducibility of the pair $[I_\chi - \zeta \mathrm{e}^{AT}, C]$ can be shown analogously.

(2)

**Theorem 5.8** *Let the conditions of Theorem 5.7 hold, and in addition assume the ILMFD and IRMFD*

$$\begin{aligned} C(I_\chi - \zeta \mathrm{e}^{AT})^{-1} &= a_l^{-1}(\zeta) b_l(\zeta), \\ (I_\chi - \zeta \mathrm{e}^{AT})^{-1} B &= b_r(\zeta) a_r^{-1}(\zeta). \end{aligned} \tag{5.108}$$

*Then the following statements are true:*

*(i) The matrices*

$$a_l(\zeta) D_W(T, \zeta, t) \stackrel{\triangle}{=} b_l(\zeta, t), \quad D_W(T, \zeta, t) a_r(\zeta) \stackrel{\triangle}{=} b_r(\zeta, t), \tag{5.109}$$

*are polynomials for $t < T$.*

*(ii) The representations derived from (5.109)*

$$D_W(T, \zeta, t) = a_l^{-1}(\zeta) b_l(\zeta, t), \quad D_W(T, \zeta, t) = b_r(\zeta, t) a_r^{-1}(\zeta), \tag{5.110}$$

*for $t < T$ are ILMDF resp. IRMDF.*

*(iii) The pseudo-equivalence relations*

$$a_l(\zeta) \stackrel{p}{\sim} a_r(\zeta) \stackrel{p}{\sim} I_\chi - \zeta \mathrm{e}^{AT} \tag{5.111}$$

*are true. Thus, in particular*

$$\det a_l(\zeta) \sim \det a_r(\zeta) \sim \det(I_\chi - \zeta \mathrm{e}^{AT}). \tag{5.112}$$

*Moreover, when*

$$\det(\lambda I_\chi - A) = (\lambda - \lambda_1)^{\mu_1} \cdots (\lambda - \lambda_q)^{\mu_q}, \tag{5.113}$$

*then*

$$\det(I_\chi - \zeta \mathrm{e}^{AT}) = (1 - \zeta \mathrm{e}^{\lambda_1 T})^{\mu_1} \cdots (1 - \zeta \mathrm{e}^{\lambda_q T})^{\mu_q}. \tag{5.114}$$

*Proof* (i) From (5.101) and (5.108) we obtain

$$
\begin{aligned}
a_l(\zeta)\mathscr{D}_W(T,\zeta,t) &= b_l(\zeta)e^{A(t-kT)}B\zeta^{-k} = b_l(\zeta,t),\\
\mathscr{D}_W(T,\zeta,t)a_r(\zeta) &= Ce^{A(t-kT)}b_r(\zeta)\zeta^{-k} = b_r(\zeta,t)
\end{aligned}
\tag{5.115}
$$

for $kT < t < (k+1)T$. Hence, for $k \le 0$ $(t < T)$ the quantities $b_l(\zeta,t)$ and $b_r(\zeta,t)$ are polynomial matrices.

(ii) This statement directly follows from the irreducibility of PMD (5.102) and Lemma 2.7.

(iii) Relations (5.111) and (5.112) are achieved from the general properties of IMFD, provided in Chap. 2. For the proof of (5.114) notice that from the properties of the characteristic polynomial of functions of matrices and (5.113), we find

$$
\det(zI_\chi - e^{AT}) = (z - e^{\lambda_1 T})^{\mu_1}\cdots(z - e^{\lambda_q T})^{\mu_q}. \tag{5.116}
$$

Hence with the help of Theorem 1.21 using the regularity of the matrix $e^{AT}$, we obtain

$$
\det(I_\chi - \zeta e^{AT}) = \operatorname{rec}\det(zI_\chi - e^{AT}) = (1 - \zeta e^{\lambda_1 T})^{\mu_1}\cdots(1 - \zeta e^{\lambda_q T})^{\mu_q}, \tag{5.117}
$$

that leads us to formula (5.114). ■

## 5.7 Parametric Discrete Models of Continuous Processes with Delay

(1) Let $W(s) \in \mathbb{R}^{nm}(s)$ be a strictly proper rational matrix. Denote

$$
W_\tau(s) \triangleq W(s)e^{-s\tau}, \tag{5.118}
$$

where $\tau > 0$ is a real constant. Further on, the matrix $W_\tau(s)$ is called the transfer matrix of the process with delay.

For the transfer matrix $W_\tau(s)$, we can build the DLT

$$
\breve{D}_{W_\tau}(T,s,t) = \frac{1}{T}\sum_{k=-\infty}^{\infty} W_\tau(s+kj\omega)e^{(s+kj\omega)t}, \quad -\infty < t < \infty, \tag{5.119}
$$

and the DPFR

$$\begin{aligned}\varphi_{W_\tau}(T,s,t) &= \breve{D}_{W_\tau}(T,s,t)\mathrm{e}^{-st} = \frac{1}{T}\sum_{k=-\infty}^{\infty} W_\tau(s+k\mathrm{j}\omega)\mathrm{e}^{k\mathrm{j}\omega t} \\ &= \mathrm{e}^{-s\tau}\varphi_W(T,\lambda,t-\tau), \quad -\infty < t < \infty.\end{aligned} \tag{5.120}$$

**Theorem 5.9** *Under the taken propositions, the following statements hold:*

*(i)*

$$\breve{D}_{W_\tau}(T,s,t) = \breve{D}_W(T,s,t-\tau). \tag{5.121}$$

*(ii) When the representation*

$$\tau = pT + \theta = (p+1)T - \gamma, \quad \gamma = T - \theta, \tag{5.122}$$

*is applied, where* $0 \le \theta < T$, $0 < \gamma \le T$ *and* $p$ *is a nonnegative integer, then*

$$\breve{D}_{W_\tau}(T,s,t) = \breve{\mathscr{D}}_{W_\tau}(T,s,t), \quad 0 < t < T, \tag{5.123}$$

*where*

$$\breve{\mathscr{D}}_{W_\tau}(T,s,t) \triangleq \begin{cases} \breve{\mathscr{D}}_W(T,s,t+\gamma)\mathrm{e}^{-(p+1)sT}, & 0 < t < \theta, \\ \breve{\mathscr{D}}_W(T,s,t-\theta)\mathrm{e}^{-psT}, & \theta < t < T. \end{cases} \tag{5.124}$$

*Moreover,*

$$\breve{D}_{W_\tau}(T,s,t) = \breve{\mathscr{D}}_{W_\tau}(T,s,t-kT)\mathrm{e}^{ksT}, \quad kT < t < (k+1)T. \tag{5.125}$$

*(iii) For* exc $W(s) > 1$, *the matrix* $\breve{D}_{W_\tau}(T,s,t)$ *depends continuously on* $t$ *for* $\infty < t < \infty$.

*Proof*

(i) Inserting (5.118) into (5.119), we find

$$\breve{D}_{W_\tau}(T,s,t) = \frac{1}{T}\sum_{k=-\infty}^{\infty} W(s+k\mathrm{j}\omega)\mathrm{e}^{(s+k\mathrm{j}\omega)(t-\tau)} = \breve{D}_W(T,s,t-\tau). \tag{5.126}$$

(ii) When (5.122) is true, we obtain

$$\begin{aligned}\breve{D}_{W_\tau}(T,s,t) &= \breve{D}_W(T,s,t-\tau) = \breve{D}_W(T,s,t-\theta-pT) \\ &= \tilde{D}_W(T,s,t-\theta)\mathrm{e}^{-psT}.\end{aligned} \tag{5.127}$$

Since for $0 < t < \theta < T$, we have $-T < t-\theta < 0$, so using (5.92), we obtain

$$\breve{D}_W(T,s,t-\theta) = \breve{\mathscr{D}}_W(T,s,t-\theta+T)e^{-sT} = \breve{\mathscr{D}}_W(T,s,t+\gamma)e^{-sT}, \quad 0 < t < \theta. \tag{5.128}$$

Analogously, we verify

$$\breve{D}_W(T,s,t-\theta) = \breve{\mathscr{D}}_W(T,s,t-\theta), \quad \theta < t < T. \tag{5.129}$$

Inserting (5.128) and (5.129) into (5.127), formula (5.124) is achieved. Formula (5.125) follows from the general properties of DLP.

(iii) This statement follows directly from (5.121) and Theorem 5.6. ∎

With the help of (5.91), (5.92) formula (5.124) can be written in the form

$$\breve{\mathscr{D}}_{W_\tau}(T,s,t) = \begin{cases} Ce^{A(t+\gamma)}(I_\chi - e^{-sT}e^{AT})^{-1}Be^{-(p+1)sT}, & 0 < t < \theta, \\ Ce^{A(t-\theta)}(I_\chi - e^{-sT}e^{AT})^{-1}Be^{-psT}, & \theta < t < T. \end{cases} \tag{5.130}$$

With the help of (5.125), formula (5.130) can be extended onto the whole axis $\infty < t < \infty$.

(2) From the above-derived relations, we conclude that for all $t$ the matrix $\breve{D}_{W_\tau}(T,s,t)$ is RP. Therefore, the connected matrix

$$D_{W_\tau}(T,\zeta,t) = \tilde{D}_{W_\tau}(T,s,t)|_{e^{-sT}=\zeta}, \tag{5.131}$$

achieved from $\breve{D}_{W_\tau}(T,s,t)$ by substituting $e^{-sT}$ by $\zeta$, is a rational function according to $\zeta$ for all $t$. Hence, from (5.124)–(5.126) and (5.130)

$$D_{W_\tau}(T,\zeta,t) = \mathscr{D}_{W_\tau}(T,\zeta,t), \quad 0 < t < T, \tag{5.132}$$

where

$$\mathscr{D}_{W_\tau}(T,\zeta,t) = \begin{cases} Ce^{A(t+\gamma)}(I_\chi - \zeta e^{AT})^{-1}B\zeta^{p+1}, & 0 < t < \theta, \\ Ce^{A(t-\theta)}(I_\chi - \zeta e^{AT})^{-1}B\zeta^{p}, & \theta < t < T. \end{cases} \tag{5.133}$$

Using the relation

$$D_{W_\tau}(T,\zeta,t+kT) = \mathscr{D}_{W_\tau}(T,\zeta,t)\zeta^{-k} \tag{5.134}$$

formula (5.133) can be extended onto the whole axis $-\infty < t < \infty$.

(3) We know from (5.102) and (5.109) that in cases, when IMFD (5.108) takes place, we can apply without changes Theorem 5.8 on processes with delay. In particular, when the conditions of this theorem are fulfilled, the following statements hold:

(i) For $t < T + \tau$ the matrices

$$\begin{aligned} a_l(\zeta)D_{W_\tau}(T,\zeta,t) &= b_l(\zeta,t-\tau), \\ D_{W_\tau}(T,\zeta,t)a_r(\zeta) &= b_r(\zeta,t-\tau) \end{aligned} \tag{5.135}$$

are polynomials.

(ii) The MFD configured in (5.135)

$$D_{W_\tau}(T,\zeta,t) = a_l^{-1}(\zeta)b_l(\zeta,t-\tau),$$
$$D_{W_\tau}(T,\zeta,t) = b_r(\zeta,t-\tau)a_r^{-1}(\zeta) \tag{5.136}$$

are IMDF.

(iii) For the matrices $a_l(\zeta)$, $a_r(\zeta)$ relations (5.111)–(5.114) are valid.

## 5.8 DLT and DPFR of Modulated Processes

(1) Let $W(s) \in \mathbb{R}^{nm}(s)$ be a strictly proper rational matrix and

$$\mu_0(s) = \int_0^T h_0(\nu)\mathrm{e}^{-s\nu}\,\mathrm{d}\nu \tag{5.137}$$

the $m \times q$ transfer matrix of the forming filter for zero-order hold (4.22). Denote

$$W_{\mu_0}(s) \triangleq W(s)\mu_0(s). \tag{5.138}$$

**Definition 5.6** Matrix (5.138) is called transfer matrix of the modulated process with zero-order hold.

In harmony with the general definitions, the DLT of matrix (5.138) $\breve{D}_{W\mu_0}(T,s,t)$ is defined by the relation

$$\breve{D}_{W\mu_0}(T,s,t) = \frac{1}{T}\sum_{k=-\infty}^{\infty} W(s+k\mathrm{j}\omega)\mu_0(s+k\mathrm{j}\omega)\mathrm{e}^{(s+k\mathrm{j}\omega)t}, \quad -\infty < t < \infty. \tag{5.139}$$

In the same way, the DPFR of matrix (5.138) is given by

$$\begin{aligned}\varphi_{W\mu_0}(T,s,t) &= \breve{D}_{W\mu_0}(T,s,t)\mathrm{e}^{-st}\\ &= \frac{1}{T}\sum_{k=-\infty}^{\infty} W(s+k\mathrm{j}\omega)\mu_0(s+k\mathrm{j}\omega)\mathrm{e}^{k\mathrm{j}\omega t}, \quad -\infty < t < \infty.\end{aligned} \tag{5.140}$$

**Theorem 5.10** *Let $W(s) \in \mathbb{R}^{nm}(s)$ be matrix (5.90). Then the following claims hold:*

*(i) For $0 < t < T$*

$$\breve{D}_{W\mu_0}(T,s,t) = \breve{\mathscr{D}}_{W\mu_0}(T,s,t), \tag{5.141}$$

*where $\breve{\mathscr{D}}_{W\mu_0}(T,s,t)$ is defined by the equivalent expressions*

$$\tilde{\mathscr{D}}_{W\mu_0}(T,s,t) = Ce^{At}(I_\chi - e^{-sT}e^{AT})^{-1}e^{-sT}e^{AT}m_0 + f_0(t),$$
$$\tilde{\mathscr{D}}_{W\mu_0}(T,s,t) = Ce^{At}(I_\chi - e^{-sT}e^{AT})^{-1}m_0 + g_0(t). \tag{5.142}$$

*Herein we used the notations*

$$\begin{aligned} m_0 &\triangleq \int_0^T e^{-A\nu}Bh_0(\nu)\,d\nu, \\ f_0(t) &\triangleq C\int_0^t e^{A(t-\nu)}Bh_0(\nu)\,d\nu, \\ g_0(t) &\triangleq -C\int_t^T e^{A(t-\nu)}Bh_0(\nu)\,d\nu. \end{aligned} \tag{5.143}$$

(ii) *Beside of (5.142) the matrix $\breve{\mathscr{D}}_{W\mu_0}(T,s,t)$ for $0 < t < T$ can be defined by the expression*

$$\breve{\mathscr{D}}_{W\mu_0}(T,s,t) = C(I_\chi - e^{-sT}e^{AT})^{-1}(\bar{\Gamma}_1(t) + e^{-sT}e^{AT}\bar{\Gamma}_2(t)), \tag{5.144}$$

*where*

$$\bar{\Gamma}_1(t) \triangleq \int_0^t e^{A(t-\nu)}Bh_0(\nu)\,d\nu, \quad \bar{\Gamma}_2(t) \triangleq \int_t^T e^{A(t-\nu)}Bh_0(\nu)\,d\nu. \tag{5.145}$$

(iii) *Formula (5.141) can be extended onto the whole axis $-\infty < t\infty$ with the help of the relation*

$$\breve{D}_{W\mu_0}(T,s,t) = \breve{\mathscr{D}}_{W\mu_0}(T,s,t-kT)e^{ksT}, \quad kT < t < (k+1)T. \tag{5.146}$$

*Proof* (i) Substitute (5.137) in (5.139), to achieve

$$\breve{D}_{W\mu_0}(T,s,t) = \frac{1}{T}\sum_{k=-\infty}^{\infty} W(s+kj\omega)\int_0^T e^{-(s+kj\omega)\nu}h_0(\nu)\,d\nu e^{(s+kj\omega)t}. \tag{5.147}$$

Exchanging here the order of the factors and integration, we obtain

$$\breve{D}_{W\mu_0}(T,s,t) = \int_0^T \breve{D}_W(T,s,t-\nu)h_0(\nu)\,d\nu, \tag{5.148}$$

where

$$\breve{D}_W(T,s,t-\nu) = \frac{1}{T}\sum_{k=-\infty}^{\infty} W(s+kj\omega)e^{(s+kj\omega)(t-\nu)}. \tag{5.149}$$

Using (5.90), the last relation can be written as

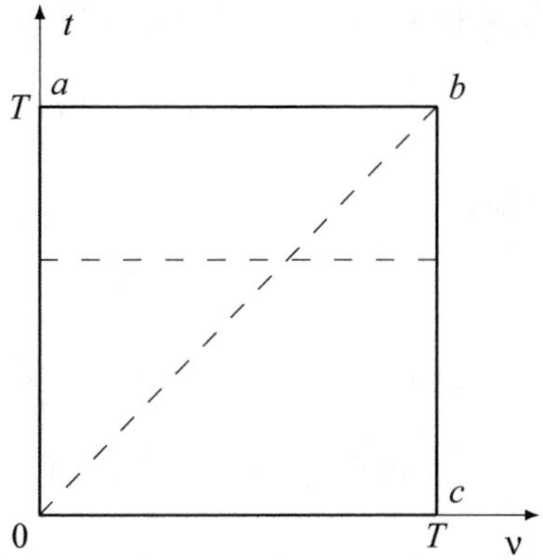


**Fig. 5.1** Integration areas

$$\breve{D}_W(T,s,t-\nu) = C\breve{D}_A(T,s,t-\nu)B, \tag{5.150}$$

where

$$\breve{D}_A(T,s,t-\nu) = \frac{1}{T}\sum_{k=-\infty}^{\infty}\left[(s+kj\omega)I_\chi - A\right]^{-1}e^{(s+kj\omega)(t-\nu)}. \tag{5.151}$$

Consider in the $(\nu, t)$ plane the square $\square_{0abc}$ in Fig. 5.1: Since in the triangle $\Delta_{0ab}$, we have $0 < t-\nu < T$, so from (5.80) we obtain

$$\begin{aligned}\breve{D}_A(T,s,t-\nu) &= \breve{\mathscr{D}}_A(T,s,t-\nu)\\ &= e^{At}(I_\chi - e^{-sT}e^{AT})^{-1}e^{-A\nu}, \quad t,\nu\in\Delta_{0ab}.\end{aligned} \tag{5.152}$$

Moreover, regarding that in the triangle $\Delta_{0bc}$ $T < t-\nu < 0$, we obtain from (5.81)

$$\breve{D}_A(T,s,t-\nu) = \breve{\mathscr{D}}_A(T,s,t-\nu+T)e^{-sT}, \quad t,\nu\in\Delta 0bc. \tag{5.153}$$

With the help of (5.82), this formula can be written in detail as

$$\breve{D}_A(T,s,t-\nu) = e^{At}e^{-sT}e^{AT}(I_\chi - e^{-sT}e^{AT})^{-1}e^{-A\nu}, \quad t,\nu\in\Delta 0bc. \tag{5.154}$$

Using the relation

$$e^{-sT}e^{AT}(I_\chi - e^{-sT}e^{AT})^{-1} = (I_\chi - e^{-sT}e^{AT})^{-1} - I_\chi, \tag{5.155}$$

the last expression takes the form

$$\breve{D}_A(T,s,t-\nu) = e^{At}(I_\chi - e^{-sT}e^{AT})^{-1}e^{-A\nu} - e^{A(t-\nu)}, \quad t,\nu\in\Delta_{0bc}. \tag{5.156}$$

Together from (5.152) and (5.156), according to Fig. 5.1, we find

$$\breve{D}_A(T, s, t-\nu) = e^{At}(I_\chi - e^{-sT}e^{AT})^{-1}e^{-A\nu}B + r(t, \nu), \tag{5.157}$$

where

$$r(t, \nu) = \begin{cases} 0_{\chi\chi}, & 0 < \nu < t < T, \\ -e^{A(t-\nu)}, & 0 < t < \nu < T. \end{cases} \tag{5.158}$$

Substituting (5.157) in (5.150), we achieve

$$\breve{D}_W(T, s, t-\nu) = Ce^{At}(I_\chi - e^{-sT}e^{AT})^{-1}e^{-A\nu}B + Cr(t, \nu)B. \tag{5.159}$$

From this relation and (5.148), we find

$$\breve{D}_{W\mu_0}(T, s, t) = \\ Ce^{At}(I_\chi - e^{-sT}e^{AT})^{-1}\int_0^T e^{-A\nu}Bh_0(\nu)\,d\nu - C\int_t^T e^{A(t-\nu)}Bh_0(\nu)\,d\nu, \tag{5.160}$$

that coincides with the second formula in (5.142). The first relation in (5.142) is achieved from (5.160) with the help of (5.155).
(ii) Applying expressions (5.157), (5.158), (5.155), we obtain

$$\breve{D}_A(T, s, t-\nu) = \begin{cases} (I_\chi - e^{-sT}e^{AT})^{-1}e^{A(t-\nu)}, & t, \nu \in \Delta_{0ab}, \\ (I_\chi - e^{-sT}e^{AT})^{-1}e^{-sT}e^{AT}e^{A(t-\nu)}, & t, \nu \in \Delta_{0bc}. \end{cases} \tag{5.161}$$

Inserting these expressions into (5.148) leads to formulae (5.144), (5.145).
(iii) Relation (5.146) is achieved from the general properties of DLT. ■

**Theorem 5.11** *For a strictly proper matrix $W(s)$, the matrix $\breve{D}_{W\mu_0}(T, s, t)$ depends continuously on $t$ for $-\infty < t < \infty$.*

*Proof* Assume (5.90). Consider the DPFR of matrix (5.138)

$$\varphi_{W\mu_0}(T, s, t) = \breve{D}_{W\mu_0}(T, s, t)e^{-st} = \varphi_{W\mu_0}(T, s, t+T), \quad \infty < t < \infty. \tag{5.162}$$

It follows from (5.162) and (5.142), (5.143) that the $T$-periodic matrix $\varphi_{W\mu_0}(T, s, t)$ at the interval $0 < t < T$ can be defined by the expression

$$\varphi_{W\mu_0}(T, s, t) = \\ Ce^{-st}e^{At}(I_\chi - e^{-sT}e^{AT})^{-1}\int_0^T e^{-A\nu}Bh_0(\nu)\,d\nu - Ce^{-st}\int_t^T e^{A(t-\nu)}Bh_0(\nu)\,d\nu. \tag{5.163}$$

Hence, for $t = +0$ and $t = T - 0$

$$\varphi_{W\mu_0}(T,s,+0) = C\left[(I_\chi - \mathrm{e}^{-sT}\mathrm{e}^{AT})^{-1} - I_\chi\right]\int_0^T \mathrm{e}^{-A\nu}Bh_0(\nu)\,\mathrm{d}\nu,$$
$$\varphi_{W\mu_0}(T,s,T-0) = C\mathrm{e}^{-sT}\mathrm{e}^{AT}(I_\chi - \mathrm{e}^{-sT}\mathrm{e}^{AT})^{-1}\int_0^T \mathrm{e}^{-A\nu}Bh_0(\nu)\,\mathrm{d}\nu. \tag{5.164}$$

It can be seen from (5.164), (5.155) that $\varphi_{W\mu_0}(T,s,T+0) = \varphi_{W\mu_0}(T,s,T-0)$, i.e., the matrix $\varphi_{W\mu_0}(T,s,t)$, is continuous according to $t$. Therefore, due to (5.162), we realize that the matrix $\tilde{D}_{W\mu_0}(T,s,t)$ also depends continuously on $t$. ∎

(2) We will show that series (5.139) can converge also in the case, when the matrix $W(s)$ is only proper (but not strictly proper). We will consider this special case now. Assume

$$W(s) = W_1(s) + W_2, \tag{5.165}$$

where $W_1(s)$ is a strictly proper matrix, and $W_2$ is a constant matrix different from zero. Inserting (5.165) into (5.139), we find

$$\breve{D}_{W\mu_0}(T,s,t) = \breve{D}_{W_1\mu_0}(T,s,t) + W_2\breve{D}_{\mu_0}(T,s,t), \tag{5.166}$$

where

$$\breve{D}_{\mu_0}(T,s,t) = \frac{1}{T}\sum_{k=-\infty}^{\infty}\mu_0(s+kj\omega)\mathrm{e}^{(s+kj\omega)t}. \tag{5.167}$$

The sum of series (5.164) can be written in closed form. For that reason consider the periodic in $t$ matrix

$$\varphi_{\mu_0}(T,s,t) \stackrel{\triangle}{=} \mathrm{e}^{-st}h_0(t), \quad 0<t<T,$$
$$\varphi_{\mu_0}(T,s,t) = \varphi_{\mu_0}(T,s,t+T) = \mathrm{e}^{-s(t-kT)}h_0(t-kT), \quad kT<t<(k+1)T. \tag{5.168}$$

If the propositions according to the properties of matrix $h_0(t)$ in Chap. 4 are fulfilled, then the matrix $\varphi_{\mu_0}(T,s,t)$ is of bounded variation on the interval $0<t<T$, and hence it can be presented by the convergent Fourier series

$$\varphi_{\mu_0}(T,s,t) = \sum_{k=-\infty}^{\infty}\varphi_k(s)\mathrm{e}^{kj\omega t}, \quad \varphi_k(s) = \frac{1}{T}\int_0^T \varphi_{\mu_0}(T,s,\nu)\mathrm{e}^{-kj\omega\nu}\,\mathrm{d}\nu. \tag{5.169}$$

With the help of (5.168) and (5.137), we find

$$\varphi_k(s) = \frac{1}{T}\int_0^T h_o(\nu)\mathrm{e}^{-(s+kj\omega)\nu}\,\mathrm{d}\nu = \frac{1}{T}\mu_0(s+kj\omega), \tag{5.170}$$

and together with (5.169), we obtain

$$\varphi_{\mu_0}(T,s,t)=\frac{1}{T}\sum_{k=-\infty}^{\infty}\mu_0(s+kj\omega)e^{kj\omega t}=\breve{D}_{\mu_0}(T,s,t)e^{-st}. \tag{5.171}$$

Therefore, using (5.165), we arrive at

$$\breve{D}_{\mu_0}(T,s,t)=\varphi_{\mu_0}(T,s,t)e^{st}=\begin{cases}h_0(t), & 0<t<T,\\ e^{ksT}h_0(t-kT), & kT<t<(k+1)T.\end{cases} \tag{5.172}$$

This equation holds for all $t$, where the right side of (5.172) is continuous.

(3) The last equation can be extended onto the case of generalized hold (4.18), (4.26). Denote by $\mu(s)$ the transfer function (matrix) of the forming element

$$\mu(s)=\sum_{i=0}^{P}e^{-isT}\mu_i(s),\quad \mu_i(s)=\int_0^T h_i(t)e^{-st}\,dt. \tag{5.173}$$

Moreover, denote by $W_\mu(s)$ the matrix

$$W_\mu(s)\stackrel{\triangle}{=}W(s)\mu(s)=W(s)\sum_{i=0}^{P}e^{-isT}\mu_i(s)\stackrel{\triangle}{=}\sum_{i=0}^{P}e^{-isT}W_{\mu_i}(s), \tag{5.174}$$

which is called the transfer matrix of the associated modulated process. The DLT of matrix (5.174) is determined by the sum of the series

$$\begin{aligned}\breve{D}_{\mu_i}(T,s,t)&=\frac{1}{T}\sum_{k=-\infty}^{\infty}W_\mu(s+kj\omega)e^{(s+kj\omega)t}\\ &=\frac{1}{T}\sum_{k=-\infty}^{\infty}W(s+kj\omega)\mu(s+kj\omega)e^{(s+kj\omega)t}.\end{aligned} \tag{5.175}$$

Substituting (5.173) in (5.175), we find

$$\breve{D}_\mu(T,s,t)=\sum_{i=0}^{P}\breve{D}_{W_{\mu_i}}(T,s,t)e^{-isT}, \tag{5.176}$$

where

$$\breve{D}_{W_{\mu_i}}(T,s,t)=\frac{1}{T}\sum_{k=-\infty}^{\infty}W(s+kj\omega)\mu_i(s+kj\omega)e^{(s+kj\omega)t}.$$

With the help of the above-derived formulae, it is possible to obtain closed formulae for the sum of series (5.175). Indeed, assume in (5.174) for the matrix $W(s)$ the form

$$W(s)=C(sI_\chi-A)^{-1}B+W_2. \tag{5.177}$$

Then, using (5.144) and (5.145), we find for $0 < t < T$

$$\breve{D}_{W_{\mu_i}}(T, s, t) = \breve{\mathscr{D}}_{W\mu_i}(T, s, t) = Ce^{At}(I_\chi - e^{-sT}e^{AT})^{-1}n_i(s, t) + W_2h_i(t), \tag{5.178}$$

where

$$\begin{aligned} n_i(t) &\triangleq \int_0^t e^{-A\nu}Bh_i(\nu)\,d\nu + e^{-sT}e^{AT}\int_t^T e^{-A\nu}Bh_i(\nu)\,d\nu \\ &= \bar{\Gamma}_{1i}(t) + e^{-sT}e^{AT}\bar{\Gamma}_{2i}(t). \end{aligned} \tag{5.179}$$

Inserting (5.178) and (5.179) into (5.176), we find for $0 < t < T$

$$\breve{D}_{W_\mu}(T, s, t) = \breve{\mathscr{D}}_{W_\mu}(T, s, t) = Ce^{At}(I_\chi - e^{-sT}e^{AT})^{-1}\breve{n}(s, t) + W_2\breve{h}(s, t). \tag{5.180}$$

Herein

$$\begin{aligned} \breve{n}(s, t) &\triangleq \sum_{i=0}^{P} e^{-sT}n_i(s, t) = \int_0^t e^{-A\nu}B\breve{h}(s, \nu)\,d\nu + e^{-sT}e^{AT}\int_t^T e^{-A\nu}B\breve{h}(s, \nu)\,d\nu, \\ \breve{h}(s, t) &\triangleq \sum_{i=0}^{P} e^{-isT}h_i(t). \end{aligned} \tag{5.181}$$

Formula (5.180) can be extended onto the whole axis $-\infty < t < \infty$ by applying the relation

$$\breve{D}_{W_\mu}(T, s, t) = \breve{\mathscr{D}}_{W_\mu}(T, s, t - kT)e^{ksT}, \quad kT < t < (k+1)T. \tag{5.182}$$

## 5.9 Parametric Discrete Models for Modulated Processes

(1) As it follows from the formulae of the preceding section, the matrix $\breve{D}_{W_\mu}(T, s, t)$ is RP for all $t$. Therefore, substituting in the matrix $\breve{D}_{W_\mu}(T, s, t)$ the argument $e^{-sT}$ by $\zeta$, we obtain the for all $t$ rational matrix

$$D_{W_\mu}(T, \zeta, t) = \breve{D}_{W_\mu}(T, s, t)|_{e^{-sT}=\zeta}. \tag{5.183}$$

**Definition 5.7** According to $\zeta$ rational matrix $D_{W_\mu}(T, \zeta, t)$ is called parametric discrete model of the modulated process $W(s)\mu(s)$.

Applying formulae (5.142), (5.176), by substituting in (5.180), (5.181) $e^{-sT}$ for $\zeta$, we obtain for $0 < t < T$

$$D_{W_\mu}(T, \zeta, t) = \mathscr{D}_{W_\mu}(T, \zeta, t) = \breve{\mathscr{D}}_{W_\mu}(T, s, t)|_{e^{-sT}=\zeta}. \tag{5.184}$$

Here

$$\mathscr{D}_{W_\mu}(T, \zeta, t) = C e^{At}(I_\chi - \zeta e^{AT})^{-1} m(\zeta) + g(\zeta, t) + W_2 h(\zeta, t), \tag{5.185}$$

where

$$\begin{aligned} m_P(\zeta) &\triangleq \int_0^T e^{-A\nu} B h_P(\zeta, \nu)\, d\nu, \\ g_P(\zeta, t) &\triangleq -C \int_t^T e^{A(t-\nu)} B h_P(\zeta, \nu)\, d\nu, \\ h_P(\zeta, t) &\triangleq \sum_{i=0}^P \zeta^i h_i(t). \end{aligned} \tag{5.186}$$

All matrices arising in (5.186) are polynomial matrices in $\zeta$. Moreover, from (5.182), after exchanging $\zeta$ for $e^{-sT}$, we find

$$D_{W_\mu}(T, \zeta, t) = \mathscr{D}_{W_\mu}(T, \zeta, t - kT)\zeta^{-k}, \quad kT < t < (k+1)T, \tag{5.187}$$

that for $kT < t < (k+1)T$ can be written in detailed form as

$$\begin{aligned} &D_{W_\mu}(T, \zeta, t) \\ &\quad = \left[C e^{(t-kT)}(I_\chi - \zeta e^{AT})^{-1} m_P(\zeta) + g_P(\zeta, t - kT) + W_2 h_P(\zeta, t - kT)\right]\zeta^{-k}. \end{aligned} \tag{5.188}$$

(2) From the last formulae, it follows that for $k \le 0$ and $kT < t < (k+1)T$ the matrix $D_{W_\mu}(T, \zeta, t)$ can be considered as transfer matrix of the inhomogeneous PMD

$$\Pi_k(\zeta) \triangleq (a(\zeta), b_k(\zeta, t), c(\zeta), d_k(\zeta, t)), \tag{5.189}$$

where

$$\begin{aligned} c(\zeta) = C, \quad a(\zeta) = I_\chi - \zeta e^{AT}, \quad b_k(\zeta, t) = e^{A(t-kT)} m_P(\zeta)\zeta^{-k}, \\ d_k(\zeta, t) = [g(\zeta, t - kT) + W_2 h_P(\zeta, t - kT)]\,\zeta^{-k}, \\ m_P(\zeta) = \sum_{i=0}^P \zeta^i \int_0^T e^{-A\nu} B h_i(\nu)\, d\nu. \end{aligned} \tag{5.190}$$

Introduce

$$m_\ell(A) \triangleq \sum_{i=0}^P e^{-iAT} \int_0^T e^{-A\nu} B h_i(\nu) d\nu. \tag{5.191}$$

**Theorem 5.12** *Let the period $T$ be non-pathological, and the pairs $[A, C]$ and $(e^{-AT}, m_\ell(A))$ be minimal. Then the following statements hold:*

*(i) PMD (5.189) is irreducible (minimal).*
*(ii) For the controllability of the pair* $(\mathrm{e}^{-AT}, m_\ell(A))$ *it is necessary that the pair* $(A, B)$ *is controllable.*

*Proof* (i) As follows from the definition in Sect. 3.12, PMD (5.189) is irreducible, if and only if the polynomial pairs $(a(\zeta), b_k(\zeta, t))$ and $[a(\zeta), c(\zeta)]$ are irreducible. Since the matrix $\mathrm{e}^{AT}$ is non-singular, the pair $(a(\zeta, b_k(\zeta, t))$ is irreducible in the case and only in the case, when the pair $(I_\chi - \zeta \mathrm{e}^{AT}, \mathrm{e}^{A(t-kT)} m_P(\zeta))$ is irreducible. It follows from Corollary 2.5 that this pair is irreducible exactly in the case, when the pair $(I_\chi - \zeta \mathrm{e}^{AT}, \mathrm{e}^{A(t-kT)} m_\ell(A))$ is irreducible, where $m_\ell(A)$ is matrix (5.191). On the other side, due to

$$\mathrm{e}^{A(t-kT)} \left[ I_\chi - \zeta \mathrm{e}^{AT} \;\; m_\ell(A) \right] \begin{bmatrix} \mathrm{e}^{-A(t-kT)} & 0 \\ 0 & I_\chi \end{bmatrix} = \left[ I_\chi - \zeta \mathrm{e}^{AT} \;\; \mathrm{e}^{A(t-kT)} m_\ell(A) \right],$$

the pair $(I_\chi - \zeta \mathrm{e}^{AT}, \mathrm{e}^{A(t-kT)} m_\ell(A))$ is reducible or irreducible at the same time as the pair $(I_\chi - \zeta \mathrm{e}^{AT}, m_\ell(A))$. The irreducibility of the pair $[a(\zeta), c(\zeta)]$ under the taken propositions was already shown above. So claim (i) is shown.

(ii) It is enough to show that the uncontrollability of the pair $(A, B)$ implies the uncontrollability of the pair $(\mathrm{e}^{-AT}, m_\ell(A))$. Notice that for an uncontrollable pair $(A, B)$, there exists an eigenvalue $\lambda_0$ of the matrix $A$ and a constant vector $x$ of size $1 \times \chi$, such that

$$xA = \lambda_0 x, \quad xB = 0. \tag{5.192}$$

From the first relation in (5.192) we realize with [5] that for all $t$

$$x\mathrm{e}^{At} = \mathrm{e}^{\lambda_0 t} x. \tag{5.193}$$

Hence

$$xm_\ell(A) = \sum_{i=0}^{P} \int_0^T x\mathrm{e}^{-A(iT+\nu)} Bh_i(\nu)\, \mathrm{d}\nu = \sum_{i=1}^{P} \int_0^T \mathrm{e}^{-\lambda_0(iT+\nu)} xBh_i(\nu)\, \mathrm{d}\nu = 0. \tag{5.194}$$

Then, from (5.192)–(5.194) we obtain

$$x\mathrm{e}^{-AT} = \mathrm{e}^{-\lambda_0 T} x, \quad xm_\ell(A) = 0. \tag{5.195}$$

The last equation means that the matrix $\left[ I_\chi - \zeta \mathrm{e}^{AT} \;\; m_\ell(A) \right]$ is latent, and thus the pair $(\mathrm{e}^{-AT}, m_\ell(A))$ is uncontrollable. ■

**Theorem 5.13** *Under the suppositions of Theorem 5.12, let us have the IMFD*

$$C(I_\chi - \zeta \mathrm{e}^{AT})^{-1} = a_l^{-1}(\zeta) b_l(\zeta), \quad (I_\chi - \zeta \mathrm{e}^{AT})^{-1} m_\ell(A) = b_r(\zeta) a_r^{-1}(\zeta). \tag{5.196}$$

*Then the following statements hold:*

(i) *The matrices*

$$b_l(\zeta,t) \stackrel{\triangle}{=} a_l(\zeta)D_{W_\mu}(T,\zeta,t), \quad b_r(\zeta,t) \stackrel{\triangle}{=} D_{W_\mu}(T,\zeta,t)a_r(\zeta) \tag{5.197}$$

*for* $t < T$ $(k \le 0)$ *are polynomial matrices.*

(ii) *The MFD arising from (5.197)*

$$D_{W_\mu}(T,\zeta,t) = a_l^{-1}(\zeta)b_l(\zeta,t), \quad D_{W_\mu}(T,\zeta,t) = b_r(\zeta,t)a_r^{-1}(\zeta) \tag{5.198}$$

*are IMDF for* $t < T$.

(iii) *For the matrices* $a_l(\zeta)$ *and* $a_r(\zeta)$ *relations (5.111)–(5.114) are true.*

The proof is not provided, because it repeats the arguments of the proof for Theorem 5.8. ■

## 5.10 Parametric Discrete Models of Modulated Processes with Delay

(1) Further on the matrix

$$W_{\mu\tau}(s) \stackrel{\triangle}{=} W_\tau(s)\mu(s) = W(s)\mu(s)e^{-s\tau} \tag{5.199}$$

is called the transfer matrix of the modulated process with delay. Here $W(s)$ is rational matrix (5.177, $\mu(s)$ is transfer matrix of the forming element for general hold of order $P$ (5.173), and $\tau > 0$ is a real constant.

Matrix (5.199) is related to the DLT

$$\breve{D}_{W_{\mu\tau}}(T,s,t) \stackrel{\triangle}{=} \frac{1}{T}\sum_{k=-\infty}^{\infty} W_{\mu\tau}(s+kj\omega)e^{(s+kj\omega)t}. \tag{5.200}$$

**Theorem 5.14** *Under the taken propositions the following statements hold:*

(i) *The formula is valid:*

$$\breve{D}_{W_{\mu\tau}}(T,s,t) = \breve{D}_\mu(T,s,t-\tau). \tag{5.201}$$

(ii) *If proposition (5.122) is true, then*

$$\breve{D}_{W_{\mu\tau}}(T,s,t) = \breve{\mathscr{D}}_{W_{\mu\tau}}(T,s,t), \quad 0 < t < T. \tag{5.202}$$

*Here*

$$\breve{\mathscr{D}}_{W_{\mu\tau}}(T,s,t)=\begin{cases}\breve{\mathscr{D}}_{W_\mu}(T,s,t+\gamma)\mathrm{e}^{-(p+1)sT}, & 0<t<\theta,\\ \breve{\mathscr{D}}_{W_\mu}(T,s,t-\theta)\mathrm{e}^{-psT}, & \theta<t<T,\end{cases} \tag{5.203}$$

*and* $\breve{\mathscr{D}}_{W_\mu}(T,s,t)$ *is matrix (5.180) and fulfills relation (5.122). Formula (5.202) can be extended onto the whole axis* $-\infty<t<\infty$ *with the help of*

$$\breve{D}_{W_{\mu\tau}}(T,s,t)=\breve{\mathscr{D}}_{W_{\mu\tau}}(t,s,t-kT)\mathrm{e}^{ksT},\quad kT<t<(k+1)T. \tag{5.204}$$

The proof of Theorem 5.14 is not provided, because it deviates only in some few details from the proof of Theorem 5.9. ■

Substituting in (5.185) $\zeta$ by $\mathrm{e}^{-sT}$, formulae (5.203) can be written in closed form

$$\breve{\mathscr{D}}_{W_{\mu\tau}}(T,s,t)= \tag{5.205}$$
$$\begin{cases}\Big[C\mathrm{e}^{A(t+\gamma)}(I_\chi-\mathrm{e}^{-sT}\mathrm{e}^{AT})^{-1}\breve{m}_P(s)+\breve{g}_P(s,t+\gamma)+W_2\breve{h}(s,t+\gamma)\Big]\mathrm{e}^{-(p+1)sT}, & 0<t<\theta,\\ \Big[C\mathrm{e}^{A(t-\theta)}(I_\chi-\mathrm{e}^{-sT}\mathrm{e}^{AT})\breve{m}_P(s)+\breve{g}_P(s,t-\theta)+W_2\breve{h}(s,t-\theta)\Big]\mathrm{e}^{-psT}, & \theta<t<T.\end{cases}$$

(2) From the above relations, it is achieved that for all $t$ the matrix $\breve{\mathscr{D}}_{W_{\mu\tau}}(T,s,t)$ is RP. Therefore, the matrix

$$\mathscr{D}_{W_{\mu\tau}}(T,\zeta,t)=\breve{\mathscr{D}}_{W_{\mu\tau}}(T,s,t)|_{\mathrm{e}^{-sT}=\zeta} \tag{5.206}$$

is rational in $\zeta$ for all $t$. Hence, from (5.205) we find that for $0<t<T$

$$D_{W_{\mu\tau}}(T,\zeta,t)=\mathscr{D}_{W_{\mu\tau}}(T,\zeta,t)= \tag{5.207}$$
$$\begin{cases}\big[C\mathrm{e}^{A(t+\gamma)}(I_\chi-\zeta\mathrm{e}^{AT})^{-1}m_P(\zeta)+g_P(\zeta,t+\gamma)+W_2h(\zeta,t+\gamma)\big]\zeta^{p+1}, & 0<t<\theta,\\ \big[C\mathrm{e}^{A(t-\theta)}(I_\chi-\zeta\mathrm{e}^{AT})^{-1}m_P(\zeta)+g_P(\zeta,t-\theta)+W_2h(\zeta,t-\theta),\big], & \theta<t<T.\end{cases}$$

Moreover, as usual

$$D_{W_{\mu\tau}}(T,\zeta,t)=\breve{\mathscr{D}}_{W_{\mu\tau}}(T,\zeta,t-kT)\zeta^{-k},\quad kT<t<(k+1)T. \tag{5.208}$$

From (5.207), (5.208) we realize that without essential changes, we can apply Theorem 5.13 on modulated processes with delay.

**Theorem 5.15** *Let the propositions of Theorem 5.12 hold, and assume IMFD (5.196). Then*

*(i) For* $t<T+\tau$ *the matrices*

$$b_{l\tau}(\zeta,t)\triangleq a_l(\zeta)D_{W_{\mu\tau}}(T,\zeta,t),\quad b_{r\tau}(\zeta,t)\triangleq D_{W_{\mu\tau}}(T,\zeta,t)a_r(\zeta) \tag{5.209}$$

*are polynomial matrices. Moreover,*

$$b_{l\tau}(\zeta, t) = b_l(\zeta, t - \tau), \quad b_{r\tau}(\zeta, t) = b_r(\zeta, t - \tau), \tag{5.210}$$

*where $b_l(\zeta, t)$, $b_r(\zeta, t)$ are the polynomial matrices in (5.197).*

*(ii) The MFD arising from (5.209), (5.210)*

$$\begin{aligned} D_{W_{\mu\tau}}(T, \zeta, t) &= a_l^{-1}(\zeta) b_l(\zeta, t - \tau), \\ D_{W_{\mu\tau}}(T, \zeta, t) &= b_r(\zeta, t - \tau) a_r^{-1}(\zeta) \end{aligned} \tag{5.211}$$

*are IMFD for $t < T + \tau$.*

*(iii) The matrices $a_l(\zeta)$, $a_r(\zeta)$ fulfill relations analog to (5.111)–(5.114).* ■

**Corollary 5.3** *It is important to remark that under the conditions of Theorem 5.15, polynomial matrices $a_l(\zeta)$ and $a_r(\zeta)$ do not depend on the value of the delay $\tau$ or the concrete form of the hold.*

## 5.11 Response of LTI System with Delay to Exponential-Periodic Inputs

(1) Assume a linear time invariant (LTI) system with delay, described by the operator equation

$$y(t) = W(p)\mathrm{e}^{-p\tau} x(t) = W_\tau(p) x(t), \tag{5.212}$$

where $p = \frac{\mathrm{d}}{\mathrm{d}t}$ is the differential operator, $\tau > 0$ is the value of the pure delay, $W(p) \in \mathbb{R}^{nm}(p)$ is a rational matrix, and $x(t)$, $y(t)$ are vectors of dimensions $m \times 1$ and $n \times 1$, respectively.

The vector signal $x(t)$ is called exponential-periodic (EP) if it has the form

$$x(t) = \mathrm{e}^{\lambda t} x_T(t), \quad x_T(t) = x_T(t + T). \tag{5.213}$$

Here $\lambda$ is a constant, which is called index of the EP function $x(t)$, and $T > 0$ is a constant, called its period. Below, we will always assume that the elements of the vector $x_T(t)$ are of bounded variation. The problem in the actual section consists in determining the steady EP response $y(\lambda, t)$ of system (5.212) to the EP input signal $x(t)$. This response has the form

$$y(\lambda, t) = \mathrm{e}^{\lambda t} y_T(\lambda, t), \quad y_T(\lambda, t) = y_T(\lambda, t + T), \tag{5.214}$$

i.e., it is an EP function with the same index and period as the input.

(2)

**Lemma 5.1** *Assume in (5.212) the standard form of the matrix $W(p)$ in the form*

$$W(p) = \frac{M(p)}{d(p)}, \tag{5.215}$$

*where $M(s) \in \mathbb{R}^{nm}[s]$ and $d(s)$ is a scalar polynomial of the form*

$$d(p) = (p - p_1)^{\mu_1} \cdots (p - p_q)^{\mu_q}, \tag{5.216}$$

*where all numbers $p_i$ are different. Assume the input vector*

$$x(t) = X\mathrm{e}^{\lambda t}, \tag{5.217}$$

*where $X$ is a constant vector and $\lambda$ is a constant satisfying*

$$\lambda \neq p_i, \quad (i = 1, \ldots, q). \tag{5.218}$$

*Then, there exists a unique EP output signal of system (5.212) of the form*

$$y(\lambda, t) = Y(\lambda)\mathrm{e}^{\lambda t}, \tag{5.219}$$

*where $Y(\lambda)$ is a constant vector, which satisfies the relation*

$$Y(\lambda) = W(\lambda)X\mathrm{e}^{-\lambda\tau} = W_\tau(\lambda)X. \tag{5.220}$$

*Proof* Assume the ILMDF

$$W(p) = a^{-1}(p)b(p). \tag{5.221}$$

Then, system (5.212) can be represented by the equivalent differential equation

$$a\left(\frac{\mathrm{d}}{\mathrm{d}t}\right) y(\lambda, t) = b\left(\frac{\mathrm{d}}{\mathrm{d}t}\right) x(t - \tau). \tag{5.222}$$

Due to (5.217) and (5.219),

$$\begin{aligned} a\left(\frac{\mathrm{d}}{\mathrm{d}t}\right) y(\lambda, t) &= a(\lambda)Y(\lambda)\mathrm{e}^{\lambda t}, \\ b\left(\frac{\mathrm{d}}{\mathrm{d}t}\right) x(t - \tau) &= b(\lambda)X\mathrm{e}^{\lambda(t-\tau)}. \end{aligned} \tag{5.223}$$

Inserting these relations into (5.222), we find

$$a(\lambda)Y(\lambda) = b(\lambda)\mathrm{e}^{-\lambda\tau}X. \tag{5.224}$$

From (5.218) and the general properties of ILMFD emerge, that $\det a(\lambda) \neq 0$. Therefore (5.224) yields

$$Y(\lambda) = a^{-1}(\lambda)b(\lambda)\mathrm{e}^{-\lambda\tau}X \tag{5.225}$$

that coincides with (5.220). Moreover, it follows from (5.218) that the achieved solution of form (5.219) is unique, because the homogeneous equation

$$a\left(\frac{\mathrm{d}}{\mathrm{d}t}\right)y(t) = 0_{n1} \tag{5.226}$$

does not possess nontrivial solutions. ■

(3) On basis of Lemma 5.1 we derive some important more general results.

**Theorem 5.16** *Let in (5.212) the transfer matrix $W(p)$ be strictly proper, and the EP input have form (5.213). Assume that for all $k = 0, \pm 1, \dots$ the index $\lambda$ satisfies*

$$\lambda + k\mathrm{j}\omega \neq p_i, \quad (i = 1, \dots, \rho). \tag{5.227}$$

*Then Eq. (5.222) has a unique EP solution of form (5.214). Hereby, the function $y_T(\lambda, t)$ is determined by*

$$y_T(\lambda, t) = \int_0^T \varphi_{W_\tau}(T, \lambda, t - \nu)x_T(\nu)\,\mathrm{d}\nu = y_T(\lambda, t + T), \tag{5.228}$$

*where*

$$\varphi_{W_\tau}(T, \lambda, t) = \mathrm{e}^{-\lambda\tau}\frac{1}{T}\sum_{k=-\infty}^{\infty} W(\lambda + k\mathrm{j}\omega)\mathrm{e}^{k\mathrm{j}\omega(t-\tau)} \tag{5.229}$$

*is DPFR (5.120).*

*Proof* Under the taken propositions, the periodic vector $x_T(t)$ can be represented by the convergent Fourier series

$$x_T(t) = \sum_{k=-\infty}^{\infty} x_k \mathrm{e}^{k\mathrm{j}\omega t}, \tag{5.230}$$

where

$$x_k = \frac{1}{T}\int_0^T x_T(\nu)\mathrm{e}^{-k\mathrm{j}\omega\nu}\,\mathrm{d}\nu. \tag{5.231}$$

Hence, EP input (5.213) relates to the convergent series

$$x(t) = \sum_{k=-\infty}^{\infty} x_k \mathrm{e}^{(\lambda+k\mathrm{j}\omega)t}. \tag{5.232}$$

Therefore,

$$x(t-\tau) = \sum_{k=-\infty}^{\infty} x_k e^{(\lambda+kj\omega)(t-\tau)}. \tag{5.233}$$

From (5.227) and Lemma 5.1, we find that EP response (5.214) in the actual case becomes

$$y(\lambda, t) = \sum_{k=-\infty}^{\infty} W(\lambda + kj\omega) x_k e^{(\lambda+kj\omega)t} e^{-(\lambda+kj\omega)\tau}. \tag{5.234}$$

Thus

$$y_T(\lambda, t) = y_T(\lambda, t+T) = e^{-\lambda\tau} \sum_{k=-\infty}^{\infty} W(\lambda + kj\omega) x_k e^{kj\omega(t-\tau)}. \tag{5.235}$$

Inserting here (5.231), we arrive at the expression

$$y_T(\lambda, t) = e^{-\lambda\tau} \sum_{k=-\infty}^{\infty} W(\lambda + kj\omega) \frac{1}{T} \int_0^T x_T(\nu) e^{-kj\omega\nu} \, d\nu \; e^{kj\omega(t-\tau)},$$

which, after exchanging the order of summation and integration, takes the form

$$y_T(\lambda, t) = e^{-\lambda\tau} \int_0^T \left[ \frac{1}{T} \sum_{k=-\infty}^{\infty} W(\lambda + kj\omega) e^{kj\omega(t-\tau-\nu)} \right] x_T(\nu) \, d\nu, \tag{5.236}$$

which is equivalent to (5.228). The uniqueness of the constructed EP solution arises from the fact that, when (5.227) is fulfilled, homogeneous Eq. (5.226) does not possess a solution of form (5.214). ■

## 5.12 Response of Open Sampled-Data System with Delay to EP Input

(1) In this section, we study the EP response of an open sampled-data system with delay to EP input (5.213), where the period $T$ coincides with the sampling period. The structure of the system is shown in Fig. 5.2.

In harmony with the expositions in Chap. 4, in Fig. 5.2 ADC is the analog to digital converter described by

$$\xi_k = x(kT) \triangleq x_k, \quad (k = 0, \pm 1, \ldots). \tag{5.237}$$

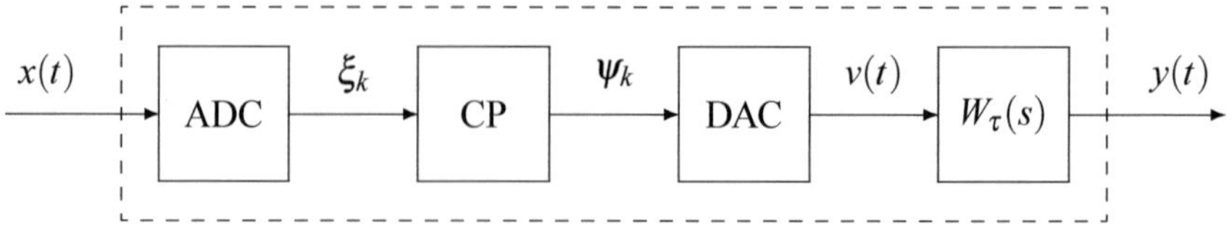


**Fig. 5.2** Structure of open sampled-data system with delay

Here and later on, we propose that the input $x(t)$ is continuous at the sampling instants $t_k = kT$. We also assume that the computer program CP is described by the causal backward model

$$\alpha(\zeta)\psi_k = \beta(\zeta)\xi_k, \tag{5.238}$$

where

$$\begin{aligned}\alpha(\zeta) &= \alpha_0 + \alpha_1\zeta + \cdots + \alpha_\rho\zeta^\rho,\\ \beta(\zeta) &= \beta_0 + \beta_1\zeta + \cdots + \beta_\rho\zeta^\rho\end{aligned} \tag{5.239}$$

are polynomial matrices, and $\det \alpha_0 \neq 0$.

Moreover, in Fig. 5.2 DAC is the digital to analog converter of order $P$, described by

$$u(t) = \sum_{i=0}^{P} h_i(t - kT)\psi_{k-i}, \quad (kT < t < (k+1)T). \tag{5.240}$$

With respect to the transfer matrix of the continuous element with delay $W_\tau(p)$, we assume form (5.118), where the rational matrix $W(p)$ is at least proper.

(2) Let the EP input in Fig. 5.2 have form (5.213). The problem consists in determining the EP output $y(\lambda, t)$ of form (5.214). From (5.213) and (5.237), we obtain

$$\xi_k = \mathrm{e}^{k\lambda T}x_0, \quad x_0 \overset{\triangle}{=} x_T(0). \tag{5.241}$$

**Definition 5.8** The property of the ADC, expressed by relations (5.241), is called stroboscopic property.

In the steady-state EP regime at the output of the ALG should be

$$\psi_k(\lambda) = \mathrm{e}^{k\lambda T}\psi_0(\lambda), \quad \xi_k(\lambda) = \mathrm{e}^{k\lambda T}\xi_0(\lambda), \tag{5.242}$$

where $\psi_0(\lambda)$, $\xi_0(\lambda)$ are constant vectors. Inserting (5.242) into (5.238) yields

$$\breve{\alpha}(\lambda)\psi_0(\lambda) = \breve{\beta}(\lambda)\xi_0, \tag{5.243}$$

where

$$\begin{aligned}\breve{\alpha}(\lambda) &= \alpha(\zeta)|_{\zeta=e^{-\lambda T}} = \alpha_0 + \alpha_1 e^{-\lambda T} + \cdots + \alpha_\rho e^{-\rho\lambda T},\\ \breve{\beta}(\lambda) &= \beta(\zeta)|_{\zeta=e^{-\lambda T}} = \beta_0 + \beta_1 e^{-\lambda T} + \cdots + \beta_\rho e^{-\rho\lambda T}.\end{aligned} \tag{5.244}$$

Further on, it is always assumed that the number $\zeta = e^{-\lambda T}$ is not an eigenvalue of the polynomial matrix $\alpha(\zeta)$ in (5.239). Then, from (5.243) we find

$$\psi_0(\lambda) = \breve{W}_d(\lambda)\xi_0 = \breve{W}_d(\lambda)x_0, \tag{5.245}$$

where

$$\breve{W}_d(\lambda) = W_d(\zeta)|_{\zeta=e^{-\lambda T}} = \breve{\alpha}^{-1}(\lambda)\breve{\beta}(\lambda) \tag{5.246}$$

and $W_d(\zeta)$ is the backward transfer matrix of the control algorithm.

(3) Find the response $u(\lambda, t)$ of the DAC to input sequence (5.242), when a zero-order hold is applied. By definition, in this case we obtain

$$u(\lambda, t) = u_0(\lambda, t) = h_0(t - kT)\psi_k(\lambda), \quad kT < t < (k+1)T, \tag{5.247}$$

that can be written in the form

$$u_0(\lambda, t) = e^{\lambda t} u_{T0}(\lambda, t)\psi_0(\lambda), \tag{5.248}$$

where

$$u_{T0}(\lambda, t) = e^{-\lambda(t-kT)} h_0(t - kT), \quad kT < t < (k+1)T. \tag{5.249}$$

Comparing (5.249) with (5.168), we find

$$u_{T0}(\lambda, t) = \varphi_{\mu_0}(T, \lambda, t) = \frac{1}{T}\sum_{k=-\infty}^{\infty} \mu_0(s + kj\omega)e^{kj\omega t}. \tag{5.250}$$

This relation proves that the input to the continuous element $u_0(\lambda, t)$ is an EP signal, and it can be presented in the form

$$u_0(\lambda, t) = e^{\lambda t}\varphi_{\mu_0}(T, \lambda, t)\psi_0(\lambda). \tag{5.251}$$

The achieved result can be extended easily to the case, when the DAC in Fig. 5.2 is a general hold of order $P$. Indeed, looking back to the above considerations and using instead of (5.247)

$$u(\lambda, t) = u_i(\lambda, t) = h_i(t - kT)\psi_{k-i}(\lambda), \quad kT < t < (k+1)T, \tag{5.252}$$

then under condition (5.242), we obtain

$$u_i(\lambda, t) = \mathrm{e}^{\lambda t} \varphi_{\mu_i}(T, \lambda, t) \psi_0(\lambda) \mathrm{e}^{-i\lambda T}. \tag{5.253}$$

Therefore, when the DAC has general form (5.240), then we achieve

$$u(t) = \mathrm{e}^{\lambda t} \varphi_\mu(T, \lambda, t) \psi_0(\lambda), \tag{5.254}$$

where

$$\varphi_\mu(T, \lambda, t) = \frac{1}{T} \sum_{k=-\infty}^{\infty} \mu(\lambda + kj\omega) \mathrm{e}^{kj\omega t} = \sum_{i=0}^{P} \varphi_{\mu_i}(T, \lambda, t) \mathrm{e}^{-i\lambda T}, \tag{5.255}$$

and the vector $\psi_0(\lambda)$ is determined by relation (5.245).

(4) Thus, if at the input of the investigated open sampled-data system, EP signal (5.213) is acting, then under the taken propositions, in the steady-state EP regime, the input of the LTI element with delay is the EP signal

$$u(\lambda, t) = \mathrm{e}^{\lambda t} \varphi_\mu(T, \lambda, t) \breve{W}_d(\lambda) x_0. \tag{5.256}$$

The response of the LTI block to input (5.256) can be found by applying Lemma 5.1. Hence, when condition (5.227) is valid, we obtain for the EP output of the system $y(\lambda, t)$ expression (5.214), wherein

$$y_T(\lambda, t) = \int_0^T \varphi_{W_\tau}(T, \lambda, t - \nu) \varphi_\mu(T, \lambda, \nu) \, \mathrm{d}\nu \; \breve{W}_d(\lambda) x_0. \tag{5.257}$$

Noticing that

$$\varphi_{W_\tau}(T, \lambda, t - \nu) = \mathrm{e}^{-\lambda \tau} \varphi_W(T, \lambda, t - \tau - \nu) \tag{5.258}$$

and using (5.255), from (5.257) after integration, we obtain

$$y_T(\lambda, t) = \mathrm{e}^{-\lambda \tau} \varphi_{W\mu}(T, \lambda, t - \tau) \breve{W}_d(\lambda) x_0, \tag{5.259}$$

that with the help of (5.214) yields the final result

$$y(\lambda, t) = \mathrm{e}^{-\lambda(t-\tau)} \varphi_{W_\mu}(T, \lambda, t - \tau) \breve{W}_d(\lambda) x_0 = \breve{D}_{W_\mu}(T, \lambda, t - \tau) \breve{W}_d(\lambda) x_0. \tag{5.260}$$

## References

1. E.N. Rosenwasser, B.P. Lampe, *Computer Controlled Systems: Analysis and Design with Process-orientated Models* (Springer, Berlin, 2000)
2. F.R. Gantmacher, *The Theory of Matrices* (Chelsea, New York, 1959)
3. T. Kailath, *Linear Systems* (Prentice Hall, Englewood Cliffs, 1980)

4. K.J. Åström, B. Wittenmark, *Computer Controlled Systems: Theory and Design*, 3rd edn. (Prentice-Hall, Englewood Cliffs, 1997)
5. T. Chen, B.A. Francis, *Optimal Sampled-Data Control Systems* (Springer, Berlin, 1995)
6. R.E. Kalman, Y.C. Ho, K. Narendra, Controllabiltiy of linear dynamical systems. Contrib. Theory Differ. Equ. **1**, 189–213 (1963)
7. B. van der Pol, H. Bremmer, *Operational Calculus Based on the Two-Sided Laplace Integral* (University Press, Cambridge, 1959)
8. E.N. Rosenwasser, *Linear Theory of Digital Control in Continuous Time* (Nauka, Moscow, 1994). (in Russian)
9. E.C. Titchmarsh, *The Theory of Functions*, 2nd edn. (Oxford University Press, Oxford, 1997)

# Chapter 6
# Mathematical Description of Standard Sampled-Data Systems with Delay

**Abstract** This chapter introduces a system of differential–difference equations consisting of several components as the standard model of the multidimensional SD system with delay. This standard model is associated with a generalized structure called the standard structure. The parametric transmission matrix (PTM) $W_{zx}(\lambda, t)$, which corresponds to this standard structure with input $x(t)$ and output $z(t)$, is constructed and its properties are investigated. The chapter also provides the solutions for modal control and stabilization problems of the standard model. A parameterization of the set of causal stabilizing regulators is generated. On this basis, a representation of the PTM is constructed using the system parameter matrix $\theta(s)$. At the end of the chapter necessary and sufficient conditions under which the standard PTM $W_{zx}(\lambda, t)$ is generated by the standard model are given.

## 6.1 Standard Model of SD Systems with Delay

(1) In this chapter, we consider sampled-data (SD) systems in which the controlled LTI process is described in state space

$$\begin{aligned} \frac{\mathrm{d}v(t)}{\mathrm{d}t} &= Av(t) + B_1 x(t-\tilde{\mu}_1) + B_2 u(t-\tilde{\mu}_2), \\ y(t) &= C_2 v(t-\tilde{\mu}_3). \end{aligned} \tag{6.1}$$

Here, $v(t)$ is the $\chi \times 1$ state vector, $x(t)$ is the $\ell \times 1$ input vector, y(t) is the $n \times 1$ output vector of the LTI process, and $u(t)$ is the $m \times 1$ control vector. Moreover, $A, B_1, B_2, C_2$ are constant matrices of appropriate size, and $\tilde{\mu}_1, \tilde{\mu}_2, \tilde{\mu}_3$ are nonnegative constants, characterizing the pure delay at the input and output of the process.

Passing in (6.1) to the Laplace transforms under zero initial conditions, we obtain the corresponding images

$$V(s) = F(s)\mathrm{e}^{-s\tilde{\mu}_1}X(s) + G(s)\mathrm{e}^{-s\tilde{\mu}_2}U(s), \tag{6.2}$$

and

$$Y(s) = M(s)\mathrm{e}^{-s(\tilde{\mu}_1+\tilde{\mu}_3)}X(s) + N(s)\mathrm{e}^{-s(\tilde{\mu}_2+\tilde{\mu}_3)}U(s). \tag{6.3}$$



E. N. Rosenwasser et al., *Computer-Controlled Systems with Delay*,
https://doi.org/10.1007/978-3-030-15042-6_6

In these formulae

$$F(s) \stackrel{\triangle}{=} (sI_\chi - A)^{-1}B_1, \quad G(s) \stackrel{\triangle}{=} (sI_\chi - A)^{-1}B_2, \tag{6.4}$$

and also

$$M(s) \stackrel{\triangle}{=} C_2F(s) = C_2(sI_\chi - A)^{-1}B_1, \quad N(s) \stackrel{\triangle}{=} C_2G(s) = C_2(sI_\chi - A)^{-1}B \tag{6.5}$$

are rational matrices.

Notice that without loss of generality, it can be supposed in (6.1) that $\tilde{\mu}_3 = 0$, i.e., the delay at the output of the LTI process is omitted. This assumption will essentially simplify our derivation. Indeed, if we consider the state-space equations in the form

$$\begin{aligned}\frac{\mathrm{d}v_1(t)}{\mathrm{d}t} &= Av_1(t) + B_1x(t - \tilde{\mu}_1 - \tilde{\mu}_3) + B_2u(t - \tilde{\mu}_2 - \tilde{\mu}_3),\\ y_1(t) &= C_2v_1(t),\end{aligned}$$

then for the output image $Y_1(s)$, we obtain an expression matching with (6.3). Therefore, in what follows, we assume that the equations for the LTI process have the form

$$\begin{aligned}\frac{\mathrm{d}v(t)}{\mathrm{d}t} &= Av(t) + B_1x(t - \tau_1) + B_2u(t - \tau_2),\\ y(t) &= C_2v(t),\end{aligned} \tag{6.6}$$

where we used the notation

$$\tau_1 = \tilde{\mu}_1 + \tilde{\mu}_3, \quad \tau_2 = \tilde{\mu}_2 + \tilde{\mu}_3 \tag{6.7}$$

and the constants $\tau_1$ and $\tau_2$ include the delay at the output of process (6.1).

(2) Below we suppose that the LTI process (6.5) is controlled with the help of a digital controller (DC), in which the ADC is described by the equation

$$\xi_k = y(kT), \quad (k = 0, \pm 1, \ldots). \tag{6.8}$$

The control algorithm is given by the causal backward model

$$\alpha(\zeta)\psi_k = \beta(\zeta)\xi_k, \tag{6.9}$$

where $\alpha(\zeta)$, $\beta(\zeta)$ are polynomial $q \times q$, respectively, $q \times n$ matrices. The control vector $u(t)$ is generated with the help of a multivariable generalized DAC of order $P$, i.e.,

$$u(t) = \sum_{i=0}^{P} h_i(t - kT)\psi_{k-i}, \quad kT < t < (k+1)T, \tag{6.10}$$

where $h_i(t)$ is a $m \times q$ matrix. Below, we will always propose that the matrices $h_i(t)$ are generalized linear independent, i.e., the equality

$$\sum_{i=0}^{P} h_i(t)C_i = 0_{mq},$$

where $C_i$ are constant matrices, implies $C_i = 0_{qq}$ for all $i$.

Equations (6.6)–(6.10) define the closed loop of the digitally controlled LTI process (6.6).

(3) As output of the closed SD system, we consider the $r \times 1$ vector $z(t)$, given by the relation

$$z(t) = C_1 v(t - \tau_3) + Du(t - \tau_2 - \tau_3), \tag{6.11}$$

where $C_1$ and $D$ are constant matrices of size $r \times \chi$ and $r \times m$, respectively. Moreover, $\tau_3$ is a constant. It is important to notice that in contrast to $\tau_1$ and $\tau_2$,the constant $\tau_3$ can be negative. As it will be shown below by an example, such a situation can happen in those cases, when the output $z(t)$ contains contributions of some internal variables. Together with equations of the closed loop (6.6)–(6.10), the output Eq. (6.11) leads to the system of differential–difference equations

$$\begin{aligned}
\frac{dv(t)}{dt} &= Av(t) + B_1 x(t - \tau_1) + B_2 u(t - \tau_2),\\
y(t) &= C_2 v(t),\\
z(t) &= C_1 v(t - \tau_3) + Du(t - \tau_2 - \tau_3),\\
\xi_k &= y(kT), \quad (k = 0, \pm 1, \ldots),\\
\alpha(\zeta)\psi_k &= \beta(\zeta)\xi_k,\\
u(t) &= \sum_{i=0}^{P} h_i(t - kT)\psi_{k-i}, \quad kT < t < (k+1)T.
\end{aligned} \tag{6.12}$$

Below, we will always assume, if not explicitly said otherwise, that in Eq. (6.12) the pair $(A, B_2)$ is controllable, and the pair $[A, C_2]$ is observable. This requirement is essential, because it implies the controllability and observability of the LTI process in the closed loop.

**Definition 6.1** When the condition on minimality of the pairs $(A, B_2)$ and $[A, C_2]$ is fulfilled, then system of Eq. (6.12) is called standard model of the sampled-data system with delay, and is denoted by $\mathscr{S}_\tau$.

As a solution of the model, we understand the totality of vectors $v(t)$, $y(t)$, $z(t)$, $u(t)$ and sequences $\{\psi_k\}$, $\{\xi_k\}$, which for all $t$ and $k$ satisfy Eq. (6.12). Hereby, the vectors $v(t)$, $y(t)$ are supposed to be continuous.

## 6.2 Standard Structure of SD Systems with Delay

(1) The transfer to images in (6.11) generates

$$Z(s) = C_1 e^{-s\tau_3} V(p) + De^{-s(\tau_2-\tau_3)} U(s). \tag{6.13}$$

Using that, due to (6.2),

$$V(s) = F(s)e^{-s\tau_1} X(s) + G(s)e^{-s\tau_3} U(s),$$

from (6.13), we find

$$Z(s) = K(s)e^{-s(\tau_1+\tau_3)} X(s) + L(s)e^{-s(\tau_2+\tau_3)} U(s), \tag{6.14}$$

where

$$K(s) = C_1(sI_\chi - A)^{-1} B_1; \quad L(s) = C_1(sI_\chi - A)^{-1} B_2 + D. \tag{6.15}$$

Exchanging the complex variable $s$ with the differential operator $\mathrm{d}/\mathrm{d}t$ in the above relations, the standard model of SD system (6.12) can be written as a system of operator equations

$$\begin{aligned}
z(t) &= K_\tau(p)x(t) + L_\tau(p)u(t),\\
y(t) &= M_\tau(p)x(t) + N_\tau(p)u(t),\\
\xi_k &= y(kT), \quad (k = 0, \pm 1, \ldots),\\
\alpha(\zeta)\psi_k &= \beta(\zeta)\xi_k,\\
u(t) &= \sum_{i=0}^{P} h_i(t - kT)\psi_{k-i}, \quad kT < t < (k+1)T.
\end{aligned} \tag{6.16}$$

In (6.16), we used the notation

$$\begin{aligned}
K_\tau(p) &\stackrel{\triangle}{=} K(p)e^{-p\tau_K}, \quad L_\tau(p) \stackrel{\triangle}{=} L(p)e^{-p\tau_L},\\
M_\tau(p) &\stackrel{\triangle}{=} M(p)e^{-p\tau_M}, \quad N_\tau(p) \stackrel{\triangle}{=} N(p)e^{-p\tau_N}.
\end{aligned} \tag{6.17}$$

Here, in consistency with earlier formulae

$$\begin{aligned}
K(p) &= C_1(pI_\chi - A)^{-1} B_1, \quad L(p) = C_1(pI_\chi - A)^{-1} B_2 + D,\\
M(p) &= C_2(pI_\chi - A)^{-1} B_2, \quad N(p) = C_2(pI_\chi - A)^{-1} B_2,
\end{aligned} \tag{6.18}$$

where

$$\tau_K = \tau_1 + \tau_3, \quad \tau_L = \tau_2 + \tau_3,$$
$$\tau_M = \tau_1, \quad \tau_N = \tau_2. \tag{6.19}$$

The matrices $K_\tau(p), L_\tau(p), M_\tau(p), N_\tau(p)$ can be considered as transfer matrices from inputs $x(t)$ and $u(t)$ to the respective outputs $z(t)$ and $y(t)$.

(2) The first two equations in (6.16) can be understood as an extended LTI process $\mathscr{L}_\tau$ from input $\bar{x}(t)$ to output $\bar{y}(t)$, where

$$\bar{x}(t) = \begin{bmatrix} x(t) \\ u(t) \end{bmatrix}, \quad \bar{y}(t) = \begin{bmatrix} z(t) \\ y(t) \end{bmatrix} \tag{6.20}$$

are vectors of dimensions $(\ell + m) \times 1$ and $(r + n) \times 1$, respectively. The operational description of the extended LTI process takes the form

$$\bar{y}(t) = \bar{W}_\tau(p)\bar{x}(t), \tag{6.21}$$

where $\bar{W}_\tau(p)$ is the extended transfer matrix of size $(r + n) \times (\ell + m)$ defined by

$$\bar{W}_\tau(p) = \begin{bmatrix} K_\tau(p) & L_\tau(p) \\ M_\tau(p) & N_\tau(p) \end{bmatrix}. \tag{6.22}$$

Moreover, Eq. (6.16) is related to the generalized structure shown in Fig. 6.1 In detail, the structure of Fig. 6.1 can be represented as shown in Fig. 6.2.

**Definition 6.2** The structure shown in Fig. 6.2 is called standard structure of the SD system with delay.

*Remark 6.1* In Definition 6.2, it is not assumed that relations (6.19) are fulfilled.

(3) We will list some special properties of the standard structure for the case when it is defined by the standard model (6.12).

(a) The rational matrices $K(p), L(p), M(p), N(p)$ are connected by relations (6.18).
(b) The matrices $K(p), M(p), N(p)$ are strictly proper, but the matrix $L(p)$ is at least proper.

**Fig. 6.1** Structure of standard SD control system with delay

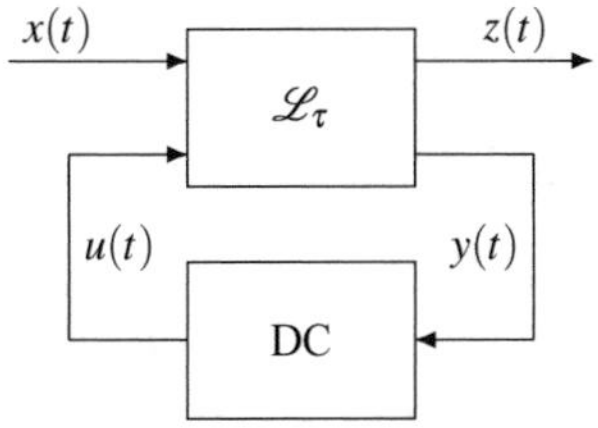

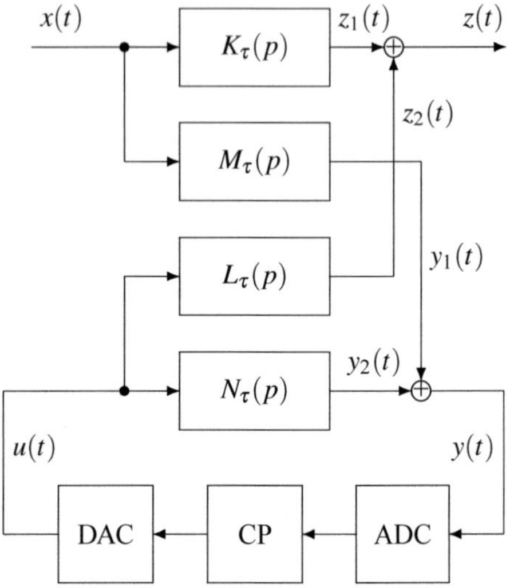


**Fig. 6.2** Detailed structure of a SD control system with delay

(c) Condition (6.19) is true.

For $\tau_1 = \tau_2 = \tau_3 = 0$ matrix (6.22) establishes as

$$\bar{W}_0(p) = \begin{bmatrix} K(p) & L(p) \\ M(p) & N(p) \end{bmatrix}. \tag{6.23}$$

**Theorem 6.1** *In Eq.* (6.12), *let the pair* $(A, B_2)$ *be controllable and the pair* $[A, C_2]$ *be observable. Then the matrix* $N(p)$ *dominates in matrix* (6.23), *i.e., it satisfies*

$$\operatorname{Mind} \bar{W}_0(p) = \operatorname{Mind} N(p),$$

*where, as earlier,* Mind *is the McMillan index of the relevant matrix.*

*Proof* Using relation (6.18), from (6.23), we obtain

$$\bar{W}_0(p) = \begin{bmatrix} C_1(pI_\chi - A)^{-1}B_1 & C_1(pI_\chi - A)^{-1}B_2 + D \\ C_2(pI_\chi - A)^{-1}B_1 & C_2(pI_\chi - A)^{-1}B_2 \end{bmatrix}, \tag{6.24}$$

which can be written in the form

$$\bar{W}_0(p) = \bar{W}_1(p) + \bar{W}_2, \tag{6.25}$$

where $\bar{W}_1(p)$ is the strictly proper rational matrix

$$\bar{W}_1(p) = \begin{bmatrix} C_1(pI_\chi - A)^{-1}B_1 & C_1(pI_\chi - A)^{-1}B_2 \\ C_2(pI_\chi - A)^{-1}B_1 & C_2(pI_\chi - A)^{-1}B_2 \end{bmatrix} \tag{6.26}$$

and $\bar{W}_2$ is the constant matrix

$$\bar{W}_2 = \begin{bmatrix} O_{r\ell} & D \\ O_{n\ell} & O_{nm} \end{bmatrix}. \tag{6.27}$$

Lemma 2.11 says that

$$\text{Mind}\, \bar{W}_0(p) = \text{Mind}\, \bar{W}_1(p). \tag{6.28}$$

Moreover, notice that the matrix $\bar{W}_1(p)$ can be represented in the form

$$\bar{W}_1(p) = C(pI_\chi - A)^{-1}B, \tag{6.29}$$

where

$$C \triangleq \begin{bmatrix} C_1 \\ C_2 \end{bmatrix}, \quad B \triangleq \begin{bmatrix} B_1 & B_2 \end{bmatrix}. \tag{6.30}$$

Equation (6.29) shows that the matrix $\bar{W}_1(p)$ is the transfer matrix of the realization $(A, B, C)$. Hence

$$\text{Mind}\, W_1(p) \le \chi. \tag{6.31}$$

On the other side, since the pairs $(A, B_2)$ and $[A, C_2]$ are minimal, so

$$\text{Mind}\, N(p) = \text{Mind}\, C_2(pI_\chi - A)^{-1}B_2 = \chi.$$

Therefore, due to Corollary 2.6

$$\text{Mind}\, \bar{W}_0(p) = \text{Mind}\, \bar{W}_1(p) \ge \chi. \tag{6.32}$$

Comparing relations (6.31) and (6.32), we realize that Mind $\bar{W}_1(p) =$ Mind $\bar{W}_0(p) = \chi$, i.e., the matrix $N(p)$ dominates in the matrix $\bar{W}_0(p)$. ■

## 6.3 Parametric Transfer Matrices (PTM) for the Standard Structure of SD Systems with Delay

(1) The standard structure for SD systems with delay can be interpreted for matrix input signals. That can be done in the following way: Let the $\ell \times \ell$ matrix $x(t)$ have the form

$$x(t) = \begin{bmatrix} x_1(t) & \dots & x_\ell(t) \end{bmatrix}, \tag{6.33}$$

where $x_i(t)$ is an $\ell \times 1$ vector. Let the vectors $v_i(t)$, $y_i(t)$, $u_i(t)$, $z_i(t)$, $\{\psi_{ki}\}$ define the corresponding responses of system (6.12) to the vector $x_i(t)$. Then, the matrices

$$\begin{aligned} \upsilon(t) &= \begin{bmatrix} \upsilon_1(t)\ \upsilon_2(t)\ \dots\ \upsilon_\ell(t) \end{bmatrix}, \\ y(t) &= \begin{bmatrix} y_1(t)\ y_2(t)\ \dots\ y_\ell(t) \end{bmatrix}, \\ u(t) &= \begin{bmatrix} u_1(t)\ u_2(t)\ \dots\ u_\ell(t) \end{bmatrix}, \\ z(t) &= \begin{bmatrix} z_1(t)\ z_2(t)\ \dots\ z_\ell(t) \end{bmatrix}, \\ \{\psi_k\} &= \begin{bmatrix} \{\psi_{k1}\}\ \{\psi_{k2}\}\ \dots\ \{\psi_{k\ell}\} \end{bmatrix}. \end{aligned} \tag{6.34}$$

are called the response of system (6.16) to the matrix input (6.33).

(2) Assume that the matrix input $x(t)$ is an EP function and has the form

$$x(\lambda, t) = \mathrm{e}^{\lambda t} x_T(t), \quad x_T(t) = x_T(t+T), \tag{6.35}$$

where $\lambda$ is a scalar parameter, $x_T(t)$ is a matrix of form (6.33), and $T$ is the sampling period of the digital controller.

**Definition 6.3** The steady-state EP response of the standard structure to the matrix EP input (6.35) is called matrix solution of Eq. (6.12), where

$$\begin{aligned} \upsilon(\lambda, t) &= \mathrm{e}^{\lambda t} \upsilon_T(\lambda, t),\ \upsilon_T(\lambda, t) = \upsilon_T(\lambda, t+T), \\ y(\lambda, t) &= \mathrm{e}^{\lambda t} y_T(\lambda, t),\ y_T(\lambda, t) = y_T(\lambda, t+T), \\ u(\lambda, t) &= \mathrm{e}^{\lambda t} u_T(\lambda, t),\ u_T(\lambda, t) = u_T(\lambda, t+T). \end{aligned} \tag{6.36}$$

If (6.36) is fulfilled, an analog representation is also possible for the output $z(\lambda, t)$

$$z(\lambda, t) = \mathrm{e}^{\lambda t} z_T(\lambda, t), \quad z_T(\lambda, t) = z_T(\lambda, t+T). \tag{6.37}$$

(3) Assume that the totality of matrices

$$\upsilon(\lambda, t), y(\lambda, t), u(\lambda, t), \{\psi_k(\lambda)\} \tag{6.38}$$

define a particular solution of the closed-loop equations of the standard model (6.12) for the EP input (6.35). Since by definition the matrices $\upsilon(\lambda, t)$ and $y(\lambda, t)$ depend continuously on $t$, so the matrix sequences

$$\{\upsilon_k(\lambda)\} \triangleq \{\upsilon(\lambda, kT)\}, \quad \{y_k(\lambda)\} \triangleq \{y(\lambda, kT)\}. \tag{6.39}$$

are defined.

**Theorem 6.2** *For the EP input* (6.35), *the matrices* $\upsilon(\lambda, t), y(\lambda, t), u(\lambda, t)$, *and* $z(\lambda, t)$ *are EP, if and only if the elements of the matrix sequences* $\{\upsilon_k(\lambda)\}$ *and* $\{\psi_k(\lambda)\}$ *satisfy*

$$\upsilon_k(\lambda) = \mathrm{e}^{\lambda T} \upsilon_{k-1}(\lambda), \quad \psi_k(\lambda) = \mathrm{e}^{\lambda T} \psi_{k-1}(\lambda). \tag{6.40}$$

*Proof* Necessity: Assume matrices (6.34) to be EP. Then obviously the first relation in (6.40) is fulfilled, because $\upsilon_k(\lambda) = \mathrm{e}^{k\lambda T} \upsilon_T(\lambda, 0) = \mathrm{e}^{k\lambda T} \upsilon_0(\lambda)$, and the first relation in (6.40) is true. For the necessity of the second relation in (6.40), introduce the periodic functions $h_{Ti}(t)$, defined by the relations

$$h_{Ti}(t) = h_i(t - kT), \quad kT < t < (k+1)T. \tag{6.41}$$

Then the equation of the generalized DAC (6.10) can be written in the form

$$u(\lambda, t) = \sum_{i=0}^{P} h_{Ti}(t)\psi_{k-i}(\lambda), \quad kT < t < (k+1)T. \tag{6.42}$$

Hereby,

$$u_T(\lambda, t) = \mathrm{e}^{-\lambda t} u(\lambda, t) = \mathrm{e}^{-\lambda t} \sum_{i=0}^{P} h_{Ti}(t)\psi_{k-i}(\lambda), \quad kT < t < (k+1)T. \tag{6.43}$$

On the other side, introduce

$$u_{T0}(t) \triangleq \mathrm{e}^{-\lambda t} \sum_{i=0}^{P} h_{Ti}(t)\psi_{-i}(\lambda), \quad 0 < t < T. \tag{6.44}$$

Then, due to $u_T(\lambda, t) = u_T(\lambda, t + T)$, we obtain

$$u_T(\lambda, t) = u_{T0}(\lambda, t - kT), \quad kT < t < (k+1)T, \tag{6.45}$$

which with the help of (6.44) can be written as

$$u_T(\lambda, t) = \mathrm{e}^{-\lambda(t-kT)} \sum_{i=0}^{P} h_{Ti}(t)\psi_{-i}(\lambda), \quad kT < t < (k+1)T. \tag{6.46}$$

Comparing (6.43) and (6.46), we arrive at

$$\sum_{i=0}^{P} h_{Ti}(t)(\mathrm{e}^{\lambda kT}\psi_{-i}(\lambda) - \psi_{k-i}(\lambda)) = 0_{mq}, \quad kT < t < (k+1)T. \tag{6.47}$$

Under the supposition of generalized linear independence of the matrices $h_i(t)$, from (6.47) we find that, for all integers $k$, we obtain $\mathrm{e}^{\lambda kT}\psi_{-i}(\lambda) = \psi_{k-i}(\lambda)$, which is equivalent to the second relation in (6.40). Thus, the necessity is shown.
Sufficiency: Let relations (6.40) be fulfilled for input (6.35). We demonstrate that in this case the matrix

$$u(\lambda, t) = \sum_{i=0}^{P} h_{Ti}(t)\psi_{k-i}, \quad kT < t < (k+1)T \tag{6.48}$$

is EP. Indeed, the matrix

$$u_1(\lambda, t) \triangleq e^{-\lambda t} u(\lambda, t), \tag{6.49}$$

under conditions (6.40), takes the form

$$u_1(\lambda, t) = e^{-\lambda t} \sum_{i=0}^{P} h_{Ti}(t) e^{(k-i)\lambda T} \psi_0 = e^{-\lambda(t-kT)} \sum_{i=0}^{P} h_i(t-kT) e^{-i\lambda T} \psi_0,$$
$$kT < t < (k+1)T,$$

which is equivalent to the expression

$$u_1(\lambda, t) = u_{10}(\lambda, t-kT), \quad kT < t < (k+1)T, \tag{6.50}$$

where

$$u_{10}(\lambda, t) = e^{-\lambda t} \breve{h}(\lambda, t) \psi_0, \quad 0 < t < T$$

and

$$\breve{h}(\lambda, t) = \sum_{i=0}^{P} h_i(t) e^{-i\lambda T}.$$

Obviously, matrix (6.50) is $T$-periodic, and hence matrix (6.48) is EP. Further on, we will show that under condition (6.40) the matrix $v(\lambda, t)$ is also EP. For that, we integrate the first equation in (6.12), for the initial values $v(\lambda, kT) = v_k(\lambda)$

$$\begin{aligned} v(\lambda, t) = e^{A(t-kT)} v_k(\lambda) &+ \int_{kT}^{t} e^{A(t-\nu)} B_1 x(\lambda, \nu - \tau_1)\, d\nu \\ &+ \int_{kT}^{t} e^{A(t-\nu)} B_2 u(\lambda, \nu - \tau_2)\, d\nu. \end{aligned} \tag{6.51}$$

Applying here $t = kT + t_1$, we find

$$\begin{aligned} v(\lambda, kT + t_1) = e^{At_1} v_k(\lambda) &+ \int_{kT}^{kT+t_1} e^{A(kT+t_1-\nu)} B_1 x(\lambda, \nu - \tau_1)\, d\nu \\ &+ \int_{kT}^{kT+t_1} e^{A(kT+t_1-\nu)} B_2 u(\lambda, \nu - \tau_2)\, d\nu. \end{aligned} \tag{6.52}$$

After transfer to the integration variable $\nu = kT + \mu$, we obtain

$$\begin{aligned} v(\lambda, kT + t_1) = e^{At_1} v_k(\lambda) &+ \int_{0}^{t_1} e^{A(t_1-\mu)} B_1 x(\lambda, kT + \mu - \tau_1)\, d\mu \\ &+ \int_{0}^{t_1} e^{A(t_1-\mu)} B_2 u(\lambda, kT + \mu - \tau_1)\, d\mu. \end{aligned} \tag{6.53}$$

Since the matrices $x(\lambda, t)$ and $u(\lambda, t)$ are EP, we find

$$\begin{aligned} x(\lambda, kT + \mu - \tau_1) &= e^{k\lambda T} x(\lambda, \mu - \tau_1), \\ u(\lambda, kT + \mu - \tau_2) &= e^{k\lambda T} u(\lambda, \mu - \tau_2). \end{aligned} \tag{6.54}$$

Using this relation and (6.40), relation (6.53) can be written in the form

$$\begin{aligned} &\upsilon(\lambda, kT + t_1) \\ &= e^{k\lambda T} \left[ e^{At_1} \upsilon_0(\lambda) + \int_0^{t_1} e^{A(t_1-\mu)} B_1 x(\mu - \tau_1)\, d\mu + \int_0^{t_1} e^{A(t_1-\mu)} B_2 u(\mu - \tau_2)\, d\mu \right], \end{aligned} \tag{6.55}$$

which for $k = 1, t_1 = t$ yields

$$\upsilon(\lambda, t + T) = e^{\lambda T} \upsilon(\lambda, t). \tag{6.56}$$

When (6.56) is true, the matrix $\upsilon_1(\lambda, t) = e^{-\lambda t}\upsilon(\lambda, t)$ is $T$-periodic. This fact emerges from

$$\upsilon_1(\lambda, t + T) = e^{-\lambda(t+T)} \upsilon(\lambda, t + T) = e^{-\lambda t} \upsilon(\lambda, t) = \upsilon_1(\lambda, t). \tag{6.57}$$

■

(4) If, in particular, the matrix EP input (6.35) takes the form

$$x(\lambda, t) = e^{\lambda t} I_\ell, \tag{6.58}$$

then the steady-state EP response (6.36) is indicated by the special notation

$$\begin{aligned} \upsilon_T(\lambda, t) &\stackrel{\triangle}{=} W_{\upsilon x}(\lambda, t), \quad y_T(\lambda, t) \stackrel{\triangle}{=} W_{yx}(\lambda, t), \\ u_T(\lambda, t) &\stackrel{\triangle}{=} W_{ux}(\lambda, t), \quad z_T(\lambda, t) \stackrel{\triangle}{=} W_{zx}(\lambda, t). \end{aligned} \tag{6.59}$$

**Definition 6.4** Matrices (6.59), periodic in $t$, are called parametric transfer matrices (PTM) of the standard structure (6.12) from input $x(t)$ to the respective outputs.

Our next goal is to find closed expressions for the PTM (6.59).

**Theorem 6.3** *Let the matrix $N(p)$ in Fig. 6.2 be strictly proper, and the matrix $L(p)$ at least proper. Then the following formulae are valid:*

*(i) The PTM $W_{yx}(\lambda, t)$ is determined by*

$$W_{yx}(\lambda, t) = \varphi_{N_\tau \mu}(T, \lambda, t) \check{R}_N(\lambda) M_\tau(\lambda) + M_\tau(\lambda). \tag{6.60}$$

*Herein*

$$\varphi_{N_\tau\mu}(T,\lambda,t)=\frac{1}{T}\sum_{k=-\infty}^{\infty}N_\tau(\lambda+k\mathrm{j}\omega)\mu(\lambda+k\mathrm{j}\omega)\mathrm{e}^{k\mathrm{j}\omega t}=\mathrm{e}^{-\lambda\tau_2}\varphi_{N\mu}(T,\lambda,t-\tau_2), \tag{6.61}$$

*and*

$$\begin{aligned}\breve{R}_N(\lambda)&=\tilde{W}_d(\lambda)\left[I_n-\tilde{D}_{N\mu}(T,\lambda,-\tau_2)\breve{W}_d(\lambda)\right]^{-1},\\ \breve{W}_d(\lambda)&=\left[\alpha^{-1}(\zeta)\beta(\zeta)\right]_{\zeta=\mathrm{e}^{-\lambda T}},\end{aligned} \tag{6.62}$$

*and finally*

$$\mu(\lambda)=\sum_{i=0}^{P}\mathrm{e}^{-i\lambda T}\mu_i(\lambda),\quad \mu_i(\lambda)=\int_0^T h_i(t)\mathrm{e}^{-\lambda t}\,\mathrm{d}t. \tag{6.63}$$

*(ii) The PTM $W_{ux}(\lambda,t)$ is determined by*

$$W_{ux}(\lambda,t)=\varphi_\mu(T,\lambda,t)\tilde{R}_N(\lambda)M_\tau(\lambda), \tag{6.64}$$

*where*

$$\varphi_\mu(T,\lambda,t)=\frac{1}{T}\sum_{k=-\infty}^{\infty}\mu(\lambda+k\mathrm{j}\omega)\mathrm{e}^{k\mathrm{j}\omega t}, \tag{6.65}$$

*or equivalently*

$$\begin{aligned}\varphi_\mu(T,\lambda,t)&=\mathrm{e}^{-\lambda t}\sum_{i=0}^{P}\mathrm{e}^{-i\lambda T}h_i(t)=\mathrm{e}^{-\lambda t}\breve{h}(\lambda,t),\quad 0<t<T\\ \varphi_\mu(T,\lambda,t)&=\varphi_\mu(T,\lambda,t+T).\end{aligned} \tag{6.66}$$

*(iii) The PTM $W_{zx}(\lambda,t)$ has the form*

$$W_{zx}(\lambda,t)=\varphi_{L_\tau\mu}(T,\lambda,t)\breve{R}_N(\lambda)M_\tau(\lambda)+K_\tau(\lambda), \tag{6.67}$$

*where*

$$\begin{aligned}\varphi_{L_\tau\mu}(T,\lambda,t)&=\frac{1}{T}\sum_{k=-\infty}^{\infty}L_\tau(\lambda+k\mathrm{j}\omega)\mu(\lambda+k\mathrm{j}\omega)\mathrm{e}^{k\mathrm{j}\omega t}\\ &=\mathrm{e}^{-\lambda(\tau_2+\tau_3)}\varphi_{L\mu}(T,\lambda,t-\tau_2-\tau_3).\end{aligned} \tag{6.68}$$

*Below the right side of formula* (6.67) *is called standard form of the PTM to the output $z(t)$.*

*(iv) The PTM $W_{vx}(\lambda, t)$ is determined by the expression*

$$W_{vx}(\lambda, t) = \varphi_{G_\tau \mu}(T, \lambda, t)\breve{R}_N(\lambda)M_\tau(\lambda) + F_\tau(\lambda), \tag{6.69}$$

*where*

$$\varphi_{G_\tau \mu}(T, \lambda, t) = \mathrm{e}^{-\lambda\tau_2}\varphi_{G\mu}(T, \lambda, t - \tau_2) \tag{6.70}$$

*and*

$$\begin{aligned} G_\tau(\lambda) &= G(\lambda)\mathrm{e}^{-\lambda\tau_2} = (\lambda I_\chi - A)^{-1}B_2\mathrm{e}^{-\lambda\tau_2}, \\ F_\tau(\lambda) &= F(\lambda)\mathrm{e}^{-\lambda\tau_1} = (\lambda I_\chi - A)^{-1}B_1\mathrm{e}^{-\lambda\tau_1}. \end{aligned} \tag{6.71}$$

*Proof* (i) It follows from Definition 6.4 that the PTM (6.59) can be defined by calculating the EP responses of the outputs $y(\lambda, t)$, $u(\lambda, t)$, $z(\lambda, t)$, $v(\lambda, t)$, when the input of the standard structure is excited by the matrix EP signal (6.58). In particular, by definition, we obtain in the steady-state EP regime

$$y(\lambda, t) = \mathrm{e}^{\lambda t}W_{yx}(\lambda, t), \quad W_{yx}(\lambda, t) = W_{yx}(\lambda, t + T). \tag{6.72}$$

Moreover, since the matrix $N(\lambda)$ is strictly proper, so the matrix $y(\lambda, t)$ continuously depends on $t$. The same dependence holds for the PTM $W_{yx}(\lambda, t)$. In the steady EP regime, as it can be seen from Fig. 6.2, the matrix $y_2(\lambda, t)$ can be considered as the EP output of the open sampled-data system shown in Fig. 6.3 under the input excitation by an EP signal (6.72). Using the continuity of the matrix $W_{yx}(\lambda, t)$ with respect to $t$, and applying the stroboscopic property of the ADC, given in Definition 5.8, we can state that the output $y_2(\lambda, t)$ of the system in Fig. 6.3 coincides with the output of the same system with the matrix input $\mathrm{e}^{\lambda t}W_{yx}(\lambda, 0)$. So the equivalent structure shown in Fig. 6.4 is achieved. From Fig. 6.4, using formula (5.256), we obtain

$$y_2(\lambda, t) = \mathrm{e}^{\lambda(t-\tau_2)}\varphi_{N\mu}(T, \lambda, t - \tau_2)\breve{W}_d(\lambda)W_{yx}(\lambda, 0). \tag{6.73}$$

Here

$$\varphi_{N\mu}(T, \lambda, t) \triangleq \frac{1}{T}\sum_{k=-\infty}^{\infty} N(\lambda + kj\omega)\mu(\lambda + kj\omega)\mathrm{e}^{kj\omega t}, \tag{6.74}$$

where, as earlier,

$$\mu(\lambda) = \sum_{i=0}^{P} \mathrm{e}^{-i\lambda T}\mu_i(\lambda), \quad \mu_i(\lambda) = \int_0^T h_i(t)\mathrm{e}^{-\lambda t}\,\mathrm{d}t. \tag{6.75}$$

Moreover, in (6.73)

$$\breve{W}_d(\lambda) = \breve{\alpha}^{-1}(\lambda)\breve{\beta}(\lambda), \tag{6.76}$$

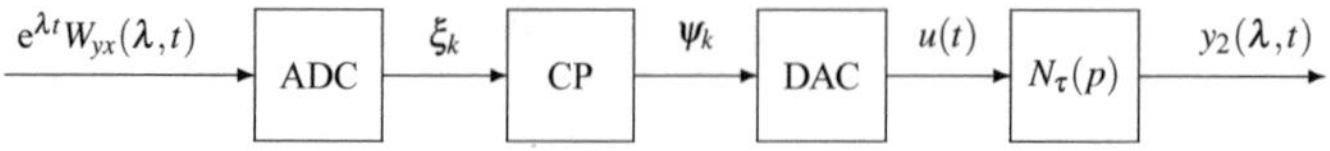


**Fig. 6.3** Structure for steady-state regime

where

$$\begin{aligned}\breve{\alpha}(\lambda) &= \alpha_0 + \alpha_1 e^{-\lambda T} + \cdots + \alpha_\rho e^{-\rho\lambda T},\\ \breve{\beta}(\lambda) &= \beta_0 + \beta_1 e^{-\lambda T} + \cdots + \beta_\rho e^{-\rho\lambda T}.\end{aligned} \tag{6.77}$$

Further, notice that under the conditions of Theorem 6.3, we obtain

$$y_1(\lambda, t) = M_\tau(\lambda) e^{\lambda t}. \tag{6.78}$$

From Fig. 6.2, we also read that

$$y(\lambda, t) = e^{\lambda t} W_{yx}(\lambda, t) = y_1(\lambda, t) + y_2(\lambda, t). \tag{6.79}$$

Collecting (6.73) and (6.78), after cancelation of $e^{\lambda t}$, we arrive at

$$W_{yx}(\lambda, t) = e^{-\lambda\tau_2} \varphi_{N\mu}(T, \lambda, t - \tau_2) \breve{W}_d(\lambda) W_{yx}(\lambda, 0) + M_\tau(\lambda). \tag{6.80}$$

Thanks to the earlier derived results, the left and right parts of this equation depend continuously on $t$. Therefore, in (6.80), we can set $t = 0$, and so we obtain

$$W_{yx}(\lambda, 0) = e^{-\lambda\tau_2} \varphi_{N\mu}(T, \lambda, -\tau_2) \breve{W}_d(\lambda) W_{yx}(\lambda, 0) + M_\tau(\lambda). \tag{6.81}$$

Taking into account

$$e^{-\lambda\tau_2} \varphi_{N\mu}(T, \lambda, -\tau_2) = \breve{D}_{N\mu}(T, \lambda, -\tau_2), \tag{6.82}$$

Equation (6.81) can be written as

$$W_{yx}(\lambda, 0) = \breve{D}_{N\mu}(T, \lambda, -\tau_2) \breve{W}_d(\lambda) W_{yx}(\lambda, 0) + M_\tau(\lambda). \tag{6.83}$$

Hence

$$W_{yx}(\lambda, 0) = \left[ I_n - \breve{D}_{N\mu}(T, \lambda, -\tau_2) \breve{W}_d(\lambda) \right]^{-1} M_\tau(\lambda). \tag{6.84}$$

Inserting this expression into (6.80), we establish (6.60)–(6.63).

(ii) From Fig. 6.4, we directly realize that in the steady-state EP regime

$$u(\lambda, t) = e^{\lambda t} W_{ux}(\lambda, t) = e^{\lambda t} \varphi_\mu(T, \lambda, t) \breve{W}_d(\lambda) W_{yx}(\lambda, 0). \tag{6.85}$$

Inserting here (6.84), we arrive at (6.64)–(6.66).

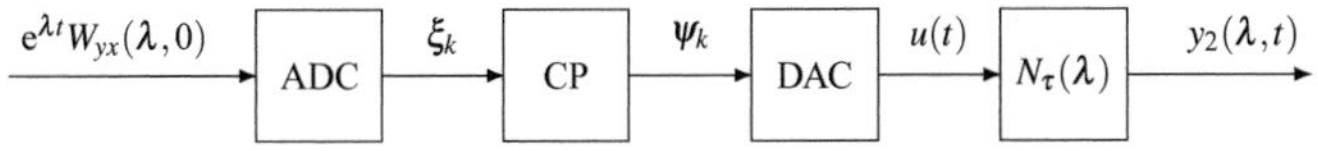


**Fig. 6.4** Equivalent structure due to the stroboscopic property of ADC

**Fig. 6.5** Ancillary structure

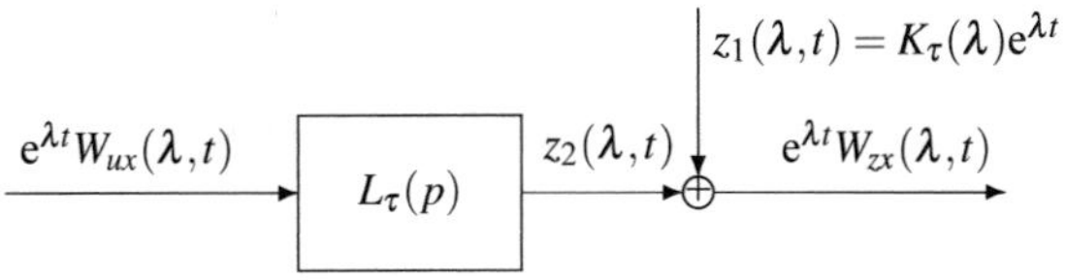


(iii) For the proof, on basis of Fig. 6.2, we consider the open structure in Fig. 6.5. From (6.64), (6.65), we derive

$$e^{\lambda t}W_{ux}(\lambda,t)=\left[\frac{1}{T}\sum_{k=-\infty}^{\infty}\mu(\lambda+kj\omega)e^{(\lambda+kj\omega)t}\right]\breve{R}_N(\lambda)M_\tau(\lambda). \tag{6.86}$$

From Fig. 6.5, on basis of Lemma 5.1, it follows that the EP response $z_2(\lambda,t)$ to EP input (6.86) is determined by the relations

$$\begin{aligned} z_2(\lambda,t)&=\left[\frac{1}{T}\sum_{k=-\infty}^{\infty}L_\tau(\lambda+kj\omega)\mu(\lambda+kj\omega)e^{(\lambda+kj\omega)t}\right]\breve{R}_N(\lambda)M_\tau(\lambda)\\ &=\tilde{D}_{L_\tau\mu}(T,\lambda,t)\breve{R}_N(\lambda)M_\tau(\lambda), \end{aligned} \tag{6.87}$$

which due to (5.201), can be presented in the form

$$z_2(\lambda,t)=e^{\lambda(t-\tau_L)}\varphi_{L\mu}(T,\lambda,t-\tau_L)\breve{R}_N(\lambda)M_\tau(\lambda). \tag{6.88}$$

Since from Fig. 6.5, it follows

$$W_{zx}(\lambda,t)=e^{-\lambda t}z_2(\lambda,t)+K_\tau(\lambda), \tag{6.89}$$

so we arrive at formulae (6.67), (6.68).

(iv) For the proof, we use the formula

$$V(s)=F(s)e^{-s\tau_1}X(s)+G(s)e^{-s\tau_2}U(s), \tag{6.90}$$

which arises from the first equation in (6.12), after transfer to the images and applying notations (6.4) and (6.19). This formula, in the stationary EP regime, can be configured to Fig. 6.6.

**Fig. 6.6** Ancillary structure for the proof (iv)

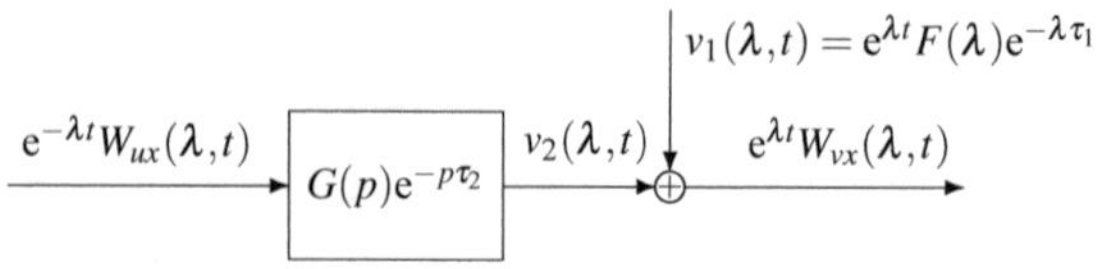


In the given case, in analogy to (6.87), we obtain

$$v_2(\lambda, t) = \left[\frac{1}{T}\sum_{k=-\infty}^{\infty} G(\lambda + kj\omega)\mu(\lambda + kj\omega)e^{(\lambda+kj\omega)(t-\tau_2)}\right]\breve{R}_N(\lambda)M_\tau(\lambda). \tag{6.91}$$

Hence, from Fig. 6.6

$$W_{vx}(\lambda, t) = e^{-\lambda t}v_2(\lambda, t) + F(\lambda)e^{-\lambda\tau_1}, \tag{6.92}$$

so applying (6.91), we establish formulae (6.69)–(6.71). ■

Analyzing the relations presented in Sect. 6.3, we realize that these relations do not depend on the validity of condition (6.19). Therefore, we can claim that the formulae achieved for the PTM $W_{yx}(\lambda, t)$, $W_{ux}(\lambda, t)$, $W_{zx}(\lambda, t)$ are valid for any standard structure when the conditions of Theorem 6.3 are fulfilled.

## 6.4 PTM of the Standard Model as a Function of Arguments $t$ and $\lambda$

As it emerges from the relations of Sect. 6.3, the PTM $W_{vx}(\lambda, t)$ and $W_{yx}(\lambda, t) = C_2W_{vx}(\lambda, t)$ depend continuously on $t$ for $-\infty < t < \infty$. The PTM $W_{ux}(\lambda, t)$, and also $W_{zx}(\lambda, t)$ for $D \neq 0_{r\ell}$, possess finite jump discontinuities.

For the study of the properties of PTM (6.59) as a function of the argument $\lambda$, we assume, that in Eq. (6.12),

$$\tau_2 = m_2T + \theta_2 = (m_2 + 1)T - \gamma_2, \tag{6.93}$$

where $m_2 \geq 0$ is an integer, and

$$0 \leq \theta_2 < T, \quad 0 < \gamma_2 = T - \theta_2 \leq T. \tag{6.94}$$

Introduce the matrix

$$\breve{Q}(\lambda, \breve{\alpha}, \breve{\beta}) = \begin{bmatrix} I_\chi - e^{-\lambda T}e^{AT} & 0_{\chi n} & -e^{-(m_2+1)\lambda T}\breve{b}(\lambda) \\ -C_2 & I_n & 0_{nq} \\ 0_{q\chi} & -\breve{\beta}(\lambda) & \breve{\alpha}(\lambda) \end{bmatrix}, \tag{6.95}$$

where

$$\breve{b}(\lambda) = \sum_{i=0}^{P} \left( \varGamma_{1i}(T) + \mathrm{e}^{-\lambda T} \varGamma_{2i}(T) \right) \mathrm{e}^{-i\lambda T} \tag{6.96}$$

and the matrices $\varGamma_{1i}(T)$ and $\varGamma_{2i}(T)$ are determined by relations (4.72), which in the actual case have the form

$$\begin{aligned} \varGamma_{1i}(T) &= \mathrm{e}^{A\gamma_2} \int_0^{\gamma_2} \mathrm{e}^{-A\nu} B_2 h_i(\nu)\, \mathrm{d}\nu, \\ \varGamma_{2i}(T) &= \mathrm{e}^{A(T+\gamma_2)} \int_{\gamma_2}^{T} \mathrm{e}^{-A\nu} B_2 h_i(\nu)\, \mathrm{d}\nu. \end{aligned} \tag{6.97}$$

**Theorem 6.4** *For the PTM of the standard model* (6.12), *defined by the relations of Sect. 6.3, the following representations are valid:*

$$\begin{aligned} W_{\upsilon x}(\lambda, t) &= \frac{L_{\upsilon x}(\lambda, t)}{\det \breve{Q}(\lambda, \tilde{\alpha}, \tilde{\beta})}, \quad W_{yx}(\lambda, t) = \frac{L_{yx}(\lambda, t)}{\det \breve{Q}(\tilde{\lambda}, \tilde{\alpha}, \tilde{\beta})}, \\ W_{ux}(\lambda, t) &= \frac{L_{ux}(\lambda, t)}{\det \breve{Q}(\lambda, \tilde{\alpha}, \tilde{\beta})}, \quad W_{zx}(\lambda, t) = \frac{L_{zx}(\lambda, t)}{\det \breve{Q}(\tilde{\lambda}, \alpha, \tilde{\beta})}, \end{aligned} \tag{6.98}$$

*where the matrices* $L_{\upsilon x}(\lambda, t) = L_{\upsilon x}(\lambda, t+T), L_{yx}(\lambda, t) = L_{yx}(\lambda, t+T)$, $L_{ux}(\lambda, t) = L_{ux}(\lambda, t+T), L_{zx}(\lambda, t) = L_{zx}(\lambda, t+T)$ *are integral functions of the argument* $\lambda$ *for all* $t$ .

*Proof* Since

$$u(t) = \sum_{i=0}^{P} h_i(t - kT)\psi_{k-i}, \quad kT < t < (k+1)T, \tag{6.99}$$

then, as it follows from (4.62) and (6.93), we obtain

$$u(t-\tau_2) = \begin{cases} \sum_{i=0}^{P} h_i(t - kT + \gamma_2)\psi_{k-m_2-i-1}, & kT < t < kT + \theta_2, \\ \sum_{i=0}^{P} h_i(t - kT - \theta_2)\psi_{k-m_2-i}, & kT + \theta_2 < t < (k+1)T. \end{cases} \tag{6.100}$$

Substituting (6.100) in (6.51), we find

$$\upsilon(t) = \begin{cases} \mathrm{e}^{A(t-kT)}\upsilon_k + \sum_{i=0}^{P} \int_{kT}^{t} \mathrm{e}^{A(t-\nu)} B_2 h_i(\nu - kT + \gamma_2)\, \mathrm{d}\nu \psi_{k-m_2-i-1} \\ \quad + \int_{kT}^{t} \mathrm{e}^{A(t-\nu)} B_1 x(\nu - \tau_1)\, \mathrm{d}\nu, \qquad kT \le t \le kT + \theta_2, \\ \mathrm{e}^{A(t-kT)}\upsilon_k + \sum_{i=0}^{P} \int_{kT}^{kT+\theta_2} \mathrm{e}^{A(t-\nu)} B_2 h_i(\nu - kT + \gamma_2)\, \mathrm{d}\nu \psi_{k-m_2-i-1} \\ \quad + \sum_{i=0}^{P} \int_{kT+\theta_2}^{t} \mathrm{e}^{A(t-\nu)} B_2 h_i(\nu - kT - \theta_2)\, \mathrm{d}\nu \psi_{k-m_2-i} \\ \quad + \int_{kT}^{t} \mathrm{e}^{A(t-\nu)} B_1 x(\nu - \tau_1)\, \mathrm{d}\nu, \qquad kT + \theta_2 \le t \le (k+1)T. \end{cases} \tag{6.101}$$

For $t = (k+1)T$, we arrive at

$$\begin{aligned} \upsilon_{k+1} = \mathrm{e}^{AT}\upsilon_k &+ \sum_{i=0}^{P}\int_{kT}^{kT+\theta_2} \mathrm{e}^{A(kT+T-\nu)}B_2 h_i(\nu - kT + \gamma_2)\,\mathrm{d}\nu\psi_{k-m_2-i-1} \\ &+ \sum_{i=0}^{P}\int_{kT+\theta_2}^{(k+1)T} \mathrm{e}^{A(kT+T-\nu)}B_2 h_i(\nu - kT - \theta_2)\,\mathrm{d}\nu\psi_{k-m_2-i} \\ &+ \int_{kT}^{(k+1)T} \mathrm{e}^{A(kT+T-\nu)}B_1 x(\nu - \tau_1)\,\mathrm{d}\nu. \end{aligned} \tag{6.102}$$

After redefinition of the integration variable $\nu - kT + \gamma_2 = u$, we verify

$$\int_{kT}^{kT+\theta_2} \mathrm{e}^{A(kT+T-\nu)}B_2 h_i(\nu - kT + \gamma_2)\,\mathrm{d}\nu = \mathrm{e}^{A(T+\gamma_2)}\int_{\gamma_2}^{T} \mathrm{e}^{u}B_2 h_i(u)\,\mathrm{d}u = \Gamma_{2i}(T), \tag{6.103}$$

wherein we used notation (6.97). Analogously, by applying $u = \nu - kT - \theta_2$, we obtain

$$\int_{kT+\theta_2}^{(k+1)T} \mathrm{e}^{A(kT+T-\nu)}B_2 h_i(\nu - kT - \theta_2)\,\mathrm{d}\nu = \mathrm{e}^{A\gamma_2}\int_{0}^{\gamma_2} \mathrm{e}^{-Au}B_2 h_i(u)\,\mathrm{d}u = \Gamma_{1i}(T), \tag{6.104}$$

where in we also used notation (6.97). For the further calculation notice that under condition (6.58)

$$g_k(\lambda) \stackrel{\triangle}{=} \int_{kT}^{(k+1)T} \mathrm{e}^{A(kT+T-\nu)}B_1 x(\nu - \tau_1)\,\mathrm{d}\nu = \mathrm{e}^{-\lambda\tau_1}\mathrm{e}^{A(kT+T)}\int_{kT}^{(k+1)T} \mathrm{e}^{(\lambda I_\chi - A)\nu}\,d\nu B_1. \tag{6.105}$$

Applying the rule for the integration of the index function of a matrix, we find

$$\begin{aligned} \int_{kT}^{(k+1)T} \mathrm{e}^{(\lambda I_\chi - A)\nu}\,\mathrm{d}\nu &= (\lambda I_\chi - A)^{-1}\left[\mathrm{e}^{(\lambda I_\chi - A)(kT+T)} - \mathrm{e}^{(\lambda I_\chi - A)kT}\right] \\ &= \mathrm{e}^{k\lambda T}\mathrm{e}^{-kAT}(\lambda I_\chi - A)^{-1}\left[\mathrm{e}^{\lambda T}\mathrm{e}^{-AT} - I_\chi\right]. \end{aligned} \tag{6.106}$$

Insert this expression into (6.105), so

$$g_k(\lambda) = \mathrm{e}^{-\lambda\tau_1}\mathrm{e}^{k\lambda T}(\lambda I_\chi - A)^{-1}(\mathrm{e}^{\lambda T}I_\chi - \mathrm{e}^{AT})B_1. \tag{6.107}$$

With the help of relations (6.103), (6.104), and (6.107), formula (6.102) can be written in the form

$$\begin{aligned} \upsilon_{k+1} = \mathrm{e}^{AT}\upsilon_k &+ \sum_{i=0}^{P}\Gamma_{1i}(T)\psi_{k-m_2-i} + \sum_{i=0}^{P}\Gamma_{2i}(T)\psi_{k-m_2-i-1} \\ &+ \mathrm{e}^{-\lambda\tau_1}\mathrm{e}^{k\lambda T}(\lambda I_\chi - A)^{-1})(\mathrm{e}^{\lambda T}I_\chi - \mathrm{e}^{AT})B_1. \end{aligned} \tag{6.108}$$

Rename herein $k$ by $k-1$ and add to the achieved equations relations (6.8) and (6.9), we arrive at the system matrix difference equations

$$\begin{aligned} \upsilon_k &= e^{AT}\upsilon_{k-1} + \sum_{i=0}^{P} \Gamma_{1i}(T)\psi_{k-m_2-i-1} + \sum_{i=0}^{P} \Gamma_{2i}(T)\psi_{k-m_2-i-2} \\ &\quad + e^{-\lambda\tau_1} e^{k\lambda T}(\lambda I_\chi - A)^{-1}(I_\chi - e^{-\lambda T}e^{AT})B_1, \\ \xi_k &= C_2\upsilon_k, \\ \alpha_0\psi_k &+ \alpha_1\psi_{k-1} + \cdots + \alpha_\rho\psi_{k-\rho} = \beta_0\xi_k + \beta_1\xi_{k-1} + \cdots + \beta_\rho\xi_{k-\rho}. \end{aligned} \tag{6.109}$$

Equation (6.109) defines a discrete matrix model of the closed-loop standard model $\mathscr{S}_\tau$ under the input excitation (6.58). Now, on basis of Theorem 6.2, we search for a solution of (6.109) in the form of the two-sided infinite matrix sequences

$$\begin{aligned} \{\upsilon_k(\lambda)\} &= \{\ldots \quad \upsilon_{-1}(\lambda) \quad \upsilon_0(\lambda) \quad \upsilon_1(\lambda) \quad \ldots\}, \\ \{\xi_k(\lambda)\} &= \{\ldots \quad \xi_{-1}(\lambda) \quad \xi_0(\lambda) \quad \xi_1(\lambda) \quad \ldots\}, \\ \{\psi_k(\lambda)\} &= \{\ldots \quad \psi_{-1}(\lambda) \quad \psi_0(\lambda) \quad \psi_1(\lambda) \quad \ldots\}, \end{aligned} \tag{6.110}$$

and their elements are connected by the relations

$$\upsilon_k(\lambda) = e^{\lambda T}\upsilon_{k-1}(\lambda), \quad \xi_k(\lambda) = e^{\lambda T}\xi_{k-1}(\lambda), \quad \psi_k(\lambda) = e^{\lambda T}\psi_{k-1}(\lambda). \tag{6.111}$$

For the matrices $\upsilon_0(\lambda)$, $\xi_0(\lambda)$, $\psi_0(\lambda)$, we can gain closed expressions. For that, we use relations (6.111). From the first equation in (6.109), we find

$$\begin{aligned} e^{k\lambda T}(I_\chi - e^{-\lambda T}e^{AT})\upsilon_0(\lambda) &= \sum_{i=0}^{P}\left[\Gamma_{1i}(T)e^{(k-m_2-i-1)\lambda T} + \Gamma_{2i}(T)e^{(k-m_2-i-2)\lambda T}\right]\psi_0(\lambda) \\ &\quad + e^{-\lambda\tau_1}e^{k\lambda T}(I_\chi - e^{-\lambda T}e^{AT})(\lambda I_\chi - A)^{-1}B_1. \end{aligned}$$

After cancelation, this expression can be represented in the form

$$\begin{aligned} \upsilon_0(\lambda) &= e^{-(m_2+1)\lambda T}(I_\chi - e^{-\lambda T}e^{AT})^{-1}\sum_{i=0}^{P}\left(\Gamma_{1i}(T) + e^{-\lambda T}\Gamma_{2i}(T)\right)e^{-i\lambda T}\psi_0(\lambda) \\ &\quad + e^{-\lambda\tau_1}(\lambda I_\chi - A)^{-1}B_1. \end{aligned} \tag{6.112}$$

For the next transforms, we prove the auxiliary formula

$$e^{-(m_2+1)\lambda T}(I_\chi - e^{-\lambda T}e^{AT})^{-1}\sum_{i=0}^{P}[\Gamma_{1i}(T) + e^{-\lambda T}\Gamma_{2i}(T)]e^{-i\lambda T} = \breve{D}_{G\mu}(T,\lambda,-\tau_2), \tag{6.113}$$

where the matrix $G(p)$ is determined in (6.4). For this reason, notice that

$$\breve{D}_{G\mu}(T,\lambda,-\tau_2)=\sum_{i=0}^{P}\breve{D}_{G\mu_i}(T,\lambda,-\tau_2)\mathrm{e}^{-i\lambda T}. \tag{6.114}$$

Moreover, using (6.93), we obtain

$$\begin{aligned}\breve{D}_{G\mu_i}(T,\lambda,-\tau_2)&=\breve{D}_{G\mu_i}(T,\lambda,-m_2T-\theta_2)=\breve{\mathscr{D}}_{G\mu_i}(T,\lambda,T-\theta_2)\mathrm{e}^{-(m_2+1)\lambda T}\\&=\breve{\mathscr{D}}_{G\mu_i}(T,\lambda,\gamma_2)\mathrm{e}^{-(m_2+1)\lambda T}.\end{aligned} \tag{6.115}$$

But (5.144), (5.145) for $t=\gamma_2$ and (6.97) yield

$$\breve{\mathscr{D}}_{G\mu_i}(T,\lambda,\gamma_2)=(I_\chi-\mathrm{e}^{-\lambda T}\mathrm{e}^{AT})^{-1}[\Gamma_{1i}(T)+\mathrm{e}^{-\lambda T}\Gamma_{2i}(T)]. \tag{6.116}$$

By inserting (6.115) and (6.116) into (6.114), formula (6.113) is achieved. With the help of (6.113) and (6.4), formula (6.112) takes the form

$$v_0(\lambda)=\tilde{D}_{G\mu}(T,\lambda,-\tau_2)\psi_0(\lambda)+F(\lambda)\mathrm{e}^{-\lambda\tau_1}. \tag{6.117}$$

Proceeding further, notice that under condition (6.111), from (6.9), we obtain

$$\psi_0(\lambda)=\breve{W}_d(\lambda)\xi_0(\lambda). \tag{6.118}$$

Hence, formula (6.117) takes the form

$$v_0(\lambda)=\breve{D}_{G\mu}(T,\lambda,-\tau_2)\breve{W}_d(\lambda)\xi_0(\lambda)+F(\lambda)\mathrm{e}^{-\lambda\tau_1}. \tag{6.119}$$

Multiplying the last equation from left by the matrix $C_2$, we find

$$C_2v_0(\lambda)=C_2\breve{D}_{G\mu}(T,\lambda,-\tau_2)\breve{W}_d(\lambda)\xi_0(\lambda)+C_2F(\lambda)\mathrm{e}^{-\lambda\tau_1}. \tag{6.120}$$

Further realize that due to (6.6) and (6.8)

$$C_2v_0(\lambda)=y(\lambda,0)=\xi_0(\lambda), \tag{6.121}$$

and from (6.4) and (6.5), it follows that

$$C_2F(s)\mathrm{e}^{-s\tau_1}=M(s)\mathrm{e}^{-s\tau_1}. \tag{6.122}$$

Moreover,

$$\begin{aligned}C_2\breve{D}_{G\mu}(T,\lambda,-\tau_2)&=\frac{1}{T}\sum_{k=-\infty}^{\infty}C_2G(\lambda+k\mathrm{j}\omega)\mu(\lambda+k\mathrm{j}\omega)\mathrm{e}^{-(\lambda+k\mathrm{j}\omega)\tau_2}\\&=\breve{D}_{N\mu}(T,\lambda,-\tau_2).\end{aligned} \tag{6.123}$$

Therefore, with the help of (6.121)–(6.123), relation (6.120) can be written as

$$\xi_0(\lambda) = \breve{D}_{N\mu}(T, \lambda, -\tau_2)\breve{W}_d(\lambda)\xi_0(\lambda) + M(\lambda)\mathrm{e}^{-\lambda\tau_1}. \tag{6.124}$$

Thus

$$\xi_0(\lambda) = \left[I_n - \breve{D}_{N\mu}(T, \lambda, -\tau_2)\breve{W}_d(\lambda)\right]^{-1} M(\lambda)\mathrm{e}^{-\lambda\tau_1}. \tag{6.125}$$

Inserting the last expression into (6.118) and applying notation (6.62), we obtain

$$\begin{aligned}\psi_0(\lambda) &= \breve{W}_d(\lambda)\left[I_n - \breve{D}_{N\mu}(T, \lambda, -\tau_2)\breve{W}_d(\lambda)\right]^{-1} M(\lambda)\mathrm{e}^{-\lambda\tau_1} \\ &= \breve{R}_N(\lambda)M(\lambda)\mathrm{e}^{-\lambda\tau_1}.\end{aligned} \tag{6.126}$$

Moreover, inserting (6.125) into (6.119), we arrive at the formula

$$\upsilon_0(\lambda) = \breve{D}_{G\mu}(T, \lambda, t - \tau_2)\breve{R}_N(\lambda)M(\lambda)\mathrm{e}^{-\lambda\tau_1} + F(\lambda)\mathrm{e}^{-\lambda\tau_1}. \tag{6.127}$$

Now, we will show that the so constructed matrices $\upsilon_0(\lambda)$, $\xi_0(\lambda)$, $\psi_0(\lambda)$ are meromorphic functions of the argument $\lambda$, and they allow representations of the form

$$\upsilon_0(\lambda) = \frac{A_\upsilon(\lambda)}{\det \breve{Q}(\lambda, \breve{\alpha}, \breve{\beta})}, \quad \xi_0(\lambda) = \frac{A_\xi(\lambda)}{\det \breve{Q}(\lambda, \breve{\alpha}, \breve{\beta})}, \quad \psi_0(\lambda) = \frac{A_\psi(\lambda)}{\det \breve{Q}(\lambda, \breve{\alpha}, \breve{\beta})}, \tag{6.128}$$

where $\breve{Q}(\lambda, \breve{\alpha}, \breve{\beta})$ is the block matrix (6.95). Indeed, by construction, the matrices $\upsilon_0(\lambda)$, $\xi_0(\lambda)$, $\psi_0(\lambda)$ are solutions of a linear system of equations, which is obtained from (6.109) by using relation (6.111), and it has the form

$$\begin{aligned}(I_\chi - \mathrm{e}^{-\lambda T}\mathrm{e}^{AT})\upsilon_0(\lambda) - \mathrm{e}^{-(m_2+1)\lambda T}\sum_{i=0}^{P}[\Gamma_{1i}(T) + \mathrm{e}^{-\lambda T}\Gamma_{2i}(T)]\mathrm{e}^{-i\lambda T}\psi_0(\lambda) & \\ = \mathrm{e}^{-\lambda\tau_1}(\lambda I_\chi - A)^{-1}(I_\chi - \mathrm{e}^{-\lambda T}\mathrm{e}^{AT})B_1, & \\ -C_2\upsilon_0(\lambda) + \xi_0(\lambda) \quad = 0_{n\ell}, & \\ -\breve{\beta}(\lambda)\xi_0(\lambda) + \breve{\alpha}(\lambda)\psi_0(\lambda) \quad = 0_{m\ell}. &\end{aligned} \tag{6.129}$$

Introducing the corresponding block matrices, this system of equations can be written in the form

$$\breve{Q}(\lambda, \breve{\alpha}, \breve{\beta})\begin{bmatrix}\upsilon_0(\lambda) \\ \xi_0(\lambda) \\ \psi_0(\lambda)\end{bmatrix} = P(\lambda), \tag{6.130}$$

where

$$P(\lambda) = \begin{bmatrix}\mathrm{e}^{-\lambda\tau_1}(\lambda I_\chi - A)^{-1}(I_\chi - \mathrm{e}^{-\lambda T}\mathrm{e}^{AT})B_1 \\ 0_{n\ell} \\ 0_{m\ell}\end{bmatrix}. \tag{6.131}$$

From the equality

$$(\lambda I_\chi - A)^{-1}(I_\chi - e^{-\lambda T}e^{AT}) = e^{-\lambda T}e^{AT}\int_0^T e^{(\lambda I_\chi - A)\nu}\, d\nu$$

it turns out that the matrix $P(\lambda)$ does not possess poles, i.e., it is an integral function of the argument $\lambda$. From (6.130), we obtain

$$\begin{bmatrix} \upsilon_0(\lambda) \\ \xi_0(\lambda) \\ \psi_0(\lambda) \end{bmatrix} = \breve{Q}^{-1}(\lambda, \breve{\alpha}, \breve{\beta})P(\lambda). \tag{6.132}$$

Since the matrix $P(\lambda)$ does not posses poles, so from (6.132) it directly emerges that the matrices $\upsilon_0(\lambda)$, $\xi_0(\lambda)$, $\psi_0(\lambda)$ are meromorphic functions of the argument $\lambda$, for which relation (6.112) holds.

On the basis of the prepared relations, we are ready to prove formula (6.98). At first notice that, under conditions (6.58) and (6.111), from (6.101), we derive

$$\begin{aligned} W_{\upsilon x}(\lambda, t) &= \upsilon(\lambda, t)e^{-\lambda t} \\ &= \begin{cases} e^{-\lambda t}\Big[e^{A(t-kT)}e^{\lambda kT}\upsilon_0 \\ \quad + \sum_{i=0}^{P}\int_{kT}^{t} e^{A(t-\nu)}B_2 h_i(\nu - kT + \gamma_2)\, d\nu e^{(k-m_2-i-1)\lambda T}\psi_0(\lambda) \\ \quad + e^{-\lambda\tau_1}\int_{kT}^{t} e^{A(t-\nu)}e^{\lambda\nu}\, d\nu B_1\Big], \qquad kT \le t \le kT + \theta_2, \\ e^{-\lambda t}\Big[e^{A(t-kT)}e^{k\lambda T}\upsilon_0(\lambda) \\ \quad + \sum_{i=0}^{P}\int_{kT}^{kT+\theta_2} e^{A(t-\nu)}B_2 h_i(\nu - kT + \gamma_2)\, d\nu e^{(k-m_2-i-1)\lambda T}\psi_0(\lambda) \\ \quad + \sum_{i=0}^{P}\int_{kT+\theta_2}^{t} e^{A(t-\nu)}B_2 h_i(\nu - kT - \theta_2)\, d\nu e^{(k-m_2-i)\lambda T}\psi_0(\lambda) \\ \quad + e^{-\lambda\tau_1}\int_{kT}^{t} e^{A(t-\nu)}e^{\lambda\nu}\, d\nu B_1\Big], \qquad kT + \theta_2 \le t \le (k+1)T. \end{cases} \end{aligned} \tag{6.133}$$

Analog as above, we state that the matrix

$$\int_{kT}^{t} e^{(\lambda I_\chi - A)\nu}\, d\nu = (\lambda I_\chi - A)^{-1}(e^{\lambda t}e^{-At} - e^{k\lambda T}e^{-kAT}) \tag{6.134}$$

is an integral function of the argument $\lambda$. Therefore, from (6.128) and (6.133), the validity of representation (6.98) for the PTM $W_{\upsilon x}(\lambda, t)$ emerges.

Due to $y(t) = C_2\upsilon(t)$, the same is true for the PTM $W_{yx}(\lambda, t)$. Moreover, from (6.111) and (6.10)

$$W_{ux}(\lambda, t) = \left[e^{-\lambda(t-kT)}\sum_{i=0}^{P} h_i(t - kT)e^{-i\lambda T}\right]\psi_0(\lambda), \quad kT < t < (k+1)T. \tag{6.135}$$

This relation proves the statement of Theorem 6.3 regarding the PTM $W_{ux}(\lambda, t)$. It only remains to show the validity of the representation (6.98) for the PTM $W_{zx}(\lambda, t)$. For this purpose, we consider the PTM $W_{gf}(\lambda, t)$ of the series connection shown in

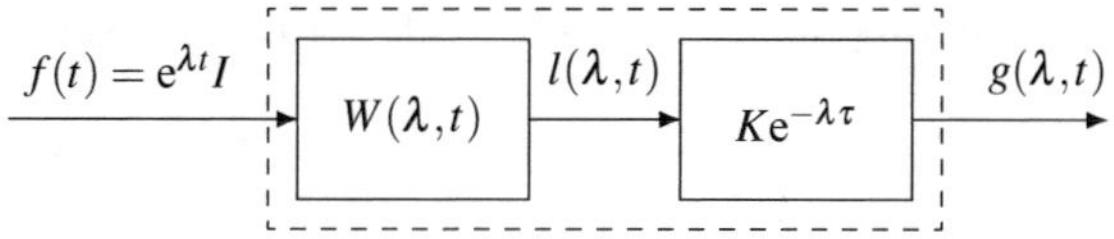


**Fig. 6.7** Series connection of PTM and delay

Fig. 6.7, where $W(\lambda, t) = W(\lambda, t+T)$ is a known PTM, $K$ is a constant matrix of appropriate size, and $\tau$ is a real constant. If we assume $f(t) = \mathrm{e}^{\lambda t}I$, then by definition the PTM in the stationary EP regime becomes

$$l(\lambda, t) = \mathrm{e}^{\lambda t}W(\lambda, t). \tag{6.136}$$

Assume the Fourier series

$$W(\lambda, t) = \sum_{k=-\infty}^{\infty} W_k(\lambda)\mathrm{e}^{kj\omega t}. \tag{6.137}$$

Then

$$l(\lambda, t) = \sum_{k=-\infty}^{\infty} W_k(\lambda)\mathrm{e}^{(\lambda+kj\omega)t}. \tag{6.138}$$

From Fig. 6.7, we read

$$g(\lambda, t) = K\sum_{k=-\infty}^{\infty} W_k(\lambda)\mathrm{e}^{(\lambda+kj\omega)(t-\tau)}. \tag{6.139}$$

Multiplying the last expression by $\mathrm{e}^{-\lambda t}$, we arrive at the PTM $W_{gf}(\lambda, t)$ of the system in Fig. 6.7

$$W_{gf}(\lambda, t) = K\mathrm{e}^{-\lambda\tau}\sum_{k=-\infty}^{\infty} W_k(\lambda)\mathrm{e}^{kj\omega(t-\tau)}. \tag{6.140}$$

With the help of (6.137), this relation can be written as

$$W_{gf}(\lambda, t) = \mathrm{e}^{-\lambda\tau}KW(\lambda, t-\tau). \tag{6.141}$$

Further notice that formula (6.11) can be configured to the structure in Fig. 6.8.

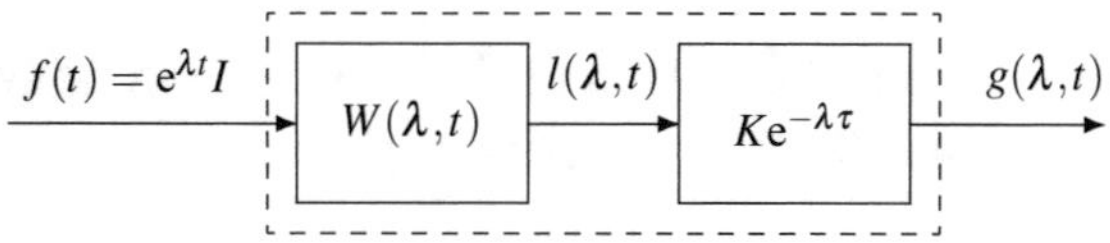


**Fig. 6.8** Superposition of delayed outputs

Applying formula (6.141) to the system in Fig. 6.8, we can write

$$W_{zx}(\lambda,t)=C_1\mathrm{e}^{-\lambda\tau_3}W_{vx}(\lambda,t-\tau_3)+D\mathrm{e}^{-\lambda(\tau_2+\tau_3)}W_{ux}(\lambda,t-\tau_2-\tau_3). \quad (6.142)$$

Since representation (6.98) is valid for both PTM $W_{vx}(\lambda,t)$ and $W_{ux}(\lambda,t)$ for all $t$, it follows from (6.142) that the analog property holds for the PTM $W_{zx}(\lambda,t)$.

For computational reasons, we show that formula (6.142) is equivalent to (6.67), (6.68). Notice that due to

$$L(p)=C_1G(p)+D, \quad (6.143)$$

formula (6.67) can be written in the form

$$\begin{aligned}W_{zx}(\lambda,t)&=\mathrm{e}^{-\lambda(\tau_2+\tau_3)}C_1\varphi_{G\mu}(T,\lambda,t-\tau_2-\tau_3)\breve{R}_N(\lambda)M(\lambda)\mathrm{e}^{-\lambda\tau_1}\\&\quad+C_1F(\lambda)\mathrm{e}^{-\lambda(\tau_1+\tau_3)}+D\mathrm{e}^{-\lambda(\tau_2+\tau_3)}\varphi_\mu(T,\lambda,t-\tau_2-\tau_3)\breve{R}_N(\lambda)M(\lambda)\mathrm{e}^{-\lambda\tau_1}.\end{aligned} \quad (6.144)$$

On the other side, from (6.69), (6.70), we obtain

$$W_{vx}(\lambda,t)=\mathrm{e}^{-\lambda\tau_2}\varphi_{G\mu}(T,\lambda,t-\tau_2)\breve{R}_N(\lambda)M(\lambda)\mathrm{e}^{-\lambda\tau_1}+F(\lambda)\mathrm{e}^{-\lambda\tau_1}. \quad (6.145)$$

Hence

$$W_{vx}(\lambda,t-\tau_3)=\mathrm{e}^{-\lambda\tau_2}\varphi_{G\mu}(T,\lambda,t-\tau_2-\tau_3)\breve{R}_N(\lambda)M(\lambda)\mathrm{e}^{-\lambda\tau_1}+F(\lambda)\mathrm{e}^{-\lambda\tau_1} \quad (6.146)$$

and

$$\begin{aligned}\mathrm{e}^{-\lambda\tau_3}C_1W_{vx}(\lambda,t-\tau_3)&=\mathrm{e}^{-\lambda(\tau_2+\tau_3)}C_1\varphi_{G\mu}(T,\lambda,t-\tau_2-\tau_3)\breve{R}_N(\lambda)M(\lambda)\mathrm{e}^{-\lambda\tau_1}\\&\quad+C_1F(\lambda)\mathrm{e}^{-\lambda(\tau_1+\tau_3)}.\end{aligned} \quad (6.147)$$

Moreover, using (6.64), we find

$$\begin{aligned}&\mathrm{e}^{-\lambda(\tau_2+\tau_3)}DW_{ux}(\lambda,t-\tau_2-\tau_3)\\&\qquad=\mathrm{e}^{-\lambda(\tau_2+\tau_3)}D\varphi_\mu(T,\lambda,t-\tau_2-\tau_3)\breve{R}_N(\lambda)M(\lambda)\mathrm{e}^{-\lambda\tau_1}.\end{aligned} \quad (6.148)$$

Substituting (6.147) and (6.148) in (6.142), formula (6.144) is achieved.

## 6.5 Modal Control and Stabilization of the Standard Model for SD System with Delay

(1) In the following considerations, the poles of the standard model for SD systems with delay $\mathscr{S}_\tau$ are called roots of the characteristic equation of the closed loop shown

in Fig. 6.2. The equations of this configuration in consensus with (6.6)–(6.10) take the form

$$
\begin{aligned}
\frac{\mathrm{d}v(t)}{\mathrm{d}t} &= Av(t) + B_1 x(t-\tau_1) + B_2 u(t-\tau_2), \\
y(t) &= C_2 v(t), \\
\xi_k &= y(kT), \quad (k = 0, \pm 1, \ldots), \\
\alpha(\zeta)\psi_k &= \beta(\zeta)\xi_k, \\
u(t) &= \sum_{i=0}^{P} h_i(t-kT)\psi_{k-i}, \quad kT < t < (k+1)T.
\end{aligned}
\tag{6.149}
$$

These equations are a special case of the general Eq. (4.40). Therefore, the solution of the modal control and the stabilization problems for the system (6.149) can be achieved with the help of the general formulae in Chap. 4. However, the concrete form of Eq. (6.149) allows to derive additional relations, which are convenient for applications. Therefore, the modal control and stabilization problems for the system (6.149) are considered separately.

(2) In the absence of external excitation, the equations of the discrete backward model of the closed loop, derived from (6.109), take the form

$$
\begin{aligned}
v_k &= \mathrm{e}^{AT} v_{k-1} + \sum_{i=0}^{P} \Gamma_{1i}(T)\psi_{k-m_2-i-1} + \sum_{i=0}^{P} \Gamma_{2i}(T)\psi_{k-m_2-i-2}, \\
\xi_k &= C_2 v_k, \\
\alpha_0\psi_k &+ \alpha_1\psi_{k-1} + \cdots + \alpha_\rho\psi_{k-\rho} = \beta_0\xi_k + \beta_1\xi_{k-1} + \cdots + \beta_\rho\xi_{k-\rho},
\end{aligned}
\tag{6.150}
$$

which, using the backward shift operator $\zeta$, can be written in polynomial form

$$
\begin{aligned}
a(\zeta)v_k - \zeta^{m_2+1} b(\zeta)\psi_k &= 0, \\
-C_2 v_k + \xi_k &= 0, \\
-\beta(\zeta)\xi_k + \alpha(\zeta)\psi_k &= 0,
\end{aligned}
\tag{6.151}
$$

where

$$
a(\zeta) = I_\chi - \zeta \mathrm{e}^{AT}, \quad b(\zeta) = \sum_{i=0}^{P} [\Gamma_{1i}(T) + \zeta \Gamma_{2i}(T)]\zeta^i. \tag{6.152}
$$

The characteristic matrix for the system (6.151) emerges as

$$
Q(\zeta, \alpha, \beta) = \begin{bmatrix} I_\chi - \zeta \mathrm{e}^{AT} & 0_{\chi n} & -\zeta^{m_2+1} b(\zeta) \\ -C_2 & I_n & 0_{nq} \\ 0_{q\chi} & -\beta(\zeta) & \alpha(\zeta) \end{bmatrix} \tag{6.153}
$$

and the modal control problem for the standard model $\mathcal{S}_\tau$ comes down to the solution of the DPE

$$\det Q(\zeta,\alpha,\beta) = \det \begin{bmatrix} I_\chi - \zeta e^{AT} & 0_{\chi n} & -\zeta^{m_2+1} b(\zeta) \\ -C_2 & I_n & 0_{nq} \\ 0_{q\chi} & -\beta(\zeta) & \alpha(\zeta) \end{bmatrix} \sim f(\zeta), \qquad (6.154)$$

where $f(\zeta)$ is a given polynomial.

(3) Introduce the constant matrix $m_\ell(A)$, defined by relations (5.191) for $B = B_2$

$$m_\ell(A) = \sum_{i=0}^{P} e^{-iAT} \int_0^T e^{-A\nu} B_2 h_i(\nu)\, d\nu. \qquad (6.155)$$

**Theorem 6.5** *Let the period T be non-pathological, and the pairs* $(e^{-AT}, m_\ell(A))$ *and* $[A, C_2]$ *be minimal. Then the following claims hold:*

*(i) The system* $\mathcal{S}_\tau$ *is modal controllable.*
*(ii) The system* $\mathcal{S}_\tau$ *is modal controllable, if and only if the corresponding standard model without delay possesses this property.*
*(iii) For the modal control of the standard model* $\mathcal{S}_\tau$*, it is necessary that the pairs* $(A, B_2)$ *and* $[A, C_2]$ *are minimal.*

*Proof* (i) From the results of Sect. 3.12, we know that for the modal controllability of system $\mathcal{S}_\tau$, it is necessary and sufficient that the polynomial pairs $(I_\chi - \zeta e^{AT}, \zeta^{m_2+1} b(\zeta))$, and $[I_\chi - \zeta e^{AT}, C_2]$ are irreducible. Thanks to the results in Sects. 5.4 and 5.9, this takes place, if and only if the pairs of constant matrices $(e^{-AT}, m_\ell(A))$ and $[e^{-AT}, C_2]$ are minimal.

(ii) This claim follows from the fact that the matrices $e^{-AT}$, $m_\ell(A)$, and $C_2$ do not depend on the delay.

(iii) For the proof of this claim, we use the ideas applied in the proof of Theorem 5.12. We will show that, when at least one of the pairs $(A, B_2)$ or $[A, C_2]$ is not minimal, then the standard model (6.12) is not modal controllable. Let, for instance, the pair $(A, B_2)$ be not minimal. Then, there exists an eigenvalue $\lambda_0$ of the matrix $A$ and a constant vector $x$ of size $1 \times \chi$ such that

$$xA = \lambda_0 x, \quad xB_2 = 0_{1m}. \qquad (6.156)$$

Hereby, as was shown earlier, we obtain

$$xe^{AT} = e^{\lambda_0 T} x, \quad xB_2 = 0_{1m}, \qquad (6.157)$$

and hence

$$x\left[ I_\chi - e^{-\lambda_0 T} e^{AT} \;\; m_\ell(A) \right] = 0_{1,\chi+m}. \qquad (6.158)$$

From the last equation, we see that the pair $(I_\chi - \zeta e^{AT}, \zeta^{m_2+1}b(\zeta))$ is not irreducible. Analogously, if the pair $[A, C_2]$ is not minimal, then the polynomial pair $[I_\chi - \zeta e^{AT}, C_2]$ is not irreducible. ∎

(4) With the thoughts used for the proof of Theorem 6.5, we also can solve the stabilization problem for the system $\mathscr{S}_\tau$. From the structure of the system $\mathscr{S}_\tau$, provided in Fig. 6.2, we recognize that this system is stable, if and only if the contained closed loop of the system is stable. Therefore, the stabilization problem for the standard model $\mathscr{S}_\tau$ is equivalent to the stabilization problem of the discrete closed loop, described by Eq. (6.150).

**Theorem 6.6** *The following statements are true:*

*(i) For the stabilizability of the standard model $\mathscr{S}_\tau$, it is necessary and sufficient that all latent numbers $\zeta_i$ of the matrices*

$$J_1(\zeta) = |I_\chi - \zeta e^{AT} \quad m_\ell(A)|, \quad J_2(\zeta) = \begin{bmatrix} I_\chi - \zeta e^{AT} \\ C_2 \end{bmatrix}, \tag{6.159}$$

*satisfy the condition*

$$|\zeta_i| > 1. \tag{6.160}$$

*(ii) The system $\mathscr{S}_\tau$ is stabilizable, if and only if the corresponding system without delay is stabilizable.*

*(iii) For the stabilizability of the system $\mathscr{S}_\tau$, it is necessary that the continuous LTI process (6.6) is stabilizable for $\tau_2 = 0$, i.e., the latent numbers $s_i$ of the polynomial matrices*

$$\begin{bmatrix} pI_\chi - A & B_2 \end{bmatrix}, \quad \begin{bmatrix} pI_\chi - A \\ C_2 \end{bmatrix} \tag{6.161}$$

*satisfy*

$$\operatorname{Re} s_i < 0. \tag{6.162}$$

*Proof* (i) In consistence with Theorem 3.39, for the stabilizability of system (6.150), it is necessary and sufficient that the latent numbers of the matrices

$$J_3(\zeta) \triangleq \begin{bmatrix} I_\chi - \zeta e^{AT} & \zeta^{m_2+1}b(\zeta) \end{bmatrix}, \quad J_4(\zeta) \triangleq \begin{bmatrix} I_\chi - \zeta e^{AT} \\ C_2 \end{bmatrix} \tag{6.163}$$

fulfill condition (6.160). Therefore, the equivalence condition

$$\begin{bmatrix} I_\chi - \zeta e^{AT} & \zeta^{m+1}b(\zeta) \end{bmatrix} \sim \begin{bmatrix} I_\chi - \zeta e^{AT} & m_\ell(A) \end{bmatrix} \tag{6.164}$$

proves the first claim.

(ii) This claim follows from the circumstance that matrices (6.159) do not depend on the delay.

(iii) Assume, for instance, that the first matrix in (6.161) has a latent number $\lambda_0$ with $\mathrm{Re}\lambda_0 \geq 0$. Then, there exists a $1 \times \chi$ vector $x$, such that

$$xA = \lambda_0 x, \quad xB_2 = 0_{1,m}, \tag{6.165}$$

and in the same way as above, we conclude

$$x\left[I_\chi - \mathrm{e}^{-\lambda_0 T}\mathrm{e}^{AT}\; m_\ell(A)\right] = 0_{1,\chi+q}, \tag{6.166}$$

i.e., the matrix $J_1(\zeta)$ in (6.159) has a latent number $\zeta_0 = \mathrm{e}^{-\lambda_0 T}$, where $|\zeta_0| \leq 1$. This means that matrix (6.153) is not stabilizable. The second condition in (iii) can be proven analogously. ■

(5) In the following considerations, we will assume that the conditions of Theorem 6.5 are fulfilled. In this case, the closed loop of system $\mathscr{S}_\tau$ is modal controllable, and thus stabilizable. Hereby, the stabilization problem comes down to the stabilization problem for the strictly causal irreducible homogeneous PMD

$$\Pi_N(\zeta) \stackrel{\triangle}{=} (I_\chi - \zeta\mathrm{e}^{AT}, \zeta^{m_2+1}b(\zeta), C_2). \tag{6.167}$$

As it was shown earlier, the transfer matrix of this PMD is given by

$$W_N(\zeta) \stackrel{\triangle}{=} \zeta^{m_2+1}C_2(I_\chi - \zeta\mathrm{e}^{AT})^{-1}b(\zeta), \tag{6.168}$$

which, with the help of (6.123), leads to

$$W_N(\zeta) = D_{N\mu}(T, \zeta, -\tau_2), \tag{6.169}$$

where we used that, for the standard model, $\tau_N = \tau_2$. On basis of the relations that already hold, we can formulate the following.

**Theorem 6.7** *Let the period $T$ be non-pathological, and the pairs $(A, B_2)$, $[A, C_2]$ be minimal. Then, the following claims hold:*

*(i) Assume the ILMFD of form (3.278)*

$$C_2(I_\chi - \zeta\mathrm{e}^{AT})^{-1} = a_l^{-1}(\zeta)b_l(\zeta), \tag{6.170}$$

*then*

$$b_l(\zeta, -\tau_2) \stackrel{\triangle}{=} a_l(\zeta)W_N(\zeta) = \zeta^{m_2+1}b_l(\zeta)\sum_{i=0}^{P}[\Gamma_{1i}(T) + \zeta\Gamma_{2i}(T)]\zeta^i \tag{6.171}$$

*is a polynomial matrix, the representation*

$$W_N(\zeta) = a_l(\zeta)^{-1}b_l(\zeta, -\tau_2) \tag{6.172}$$

*is an ILMFD, and the pair* $(a_l(\zeta), b_l(\zeta, -\tau_2))$ *is irreducible.*

*(ii) Let* $(\alpha_{l0}(\zeta), \beta_{l0}(\zeta))$ *be a basic controller for the irreducible pair* $(a_l(\zeta), b_l(\zeta, -\tau_2))$*. Then, owing to Theorem 3.35, the set of stabilizing controllers* $(\alpha_l(\zeta), \beta_l(\zeta))$ *for the standard model* $\mathscr{S}_\tau$ *is determined by the parametrization*

$$\begin{aligned} \alpha_l(\zeta) &= D_l(\zeta)\alpha_{l0}(\zeta) - M_l(\zeta)b_l(\zeta, -\tau_2), \\ \beta_l(\zeta) &= D_l(\zeta)\beta_{l0}(\zeta) - M_l(\zeta)a_l(\zeta), \end{aligned} \tag{6.173}$$

*where the polynomial matrix* $D_\ell(\zeta)$ *is stable, and the polynomial matrix* $M_\ell(\zeta)$ *is arbitrary. All controllers* (6.175) *are causal.*

*Proof* (i) This claim is a corollary of Theorem 5.15.

(ii) This claim follows from Theorem 3.20 when considering that the process $(a_l(\zeta), b_l(\zeta, -\tau_2))$ is strictly causal. ∎

(6) We derive, in addition to Chap. 5, some relations that are necessary for the further investigations. Introduce, in correspondence to (3.28), the matrix

$$Q_l(\zeta, \alpha_l, \beta_l) \triangleq \begin{bmatrix} a_l(\zeta) & -b_l(\zeta, -\tau_2) \\ -\beta_l(\zeta) & \alpha_l(\zeta) \end{bmatrix}, \tag{6.174}$$

where $\alpha_l(\zeta), \beta_l(\zeta)$ are matrices (6.173). From (6.173), (6.174), we obtain

$$Q_l(\zeta, \alpha_l, \beta_l) = \begin{bmatrix} I_n & 0_{nq} \\ M_l(\zeta) & D_l(\zeta) \end{bmatrix} Q_l(\zeta, \alpha_{l0}, \beta_{l0}), \tag{6.175}$$

where

$$Q_l(\zeta, \alpha_{l0}, \beta_{l0}) = \begin{matrix} \\ n \\ q \end{matrix} \begin{matrix} \quad n \qquad\qquad q \\ \begin{bmatrix} a_l(\zeta) & -b_l(\zeta, -\tau_2) \\ -\beta_{l0}(\zeta) & \alpha_{l0}(\zeta) \end{bmatrix} \end{matrix} \tag{6.176}$$

is an unimodular matrix. Let the matrix $Q_r(\zeta, \alpha_{r0}, \beta_{r0})$ be defined by the relation

$$Q_r(\zeta, \alpha_{r0}, \beta_{r0}) \triangleq Q_l^{-1}(\zeta, \alpha_{l0}, \beta_{l0}) \triangleq \begin{matrix} \\ n \\ q \end{matrix} \begin{matrix} \quad n \qquad\qquad q \\ \begin{bmatrix} \alpha_{r0}(\zeta) & b_r(\zeta, -\tau_2) \\ \beta_{r0}(\zeta) & a_r(\zeta) \end{bmatrix} \end{matrix}. \tag{6.177}$$

Since by construction

$$Q_l(\zeta, \alpha_{l0}, \beta_{l0})Q_r(\zeta, \alpha_{r0}, \beta_{r0}) = Q_r(\zeta, \alpha_{r0}, \beta_{r0})Q_l(\zeta, \alpha_{l0}, \beta_{l0}) = I_{n+q}, \tag{6.178}$$

we find

$$\begin{aligned} a_l(\zeta)b_r(\zeta, -\tau_2) &= b_l(\zeta, -\tau_2)a_r(\zeta), \\ \beta_{l0}(\zeta)\alpha_{r0}(\zeta) &= \alpha_{l0}(\zeta)\beta_{r0}(\zeta), \end{aligned} \tag{6.179}$$

or what is equivalent

$$\begin{aligned} D_{N\mu}(T,\zeta,-\tau_2) &= a_l^{-1}(\zeta)b_l(\zeta,-\tau_2) = b_r(\zeta,-\tau_2)a_r^{-1}(\zeta),\\ \alpha_{l0}^{-1}(\zeta)\beta_{l0}(\zeta) &= \beta_{r0}(\zeta)\alpha_{r0}^{-1}(\zeta), \end{aligned} \tag{6.180}$$

where the right sides are also IRMFD. We observe from (6.180) that the pair $(a_r(\zeta), b_r(\zeta,-\tau_2))$ defines a dual right process, and the pair $[\alpha_{r0}(\zeta), \beta_{r0}(\zeta)]$ a dual basic controller.

The inverse matrices of the left and right sides of Eq. (6.175) can be built with the help of (6.177)

$$Q_l^{-1}(\zeta,\alpha,\beta) = \begin{bmatrix} \alpha_{r0}(\zeta) & b_r(\zeta,-\tau_2) \\ \beta_r(\zeta) & a_r(\zeta) \end{bmatrix} \begin{bmatrix} I_n & 0_{nq} \\ M_l(\zeta) & D_l(\zeta) \end{bmatrix}^{-1}. \tag{6.181}$$

Since

$$\begin{bmatrix} I_n & 0_{nq} \\ M_l(\zeta) & D_l(\zeta) \end{bmatrix}^{-1} = \begin{bmatrix} I_n & 0_{nq} \\ -D_l^{-1}(\zeta)M_l(\zeta) & D_l^{-1}(\zeta) \end{bmatrix}, \tag{6.182}$$

so from (6.181), we obtain

$$Q_l^{-1}(\zeta,\alpha,\beta) = \begin{matrix} \\ n \\ q \end{matrix} \begin{matrix} \quad n \qquad\qquad\qquad\qquad q \\ \begin{bmatrix} \alpha_{r0}(\zeta) - b_r(\zeta,-\tau_2)\theta(\zeta) & q_{12}(\zeta) \\ \beta_{r0}(\zeta) - a_r(\zeta)\theta(\zeta) & q_{22}(\zeta) \end{bmatrix} \end{matrix}, \tag{6.183}$$

where, as already done in Chap. 3, we utilize the notation of the system function

$$\theta(\zeta) = D_l^{-1}(\zeta)M_l(\zeta). \tag{6.184}$$

On the other side, applying formulae (3.164) and (3.171) to matrix (6.175), we arrive at

$$Q_l^{-1}(\zeta,\alpha,\beta) = \begin{bmatrix} \upsilon_1(\zeta) & q_{12}(\zeta) \\ \upsilon_2(\zeta) & q_{22}(\zeta) \end{bmatrix}. \tag{6.185}$$

Herein

$$\begin{aligned} \upsilon_1(\zeta) &= \left[a_l(\zeta) - b_l(\zeta,-\tau_2)W_d(\zeta)\right]^{-1},\\ \upsilon_2(\zeta) &= W_d(\zeta)\left[a_l(\zeta) - b_l(\zeta,-\tau_2)W_d(\zeta)\right]^{-1}, \end{aligned} \tag{6.186}$$

where, as before,

$$W_d(\zeta) = \alpha^{-1}(\zeta)\beta(\zeta) \tag{6.187}$$

is the transfer matrix of the discrete controller.

Comparing Eqs. (6.183) and (6.185), due to the uniqueness of the inverse matrices, we find in addition to (6.184)

$$\begin{aligned} \upsilon_1(\zeta) &= \alpha_{r0}(\zeta) - b_r(\zeta, -\tau_2)\theta(\zeta), \\ \upsilon_2(\zeta) &= \beta_{r0}(\zeta) - a_r(\zeta)\theta(\zeta). \end{aligned} \tag{6.188}$$

These relations can be represented in the form

$$\begin{bmatrix} \upsilon_1(\zeta) \\ \upsilon_2(\zeta) \end{bmatrix} = Q_r(\zeta, \alpha_{r0}, \beta_{r0}) \begin{bmatrix} I_n \\ -\theta(\zeta) \end{bmatrix}, \tag{6.189}$$

from which, we obtain

$$\begin{bmatrix} I_n \\ -\theta(\zeta) \end{bmatrix} = Q_r^{-1}(\zeta, \alpha_{r0}, \beta_{r0}) \begin{bmatrix} \upsilon_1(\zeta) \\ \upsilon_2(\zeta) \end{bmatrix} = Q_l(\zeta, \alpha_{l0}, \beta_{l0}) \begin{bmatrix} \upsilon_1(\zeta) \\ \upsilon_2(\zeta) \end{bmatrix}. \tag{6.190}$$

From (6.190), we find a representation of the system function by the matrices $\upsilon_1(\zeta)$ and $\upsilon_2(\zeta)$

$$\theta(\zeta) = \beta_{l0}(\zeta)\upsilon_1(\zeta) - \alpha_{l0}(\zeta)\upsilon_2(\zeta). \tag{6.191}$$

With the above preparations, we are able to formulate an important result for further developments.

**Theorem 6.8** *For the stability of the standard model $\mathscr{S}_\tau$, it is necessary and sufficient that the system function $\theta(\zeta)$ is stable.*

*Proof* Necessity: Let the standard model $\mathscr{S}_\tau$ be stable. Then, due to Theorem 3.24, the matrices $\upsilon_1(\zeta)$ and $\upsilon_2(\zeta)$ are stable. Then from (6.191) we conclude that the matrix $\theta(\zeta)$ is stable too.

Sufficiency: Let the matrix $\theta(\zeta)$ be stable. Then from (6.188) we recognize that the matrices $\upsilon_1(\zeta)$ and $\upsilon_2(\zeta)$ are stable and thus the system $\mathscr{S}_\tau$ is stable thanks to Theorem 3.24. ■

## 6.6 Representing PTM $W_{zx}(\lambda, t)$ of the Standard Model of SD System with Delay by a System Function

(1) Denote in the expression for matrix $\breve{R}_N(\lambda)$ (6.62) the variable $e^{-\lambda T}$ by $\zeta$, we find the associated rational matrix

$$R_N(\zeta) = \breve{R}_N(\lambda)|_{e^{-\lambda T}=\zeta} = W_d(\zeta)\left[I_\chi - D_{N\mu}(T, \zeta, -\tau_2)W_d(\zeta)\right]^{-1}, \tag{6.192}$$

which with the help of (6.180) can be represented in the form

$$
\begin{aligned}
R_N(\zeta) &= W_d(\zeta)\left[I_\chi - a_l^{-1}(\zeta)b_l(\zeta,-\tau_2)W_d(\zeta)\right]^{-1}\\
&= W_d(\zeta)\left[a_l(\zeta) - b_l(\zeta,-\tau_2)W_d(\zeta)\right]^{-1}a_l(\zeta).
\end{aligned}
\tag{6.193}
$$

Using (6.186), we obtain

$$
R_N(\zeta) = \upsilon_2(\zeta)a_l(\zeta), \tag{6.194}
$$

which, according to (6.190), is equivalent to

$$
R_N(\zeta) = \beta_{r0}(\zeta)a_l(\zeta) - a_r(\zeta)\theta(\zeta)a_l(\zeta). \tag{6.195}
$$

Rename here $\zeta$ by $\mathrm{e}^{-\lambda T}$. We arrive at

$$
\breve{R}_N(\lambda) = \breve{\beta}_{r0}(\lambda)\breve{a}_l(\lambda) - \breve{a}_r(\lambda)\breve{\theta}(\lambda)\breve{a}_l(\lambda), \tag{6.196}
$$

where

$$
\begin{aligned}
\breve{\beta}_{r0}(\lambda) &= \beta_{r0}(\zeta)|_{\zeta=\mathrm{e}^{-\lambda T}}, \quad & \breve{a}_l(\lambda) &= a_l(\zeta)|_{\zeta=\mathrm{e}^{-\lambda T}},\\
\breve{a}_r(\lambda) &= a_r(\zeta)|_{\zeta=\mathrm{e}^{-\lambda T}}, \quad & \breve{\theta}(\lambda) &= \theta(\zeta)|_{\zeta=\mathrm{e}^{-\lambda T}}.
\end{aligned}
\tag{6.197}
$$

(2) With the help of the notation (6.17)–(6.19), the expression for the PTM $W_{zx}(\lambda, t)$, defined by formula (6.67), can be written as

$$
W_{zx}(\lambda, t) = \mathrm{e}^{-\lambda(\tau_2+\tau_3)}\varphi_{L\mu}(T,\lambda,t-\tau_2-\tau_3)\breve{R}_N(\lambda)M(\lambda)\mathrm{e}^{-\lambda\tau_1} + K(\lambda)\mathrm{e}^{-\lambda(\tau_1+\tau_3)}. \tag{6.198}
$$

Inserting here (6.196), we find

$$
W_{zx}(\lambda, t) = \psi(\lambda, t)\breve{\theta}(\lambda)\xi(\lambda) + \eta(\lambda, t), \tag{6.199}
$$

where we applied the notation

$$
\begin{aligned}
\psi(\lambda, t) &= -\mathrm{e}^{-\lambda(\tau_2+\tau_3)}\varphi_{L\mu}(T,\lambda,t-\tau_2-\tau_3)\breve{a}_r(\lambda) = -\varphi_{L_\tau\mu}(T,\lambda,t)\breve{a}_r(\lambda),\\
\xi(\lambda) &= \breve{a}_l(\lambda)M(\lambda)\mathrm{e}^{-\lambda\tau_1},\\
\eta(\lambda, t) &= \mathrm{e}^{-\lambda(\tau_2+\tau_3)}\varphi_{L\mu}(T,\lambda,t-\tau_2-\tau_3)\breve{\beta}_{r0}(\lambda)\breve{a}_l(\lambda)M(\lambda)\mathrm{e}^{-\lambda\tau_1}\\
&\quad + K(\lambda)\mathrm{e}^{-\lambda(\tau_1+\tau_3)}.
\end{aligned}
\tag{6.200}
$$

**Definition 6.5** Further on, formula (6.199) is called the representation of the PTM $W_{zx}(\lambda, t)$ by the system function, and matrices (6.200) are called the coefficients of this representation.

**Theorem 6.9** *Matrices* (6.200) *do not possess poles, and they are integral functions of the argument* $\lambda$.

*Proof* (a) Let us start with the proof of Theorem 6.9 for $\psi(\lambda, t)$. For this consider the matrix

$$\psi_1(\lambda, t) \stackrel{\triangle}{=} -e^{\lambda t}\psi(\lambda, t) = -\breve{D}_{L\mu}(T, \lambda, t - \tau_L)\breve{a}_r(\lambda), \tag{6.201}$$

where, as before, $\tau_L = \tau_2 + \tau_3$. For $mT < t - \tau_L < (m+1)T$, we obtain

$$\breve{D}_{L\mu}(T, \lambda, t - \tau_L) = \breve{\mathscr{D}}_{L\mu}(T, \lambda, t - \tau_L - mT)e^{m\lambda T}, \tag{6.202}$$

where, in agreement with (5.180), (5.181), and (6.15),

$$\breve{\mathscr{D}}_{L\mu}(T, \lambda, t) = C_1 e^{At}(I_\chi - e^{-\lambda T}e^{AT})^{-1}\breve{n}(\lambda, t) + D\breve{h}(\lambda, t), \tag{6.203}$$

and, as earlier,

$$\begin{aligned}\breve{n}(\lambda, t) &= \sum_{i=0}^{P} e^{-i\lambda T}\left[\int_0^t e^{-A\nu}B_2 h_i(\nu)d\nu + e^{-\lambda T}e^{AT}\int_t^T e^{-A\nu}B_2 h_i(\nu)d\nu\right],\\ \breve{h}(\lambda, t) &= \sum_{i=0}^{P} e^{-i\lambda T}h_i(t).\end{aligned} \tag{6.204}$$

Rename in (6.203), (6.204) $e^{-\lambda T}$ by $\zeta$. Then

$$\mathscr{D}_{L\mu}(T, \zeta, t) = C_1 e^{-At}(I_\chi - \zeta e^{AT})^{-1}n(\zeta, t) + Dh(\zeta, t), \tag{6.205}$$

where

$$\begin{aligned}n(\zeta, t) &= \sum_{i=0}^{P} \zeta^i\left[\int_0^t e^{-A\nu}B_2 h_i(\nu)d\nu + \zeta e^{AT}\int_t^T e^{-A\nu}B_2 h_i(\nu)d\nu,\right],\\ h(\zeta, t) &= \sum_{i=0}^{P} \zeta^i h_i(t)\end{aligned} \tag{6.206}$$

are matrices, which polynomially depend on $\zeta$.

Consider the result of the left division

$$n(\zeta, t) = (I_\chi - \zeta e^{AT})q(\zeta, t) + p(\zeta, t), \tag{6.207}$$

where $q(\zeta, t)$, $p(\zeta, t)$ are matrices which depend polynomially on $\zeta$. Inserting here $\zeta = e^{-AT}$, we obtain that the remainder $p(\zeta, t)$ does not depend on $t$. Indeed, we have

$$p(\zeta, t) = n(\zeta, t)|_{\zeta = e^{-AT}}, \tag{6.208}$$

which according to (6.206) yields

$$p(\mathrm{e}^{-AT}, t) = \sum_{i=0}^{P} \mathrm{e}^{-iAT} \int_0^T \mathrm{e}^{-A\nu} B_2 h_i(\nu)\,\mathrm{d}\nu = m_\ell(A), \tag{6.209}$$

where $m_\ell(A)$ is the matrix (6.155).

Considering (6.207)–(6.209), relation (6.205) can be written in the form

$$\mathscr{D}_{L\mu}(T, \zeta, t) = C_1 \mathrm{e}^{At}(I_\chi - \zeta \mathrm{e}^{AT})^{-1} m_\ell(A) + r(\zeta, t), \tag{6.210}$$

where $r(\zeta, t)$ is a matrix depending polynomially on $\zeta$ and the pair $(\mathrm{e}^{-AT}, m_\ell(A))$ was proposed to be controllable. Using the IRMFD

$$(I_\chi - \zeta \mathrm{e}^{AT})^{-1} m_\ell(A) = b_r(\zeta) a_r^{-1}(\zeta), \tag{6.211}$$

from (6.210), we find

$$\mathscr{D}_{L\mu}(T, \zeta, t) = \left[C_1 \mathrm{e}^{At} b_r(\zeta) + r(\zeta, t) a_r(\zeta)\right] a_r^{-1}(\zeta), \tag{6.212}$$

which allows the conclusion that the product

$$\mathscr{D}_{L\mu}(T, \zeta, t) a_r(\zeta) = \left[C_1 \mathrm{e}^{At} b_r(\zeta) + r(\zeta, t) a_r(\zeta)\right] \tag{6.213}$$

depends polynomially on $\zeta$.

Substituting in (6.213) $\zeta$ by $\mathrm{e}^{-\lambda T}$, we find that the matrix

$$\breve{\mathscr{D}}_{L\mu}(T, \lambda, t) \breve{a}_r(\lambda) = \left[C_1 \mathrm{e}^{At} \breve{b}_r(\lambda) + \breve{r}(\lambda, t) \breve{a}_r(\lambda)\right] \tag{6.214}$$

for all $t$ is an integral function of the argument $\lambda$.

Since in any interval $mT < t - \tau_L < (m+1)T$, we have (6.202), so with the help of (6.212), we realize that the product

$$\begin{aligned} \psi_1(\lambda, t) &= \breve{D}_{L\mu}(T, \lambda, t - \tau_L) \tilde{a}_r(\lambda) = \breve{\mathscr{D}}_{L\mu}(T, \lambda, t - \tau_L - mT) \breve{a}_r(\lambda) \mathrm{e}^{m\lambda T} \\ &= \left[C_1 \mathrm{e}^{A(t-\tau_L-mT)} \tilde{b}_r(\lambda) + \breve{r}(\lambda, t - \tau_L - mT) \breve{a}_r(\lambda)\right] \mathrm{e}^{m\lambda T} \end{aligned} \tag{6.215}$$

is an integral function of the argument $\lambda$ for all $t$. Obviously, this claim is also true for the matrix $\psi(\lambda, t)$.

(b) Now we will prove that the matrix $\xi(\lambda)$ does not possess poles. For this purpose, consider the series

$$\breve{D}_\xi(T, \lambda, t) = \frac{1}{T} \sum_{k=-\infty}^{\infty} \xi(\lambda + k\mathrm{j}\omega) \mathrm{e}^{(\lambda + k\mathrm{j}\omega)t}. \tag{6.216}$$

Applying (6.200), considering $\breve{a}_l(\lambda + kj\omega) = \breve{a}_l(\lambda)$, we find

$$\begin{aligned}\breve{D}_\xi(T, \lambda, t) &= \breve{a}_l(\lambda)\frac{1}{T}\sum_{k=-\infty}^{\infty} M(\lambda + kj\omega)e^{(\lambda+kj\omega)(t-\tau_1)} \\ &= \breve{a}_l(\lambda)\breve{D}_M(T, \lambda, t - \tau_1).\end{aligned} \tag{6.217}$$

Substituting herein $e^{-\lambda T}$ by $\zeta$, we obtain

$$D_\xi(T, \zeta, t) = a_l(\zeta)D_M(T, \zeta, t - \tau_1). \tag{6.218}$$

With the help of (5.101), for $mT < t - \tau_1 < (m + 1)T$, we immediately find

$$D_\xi(T, \zeta, t) = a_l(\zeta)\left[C_2(I_\chi - \zeta e^{AT})^{-1}e^{A(t-\tau_1-mT)})B_1\right]\zeta^{-m}. \tag{6.219}$$

Using ILMFD (6.170), from (6.219), we come to the expression

$$D_\xi(T, \zeta, t) = b_l(\zeta)e^{A(t-\tau_1-mT)}B_1\zeta^m, \quad mT < t - \tau_1 < (m + 1)T. \tag{6.220}$$

From (6.220), we immediately find out that the matrix

$$\breve{D}_\xi(T, \lambda, t) = \breve{D}_\xi(T, \zeta, t)|_{\zeta=e^{-\lambda T}} = \breve{b}_l(\lambda)e^{A(t-\tau_1-mT)}B_1e^{m\lambda T} \tag{6.221}$$

does not possess poles for any $m$. Therefore, due to

$$\xi(\lambda) = \int_0^T \breve{D}_\xi(T, \lambda, t)e^{-\lambda t}\,dt, \tag{6.222}$$

we conclude that the matrix $\xi(\lambda)$ is free of poles.

(c) It remains to show that the matrix $\eta(\lambda, t)$ in (6.200) also does not possess poles. To that, notice that, thanks to (6.199),

$$\eta(\lambda, t) = W_{zx}(\lambda, t)|_{\breve{\theta}(\lambda)=0_{mn}}. \tag{6.223}$$

For $\breve{\theta}(\lambda) = \theta(\zeta) = 0_{qm}$, we obtain $M_l(\zeta) = 0$, and from (6.173), we find the ILMFD

$$W_d(\zeta) = \alpha_l^{-1}(\zeta)\beta_l(\zeta) = \alpha_{l0}^{-1}(\zeta)\beta_{l0}(\zeta),$$

where $(\alpha_{l0}(\zeta), \beta_{l0}(\zeta))$ is any basic controller. Moreover, the controller $(\alpha_{l0}(\zeta), [0_{m\chi}\,\beta_{l0}(\zeta)])$ is basic for matrix (6.153). Therefore, in this case for matrix (6.95), we obtain

$$\det \breve{Q}(\lambda, \breve{\alpha}, \breve{\beta}) = \text{const.} \neq 0,$$

and due to Theorem 6.4, the PTM $W_{zx}(\lambda, t)$ for $\breve{\theta}(\lambda) = 0_{qn}$ does not possess poles. But the same, with the help of (6.223), is true for the matrix $\eta(\lambda, t)$. ■

## 6.7 Structural Standardizable SD Systems with Delay

(1) The approach chosen in Sect. 6.3 for constructing the PTM for the standard structure of SD systems with delay is based on the application of the stroboscopic property of the ADC. This approach is of universal nature, and in principle, it can be applied to SD systems of arbitrary structure, as it was shown in [1, 2] for SISO SD systems. Hereby, in many cases, it happens that the PTM of the investigated SD system for the chosen input and output takes the standard form, despite the fact that the structure of this system deviates from the standard one. We will show that by examples.

*Example 6.1* Consider the single-loop SD system shown in Fig. 6.9. Herein, the $W_i(p)$ $(i = 1, 2, 3)$ are rational matrices, but their properties will be described later. Moreover, the $\nu_i$, $(i = 1, 2, 3)$ are nonnegative constants, and, as earlier, the computer is described by its linear model given by the equations

$$\begin{aligned} \xi_k &= y(kT), \quad (k = 0, \pm 1, \ldots), \\ \alpha(\zeta)\psi_k &= \beta(\zeta)\xi_k, \\ u(t) &= \sum_{i=0}^{P} h_i(t - kT)\psi_{k-i}, \quad kT < t < (k+1)T. \end{aligned} \tag{6.224}$$

In order to give sense to the first equation in (6.224), we assume that vector $y(t)$ is continuous at the sampling instants $t_k = kT$. This condition is fulfilled when the matrix

$$N(p) \triangleq W_3(p)W_2(p)W_1(p) \tag{6.225}$$

is strictly proper. Below, we will always assume that this condition holds. Then, for the determination of the PTM $W_{z_3x}(\lambda, t)$ from the input $x(t)$ to the output $z_3(t)$, we

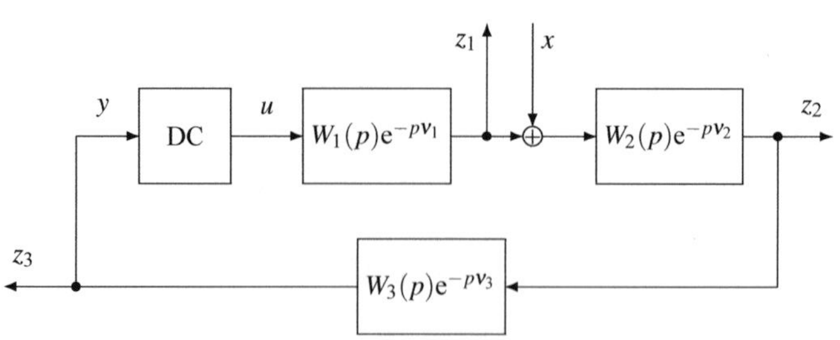


**Fig. 6.9** SD system with delay in nonstandard structure

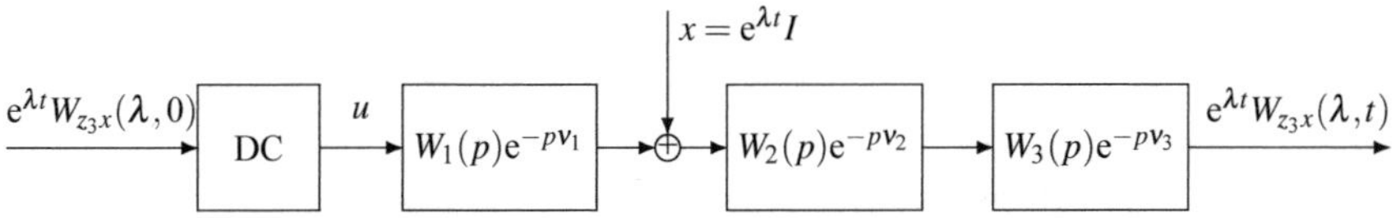


**Fig. 6.10** Open chain as part of the SD system in Fig. 6.9

attempt

$$x(\lambda, t) = e^{\lambda t} I, \quad z_3(\lambda, t) = e^{\lambda t} W_{z_3 x}(\lambda, t), \tag{6.226}$$

where $I$ is an identity matrix of appropriate size and the PTM $W_{z_3 x}(\lambda, t) = W_{z_3 x}(\lambda, t + T)$ is continuous with respect to $t$. Under the taken propositions, using the stroboscopic property of the ADC, we can assume that the exponential signal $e^{\lambda t} W_{z_3 x}(\lambda, 0)$ acts as input to the computer. With regard to this fact, we consider the open SD system shown in Fig. 6.10. As above, from Fig. 6.10 and (5.260), it follows directly that

$$W_{z_3 x}(\lambda, t) = e^{-\lambda \tau_N} \varphi_{N\mu}(T, \lambda, t - \tau_N) \breve{W}_d(\lambda) W_{z_3 x}(\lambda, 0) + W_3(\lambda) W_2(\lambda) e^{-\lambda(\nu_3 + \nu_2)}, \tag{6.227}$$

where

$$\tau_N \triangleq \nu_3 + \nu_2 + \nu_1 \,. \tag{6.228}$$

Since the left and right sides of formula (6.227) continuously depend on $t$, so for $t = 0$, we obtain

$$W_{z_3 x}(\lambda, 0) = e^{-\lambda \tau_N} \varphi_{N\mu}(T, \lambda, -\tau_N) \breve{W}_d(\lambda) W_{z_3 x}(\lambda, 0) + W_3(\lambda) W_2(\lambda) e^{-\lambda(\nu_3 + \nu_2)}, \tag{6.229}$$

or, equivalently,

$$W_{z_3 x}(\lambda, 0) = \breve{D}_{N\mu}(T, \lambda, -\tau_N) \breve{W}_d(\lambda) W_{z_3 x}(\lambda, 0) + W_3(\lambda) W_2(\lambda) e^{-\lambda(\nu_3 + \nu_2)}. \tag{6.230}$$

From this relation, we easily derive

$$W_{z_3 x}(\lambda, 0) = \left[ I - \breve{D}_{N\mu}(T, \lambda, -\tau_N) \breve{W}_d(\lambda) \right]^{-1} W_3(\lambda) W_2(\lambda) e^{-\lambda(\nu_3 + \nu_2)}. \tag{6.231}$$

Substituting this expression in (6.227) yields the desired expression for the PTM

$$\begin{aligned} W_{z_3 x}(\lambda, t) = e^{-\lambda \tau_N} \varphi_{N\mu}(T, \lambda, t - \tau_N) \breve{W}_d(\lambda) \left[ I - \breve{D}_{N\mu}(T, \lambda, -\tau_N) \breve{W}_d(\lambda) \right]^{-1} \times \\ \times W_3(\lambda) W_2(\lambda) e^{-\lambda(\nu_3 + \nu_2)} + W_3(\lambda) W_2(\lambda) e^{-\lambda(\nu_3 + \nu_2)}. \end{aligned} \tag{6.232}$$

Comparing (6.232) and (6.67), we find out that the PTM $W_{z_3 x}(\lambda, t)$ can be presented in the standard form, where

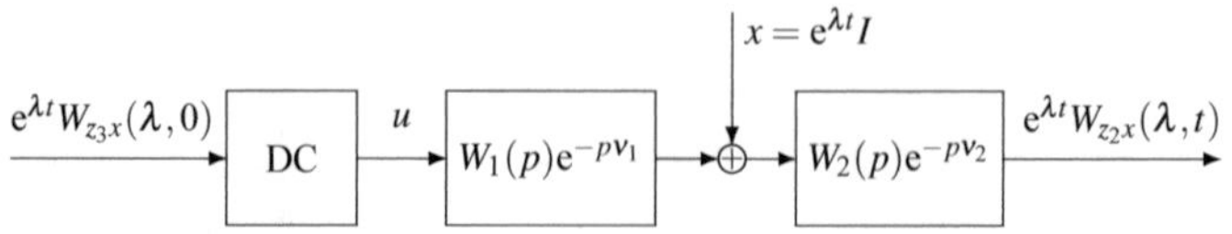


**Fig. 6.11** Open chain for Example 6.2

$$
\begin{aligned}
K(\lambda) &= W_3(\lambda)W_2(\lambda), \quad L(\lambda) = W_3(\lambda)W_2(\lambda)W_1(\lambda),\\
M(\lambda) &= W_3(\lambda)W_2(\lambda), \quad N(\lambda) = W_3(\lambda)W_2(\lambda)W_1(\lambda),
\end{aligned}
\tag{6.233}
$$

and, moreover

$$
\begin{aligned}
\tau_K &= \nu_3 + \nu_2, \quad \tau_L = \nu_3 + \nu_2 + \nu_1,\\
\tau_M &= \nu_3 + \nu_2, \quad \tau_N = \nu_3 + \nu_2 + \nu_1.
\end{aligned}
\tag{6.234}
$$

*Example 6.2* In order to construct the PTM $W_{z_2x}(\lambda, t)$ for the system shown in Fig. 6.9, consider the open SD chain shown in Fig. 6.11. Obviously,

$$
W_{z_2x}(\lambda, t) = e^{-\lambda(\nu_2+\nu_1)} \varphi_{W_2W_1\mu}(T, \lambda, t - \nu_2 - \nu_1) W_{z_3x}(\lambda, 0) + W_2(\lambda)e^{-\lambda\nu_2}.
\tag{6.235}
$$

After inserting expression (6.231), we immediately obtain

$$
\begin{aligned}
W_{z_2x}(\lambda, t) = e^{-\lambda(\nu_2+\nu_1)} \varphi_{W_2W_1\mu}(T, \lambda, t - \nu_2 - \nu_1)\tilde{W}_d(\lambda)\left[I - D_{N\mu}(T, \lambda, -\tau_N)\tilde{W}_d(\lambda)\right]^{-1} \times\\
\times W_3(\lambda)W_2(\lambda)e^{-\lambda(\nu_3+\nu_2)} + W_2(\lambda)e^{-\lambda\nu_2}.
\end{aligned}
\tag{6.236}
$$

We easily realize that the obtained PTM has standard form (6.67), where

$$
\begin{aligned}
K(p) &= W_2(p), \quad L(p) = W_2(p)W_1(p),\\
M(p) &= W_3(p)W_2(p), \quad N(p) = W_3(p)W_2(p)W_1(p),
\end{aligned}
\tag{6.237}
$$

and the corresponding delays are

$$
\begin{aligned}
\tau_K &= \nu_2, \quad \tau_L = \nu_2 + \nu_1,\\
\tau_M &= \nu_3 + \nu_2, \quad \tau_N = \nu_3 + \nu_2 + \nu_1.
\end{aligned}
\tag{6.238}
$$

*Example 6.3* Following the same ideas, we can build the PTM $W_{z_1,x}(\lambda, t)$. For that reason, we investigate the structure in Fig. 6.12. In analogy to above, we obtain

$$
W_{z_1x}(\lambda, t) = e^{-\lambda\nu_1} \varphi_{W_1\mu}(T, \lambda, t - \nu_1)\breve{W}_d(\lambda)W_{z_3x}(\lambda, 0),
\tag{6.239}
$$

which, with the help of (6.231), leads to the expression

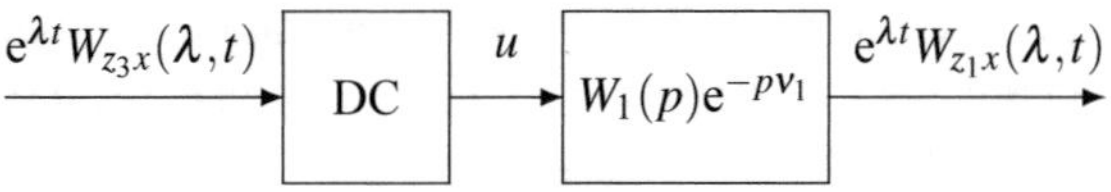


**Fig. 6.12** Open chain for Example 6.3

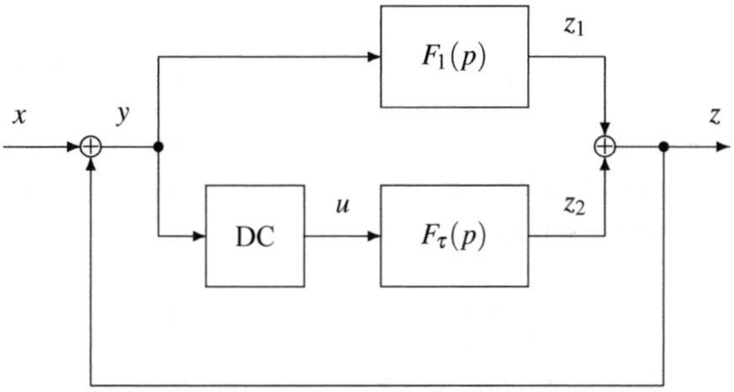


**Fig. 6.13** SD system with delay for Example 6.4

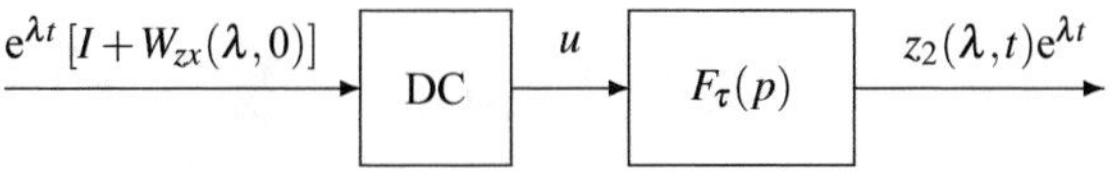


**Fig. 6.14** Open chain for Example 6.4

$$W_{z_1x}(\lambda,t) = e^{-\lambda\nu_1}\varphi_{W_1\mu}(T,\lambda,t-\nu_1)\tilde{W}_d(\lambda)\left[I - \breve{W}_{N\mu}(T,\lambda,-\tau_N)\breve{W}_d(\lambda)\right]^{-1}\times$$
$$\times e^{-\lambda(\nu_3+\nu_1)}W_3(\lambda)W_L(\lambda). \tag{6.240}$$

The PTM obtained has the standard form (6.67), where

$$K(p) = 0,\;\; L(p) = W_1(p),$$
$$M(p) = W_3(p)W_2(p),\;\; N(p) = W_3(p)W_2(p)W_1(p), \tag{6.241}$$

and moreover,

$$\tau_K = 0,\;\; \tau_L = \nu_1,$$
$$\tau_M = \nu_3 + \nu_2,\;\; \tau_N = \nu_3 + \nu_2 + \nu_1. \tag{6.242}$$

(2)

*Example 6.4* As a more complicated example, we consider the SD system shown in Fig. 6.13, where $F_\tau(p) = F(p)e^{-p\tau}$. Moreover, $F_1(p)$, $F(p)$ are rational matrices of appropriate size, and $\tau > 0$ is a real constant. For the determination of the PTM $W_{zx}(\lambda,t)$, as earlier, we assume

$$x(t) = e^{\lambda t}I,\quad z(t) = e^{\lambda t}W_{zx}(\lambda,t),\quad W_{zx}(\lambda,t) = W_{zx}(\lambda,t+T). \tag{6.243}$$

For correctness in the further calculations, it is necessary that the matrix $W_{zx}(\lambda, t)$ is continuous with respect to $t$. This condition holds if the matrices $F_1(p)$ and $F(p)$ are strictly proper, which is assumed.

Under assumption (6.243), we obtain from Fig. 6.13

$$y(\lambda, t) = \mathrm{e}^{\lambda t}\left[I + W_{zx}(\lambda, t)\right]. \tag{6.244}$$

Therefore, using the stroboscopic property of the ADC, the matrix $z_2(\lambda, t)$ can be generated by investigation of the open chain in Fig. 6.14. From this figure, in analogy to above, we find

$$z_2(\lambda, t) = \varphi_{F_\tau \mu}(T, \lambda, t)\breve{W}_d(\lambda)\left[I + W_{zx}(\lambda, 0)\right]\mathrm{e}^{\lambda t}. \tag{6.245}$$

For determining the matrix $z_1(\lambda, t)$, we suppose the Fourier series

$$W_{zx}(\lambda, t) = \sum_{k=-\infty}^{\infty} W_k(\lambda)\mathrm{e}^{kj\omega t}. \tag{6.246}$$

Then, from (6.244), we obtain

$$y(\lambda, t) = \mathrm{e}^{\lambda t} I + \sum_{k=-\infty}^{\infty} W_k(\lambda)\mathrm{e}^{(\lambda + kj\omega)t}. \tag{6.247}$$

The EP response of the LTI element with transfer matrix $F_1(p)$ to the EP input (6.247) takes the form

$$z_1(\lambda, t) = \mathrm{e}^{\lambda t} F_1(\lambda) + \sum_{k=-\infty}^{\infty} F_1(\lambda + kj\omega)W_k(\lambda)\mathrm{e}^{(\lambda + kj\omega)t}. \tag{6.248}$$

From Fig. 6.13, we find

$$W_{zx}(\lambda, t)\mathrm{e}^{\lambda t} = z_1(\lambda, t) + z_2(\lambda, t). \tag{6.249}$$

Substituting herein (6.245) and (6.248), after cancelation of $\mathrm{e}^{\lambda t}$, we obtain

$$\begin{aligned} W_{zx}(\lambda, t) = {} & \varphi_{F_\tau \mu}(T, \lambda, t)\breve{W}_d(\lambda)\left[I + W_{zx}(\lambda, 0)\right] + F_1(\lambda) \\ & + \sum_{k=-\infty}^{\infty} F_1(\lambda + kj\omega)W_k(\lambda)\mathrm{e}^{kj\omega t}. \end{aligned} \tag{6.250}$$

Owing to (6.246), (6.248), the last expression can be written in the form

$$\sum_{k=-\infty}^{\infty} W_k(\lambda)\mathrm{e}^{kj\omega t} = \frac{1}{T}\sum_{k=-\infty}^{\infty} F_\tau(\lambda+kj\omega)\mu(\lambda+kj\omega)\mathrm{e}^{kj\omega t}\breve{W}_d(\lambda)B(\lambda)$$

$$+ F_1(\lambda) + \sum_{k=-\infty}^{\infty} F_1(\lambda+kj\omega)W_k(\lambda)\mathrm{e}^{kj\omega t}, \qquad (6.251)$$

where

$$B(\lambda) \triangleq I + W_{zx}(\lambda, 0). \qquad (6.252)$$

Comparing the coefficients of equal harmonics on the left and right sides of (6.251), we obtain for $k = 0$

$$W_0(\lambda) = \frac{1}{T}F_\tau(\lambda)\mu(\lambda)\breve{W}_d(\lambda)B(\lambda) + F_1(\lambda)W_0(\lambda) + F_1(\lambda), \qquad (6.253)$$

and for $k \neq 0$

$$W_k(\lambda) = \frac{1}{T}F_\tau(\lambda+kj\omega)\mu(\lambda+kj\omega)\breve{W}_d(\lambda)B(\lambda) + F_1(\lambda+kj\omega)W_k(\lambda). \qquad (6.254)$$

From (6.253) and (6.254), we obtain

$$\begin{aligned} W_0(\lambda) &= \frac{1}{T}\left[I - F_1(\lambda)\right]^{-1} F(\lambda)\mathrm{e}^{-\lambda\tau}\mu(\lambda)\breve{W}_d(\lambda)B(\lambda) + \left[I - F_1(\lambda)\right]^{-1} F_1(\lambda), \\ W_k(\lambda) &= \frac{1}{T}\left[I - F_1(\lambda+kj\omega)\right]^{-1} F(\lambda+kj\omega)\mathrm{e}^{-(\lambda+kj\omega)\tau}\mu(\lambda+kj\omega)\breve{W}_d(\lambda)B(\lambda). \end{aligned} \qquad (6.255)$$

Introduce the strictly proper rational matrices

$$F_2(\lambda) \triangleq \left[I - F_1(\lambda)\right]^{-1} F_1(\lambda), \quad F_3(\lambda) \triangleq \left[I - F_1(\lambda)\right]^{-1} F(\lambda). \qquad (6.256)$$

Then, from (6.255), we find

$$\begin{aligned} W_0(\lambda) &= F_2(\lambda) + \frac{1}{T}\mathrm{e}^{-\lambda\tau}F_3(\lambda)\mu(\lambda)\breve{W}_d(\lambda)B(\lambda), \\ W_k(\lambda) &= \frac{1}{T}\mathrm{e}^{-(\lambda+kj\omega)\tau}F_3(\lambda+kj\omega)\mu(\lambda+kj\omega)\breve{W}_d(\lambda)B(\lambda), \end{aligned} \qquad (6.257)$$

from which we derive

$$\begin{aligned}
W_{zx}(\lambda, t) &= \sum_{k=-\infty}^{\infty} W_k(\lambda)\mathrm{e}^{kj\omega t} \\
&= \mathrm{e}^{-\lambda\tau}\frac{1}{T}\sum_{k=-\infty}^{\infty} F_3(\lambda + kj\omega)\mu(\lambda + kj\omega)\mathrm{e}^{kj\omega(t-\tau)}\breve{W}_d(\lambda)B(\lambda) + F_2(\lambda) \\
&= \mathrm{e}^{-\lambda\tau}\varphi_{F_3\mu}(T, \lambda, t-\tau)\breve{W}_d(\lambda)B(\lambda) + F_2(\lambda). \qquad (6.258)
\end{aligned}$$

For $t = 0$, this results in

$$W_{zx}(\lambda, 0) = \breve{D}_{F_3\mu}(T, \lambda, -\tau)\breve{W}_d(\lambda)B(\lambda) + F_2(\lambda), \qquad (6.259)$$

which, owing to (6.252), can be written in the form

$$B(\lambda) = \breve{D}_{F_3\mu}(T, \lambda, -\tau)\breve{W}_d(\lambda)B(\lambda) + I + F_2(\lambda). \qquad (6.260)$$

Since $I + F_2(\lambda) = (I - F_1(\lambda))^{-1}$, we find

$$B(\lambda) = \left[I - \breve{D}_{F_3\mu}(T, \lambda, -\tau)\breve{W}_d(\lambda)\right]^{-1}[I + F_2(\lambda)]. \qquad (6.261)$$

With the help of this relation, from (6.258), we find the required PTM

$$\begin{aligned}
W_{zx}(\lambda, t) = \mathrm{e}^{-\lambda\tau}\varphi_{F_3\mu}(T, \lambda, t-\tau)\breve{W}_d(\lambda)\left[I - \tilde{D}_{F_3\mu}(T, \lambda, -\tau)\breve{W}_d(\lambda)\right]^{-1} \times \\
\times [I + F_2(\lambda) + F_2(\lambda)]. \qquad (6.262)
\end{aligned}$$

The obtained PTM possesses the standard form, where

$$\begin{aligned}
K(p) = F_2(p) = [I - F_1(p)]^{-1} F_1(p), \quad L(p) = F_3(p) = [I - F_1(p)]^{-1} F(p), \\
M(p) = I + F_2(p) = [I - F_1(p)]^{-1}, \quad N(p) = F_3(p) = [I - F_1(p)]^{-1} F(p). \\
(6.263)
\end{aligned}$$

Moreover, we get

$$\tau_L = \tau_N = \tau, \quad \tau_K = \tau_M = 0. \qquad (6.264)$$

(3)

**Definition 6.6** Sampled-data systems possessing the same PTM with respect to input $x(t)$ and output $z(t)$ are called similar.

Examples 6.1–6.4 show that the SD systems with delay investigated in these examples are similar to a standard structure with appropriate delays and matrices $K(p)$, $L(p)$, $M(p), andN(p)$.

**Definition 6.7** Similar SD systems with delay are called structural standardizable, if their PTM has standard form.

The fact that an SD system with delay and multivariable controlling computer is structural standardizable, it can be ascertained directly by its block structure, without calculation of the corresponding PTM. Thereto, we have to consider the LTI systems, which we obtain from the SD system, when the controlling computer is removed. Hereby, denote by $z(t)$ the selected output of the SD system, by $y(t)$ the input to the computer, and by $u(t)$ its output. Then, we have to find the transfer matrices

$$\begin{aligned} K_\tau(p) &= W_{zx}(p), \quad L_\tau(p) = W_{zu}(p), \\ M_\tau(p) &= W_{yx}(p), \quad N_\tau(p) = W_{yu}. \end{aligned} \tag{6.265}$$

Hereby, if it turns out that

$$\begin{aligned} K_\tau(p) &= K(p)\mathrm{e}^{-p\tau_K}, \quad L_\tau(p) = L(p)\mathrm{e}^{-p\tau_L}, \\ M_\tau(p) &= M(p)\mathrm{e}^{-p\tau_M}, \quad N_\tau(p) = N(p)\mathrm{e}^{-p\tau_N}, \end{aligned} \tag{6.266}$$

where $K(p), L(p), M(p), and N(p)$ are rational matrices, and $\tau_K, \tau_L, \tau_M, and \tau_N$ are nonnegative constants, then the considered SD system with delay is structural standardizable if its PTM exists.

*Example 6.5* After removing the computer in the structure of Fig. 6.9, the structure of Fig. 6.15 is achieved. For the output $z_3 = y$, from Fig. 6.15, we read

$$\begin{aligned} W_{z_3x}(p) = W_3(p)W_2(p)\mathrm{e}^{-p(\nu_3+\nu_2)}, \quad W_{z_3u}(p) = W_3(p)W_2(p)W_1(p)\mathrm{e}^{-p(\nu_3+\nu_2+\nu_1)}, \\ W_{yx}(p) = W_{z_3x}(p), \quad W_{yu}(p) = W_{z_3u}(p). \end{aligned} \tag{6.267}$$

Thus, in this case,

$$\begin{aligned} K(p) &= W_3(p)W_2(p), \quad L(p) = W_3(p)W_2(p)W_1(p), \\ M(p) &= W_3(p)W_2(p), \quad N(p) = W_3(p)W_2(p)W_1(p), \end{aligned} \tag{6.268}$$

and, in harmony with Example 6.1.

$$\tau_K = \tau_M = \nu_3 + \nu_2, \quad \tau_L = \tau_N = \nu_3 + \nu_2 + \nu_1. \tag{6.269}$$

In analogy, for the output $z_2$, we obtain

$$\begin{aligned} W_{z_2x}(p) = W_2(p)\mathrm{e}^{-p\nu_2}, \quad W_{z_2u}(p) = W_2(p)W_1(p)\mathrm{e}^{-p(\nu_2+\nu_1)}, \\ W_{yx}(p) = W_3(p)W_2(p)\mathrm{e}^{-p(\nu_3+\nu_2)}, \quad W_{yu}(p) = W_3(p)W_2(p)W_1(p)\mathrm{e}^{-p(\nu_3+\nu_2+\nu_1)}. \end{aligned} \tag{6.270}$$

Hence

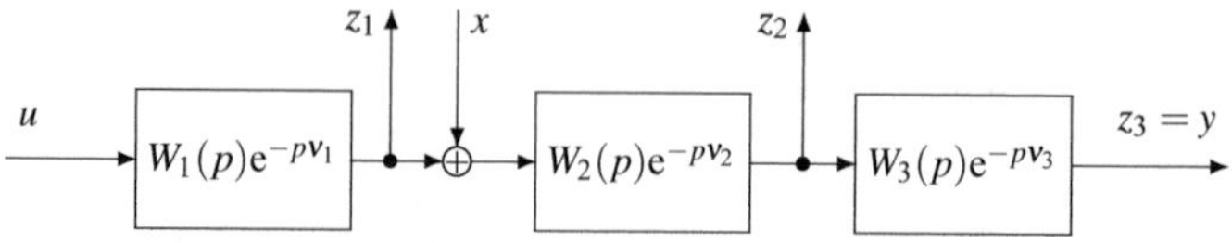


**Fig. 6.15** Open chain of Fig. 6.9 after removing the computer

**Fig. 6.16** LTI system for Example 6.3 after removing the computer

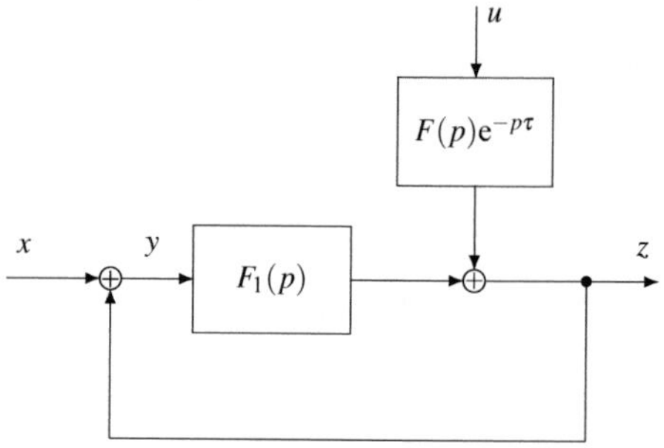


$$
\begin{aligned}
K(p) &= W_2(p), \quad L(p) = W_2(p)W_1(p), \\
M(p) &= W_3(p)W_2(p), \quad N(p) = W_3(p)W_2(p)W_1(p),
\end{aligned} \tag{6.271}
$$

and

$$
\begin{aligned}
\tau_K &= \nu_2, \quad \tau_L = \nu_2 + \nu_1, \\
\tau_M &= \nu_3 + \nu_2, \quad \tau_N = \nu_3 + \nu_2 + \nu_1.
\end{aligned} \tag{6.272}
$$

Finally, for the output $z_1$, from Fig. 6.15, we find

$$
\begin{aligned}
W_{z_1x}(p) &= 0, \quad W_{z_1u} = W_1(p)e^{-p\nu_1}, \\
W_{yx}(p) &= W_3(p)W_2(p)e^{-p(\nu_3+\nu_2)}, \quad W_{yu}(p) = W_3(p)W_2(p)W_1(p)e^{-p(\nu_3+\nu_2+\nu_1)}.
\end{aligned} \tag{6.273}
$$

Moreover, using $\tau_K = 0$, we obtain

$$
\begin{aligned}
\tau_K &= 0, \quad \tau_L = \nu_1, \\
\tau_M &= \nu_3 + \nu_2, \quad \tau_N = \nu_3 + \nu_2 + \nu_1.
\end{aligned} \tag{6.274}
$$

From (6.267)–(6.274), we realize that the system in Fig. 6.9 is structural standardizable according to the outputs $z_1, z_2, z_3$.

(4)

*Example 6.6* After removing the computer from the structure in Fig. 6.13, the closed LTI system in Fig. 6.16 is achieved. A direct calculation proves that, in this case,

$$W_{zx}(p) = \left[I - F_1(p)\right]^{-1} F_1(p), \quad W_{zu}(p) = \left[I - F_1(p)\right]^{-1} F(p)e^{-p\tau},$$
$$W_{yx}(p) = \left[I - F_1(p)\right]^{-1} \quad W_{yu} = \left[I - F_1(p)\right]^{-1} F(p)e^{-p\tau}. \tag{6.275}$$

Hence

$$K(p) = \left[I - F_1(p)\right]^{-1} F_1(p), \quad L(p) = \left[I - F_1(p)\right]^{-1} F(p),$$
$$M(p) = \left[I - F_1(p)\right]^{-1}, \quad N(p) = \left[I - F_1(p)\right]^{-1} F(p). \tag{6.276}$$

Moreover,

$$\tau_K = 0, \quad \tau_L = \tau,$$
$$\tau_M = 0, \quad \tau_N = \tau. \tag{6.277}$$

Owing to (6.275), we conclude that the continuous–discrete SD system with delay in Fig. 6.13 is structural standardizable.

## 6.8 Model-Standardizable SD Systems with Delay

(1) According to the definition in Sect. 6.7, an SD system with delay for the selected input vector $x(t)$ and output vector $z(t)$ is structural standardizable, if the corresponding PTM can be represented in the standard form of shape

$$W_{zx}(\lambda, t) = e^{-\lambda\tau_L}\varphi_{L\mu}(T, \lambda, t - \tau_L)\breve{W}_d(\lambda)\left[I - \breve{D}_{N\mu}(T, \lambda, -\tau_N)\breve{W}_d(\lambda)\right]^{-1} \times$$
$$\times M(\lambda)e^{-\lambda\tau_M} + K(\lambda)e^{-\lambda\tau_K}, \tag{6.278}$$

where $K(\lambda), L(\lambda), M(\lambda),$ *and* $N(\lambda)$ are rational matrices, and $\tau_K, \tau_L, \tau_M,$ *and* $\tau_N$ are nonnegative real constants.

**Definition 6.8** A structural standardizable SD system with delay, possessing PTM (6.278), is called model standardizable, if there exists a similar standard model of form (6.12) and its PTM coincides with matrices (6.278).

The statement of the fact that an SD system with delay is model standardizable is very important for applications. Indeed, owing to the statements in Sects. 6.6–6.8, the properties of the PTM of the standard model do not depend on the form of the matrices $A, B_1, B_2, C_1, C_2$ in (6.12), but are only determined by the properties of the matrices $K(p), L(p), M(p), N(p)$ and the corresponding $\tau_K, \tau_L, \tau_M, \tau_N$. Therefore, the PTM (6.278) of a model-standardizable system disposes of all properties of the PTM of the equivalent standard model. In particular, this includes the properties of the poles of the corresponding PTM, the solution of the modal control and the stabilization problem, the representation of the PTM by its system function, etc.

The next theorem provides necessary and sufficient conditions for model standardizability.

**Theorem 6.10** *For the structural standardizable SD system with delay, possessing the standard PTM* (6.280), *to be model standardizable, it is necessary and sufficient that the following conditions are fulfilled:*

*(a) The rational matrices* $K(p), M(p), N(p)$ *are strictly proper.*
*(b) The rational matrix* $L(p)$ *is at least proper.*
*(c) The matrix* $N(p)$ *dominates in the matrix*

$$\bar{W}_0(p) = \begin{bmatrix} K(p) & L(p) \\ M(p) & N(p) \end{bmatrix}, \tag{6.279}$$

*i.e.,*

$$\text{Mind}\,\bar{W}_0(p) = \text{Mind}\,N(p). \tag{6.280}$$

*(d) The delay constants satisfy*

$$\tau_K + \tau_N = \tau_M + \tau_L. \tag{6.281}$$

*Proof* Necessity: Assume that there exists a standard model (6.12), and its PTM $W_{zx}(\lambda, t)$ coincides with matrix (6.278). Then conditions (a)–(c) are fulfilled, as it was shown in Sect. 6.2. Moreover, from (6.19), it follows that condition (6.281) also is true. Therefore, conditions (a)–(d) are necessary.

Sufficiency: Let conditions (a)–(d) be fulfilled. Then the matrices $K(p), M(p), N(p)$ are strictly proper, and the matrix $L(p)$ is at least proper. Assume

$$L(\lambda) = L_1(\lambda) + D, \tag{6.282}$$

where the matrix $L_1(\lambda)$ is strictly proper and the matrix $D$ is constant. Then

$$\bar{W}_0(p) = \bar{W}_1(p) + \bar{D}_1, \tag{6.283}$$

where

$$\bar{W}_1(p) = \begin{bmatrix} K(p) & L_1(p) \\ M(p) & N(p) \end{bmatrix}, \quad \bar{D}_1 = \begin{bmatrix} 0 & D \\ 0 & 0 \end{bmatrix}, \tag{6.284}$$

and herein the matrix $\bar{W}_1(p)$ is strictly proper. From Lemma 2.11, it follows that

$$\text{Mind}\,\bar{W}_0(p) = \text{Mind}\,\bar{W}_1(p). \tag{6.285}$$

Since by assumption the matrix $N(p)$ dominates in matrix (6.279), i.e.,

$$\text{Mind}\,\bar{W}_0(p) = \text{Mind}\,N(p), \tag{6.286}$$

so (6.285) implies

$$\operatorname{Mind} \bar{W}_1(p) = \operatorname{Mind} N(p).$$

This relation means that the matrix $N(p)$ dominates in the strictly proper matrix $\bar{W}_1(p)$. Assume the minimal representation

$$N(p) = C_2(pI_\chi - A)^{-1}B_2. \tag{6.287}$$

Then, due to the strict properness of the matrix $\bar{W}_1(p)$, Theorem 2.12 yields

$$\begin{aligned} K(p) &= C_1(pI_\chi - A)^{-1}B_1, \;\; L_1(p) = C_1(pI_\chi - A)^{-1}B_2, \\ M(p) &= C_2(pI_\chi - A)^{-1}B_1, \;\; N(p) = C_2(pI_\chi - A)^{-1}B_2, \end{aligned} \tag{6.288}$$

where $C_1, B_1$ are constant matrices. Denote

$$\tau_1 \stackrel{\triangle}{=} \tau_M, \quad \tau_2 \stackrel{\triangle}{=} \tau_N. \tag{6.289}$$

Moreover, using (6.281), name

$$\tau_3 \stackrel{\triangle}{=} \tau_K - \tau_M = \tau_L - \tau_N. \tag{6.290}$$

Consider the standard model of an SD system with delay, where its continuous part is described by

$$\begin{aligned} \frac{dv(t)}{dt} &= Av(t) + B_1x(t-\tau_1) + B_2u(t-\tau_2), \\ y(t) &= C_2v(t), \\ z(t) &= C_1v(t-\tau_3) + Du(t-\tau_2-\tau_3). \end{aligned} \tag{6.291}$$

After transfer to the images, these relations appear as

$$V(s) = (sI_\chi - A)^{-1}B_1\mathrm{e}^{-s\tau_1}X(s) + (sI_\chi - A)^{-1}B_2\mathrm{e}^{-s\tau_2}U(s), \tag{6.292}$$

and, further considering (6.288), we obtain

$$Y(s) = C_2V(s) = M(s)\mathrm{e}^{-s\tau_M}X(s) + N(s)\mathrm{e}^{-s\tau_N}U(s). \tag{6.293}$$

Moreover, from (6.291), we find

$$Z(s) = C_1\mathrm{e}^{-s\tau_3}V(s) + D\mathrm{e}^{-s(\tau_2+\tau_3)}U(s). \tag{6.294}$$

Substituting this relation in (6.292), and using (6.289), (6.290) yield

$$Z(s) = K(s)\mathrm{e}^{-s\tau_K}X(s) + L(s)\mathrm{e}^{-s\tau_L}U(s). \tag{6.295}$$

Adding to (6.293), (6.295) the equations of the digital controller, we obtain a system of form (6.16). Therefore, the PTM of the corresponding standard model coincides with the PTM (6.278). The sufficiency is proven. ■

(2) Let us consider some examples for the application of Theorem 6.10.

*Example 6.7* Under the condition of Example 6.1, from (6.233), we obtain

$$\bar{W}_0(p) = \begin{bmatrix} W_3(p)W_2(p) & W_3(p)W_2(p)W_1(p) \\ W_3(p)W_2(p) & W_3(p)W_2(p)W_1(p) \end{bmatrix}. \tag{6.296}$$

For conditions (a), (b) to be fulfilled, Theorem 6.10 says that a necessary condition is that the matrices $W_3(p)W_2(p)$ and $W_3(p)W_2(p)W_1(p)$ must be strictly proper. Condition (d), as it follows from (6.234), is always true. It remains to find a condition under which the matrix $N(p) = W_3(p)W_2(p)W_1(p)$ becomes dominant in matrix (6.296). We will show that this takes place, when the matrices $W_3(p)$, $W_2(p)$ and $W_1(p)$ are independent, i.e.,

$$\operatorname{Mind} N(p) = \operatorname{Mind} W_3(p) + \operatorname{Mind} W_2(p) + \operatorname{Mind} W_1(p). \tag{6.297}$$

Hereto, notice that

$$\bar{W}_0(p) = \bar{W}_2(p)\bar{W}_3(p), \tag{6.298}$$

where

$$\bar{W}_2(p) \triangleq \begin{bmatrix} W_3(p)W_2(p) & W_3(p)W_2(p) \\ W_3(p)W_2(p) & W_3(p)W_2(p) \end{bmatrix}, \quad \bar{W}_3(p) \triangleq \begin{bmatrix} I & 0 \\ 0 & W_1(p) \end{bmatrix}. \tag{6.299}$$

We will prove that under condition (6.297), we obtain

$$\operatorname{Mind}\left[W_3(p)W_2(p)\right] = \operatorname{Mind} W_3(p) + \operatorname{Mind} W_2(p) \tag{6.300}$$

and

$$\operatorname{Mind} \bar{W}_2(p) = \operatorname{Mind}\left[W_3(p)W_2(p)\right] = \operatorname{Mind} W_3(p) + \operatorname{Mind} W_2(p). \tag{6.301}$$

Formula (6.300) is directly achieved from independence condition (6.297). For the proof of formula (6.301), we use the relation

$$\bar{W}_2(p) = \begin{bmatrix} W_3(p)W_2(p) & W_3(p)W_2(p) \\ W_3(p)W_2(p) & W_3(p)W_2(p) \end{bmatrix} = \begin{bmatrix} I & I \\ 0 & I \end{bmatrix} \bar{W}_4(p) \begin{bmatrix} I & 0 \\ I & I \end{bmatrix}, \tag{6.302}$$

where

$$\bar{W}_4(p) \triangleq \begin{bmatrix} 0 & 0 \\ 0 & W_3(p)W_2(p) \end{bmatrix}. \tag{6.303}$$

Now (6.304) yields

$$\text{Mind}\, \bar{W}_2(p) = \text{Mind}\, \bar{W}_4(p), \tag{6.304}$$

because the first and last factors on the right side of (6.302) are independent matrices. We prove now that

$$\text{Mind}\, \bar{W}_4(p) = \text{Mind}\left[W_3(p)W_2(p)\right]. \tag{6.305}$$

Indeed, (6.303) and Corollary 2.6 yield

$$\text{Mind}\, \bar{W}_4(p) \geq \text{Mind}\left[W_3(p)W_2(p)\right]. \tag{6.306}$$

On the other side, assume the ILMFD

$$W_3(p)W_2(p) = a^{-1}(p)b(p), \tag{6.307}$$

where, due to the general properties of ILMFD

$$\det a(p) = \text{Mind}\left[W_3(p)W_2(p)\right]. \tag{6.308}$$

Hence

$$\bar{W}_4(p) = \begin{bmatrix} I & 0 \\ 0 & a(p) \end{bmatrix}^{-1} \begin{bmatrix} 0 & 0 \\ 0 & b(p) \end{bmatrix} \tag{6.309}$$

and, from Lemma 2.10, we find

$$\text{Mind}\, \bar{W}_4(p) \leq \deg\det \begin{bmatrix} I & 0 \\ 0 & a(p) \end{bmatrix} = \deg\det a(p) = \text{Mind}\left[W_3(p)W_2(p)\right]. \tag{6.310}$$

Comparing inequalities (6.306) and (6.310), we find (6.305), which with the help of (6.304) yields (6.301).

Further notice that

$$\text{Mind}\, \bar{W}_3(p) = \text{Mind} \begin{bmatrix} I & 0 \\ 0 & W_1(p) \end{bmatrix} = \text{Mind}\, W_1(p). \tag{6.311}$$

Indeed, on one side from (6.299) directly follows

$$\text{Mind}\, \bar{W}_3(p) \geq \text{Mind}\, W_1(p). \tag{6.312}$$

But on the other side, when we have the ILMFD $W_1(p) = a_1{}^{-1}(p)b_1(p)$, so in analogy to above, it is easy to show that from the existence of the LMFD

$$\bar{W}_3(p) = \begin{bmatrix} I & 0 \\ 0 & a_1(p) \end{bmatrix}^{-1} \begin{bmatrix} I & 0 \\ 0 & b_1(p) \end{bmatrix} \tag{6.313}$$

we realize

$$\text{Mind}\, \bar{W}_3(p) \leq \deg\det a_1(p) = \text{Mind}\, W_1(p). \tag{6.314}$$

Comparing (6.312) and (6.314), we arrive at (6.311).

Now, we are ready to prove that, under condition (6.297), the matrix $N(p) = W_3(p)W_2(p)W_1(p)$ dominates in matrix (6.296). Indeed, from (6.296), (6.297), we obtain

$$\operatorname{Mind} \bar{W}_0(p) \geq \operatorname{Mind} N(p) = \operatorname{Mind} W_3(p) + \operatorname{Mind} W_2(p) + \operatorname{Mind} W_1(p). \tag{6.315}$$

At the same time, from (6.298), (6.301), and (6.311), it follows that

$$\begin{aligned} \operatorname{Mind} \bar{W}_0(p) &\leq \operatorname{Mind} \bar{W}_2(p) + \operatorname{Mind} \bar{W}_3(p) \\ &= \operatorname{Mind} W_3(p) + \operatorname{Mind} W_2(p) + Mind\, W_1(p). \end{aligned} \tag{6.316}$$

Comparison of the last two inequalities allows

$$\operatorname{Mind} \bar{W}_0(p) = \operatorname{Mind} N(p),$$

i.e., the matrix $N(p)$ dominates in the matrix $\bar{W}_0(p)$.

Since it follows from (6.269) that condition (6.290) in the actual case is always fulfilled, so, summing up the above said, we can claim, using (6.234), that the system in Fig. 6.9 is model standardizable for the output vector $z_3(t)$, if the matrices, configured in (6.296), are strictly proper and satisfy condition (6.297). Notice also that from (6.234) and (6.289), (6.290), we find out that in the actual example

$$\begin{aligned} \tau_1 = \tau_M &= \nu_3 + \nu_2,\ \tau_2 = \tau_L = \tau_N = \nu_3 + \nu_2 + \nu_1, \\ \tau_3 &= \tau_K - \tau_M = \tau_L - \tau_N = 0. \end{aligned} \tag{6.317}$$

Moreover, when (6.295) and (6.317) are valid, the equations of the LTI part of the corresponding standard model (6.291) take the form

$$\begin{aligned} \frac{dv(t)}{dt} &= Av(t) + B_1x(t - \nu_3 - \nu_2) + B_2u(t - \nu_3 - \nu_2 - \nu_1), \\ y(t) &= C_2v(t), \\ z(t) &= C_1v(t) + Du(t - \nu_3 - \nu_2 - \nu_1). \end{aligned} \tag{6.318}$$

(3)

*Example 6.8* Under the conditions of Example 6.2, it follows from (6.238) that condition (6.281) is always fulfilled. The matrix $\bar{W}_0(p)$ in the given case, as it follows from (6.237), takes the form

$$\bar{W}_0(p) = \begin{bmatrix} W_2(p) & W_2(p)W_1(p) \\ W_3(p)W_2(p) & W_3(p)W_2(p)W_1(p) \end{bmatrix}. \tag{6.319}$$

This relation can be presented in the form

$$\bar{W}_0(p) = \bar{W}_1(p)\bar{W}_2(p)\bar{W}_3(p), \tag{6.320}$$

where

$$\bar{W}_1(p) \triangleq \begin{bmatrix} I & 0 \\ 0 & W_3(p) \end{bmatrix}, \quad \bar{W}_2(p) \triangleq \begin{bmatrix} W_2(p) & W_2(p) \\ W_2(p) & W_2(p) \end{bmatrix},$$
$$\bar{W}_3(p) \triangleq \begin{bmatrix} I & 0 \\ 0 & W_1(p) \end{bmatrix}. \tag{6.321}$$

In analogy to Example 6.7, we state

$$\text{Mind}\,\bar{W}_1(p) = \text{Mind}\,W_3(p), \quad \text{Mind}\,\bar{W}_2(p) = \text{Mind}\,W_2(p),$$
$$\text{Mind}\,\bar{W}_3(p) = \text{Mind}\,W_1(p). \tag{6.322}$$

Hence, we conclude that, under condition (6.297), the matrix $N(p) = W_3(p)$ $W_2(p)W_1(p)$ dominates in the matrix (6.319). Indeed, from (6.319), we directly obtain

$$\text{Mind}\,\bar{W}_0(p) \geq \text{Mind}\,N(p). \tag{6.323}$$

On the other side, from (6.320), (6.322), it emerges that

$$\text{Mind}\,\bar{W}_0(p) \leq \text{Mind}\,N(p). \tag{6.324}$$

Thus

$$\text{Mind}\,\bar{W}_0(p) = \text{Mind}\,N(p)$$

Summing up the obtained results, on basis of Theorem 6.10, we can state that, under the conditions of Example 6.2, the structural standardizable SD system with delay is model standardizable, if condition (6.297) is fulfilled and, moreover, the matrices $W_2(p)$, $W_3(p)W_2(p)$, $W_3(p)W_2(p)W_1(p)$ are strictly proper, and the matrix $W_2(p)W_1(p)$ is at least proper. Hereby, for the corresponding standard model from (6.238), we find

$$\tau_1 = \tau_M = \nu_3 + \nu_2, \quad \tau_2 = \tau_N = \nu_3 + \nu_2 + \nu_1,$$
$$\tau_3 = \tau_K - \tau_M = \tau_L - \tau_N = -\nu_3. \tag{6.325}$$

Hence, in the given case, the quantity $\tau_3$ becomes nonpositive.

(4)

*Example 6.9* For Example 6.3, due to (6.241), the matrix $\bar{W}_0(p)$ can be written in the form

$$\bar{W}_0(p) = \begin{bmatrix} 0 & W_1(p) \\ W_3(p)W_2(p) & W_3(p)W_2(p)W_1(p) \end{bmatrix}, \tag{6.326}$$

which can be presented as the product

$$\bar{W}_0(p) = \bar{W}_1(p)\bar{W}_2(p)\bar{W}_3(p), \tag{6.327}$$

where

$$\bar{W}_1(p) \triangleq \begin{bmatrix} I & 0 \\ 0 & W_3(p) \end{bmatrix}, \quad \bar{W}_2(p) \triangleq \begin{bmatrix} 0 & I \\ W_2(p) & W_2(p) \end{bmatrix},$$
$$\bar{W}_3(p) \triangleq \begin{bmatrix} I & 0 \\ 0 & W_1(p) \end{bmatrix}. \tag{6.328}$$

The calculations in Examples 6.6 and 6.7 yield

$$\operatorname{Mind} \bar{W}_1(p) = \operatorname{Mind} W_3(p), \quad \operatorname{Mind} \bar{W}_3(p) = \operatorname{Mind} W_1(p). \tag{6.329}$$

We prove that

$$\operatorname{Mind} \bar{W}_2(p) = \operatorname{Mind} W_2(p). \tag{6.330}$$

Indeed, from (6.328), we find

$$\operatorname{Mind} \bar{W}_2(p) \geq \operatorname{Mind} W_2(p). \tag{6.331}$$

On the other side, assume the ILMFD

$$W_2(p) = a_2^{-1}(p)b_2(p). \tag{6.332}$$

Then, there exists the LMFD

$$\bar{W}_2(p) = \begin{bmatrix} I & 0 \\ 0 & a_2(p) \end{bmatrix}^{-1} \begin{bmatrix} 0 & I \\ b_2(p) & b_2(p) \end{bmatrix}. \tag{6.333}$$

Hence

$$\operatorname{Mind} \bar{W}_2(p) \leq \deg\det \begin{bmatrix} I & 0 \\ 0 & a_2(p) \end{bmatrix} = \deg\det a(p) = \operatorname{Mind} W_2(p). \tag{6.334}$$

Comparing inequalities (6.331) and (6.334), the relation (6.330) is achieved. Since, in the present case, condition (6.281) is fulfilled, we can state that for the output $z_1(t)$ the system in Fig. 6.9 is model standardizable if the matrices $W_3(p)W_2(p)$ and $W_3(p)W_2(p)W_1(p)$ are strictly proper, and the matrix $W_1(p)$ is at least proper, and in addition the matrices $W_3(p)$, $W_2(p)$, $W_1(p)$ are independent and satisfy (6.297).

(5)

*Example 6.10* For the structural standardizable system shown in Fig. 6.13, from (6.265), we obtain

$$M(p) = \left[I - F_1(p)\right]^{-1}. \tag{6.335}$$

Since by supposition the matrix $F_1(p)$ is strictly proper, so we obtain

$$\lim_{p\to\infty} M(p) = \lim_{p\to\infty} \left[I - F_1(p)\right]^{-1} = I,$$

i.e., the matrix $M(p)$ is proper. This means that condition (a) in Theorem 6.10 is not fulfilled, and consequently, the structural standardizable system in Fig. 6.13 is not model standardizable.

## References

1. E.N. Rosenwasser, *Linear Theory of Digital Control in Continuous Time* (Nauka, Moscow, 1994). (in Russian)
2. E.N. Rosenwasser, B.P. Lampe, *Computer Controlled Systems: Analysis and Design with Process-orientated Models* (Springer, Berlin, 2000)

# Part III
# Optimal SD Control with Delay

# Chapter 7
# $\mathscr{H}_2$ Optimization and Fixed Poles of the Standard Model for SD Systems with Delay

**Abstract** Based on the results of Chap. 6, a square function is constructed as $\mathscr{H}_2$ norm of the SD standard model depending on the choice of the system parameters. Using the Wiener–Hopf minimization method, the optimal value of a stable system function is determined and a $\mathscr{H}_2$-optimal causal control program is generated from it. The application of the described method can provide some important facts that are useful for the application: 1. The characteristic polynomial of the $\mathscr{H}_2$-optimal system is a divisor of a polynomial independent of delays. 2. A part of the poles of a regulated $\mathscr{H}_2$ optimal system does not depend on the delays and the choice of the generalized digital-to-analog converter and is completely determined by the structure of the continuous part of the system and the properties of the elements it contains. These poles are called fixed. The existence of elements with unfavorable fixed poles can lead to the synthesis of a $\mathscr{H}_2$-optimal system with unsatisfactory quality. As an example, the construction of a set of fixed poles for a multidimensional single-loop SD system with a different selection of inputs and outputs is considered.

## 7.1 Advanced Statistical Analysis for the Standard Model of SD Systems with Delay

(1) Assume that as input of the standard model (6.12) acts a vector signal $x(t)$, proving to be a stationary stochastic process with covariance matrix

$$K_x(\nu) = \mathscr{E}\left[x(t)x'(t+\nu)\right], \tag{7.1}$$

where $\mathscr{E}\,[\,\cdot\,]$ denotes the operator of mathematical expectation. Hereby, the integral (two-sided Laplace transform)

$$\Phi_x(s) = \int_{-\infty}^{\infty} K_x(\nu)\mathrm{e}^{-s\nu}\,\mathrm{d}\nu \tag{7.2}$$

is called the spectral density of the input. It is supposed that integral (7.2) converges absolutely in the stripe $-\alpha_0 \leq \operatorname{Re} s \leq \alpha_0$, where $\alpha_0$ is a positive number.



E. N. Rosenwasser et al., *Computer-Controlled Systems with Delay*,
https://doi.org/10.1007/978-3-030-15042-6_7

(2) As it is known from [1–3], if the standard system (6.12) is stable, then after the transient process, at the output, $z(t)$ establishes a quasi-stationary process $z_\infty(t)$, and its covariance matrix is determined by the formula

$$K_z(t_1, t_2) = \frac{1}{2\pi \mathrm{j}} \int_{-\mathrm{j}\infty}^{\mathrm{j}\infty} W'_{zx}(-s, t_1)\Phi_x(s)W_{zx}(s, t_2)\mathrm{e}^{s(t_2-t_1)}\, \mathrm{d}s, \tag{7.3}$$

where $W_{zx}(s, t)$ is the PTM (6.67). It follows from (7.3) that the covariance matrix of the output $K_z(t_1, t_2)$ depends on each of the arguments $t_1$ and $t_2$ separately. This means that the stochastic vector output $z_\infty(t)$ establishes itself as nonstationary stochastic process.

Since from the above-stated properties of the PTM

$$W_{zx}(s, t) = W_{zx}(s, t + T), \quad W_{zx}(-s, t) = W_{zx}(-s, t + T), \tag{7.4}$$

due to (7.3), we find

$$K_z(t_1, t_2) = K_z(t_1 + T, t_2 + T). \tag{7.5}$$

When the covariance matrix of a stochastic process satisfies (7.5), it is called periodic nonstationary. Thus, the quasi-stationary process at the output of the stable standard model $\mathscr{S}_\tau$ is a periodic nonstationary stochastic process.

(3) The scalar function

$$d_z(t) = \text{trace}\, K_z(t, t) \tag{7.6}$$

is called the variance of the quasi-stationary output. Here, the symbol trace, as usual, denotes the trace of the matrix. For $t_1 = t_2 = t$, formula (7.3) gives

$$d_z(t) = \frac{1}{2\pi \mathrm{j}} \int_{-\mathrm{j}\infty}^{\mathrm{j}\infty} \text{trace}\left[W'(-s, t)\Phi_x(s)W(s, t)\right] \mathrm{d}s. \tag{7.7}$$

From (7.4) and (7.7), the important relation

$$d_z(t) = d_z(t + T) \tag{7.8}$$

comes out which means that the variance of the quasi-stationary output of the standard model of a SD system with delay is periodic, and the period coincides with the sampling period of the digital controller. This is a deduction from the fact that in this situation, the SD system, when considered in continuous time, behaves as linear periodic system.

Using that for matrices $A$ and $B$ of appropriate size

$$\text{trace}(AB) = \text{trace}(BA), \tag{7.9}$$

formula (7.7) can be represented by the two equivalent formulae

$$
\begin{aligned}
d_z(t) &= \frac{1}{2\pi \mathrm{j}} \int_{-\mathrm{j}\infty}^{\mathrm{j}\infty} \operatorname{trace}\left[W_{zx}(s,t)W'_{zx}(-s,t)\Phi_x(s)\right] \mathrm{d}s, \\
d_z(t) &= \frac{1}{2\pi \mathrm{j}} \int_{-\mathrm{j}\infty}^{\mathrm{j}\infty} \operatorname{trace}\left[\Phi_x(p)W_{zx}(s,t)W'_{zx}(-s,t)\right] \mathrm{d}s.
\end{aligned}
\tag{7.10}
$$

(4) Consider the important special case, when $\Phi_x(p) = I_l$, i.e., the input signal is uncorrelated vectorial white noise. Then using for this case the notation

$$
d_z(t) \triangleq d_z^0(t), \tag{7.11}
$$

from (7.7) and (7.10), we obtain

$$
\begin{aligned}
d_z^0(t) &= \frac{1}{2\pi \mathrm{j}} \int_{-\mathrm{j}\infty}^{\mathrm{j}\infty} \operatorname{trace}\left[W'_{zx}(-s,t)W_{zx}(s,t)\right] \mathrm{d}s \\
&= \frac{1}{2\pi \mathrm{j}} \int_{-\mathrm{j}\infty}^{\mathrm{j}\infty} \operatorname{trace}\left[W_{zx}(s,t)W'_{zx}(-s,t)\right] \mathrm{d}s.
\end{aligned}
\tag{7.12}
$$

Interchanging herein $s$ and $-s$, we also find

$$
\begin{aligned}
d_z^0(t) &= \frac{1}{2\pi \mathrm{j}} \int_{-\mathrm{j}\infty}^{\mathrm{j}\infty} \operatorname{trace}\left[W'_{zx}(s,t)W_{zx}(-s,t)\right] \mathrm{d}s \\
&= \frac{1}{2\pi \mathrm{j}} \int_{-\mathrm{j}\infty}^{\mathrm{j}\infty} \operatorname{trace}\left[W_{zx}(-s,t)W'_{zx}(s,t)\right] \mathrm{d}s.
\end{aligned}
\tag{7.13}
$$

(5) As examples show, [4], the variation of the variance $d_z(t)$ inside a period for some linear periodic systems can take substantial values. Therefore, in principle, for a complete performance estimation of stable SD systems, when a stationary stochastic signal acts at the input, we have to calculate the function $d_z(t)$.

**Definition 7.1** The problem of calculating the variance $d_z(t)$ as function of the time is called the advanced statistical analysis problem for the SD system.

The direct application of formulae (7.7), (7.10) for the solution of the advanced statistical analysis problem is connected with technical difficulties, because the integrands in formulae (7.7) and (7.10) are transcendent functions of the argument $s$. Below, we describe how to surmount these difficulties. Let the matrix $\Phi_x(s)$ be rational and proper. Then, knowing that for a stable standard model the poles of the PTM $W_{zx}(s,t)$ are located in the left half-plane $\operatorname{Re} s < 0$, from (6.67) we conclude that integrals (7.10) for the standard SD system converge absolutely, and they can be transformed into integrals with finite limits

$$d_z(t) = \frac{T}{2\pi \mathrm{j}} \int_{-\mathrm{j}\frac{\omega}{2}}^{\mathrm{j}\frac{\omega}{2}} \operatorname{trace} \breve{U}_1(T, s, t)\,\mathrm{d}s = \frac{T}{2\pi \mathrm{j}} \int_{-\mathrm{j}\frac{\omega}{2}}^{\mathrm{j}\frac{\omega}{2}} \operatorname{trace} \breve{U}_2(T, s, t)\,\mathrm{d}s, \tag{7.14}$$

where $\omega = \frac{2\pi}{T}$ and

$$\breve{U}_1(T, s, t) \triangleq \frac{1}{T} \sum_{k=-\infty}^{\infty} \Phi(s + k\mathrm{j}\omega) W'_{zx}(-s - k\mathrm{j}\omega, t) W_{zx}(s + k\mathrm{j}\omega, t),$$

$$\breve{U}_2(T, s, t) \triangleq \frac{1}{T} \sum_{k=-\infty}^{\infty} W'_{zx}(-s - k\mathrm{j}\omega, t) W_{zx}(s + k\mathrm{j}\omega, t) \Phi_x(s + k\mathrm{j}\omega). \tag{7.15}$$

As follows from [2, 5], since the matrix $\Phi_x(s)$ is rational, so the matrices (7.15) are rational periodic (RP) in the sense of [3] according to the argument $s$, and de facto they are functions of the argument $\zeta = \mathrm{e}^{-sT}$. Therefore, substituting in integrals (7.14) $\mathrm{e}^{-sT} = \zeta$, formula (7.14) takes the form

$$d_z(t) = \frac{1}{2\pi \mathrm{j}} \oint \operatorname{trace} U_1(T, \zeta, t) \frac{\mathrm{d}\zeta}{\zeta} = \frac{1}{2\pi \mathrm{j}} \oint \operatorname{trace} U_2(T, \zeta, t) \frac{\mathrm{d}\zeta}{\zeta}, \tag{7.16}$$

where the integrals are running over the unit circle in positive direction (counterclockwise), and the following notation was used:

$$U_i(T, \zeta, t) = \breve{U}_i(T, s, t)\,\big|_{\mathrm{e}^{-sT} = \zeta}, \quad (i = 1, 2). \tag{7.17}$$

Integrals (7.16) are elementarily calculated with the help of the residues theorem. Examples for the application of formulae (7.16) for the solution of advanced statistical analysis problems can be found in [4, 6].

## 7.2 Mean Variance and $\mathscr{H}_2$ Norm

(1) Let $d_z(t)$ be the variance of the quasi-stationary output of the stable standard model, calculated with the help of formulae in Sect. 7.1.

**Definition 7.2** The number

$$\bar{d}_z \triangleq \bar{d}_z^0 = \frac{1}{T} \int_0^T d_z(t)\,\mathrm{d}t \tag{7.18}$$

is called the mean variance of the quasi-stationary output.

From (7.18) using (7.10), we find

$$\bar{d}_z = \frac{1}{2\pi \mathrm{j}} \int_{-\mathrm{j}\infty}^{\mathrm{j}\infty} \operatorname{trace}\left[\Phi_x(s) A_{zx}(s)\right] \mathrm{d}s = \frac{1}{2\pi \mathrm{j}} \int_{-\mathrm{j}\infty}^{\mathrm{j}\infty} \operatorname{trace}\left[A_{zx}(s) \Phi_x(s)\right] \mathrm{d}s, \tag{7.19}$$

where

$$A_{zx}(s) = \frac{1}{T} \int_0^T W'_{zx}(-s, t) W_{zx}(s, t)\, \mathrm{d}t. \tag{7.20}$$

In particular, for $\Phi_x(s) = I_l$ from (7.18) and (7.13), we achieve

$$\bar{d}_z^0 \stackrel{\triangle}{=} \frac{1}{2\pi \mathrm{j}} \int_{-\mathrm{j}\infty}^{\mathrm{j}\infty} \operatorname{trace} A_{zx}(s)\, \mathrm{d}s. \tag{7.21}$$

**Definition 7.3** The number

$$\|\mathscr{S}_\tau\|_2 \stackrel{\triangle}{=} +\sqrt{d_z^0} \tag{7.22}$$

is called the $\mathscr{H}_2$ norm of the standard model of the SD system with delay $\mathscr{S}_\tau$.

From (7.21) and (7.22), we find

$$\|\mathscr{S}_\tau\|_2^2 = \frac{1}{2\pi \mathrm{j}} \int_{-\mathrm{j}\infty}^{\mathrm{j}\infty} \operatorname{trace} A_{zx}(s)\, \mathrm{d}s. \tag{7.23}$$

The derived formulae allow to express the $\mathscr{H}_2$ norm by the PTM $W_{zx}(s, t)$. From (6.67), we obtain for $\lambda = s$

$$W_{zx}(s, t) = \varphi_{L_\tau \mu}(T, s, t) \breve{R}_N(s) M_\tau(s) + K_\tau(s). \tag{7.24}$$

Denote

$$\underline{W}_{zx}(s, t) \stackrel{\triangle}{=} W_{zx}(-s, t) = \underline{\varphi}_{L_\tau \mu}(T, s, t) \underline{\breve{R}}_N(s) \underline{M}_\tau(s) + \underline{K}_\tau(s). \tag{7.25}$$

In (7.25) and further on, in order to shorten the notation, we will write for any function $f(s)$

$$\underline{f}(s) \stackrel{\triangle}{=} f(-s). \tag{7.26}$$

Transposing both sides of (7.25), we find

$$\underline{W}'_{zx}(s, t) = \underline{M}'_\tau(s) \underline{\breve{R}}'_N(s) \underline{\varphi}'_{L_\tau \mu}(T, s, t) + \underline{K}'_\tau(s). \tag{7.27}$$

Multiplying relations (7.24) and (7.27), we obtain

$$
\begin{aligned}
W'_{zx}(-s,t)\ \ W_{zx}(s,t) &= \underline{K'_\tau}(s)K_\tau(s) \\
&+ \underline{M'_\tau}(s)\underline{\check{R}'_N}(s)\varphi'_{L_\tau\mu}(T,-s,t)\varphi_{L_\tau\mu}(T,s,t)\check{R}_N(s)M_\tau(s) \\
&+ \underline{M'_\tau}(s)\underline{\check{R}'_N}(s)\varphi'_{L_\tau\mu}(T,-s,t)K_\tau(s) + \underline{K'_\tau}(s)\varphi_{L_\tau\mu}(T,s,t)\check{R}_N(s)M_\tau(s).
\end{aligned}
\tag{7.28}
$$

Substituting (7.28) in (7.20) results in

$$
\begin{aligned}
A_{zx}(s) = {} & \underline{M'_\tau}(s)\underline{\check{R}'_N}(s)\check{D}_L(s)\check{R}_N(s)M_\tau(s) + \underline{K'_\tau}(s)K(s) \\
& + \underline{M'_\tau}(s)\underline{\check{R}'_N}(s)\underline{Q'_L}(s)K_\tau(s) + \underline{K'_\tau}(s)Q_L(s)\check{R}_N(s)M_\tau(s),
\end{aligned}
\tag{7.29}
$$

where

$$
\begin{aligned}
\check{D}_L(s) &\stackrel{\triangle}{=} \frac{1}{T}\int_0^T \varphi'_{L_\tau\mu}(T,-s,t)\varphi_{L_\tau\mu}(T,s,t)\,\mathrm{d}t, \\
Q_L(s) &\stackrel{\triangle}{=} \frac{1}{T}\int_0^T \varphi_{L_\tau\mu}(T,s,t)\,\mathrm{d}t, \\
\underline{Q'_L}(s) &\stackrel{\triangle}{=} \frac{1}{T}\int_0^T \varphi'_{L_\tau\mu}(T,-s,t)\,\mathrm{d}t.
\end{aligned}
\tag{7.30}
$$

From (7.29) and (7.23), we achieve the wanted expressions for the quantity $\|\mathscr{S}_\tau\|_2^2$

$$
\begin{aligned}
\|\mathscr{S}_\tau\|_2^2 = \frac{1}{2\pi \mathrm{j}}\int_{-\mathrm{j}\infty}^{\mathrm{j}\infty} & \operatorname{trace}\Big[\underline{M'_\tau}(s)\underline{\check{R}'_N}(s)\check{D}_L(s)\check{R}_N(s)M_\tau(s) \\
& + \underline{M'_\tau}(s)\underline{\check{R}'_N}(s)\underline{Q'_L}(s)K_\tau(s) + \underline{K'_\tau}(s)Q_L(s)\check{R}_N(s)M_\tau(s) \\
& + \underline{K'_\tau}(s)K_\tau(s)\Big]\,\mathrm{d}s.
\end{aligned}
\tag{7.31}
$$

Applying the relations from Chap. 6, the achieved formula can be essentially simplified. Indeed, since, due to (6.17),

$$
\begin{aligned}
K_\tau(s) &= K(s)\mathrm{e}^{-s\tau_K}, \quad L_\tau(s) = L(s)\mathrm{e}^{-s\tau_L}, \\
M_\tau(s) &= M(s)\mathrm{e}^{-s\tau_M}, \quad N_\tau(s) = N(s)\mathrm{e}^{-s\tau_N},
\end{aligned}
\tag{7.32}
$$

so we find

$$
\underline{K'_\tau}(s) = \underline{K'}(s)\mathrm{e}^{s\tau_K}, \quad \underline{M'_\tau}(s) = \underline{M'}(s)\mathrm{e}^{s\tau_M}. \tag{7.33}
$$

Using

$$
\begin{aligned}
\varphi_{L_\tau\mu}(T,s,t) &= \frac{\mathrm{e}^{-s\tau_L}}{T}\sum_{k=-\infty}^{\infty} L(s+kj\omega)\mu(s+kj\omega)\mathrm{e}^{kj\omega(t-\tau_L)},\\
\varphi'_{L_\tau\mu}(T,-s,t) &= \frac{\mathrm{e}^{s\tau_L}}{T}\sum_{k=-\infty}^{\infty} \mu'(-s+kj\omega)L'(-s+kj\omega)\mathrm{e}^{kj\omega(t-\tau_L)},
\end{aligned}
\tag{7.34}
$$

we obtain after termwise integration

$$
\breve{D}_L(s) \triangleq \frac{1}{T}\int_0^T \varphi'_{L_\tau\mu}(T,-s,t)\varphi_{L_\tau\mu}(T,s,t)\,\mathrm{d}t = \frac{1}{T}\breve{D}_{\underline{\mu'L'}L\mu}(T,s,0), \tag{7.35}
$$

where

$$
\breve{D}_{\underline{\mu'L'}L\mu}(T,s,0) \triangleq \frac{1}{T}\sum_{k=-\infty}^{\infty} \mu'(-s-kj\omega)L'(-s-kj\omega)L(s+kj\omega)\mu(s+kj\omega). \tag{7.36}
$$

Moreover, applying (7.30) and (7.34), we obtain

$$
Q_L(s) = \frac{1}{T}\mathrm{e}^{-s\tau_L}L(s)\mu(s), \quad Q'_L(-s) = \frac{1}{T}\mathrm{e}^{s\tau_L}\mu'(-s)L'(-s). \tag{7.37}
$$

Considering (7.32)–(7.36) and (6.19), from (7.31), we find

$$
\begin{aligned}
T\|\mathscr{S}_\tau\|_2^2 = \frac{1}{2\pi j}\int_{-j\infty}^{j\infty} &\operatorname{trace}\Big[M'(-s)\breve{R}'_N(-s)\breve{D}_{\underline{\mu'L'}L\mu}(T,s,0)\breve{R}_N(s)M(s) +\\
&+ M'(-s)\breve{R}'_N(-s)\mu'(-s)L'(-s)K(s)\mathrm{e}^{s\tau_N} +\\
&+ K'(-s)L(s)\mu(s)\breve{R}_N(s)M(s)\mathrm{e}^{-s\tau_N} + TK'(-s)K(s)\Big]\,\mathrm{d}s.
\end{aligned}
\tag{7.38}
$$

Formula (7.38) suggests that the $\mathscr{H}_2$ norm of the standard model (6.12) does only depend on the delay $\tau_2 = \tau_N$ in the closed loop, and it does not depend on the delays $\tau_1$ and $\tau_3$. Notice, however, that all matrices appearing in (7.38), including the matrix $\breve{R}_N(s)$, do not depend on the choice of the digital controller, and they are exclusively determined by the properties of matrix $\bar{W}_0(s)$ (6.23). For a given matrix $\bar{W}_0(s)$ and delay $\tau_2 = \tau_N$, and also Eqs. (6.8)–(6.10) of the digital controller, expression (7.38) can be considered as a functional defined over the set of transfer matrices of the discrete stabilizing controllers $W_d(\zeta)$. In this connection, the following $\mathscr{H}_2$ optimization problem can be formulated.

**$\mathscr{H}_2$ optimization problem**: Let the matrix $\bar{W}_0(s)$ (6.23), the delay $\tau_2$, the sampling period $T$, and the equations of the digital controller (6.10) be given. Find the transfer matrix of a causal stabilizing controller yielding the minimum of the functional (7.38).

## 7.3 Representation of the $\mathscr{H}_2$ Norm by a System Function

(1) In this section, we construct a representation of the $\mathscr{H}_2$ norm of the standard model by the system function, which can be held from expression (6.199), when $\lambda$ is substituted by $s$

$$W_{zx}(s,t) = \psi(s,t)\breve{\theta}(s)\xi(s) + \eta(s,t), \tag{7.39}$$

from which we obtain

$$W'_{zx}(-s,t) = \xi'(-s)\breve{\theta}'(-s)\psi'(-s,t) + \eta'(-s,t). \tag{7.40}$$

Multiplying these expressions, we find

$$\begin{aligned} W'_{zx}(-s,t)W_{zx}(s,t) = \\ \eta'(-s,t)\eta(s,t) + \xi'(-s)\breve{\theta}'(-s)\psi'(-s,t)\psi(s,t)\breve{\theta}(s)\xi(s) \\ + \xi'(-s)\breve{\theta}'(-s)\psi'(-s,t)\eta(s,t) + \eta'(-s,t)\psi(s,t)\breve{\theta}(s)\xi(s). \end{aligned} \tag{7.41}$$

Inserting this expression into (7.20) yields

$$TA_{zx}(s) = \int_0^T W'_{zx}(-s,t)W_{zx}(s,t)\,\mathrm{d}t = g_1(s) - g_2(s) - g_3(s) + g_4(s), \tag{7.42}$$

where

$$\begin{aligned} g_1(s) &\triangleq \xi'(-s)\breve{\theta}'(-s)\int_0^T \psi'(-s,t)\psi(s,t)\,\mathrm{d}t\,\breve{\theta}(s)\xi(s), \\ g_2(s) &\triangleq -\xi'(-s)\breve{\theta}'(-s)\int_0^T \psi'(-s,t)\eta(s,t)\,\mathrm{d}t, \\ g_3(s) &\triangleq -\int_0^T \eta'(-s,t)\psi(s,t)\,\mathrm{d}t\,\breve{\theta}(s)\xi(s), \\ g_4(s) &\triangleq \int_0^T \eta'(-s,t)\eta(s,t)\,\mathrm{d}t. \end{aligned} \tag{7.43}$$

(2) Let us calculate the matrix (7.43).

Introduce

$$\breve{A}_L(s) \triangleq \int_0^T \psi'(-s,t)\psi(s,t)\,\mathrm{d}t. \tag{7.44}$$

Since from (6.202), using $\tau_2 + \tau_3 = \tau_L$, we obtain

$$\psi(s,t) = -\mathrm{e}^{-s\tau_L}\varphi_{L\mu}(T,s,t-\tau_L)\breve{a}_r(s),$$
$$\psi'(-s,t) = -\mathrm{e}^{s\tau_L}\breve{a}_r'(-s)\varphi_{L\mu}'(T,-s,t-\tau_L), \tag{7.45}$$

so from (7.44), we arrive at

$$\breve{A}_L(s) = \breve{a}_r'(-s)\int_0^T \varphi_{L\mu}'(T,-s,t-\tau_L)\varphi_{L\mu}(T,s,t-\tau_L)\,\mathrm{d}t\,\breve{a}_r(s). \tag{7.46}$$

Notice, that

$$\varphi_{L\mu}(T,s,t-\tau_L) = \frac{1}{T}\sum_{k=-\infty}^{\infty} L(s+k\mathrm{j}\omega)\mu(s+k\mathrm{j}\omega)\mathrm{e}^{k\mathrm{j}\omega(t-\tau_L)},$$
$$\varphi_{L\mu}'(T,-s,t-\tau_L) = \frac{1}{T}\sum_{k=-\infty}^{\infty} \mu'(-s+k\mathrm{j}\omega)L'(-s+k\mathrm{j}\omega)\mathrm{e}^{k\mathrm{j}\omega(t-\tau_L)}, \tag{7.47}$$

so, after termwise integration, we obtain

$$\int_0^T \varphi_{L\mu}'(T,-s,t-\tau_L)\varphi_{L\mu}(T,s,t-\tau_L)\,\mathrm{d}t = \breve{D}_{\underline{\mu'L'}L\mu}(T,s,0), \tag{7.48}$$

where as earlier

$$\breve{D}_{\underline{\mu'L'}L\mu}(T,s,0) = \frac{1}{T}\sum_{k=-\infty}^{\infty} \mu'(-s-k\mathrm{j}\omega)L'(-s-k\mathrm{j}\omega)L(s+k\mathrm{j}\omega)\mu(s+k\mathrm{j}\omega). \tag{7.49}$$

With the help of (7.48), relation (7.46) takes the form

$$\breve{A}_L(s) = \breve{a}_r'(-s)\breve{D}_{\underline{\mu'L'}L\mu}(T,s,0)\breve{a}_r(s). \tag{7.50}$$

Applying this relation to the first formula of (7.43), we find

$$g_1(s) = \xi'(-s)\breve{\theta}'(-s)\breve{a}_r'(-s)\breve{D}_{\underline{\mu'L'}L\mu}(T,s,0)\breve{a}_r(s)\breve{\theta}(s)\xi(s). \tag{7.51}$$

(3) In order to calculate the matrices $g_2(s)$ and $g_3(s)$, denote

$$Q(s) \stackrel{\triangle}{=} -\int_0^T \eta'(-s,t)\psi(s,t)\,\mathrm{d}t. \tag{7.52}$$

Then

$$Q'(-s) = -\int_0^T \psi'(-s,t)\eta(s,t)\,\mathrm{d}t. \tag{7.53}$$

Using relations (6.200), we obtain

$$-\psi'(-s,t)\eta(s,t) = \left[\mathrm{e}^{s\tau_L}\breve{a}_r'(-s)\varphi_{L\mu}'(T,-s,t-\tau_L)\right]\times$$
$$\left[\mathrm{e}^{-s\tau_L}\varphi_{L\mu}(T,s,t-\tau_L)\breve{\beta}_{r0}(s)\breve{a}_l(s)M(s)\mathrm{e}^{-s\tau_M} + K(s)\mathrm{e}^{-s\tau_K}\right], \tag{7.54}$$

and after expanding, we achieve

$$\begin{aligned}-\psi'(-s,t)\eta(s,t) &= \breve{a}_r'(-s)\varphi_{L\mu}'(T,-s,t-\tau_L)\varphi_{L\mu}(T,s,t-\tau_L)\breve{\beta}_{r0}(s)\breve{a}_l(s)M(s)\mathrm{e}^{-s\tau_M}\\ &\quad + \breve{a}_r'(-s)\varphi_{L\mu}'(T,-s,t-\tau_L)K(s)\mathrm{e}^{s(\tau_L-\tau_K)}.\end{aligned} \tag{7.55}$$

Inserting the last expression into the right side of (7.53), we obtain

$$\begin{aligned}Q'(-s) &= \breve{a}_r'(-s)\int_0^T \varphi_{L\mu}'(T,-s,t-\tau_L)\varphi_{L\mu}'(T,s,t-\tau_L)\,\mathrm{d}t\;\breve{\beta}_{r0}(s)\breve{a}_l(s)M(s)\mathrm{e}^{-s\tau_M}\\ &\quad +\breve{a}_r'(-s)\int_0^T \varphi_{L\mu}'(T,-s,t-\tau_L)\,\mathrm{d}t\;K(s)\mathrm{e}^{s(\tau_L-\tau_K)}.\end{aligned} \tag{7.56}$$

Using herein (7.47), (7.48), we find

$$\begin{aligned}Q'(-s) &= \breve{a}_r'(-s)\breve{D}_{\underline{\mu' L'}L\mu}(T,s,0)\breve{\beta}_{r0}(s)\breve{a}_l(s)M(s)\mathrm{e}^{-s\tau_M}\\ &\quad +\breve{a}_r'(-s)\mu'(-s)L'(-s)K(s)\mathrm{e}^{s(\tau_L-\tau_K)}.\end{aligned} \tag{7.57}$$

Substituting here $s$ by $-s$ and after transposition

$$\begin{aligned}Q(s) &= M'(-s)\breve{a}_l'(-s)\breve{\beta}_{r0}'(-s)\breve{D}_{\underline{\mu' L'}L\mu}(T,s,0)\breve{a}_r(s)\mathrm{e}^{s\tau_M}\\ &\quad + K'(-s)L(s)\mu(s)\breve{a}_r(s)\mathrm{e}^{-s(\tau_L-\tau_K)}\end{aligned} \tag{7.58}$$

is achieved, when considering that

$$\breve{D}_{\underline{\mu' L'}L\mu}'(T,-s,0) = \breve{D}_{\underline{\mu' L'}L\mu}(T,s,0). \tag{7.59}$$

Finally, (7.52), (7.53), and (7.43) yield

$$\begin{aligned}g_2(s) &= \xi'(-s)\breve{\theta}'(-s)Q'(-s),\\ g_3(s) &= Q(s)\breve{\theta}(s)\xi(s).\end{aligned} \tag{7.60}$$

(4) On the basis of the achieved formulae, a further calculation can be obtained from (7.20), (7.42)

$$\|\mathscr{S}_\tau\|_2^2 = \frac{1}{2\pi \mathrm{j}} \int_{-\mathrm{j}\infty}^{\mathrm{j}\infty} \operatorname{trace} A_{zx}(s)\, \mathrm{d}s = J_1 + J_2, \tag{7.61}$$

where

$$\begin{aligned} TJ_1 = \frac{1}{2\pi \mathrm{j}} \int_{-\mathrm{j}\infty}^{\mathrm{j}\infty} & \operatorname{trace} \Big[ \xi'(-s)\breve{\theta}'(-s)\breve{A}_L(s)\breve{\theta}(s)\xi'(s) - \\ & - \xi'(-s)\breve{\theta}'(-s)Q'(-s) - Q(s)\breve{\theta}(s)\xi(s) \Big]\, \mathrm{d}s, \end{aligned} \tag{7.62}$$

$$TJ_2 = \frac{1}{2\pi \mathrm{j}} \int_{-\mathrm{j}\infty}^{\mathrm{j}\infty} g_4(s)\, \mathrm{d}s.$$

Obviously, it is not necessary to find an expression for the integral $J_2$, because it does not depend on the selection of the controller, that's why it is not considered in the further transformations. Hence, the $\mathscr{H}_2$ optimization problem leads to the minimization of the quantity $J_1$, which using (7.9) can be written in the form

$$\begin{aligned} J_1 = \frac{1}{2\pi \mathrm{j}} \int_{-\mathrm{j}\infty}^{\mathrm{j}\infty} \frac{1}{T} & \operatorname{trace} \Big[ \breve{\theta}'(-s)\breve{A}_L(s)\breve{\theta}(s)\xi(s)\xi'(-s) - \\ & - \breve{\theta}'(-s)Q'(-s)\xi'(-s) - \xi(p)Q(s)\breve{\theta}(s) \Big]\, \mathrm{d}s, \end{aligned} \tag{7.63}$$

where the matrices $\breve{A}_L(s)$ and $\xi(s)$ under the integral satisfy

$$\begin{aligned} \breve{A}_L(s) &= \breve{a}_r'(-s)\breve{D}_{\mu'\underline{L'}L\mu}(T, s, 0)\breve{a}_r(s), \\ \xi(s) &= \breve{a}_l(s)M(s)\mathrm{e}^{-s\tau_M}, \\ \xi'(-s) &= M'(-s)\breve{a}_l'(-s)\mathrm{e}^{s\tau_M}. \end{aligned} \tag{7.64}$$

Thus, using (7.58)

$$\begin{aligned} \xi(s)Q(s) = \breve{a}_l(s)M(s)\mathrm{e}^{-s\tau_M} & \Big[ \mathrm{e}^{s\tau_M} M'(-s)\breve{a}_l'(-s)\breve{\beta}_{r0}'(-s) \\ & + \breve{D}_{\mu'\underline{L'}L\mu}(T, s, 0)\breve{a}_r(s) + K'(-s)L(s)\mu(s)\breve{a}_r(s)\mathrm{e}^{-s(\tau_L - \tau_K)} \Big]. \end{aligned} \tag{7.65}$$

Regarding that, due to (6.19),

$$\tau_M + \tau_L - \tau_K = \tau_2 = \tau_N, \tag{7.66}$$

after expanding, expression (7.65) is written in the form

$$\begin{aligned}\xi(s)Q(s) &= \breve{a}_l(s)M(s)M'(-s)\breve{a}_l'(-s)\breve{\beta}_{r0}'(-s)\breve{D}_{\mu'\underline{L'}L\mu}(T,s,0)\breve{a}_r(s)\\ &\quad+\breve{a}_l(s)M(s)K'(-s)L(s)\mu(s)\breve{a}_r(s)\mathrm{e}^{-s\tau_N},\end{aligned} \tag{7.67}$$

which with the help of (7.59) yields

$$\begin{aligned}Q'(-s)\xi'(-s) &= \breve{a}_r(-s)\breve{D}_{\mu'\underline{L'}L\mu}(T,s,0)\breve{\beta}_{r0}(s)\breve{a}_l'(s)M(s)M'(-s)\breve{a}_l'(-s)\\ &\quad+\breve{a}_r'(-s)\mu'(-s)L'(-s)K(s)M'(-s)\breve{a}_l'(-s)\mathrm{e}^{s\tau_N}.\end{aligned} \tag{7.68}$$

(5) Under the suppositions taken above, integral (7.63) converges absolutely, and by standard methods, it can be written as integral with finite integration limits

$$J_1 = \frac{1}{2\pi\mathrm{j}}\int_{-\mathrm{j}\frac{\omega}{2}}^{\mathrm{j}\frac{\omega}{2}} \operatorname{trace}\left[\breve{\theta}'(-s)\breve{A}_L(s)\breve{\theta}(s)\breve{A}_M(s) - \breve{\theta}'(-s)\breve{C}'(-s) - \breve{C}(s)\breve{\theta}(s)\right]\mathrm{d}s, \tag{7.69}$$

where in addition to (7.50)

$$\begin{aligned}\breve{A}_M(s) &\triangleq \frac{1}{T}\sum_{k=-\infty}^{\infty}\xi(s+k\mathrm{j}\omega)\xi'(-s-k\mathrm{j}\omega)\\ &= \frac{1}{T}\sum_{k=-\infty}^{\infty}\breve{a}_l(s+k\mathrm{j}\omega)M(s+k\mathrm{j}\omega)M'(-s-k\mathrm{j}\omega)\breve{a}_l'(-s-k\mathrm{j}\omega)\\ &= \breve{a}_l(s)\frac{1}{T}\sum_{k=-\infty}^{\infty}M(s+k\mathrm{j}\omega)M'(-s-k\mathrm{j}\omega)\breve{a}_l'(-s)\\ &= \breve{a}_l(s)\breve{D}_{M\underline{M'}}(T,s,0)\breve{a}_l'(-s).\end{aligned} \tag{7.70}$$

Moreover, using (7.67), we obtain

$$\begin{aligned}\breve{C}(s) &\triangleq \frac{1}{T}\sum_{k=-\infty}^{\infty}\xi(s+k\mathrm{j}\omega)Q(s+k\mathrm{j}\omega)\\ &= \breve{A}_M(p)\breve{\beta}_{r0}'(-s)\breve{D}_{\mu'\underline{L'}L\mu}(T,s,0)\breve{a}_r(s)+\breve{a}_l(s)\breve{D}_{M\underline{K'}L\mu}(T,s,-\tau_N)\breve{a}_r(s),\end{aligned} \tag{7.71}$$

where

$$\breve{D}_{M\underline{K}'L\mu}(T,s,-\tau_N) = \\ \frac{1}{T}\sum_{k=-\infty}^{\infty} M(s+k\mathrm{j}\omega)K'(-s-k\mathrm{j}\omega)L(s+k\mathrm{j}\omega)\mu(s+k\mathrm{j}\omega)\mathrm{e}^{-(s+k\mathrm{j}\omega)\tau_N}. \tag{7.72}$$

After transposition of (7.71) and substituting $s$ by $-s$, we obtain

$$\begin{aligned}\breve{C}'(-s) = \breve{a}_r'(-s)\breve{D}_{\underline{\mu}'\underline{L}'L\mu}(T,s,0)\breve{\beta}_{r0}(s)\breve{a}_l(s)\breve{A}_M(s)\breve{a}_l'(-s)\\ +\breve{a}_r'(-s)D_{\underline{\mu}'\underline{L}'K\underline{M}'}(T,s,\tau_N)\breve{a}_l'(-s).\end{aligned} \tag{7.73}$$

Notice, that

$$\breve{A}_M'(-s) = \breve{A}_M(s), \tag{7.74}$$

and, moreover,

$$\breve{D}_{M\underline{K}'L\mu}'(T,-s,-\tau_N) = \breve{D}_{\underline{\mu}'\underline{L}'K\underline{M}'}(T,s,\tau_N). \tag{7.75}$$

The last formula can be proven as follows. Let us have for the matrix $A(s)$

$$\breve{D}_A(T,s,t) = \frac{1}{T}\sum_{k=-\infty}^{\infty} A(s+k\mathrm{j}\omega)\mathrm{e}^{(s+k\mathrm{j}\omega)t}. \tag{7.76}$$

Then

$$\breve{D}'_A(T,-s,t) = \frac{1}{T}\sum_{k=-\infty}^{\infty} A'(-s+k\mathrm{j}\omega)\mathrm{e}^{(-s+k\mathrm{j}\omega)t}. \tag{7.77}$$

Substituting herein $k$ by $-k$, we directly find

$$\breve{D}_A'(T,-s,t) = \breve{D}_{\underline{A}'}(T,s,-t). \tag{7.78}$$

This relation merges into (7.75) for $A(s) = M(s)K'(-s)L(s)\mu(s),\ t = -\tau_N$.

(6) Assuming in (7.69) $\mathrm{e}^{-sT} = \zeta$, we obtain the integral in the form

$$TJ_1 = \frac{1}{2\pi\mathrm{j}}\oint \operatorname{trace}\left[\hat{\theta}(\zeta)A_L(\zeta)\theta(\zeta)A_M(\zeta) - \hat{\theta}(\zeta)\hat{C}(\zeta) - C(\zeta)\theta(\zeta)\right]\frac{\mathrm{d}\zeta}{\zeta}, \tag{7.79}$$

where the integration is taken along the circle $|\zeta| = 1$, and the notation

$$\hat{F}(\zeta) \stackrel{\triangle}{=} F'(\zeta^{-1}) \tag{7.80}$$

has been applied. The matrices $A_L(\zeta)$, $A_M(\zeta)$, and $C(\zeta)$, appearing under integral (7.79), are determined by formulae, generated from (7.64), (7.70), and (7.71), when substituting $e^{-sT}$ by $\zeta$

$$\begin{aligned} A_L(\zeta) &= \hat{a}_r(\zeta) D_{\underline{\mu' L'} L\mu}(T, \zeta, 0) a_r(\zeta), \\ A_M(\zeta) &= a_l(\zeta) D_{M\underline{M'}}(T, \zeta, 0) \hat{a}_l(\zeta), \\ C(\zeta) &= A_M(\zeta) \hat{\beta}_{r0}(\zeta) D_{\underline{\mu' L'} L\mu}(T, \zeta, 0) a_r(\zeta) + a_l(\zeta) D_{M\underline{K'} L\mu}(T, \zeta, -\tau_N) a_r(\zeta). \end{aligned} \tag{7.81}$$

## 7.4 Properties of Matrix $A_M(\zeta)$

(1) For studying the properties of the matrix $A_M(\zeta)$ in (7.81), we will state some preliminary relations, which are necessary for the following.

At first, notice that the representation of the $n \times \ell$ matrix $M(s)$

$$M(s) = C_2(sI_\chi - A)^{-1} B_1, \tag{7.82}$$

given in (6.5), normally is not minimal, e.g., when the pair $(A, B_1)$ is noncontrollable. If this happens, there exists a minimal representation

$$M(s) = C_M(sI_\alpha - A_M)^{-1} B_M, \tag{7.83}$$

where $\alpha < \chi$. In what follows, we always assume

$$n \le \alpha, \quad \ell \le \alpha, \tag{7.84}$$

which is normally fulfilled in applications.

**Lemma 7.1** *The quantity*

$$\delta_M(s) \triangleq \frac{\det(sI_\chi - A)}{\det(sI_\alpha - A_M)} \tag{7.85}$$

*is a polynomial.*

*Proof* Assume the ILMFD

$$C_2(sI_\chi - A)^{-1} = g_l^{-1}(s) d_l(s), \quad C_M(sI_\alpha - A_M)^{-1} = g_M^{-1}(s) d_M(s), \tag{7.86}$$

where

$$\det g_l(s) \sim \det(sI_\chi - A), \quad \det g_M(s) \sim \det(sI_\alpha - A_M). \tag{7.87}$$

Herein, the representation

$$M(s) = g_M^{-1}(s)\,[d_M(s)B_M] \tag{7.88}$$

is an ILMFD. From (7.82) and (7.86), we conclude that the product

$$g_l(s)M(s) = d_l(s)B_1$$

is a polynomial matrix. Since the right side of (7.88) is an ILMFD, so from the last equation, we achieve that

$$g_l(s) = g_{M1}(s)g_M(s), \tag{7.89}$$

where $g_{M1}(s)$ is a polynomial matrix. Hence, using (7.87)

$$\det g_{M1}(s) = \frac{\det g_l(s)}{\det g_M(s)} \sim \delta_M(s) = \frac{\det(sI_\chi - A)}{\det(sI_\alpha - A_M)}. \tag{7.90}$$

■

Lemma 7.1 allows us to conclude that, when we have the expansion

$$\det(sI_\chi - A) = (s - p_1)^{\nu_1} \cdots (s - p_\rho)^{\nu_\rho}, \quad \nu_1 + \cdots + \nu_\rho = \chi, \tag{7.91}$$

where all $p_i$ are different, then

$$\det(sI_\alpha - A_M) = (s - p_1)^{\mu_1} \cdots (s - p_\rho)^{\mu_\rho}, \quad \mu + \cdots + \mu_\rho = \alpha, \tag{7.92}$$

where $0 \le \mu_i \le \nu_i$. Moreover, (7.85), (7.91), and (7.92) imply

$$\delta_M(s) = (s - p_1)^{\nu_1 - \mu_1} \cdots (s - p_\rho)^{\nu_\rho - \mu_\rho}, \quad \deg \delta_M(s) = \chi - \alpha. \tag{7.93}$$

(2) Introduce the DLT of the matrix $M(s)$

$$\breve{D}_M(T, s, t) = \frac{1}{T} \sum_{k=-\infty}^{\infty} M(s + kj\omega)e^{(s+kj\omega)t}. \tag{7.94}$$

The relations of Chap. 5 provide for $0 < t < T$

$$\breve{D}_M(T, s, t) = \breve{\mathscr{D}}_M(T, s, t) = C_M(I_\alpha - e^{-sT}e^{A_M T})^{-1}e^{A_M t}B_M. \tag{7.95}$$

When substituting $e^{-sT}$ by $\zeta$, we find

$$D_M(T, \zeta, t) = \mathscr{D}_M(T, \zeta, t) = C_M\left(I_\alpha - \zeta e^{A_M T}\right)^{-1} e^{A_M t}B_M, \quad 0 < t < T. \tag{7.96}$$

**Lemma 7.2** *Let the sampling period T be non-pathological, and assume the ILMFD*

$$C_2(I_\chi - \zeta e^{AT}) = a_l^{-1}(\zeta)b_l(\zeta), \quad C_M(I_\alpha - \zeta e^{A_M T})^{-1} = a_M^{-1}(\zeta)b_M(\zeta). \tag{7.97}$$

*Then the following claims hold:*

*(i) The relation*

$$\Delta_M(\zeta) \triangleq \frac{\det(I_\chi - \zeta \mathrm{e}^{AT})}{\det(I_\alpha - \zeta \mathrm{e}^{A_M T})} \tag{7.98}$$

*defines a polynomial.*

*(ii) If expansion* (7.93) *is true, then*

$$\Delta_M(\zeta) = (1 - \zeta \mathrm{e}^{p_1 T})^{\nu_1 - \mu_1} \cdots (1 - \zeta \mathrm{e}^{p_\rho T})^{\nu_\rho - \mu_\rho}. \tag{7.99}$$

*Proof* (i) Owing to Lemma 7.1, since the sampling period $T$ is non-pathological, according to the relations between the eigenvalues of the matrix $A$, also the sampling period is non-pathological with respect to the eigenvalues of the matrix $A_M$. Due to Theorem 5.7, the realization on the right side of (7.96) is minimal for all $t$, and consequently the right side of

$$\mathscr{D}_M(T, \zeta, t) = a_M{}^{-1}(\zeta) \left[ b_M(\zeta) \mathrm{e}^{A_M t} B_M \right] \triangleq a_M{}^{-1}(\zeta) b_M(\zeta, t) \tag{7.100}$$

is an ILMFD. Hence,

$$\det a_M(\zeta) \sim \det(I_\alpha - \zeta \mathrm{e}^{A_M T}). \tag{7.101}$$

On the other side, using (7.82), we can build the representation

$$\mathscr{D}_M(T, \zeta, t) = C_2 (I_\chi - \zeta \mathrm{e}^{AT})^{-1} \mathrm{e}^{At} B_1, \tag{7.102}$$

from which, regarding the first relation in (7.97), we find the LMFD

$$\mathscr{D}_M(T, \zeta, t) = a_l{}^{-1}(\zeta) \left[ b_l(\zeta) \mathrm{e}^{At} B_1 \right], \quad 0 < t < T, \tag{7.103}$$

wherein

$$\det a_l(\zeta) \sim \det(I_\chi - \zeta \mathrm{e}^{AT}). \tag{7.104}$$

Since the right part of (7.100) is an ILMFD, so comparing (7.100) and (7.103) proves that

$$a_l(\zeta) = a_{M1}(\zeta) a_M(\zeta), \tag{7.105}$$

where $a_{M1}(\zeta)$ is a polynomial matrix. Now, owing to (7.101) and (7.104), we conclude that

$$\det a_{M1}(\zeta) = \frac{\det a_l(\zeta)}{\det a_M(\zeta)} \sim \Delta_{M1}(\zeta) = \frac{\det(I_\chi - \zeta \mathrm{e}^{AT})}{\det(I_\alpha - \zeta \mathrm{e}^{A_M T})} \tag{7.106}$$

is a polynomial.

(ii) From (5.31) and (7.92), we obtain

$$\det(zI_\alpha - e^{A_M T}) = (z - e^{p_1 T})^{\mu_1} \cdots (z - e^{p_\rho T})^{\mu_\rho}. \tag{7.107}$$

Moreover, from

$$I_\alpha - \zeta e^{A_M T} = \operatorname{rec}(zI_\alpha - e^{A_M T}), \tag{7.108}$$

owing to (7.107), it follows

$$\det(I_\alpha - \zeta e^{A_M T}) \sim rec \det(zI_\alpha - e^{A_M T}) \sim \kappa(1 - \zeta e^{p_1 T})^{\mu_1} \dots (1 - \zeta e^{p_\rho T})^{\mu_\rho}, \tag{7.109}$$

where $\kappa$ is a constant coefficient. Since from (7.98), we find

$$\Delta_M(\zeta)_{|\zeta=0} = 1, \tag{7.110}$$

so $\kappa = 1$. Thus,

$$\det(I_\alpha - \zeta e^{A_M T}) = (1 - \zeta e^{p_1 T})^{\mu_1} \dots (1 - \zeta e^{p_\rho T})^{\mu_\rho}. \tag{7.111}$$

Analogously, we state

$$\det(I_\chi - \zeta e^{AT}) = (1 - \zeta e^{p_1 T})^{\nu_1} \dots (1 - \zeta e^{p_\rho T})^{\nu_\rho}. \tag{7.112}$$

The last two formulae together with (7.106) prove (ii). ∎

(3) In the following considerations, a rational matrix $F(\zeta)$ is called quasi-polynomial, if it possesses poles only at $\zeta = 0$. A quadratic quasi-polynomial $F(\zeta)$ is called symmetric, when

$$F'(\zeta^{-1}) \stackrel{\triangle}{=} \hat{F}(\zeta) = F(\zeta). \tag{7.113}$$

A symmetric quasi-polynomial $F(\zeta)$ is called nonnegative (positive) on the circle $|\zeta| = 1$, if for $|\zeta| = 1$

$$xF(\zeta)x^* \geq 0, \quad (x^* F(\zeta)x > 0), \tag{7.114}$$

for any vector $x$ of admissible size, where

$$x^* \stackrel{\triangle}{=} \bar{x}' \tag{7.115}$$

denotes the complex conjugate transposed vector.

Introduce

$$A_{M1}(\zeta) \stackrel{\triangle}{=} a_M(\zeta) D_{M\underline{M}'}(T, \zeta, 0) \hat{a}_M(\zeta), \tag{7.116}$$

where

$$D_{M\underline{M}'}(T,\zeta,0)=\left[\frac{1}{T}\sum_{k=-\infty}^{\infty}M(s+k\mathrm{j}\omega)M'(-s-k\mathrm{j}\omega)\right]_{\mathrm{e}^{-sT}=\zeta}. \tag{7.117}$$

Expression (7.117) makes sense because $\operatorname{exc}[M(s)M'(-s)]>1$.

**Lemma 7.3** *Matrix* (7.116) *is a symmetric quasi-polynomial, which is nonnegative on the unit circle.*

*Proof* At first, we prove that

$$\breve{D}_{M\underline{M}'}(T,s,0)=\int_0^T\breve{D}_M(T,s,t)\breve{D}'_M(T,-s,t)\,\mathrm{d}t. \tag{7.118}$$

Herein

$$\begin{aligned}\breve{D}_M(T,s,t)&=\frac{1}{T}\sum_{k=-\infty}^{\infty}M(s+k\mathrm{j}\omega)\mathrm{e}^{(s+k\mathrm{j}\omega)t},\\ \breve{D}'_M(T,-s,t)&=\frac{1}{T}\sum_{k=-\infty}^{\infty}M'(-s+k\mathrm{j}\omega)\mathrm{e}^{(-s+k\mathrm{j}\omega)t}.\end{aligned} \tag{7.119}$$

Equation (7.118) can be held after inserting (7.119) into the right side of (7.118) and termwise integration. Substituting in (7.118) $\mathrm{e}^{-sT}$ by $\zeta$, we find

$$D_{M\underline{M}'}(T,\zeta,0)=\int_0^T D_M(T,\zeta,t)\hat{D}_M(T,\zeta,t)\,\mathrm{d}t. \tag{7.120}$$

Applying (7.100), we obtain for $0<t<T$

$$\begin{aligned}D_M(T,\zeta,t)&=\mathscr{D}_M(T,\zeta,t)=a_M{}^{-1}(\zeta)b_M(\zeta,t),\\ \hat{D}_M(T,\zeta,t)&=\hat{b}_M(\zeta,t)\hat{a}^{-1}{}_M(\zeta).\end{aligned} \tag{7.121}$$

Inserting this relation in (7.120), we achieve

$$\begin{aligned}D_{M\underline{M}'}(T,\zeta,0)&=\int_0^T\mathscr{D}_M(T,\zeta,t)\breve{\mathscr{D}}_M(T,\zeta,t)\,\mathrm{d}t\\ &=a_M{}^{-1}(\zeta)\int_0^T b_M(\zeta,t)\hat{b}_M(\zeta,t)\,\mathrm{d}t\hat{a}_M{}^{-1}(\zeta),\end{aligned} \tag{7.122}$$

from which, with the help of (7.116), we obtain

$$A_{M1}(\zeta) = \int_0^T b_M(\zeta, t)\hat{b}_M(\zeta, t)\,\mathrm{d}t = \int_0^T b_M(\zeta, t)b'_M(\zeta^{-1}, t)\,\mathrm{d}t. \tag{7.123}$$

From (7.123), we directly see that the matrix $A_{M1}(\zeta)$ is a quasi-polynomial. Hereby, obviously $A_{M1}(\zeta) = A'_{M1}(\zeta^{-1})$, i.e., this quasi-polynomial is symmetric. So it remains to show that the quasi-polynomial $A_{M1}(\zeta)$ is nonnegative on the circle $|\zeta| = 1$. This claim follows immediately from the inequality

$$xA_{M1}(\zeta)x^* = \int_0^T [xb_M(\zeta, t)]\left[\bar{x}b_M(\zeta^{-1}, t)\right]'\,\mathrm{d}t \geq 0,$$

that is a consequence of (7.123), regarding the fact that for $|\zeta| = 1$, the equality $\zeta^{-1} = \bar{\zeta}$ holds. ■

(4)

**Definition 7.4** As degree of the quasi-polynomial matrix $F(\zeta)$, we name the least integer $\lambda$ for which $\zeta^\lambda F(\zeta)$ becomes a polynomial matrix.

Below, the degree of the quasi-polynomial matrix $F(\zeta)$ is denoted by $\mathrm{Deg}\, F(\zeta)$.

**Lemma 7.4** *Let the period $T$ be non-pathological. Then, the matrices $a_M(\zeta)$ and $b_M(\zeta, t)$ in ILMFD* (7.121) *can be chosen in such a way that*

$$\deg a_M(\zeta) \leq \alpha, \quad \deg b_M(\zeta, t) \leq \deg a_M(\zeta) - 1 \leq \alpha - 1, \tag{7.124}$$

*and*

$$\mathrm{Deg}\, A_{M1}(\zeta) \leq \alpha - 1. \tag{7.125}$$

*Proof* Under the condition of Lemma 7.4, the representation

$$\mathscr{D}_M(T, \zeta, t) = C_M(I_\alpha - \zeta \mathrm{e}^{A_M T})^{-1}\mathrm{e}^{A_M t}B_M \tag{7.126}$$

is minimal, and consequently, $\mathrm{Mind}\, \mathscr{D}_M(T, \zeta, t) = \alpha$. Therefore, in any ILMFD (7.121),

$$\deg \det a_M(\zeta) = \alpha. \tag{7.127}$$

If we choose in ILMFD (7.121) the matrix $a_M(\zeta)$ as row reduced, then due to (1.76), we obtain $\deg a_M(\zeta) \leq \alpha$. Moreover, knowing that matrix $\mathscr{D}_M(T, \zeta, t)$ is strictly proper, we have $\deg b_M(\zeta, t) \leq \deg a_M(\zeta) - 1 \leq \alpha - 1$. Hence, inequality (7.125) follows directly from (7.123). ■

(5)

**Lemma 7.5** *Under the suppositions of Lemma 7.4, the following claims hold:*

*(i) The determinant* $\det A_{M1}(\zeta)$ *is a symmetric quasi-polynomial, which is nonnegative on the unit circle.*

*(ii) The inequality*

$$\operatorname{Deg}\det A_{M1}(\zeta) \le \alpha - n. \tag{7.128}$$

*is true.*

*Proof* (i) Since matrix (7.123) is a quasi-polynomial, also the determinant $\det A_{M1}(\zeta)$ is a quasi-polynomial. Moreover, due to $\det A_{M1}(\zeta) = \det A_{M1}(\zeta^{-1})$, the quasi-polynomial $\det A_{M1}(\zeta)$ is symmetric. The nonnegativity of the quasi-polynomial $\det A_{M1}(\zeta)$ on the circle $|\zeta| = 1$ follows from the nonnegativity of the quasi-polynomial $A_{M1}(\zeta)$ on the unit circle thanks to the Sylvester condition, because $A_{M1}(\zeta)$ is a Hermitian matrix on $|\zeta| = 1$.

(ii) At first we notice that the rational matrix $D_{M\underline{M}'}(T, \zeta, 0)$ is strictly proper. Indeed, from (7.126), we obtain

$$\hat{\mathscr{D}}_M(T, \zeta, t) = B'_M e^{A'_M t}(I_\alpha - \zeta^{-1} e^{A'_M T})^{-1} C'_M. \tag{7.129}$$

Inserting (7.126) and (7.129) into (7.120), we find

$$D_{M\underline{M}'}(T, \zeta, 0) = C_M(I_\alpha - \zeta e^{A_M T})^{-1} R_M (I_\alpha - \zeta^{-1} e^{A'_M T})^{-1} C'_M, \tag{7.130}$$

where the constant matrix $R_M$ is determined by

$$R_M \triangleq \int_0^T e^{A_M t} B_M B'_M e^{A'_M t}\, dt. \tag{7.131}$$

Applying that the matrix $I_\alpha - \zeta e^{A_M T}$ is strictly proper, we find from Eq. (7.130) that there exists the finite limit

$$\begin{aligned} \lim_{\zeta\to\infty} \zeta D_{M\underline{M}'}(T, \zeta, 0) &= \lim_{\zeta\to\infty} C_M(\zeta^{-1} I_\alpha - e^{A_M T})^{-1} R_M (I_\alpha - \zeta^{-1} e^{A'_M T})^{-1} C'_M \\ &= -C_M e^{-A_M T} R_M C'_M. \end{aligned}$$

Hence,

$$\operatorname{exc} D_{M\underline{M}'}(T, \zeta, 0) \ge 1, \tag{7.132}$$

i.e., the rational matrix $D_{M\underline{M}'}(T, \zeta, 0)$ is strictly proper.
Further, notice that due to (7.116)

$$D_{M\underline{M}'}(T, \zeta, 0) = a_M{}^{-1}(\zeta) A_{M1}(\zeta) \hat{a}_M{}^{-1}(\zeta). \tag{7.133}$$

Thus,

$$\det D_{M\underline{M}'}(T, \zeta, 0) = \frac{\det A_{M1}(\zeta)}{\det a_M(\zeta)\det a_M(\zeta^{-1})}. \tag{7.134}$$

Moreover, since the matrix $D_{M\underline{M}'}(T, \zeta, 0)$ is quadratic of size $n \times n$, so from (7.133), we derive

$$\operatorname{exc}\det D_{M\underline{M}'}(T, \zeta, 0) = \operatorname{exc}\frac{\det A_{M1}(\zeta)}{\det a_M(\zeta)\det a_M(\zeta^{-1})} \geq n. \tag{7.135}$$

Denote by $\Delta_{M\underline{M}'}(\zeta)$ the symmetric quasi-polynomial

$$\Delta_{M\underline{M}'}(\zeta) \stackrel{\triangle}{=} \det a_M(\zeta)\det a_M(\zeta^{-1}). \tag{7.136}$$

Since $\det a_M(0) \neq 0$, so from (7.127), we find

$$\operatorname{Deg}\Delta_{M\underline{M}'}(\zeta) = \alpha. \tag{7.137}$$

Hereby,

$$\Delta_{M\underline{M}'}(\zeta) = \zeta^{-\alpha} d_{M\underline{M}'}(\zeta), \tag{7.138}$$

where $d_{M\underline{M}'}(\zeta)$ is a polynomial with $\deg d_{M\underline{M}'}(\zeta) = 2\alpha$.
Assume

$$\operatorname{Deg}\det A_{M1}(\zeta) = \gamma. \tag{7.139}$$

Then

$$\det A_{M1}(\zeta) = \zeta^{-\gamma} d_{M1}(\zeta), \tag{7.140}$$

where $d_{M1}(\zeta)$ is a polynomial, and $\deg d_{M1}(\zeta) = 2\gamma$. With the help of (7.138) and (7.140), from (7.134), we find

$$\det D_{M\underline{M}'}(T, \zeta, 0) = \frac{\zeta^{\alpha} d_{M1}(\zeta)}{\zeta^{\gamma} d_{M\underline{M}'}(\zeta)}, \tag{7.141}$$

which directly implies

$$\operatorname{exc}\det D_{M\underline{M}'}(T, \zeta, 0) = \gamma + \deg d_{M\underline{M}'}(\zeta) - \alpha - \deg d_M(\zeta) = \alpha - \gamma. \tag{7.142}$$

This shows, using (7.135), that

$$\alpha - \gamma \geq n, \tag{7.143}$$

which on basis of (7.139) is equivalent to (7.128). ∎

(6) On basis of Lemmata 7.1–7.5, we can prove the fundamental statement of this section.

**Theorem 7.1** *Let the sampling period $T$ be non-pathological. Then the following claims hold:*

*(i) The matrix* $A_M(\zeta)$ *in* (7.81) *is a symmetric quasi-polynomial, which is nonnegative on the circle* $|\zeta| = 1$.

*(ii) The polynomial matrices* $a_M(\zeta)$ *and* $a_{M1}(\zeta)$ *in* (7.105) *can be chosen in such a way that*

$$A_M(\zeta) = \sum_{k=-\eta}^{\eta} a_k \zeta^k, \quad a_k = a'_k, \tag{7.144}$$

*where*

$$0 \le \eta = \operatorname{Deg} A_M(\zeta) \le \chi - 1. \tag{7.145}$$

*(iii) The determinant* det $A_M(\zeta)$ *is a symmetric quasi-polynomial, which is nonnegative on the circle* $|\zeta| = 1$.

*(iv)*

$$\operatorname{Deg} \det A_M(\zeta) \le \chi - n. \tag{7.146}$$

*Proof* (i) From (7.122), (7.81), and (7.105), we obtain

$$\begin{aligned} A_M(\zeta) &= a_l(\zeta) D_{M\underline{M}'}(T, \zeta, 0) \hat{a}_l(\zeta) = a_{M1}(\zeta) a_M(\zeta) D_{M\underline{M}'}(T, \zeta, 0) \hat{a}_M(\zeta) \hat{a}_{M1}(\zeta) \\ &= a_{M1}(\zeta) A_{M1}(\zeta) \hat{a}_{M1}(\zeta). \end{aligned} \tag{7.147}$$

Using (7.123), this relation can be written in the form

$$A_M(\zeta) = \int_0^T [a_{M1}(\zeta) b_M(\zeta, t)] \left[\widehat{a_{M1}(\zeta) b_M(\zeta, t)}\right] \mathrm{d}t. \tag{7.148}$$

Hence, with the help of the thoughts in the proof of Lemma 7.3, $A_M(\zeta)$ is a symmetric quasi-polynomial, which is nonnegative on the circle $|\zeta| = 1$.

(ii) By definition, the matrix $a_l(\zeta) = a_{M1}(\zeta) a_M(\zeta)$ is determined by ILMFD (7.97)

$$C_2 (I_\chi - \zeta e^{AT})^{-1} = a_l^{-1}(\zeta) b_l(\zeta). \tag{7.149}$$

Moreover, let $k(\zeta)$ be any unimodular $n \times n$ matrix. Then the horizontal pair $(k(\zeta) a_l(\zeta), k(\zeta) b_l(\zeta))$ also defines an ILMFD of the matrix $C_2(I_\chi - \zeta e^{AT})^{-1}$. In particular, assuming that the matrix $a_M(\zeta)$ is row reduced, we can choose the matrix $k(\zeta)$ in such a way that the matrix $k(\zeta) a_{M1}(\zeta)$ becomes also row reduced. Hence, without loss of generality, we can suppose that the matrix $a_{M1}(\zeta)$ is row reduced. Moreover, because of *deg* det $a_{M1}(\zeta) = \chi - \alpha$,

$$\deg a_{M1}(\zeta) \le \chi - \alpha.$$

Since the matrix $a_M(\zeta)$ is row reduced, so thanks to the above shown

$$\deg a_M(\zeta) \le \alpha, \quad \deg b_M(\zeta, t) \le \deg a_M(\zeta) - 1 \le \alpha - 1.$$

From the last two estimates, we see that under the taken suppositions,

$$\deg\,[a_{M1}(\zeta)b_M(\zeta,t)] \le \deg a_{M1}(\zeta) + \deg b_M(\zeta,t) \le \chi - 1. \tag{7.150}$$

This formula, together with (7.147) proves claim (ii).

(iii) This claim is a direct consequence of the same properties of the matrix $A_M(\zeta)$.

(iv) Calculating the determinants of both sides of (7.147), we obtain

$$\det A_M(\zeta) = \det a_{M1}(\zeta)\det A_{M1}(\zeta)\det a_{M1}(\zeta^{-1}). \tag{7.151}$$

Hence, using $\det a_{M1}(0) \neq 0$, we find

$$\begin{aligned}\text{Deg}\det A_M(\zeta) &= \deg\det a_{M1}(\zeta) + Deg A_{M1}(\zeta)\\ &\le \chi - \alpha + \alpha - n = \chi - n.\end{aligned} \tag{7.152}$$

■

## 7.5 Properties of Matrices $A_L(\zeta)$ and $C(\zeta)$

(1) To study the properties of the matrix $A_L(\zeta)$, we derive some preliminary relations from (7.81) in analogy to the relations in Sect. 7.4 for the matrix $A_M(\zeta)$.

At first notice that, in most applications, the representation (6.15) for the $r \times m$ matrix

$$L(s) = C_1(sI_\chi - A)^{-1}B_2 + D \tag{7.153}$$

is not minimal, because the pair $[A, C_1]$ is not observable. In this situation, there exists a minimal realization

$$L(s) = C_L(sI_\beta - A_L)^{-1}B_L + D, \tag{7.154}$$

where $\beta < \chi$. Further on, we always suppose

$$r \le \beta, \quad m \le \beta, \tag{7.155}$$

where $r$ and $m$ are the number of rows and columns of the matrix $L(s)$, respectively. In practical applications, this supposition is fulfilled.

**Lemma 7.6** *The quantity*

$$\delta_L(s) \triangleq \frac{\det(sI_\chi - A)}{\det(sI_\beta - A_L)} \tag{7.156}$$

*is a polynomial.*

*Proof* Consider the IRMFD

$$(sI_\chi - A)^{-1}B_2 = d_r(s)g_r{}^{-1}(s), \quad (sI_\beta - A_L)^{-1}B_L = d_L(s)g_L{}^{-1}(s). \tag{7.157}$$

Since by definition, the pairs $(A, B_2)$ and $(A_L, B_L)$ are minimal, we obtain

$$\det g_r(s) \sim \det(sI_\chi - A), \quad \det g_L(s) \sim \det(sI_\beta - A_L). \tag{7.158}$$

Moreover, the representation

$$L(s) = [C_L d_L(s) + Dg_L(s)]\, g_L{}^{-1}(s) \tag{7.159}$$

is an IRMFD. From (7.153) and (7.157), we find that the product

$$L(s)g_r(s) = C_1 d_r(s) + Dg_r(s) \tag{7.160}$$

is a polynomial matrix. Since the right side of (7.159) is an IRMFD, we achieve

$$g_r(s) = g_L(s)g_{L1}(s), \tag{7.161}$$

where $g_{L1}(p)$ is a polynomial matrix. Hence, using (7.158), the function

$$\det g_{L1}(s) = \frac{\det g_r(s)}{\det g_L(s)} \sim \delta_L(s) = \frac{\det(sI_\chi - A)}{\det(sI_\beta - A_L)} \tag{7.162}$$

is a polynomial. ■

As earlier, it follows from Lemma 7.6 that under (7.91), the expansion

$$\det(sI_\beta - A_L) = (s - p_1)^{\lambda_1} \cdots (s - p_\rho)^{\lambda_\rho}, \quad \lambda_1 + \cdots + \lambda_\rho = \beta, \tag{7.163}$$

takes place, where $\lambda_i \le \nu_i$. Moreover,

$$\delta_L(s) \sim (s - p_1)^{\nu_1 - \lambda_1} \cdots (s - p_\rho)^{\nu_\rho - \lambda_\rho}. \tag{7.164}$$

(2) Let, in accordance with earlier investigations, the equations of the generalized DAC possess form (6.10)

$$u(t) = \sum_{i=0}^{P} h_i(t - kT)\psi_{k-i}, \quad kT < t < (k+1)T, \tag{7.165}$$

where $h_i(t)$ are $m \times q$ matrices, and the transfer matrix of the hold is defined by the formulae

$$\mu(s) = \sum_{i=0}^{P} \mu_i(s)\mathrm{e}^{-isT}, \quad \mu_i(s) = \int_0^T h_i(t)\mathrm{e}^{-st}\, \mathrm{d}t. \tag{7.166}$$

Then, as follows from relations (6.205) and (6.206), for $0 < t < T$, with the help of (7.153), (7.154), we obtain the equivalent relations

$$\breve{D}_{L\mu}(T, s, t) = \frac{1}{T}\sum_{k=-\infty}^{\infty} L(s + k\mathrm{j}\omega)\mu(s + k\mathrm{j}\omega)\mathrm{e}^{(s+k\mathrm{j}\omega)t} = \breve{\mathscr{D}}_{L\mu}(T, s, t)$$
$$= \begin{cases} C_1\mathrm{e}^{At}(I_\chi - \mathrm{e}^{-sT}\mathrm{e}^{AT})^{-1}\breve{n}(s, t) + D\breve{h}(s, t), \\ C_L\mathrm{e}^{A_Lt}(I_\beta - \mathrm{e}^{-sT}\mathrm{e}^{A_LT})^{-1}\breve{n}_L(s, t) + D\breve{h}(s, t), \end{cases} \tag{7.167}$$

where

$$\breve{n}(s, t) = \sum_{i=0}^{P}\mathrm{e}^{-isT}\left[\int_0^t \mathrm{e}^{-A\nu}B_2h_i(\nu)\,\mathrm{d}\nu + \mathrm{e}^{-sT}\mathrm{e}^{AT}\int_t^T \mathrm{e}^{-A\nu}B_2h_i(\nu)\,\mathrm{d}\nu\right],$$
$$\breve{n}_L(s, t) = \sum_{i=0}^{P}\mathrm{e}^{-isT}\left[\int_0^t \mathrm{e}^{-A_L\nu}B_Lh_i(\nu)\,\mathrm{d}\nu + \mathrm{e}^{-sT}\mathrm{e}^{A_LT}\int_t^T \mathrm{e}^{-A_L\nu}B_Lh_i(\nu)\,\mathrm{d}\nu\right], \tag{7.168}$$
$$\breve{h}(s, t) = \sum_{i=0}^{P}\mathrm{e}^{-isT}h_i(t).$$

Substituting in (7.167) and (7.168) $\mathrm{e}^{-sT}$ by $\zeta$, we obtain

$$\mathscr{D}_{L\mu}(T, \zeta, t) = \breve{\mathscr{D}}_{L\mu}(T, s, t)\,\big|_{\mathrm{e}^{-sT}=\zeta}$$
$$= \begin{cases} C_1\mathrm{e}^{At}(I_\chi - \zeta\mathrm{e}^{AT})^{-1}n(\zeta, t) + Dh(\zeta, t), \\ C_L\mathrm{e}^{A_Lt}(I_\beta - \zeta\mathrm{e}^{A_LT})^{-1}n_L(\zeta, t) + Dh(\zeta, t), \end{cases} \tag{7.169}$$

where

$$n(\zeta, t) = \sum_{i=0}^{P}\zeta^i\left[\int_0^t \mathrm{e}^{-A\nu}B_2h_i(\nu)\,\mathrm{d}\nu + \zeta\mathrm{e}^{AT}\int_t^T \mathrm{e}^{-A\nu}B_2h_i(\nu)\,\mathrm{d}\nu\right],$$
$$n_L(\zeta, t) = \sum_{i=0}^{P}\zeta^i\left[\int_0^t \mathrm{e}^{-A_L\nu}B_Lh_i(\nu)\,\mathrm{d}\nu + \zeta\mathrm{e}^{A_LT}\int_t^T \mathrm{e}^{-A_L\nu}B_Lh_i(\nu)\,\mathrm{d}\nu\right], \tag{7.170}$$
$$h(\zeta, t) = \sum_{i=0}^{P}\zeta^ih_i(t).$$

In accordance with (6.209)–(6.211), let us consider the results of the left division

$$
\begin{aligned}
n(\zeta, t) &= (I_\chi - \zeta \mathrm{e}^{AT}) q(\zeta, t) + m_l(A), \\
n_L(\zeta, t) &= (I_\beta - \zeta \mathrm{e}^{A_L T}) q_L(\zeta, t) + m_L(A_L).
\end{aligned}
\tag{7.171}
$$

Here $q(\zeta, t), q_L(\zeta, t)$ are matrices, which depend polynomially on $\zeta$ and

$$
\begin{aligned}
m_l(A) &\triangleq \sum_{i=0}^{P} \mathrm{e}^{-iAT} \int_0^T \mathrm{e}^{-A\nu} B_2 h_i(\nu)\,\mathrm{d}\nu, \\
m_L(A_L) &\triangleq \sum_{i=0}^{P} \mathrm{e}^{-iA_L T} \int_0^T \mathrm{e}^{-A_L\nu} B_L h_i(\nu)\,\mathrm{d}\nu
\end{aligned}
\tag{7.172}
$$

are constant matrices. Applying (7.171) from (7.169), we derive for $0 < t < T$ the equivalent formulae

$$
\begin{aligned}
D_{L\mu}(T, \zeta, t) &= \mathscr{D}_{L\mu}(T, \zeta, t) \\
&= \begin{cases} C_1 \mathrm{e}^{At}(I_\chi - \zeta \mathrm{e}^{AT})^{-1} m_l(A) + r_l(\zeta, t), \\ C_L \mathrm{e}^{A_L t}(I_\beta - \zeta \mathrm{e}^{A_L T})^{-1} m_L(A_L) + r_L(\zeta, t), \end{cases}
\end{aligned}
\tag{7.173}
$$

where

$$
r_l(\zeta, t) = r_L(\zeta, t) = C_1 \mathrm{e}^{At} q(\zeta, t) + Dh(\zeta, t) = C_L \mathrm{e}^{A_L t} q_L(\zeta, t) + Dh(\zeta, t)
\tag{7.174}
$$

is a matrix depending polynomially on $\zeta$.

**Lemma 7.7** *Let the period $T$ be non-pathological, and the pairs* $(\mathrm{e}^{-AT}, m_l(A))$, $(\mathrm{e}^{-A_L T}, m_L(A_L))$ *be minimal. Assume the IRMFD*

$$
\begin{aligned}
(I_\chi - \zeta \mathrm{e}^{AT})^{-1} m_l(A) &= b_r(\zeta) a_r{}^{-1}(\zeta), \\
(I_\beta - \zeta \mathrm{e}^{A_L T})^{-1} m_L(A_L) &= b_L(\zeta) a_L{}^{-1}(\zeta).
\end{aligned}
\tag{7.175}
$$

*Then the following claims hold:*

*(i)*

$$
L_1(\zeta) \triangleq C_1 \mathrm{e}^{At}(I_\chi - \zeta \mathrm{e}^{AT})^{-1} m_l(A) = C_L \mathrm{e}^{A_L t}(I_\beta - \zeta \mathrm{e}^{A_L T})^{-1} m_L(A_L).
\tag{7.176}
$$

*(ii) The realization*

$$
L_1(\zeta) = C_L \mathrm{e}^{A_L t}(I_\beta - \mathrm{e}^{A_L T})^{-1} m_L(A)
\tag{7.177}
$$

*is minimal, and the right side of the relation*

$$\mathscr{D}_{L\mu}(T,\zeta,t)=\left[C_L\mathrm{e}^{A_Lt}b_L(\zeta)+r_L(\zeta,t)a_L(\zeta)\right]a_L{}^{-1}(\zeta)\stackrel{\triangle}{=}b_L(\zeta,t)a_L{}^{-1}(\zeta) \tag{7.178}$$

*is an IRMFD.*

*(iii) The product*

$$b_r(\zeta,t)\stackrel{\triangle}{=}\mathscr{D}_{L\mu}(T,\zeta,t)a_r(\zeta) \tag{7.179}$$

*is a matrix depending polynomially on $\zeta$, but the matrices $a_r(\zeta)$ and $a_L(\zeta)$ are connected by*

$$a_r(\zeta)=a_L(\zeta)a_{L1}(\zeta), \tag{7.180}$$

*where $a_{L1}(\zeta)$ is a polynomial matrix. Moreover,*

$$\det a_{L1}(\zeta)\sim\Delta_L(\zeta)\stackrel{\triangle}{=}\frac{\det(I_\chi-\zeta\mathrm{e}^{AT})}{\det(I_\beta-\zeta\mathrm{e}^{A_LT})}. \tag{7.181}$$

*Proof* (i) Formula (7.173) represents the separation of a rational matrix into an integral and a broken part. Since such a separation is unique, equality (7.176) is valid.

(ii) Since the period $T$ is non-pathological, and the pair $[A_L, C_L]$ is minimal by definition, so owing to Theorem 5.7, the pair $[I_\beta-\zeta\mathrm{e}^{A_LT}, C_L]$ is irreducible. Since the pair $[\mathrm{e}^{-A_LT}, m_L(A_L)]$ is irreducible too, so realization (7.177) is minimal. Hence the right side of

$$L_1(\zeta)=\left[C_L\mathrm{e}^{A_Lt}b_L(\zeta)\right]a_L{}^{-1}(\zeta) \tag{7.182}$$

is an IRMFD. Therefore, the right side of formula (7.178) is also an IRMFD.

(iii) Since at the same time with (7.154) Eq. (7.153) is true, so the product

$$C_1\mathrm{e}^{At}(I_\chi-\zeta\mathrm{e}^{AT})^{-1}m_l(A)a_r(\zeta)\stackrel{\triangle}{=}b_r(\zeta,t) \tag{7.183}$$

is a polynomial matrix in $\zeta$. Hence, the matrix

$$\mathscr{D}_{L\mu}(T,\zeta,t)a_r(\zeta)=b_r(\zeta,t)+Dh(\zeta,t)a_r(\zeta) \tag{7.184}$$

is also polynomial. Moreover, since the right part of (7.178) is an IRMFD, so relation (7.180) is valid. It remains to show formula (7.181). From (7.180), we find

$$\det a_{L1}(\zeta)=\frac{\det a_r(\zeta)}{\det a_L(\zeta)}. \tag{7.185}$$

On the other side, under the taken suppositions, from (7.175), we obtain

$$\begin{aligned}\det a_r(\zeta) &\sim \det(I_\chi - \zeta e^{AT}),\\ \det a_L(\zeta) &\sim \det(I_\beta - \zeta e^{A_L T}),\end{aligned} \tag{7.186}$$

so that (7.185) implies (7.181). ■

(3) Introduce

$$A_{L1}(\zeta) \stackrel{\triangle}{=} \hat{a}_L(\zeta) D_{\underline{\mu}'\underline{L}' L\mu}(T, \zeta, 0) a_L(\zeta), \tag{7.187}$$

where, as before,

$$D_{\underline{\mu}'\underline{L}' L\mu}(T,\zeta, 0) = \left[\frac{1}{T}\sum_{k=-\infty}^{\infty} \mu'(-s - kj\omega) L'(-s - kj\omega) L(s + kj\omega)\mu(s + kj\omega)\right]_{e^{-sT}=\zeta}. \tag{7.188}$$

**Lemma 7.8** *Matrix* (7.187) *of size* $q \times q$ *is a symmetric quasi-polynomial, which is nonnegative on the circle* $|\zeta| = 1$.

*Proof* At first, we state that

$$\begin{aligned}\breve{D}_{\underline{\mu}'\underline{L}' L\mu}(T, s, 0) &= \frac{1}{T}\sum_{k=-\infty}^{\infty} \mu'(-s - kj\omega) L'(-s - kj\omega) L(s + kj\omega)\mu(s + kj\omega)\\ &= \int_0^T \breve{D}_{\mu' L'}(T, -s, t)\breve{D}_{L\mu}(T, s, t)\, dt.\end{aligned} \tag{7.189}$$

Indeed, by definition

$$\begin{aligned}\breve{D}_{L\mu}(T, s, t) &= \frac{1}{T}\sum_{k=-\infty}^{\infty} L(s + kj\omega)\mu(s + kj\omega)e^{(s+kj\omega)t},\\ \breve{D}_{\mu' L'}(T, -s, t) &= \frac{1}{T}\sum_{k=-\infty}^{\infty} \mu'(-s + kj\omega) L'(-s + kj\omega)e^{(-s+kj\omega)t}.\end{aligned} \tag{7.190}$$

Inserting these expressions into the right part of (7.189), after termwise integration, we obtain the right part of formula (7.189). Substituting in (7.189) $e^{-sT}$ by $\zeta$, we achieve

$$D_{\underline{\mu}'\underline{L}' L\mu}(T, \zeta, 0) = \int_0^T \mathscr{D}_{\mu' L'}(T, \zeta^{-1}, t)\mathscr{D}_{L\mu}(T, \zeta, t)\, dt, \tag{7.191}$$

when considering that for $0 < t < T$, $D_{L\mu}(T, \zeta, t) = \mathscr{D}_{L\mu}(T, \zeta, t)$. Notice that

$$\mathscr{D}_{\mu' L'}(T, \zeta^{-1}, t) = \mathscr{D}'_{L\mu}(T, \zeta^{-1}, t) = \hat{\mathscr{D}}_{L\mu}(T, \zeta, t). \tag{7.192}$$

Then, relation (7.191) implies

$$D_{\underline{\mu' L'} L\mu}(T, \zeta, 0) = \int_0^T \hat{D}_{L\mu}(T, \zeta, t) D_{L\mu}(T, \zeta, t)\,\mathrm{d}t. \tag{7.193}$$

Furthermore, with regard to (7.178), we find

$$\begin{aligned} \mathscr{D}_{L\mu}(T, \zeta, t) &= b_L(\zeta, t) a_L{}^{-1}(\zeta), \\ \hat{\mathscr{D}}_{L\mu}(T, \zeta, t) &= \hat{a}_L{}^{-1}(\zeta) \hat{b}(\zeta, t), \end{aligned} \tag{7.194}$$

and with the help of these relations, from (7.193), we obtain

$$D_{\underline{\mu' L'} L\mu}(T, \zeta, 0) = \hat{a}_L{}^{-1}(\zeta) \int_0^T \hat{b}_L(\zeta, t) b_L(\zeta, t)\,\mathrm{d}t\, a_L{}^{-1}(\zeta). \tag{7.195}$$

Hence, observing (7.187), we achieve

$$A_{L1}(\zeta) = \int_0^T \hat{b}_L(\zeta, t) b_L(\zeta, t)\,\mathrm{d}t. \tag{7.196}$$

By analog reasoning as in Sect. 7.4 for the matrix $A_{M1}(\zeta)$, from (7.196), we realize that the matrix $A_{L1}(\zeta)$ is a symmetric quasi-polynomial, which is nonnegative on the circle $|\zeta| = 1$. ∎

(4)

**Lemma 7.9** *The matrix* $a_L(\zeta)$ *in IRMFD* (7.178) *can be chosen in such a way that the estimate*

$$\operatorname{Deg} A_{L1}(\zeta) \leq P + \beta \tag{7.197}$$

*becomes true, where* $P$ *is the order of the generalized ADC.*

*Proof* Let in (7.175) the matrix $a_L(\zeta)$ be column reduced. Then, in analogy to earlier, we state that in IRMFD (7.175)

$$\deg a_L(\zeta) \leq \beta, \quad \deg b_L(\zeta) \leq \deg a_L(\zeta) - 1 \leq \beta - 1. \tag{7.198}$$

Further notice that, owing to (7.178),

$$b_L(\zeta, t) = C_L \mathrm{e}^{A_L t} b_L(\zeta) + r_L(\zeta, t) a_L(\zeta), \tag{7.199}$$

where the polynomial matrix $r_L(\zeta, t)$ is determined by the formula

$$r_L(\zeta, t) = C_L e^{A_L t} q_L(\zeta, t) + Dh(\zeta, t), \tag{7.200}$$

and the matrix $q_L(\zeta, t)$ is determined as the result of division by (7.171).

From (7.170), we find

$$\deg n_L(\zeta, t) = P + 1. \tag{7.201}$$

Therefore, observing that the matrix $I_\beta - \zeta e^{A_L T}$ is regular, from (7.171), we obtain

$$\deg q_L(\zeta, t) = P. \tag{7.202}$$

Since from (7.170), we also obtain $\deg h(\zeta, t) = P$, so from (7.200), we realize that

$$\deg r_L(\zeta, t) \le P. \tag{7.203}$$

Hence, using (7.198) and (7.199), we achieve the estimate

$$\deg b_L(\zeta, t) \le P + \beta, \tag{7.204}$$

which, with regard to (7.196), is equivalent to (7.197). ■

**Lemma 7.10** *Let the period $T$ be non-pathological. Then*

*(i) The determinant* $\det A_{L1}(\zeta)$ *is a symmetric quasi-polynomial, which is nonnegative on the circle $|\zeta| = 1$.*
*(ii) The following estimate is valid:*

$$\operatorname{Deg} \det A_{L1}(\zeta) \le \beta + qP. \tag{7.205}$$

*Proof* (i) Since the matrix $A_{L1}(\zeta)$ is a symmetric quasi-polynomial, so its determinant $\det A_{L1}(\zeta)$ is also a symmetric quasi-polynomial. The nonnegativity of $\det A_{L1}(\zeta)$ on the circle $|\zeta| = 1$ follows from Lemma 7.8 on basis of the Sylvester condition.

(ii) From the above derived relations (7.173), it follows that

$$\begin{aligned} \mathscr{D}_{L\mu}(T, \zeta, t) &= C_L e^{A_L t}(I_\beta - \zeta e^{A_L T})^{-1} m_L(A_L) + r_L(\zeta, t), \\ \hat{\mathscr{D}}_{L\mu}(T, \zeta, t) &= m_L'(A_L)(I_\beta - \zeta^{-1} e^{A_L' T})^{-1} e^{A_L' t} C_L' + r_L'(\zeta^{-1}, t). \end{aligned} \tag{7.206}$$

Applying these relations, we build the product

$$
\begin{aligned}
&\hat{\mathscr{D}}_{L\mu}(T,\zeta,t)\mathscr{D}_{L\mu}(T,\zeta,t) \\
&\qquad = m_L'(A_L)(I_\chi - \zeta^{-1}\mathrm{e}^{A_L'T})^{-1}\mathrm{e}^{A_L't}C_L'C_L\mathrm{e}^{At}(I_\chi - \zeta\mathrm{e}^{AT})^{-1}m_L(A_L) \qquad (7.207)\\
&\qquad\quad + r_L'(\zeta^{-1},t)C_L\mathrm{e}^{A_Lt}(I_\chi - \zeta\mathrm{e}^{AT})^{-1}m_L(A_L)\\
&\qquad\quad + m_L'(A_L)(I_\chi - \zeta^{-1}\mathrm{e}^{A_L'T})^{-1}\mathrm{e}^{A_L't}C_L'r_L(\zeta,t) + r_L'(\zeta^{-1},t)r_L(\zeta,t).
\end{aligned}
$$

Applying estimate (7.203), on basis of (7.207), it is easy to realize that for $0 < t < T$, there exists the finite limit

$$\lim_{\zeta\to\infty}\left[\zeta^{-P}\hat{\mathscr{D}}_{L\mu}(T,\zeta,t)\mathscr{D}_{L\mu}(T,\zeta,t)\right].$$

Hence,

$$\operatorname{exc}\left[\hat{\mathscr{D}}_{L\mu}(T,\zeta,t)\mathscr{D}_{L\mu}(T,\zeta,t)\right] \geq -P. \qquad (7.208)$$

Since by integration over the parameter $t$, the excess can at most increase, so from (7.208), we achieve

$$\operatorname{exc}\int_0^T \hat{\mathscr{D}}_{L\mu}(T,\zeta,t)\mathscr{D}_{L\mu}(T,\zeta,t)\,\mathrm{d}t = \operatorname{exc} D_{\underline{\mu}'\underline{L}'L\mu}(T,\zeta,0) \geq -P. \quad (7.209)$$

Since the matrix $D_{\underline{\mu}'\underline{L}'L\mu}(T,\zeta,0)$ is of size $q\times q$, so from (7.209) directly follows that

$$\operatorname{exc}\det D_{\underline{\mu}'\underline{L}'L\mu}(T,\zeta,0) \geq -qP. \qquad (7.210)$$

From (7.187), we obtain

$$D_{\underline{\mu}'\underline{L}'L\mu}(T,\zeta,0) = \hat{a}_L{}^{-1}(\zeta)A_{L1}(\zeta)a_L{}^{-1}(\zeta). \qquad (7.211)$$

Taking the determinant of both sides of this equation, we find

$$\det D_{\underline{\mu}'\underline{L}'L\mu}(T,\zeta,0) = \frac{\det A_{L1}(\zeta)}{\det a_L(\zeta)\det a_l(\zeta^{-1})}. \qquad (7.212)$$

Due to $\deg\det a_L(\zeta) = \beta$ and $\det a_L(0) \neq 0$, from the last formula with the ideas as in Lemma 7.5, we obtain

$$\operatorname{exc}\det D_{\underline{\mu}'\underline{L}'L\mu}(T,\zeta,0) = \beta - \deg\det A_{L1}(\zeta). \qquad (7.213)$$

Thus, applying (7.210)

$$\beta - \operatorname{Deg}\det A_{L1}(\zeta) \geq -qP \qquad (7.214)$$

which is equivalent to (7.205). ■

(5) Finally, the derived relations allow us to claim the main results of this section.

**Theorem 7.2** *The following statements hold:*

*(i) The matrix* $A_L(\zeta)$ *in* (7.81) *is a symmetric quasi-polynomial, which is nonnegative on the circle* $|\zeta| = 1$.
*(ii) Let the period* $T$ *be non-pathological. Then, the matrices* $a_r(\zeta)$ *and* $a_L(\zeta)$*in IRMFD* (7.175) *can be chosen in such a way that*

$$\operatorname{Deg} A_L(\zeta) \leq \chi + P. \tag{7.215}$$

*(iii) When the sampling period is non-pathological, then*

$$\operatorname{Deg} \det A_L(\zeta) \leq \chi + qP. \tag{7.216}$$

*Proof* (i) Since, due to (7.180)

$$a_r(\zeta) = a_L(\zeta) a_{L1}(\zeta), \tag{7.217}$$

so from (7.81) and (7.211), we can derive

$$A_L(\zeta) = \hat{a}_r(\zeta) D_{\underline{\mu' L' L \mu}}(T, \zeta, 0) a_r(\zeta) = \hat{a}_{L1}(\zeta) A_{L1}(\zeta) a_{L1}(\zeta), \tag{7.218}$$

which with the help of (7.196) can be written in the form

$$A_L(\zeta) = \int_0^T \left[ \widehat{b_L(\zeta, t) a_{L1}(\zeta)} \right] \left[ b_L(\zeta, t) a_{L1}(\zeta) \right] \, \mathrm{d}t. \tag{7.219}$$

Thus, claim (i) is shown.

(ii) As in the proof of Theorem 7.1, we can show that the matrix $a_r(\zeta)$ in IRMFD (7.175) can be chosen in such a way that the matrix $a_{L1}(\zeta)$ becomes column reduced. In this case, using that $\deg \det a_{L1}(\zeta) = \chi - \beta$, we obtain

$$\deg a_{L1}(\zeta) \leq \chi - \beta. \tag{7.220}$$

Therefore, from (7.218) with the help of (7.205), we find

$$\operatorname{Deg} A_L(\zeta) \leq \operatorname{Deg} A_{L1}(\zeta) + \deg a_{L1}(\zeta) = \chi + P. \tag{7.221}$$

(iii) From (7.218), we obtain

$$\det A_L(\zeta) = \det A_{L1}(\zeta) \det a_{L1}(\zeta) \det a_{L1}(\zeta^{-1}). \tag{7.222}$$

Hence, using $\det a_{L1}(0) \neq 0$, we derive

$$\operatorname{Deg}\det A_L(\zeta) = \operatorname{Deg}\det A_{L1}(\zeta) + \deg\det a_{L1}(\zeta). \tag{7.223}$$

Owing to $\deg\det a_{L1}(\zeta) = \chi - \beta$, when using (7.205) and (7.223), estimate (7.216) is achieved. ■

(6)

**Theorem 7.3** *The matrix $C(\zeta)$ in* (7.81) *is a quasi-polynomial.*

*Proof* At first, we show that matrix $\breve{C}(s)$ (7.71) is an integral function of the argument $s$. Thereto, we apply formula (6.202), which for $\lambda = s$ is written as

$$\begin{aligned}\psi(s,t) &= e^{-s(\tau_2+\tau_3)}\varphi_{L\mu}(T,s,t-\tau_2-\tau_3)\breve{a}_r(s),\\ \xi(s) &= \breve{a}_l(s)M(s)e^{-s\tau_1},\\ \eta(s,t) &= e^{-s(\tau_2+\tau_3)}\varphi_{L\mu}(T,s,t-\tau_2-\tau_3)\breve{\beta}_{r0}(s)\breve{a}_l(s)M(s)e^{-s\tau_1} + K(s)e^{-s(\tau_1+\tau_3)},\end{aligned} \tag{7.224}$$

where the matrices $\psi(s,t)$, $\xi(s)$, $\eta(s,t)$ are integral functions of the argument $s$.

From the first formula in (7.224) and Theorem 6.9, it follows that the matrix $\psi(s,t)$ is bounded on any vertical stripe of the right complex $s$ plane. Moreover, since the matrices $M(s)$ and $K(s)$ are strictly proper, the matrices $\xi(s)$ and $\eta(s,t)$ decrease in any vertical stripe of the right plane $s$ as $|s|^{-1}$.

Therefore, matrix $Q(s)$ (7.52) is an integral function of the argument $s$, decreasing on the mentioned stripes as $|s|^{-1}$. Hence, the products $\xi(s)Q(s)$ and $Q'(-s)\xi'(-s)$ are integral functions of the argument $s$, and they decrease on those stripes for $|\Im s| \to \infty$ as $|s|^{-2}$. Thus, series (7.71)

$$\breve{C}(s) = \frac{1}{T}\sum_{k=-\infty}^{\infty} \xi(s+kj\omega)Q(s+kj\omega) \tag{7.225}$$

converges for all $s$ absolutely, and its sum is an integral function of the argument $s$. Therefore, when substituting $e^{-sT}$ by $\zeta$, the matrix $C(\zeta)$ (7.81) can possess eigenvalues only at the point $\zeta = 0$ in any bounded region. On the other side, from (7.81), it is seen that the matrix $C(\zeta)$ is rational. But a rational matrix having poles only at $\zeta = 0$ in any bounded region is a quasi-polynomial. Thus, the matrix $C(\zeta)$ is a quasi-polynomial. ■

## 7.6 Wiener–Hopf Method

(1) This section deals with the solution method for the $\mathscr{H}_2$ optimization problem of the standard model for SD systems with delay, based on the minimization of the quadratic functional (7.79). As solution technique for the minimization, the Wiener–Hopf method in the form [7] is applied.

The next statement lays the fundament for the further investigations.

**Theorem 7.4** *Let standard model (6.12) be modal controllable. Assume the IMFD, which, using (6.169), (6.172), and (6.180), is written in the form*

$$D_{N\mu}(T,\zeta,-\tau_2) = a_l^{-1}(\zeta)b_l(\zeta,-\tau_2) = b_r(\zeta,-\tau_2)a_r^{-1}(\zeta), \tag{7.226}$$

*where $b_l(\zeta,-\tau_2)$ and $b_r(\zeta,-\tau_2)$ are polynomial matrices, and the matrices $a_l(\zeta)$ and $a_r(\zeta)$ are determined from the IMFD*

$$C_2(I_\chi - \zeta e^{AT})^{-1} = a_l^{-1}(\zeta)b_l(\zeta),$$
$$(I_\chi - \zeta e^{AT})^{-1}m_l(A) = b_r(\zeta)a_r^{-1}(\zeta).$$

*Moreover, let the associated dual left and right basic controllers $(a_{l0}(\zeta), b_{l0}(\zeta))$, $[\alpha_{r0}(\zeta), \beta_{r0}(\zeta)]$ be given, which do exist thanks to Theorem 3.5. Suppose that the stable rational matrix $\theta^0(\zeta)$ minimizes functional (7.79). Then, the following statements hold:*

*(i) The transfer matrix of an optimal discrete stabilizing controller $W_d^0(\zeta)$ can be determined with the help of the formula*

$$W_d^0(\zeta) = V_{20}(\zeta)V_{10}^{-1}(\zeta), \tag{7.227}$$

*where*

$$\begin{aligned} V_{10}(\zeta) &= \alpha_{r0}(\zeta) - b_r(\zeta,-\tau_2)\theta^0(\zeta), \\ V_{20}(\zeta) &= \beta_{r0}(\zeta) - a_r(\zeta)\theta^0(\zeta). \end{aligned} \tag{7.228}$$

*(ii) The matrix $W_d^0(\zeta)$ can also be determined by the formula*

$$W_d^0(\zeta) = \alpha_l^{-1}(\zeta)\beta_l(\zeta), \tag{7.229}$$

*in which*

$$\begin{aligned} \alpha_l(\zeta) &= D_{l0}(\zeta)\alpha_{l0}(\zeta) - M_{l0}(\zeta)b_l(\zeta,-\tau_2), \\ \beta_l(\zeta) &= D_{l0}(\zeta)\beta_{l0}(\zeta) - M_{l0}(\zeta)a_l(\zeta), \end{aligned} \tag{7.230}$$

*and the polynomial matrices $D_{l0}(\zeta)$ and $M_{l0}(\zeta)$ are determined by the ILMFD*

$$\theta^0(\zeta) = D_{l0}^{-1}(\zeta)M_{l0}(\zeta). \tag{7.231}$$

*(iii) The characteristic polynomial $\Delta^0(\zeta)$ of the $\mathscr{H}_2$ optimal system satisfies*

$$\Delta^0(\zeta) \sim \det D_{l0}(\zeta). \tag{7.232}$$

*Proof* (i) Since, by proposition, the matrix $\theta^0(\zeta)$ is stable, so rational matrix (7.228) is stable too. Using ILMFD (7.229), consider the matrix

$$R^0(\zeta) \stackrel{\triangle}{=} a_l(\zeta)V_{10}(\zeta) - b_l(\zeta, -\tau_2)V_{20}(\zeta). \tag{7.233}$$

With the help of (7.228), this expression can be written in the form

$$\begin{aligned} R^0(\zeta) &\stackrel{\triangle}{=} a_l(\zeta)\left[\alpha_{r0}(\zeta) - b_r(\zeta, -\tau_2)\theta^0(\zeta)\right] - b_l(\zeta, -\tau_2)\left[\beta_{r0}(\zeta) - a_r(\zeta)\theta^0(\zeta)\right] \\ &= a_l(\zeta)\alpha_{r0}(\zeta) - b_l(\zeta, -\tau_2)\beta_{r0}(\zeta) \qquad (7.234) \\ &- \left[a_l(\zeta)b_r(\zeta, -\tau_2) - b_l(\zeta, -\tau_2)a_r(\zeta)\right]\theta^0(\zeta). \end{aligned}$$

Then, on basis of the Bezout identity [8], we obtain

$$\begin{aligned} a_l(\zeta)\alpha_{r0}(\zeta) - b_r(\zeta, -\tau_2)\beta_{r0}(\zeta) &= I_n, \\ a_l(\zeta)b_r(\zeta, -\tau_2) - b_l(\zeta, -\tau_2)\alpha_{r0}(\zeta) &= 0_{nq}. \end{aligned} \tag{7.235}$$

Hence,

$$R^0(\zeta) = I_n,$$

and statement (i) follows from Theorem 3.23.

(ii) Since the matrix $\theta^0(\zeta)$ is stable, so the matrix $D_{l0}(\zeta)$ in ILMFD (7.231) is stable too. Therefore, controller (7.230) is stabilizing, and formula (7.229) defines the transfer matrix of the $\mathscr{H}_2$ optimal controller.

(iii) Under the current suppositions, the characteristic matrix $Q_l(\zeta, \alpha_l, \beta_l)$ of the $\mathscr{H}_2$ optimal system is

$$Q_l^0(\zeta, \alpha, \beta) = \begin{bmatrix} a_l(\zeta) & -b_l(\zeta, -\tau_2) \\ -\beta_l(\zeta) & \alpha_l(\zeta) \end{bmatrix}. \tag{7.236}$$

Inserting herein (7.230), as earlier, we obtain

$$Q_l(\zeta, \alpha, \beta) = \begin{bmatrix} I_n & 0_{nm} \\ M_{l0}(\zeta) & D_{l0}(\zeta) \end{bmatrix} \begin{bmatrix} a_l(\zeta) & -b_l(\zeta, -\tau_2) \\ -\beta_{l0}(\zeta) & \alpha_{l0}(\zeta) \end{bmatrix}. \tag{7.237}$$

Hence,

$$\det Q_l^0(\zeta, \alpha_l^0, \beta_l^0) \sim \det D_{l0}(\zeta), \tag{7.238}$$

because, by construction, the second factor on the right side of (7.237) is a unimodular matrix. ■

(2) The next theorem provides a procedure for constructing the system function $\theta^0(\zeta)$.

**Theorem 7.5** *Let the period $T$ be non-pathological, and the quasi-polynomial, $A_L(\zeta)$ and $A_M(\zeta)$ in (7.81) be positive on the circle $|\zeta| = 1$. Then the following statements hold:*

*(i) There exist the factorizations*

$$A_L(\zeta) = \hat{\Pi}(\zeta)\Pi(\zeta), \quad A_M(\zeta) = \Gamma(\zeta)\hat{\Gamma}(\zeta), \tag{7.239}$$

*where $\Pi(\zeta)$ and $\Gamma(\zeta)$ are stable real polynomial matrices of size $q \times q$ and $n \times n$, respectively.*

*(ii) The optimal system function $\theta^0(\zeta)$ can be constructed by the following algorithm:*

*(a) Build the rational matrix*

$$H(\zeta) = \hat{\Pi}^{-1}(\zeta)\hat{C}(\zeta)\hat{\Gamma}^{-1}(\zeta), \tag{7.240}$$

*where due to* (7.81)

$$\hat{C}(\zeta) = \hat{a}_r(\zeta)D_{\underline{\mu}'\underline{L}'L\mu}(T,\zeta,0)\hat{\beta}_{r0}(\zeta)A_M(\zeta) + \hat{a}_r(\zeta)D_{\underline{\mu}'\underline{L}'K\underline{M}'}(T,\zeta,\tau_2)\hat{a}_l(\zeta). \tag{7.241}$$

*Here, we applied the relation*

$$\hat{D}_{M\underline{K}'L\mu}(T,\zeta,-\tau_2) = D_{\underline{\mu}'\underline{L}'K\underline{M}'}(T,\zeta,\tau_2), \tag{7.242}$$

*which emerges from* (7.75).

*(b) Separate the matrix $H(\zeta)$ into its entire and broken part*

$$H(\zeta) = H_+(\zeta) + H_-(\zeta), \tag{7.243}$$

*where $H_+(\zeta)$ is a polynomial matrix, but $H_-(\zeta)$ is a strictly proper rational matrix, and all its poles are located outside the circle $|\zeta| = 1$.*

*(c) The optimal system function $\theta^0(\zeta)$ emerges from the formula*

$$\theta^0(\zeta) = \Pi^{-1}(\zeta)H_+(\zeta)\Gamma^{-1}(\zeta). \tag{7.244}$$

*Proof* (i) This statement is a consequence of the general theorem about factorization of matrix quasi-polynomials, which are nonnegative on the unit circle [9].

(ii) Using (7.239), functional (7.79) can be represented in the form

$$TJ_1 = \frac{1}{2\pi \mathrm{j}} \oint \mathrm{trace}\left[\hat{\theta}(\zeta)\hat{\Pi}(\zeta)\Pi(\zeta)\theta(\zeta)\Gamma(\zeta)\hat{\Gamma}(\zeta)\right.$$
$$\left. -\hat{\theta}(\zeta)\hat{C}(\zeta) - C(\zeta)\theta(\zeta)\right]\frac{\mathrm{d}\zeta}{\zeta}. \tag{7.245}$$

Regarding that for matrices $A$ and $B$ of appropriate size

$$\mathrm{trace}(AB) = \mathrm{trace}(BA), \tag{7.246}$$

we obtain

$$\mathrm{trace}\left[\hat{\theta}(\zeta)\hat{\Pi}(\zeta)\Pi(\zeta)\theta(\zeta)\Gamma(\zeta)\hat{\Gamma}(\zeta)\right]$$
$$= \mathrm{trace}\left[\hat{\Gamma}(\zeta)\hat{\theta}(\zeta)\hat{\Pi}(\zeta)\Pi(\zeta)\theta(\zeta)\Gamma(\zeta)\right]. \tag{7.247}$$

Moreover,

$$\mathrm{trace}\left[C(\zeta)\theta(\zeta)\right] = \mathrm{trace}\left[C(\zeta)\theta(\zeta)\Gamma(\zeta)\Gamma^{-1}(\zeta)\right]$$
$$= \mathrm{trace}\left[\Gamma^{-1}(\zeta)C(\zeta)\theta(\zeta)\Gamma(\zeta)\right] = \mathrm{trace}\left[\Psi(\zeta)\theta(\zeta)\Gamma(\zeta)\right], \tag{7.248}$$

when we use the notation

$$\Psi(\zeta) \stackrel{\Delta}{=} \Gamma^{-1}(\zeta)C(\zeta). \tag{7.249}$$

In analogy to (7.248), we can derive the relation

$$\mathrm{trace}\left[\hat{\theta}(\zeta)\hat{C}(\zeta)\right] = \mathrm{trace}\left[\hat{\theta}(\zeta)\hat{C}(\zeta)\hat{\Gamma}^{-1}(\zeta)\hat{\Gamma}(\zeta)\right]$$
$$= \mathrm{trace}\left[\hat{\Gamma}(\zeta)\hat{\theta}(\zeta)\hat{C}(\zeta)\hat{\Gamma}^{-1}(\zeta)\right] = \mathrm{trace}\left[\hat{\Gamma}(\zeta)\hat{\theta}(\zeta)\hat{\Psi}(\zeta)\right]. \tag{7.250}$$

Applying (7.247)–(7.250), from (7.245), we obtain the equivalent expression

$$TJ_1 = \frac{1}{2\pi \mathrm{j}} \oint \mathrm{trace}\left[\hat{\Gamma}(\zeta)\hat{\theta}(\zeta)\hat{\Pi}(\zeta)\Pi(\zeta)\theta(\zeta)\Gamma(\zeta)\right.$$
$$\left. -\hat{\Gamma}(\zeta)\hat{\theta}(\zeta)\hat{\Psi}(\zeta) - \Psi(\zeta)\theta(\zeta)\Gamma(\zeta)\right]\frac{\mathrm{d}\zeta}{\zeta}. \tag{7.251}$$

For the further investigations, notice that

$$\begin{aligned}&\hat{\Gamma}(\zeta)\hat{\theta}(\zeta)\hat{\Pi}(\zeta)\Pi(\zeta)\theta(\zeta)\Gamma(\zeta) - \Gamma(\zeta)\hat{\theta}(\zeta)\hat{\Psi}(\zeta) - \Psi(\zeta)\theta(\zeta)\Gamma(\zeta)\\&\quad= [\widehat{\Pi(\zeta)\theta(\zeta)\Gamma(\zeta)} - H(\zeta)][\Pi(\zeta)\theta(\zeta)\Gamma(\zeta) - H(\zeta)] - \hat{H}(\zeta)H(\zeta).\end{aligned} \tag{7.252}$$

Indeed, the right side of (7.252) can be written in the form

$$\begin{aligned}&\left[\hat{\Gamma}(\zeta)\hat{\theta}(\zeta)\hat{\Pi}(\zeta) - \hat{H}(\zeta\right][\Pi(\zeta)\theta(\zeta)\Gamma(\zeta) - H(\zeta)] - \hat{H}(\zeta)H(\zeta)\\&\quad= \hat{\Gamma}(\zeta)\hat{\theta}(\zeta)\hat{\Pi}(\zeta)\Pi(\zeta)\theta(\zeta)\Gamma(\zeta)\\&\qquad- \hat{\Gamma}(\zeta)\hat{\theta}(\zeta)\hat{\Pi}(\zeta)H(\zeta) - \hat{H}(\zeta)\Pi(\zeta)\theta(\zeta)\Gamma(\zeta).\end{aligned} \tag{7.253}$$

When using that owing to (7.240), (7.249)

$$\begin{aligned}H(\zeta) &= \hat{\Pi}^{-1}\hat{C}(\zeta)\hat{\Gamma}^{-1}(\zeta) = \hat{\Pi}^{-1}(\zeta)\hat{\Psi}(\zeta),\\ \hat{H}(\zeta) &= \Psi(\zeta)\Pi^{-1}(\zeta),\end{aligned}$$

then we immediately obtain that the right side of formula (7.253) is equal to the left side of formula (7.252). Then applying separation (7.243) and identity (7.252), we find

$$\begin{aligned}&\hat{\Gamma}(\zeta)\hat{\theta}(\zeta)\hat{\Pi}(\zeta)\Pi(\zeta)\theta(\zeta)\Gamma(\zeta) - \hat{\Gamma}(\zeta)\hat{\theta}(\zeta)\hat{\Psi}(\zeta) - \Psi(\zeta)\theta(\zeta)\Gamma(\zeta)\\&\quad= \left[\widehat{\Pi(\zeta)\theta(\zeta)\Gamma(\zeta) - H_+(\zeta)} - H_-(\zeta)\right]\left[\Pi(\zeta)\theta(\zeta)\Gamma(\zeta) - H_+(\zeta) - H_-(\zeta)\right]\\&\qquad- \hat{H}(\zeta)H(\zeta)\\&\quad= \left[\widehat{\Pi(\zeta)\theta(\zeta)\Gamma(\zeta)} - H_+(\zeta)\right]\left[\Pi(\zeta)\theta(\zeta)\Gamma(\zeta) - H_+(\zeta)\right]\\&\qquad- \left[\widehat{\Pi(\zeta)\theta(\zeta)\Gamma(\zeta)} - H_+(\zeta)\right]H_-(\zeta) - \hat{H}_-(\zeta)\left[\Pi(\zeta)\theta(\zeta)\Gamma(\zeta) - H_+(\zeta)\right]\\&\qquad- \hat{H}_-(\zeta)H_-(\zeta) - \hat{H}(\zeta)H(\zeta).\end{aligned} \tag{7.254}$$

Taking advantage of these relations, for (7.245), we write

$$TJ_1 = J_{11} + J_{12} + J_{13} + J_{14}, \tag{7.255}$$

where

$$
\begin{aligned}
J_{11} &= \frac{1}{2\pi \mathrm{j}} \oint \operatorname{trace}\left[\widehat{\Pi(\zeta)\theta(\zeta)\Gamma(\zeta)} - H_+(\zeta)\right]\left[\Pi(\zeta)\theta(\zeta)\Gamma(\zeta) - H_+(\zeta)\right] \frac{\mathrm{d}\zeta}{\zeta}, \\
J_{12} &= -\frac{1}{2\pi \mathrm{j}} \oint \operatorname{trace}\left[(\widehat{\Pi(\zeta)\theta(\zeta)\Gamma(\zeta) - H_+(\zeta)})H_-(\zeta)\right] \frac{\mathrm{d}\zeta}{\zeta}, \\
J_{13} &= -\frac{1}{2\pi \mathrm{j}} \oint \operatorname{trace}\left[\hat{H}_-(\zeta)(\Pi(\zeta)\theta(\zeta)\Gamma(\zeta) - H_+(\zeta))\right] \frac{\mathrm{d}\zeta}{\zeta}, \\
J_{14} &= -\frac{1}{2\pi \mathrm{j}} \oint \operatorname{trace}\left[\hat{H}_-(\zeta)H_-(\zeta) - \hat{H}(\zeta)H(\zeta)\right] \frac{\mathrm{d}\zeta}{\zeta}.
\end{aligned}
\tag{7.256}
$$

We will show that the integral $J_{13}$ is equal to zero. Since the eigenvalues of the matrices $\Pi(\zeta)$ and $\Gamma(\zeta)$ are all located outside the unit circle, the poles of the matrices $\hat{\Pi}^{-1}(\zeta)$ and $\hat{\Gamma}^{-1}(\zeta)$ lie inside the open disc $|\zeta| < 1$. Moreover, since the matrix $\hat{C}(\zeta)$ is a quasi-polynomial, it has poles only at the point $\zeta = 0$. Hence, in separation (7.243), the matrix $H_+(\zeta)$ is a polynomial.
Since the matrix $H_-(\zeta)$ is strictly proper, so

$$
\lim_{\zeta \to \infty} H_-(\zeta) = 0. \tag{7.257}
$$

Thus,

$$
\lim_{\zeta \to 0} \hat{H}_-(\zeta) = 0. \tag{7.258}
$$

The last relation means that

$$
\hat{H}_-(\zeta) = \zeta \hat{H}_{1-}(\zeta), \tag{7.259}
$$

where $\hat{H}_{1-}(\zeta)$ is a rational matrix with all its poles lying outside the circle $|\zeta| = 1$. Since in (7.245), the matrix $\theta(\zeta)$ is stable by supposition, so due to the above shown, we conclude that the integrand of $J_{13}$ is analytic on the closed disc $|\zeta| \leq 1$. Therefore, the residue theorem yields

$$
J_{13} = 0. \tag{7.260}
$$

Moreover, by substituting in the integral $J_{12}$ the integration variable $\zeta$ by the variable $\zeta_1 = \zeta^{-1}$, we easily find out that

$$
J_{12} = J_{13} = 0. \tag{7.261}
$$

Thus, (7.255) means only

$$
J_1 = J_{11} + J_{14}. \tag{7.262}
$$

Since (7.256) says that the integral $J_{14}$ does not depend on the system function $\theta(\zeta)$, so the optimal matrix $\theta^0(\zeta)$ is determined only by the condition on minimality of the integral $J_{11}$. However, from the first relation of (7.256), we easily recognize that the minimal value of the integral $J_{11}$ is equal to zero, and this value holds when

$$\Pi(\zeta)\theta^0(\zeta)\Gamma(\zeta) - H_+(\zeta) = 0_{\alpha,\beta}, \tag{7.263}$$

which is equivalent to (7.244). ■

## 7.7 Characteristic Polynomial and Fixed Poles of the $\mathscr{H}_2$ Optimal System

(1)

**Lemma 7.11** *For the matrix $A_M(\zeta)$ in (7.81) to be positive at the circle $|\zeta| = 1$, it is necessary and sufficient that the following conditions (a) and (b) are fulfilled:*

*(a) The quasi-polynomial matrix $A_{M1}(\zeta)$ in* (7.123) *is positive on the circle $|\zeta| = 1$.*
*(b) The polynomial $\delta_M(p) \sim$ det $g_{M1}(s)$ in* (7.90) *and* (7.93) *does not possess roots on the imaginary axis* Re $s = 0$, *or equivalently, the polynomial* det $a_{M1}(\zeta) \sim \Delta_M(\zeta)$ *in* (7.106) *does not possess roots on the circle $|\zeta| = 1$.*

*Proof* Necessity: Let the quasi-polynomial $A_M(\zeta)$ be positive at the circle $|\zeta| = 1$. Then, as was shown in [9], there exists the factorization

$$\det A_M(\zeta) = \rho_M(\zeta)\rho_M(\zeta^{-1}), \tag{7.264}$$

where $\rho(\zeta)$ is a stable polynomial, i.e., it does not possess roots in the disc $|\zeta| \leq 1$. At the same time, from (7.147), we obtain

$$\det A_M(\zeta) = \det a_{M1}(\zeta) \det a_{M1}(\zeta^{-1}) \det A_{M1}(\zeta). \tag{7.265}$$

It is obvious, that in the case that the quasi-polynomial $A_{M1}(\zeta)$ is not positive on the circle $|\zeta| = 1$, or if condition (b) does not hold, then a factorization of the form (7.264) does not exist.

Sufficiency: Let condition (a) be fulfilled. Then there exists the factorization

$$\det A_{M1}(\zeta) = \rho_{M1}(\zeta)\rho_{M1}(\zeta^{-1}), \tag{7.266}$$

where $\rho_{M1}(\zeta)$ is a stable polynomial. Assume that, in addition, condition (b) is also true, and $p_1, \ldots, p_\eta$ be the eigenvalues of the matrix $g_{M1}(s)$ in (7.89), which coincide with the roots of polynomial (7.93). Let us have from (7.93)

$$\delta_M(s) = (s - p_1)^{\gamma_1} \cdots (s - p_\eta)^{\gamma_\eta}, \tag{7.267}$$

where $\gamma_i > 0, \quad (i = 1, \ldots, \eta)$. Assume herein

$$\operatorname{Re} p_i < 0, \quad (i = 1, \ldots, \psi), \quad \operatorname{Re} p_i > 0, \quad (i = \psi + 1, \ldots, \eta). \tag{7.268}$$

In this case

$$\delta_M(s) = \delta_M^+(s)\delta_M^-(s), \tag{7.269}$$

where

$$\begin{aligned} \delta_M^+(s) &\stackrel{\triangle}{=} (s - p_1)^{\gamma_1} \cdots (s - p_\psi)^{\gamma_\psi}, \\ \delta_M^-(s) &\stackrel{\triangle}{=} (s - p_{\psi+1})^{\gamma_{\psi+1}} \cdots (s - p_\eta)\gamma_\eta. \end{aligned} \tag{7.270}$$

Then, from (7.99), we find

$$\Delta_M(\zeta) \sim \Delta_{M1}^+(\zeta)\Delta_{M1}^-(\zeta), \tag{7.271}$$

where

$$\begin{aligned} \Delta_{M1}^+(\zeta) &\stackrel{\triangle}{=} (1 - \zeta e^{p_1 T})^{\gamma_1} \cdots (1 - \zeta e^{p_\psi T})^{\gamma_\psi}, \\ \Delta_{M1}^-(\zeta) &\stackrel{\triangle}{=} (1 - \zeta e^{p_{\psi+1} T})^{\gamma_{\psi+1}} \cdots (1 - \zeta e^{p_\eta T})^{\gamma_\eta} \end{aligned} \tag{7.272}$$

are polynomials, and the first one is stable, while the second one is antistable. Owing to $\operatorname{Re} p_i \neq 0, \ (i = 1, \ldots, \eta)$, there exists the factorization

$$\Delta_M(\zeta)\Delta_M(\zeta^{-1}) = \Delta_{M2}(\zeta)\Delta_{M2}(\zeta^{-1}), \tag{7.273}$$

where

$$\Delta_{M2}(\zeta) \sim (1 - \zeta e^{p_1 T})^{\gamma_1} \cdots (1 - \zeta e^{p_\psi T})^{\gamma_\psi} (1 - \zeta e^{-p_{\psi+1} T})^{\gamma_{\psi+1}} \cdots (1 - \zeta e^{-p_\eta T})^{\gamma_\eta} \tag{7.274}$$

is a stable polynomial. As a result, we conclude that when conditions (a) and (b) are satisfied, then there exists the factorization

$$\det A_M(\zeta) \sim \Delta_M^*(\zeta)\Delta_M^*(\zeta^{-1}), \tag{7.275}$$

where

$$\Delta_M^*(\zeta) \stackrel{\triangle}{=} \Delta_{M2}(\zeta)\rho_{M1}(\zeta) \tag{7.276}$$

is a stable polynomial.

Since the quasi-polynomial $A_{M1}(\zeta)$ is positive on the circle $|\zeta| = 1$, so from (7.275) and (7.276), it follows that $A_M(\zeta)$ is positive on this circle too. ■

**Theorem 7.6** *Let the period $T$ be non-pathological and the conditions of Lemma 7.11 be fulfilled. Then the following statements hold:*

*(i) There exists the factorization*

$$A_M(\zeta) = \Gamma(\zeta)\hat{\Gamma}(\zeta), \tag{7.277}$$

*where $\Gamma(\zeta) \in \mathbb{R}^{nn}[\zeta]$ is a stable polynomial matrix. Hereby, matrices $a_l(\zeta)$ and $a_M(\zeta)$ in ILMFD* (7.97) *can be chosen in such a way that*

$$\deg \Gamma(\zeta) \leq \chi - 1. \tag{7.278}$$

*(ii) The determinant of the matrix $\Gamma(\zeta)$ satisfies*

$$\det \Gamma(\zeta) \sim \Delta_M^*(\zeta) \sim \Delta_{M2}(\zeta)\rho_{M1}(\zeta), \tag{7.279}$$

*where the stable polynomial $\Delta_{M2}(\zeta)$ is defined by formula* (7.274) *and the stable polynomial $\rho_{M1}(\zeta)$ emerges from factorization* (7.266).

*(iii) The degree estimation holds:*

$$\deg \det \Gamma(\zeta) \leq \chi - n. \tag{7.280}$$

*Proof* (i) The existence of factorization (7.277) directly follows from the general theorem [9] about factorization of quasi-polynomial matrices, which are non-negative on the circle $|\zeta| = 1$. For the proof of estimate (7.278), notice that due to the stability of the matrix $\Gamma(\zeta)$, we have

$$\det \Gamma(0) \sim \Delta_M^*(0) \neq 0. \tag{7.281}$$

As was shown in [9], from (7.281), we obtain

$$\deg \Gamma(\zeta) = \operatorname{Deg} A_M(\zeta). \tag{7.282}$$

Hence, using (7.145), we verify (7.278).

(ii) From (7.277), we obtain

$$\det A_M(\zeta) = \det \Gamma(\zeta) \det \Gamma(\zeta^{-1}). \tag{7.283}$$

Inserting this relation into (7.275) and (7.276), formula (7.279) is achieved.

(iii) Hereby, estimate (7.280) is a consequence of estimate (7.146). ∎

(2)

**Lemma 7.12** *Let the conditions of Lemma 7.7 be fulfilled. Then, the quasi-polynomial matrix $A_L(\zeta)$ in* (7.81) *is positive on the circle $|\zeta| = 1$, if and only if it satisfies the following conditions (a) and (b).*

(a) *The quasi-polynomial* $A_{L1}(\zeta)$ *in* (7.187) *is positive on the circle* $|\zeta| = 1$.
(b) *The polynomial* $\delta_L(p)$ *from* (7.164) *does not possess roots on the imaginary axis* $\operatorname{Re} p = 0$*, or, what is equivalent, the polynomial* $\Delta_L(\zeta)$ *from* (7.181) *does not possess roots on the circle* $|\zeta| = 1$.

*Proof* Necessity: Let the quasi-polynomial matrix $A_L(\zeta)$ be positive on the circle $|\zeta| = 1$. Then, there exists the factorization

$$\det A_L(\zeta) = \rho_L(\zeta)\rho_L(\zeta^{-1}), \tag{7.284}$$

where $\rho_L(\zeta)$ is a stable polynomial. Moreover, from (7.218), we find

$$\det A_L(\zeta) = \det a_{L1}(\zeta) \det a_{L1}(\zeta^{-1}) \det A_{L1}(\zeta). \tag{7.285}$$

Hence, it is easy to see that, in the case when the quasi-polynomial $A_{L1}(\zeta)$ is not positive on the circle $|\zeta| = 1$, then factorization (7.284) is impossible. In analogy, if condition (b) does not hold, then the polynomial $\delta_L(p)$ has roots on the imaginary axis. Hereby, since from (7.164), it follows that

$$\det a_{L1}(\zeta) \sim (1 - \zeta e^{p_1 T})^{\nu_1 - \lambda_1} \cdots (1 - \zeta e^{p_\rho T})^{\nu_\rho - \lambda_\rho}, \tag{7.286}$$

so the polynomial $\det a_{L1}(\zeta)$ will possess roots on the circle $|\zeta| = 1$. In this case, factorization (7.284) is impossible too.

Sufficiency: Let condition (a) be fulfilled. Then the factorization

$$\det A_{L1}(\zeta) = \rho_{L1}(\zeta)\rho_{L1}(\zeta^{-1}) \tag{7.287}$$

is possible, where $\rho_{L1}(\zeta)$ is a stable polynomial. Represent polynomial (7.164) in the form

$$\delta_L(s) = (s - p_1)^{\varkappa_1} \cdots (s - p_\eta)^{\varkappa_\eta}, \tag{7.288}$$

where $\varkappa_i > 0$, $\eta \le \rho$. Under condition (b), the roots of polynomial $\delta_L(s)$ can be arranged in two groups, such that

$$\operatorname{Re} p_i < 0, \quad (i = 1, \ldots, \theta), \quad \operatorname{Re} p_i > 0, \quad (i = \theta + 1, \ldots, \eta). \tag{7.289}$$

Then we can write

$$\delta_L(s) = \delta_L^+(s)\delta_L^-(s), \tag{7.290}$$

where

$$\begin{aligned} \delta_L^+(s) &= (s - p_1)^{\varkappa_1} \cdots (s - p_\theta)^{\varkappa_\theta}, \\ \delta_L^-(s) &= (s - p_{\theta+1})^{\varkappa_{\theta+1}} \cdots (s - p_\eta)^{\varkappa_\eta}. \end{aligned} \tag{7.291}$$

Moreover, under the conditions of Lemma 7.7, using (7.185), we obtain

$$\det a_{L1}(\zeta) \sim (1-\zeta e^{p_1 T})^{\varkappa_1} \cdots (1-\zeta e^{p_\eta T})^{\varkappa_\eta} \stackrel{\triangle}{=} \Delta_{L1}(\zeta), \tag{7.292}$$

which can be written in the form

$$\det a_{L1}(\zeta) \sim \Delta_{L1}^{+}(\zeta)\Delta_{L1}^{-}(\zeta), \tag{7.293}$$

where

$$\begin{aligned} \Delta_{L1}^{+}(\zeta) &\stackrel{\triangle}{=} (1-e^{p_1 T})^{\varkappa_1} \cdots (1-e^{p_\theta T})^{\varkappa_\theta}, \\ \Delta_{L1}^{-}(\zeta) &\stackrel{\triangle}{=} (1-\zeta e^{p_{\theta+1} T})^{\varkappa_{\theta+1}} \cdots (1-\zeta e^{p_\eta} T)^{\varkappa_\eta}. \end{aligned} \tag{7.294}$$

Since condition (7.289) is satisfied, so from (7.292)–(7.294), we achieve that there exists the factorization

$$\det a_{L1}(\zeta) \det a_{L1}(\zeta^{-1}) = \Delta_{L2}(\zeta)\Delta_{L2}(\zeta^{-1}), \tag{7.295}$$

where $\Delta_{L2}(\zeta)$ is a stable polynomial satisfying the relation

$$\Delta_{L2}(\zeta) \sim (1-\zeta e^{p_1 T})^{\varkappa_1} \cdots (1-\zeta e^{p_\theta})^{\varkappa_\theta} (1-\zeta e^{-p_{\theta+1} T})^{\varkappa_{\theta+1}} \cdots (1-\zeta e^{p_\eta T})^{\varkappa_\eta}. \tag{7.296}$$

Applying (7.287) and (7.295), from (7.284), we obtain the factorization

$$\det A_L(\zeta) = \Delta_L(\zeta)\Delta_L(\zeta^{-1}), \tag{7.297}$$

where

$$\Delta_L(\zeta) \sim \Delta_{L2}(\zeta)\rho_{L1}(\zeta) \tag{7.298}$$

is a stable polynomial. Since the quasi-polynomial matrix $A_{L1}(\zeta)$ is positive on the circle $|\zeta| = 1$, so from (7.297), the positivity of the matrix $A_L(\zeta)$ on this circle emerges. ■

**Theorem 7.7** *Under the condition of Lemma 7.12, the following statements hold:*

*(i) There exists the factorization*

$$A_L(\zeta) = \hat{\Pi}(\zeta)\Pi(\zeta), \tag{7.299}$$

*where $\Pi(\zeta) \in \mathbb{R}^{qq}[\zeta]$ is a stable polynomial matrix. Hereby,*

$$\det \Pi(\zeta) \sim \Delta_L(\zeta) \sim \Delta_{L2}(\zeta)\rho_{L1}(\zeta), \tag{7.300}$$

*where the stable polynomial $\Delta_{L2}(\zeta)$ is defined by* (7.296)*, and $\rho_{L1}(\zeta)$ is the stable polynomial from factorization* (7.287).

*(ii) The matrices $a_r(\zeta)$ and $a_L(\zeta)$ in IRMFD* (7.175) *can be chosen in such a way that*

$$\deg \Pi(\zeta) \leq \chi + P, \tag{7.301}$$

*where $P$ is the order of the generalized hold.*

*(iii) The degree estimate holds:*

$$\deg \det \Pi(\zeta) \leq \chi + qP. \tag{7.302}$$

*Proof* (i) The existence of factorization (7.299) is a consequence of the positivity of the quasi-polynomial $A_L(\zeta)$. For the proof of formula (7.300), notice that due to (7.218),

$$\det A_L(\zeta) = \det a_{L1}(\zeta) \det a_{L1}(\zeta^{-1}) \det A_{L1}(\zeta). \tag{7.303}$$

Comparing this relation with (7.297)–(7.299), relation (7.300) is achieved.

(ii) The stability of the matrix $\Pi(\zeta)$ yields

$$\det \Pi(0) \neq 0. \tag{7.304}$$

From (7.304), in analogy to the above said, we find

$$\det \Pi(\zeta) = \operatorname{Deg} A_L(\zeta). \tag{7.305}$$

Hence, using (7.215), inequality (7.301) emerges.

(iii) Estimate (7.302) is a consequence of formula (7.216). ■

(3) As follows from the results of Sect. 7.6, under the conditions for Theorems 7.6 and 7.7, the $\mathcal{H}_2$ optimal system function $\theta^0(\zeta)$ is determined by the formula

$$\theta^0(\zeta) = \Pi^{-1}(\zeta) H_+(\zeta) \Gamma^{-1}(\zeta), \tag{7.306}$$

where $H_+(\zeta)$ is a polynomial matrix, and the stable polynomial matrices $\Pi(\zeta)$, $\Gamma(\zeta)$ are determined with the help of factorizations (7.277) and (7.299). Moreover, as was shown earlier, the characteristic polynomial of the $\mathcal{H}_2$ optimal system $\Delta^0(\zeta)$ is given by the formula

$$\Delta^0(\zeta) \sim \det D_{l0}(\zeta), \tag{7.307}$$

where the matrix $D_{l0}(\zeta)$ is assigned by the ILMFD

$$\theta^0(\zeta) = D_{l0}^{-1}(\zeta) M_{l0}(\zeta). \tag{7.308}$$

The next theorem provides some properties of the characteristic polynomial $\Delta^0(\zeta)$, which are important for practical applications.

**Theorem 7.8** *The following statements hold:*

*(i) The characteristic polynomial of the $\mathscr{H}_2$ optimal system is a divisor of the polynomial*

$$\Delta_H(\zeta) = \det \Pi(\zeta) \det \Gamma(\zeta). \tag{7.309}$$

*(ii) The polynomial* $\det \Pi(\zeta)$ *does not depend on the delays $\tau_i$ in the original standard model $\mathscr{S}_\tau$ (6.12), and the polynomial* $\det \Gamma(\zeta)$ *does not depend on the delays $\tau_i$, and also not from the concrete form of the GDAC (6.10).*

*Proof* (i) Lemma 2.6 provides the existence of the LMFD

$$H_+(\zeta)\Gamma^{-1}(\zeta) = \Gamma_1^{-1}(\zeta)H_{+1}(\zeta), \tag{7.310}$$

in which $\det \Gamma_1(\zeta) \sim \det \Gamma(\zeta)$. Hence, from (7.306), we obtain the LMFD

$$\theta^0(\zeta) = F^{-1}(\zeta)H_{+1}(\zeta), \tag{7.311}$$

where

$$F(\zeta) = \Gamma_1(\zeta)\Pi(\zeta) \tag{7.312}$$

and

$$\det F(\zeta) \sim \det \Pi(\zeta) \det \Gamma(\zeta). \tag{7.313}$$

Since the right side of (7.308) is an ILMFD, so the relation $\dfrac{\det F(\zeta)}{\det D_{l0}(\zeta)}$ is a polynomial.

(ii) This claim is an implication from the fact that the quasi-polynomials $A_M(\zeta)$ and $A_L(\zeta)$ do not depend on the delays and, moreover, the quasi-polynomial $A_M(\zeta)$ does not depend on the choice of the GDAC. ■

*Remark 7.1* In practical applications, the polynomial pair $(F(\zeta), H_+(\zeta))$ is irreducible. In that case, we obtain

$$\det D_{l0}(\zeta) \sim \det F(\zeta) \sim \det \Pi(\zeta) \det \Gamma(\zeta). \tag{7.314}$$

Below, we will suppose that this condition is fulfilled. Then, under (7.314), the formula

$$\det D_{l0}(\zeta) \sim \Delta_{M2}(\zeta)\rho_{M1}(\zeta)\Delta_{L2}(\zeta)\rho_{L1}(\zeta) \tag{7.315}$$

is true, where $\Delta_{M2}(\zeta)$, $\Delta_{L2}(\zeta)$ are the stable polynomials from (7.274) and (7.296), respectively, and the stable polynomials $\rho_{M1}(\zeta)$ and $\rho_{L1}(\zeta)$ are determined by factorizations (7.266) and (7.287), respectively. Hence, relation (7.315) advises that, under condition (7.314), the poles (roots of the characteristic polynomial) of the $\mathscr{H}_2$ optimal system should be partitioned into three types.

Type I Roots of the polynomials $\Delta_{M2}(\zeta)$ and $\Delta_{L2}(\zeta)$. As follows from formula (7.274) and (7.296), these roots can be easily expressed by the fixed poles (roots of polynomials (7.85) and (7.156)). If, however, one of the polynomials (7.85), (7.156) has roots with a real part almost zero, then at least one of factorizations (7.277) or (7.299) becomes impossible. In this situation, a formal application of the Wiener–Hopf method leads to a system at the border of stability. As was shown in [7], in similar cases, the minimum of functional (7.79) in general is not achieved on the set of stabilizing rational matrices $\theta(\zeta)$.

Type II Roots of the polynomial $\rho_{M1}(\zeta)$. The special feature of the roots of this type consists in the fact that they do not depend on the delays, and also not on the choice of GDAC (6.10). The reason for it is the fact that the matrix $A_M(\zeta)$ does not depend on the choice of the GDAC.

Type III Roots of the polynomial $\rho_{L1}(\zeta)$. These roots also do not depend on the delays, but, in contrast to the roots of the second type, these poles do depend on the choice of the GDAC (6.10).

(4)

**Definition 7.5** The poles of Type I are called fixed poles.

By utilizing the concept of fixed poles, we are able to solve the $\mathscr{H}_2$ optimization problem for the standard model for SD systems with delay, even in cases when some of the poles of the extended matrix $\breve{W}_0(s)$ (6.23) have $\operatorname{Re} s = 0$. This problem was formulated in [10] for standard SD models without delay, and using zero-order hold as special case of the GDAC. The general solution for the standard model of SD system with delay and GDAC is formulated now.

**Theorem 7.9** *Let $p_1, \dots, p_\mu$ be the eigenvalues of the matrix $A$ in the standard model (6.12) located on the imaginary axis. Then the following claims hold:*

*(i) If among the numbers $\zeta_i = \mathrm{e}^{-p_i T}$, $(i = 1, \dots, \mu)$, there exist some of them belonging to the set of fixed poles, then the $\mathscr{H}_2$ optimization problem does not possess a solution in the set of stable rational matrices $\theta(\zeta)$.*

*(ii) If there is no any $\mathrm{e}^{-p_i T}$, $(i = 1, \dots, \mu)$ belonging to the set of fixed poles, then the polynomials $\rho_{M1}(\zeta)$ and $\rho_{L1}(\zeta)$ do not possess roots on the circle $|\zeta| = 1$, and the $\mathscr{H}_2$ optimization problem has a solution in the set of stable system functions, and this solution can be obtained by the Wiener–Hopf method.*

*Proof* The proof obviously follows from the above considerations. ∎

Apart from the above said, we remark that the fact of the existence of fixed poles essentially restricts the ability and performance of $\mathscr{H}_2$ optimal systems. If among the fixed poles, there are some close to the circle $|\zeta| = 1$ or, equivalently, some of the corresponding fixed poles (roots of polynomial $\delta(s) = \delta_M(p)\delta_L(s)$) are located close to the imaginary axis, then the designed sampled-data system on basis of the $\mathscr{H}_2$ optimization is unusable.

Thus, for practical applications, the following conclusion is important: Before starting the $\mathcal{H}_2$ optimization procedure, investigate the character and properties of the set of fixed poles.

## 7.8 Examples for Constructing the Set of Fixed Poles

(1) In this section, we consider as example the problem of determining the set of fixed poles for single-loop MIMO SD systems with delay, the structure of which is shown in Fig. 6.9. We choose various output vectors $z(t)$ and the condition on model standardizability holds.

*Example 7.1* Under the conditions of Example 6.1, from (6.235) we obtain

$$M(s) = W_3(s)W_2(s) \tag{7.316}$$

and

$$L(s) = N(s) = W_3(s)W_2(s)W_1(s). \tag{7.317}$$

Let the conditions on model standardizability be fulfilled, which are obtained in Example 6.7, and assume the ILMFD

$$W_i(s) = g_i^{-1}(s)d_i(s), \quad (i = 1, 2, 3) \tag{7.318}$$

where

$$\deg\det g_i(s) = \text{Mind}\, W_i(s). \tag{7.319}$$

Then, due to the independence of the matrices $W_i(s)$ (6.299), we obtain

$$\text{Mind}\, N(s) = \chi_3 + \chi_2 + \chi_1, \tag{7.320}$$

where $\chi_i \triangleq \text{Mind}\, W_i(s), \quad (i = 1, 2, 3)$.

From (7.316) and (7.320), we find

$$\text{Mind}\, M(s) = \chi_3 + \chi_2. \tag{7.321}$$

Using (7.317) and (7.318), we obtain

$$N(s) = g_3^{-1}(s)d_3(s)g_2^{-1}(s)d_2(s)g_1^{-1}(s)d_1(s). \tag{7.322}$$

Hereby, from (7.320) we achieve that the pair $[g_2(s), d_3(s)]$ is irreducible, and thus, in the ILMFD

$$d_3(s)g_2^{-1}(s) = g_4^{-1}(s)d_4(s) \tag{7.323}$$

we obtain

$$\begin{aligned} \det g_4(s) &\sim \det g_2(s), \\ \deg\det g_4(s) &= \deg\det g_2(s) = \chi_2. \end{aligned} \tag{7.324}$$

With the help of (7.323), from (7.322), we find

$$N(s) = g_3{}^{-1}(s) g_4{}^{-1}(s) d_4(s) d_2(s) g_1{}^{-1}(s) d_1(s). \tag{7.325}$$

Since condition (7.320) is true, so in (7.325) the pair $[g_1(s), d_4(s)d_2(s)]$ is irreducible. Hence, for the ILMFD,

$$[d_4(s)d_2(s)] g_1{}^{-1}(s) = g_5{}^{-1}(s) d_5(s) \tag{7.326}$$

the relations

$$\begin{aligned} \det g_5(s) &\sim \det g_1(s), \\ \deg\det g_5(s) &= \deg\det g_1(s) = \chi_1. \end{aligned} \tag{7.327}$$

are true. From (7.325) and (7.326), we build the ILMFD

$$N(s) = g_3{}^{-1}(s) g_4{}^{-1}(s) g_5{}^{-1}(s) d_5(s) d_1(s) \stackrel{\triangle}{=} g_l{}^{-1}(s) d_l(s),$$

where

$$\begin{aligned} g_l(s) &= g_5(s) g_4(s) g_3(s), \\ d_l(s) &= d_5(s) d_1(s). \end{aligned} \tag{7.328}$$

Applying the analog transformations to matrix (7.316), we arrive at the ILMFD

$$M(s) = g_M{}^{-1}(s) d_M(s), \tag{7.329}$$

where

$$\begin{aligned} g_M(s) &= g_4(s) g_3(s), \\ d_M(s) &= d_4(s) d_2(s). \end{aligned} \tag{7.330}$$

With the help of (7.328) and (7.330), from (7.90), we find

$$\delta_M(s) \sim \det g_5(s) \sim \det g_1(s). \tag{7.331}$$

The last relation means that the polynomial $\delta_M(s)$ is equivalent to the McMillan denominator of the matrix $W_1(s)$. Moreover, owing to $L(s) = N(s)$, we state that the polynomial $\delta_L(s)$ in (7.156) appears as a constant.

Summing up, we can conclude that in the given example, when the conditions of model standardizability provided in Example 6.7 are fulfilled, the set of fixed poles is determined only by the properties of the poles of the matrix $W_1(s)$. Assume the minimal realization

$$W_1(s) = C(sI_{\chi_1} - A_1)^{-1}B, \tag{7.332}$$

then the set of fixed poles consists of the set of roots of the polynomial

$$\det g_1(s) \sim \det(sI_{\chi_1} - A_1). \tag{7.333}$$

Moreover, assume

$$\det(sI_{\chi_1} - A_1) = (s - p_1)^{\alpha_1} \cdots (s - p_m)^{\alpha_m}. \tag{7.334}$$

Then, if some of the roots $p_i$ are located on the imaginary axis, a $\mathscr{H}_2$ optimal stable system function $\theta^0(\zeta)$ does not exist. If, however, there are no roots on the imaginary axis, and

$$\operatorname{Re} p_i < 0, \quad (i = 1, \ldots, r), \quad \operatorname{Re} p_i > 0, \quad (i = r + 1, \ldots, m), \tag{7.335}$$

then the numbers

$$\zeta_1 = \mathrm{e}^{-p_1 T}, \ldots, \zeta_r = \mathrm{e}^{-p_r T}, \quad \zeta_{r+1} = \mathrm{e}^{p_{r+1} T}, \ldots, \zeta_m = \mathrm{e}^{p_m T} \tag{7.336}$$

appear as fixed poles of the $\mathscr{H}_2$ optimal system.

(2)

*Example 7.2* Choose as output of the system in Fig. 6.9 the vector $z_2$. Then, from (6.239), we obtain

$$M(p) = W_3(s)W_2(s), \quad L(s) = W_2(s)W_1(s) \tag{7.337}$$

and

$$N(s) = W_3(s)W_2(s)W_1(s). \tag{7.338}$$

The conditions on model standardizability, which are formulated in Example 6.8, imply relation (7.320). Therefore, the polynomial $\delta_M(s)$ in the given example is solely determined by relation (7.331). For the determination of the polynomial $\delta_L(s)$, we assume the IRMFD

$$W_i(s) = c_i(s) f_i^{-1}(s), \quad (i = 1, 2, 3), \tag{7.339}$$

where

$$\deg \det f_i(s) = \text{Mind}\, W_i(s) = \chi_i. \tag{7.340}$$

With the help of (7.339), relation (7.338) can be written in the form

$$N(s) = c_3(s) f_3^{-1}(s) c_2(s) f_2^{-1}(s) c_1(s) f_1^{-1}(s). \tag{7.341}$$

Independence condition (7.320) yields that the pair $(f_2(s), c_1(s))$ is irreducible. Therefore, in the IRMFD

$$f_2^{-1}(s) c_1(s) = c_4(s) f_4^{-1}(s), \tag{7.342}$$

takes place

$$\det f_4(s) \sim \det f_2(s), \deg \det f_4(s) = \deg \det f_2(s) = \chi_2. \tag{7.343}$$

Owing to (7.342), from (7.341) we find

$$N(s) = c_3(s) f_3^{-1}(s) c_2(s) c_4(s) f_4^{-1}(s) f_1^{-1}(s). \tag{7.344}$$

From the examinations, provided in Example 7.1, we realize that the pair $(f_3(s), c_2(s)c_4(s))$ is irreducible. Hence, in the IRMFD,

$$f_3^{-1}(s) \left[c_2(s) c_4(s)\right] = c_5(s) f_5^{-1}(s) \tag{7.345}$$

the relations

$$\det f_5(s) \sim \det f_3(s), \quad \deg \det f_5(s) = \deg \det f_3(s) = \chi_3 \tag{7.346}$$

are true. From (7.344) and (7.345), we obtain the IRMFD

$$N(s) = c_3(s) c_5(s) f_5^{-1}(s) f_4^{-1}(s) f_1^{-1}(s) \triangleq c_r(s) f_r^{-1}(s), \tag{7.347}$$

where

$$\begin{aligned} f_r(s) &= f_1(s) f_4(s) f_5(s), \\ c_r(s) &= c_3(s) c_5(s). \end{aligned} \tag{7.348}$$

With the help of analog transformations for the matrix $L(s) = W_2(s) W_1(s)$, we can obtain the IRMFD

$$L(s) = c_2(s)c_4(s)f_4^{-1}(s)f_1^{-1}(s) \triangleq c_L(s)f_L^{-1}(s), \tag{7.349}$$

where

$$f_L(s) = f_1(s)f_4sp), \quad c_L(s) = c_2(s)c_4(s). \tag{7.350}$$

As result of the above calculations, we find from (7.162)

$$\delta_L(s) \sim \frac{\det f_r(s)}{\det f_L(s)} = \det f_5(s) \sim \det f_3(s). \tag{7.351}$$

This relation implies that in the given example, the polynomial $\delta_L(s)$ is equivalent to the McMillan denominator of the matrix $W_3(s)$. Thus, if at least one of the matrices $W_1(s)$ or $W_3(s)$ possesses poles located on the imaginary axis, then, under the conditions of Example 7.2, the Wiener–Hopf method cannot be applied. If however, those poles do not exist, then the Wiener–Hopf method yields the wanted result, and the set of fixed poles coincides with the union of the sets of roots of the polynomials $\delta_M(s)$, $\delta_L(s)$ from (7.331), (7.351).

*Example 7.3* Under the conditions of Examples 6.3, 6.9, we obtain

$$M(s) = W_3(s)W_2(s), \quad L(s) = W_1(s) \tag{7.352}$$

and

$$N(s) = W_3(s)W_2(s)W_1(s). \tag{7.353}$$

As has been the case in the previous examples, when (7.318) is fulfilled, we obtain

$$\delta_M(s) \sim \det a_1(s). \tag{7.354}$$

Moreover

$$\delta_L(s) \sim \frac{\det f_r(s)}{\det f_1(s)} \sim \det f_2(s) \det f_3(s). \tag{7.355}$$

The last two relations mean that in the present example, the poles of all continuous elements $W_i(s)$ are fixed, and therefore, the Wiener–Hopf method yields the solution of the $\mathscr{H}_2$ optimization problem only when all continuous elements of the system do not possess poles on the imaginary axis.

## References

1. E.N. Rosenwasser, *Linear Theory of Digital Control in Continuous Time* (Nauka, Moscow, 1994). (in Russian)
2. E.N. Rosenwasser, B.P. Lampe, *Computer Controlled Systems: Analysis and Design with Process-orientated Models* (Springer, Belin, 2000)

3. E.N. Rosenwasser, B.P. Lampe, *Multivariable Computer-controlled Systems–A Transfer Function Approach* (Springer, London, 2006)
4. B.P. Lampe, M.A. Obraztsov, E.N. Rosenwasser, Statistical analysis of stable FDLCP systems by parametric transfer matrices. Int. J. Control **78**(10), 747–761 (2005)
5. E.N. Rosenwasser, *Linear Theory of Digital Control in Continuous Time* (Izd, LKI, St. Petersburg, 1989). (in Russian)
6. B.P. Lampe, M.A. Obraztsov, E.N. Rosenwasser, $\mathscr{H}_2$-norm computation for stable linear continuous-time periodic systems. Arch. Control Sci. **14**(2), 147–160 (2004)
7. V.N. Fomin, *Control Methods for Discrete Multidimensional Processes* (University Press, Leningrad, 1985). (in Russian)
8. T. Kailath, *Linear Systems* (Prentice Hall, Englewood Cliffs, NJ, 1980)
9. V.M. Popov, *Hyperstability of Control Systems* (Springer, Berlin, 1973)
10. T. Hagiwara, M. Araki, FR-operator approach to the $\mathscr{H}_2$-analysis and synthesis of sampled-data systems. IEEE Trans. Autom. Control, **AC-40**(8):1411–1421 (1995)

# Chapter 8
# $\mathscr{H}_2$ Optimization and Fixed Poles for the Standard Model of SD Systems with Delay Under Colored Noise

**Abstract** This chapter deals with the problem of $\mathscr{H}_2$-optimization of the standard model of a time-delayed SD system $\mathscr{S}_\tau$ affected on its input by a colored stochastic stationary signal. To solve the problem, at the input of the $\mathscr{S}_\tau$ system, a corresponding stable input shaping filter is added, and then the application of the optimization algorithm described in Chap. 7 is possible. This approach leads to a modified PTM, and unlike the approach currently used, is not related to the extension of the state space of a continuous object. The application of this method formally leads to the conclusion that the fixed poles of the $\mathscr{H}_2$-optimal system contain the external poles generated by the shaping filter for the input signal. A detailed examination of the solution obtained shows that these external poles can be shortened and the set of fixed poles for the problem of $\mathscr{H}_2$-optimization is the same in cases of white and colored input noise.

## 8.1 Statement of the $\mathscr{H}_2$ Optimization Problem Under Colored Noise

This chapter deals with the $\mathscr{H}_2$ optimization of the standard model of the system with delay (6.12), in the case when the input signal $x(t)$ is established as the response of a stable forming filter to a vector signal of white noise $g(t)$. In this case, according to the input $g(t)$, we obtain the system $\mathscr{S}_d$, the structure of which is shown in Fig. 8.1.

In Fig. 8.1, $\Phi(p)$ is the $l \times l$ transfer matrix of the coloring filter. Suppose that $\Phi(p)$ is a strictly proper rational matrix possessing the minimal realization

$$\Phi(p) = C_\Phi(pI_\varkappa - A_\Phi)^{-1}B_\Phi. \tag{8.1}$$

Obviously, the system $\mathscr{S}_d$ can be represented in the form of a series connection, as shown in Fig. 8.2, from input $g(t)$ to output $z(t)$.

From Fig. 8.2, and in consistence with the general rules for the calculation of the parametric transfer matrices for series-connected elements [1–3], we obtain that the PTM $W_{zg}(p, t)$ of the system $\mathscr{S}_d$ from the input $g(t)$ to the output $z(t)$ is



E. N. Rosenwasser et al., *Computer-Controlled Systems with Delay*,
https://doi.org/10.1007/978-3-030-15042-6_8

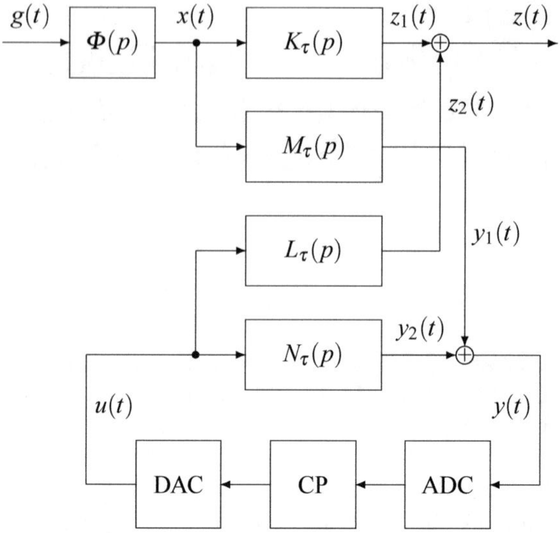


**Fig. 8.1** Standard model of SD system with delay under colored noise

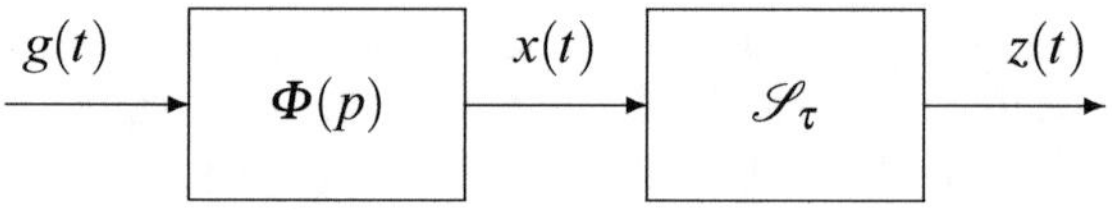


**Fig. 8.2** Equivalent structure of system $\mathscr{S}_d$

$$W_{zg}(p,t) = W_{zx}(p,t)\Phi(p), \tag{8.2}$$

where $W_{zx}(p,t)$ is the PTM of the standard model $\mathscr{S}_\tau$ from the input $x(t)$ to the output $z(t)$, which emerges from (6.67) for $\lambda = p$

$$W_{zx}(p,t) = \varphi_{L_\tau\mu}(T,p,t)\breve{R}_N(p)M_\tau(p) + K_\tau(p). \tag{8.3}$$

Thus, the investigation of the response of the model $\mathscr{S}_\tau$ to the colored noise $x(t)$ is equivalent to investigating the response of the system $\mathscr{S}_d$ to vector white noise $g(t)$. Owing to the stability of the coloring filter, the system $\mathscr{S}_d$ is stable if and only if the standard model $\mathscr{S}_\tau$ is stable. Therefore, if the standard model is stable, then in analogy to the formulae from Sect. 7.1, the solution for the advanced statistical analysis problem of the system $\mathscr{S}_d$ can be found, and it is given by the formula

$$d_z(t) = \frac{1}{2\pi \mathrm{j}} \int_{-\mathrm{j}\infty}^{\mathrm{j}\infty} \operatorname{trace}\left[W'_{zg}(-s,t)W_{zg}(s,t)\right] \mathrm{d}s. \tag{8.4}$$

According to (8.2), this formula is equivalent to

$$d_z(t) = \frac{1}{2\pi \mathrm{j}} \int_{-\mathrm{j}\infty}^{\mathrm{j}\infty} \operatorname{trace}\left[\Phi'(-s)W'_{zx}(-s,t)W_{zx}(s,t)\Phi(s)\right] \mathrm{d}s. \tag{8.5}$$

The mean variance of the output $\bar{d}_z$ is determined by the formula

$$\bar{d}_z = \frac{1}{2\pi \mathrm{j}} \int_{-\mathrm{j}\omega}^{\mathrm{j}\omega} \Phi'(-s) A_{zx}(s) \Phi(s)\, \mathrm{d}s, \tag{8.6}$$

where, as in Chap. 7,

$$A_{zx}(s) = \frac{1}{T} \int_0^T W'_{zx}(-s,t) W_{zx}(s,t)\, \mathrm{d}t. \tag{8.7}$$

The number

$$\|\mathscr{S}_d\|_2 \triangleq +\sqrt{\bar{d}_z} \tag{8.8}$$

is called the $\mathscr{H}_2$ norm of the standard model $\mathscr{S}_\tau$ under the colored noise excitation $x(t)$. Thus, the $\mathscr{H}_2$ optimization problem for the system $\mathscr{S}_d$ can be formulated as follows:

**$\mathscr{H}_2$ optimal control problem for $\mathscr{S}_d$**: For a given standard model (6.12) and given matrices (8.1), find the transfer matrix $W_d(\zeta)$ of a causal discrete controller which stabilizes the standard model and yields the minimal norm (8.8).

Notice that in the formulated $\mathscr{H}_2$ optimization problem, the system $\mathscr{S}_d$ is stable if and only if the standard model $\mathscr{S}_\tau$ is stable.

## 8.2 PTM and $\mathscr{H}_2$ Norm of the System $\mathscr{S}_d$

(1) Applying formulae (8.2) and (8.3), and also the relations of Chap. 6, we are able to formulate some general properties of the PTM $W_{zg}(s,t)$, which are provided now:

(a) The set of poles of the PTM $W_{zg}(s,t)$ is built by the unification of the sets of poles of the PTM $W_{zx}(s,t)$ and the matrix $\Phi(s)$.
(b) Since the coloring filter is by supposition stable, so the set of stabilizing controllers for the system $\mathscr{S}_d$ coincides with the set of stabilizing controllers for the standard model $\mathscr{S}_\tau$.
(c) Below, we will always assume that for the standard model $\mathscr{S}_\tau$ the conditions of Theorem 6.7 are fulfilled. Then, the set of causal stabilizing controllers $(\alpha_l(\zeta), \beta_l(\zeta))$ for the system $\mathscr{S}_d$ is determined by formulae (6.173), independently on the form of the transfer matrix of the coloring filter:

$$\begin{aligned}\alpha_l(\zeta) &= D_l(\zeta)\alpha_{l0}(\zeta) - M_l(\zeta) b_l(\zeta, -\tau_2),\\ \beta_l(\zeta) &= D_l(\zeta)\beta_{l0}(\zeta) - M_l(\zeta) a_l(\zeta),\end{aligned} \tag{8.9}$$

where the notations of Sect. 7.5 are applied.

(2) From (8.2) and (8.3), we generate simple expressions for the PTM $W_{zg}(s,t)$

$$W_{zg}(s,t) = \varphi_{L_\tau\mu}(T,s,t)\breve{R}_N(s)M_\tau(s)\Phi(s) + K_\tau(s)\Phi(s). \tag{8.10}$$

Hence, using formula (6.195), we find analogously to (6.199), the representation of the PTM $W_{zg}(s,t)$ by the system function $\breve{\theta}(s)$

$$W_{zg}(s,t) = \psi(s,t)\breve{\theta}(s)\xi_\Phi(s) + \eta_\Phi(s,t), \tag{8.11}$$

where the matrix $\psi(s,t)$ is as in (6.200) and the matrices $\xi_\Phi(s)$, $\eta_\Phi(s,t)$ are connected with the matrices $\xi(s)$, $\eta(s,t)$ from (6.200) by the relations

$$\xi_\Phi(s) = \xi(s)\Phi(s), \quad \eta_\Phi(s,t) = \eta(s,t)\Phi(s). \tag{8.12}$$

From formula (8.12) and Theorem 6.9, we conclude that the sets of poles for the matrices $\xi_\Phi(s)$ and $\eta_\Phi(s,t)$ coincide with the set of poles of the matrix $\Phi(s)$.

(3) Using (8.11) and (8.4), we can build closed expressions for the $\mathscr{H}_2$ norm for the system $\mathscr{S}_d$. Hereby, notice that the matrix $W_{zg}(s,t)$ is obtained from the matrix $W_{zx}(s,t)$, when the matrix $M(s)$ is substituted by the matrix $M(s)\Phi(s)$, and the matrix $K(s)$ by $K(s)\Phi(s)$. Therefore, the closed expression for the $\mathscr{H}_2$ norm of the system $\mathscr{S}_d$ can be held from the formulae in Sects. 7.2, 7.3 when substituting $M(p)$ s by $M(s)\Phi(s)$ and $K(s)$ by $K(s)\Phi(s)$. As result of these substitutions, we obtain, in analogy to (7.79), the quadratic functional

$$TJ_\Phi = \frac{1}{2\pi j}\oint \operatorname{trace}\left[\hat{\theta}(\zeta)A_L(\zeta)\theta(\zeta)A_{M\Phi}(\zeta) - \hat{\theta}(\zeta)\hat{C}_\Phi(\zeta) - C_\Phi(\zeta)\theta(\zeta)\right]\frac{d\zeta}{\zeta}, \tag{8.13}$$

where the matrix $A_L(\zeta)$ is determined by the first formula in (7.81), and the matrices $A_{M\Phi}(\zeta)$ and $C_\Phi(\zeta)$ have the form

$$\begin{aligned} A_{M\Phi}(\zeta) &= a_l(\zeta)D_{M\Phi\underline{\Phi'}\underline{M'}}(T,\zeta,0)\hat{a}_l(\zeta),\\ C_\Phi(\zeta) &= a_l(\zeta)D_{M\Phi\underline{\Phi'}\underline{M'}}(T,\zeta,0)\hat{a}_l(\zeta)\hat{\beta}_{r0}(\zeta)D_{\underline{\mu'}\underline{L'}L\mu}(T,\zeta,0)a_r(\zeta)\\ &\quad + a_l(\zeta)D_{M\Phi\underline{\Phi'}\underline{K'}L\mu}(T,\zeta,-\tau_2)a_r(\zeta), \end{aligned} \tag{8.14}$$

when we consider $\tau_2 = \tau_N$. Herein, as earlier, we used the notations

$$D_{M\Phi\underline{\Phi'}\underline{M'}}(T,\zeta,0) = \left[\frac{1}{T}\sum_{k=-\infty}^{\infty} M(s+kj\omega)\Phi(s+kj\omega)\Phi'(-s-kj\omega)M'(-s-kj\omega)\right]_{e^{-sT}=\zeta},$$

$$D_{M\Phi\underline{\Phi'}\underline{K'}L\mu}(T,\zeta,t) \tag{8.15}$$

$$
= \left[ \frac{1}{T} \sum_{k=-\infty}^{\infty} M(s+kj\omega)\Phi(s+kj\omega)\Phi'(-s-kj\omega)\times \right.
$$
$$
\left. \times\ K'(-s-kj\omega)L(s+kj\omega)\mu(s+kj\omega)\mathrm{e}^{(s+kj\omega)t} \right]_{\mathrm{e}^{-sT}=\zeta} .
$$

From the achieved relations, it follows that the $\mathscr{H}_2$ optimization problem for the system $\mathscr{S}_d$ leads to the minimization problem of functional (8.13) over the set of stable rational matrices $\theta(\zeta)$.

## 8.3 Properties of Matrices $A_{M\Phi}(\zeta)$ and $C_\Phi(\zeta)$

(1)

**Lemma 8.1** *Let the matrices $M(s)$ and $\Phi(s)$ be independent, not possess common poles, and the conditions on non-pathology of the sampling period $T$ be true for the poles of the matrix $M(s)\Phi(s)$. Moreover, let the minimal realizations*

$$
\begin{aligned}
M(s) &= C_M(sI_\alpha - A_M)^{-1}B_M, \\
\Phi(s) &= C_\Phi(sI_\varkappa - A_\Phi)^{-1}B_\Phi.
\end{aligned} \tag{8.16}
$$

*be given. Then the following statements are true.*

*(i) There exists the minimal realization*

$$
M(s)\Phi(s) = C_{M\Phi}(sI_{\alpha+\varkappa} - A_{M\Phi})^{-1}B_{M\Phi}, \tag{8.17}
$$

*i.e.,* $\mathrm{Mind}[M(s)\Phi(s)] = \alpha + \varkappa$.

*(ii) In the ILMFD*

$$
\begin{aligned}
C_{M\Phi}(I_{\alpha+\varkappa} - \zeta\mathrm{e}^{A_{M\Phi}T})^{-1} &= a_{M\Phi}^{-1}(\zeta)b_{M\Phi}(\zeta), \\
C_M(I_\alpha - \zeta\mathrm{e}^{A_M T})^{-1} &= a_M^{-1}(\zeta)b_M(\zeta),
\end{aligned} \tag{8.18}
$$

*the matrices $a_{M\Phi}(\zeta)$ and $a_M(\zeta)$ are connected by the equivalence relation*

$$
a_{M\Phi}(\zeta) \sim a_\Phi(\zeta)a_M(\zeta), \tag{8.19}
$$

*where the $n \times n$ matrix $a_\Phi(\zeta)$ is pseudo-equivalent to the matrix $I_\varkappa - \zeta\mathrm{e}^{A_\Phi T}$, i.e.,*

$$
a_\Phi(\zeta) \overset{p}{\sim} I_\varkappa - \zeta\mathrm{e}^{A_\Phi T}. \tag{8.20}
$$

*(iii) For all $t$, the ILMFD*

$$
\mathscr{D}_{M\Phi}(T,\zeta,t) = [a_\Phi(\zeta)a_M(\zeta)]^{-1}b_{M\Phi}(\zeta,t) \tag{8.21}
$$

*is valid, where $b_{M\Phi}(\zeta, t)$ depends polynomially on $\zeta$ for all $t$.*

*Proof* (i) Formula (8.17) is a consequence of the independence of the matrices $M(s)$ and $\Phi(s)$.

(ii) For the proof, we use the auxiliary formula

$$\breve{D}_{M\Phi}(T, s, 0) = \breve{\mathscr{D}}_{M\Phi}(T, s, 0) = \int_0^T \breve{D}_M(T, s, t)\breve{D}_\Phi(T, s, -t)\,\mathrm{d}t, \tag{8.22}$$

which can be seen from the chain rule

$$\int_0^T \left[\frac{1}{T}\sum_{k=-\infty}^{\infty} M(s+k\mathrm{j}\omega)\mathrm{e}^{(s+k\mathrm{j}\omega)t}\right]\left[\frac{1}{T}\sum_{k=-\infty}^{\infty} \Phi(s+k\mathrm{j}\omega)\mathrm{e}^{-(s+k\mathrm{j}\omega)t}\right]\mathrm{d}t$$
$$= \frac{1}{T}\sum_{k=-\infty}^{\infty} M(s+k\mathrm{j}\omega)\Phi(s+k\mathrm{j}\omega) = \breve{D}_{M\Phi}(T, s, 0). \tag{8.23}$$

From relations (5.91), (5.92), we know that for $0 < t < T$,

$$\breve{D}_M(T, s, t) = \breve{\mathscr{D}}_M(T, s, t)$$
$$\breve{D}_\Phi(T, s, -t) = \breve{\mathscr{D}}_\Phi(T, s, T-t)\mathrm{e}^{-sT}.$$

Substituting these relations in (8.22), we find

$$\breve{\mathscr{D}}_{M\Phi}(T, s, 0) = \mathrm{e}^{-sT}\int_0^T \breve{\mathscr{D}}_M(T, s, t)\breve{\mathscr{D}}_\Phi(T, s, T-t)\,\mathrm{d}t. \tag{8.24}$$

However, in the given case

$$\breve{\mathscr{D}}_M(T, s, t) = C_M(I_\alpha - \mathrm{e}^{-sT}\mathrm{e}^{A_M T})^{-1}\mathrm{e}^{A_M t}B_M,$$
$$\breve{\mathscr{D}}_\Phi(T, s, t) = C_\Phi \mathrm{e}^{A_\Phi t}(I_\varkappa - \mathrm{e}^{-sT}\mathrm{e}^{A_\Phi T})^{-1}B_\Phi,$$

so that from (8.24) we achieve

$$\breve{\mathscr{D}}_{M\Phi}(T, s, 0) = \mathrm{e}^{-sT}C_M(I_\alpha - \mathrm{e}^{-sT}\mathrm{e}^{A_M T})^{-1}N(I_\varkappa - \mathrm{e}^{-sT}\mathrm{e}^{A_\Phi T})^{-1}B_\Phi, \tag{8.25}$$

where

$$N \triangleq \int_0^T \mathrm{e}^{A_M t}B_M C_\Phi \mathrm{e}^{A_\Phi(T-t)}\,\mathrm{d}t \tag{8.26}$$

is a constant matrix. Rename in (8.25) $\mathrm{e}^{-sT}$ by $\zeta$, then we obtain the rational matrix

$$\mathscr{D}_{M\Phi}(T,\zeta,0) = \zeta C_M(I_\alpha - \zeta e^{A_M T})^{-1} N (I_\varkappa - \zeta e^{A_\Phi T})^{-1} B_\Phi. \tag{8.27}$$

Using the second formula in (8.18), the last relation is written in the form

$$\mathscr{D}_{M\Phi}(T,\zeta,0) = \zeta a_M^{-1}(\zeta) N_1(\zeta)(I_\varkappa - \zeta e^{A_\Phi T})^{-1} B_\Phi, \tag{8.28}$$

where $N_1(\zeta) \stackrel{\triangle}{=} b_M(\zeta)N$ is a polynomial matrix. Since by supposition, the pair $(A_\Phi, B_\Phi)$ is controllable, and the sampling period $T$ is non-pathological, so in the IRMFD

$$(I_\varkappa - \zeta e^{A_\Phi T})^{-1} B_\Phi = b_{r\Phi}(\zeta) a_{r\Phi}^{-1}(\zeta) \tag{8.29}$$

we have

$$a_{r\Phi}(\zeta) \stackrel{p}{\sim} (I_\varkappa - \zeta e^{A_\Phi T}). \tag{8.30}$$

With regard to (8.29), from (8.28) we obtain

$$\mathscr{D}_{M\Phi}(T,\zeta,0) = \zeta a_M^{-1}(\zeta) N_2(\zeta) a_{r\Phi}^{-1}(\zeta), \tag{8.31}$$

where $N_2(\zeta) \stackrel{\triangle}{=} N_1(\zeta) b_{r\Phi}(\zeta)$ is a polynomial matrix. Moreover, the pair $[a_{r\Phi}(\zeta), N_2(\zeta)]$ is irreducible. If we assumed the contrary, such that the pair would be reducible, then we would obtain the IRMFD

$$N_2(\zeta) a_{r\Phi}^{-1}(\zeta) = N_3(\zeta) d_{r\Phi}^{-1}(\zeta), \tag{8.32}$$

where $\deg\det d_{r\Phi}(\zeta) < \varkappa$. Then Lemma 2.6 guarantees the existence of the ILMFD

$$N_2(\zeta) a_{r\Phi}^{-1}(\zeta) = d_{l\Phi}^{-1}(\zeta) N_4(\zeta), \tag{8.33}$$

in which $\deg\det d_{l\Phi}(\zeta) < \varkappa$. When (8.33) is true, then from (8.31) we obtain the LMFD

$$\mathscr{D}_{M\Phi}(T,\zeta,0) = \zeta \left[d_{l\Phi}(\zeta) a_M(\zeta)\right]^{-1} N_4(\zeta), \tag{8.34}$$

in which $\deg\det[d_{l\Phi}(\zeta)a_M(\zeta)] < \alpha + \chi$, which is impossible, because under the taken suppositions Mind $\mathscr{D}_{M\Phi}(T,\zeta,t) = \alpha + \chi$.
From (8.31), due to the irreducibility of the pair $[a_{r\Phi}(\zeta), N_2(\zeta)]$, it follows the existence of the ILMFD

$$N_2(\zeta) a_{r\Phi}^{-1}(\zeta) = a_\Phi^{-1}(\zeta) N_5(\zeta), \tag{8.35}$$

in which the matrices $a_{r\Phi}(\zeta)$ and $a_\Phi(\zeta)$ are related by pseudo-equivalence $a_\Phi(\zeta) \stackrel{p}{\sim} a_{r\Phi}(\zeta)$. Moreover, from (8.29) we find (8.20). Further, from (8.31) and (8.35) we obtain

$$\mathscr{D}_{M\Phi}(T,\zeta,0) = \left[a_\Phi(\zeta) a_M(\zeta)\right]^{-1} \zeta N_3(\zeta). \tag{8.36}$$

The right side of this formula is an ILMFD, so that

$$\deg\det\left[a_\Phi(\zeta)a_M(\zeta)\right] = \alpha + \varkappa = \operatorname{Mind} \mathscr{D}_{M\Phi}(T,\zeta,0). \tag{8.37}$$

It follows from (8.17), (8.18) that the representation

$$\mathscr{D}_{M\Phi}(T,\zeta,0) = C_{M\Phi}(I_{\alpha+\chi} - \zeta e^{A_{M\Phi}T})^{-1}B_{M\Phi}$$

is minimal, and the matrix relation

$$\mathscr{D}_{M\Phi}(T,\zeta,0) = a_{M\Phi}^{-1}[b_{M\Phi}(\zeta)B_{M\Phi}]$$

is an ILMFD. Comparing the last formula with (8.36) proves (ii).

(iii) From (8.17) we realize that, under the taken propositions, the minimal realization

$$\mathscr{D}_{M\Phi}(T,\zeta,t) = C_{M\Phi}(I_{\alpha+\varkappa} - \zeta e^{A_{M\Phi}T})^{-1}e^{A_{M\Phi}t}B_{M\Phi} \tag{8.38}$$

takes place. Hereby, from the first formula in (8.18), we build the ILMFD

$$\mathscr{D}_{M\Phi}(T,\zeta,t) = a_{M\Phi}^{-1}(\zeta)\left[b_{M\Phi}(\zeta)e^{A_{M\Phi}t}B_{M\Phi}\right]. \tag{8.39}$$

Since the right side of (8.36) is an ILMFD too, we find

$$a_{M\Phi}(\zeta) = k(\zeta)a_\Phi(\zeta)a_M(\zeta), \tag{8.40}$$

where $k(\zeta)$ is an unimodular matrix, and the product

$$a_\Phi(\zeta)a_M(\zeta)\mathscr{D}_{M\Phi}(T,\zeta,t) = k^{-1}(\zeta)[b_{M\Phi}(\zeta)e^{A_M t}B_{M\Phi}] \stackrel{\triangle}{=} b_{M\Phi}(\zeta,t) \tag{8.41}$$

is a polynomial matrix. Herein, owing to

$$\deg\det a_{M\Phi}(\zeta) = \deg\det a_M(\zeta) + \deg\det a_\Phi(\zeta) = \alpha + \varkappa, \tag{8.42}$$

the right side of (8.21) is an ILMFD. ■

(2)

**Theorem 8.1** *Let the conditions of Lemma 8.1 be fulfilled, and assume the ILMFD*

$$C_\Phi(I_\varkappa - \zeta e^{A_\Phi T})^{-1} = a_{\Phi 1}^{-1}(\zeta)b_{\Phi 1}(\zeta). \tag{8.43}$$

*Then the matrix $A_{M\Phi}(\zeta)$ in (8.14) can be represented in the form*

$$A_{M\Phi}(\zeta) = a_{\Phi 2}^{-1}(\zeta)K_\Phi(\zeta)\hat{a}_{\Phi 2}^{-1}(\zeta), \tag{8.44}$$

*where* $K_{\Phi}(\zeta)$ *is a symmetric quasi-polynomial, which is nonnegative on the circle* $|\zeta| = 1$. *The matrix* $a_{\Phi 2}(\zeta) \in \mathbb{R}^{nn}[\zeta]$ *satisfies the equivalence relation*

$$a_{\Phi 2}(\zeta) \sim a_{\Phi 1}(\zeta) \overset{p}{\sim} I_{\varkappa} - \zeta \mathrm{e}^{A_{\Phi}T}. \tag{8.45}$$

*Moreover, the quasi-polynomial* $K_{\Phi}(\zeta)$ *can be represented in the form*

$$K_{\Phi}(\zeta) = a_{M2}(\zeta) K_{\Phi 1}(\zeta) \hat{a}_{M2}(\zeta), \tag{8.46}$$

*where* $K_{\Phi 1}(\zeta)$ *is a quasi-polynomial, which is nonnegative on the circle* $|\zeta| = 1$, *and the matrix* $a_{M2}(\zeta) \in \mathbb{R}^{nn}[\zeta]$ *satisfies*

$$a_{M2}(\zeta) \sim a_{M1}(\zeta), \tag{8.47}$$

*where* $a_{M1}(\zeta) \in \mathbb{R}^{nn}[\zeta]$ *is the polynomial matrix determined by formula (7.105).*

*Proof* For the proof, we use the equalities

$$\begin{aligned}\breve{D}_{M\Phi\underline{\Phi}'\underline{M}'}(T, s, 0) = \breve{\mathscr{D}}_{M\Phi\underline{\Phi}'\underline{M}'}(T, s, 0) &= \int_0^T \breve{D}_{M\Phi}(T, s, t)\breve{D}'_{\Phi M}(T, -s, t)\,\mathrm{d}t \\ &= \int_0^T \breve{\mathscr{D}}_{M\Phi}(T, s, t)\breve{\mathscr{D}}_{\Phi' M'}(T, -s, t)\,\mathrm{d}t \end{aligned} \tag{8.48}$$

which established analogously, to (7.118) when substituting the matrix $M(s)$ by the matrix $M(s)\Phi(s)$. Thus

$$\breve{D}_{\Phi' M'}(T, -s, t) = \frac{1}{T}\sum_{k=-\infty}^{\infty} \Phi'(-s + k\mathrm{j}\omega)M'(-s + k\mathrm{j}\omega)\mathrm{e}^{(-s+k\mathrm{j}\omega)t}. \tag{8.49}$$

Substituting in (8.48) the variable $\mathrm{e}^{-sT}$ by $\zeta$, we find, in analogy to (7.120),

$$D_{M\Phi\underline{\Phi}'\underline{M}'}(T, \zeta, 0) = \breve{D}_{M\Phi\underline{\Phi}'\underline{M}'}(T, s, 0)|_{\mathrm{e}^{-sT}=\zeta} = \int_0^T \mathscr{D}_{M\Phi}(T, \zeta, t)\hat{\mathscr{D}}_{M\Phi}(T, \zeta, t)\,\mathrm{d}t. \tag{8.50}$$

Write ILMFD (8.39) in the form

$$\mathscr{D}_{M\Phi}(T, \zeta, t) = a_{M\Phi}^{-1}(\zeta) b_{M\Phi}(\zeta, t), \tag{8.51}$$

which with regard to (8.40) is equivalent to

$$\mathscr{D}_{M\Phi}(T, \zeta, t) = a_M^{-1}(\zeta) a_{\Phi 3}^{-1}(\zeta) b_{M\Phi}(\zeta, t), \tag{8.52}$$

where

$$a_{\Phi 3}(\zeta) = k(\zeta) a_{\Phi}(\zeta), \tag{8.53}$$

while

$$a_{\Phi 3}(\zeta) \sim a_{\Phi}(\zeta) \stackrel{p}{\sim} (I_\varkappa - \zeta e^{A_\Phi T}),$$

i.e., the matrix $k(\zeta)$ is unimodular. Using (8.52), relation (8.50) can be written in the form

$$D_{M\Phi\underline{\Phi}'M'}(T,\zeta,0) = a_M^{-1}(\zeta)a_{\Phi 3}^{-1}(\zeta)K_{\Phi 1}(\zeta)\hat{a}_{\Phi 3}^{-1}(\zeta)\hat{a}_M^{-1}(\zeta), \tag{8.54}$$

where

$$K_{\Phi 1}(\zeta) \stackrel{\triangle}{=} \int_0^T b_{M\Phi}(\zeta,t)\hat{b}_{M\Phi}(\zeta,t)\,dt \tag{8.55}$$

is a quasi-polynomial, which is nonnegative on the circle $|\zeta| = 1$. Applying (8.54) and (7.105), we find

$$A_{M\Phi}(\zeta) = a_l(\zeta)D_{M\Phi\underline{\Phi}'\underline{M}'}(T,\zeta,0)\hat{a}_l(\zeta) = a_{M1}(\zeta)a_{\Phi 3}^{-1}(\zeta)K_{\Phi 1}(\zeta)\hat{a}_{\Phi 3}^{-1}(\zeta)\hat{a}_{M1}(\zeta). \tag{8.56}$$

Since, owing to (8.53) and (7.104), (7.106), the matrices $a_{M1}(\zeta)$ and $a_{\Phi 3}(\zeta)$ have disjunct eigenvalues, so the matrix relation $a_{M1}(\zeta)a_{\Phi 3}^{-1}(\zeta)$ is an IRMFD. Therefore, the ILMFD

$$a_{\Phi 2}^{-1}(\zeta)a_{M2}(\zeta) = a_{M1}(\zeta)a_{\Phi 3}^{-1}(\zeta) \tag{8.57}$$

satisfies

$$a_{\Phi 2}(\zeta) \sim a_{\Phi 3}(\zeta), \quad a_{M2}(\zeta) \sim a_{M1}(\zeta). \tag{8.58}$$

Inserting (8.57) into (8.56), with the help of earlier results, formulae (8.44)–(8.47) can be achieved. ∎

(3)

**Theorem 8.2** *The matrix $C_\Phi(\zeta)$ in (8.14) is a rational matrix, which can possess poles at the origin $\zeta = 0$ and also at points being roots of the polynomials*

$$\Delta_\Phi(\zeta) \stackrel{\triangle}{=} \det(I_\varkappa - \zeta e^{A_\Phi T}), \quad \Delta_{\underline{\Phi}}(\zeta) \stackrel{\triangle}{=} \det(I_\varkappa - \zeta e^{-A_\Phi T}). \tag{8.59}$$

*Proof* Introduce the matrix

$$Q_\Phi(s) \stackrel{\triangle}{=} -\int_0^T \eta'_\Phi(-s,t)\psi(s,t)\,dt, \tag{8.60}$$

which with regard to (8.12) can be expressed equivalently by

$$Q_\Phi(s) = -\Phi'(-s)\int_0^T \eta'(-s,t)\psi(s,t)\,dt = \Phi'(-s)Q(s), \tag{8.61}$$

where $Q(s)$ is the matrix defined in (7.52). From (8.61) and (8.12) we obtain

$$\xi_{\Phi}(s)Q_{\Phi}(s) = \xi(s)\Phi(s)\Phi'(-s)Q(s). \tag{8.62}$$

Since the matrices $\xi(s)$ and $Q(s)$ are integral functions of the argument $s$, so the matrix (8.62) can possess poles only at the points that are roots of the polynomials

$$d_{\Phi}(s) = \det(sI_{\varkappa} - A_{\Phi}), \quad d_{\Phi 1}(s) = \det(sI_{\varkappa} + A_{\Phi}). \tag{8.63}$$

Owing to the absolute convergence of the series

$$\breve{C}_{\Phi}(s) = \frac{1}{T}\sum_{k=-\infty}^{\infty} \xi_{\Phi}(s + kj\omega)Q_{\Phi}(s + kj\omega) \tag{8.64}$$

for all $s$ and because its elements do not have poles, so the matrix $\breve{C}_{\Phi}(s)$ can have poles only at points that are roots of the functions

$$\breve{\Delta}_{\Phi}(s) \triangleq \det(I_{\varkappa} - e^{-sT}e^{A_{\Phi}T}), \quad \breve{\Delta}_{\underline{\Phi}}(s) \triangleq \det(I_{\varkappa} - e^{-sT}e^{-A_{\Phi}T}). \tag{8.65}$$

Hence, after substituting $e^{-sT} = \zeta$, the matrix

$$C_{\Phi}(\zeta) = \breve{C}_{\Phi}(s)|_{e^{-sT}=\zeta} \tag{8.66}$$

can possess poles only at the point $\zeta = 0$ or at the roots of polynomials (8.59). Moreover, it follows from (8.14) that the matrix $C_{\Phi}(\zeta)$ is rational, which completes the proof. ∎

## 8.4 Wiener–Hopf Method

(1) In order to apply the Wiener–Hopf method to the minimization of functional (8.13), it is necessary to modify Theorem 7.5, because in functional (8.13), in contrast to functional (7.79), the matrices $A_{M\Phi}(\zeta)$ and $C_{M\Phi}(\zeta)$ are no longer quasi-polynomials.

**Theorem 8.3** *Let the quasi-polynomials $A_L(\zeta)$ in (7.81) and $K_{\Phi_1}(\zeta)$ in (8.46), (8.55) be positive on the circle $|\zeta| = 1$. Assume also that the polynomial $\delta_M(s)$ in (7.93) does not possess roots on the imaginary axis. Then the following statements hold:*

*(i) There exists the factorization*

$$A_L(\zeta) = \hat{\Pi}(\zeta)\Pi(\zeta), \tag{8.67}$$

*where $\Pi(\zeta)$ is a stable $q \times q$ polynomial matrix.*
*Moreover, the rational matrix $A_{M\Phi}(\zeta)$ in (8.44) allows the factorization*

$$A_{M\Phi}(\zeta) = a_l(\zeta) D_{M\Phi\underline{\Phi}'\underline{M}'}(T, \zeta, 0)\hat{a}_l(\zeta) = V(\zeta)\hat{V}(\zeta), \tag{8.68}$$

*where the rational matrix* $V(\zeta) \in \mathbb{R}^{nn}(\zeta)$ *is stable together with the inverse.*

*(ii) The optimal system function* $\theta^0(\zeta)$ *can be constructed with the help of the following algorithm:*

*(a) For the rational matrix*

$$H_\Phi(\zeta) \triangleq \hat{\Pi}^{-1}\hat{C}(\zeta)\hat{V}^{-1}(\zeta) \tag{8.69}$$

*perform the separation*

$$H_\Phi(\zeta) = H_\Phi^+(\zeta) + H_\Phi^-(\zeta), \tag{8.70}$$

*where the rational matrix* $H_\Phi^-(\zeta)$ *is strictly proper and has as poles, all the poles of the matrix* $H_\Phi(\zeta)$ *located inside the disc* $|\zeta| < 1$.

*(b) The optimal system function* $\theta^0(\zeta)$ *is determined by the formula*

$$\theta^0(\zeta) = \Pi^{-1}(\zeta)H_\Phi^+(\zeta)V^{-1}(\zeta). \tag{8.71}$$

*Proof* (i) The existence of the factorization (8.67) under the taken propositions is stated in Chap. 7. For the proof of the existence of factorization (8.68), it is essential that we propose the positivity of the quasi-polynomial $K_{\Phi 1}(\zeta)$ on the circle $|\zeta| = 1$. This ensures the existence of the factorization

$$K_{\Phi 1}(\zeta) = \Gamma_{\Phi 1}(\zeta)\hat{\Gamma}_{\Phi 1}(\zeta), \tag{8.72}$$

where $\Gamma_{\Phi_1}(\zeta) \in \mathbb{R}^{nn}[\zeta]$ is a stable polynomial matrix. Moreover, from (8.58) we obtain

$$\det a_{M2}(\zeta) \sim \det a_{M1}(\zeta). \tag{8.73}$$

Therefore, because the polynomial $\delta_M(p)$ (7.93) does not possess roots on the imaginary axis, the polynomial $\det a_{M1}(\zeta)$ does not possess roots on the circle $|\zeta| = 1$, and the same is valid for the polynomial $\det a_{M2}(\zeta)$. Hence, due to (8.46), there exists the factorization

$$K_\Phi(\zeta) = \Gamma_\Phi(\zeta)\hat{\Gamma}_\Phi(\zeta), \tag{8.74}$$

which can be constructed analogously to factorization (7.277). Inserting (8.74) into (8.44), and observing that the matrix $a_{\Phi 2}(\zeta)$ is stable, we obtain a factorization of the form (8.68), wherein

$$V(\zeta) = a_{\Phi 2}^{-1}(\zeta)\Gamma_\Phi(\zeta). \tag{8.75}$$

(ii) If (i) is fulfilled, functional (8.13) can be written in the form

$$TJ_{\Phi} = \frac{1}{2\pi \mathrm{j}} \oint \operatorname{trace} \Big[ \hat{\theta}_{\Phi}(\zeta)\Pi(\zeta)\hat{\Pi}(\zeta)\theta_{\Phi}(\zeta)V(\zeta)\hat{V}(\zeta) \\ - \theta_{\Phi}(\zeta)C_{\Phi}(\zeta) - \hat{C}_{\Phi}(\zeta)\hat{\theta}_{\Phi}(\zeta) \Big] \frac{\mathrm{d}\zeta}{\zeta}, \tag{8.76}$$

which can be presented in a form like (7.251)

$$TJ_{\Phi} = \frac{1}{2\pi \mathrm{j}} \oint \operatorname{trace} \Big[ \hat{V}(\zeta)\hat{\theta}_{\Phi}(\zeta)\hat{\Pi}(\zeta)\Pi(\zeta)\theta_{\Phi}(\zeta)V(\zeta) \\ - \hat{V}(\zeta)\hat{\theta}_{\Phi}(\zeta)\hat{\Psi}_{\Phi}(\zeta) - \Psi_{\Phi}(\zeta)\theta_{\Phi}(\zeta)V(\zeta) \Big] \frac{\mathrm{d}\zeta}{\zeta}, \tag{8.77}$$

with the notation

$$\Psi_{\Phi}(\zeta) = V^{-1}(\zeta)C_{\Phi}(\zeta) = \Gamma_{\Phi}^{-1}(\zeta)a_{\Phi 2}(\zeta)C_{\Phi}(\zeta). \tag{8.78}$$

The further proof runs in the same way as the proof of Theorem 7.5. ■

## 8.5 Cancellation of External Poles

(1) Using the achieved optimal matrix $\theta^0(\zeta)$, the transfer matrix $W_d(\zeta)$ of the optimal controller can be constructed on basis of the results provided by Theorem 7.4.

In particular, if we have the ILMFD

$$\theta^0(\zeta) = D_{\Phi 0}^{-1}(\zeta)M_{\Phi 0}(\zeta), \tag{8.79}$$

then the characteristic polynomial of the $\mathscr{H}_2$ optimal system $\Delta_{\Phi 0}(\zeta)$ is related to the matrix $D_{\Phi 0}(\zeta)$ by equivalence

$$\Delta_{\Phi 0}(\zeta) \sim \det D_{\Phi 0}(\zeta). \tag{8.80}$$

(2) Assume polynomials (8.59)

$$\Delta_{\Phi}(\zeta) = \det(I_{\varkappa} - \zeta \mathrm{e}^{A_{\Phi}T}), \quad \Delta_{\underline{\Phi}}(\zeta) = \det(I_{\varkappa} - \zeta \mathrm{e}^{-A_{\Phi}T}).$$

Further on, denote the set of roots of the polynomial $\Delta_{\Phi}(\zeta)$ by $\mathscr{M}_{\Phi}$, and the set of roots of the polynomial $\Delta_{\underline{\Phi}}$ by $\mathscr{M}_{\underline{\Phi}}$. Let $f_1, \ldots, f_{\mu}$ be all the different eigenvalues of the matrix $\mathrm{e}^{A_{\Phi}T}$. Then

$$\begin{aligned} \Delta_{\Phi}(\zeta) &= (1 - \zeta f_1)^{\theta_1} \cdots (1 - \zeta f_{\mu})^{\theta_{\mu}}, \quad \theta_1 + \cdots + \theta_{\mu} = \varkappa \\ \Delta_{\underline{\Phi}}(\zeta) &= (1 - \zeta {f_1}^{-1})^{\theta_1} \cdots (1 - \zeta {f_{\mu}}^{-1})^{\theta_{\mu}}, \end{aligned} \tag{8.81}$$

so that owing to the proposition about the stability of the coloring filter, we obtain

$$|f_i| < 1, \quad |f_i^{-1}| > 1 \tag{8.82}$$

i.e., the sets $\mathcal{M}_\Phi$ and $\mathcal{M}_{\underline{\Phi}}$ are disjunct.

**Definition 8.1** The roots $\zeta_i = f_i^{-1}$ of the polynomial $\Delta_\Phi(\zeta)$ are called external poles of the system $\mathcal{S}_d$.

After a first view on formulae (8.69)–(8.71), due to the fact that the matrix $H_\Phi^+(\zeta)$ possesses poles at the points $\zeta = f_i^{-1}$, which are also roots of the polynomial $\Delta_\Phi(\zeta)$, one might assume that the optimal system function $\theta_\Phi^0(\zeta)$ has poles at the points $\zeta = f_i^{-1}$. However, would it really happen that the numbers $\zeta_i = f_i^{-1}$ enter into the set of eigenvalues of the matrix $D_{\Phi 0}(\zeta)$ in ILMFD (8.79)? In this situation, from (8.80) it would follow, that independent on the properties of the standard model, the numbers $\zeta_i = f_i^{-1}$ would become roots of the $\mathcal{H}_2$ optimal system $\mathcal{S}_d$. In this section we will show, that under conditions fulfilled practically always, this situation does not take place, because the external poles $\zeta_i$ in relation (8.71), which determine the optimal system function, are completely canceled. The next theorem provides the accurate result.

**Theorem 8.4** *Let, in addition to the above propositions, the conditions (a)–(d) be fulfilled:*

*(a) The matrices $M(s)$, $\Phi(s)$, $\Phi'(-s)$, $M'(-s)$ are independent, i.e.,*

$$\begin{aligned}\operatorname{Mind}\left[M(s)\Phi(s)\Phi'(-s)M'(s-)\right] \\ = \operatorname{Mind} M(s) + \operatorname{Mind}\Phi(s) + \operatorname{Mind}\Phi'(-s) + \operatorname{Mind} M'(-s).\end{aligned} \tag{8.83}$$

*(b) The unification of the sets $\mathcal{M}_\Phi$ and $\mathcal{M}_{\underline{\Phi}}$ is disjunct to the union of the sets of roots of the polynomials*

$$\Delta_M(\zeta) = \det(I_\alpha - \zeta e^{A_M T}), \quad \Delta_{\underline{M}} = \det(I_\alpha - \zeta e^{-A_M T}). \tag{8.84}$$

*Below, these sets will be referred to as $\mathcal{M}_M$ and $\mathcal{M}_{\underline{M}}$, respectively.*

*(c) The period $T$ is non-pathological with respect to the union of the sets $\mathcal{M}_M$, $\mathcal{M}_{\underline{M}}$, $\mathcal{M}_\Phi$, $\mathcal{M}_{\underline{\Phi}}$*

*(d) The sets of eigenvalues $\mathcal{M}_\Pi$ and $\mathcal{M}_\Gamma$ of the stable matrices $\Pi(\zeta)$ and $\Gamma_\Phi(\zeta)$, configured in factorizations (8.67) and (8.74), are disjunct to the set $\mathcal{M}_\Phi$.*

*Then the numbers $\zeta_i = f_i^{-1}$ do not contribute to the set of poles of the matrix $\theta_\Phi^0(\zeta)$ and which is equivalent, the set of roots of the characteristic polynomial of the $\mathcal{H}_2$ optimal system $\Delta_0(\zeta)$ does not contain any of the external poles $\zeta_i$.*

(3) For the proof of Theorem 8.4, we provide first some preliminary results.

Let the irreducible rational matrix $W(s) \in \mathbb{R}^{nm}(s)$ have the form

$$W(s) = \frac{M(s)}{d_1(s)d_2(s)}, \tag{8.85}$$

where $d_1(s)$ and $d_2(s)$ are coprime scalar polynomials. Denote by $\breve{\mathscr{M}}_i$, $(i = 1, 2)$ the sets of roots of the polynomials $d_1(s)$ and $d_2(s)$, respectively.

**Definition 8.2** We will say that the non-singular matrix $k_i(s) \in \mathbb{R}^{nn}[s]$ cancels the polynomial $d_i(s)$ from the left, if the product

$$L_i(s) \overset{\triangle}{=} k_i(s)W(s), \quad (i = 1, 2) \tag{8.86}$$

does not possess poles in the set $\breve{\mathscr{M}}_i$.

**Lemma 8.2** *Assume for matrix (8.85) the separation*

$$W(s) = Q_1(s) + Q_2(s), \tag{8.87}$$

*where the matrix $Q_1(s)$ does not possess poles in the set $\breve{\mathscr{M}}_2$, and the matrix $Q_2(s)$ does not possess poles in the set $\breve{\mathscr{M}}_1$. Denote*

$$W_1(s) \overset{\triangle}{=} \frac{M(s)}{d_1(s)}, \quad W_2(s) \overset{\triangle}{=} \frac{M(s)}{d_2(s)}. \tag{8.88}$$

*Then,*

*(i)*

$$\text{Mind}\, W_i(s) = \text{Mind}\, Q_i(s), \quad (i = 1, 2). \tag{8.89}$$

*(ii) Assume the ILMFD*

$$Q_i(s) = a_i^{-1}(s)b_i(s), \quad (i = 1, 2). \tag{8.90}$$

*Then the matrix $k_i(s)$ cancels the polynomial $d_i(\zeta)$ from left, if and only if*

$$k_i(s) = n_i(s)a_i(s), \tag{8.91}$$

*where $n_i(s) \in \mathbb{R}^{nn}[s]$ is non-singular.*

*Proof* (i) Assume the ILMFD

$$W_i(s) = c_i^{-1}(s)g_i(s), \quad (i = 1, 2). \tag{8.92}$$

Due to the properties of ILMFD

$$\text{Mind}\, W_i(s) = \deg\det c_i(s). \tag{8.93}$$

Moreover, the product

$$c_i(s)W(s) = c_i(s)(Q_1(s) + Q_2(s)), \quad (i = 1, 2) \tag{8.94}$$

does not possess poles in the set $\breve{\mathscr{M}}_i$, and the matrices $c_i(s)Q_i(s), \ (i = 1, 2)$ are polynomials. Therefore,

$$\text{Mind } Q_i(s) \leq \deg\det c_i(s) = \text{Mind } W_i(s). \tag{8.95}$$

On the other side, when (8.90) is fulfilled, then the matrices

$$\begin{aligned} a_1(s)W(s) &= a_1(s)W_1(s)\frac{1}{d_2(s)}, \\ a_2(s)W(s) &= a_2(s)W_2(s)\frac{1}{d_1(s)} \end{aligned} \tag{8.96}$$

do not possess poles in the sets $\breve{\mathscr{M}}_1$ or $\breve{\mathscr{M}}_2$, respectively. Hence, $a_i(s)W_i(s), \ (i = 1, 2)$ are polynomial matrices. Thus

$$\text{Mind } W_i(s) \leq \deg\det a_i(s) = \text{Mind } Q_i(s), \quad (i = 1, 2). \tag{8.97}$$

Comparing Eqs. (8.95) and (8.97) proves Eq. (8.89).

(ii) From the proof of statement (i), we achieve that the set of matrices $k_i(s)$, which cancel the polynomial $d_i(s)$ from left, coincides with the set of matrices, for which the product $k_i(s)Q_i(s)$ becomes a polynomial matrix. With the help of (8.90), this fact proves (ii). ■

(4) Let the matrix $W(s) \in \mathbb{R}^{nm}(s)$ be given by the relation

$$W(s) = a_1^{-1}(s)b(s)a_2^{-1}(s), \tag{8.98}$$

where $a_1(s) \in \mathbb{R}^{nn}[s]$, $b(s) \in \mathbb{R}^{nm}[s]$, $a_2(s) \in \mathbb{R}^{mm}[s]$, and in addition

$$\det a_1(s) = d_1(s), \quad \det a_2(s) = d_2(s), \tag{8.99}$$

where the polynomials $d_1(s)$ and $d_2(s)$ are coprime. Then the matrix $W(s)$ can be written in form (8.85).

**Lemma 8.3** *Let the pair $(a_1(s), b(s))$ in (8.98) be irreducible. Then the non-singular matrix $k(s)$ cancels in (8.98) the polynomial $d_1(s)$ from left, if and only if*

$$k(s) = n(s)a_1(s), \tag{8.100}$$

*where $n(s) \in \mathbb{R}^{nn}[s]$ is a non-singular polynomial matrix.*

*Proof* Necessity: Let the conditions of Lemma 8.3 be fulfilled. Then both sides of the equation

$$k(s)W(s) = k(s)a_1^{-1}(s)b(s)a_2^{-1}(s) \tag{8.101}$$

do not possess poles in the set $\mathscr{M}_1$. Moreover, the matrix

$$k(s)W(s)a_2(s) = k(s)a_1^{-1}(s)b(s). \tag{8.102}$$

has the same property. Since the pair $(a_1(s), b(s))$ is irreducible, so by the properties of ILMFD, relation (8.100) is true.
Sufficiency: Assume (8.100). Then, the product

$$k(s)W(s) = n(s)b(s)a_2^{-1}(s) \tag{8.103}$$

does not possess poles located in the set of roots of the polynomial $d_1(s)$. ∎

(5) Suppose that the rational matrices $W_i(s)$, $(i = 1, 2, 3)$ have the following properties (a)–(e)

(a) The following product is defined

$$W(s) \stackrel{\triangle}{=} W_1(s)W_2(s)W_3(s). \tag{8.104}$$

(b) The matrices $W_i(s)$ are strictly proper, and possess the minimal realizations

$$W_i(s) = C_i(sI_{\gamma_i} - A_i)^{-1}B_i, \tag{8.105}$$

where $\gamma_i \stackrel{\triangle}{=} \text{Mind } W_i(s)$, $(i = 1, 2, 3)$.
(c) The matrices $W_i(s)$ are independent, i.e.,

$$\text{Mind } W(s) = \text{Mind } W_1(s) + \text{Mind } W_2(s) + \text{Mind } W_3(s) = \gamma_1 + \gamma_2 + \gamma_3. \tag{8.106}$$

(d) The sets of roots of the polynomials

$$d_i(s) \stackrel{\triangle}{=} \det(sI_{\gamma_i} - A_i), \quad (i = 1, 2, 3) \tag{8.107}$$

do not intersect.
(e) The sampling period $T$ is non-pathological for the union of the sets of roots of the polynomials $d_i(s)$, $(i = 1, 2, 3)$.

Under the above restrictions, the following rational matrices in $\zeta$ are defined:

$$D_{123}(T, \zeta, t) \stackrel{\triangle}{=} D_W(T, \zeta, t) = \left[\frac{1}{T}\sum_{k=-\infty}^{\infty} W(s + kj\omega)e^{(s+kj\omega)t}\right]_{e^{-sT}=\zeta}$$

$$= \left[\frac{1}{T}\sum_{k=-\infty}^{\infty} W_1(s+kj\omega)W_2(s+kj\omega)W_3(s+kj\omega)e^{(s+kj\omega)t}\right]_{e^{-sT}=\zeta},$$

$$D_{12}(T,\zeta,t) \overset{\triangle}{=} D_{W_{12}}(T,\zeta,t) = \left[\frac{1}{T}\sum_{k=-\infty}^{\infty} W_{12}(s+kj\omega)e^{(s+kj\omega t)}\right]_{e^{-sT}=\zeta} \tag{8.108}$$

$$= \left[\frac{1}{T}\sum_{k=-\infty}^{\infty} W_1(s+kj\omega)W_2(s+kj\omega)e^{(s+kj\omega)t}\right]_{e^{-sT}=\zeta},$$

where

$$W_{12}(s) \overset{\triangle}{=} W_1(s)W_2(s). \tag{8.109}$$

It follows from the results of Sect. 5.6 that, under the taken propositions, for $0 < t < T$, the rational matrices (8.108) are strictly proper with respect to $\zeta$ and continuously depending on $t$, i.e., the pole excess of the matrices $W(s)$ and $W_{12}(s)$ is greater than one.

For $0 < t < T$, matrices (8.108) allow the representations

$$D_{123}(T,\zeta,t) = \mathscr{D}_{123}(T,\zeta,t) = \frac{n_{123}(\zeta,t)}{\Delta_1(\zeta)\Delta_2(\zeta)\Delta_3(\zeta)},$$
$$D_{12}(T,\zeta,t) = \mathscr{D}_{12}(T,\zeta,t) = \frac{n_{12}(\zeta,t)}{\Delta_1(\zeta)\Delta_2(\zeta)}, \tag{8.110}$$

where $n_{123}(\zeta,t)$ and $n_{12}(\zeta,t)$ are polynomial matrices with respect to $\zeta$ and continuously depending on $t$. Moreover, $\Delta_i(\zeta)$ are the polynomials

$$\Delta_i(\zeta) \overset{\triangle}{=} \det(I_{\gamma_i} - \zeta e^{A_i T}), \quad (i = 1, 2, 3) \tag{8.111}$$

Denote the set of roots of the polynomials $\Delta_i(\zeta)$ by $\bar{\mathscr{M}}_i$, $(i = 1, 2, 3)$. Hereby, due to the propositions (d) and (e), the sets $\bar{\mathscr{M}}_i$ are disjunct.

The continuity of the matrix $D_{123}(T,\zeta,t)$ with respect to $t$ ensures the existence of the matrix

$$D_{123}(T,\zeta,0) = D_{123}(T,\zeta,t)|_{t=0} = \mathscr{D}_{123}(T,\zeta,0). \tag{8.112}$$

(6)

**Lemma 8.4** *Under the conditions (a)–(e), let the non-singular matrix* $k(\zeta) \in \mathbb{R}^{nn}[\zeta]$ *cancel the polynomial* $\Delta_2(\zeta)$ *from left in the matrix*

$$\mathscr{D}_{123}(T,\zeta,0) = \frac{n_{123}(\zeta,0)}{\Delta_1(\zeta)\Delta_2(\zeta)\Delta_3(\zeta)}, \tag{8.113}$$

*i.e., the product*

$$\delta_{123}(\zeta) \stackrel{\triangle}{=} k(\zeta)\mathscr{D}_{123}(T,\zeta,0) = \frac{k(\zeta)n_{123}(\zeta,0)}{\Delta_1(\zeta)\Delta_2(\zeta)\Delta_3(\zeta)} \tag{8.114}$$

*does not possess poles in the set* $\bar{\mathscr{M}}_2$. *Then the matrix* $k(\zeta)$ *cancels, for* $0 < t < T$, *the polynomial* $\Delta_2(\zeta)$ *from left in the matrix* $\mathscr{D}_{12}(T,\zeta,t)$, *i.e., the product*

$$\delta_{12}(\zeta) \stackrel{\triangle}{=} k(\zeta)\mathscr{D}_{12}(T,\zeta,t) \tag{8.115}$$

*does not possess poles in the set* $\bar{\mathscr{M}}_2$ *too.*

*Proof* Build the separation of form (8.87)

$$W_{12}(s) = Q_1(s) + Q_2(s), \tag{8.116}$$

where $Q_1(s)$ and $Q_2(s)$ are strictly proper rational matrices, where the first does not posses poles in the set $\mathscr{M}_2$, and the second not in $\mathscr{M}_1$. From Lemma 8.2 we know that

$$\text{Mind}\, Q_i(s) = \gamma_i, \quad (i = 1, 2). \tag{8.117}$$

Perform the DLT of both sides of Eq. (8.116)

$$\breve{D}_{12}(T,s,t) = \breve{D}_1(T,s,t) + \breve{D}_2(T,s,t). \tag{8.118}$$

Herein, and later

$$\breve{D}_i(T,s,t) \stackrel{\triangle}{=} \frac{1}{T}\sum_{k=-\infty}^{\infty} Q_i(s + kj\omega)e^{(s+kj\omega)t}, \quad (i = 1, 2). \tag{8.119}$$

Substitute in (8.118) the variable $e^{-pT}$ by $\zeta$, then

$$D_{12}(T,\zeta,t) = D_1(T,\zeta,t) + D_2(T,\zeta,t), \tag{8.120}$$

which for $0 < t < T$ can be written in the form

$$D_{12}(T,\zeta,t) = \mathscr{D}_{12}(T,\zeta,t) = \mathscr{D}_1(T,\zeta,t) + \mathscr{D}_2(T,\zeta,t). \tag{8.121}$$

Here, the poles of the first term on the right side are located in the set $\bar{\mathscr{M}}_1$, and of the second term in $\bar{\mathscr{M}}_2$. Hence, Theorem 5.8 implies that, for $0 < t < T$,

$$\text{Mind}\, \mathscr{D}_i(T,\zeta,t) = \gamma_i, \quad (i = 1, 2) \tag{8.122}$$

and

$$\text{Mind}\, \mathscr{D}_{12}(T,\zeta,t) = \gamma_1 + \gamma_2. \tag{8.123}$$

Hereby, the ILMFD of the form

$$\mathscr{D}_i(T,\zeta,t) = c_i^{-1}(\zeta)d_i(\zeta,t), \quad \det c_i(\zeta) \sim \det(I_{\gamma_i} - \zeta e^{A_i T}), \ (i = 1,2). \quad (8.124)$$

takes place. Further, notice that

$$\breve{D}_{123}(T,s,0) = e^{-sT} \int_0^T \breve{\mathscr{D}}_{12}(T,s,t)\breve{\mathscr{D}}_3(T,s,T-t)\,dt, \quad (8.125)$$

which follows from formula (8.24) with the help of

$$\breve{D}_3(T,s,-t) = e^{-sT}\breve{\mathscr{D}}_3(T,s,T-t),$$

which is valid for $0 < t < T$.

Rename again in (8.125) $e^{-sT}$ by $\zeta$, then

$$D_{123}(T,\zeta,0) = \zeta \int_0^T \mathscr{D}_{12}(T,\zeta,t)\mathscr{D}_3(T,\zeta,T-t)\,dt. \quad (8.126)$$

Inserting here (8.121), we find

$$D_{123}(T,\zeta,0) = \zeta \int_0^T \breve{\mathscr{D}}_2(T,\zeta,t)\breve{\mathscr{D}}_3(T,\zeta,T-t)\,dt + \zeta \int_0^T \breve{\mathscr{D}}_1(T,\zeta,t)\breve{\mathscr{D}}_3(T,\zeta,T-t)\,dt. \quad (8.127)$$

Apply the IRMFD

$$\mathscr{D}_3(T,\zeta,t) = d_3(\zeta,t)c_3^{-1}(\zeta), \quad \det c_3(\zeta) \sim \det(I_{\gamma_3} - \zeta e^{A_3 T}), \quad (8.128)$$

then with the help of (8.124) formula (8.127) can be represented by the expression

$$D_{123}(T,\zeta,0) = c_2^{-1}(\zeta)K_{23}(\zeta)c_3^{-1}(\zeta) + c_1^{-1}(\zeta)K_{13}(\zeta)c_3^{-1}(\zeta), \quad (8.129)$$

where

$$\begin{aligned} K_{23} &\stackrel{\triangle}{=} \zeta \int_0^T d_2(\zeta,t)d_3(\zeta,T-t)\,dt, \\ K_{13}(\zeta) &\stackrel{\triangle}{=} \zeta \int_0^T d_1(\zeta,t)d_3(\zeta,T-t)\,dt \end{aligned} \quad (8.130)$$

are polynomial matrices. Since the second term on the right side of (8.129) does not possess poles in the set $\mathscr{M}_2$, so the matrix $\delta_{123}(\zeta)$ in (8.114) does not possess poles in $\mathscr{M}_2$ if and only if the product

$$\delta_{23}(\zeta) \stackrel{\triangle}{=} k(\zeta)c_2^{-1}(\zeta)K_{23}c_3^{-1}(\zeta) \tag{8.131}$$

has this property too. Notice further that the pair $(c_2(\zeta), K_{23}(\zeta))$ in (8.131) is irreducible, because in the contrary condition (c) (8.106) would be violated. Moreover, due to

$$\det c_2(\zeta) \sim \det(I_{\chi_2} - \zeta e^{A_2 T}), \quad \det c_3(\zeta) \sim \det(I_{\chi_3} - \zeta e^{A_3 T}), \tag{8.132}$$

the sets of eigenvalues of the matrices $c_2(\zeta)$ and $c_3(\zeta)$ are disjunct. This fact means that the right part of relation (8.131) satisfies the conditions of Lemma 8.3. Thus,

$$k(\zeta) = n(\zeta)c_2(\zeta), \tag{8.133}$$

where $n(\zeta)$ is a non-singular polynomial matrix. Finally, on basis of (8.133), (8.124), and (8.121), we obtain that the matrix

$$\begin{aligned} k(\zeta)\mathscr{D}_{12}(T,\zeta,t) &= n(\zeta)c_2(\zeta)\left[c_2^{-1}(\zeta)b_2(\zeta,t) + c_1^{-1}(\zeta)b_1(\zeta,t)\right] \\ &= n(\zeta)\left[b_2(\zeta,t) + c_2(\zeta)c_1^{-1}(\zeta)b_1(\zeta,t)\right] \end{aligned} \tag{8.134}$$

does not possess poles in the set $\bar{\mathscr{M}}_2$. ■

(7)

**Lemma 8.5** *Assume, in harmony with (8.75), as result of factorization (8.68)*

$$V(\zeta) = a_{\Phi 2}^{-1}(\zeta)\Gamma_\Phi(\zeta), \quad \hat{V}(\zeta) = \hat{\Gamma}_\Phi(\zeta)\hat{a}_{\Phi 2}^{-1}(\zeta), \tag{8.135}$$

*where the matrices $a_{\Phi 2}(\zeta)$ and $\Gamma_\Phi(\zeta)$ are defined from (8.44), (8.46), (8.72) and (8.74). Moreover, let the conditions of Theorem 8.4 hold. Then the following claims are true:*

*(i) The set of poles of the matrix*

$$V_1(\zeta,t) \stackrel{\triangle}{=} V^{-1}(\zeta)a_l(\zeta)\mathscr{D}_{M\Phi\underline{\Phi}'}(T,\zeta,t) \tag{8.136}$$

*does not contain poles of the set $\mathscr{M}_\Phi$, and the set of poles of the matrix*

$$V_2(\zeta,t) \stackrel{\triangle}{=} \hat{V}^{-1}(\zeta)V_1(\zeta,t) = \hat{V}^{-1}(\zeta)V^{-1}(\zeta)a_l(\zeta)\mathscr{D}_{M\Phi\underline{\Phi}'}(T,\zeta,t) \tag{8.137}$$

*does not contain poles of the sets $\mathscr{M}_\Phi$ and $\mathscr{M}_{\underline{\Phi}}$.*

*(ii) The set of poles of the matrix*

$$\hat{V}_1(\zeta,t) \stackrel{\triangle}{=} \hat{\mathscr{D}}_{M\Phi\underline{\Phi}'}(T,\zeta,t)\hat{a}_l(\zeta)\hat{V}^{-1}(\zeta) \tag{8.138}$$

*does not contain poles in the set $\mathscr{M}_{\underline{\Phi}}$, and the matrix*

$$\hat{V}_2(\zeta,t) \overset{\triangle}{=} \hat{V}_1(\zeta,t)V^{-1}(\zeta) = \hat{\mathscr{D}}_{M\Phi\underline{\Phi}'}(T,\zeta,t)\hat{a}_l(\zeta)\hat{V}^{-1}(\zeta)V^{-1}(\zeta) \quad (8.139)$$

*does not contain poles of the sets $\mathscr{M}_\Phi$ and $\mathscr{M}_{\underline{\Phi}}$.*

*Proof* (i) By construction

$$a_l(\zeta)\mathscr{D}_{M\Phi\underline{\Phi}'\underline{M}'}(T,\zeta,0)\hat{a}_l(\zeta) = V(\zeta)\hat{V}(\zeta), \quad (8.140)$$

so, using (8.135) we obtain

$$a_{\Phi 2}(\zeta)a_l(\zeta)\mathscr{D}_{M\Phi\underline{\Phi}'\underline{M}'}(T,\zeta,0) = \Gamma_\Phi(\zeta)\hat{\Gamma}_\Phi(\zeta)\hat{a}_{\Phi 2}^{-1}(\zeta)\hat{a}_l^{-1}(\zeta). \quad (8.141)$$

We immediately verify that under the taken propositions, the matrix on the right side of the last equation does not possess poles in the set $\mathscr{M}_\Phi$. Therefore, choosing

$$W_1(s) = M(s), \quad W_2(s) = \Phi(s), \quad W_3(s) = \Phi'(-s)M'(-s), \quad (8.142)$$

and applying Lemma 8.4, we find that the matrix

$$a_{\Phi_2}(\zeta)a_l(\zeta)\mathscr{D}_{M\Phi\underline{\Phi}'}(T,\zeta,t) \quad (8.143)$$

has an analog property, i.e., it has no poles in $\mathscr{M}_\Phi$. So the same property holds for the matrix

$$\Gamma_\Phi^{-1}(\zeta)a_{\Phi_2}(\zeta)a_l(\zeta)\mathscr{D}_{M\Phi\Phi'}(T,\zeta,t) = V^{-1}(\zeta)a_l(\zeta)\mathscr{D}_{M\Phi\Phi'}(T,\zeta,t), \quad (8.144)$$

and this fact proves statement (i) for the matrix $V_1(\zeta,t)$. Further notice, that owing to (8.68) and (8.135), we obtain

$$V(\zeta)\hat{V}(\zeta) = a_{\Phi 2}^{-1}(\zeta)\Gamma_\Phi(\zeta)\hat{\Gamma}_\Phi(\zeta)\hat{a}_{\Phi 2}^{-1}(\zeta). \quad (8.145)$$

The set of poles of the matrix on the right side is located in the union of the sets $\mathscr{M}_\Phi$ and $\mathscr{M}_{\underline{\Phi}}$, and perhaps contains $\zeta = 0$. Therefore, there exists a representation of the form

$$V(\zeta)\hat{V}(\zeta) = a_{\Phi\underline{\Phi}}^{-1}(\zeta)K_V(\zeta)a_0^{-1}(\zeta), \quad (8.146)$$

where $K_V(\zeta)$ is a polynomial matrix, and the eigenvalues of the matrix $a_{\Phi\underline{\Phi}}(\zeta)$ are located in the set $\mathscr{M}_\Phi \cup \mathscr{M}_{\underline{\Phi}}$, and the matrix $a_0(\zeta)$ is nilpotent (i.e., it has only the single eigenvalue $\zeta = 0$). Moreover, since by supposition the matrix $\Gamma_\Phi(\zeta)$ is stable and does not possess eigenvalues in $\mathscr{M}_{\underline{\Phi}}$, so the matrix $K_V(\zeta)$ does not contain eigenvalues in $\mathscr{M}_\Phi \cup \mathscr{M}_{\underline{\Phi}}$. Moreover, by construction the pairs $(a_{\Phi\underline{\Phi}}(\zeta), K_V(\zeta))$ and $[a_0(\zeta), K_V(\zeta)]$ are irreducible.
Using (8.146) and (8.140), we can write

$$a_l(\zeta)\mathscr{D}_{M\Phi\underline{\Phi}'\underline{M}'}(T,\zeta,0)\hat{a}_l(\zeta) = a_{\Phi\underline{\Phi}}^{-1}(\zeta)K_V(\zeta)a_0^{-1}(\zeta), \quad (8.147)$$

which is equivalent to the relation

$$a_{\Phi\Phi}(\zeta)a_l(\zeta)\mathscr{D}_{M\Phi\underline{\Phi}'\underline{M}'}(T,\zeta,0) = K_V(\zeta)a_0^{-1}(\zeta)\hat{a}_l^{-1}(\zeta). \tag{8.148}$$

Since the matrix on the right side does not possess poles in the set $\mathscr{M}_\Phi \cup \mathscr{M}_{\underline{\Phi}}$, so for

$$W_1(s) = M(s), \quad W_2(s) = \Phi(s)\Phi'(-s), \quad W_3(s) = M'(-s)$$

it follows from Lemma 8.4 that this property holds for the product

$$a_{\Phi\Phi}(\zeta)a_l(\zeta)\mathscr{D}_{M\Phi\underline{\Phi}'}(T,\zeta,t). \tag{8.149}$$

Hence, using (8.146), the chain

$$\begin{aligned}\hat{V}^{-1}(\zeta)V^{-1}(\zeta)a_l(\zeta)\mathscr{D}_{M\Phi\underline{\Phi}'}(T,\zeta,t) &= \left[V(\zeta)\hat{V}(\zeta)\right]^{-1} a_l(\zeta)\mathscr{D}_{M\Phi\underline{\Phi}'}(T,\zeta,t)\\ &= a_0(\zeta)K_V^{-1}(\zeta)a_{\Phi\Phi}(\zeta)a_l(\zeta)\mathscr{D}_{M\Phi\underline{\Phi}'}(T,\zeta,t)\end{aligned} \tag{8.150}$$

proves the second part of claim i), because the matrix on the right side of (8.150) does not possess poles in $\mathscr{M}_\Phi \cup \mathscr{M}_{\underline{\Phi}}$ .

(ii) This claim is generated from (i) by transposition and substituting $\zeta$ by $\zeta^{-1}$. ■

(8)

**Lemma 8.6** *Let $W_i(s)$, $(i = 1, 2)$ be strictly proper rational matrices, and*

$$\varphi_i(T,s,t) = \frac{1}{T}\sum_{k=-\infty}^{\infty} W_i(s+kj\omega)e^{kj\omega t}, \quad (i=1,2) \tag{8.151}$$

*be their DPFR, and*

$$\breve{D}_i(T,s,t) = \varphi_i(T,s,t)e^{st}, \quad (i=1,2) \tag{8.152}$$

*the associated DLT. Then, if the product $W_{12}(s) = W_1(s)W_2(s)$ is defined, we obtain*

$$\breve{D}_{W_1W_2}(T,s,t) = \int_0^T \breve{D}_1(T,s,\nu)\breve{D}_2(T,s,t-\nu)\,d\nu. \tag{8.153}$$

*Proof* We have

$$\begin{aligned}\varphi_{12}(T,s,t) &= \frac{1}{T}\sum_{k=-\infty}^{\infty} W_1(s+kj\omega)W_2(s+kj\omega)e^{kj\omega t}\\ &= \int_0^T \varphi_1(T,s,\nu)\varphi_2(T,s,t-\nu)\,d\nu,\end{aligned} \tag{8.154}$$

which can be proved inserting series (8.151) into the right side of (8.154), and termwise integration. With the help of (8.152) and (8.154), we find

$$\begin{aligned}\breve{D}_{12}(T,s,t) &= \varphi_{12}(T,s,t)\mathrm{e}^{st} \\ &= \mathrm{e}^{st}\int_0^T \left[\breve{D}_1(T,s,\nu)\mathrm{e}^{-s\nu}\right]\left[\breve{D}_2(T,s,t-\nu)\mathrm{e}^{-s(t-\nu)}\right]\mathrm{d}\nu \\ &= \int_0^T \breve{D}_1(T,s,\nu)\breve{D}_2(T,s,t-\nu)\,\mathrm{d}\nu.\end{aligned} \tag{8.155}$$

■

(9)

*Proof* [of Theorem 8.4] In analogy to (8.126), we obtain

$$D_{M\Phi\underline{\Phi}'\underline{M}'}(T,\zeta,0) = \zeta\int_0^T \mathscr{D}_{M\Phi\underline{\Phi}'}(T,\zeta,\nu)\mathscr{D}_{\underline{M}'}(T,\zeta,T-\nu)\,\mathrm{d}\nu. \tag{8.156}$$

Moreover, applying (8.153), we find for $W_1(s) = M(s)\Phi(p)\Phi'(-s)$, $W_2(s) = K'(-s)L(s)\mu(s)$

$$\breve{D}_{M\Phi\underline{\Phi}\underline{K}'L\mu}(T,s,-\tau_2) = \int_0^T \breve{D}_{M\Phi\underline{\Phi}'}(T,s,\nu)\breve{D}_{\underline{K}'L\mu}(T,s,-\tau_2-\nu)\,\mathrm{d}\nu. \tag{8.157}$$

Observing that, in coincidence with (6.93),

$$\tau_2 = m_2T + \theta_2 = (m_2+1)T - \gamma_2, \quad 0 \le \theta_2 < T, \quad 0 < \gamma_2 \le T, \tag{8.158}$$

we arrive at

$$\begin{aligned}\breve{D}_{\underline{K}'L\mu}(T,s,-\tau_2-\nu) &= \breve{D}_{\underline{K}'L\mu}(T,s,-m_2T-T+\gamma_2-\nu) \\ &= \mathrm{e}^{-(m_2+1)pT}\,\breve{D}_{\underline{K}'L\mu}(T,s,\gamma_2-\nu).\end{aligned} \tag{8.159}$$

Herein,

$$\breve{D}_{\underline{K}'L\mu}(T,s,\gamma_2-\nu) = \begin{cases}\breve{\mathscr{D}}_{\underline{K}'L\mu}(T,s,\gamma_2-\nu), & 0 \le \nu \le \gamma_2, \\ \breve{\mathscr{D}}_{\underline{K}'L\mu}(T,s,T+\gamma_2-\nu)\mathrm{e}^{-sT}, & \gamma_2 \le \nu \le T.\end{cases} \tag{8.160}$$

After renaming $\mathrm{e}^{-pT}$ by $\zeta$ from (8.159), (8.160), we obtain

$$D_{\underline{K}'L\mu}(T,\zeta,-\tau_2-\nu) = \zeta^{(m_2+1)}\begin{cases}\mathscr{D}_{\underline{K}'L\mu}(T,\zeta,\gamma_2-\nu), & 0 \le \nu \le \gamma_2, \\ \zeta\mathscr{D}_{\underline{K}'L\mu}(T,\zeta,T+\gamma_2-\nu), & \gamma_2 \le \nu \le T.\end{cases} \tag{8.161}$$

Obviously, matrices (8.161) do not possess poles in the set $\mathscr{M}_\Phi \cup \mathscr{M}_{\underline{\Phi}}$. Moreover, from (8.157), after substituting $\mathrm{e}^{-pT}$ by $\zeta$, we obtain

$$D_{M\Phi\underline{\Phi}'\underline{K}'L\mu}(T,\zeta,-\tau_2) = \int_0^T \mathscr{D}_{M\Phi\underline{\Phi}'}(T,\zeta,\nu)D_{\underline{K}'L\mu}(T,\zeta,-\tau_2-\nu)\,\mathrm{d}\nu. \quad (8.162)$$

According to (8.156) and (8.162), the matrix $C_\Phi(\zeta)$ in (8.14) can be represented in the form

$$C_\Phi(\zeta) = \int_0^T a_l(\zeta)\mathscr{D}_{M\Phi\underline{\Phi}'}(T,\zeta,\nu)D_1(\zeta,\nu)\,\mathrm{d}\nu, \quad (8.163)$$

where $D_1(\zeta,\nu)$ is rational with respect to $\zeta$, and its poles neither depend on $\nu$ nor are they located in $\mathscr{M}_\Phi \cup \mathscr{M}_{\underline{\Phi}}$. Closed expressions for the matrix $D_1(\zeta,\nu)$ are not provided, because they are not needed in further considerations. From (8.163), we also find

$$\hat{C}_\Phi(\zeta) = \int_0^T \hat{D}_1(\zeta,\nu)\hat{\mathscr{D}}_{M\Phi\underline{\Phi}'}(T,\zeta,\nu)\hat{a}_l(\nu)\,\mathrm{d}\nu, \quad (8.164)$$

where the matrix $\hat{D}_1(\zeta,\nu)$ also does not possess poles in $\mathscr{M}_\Phi \cup \mathscr{M}_{\underline{\Phi}}$. Therefore, matrix (8.69) can be written in the form

$$\begin{aligned} H_\Phi(\zeta) &\triangleq \hat{\Pi}^{-1}(\zeta)\hat{C}_\Phi(\zeta)\hat{V}^{-1}(\zeta) \\ &= \hat{\Pi}^{-1}(\zeta)\int_0^T \hat{D}_1(\zeta,\nu)\hat{\mathscr{D}}_{M\Phi\underline{\Phi}'}(T,\zeta,\nu)\hat{a}_l(\zeta)\hat{V}^{-1}(\zeta)\,\mathrm{d}\nu. \quad (8.165)\end{aligned}$$

Hence, using the first claim of (ii) in Lemma 8.5, we come to the conclusion, that the rational matrix $H_\Phi(\zeta)$ does not possess poles in the set $\mathscr{M}_{\underline{\Phi}}$, and the rational matrix

$$\begin{aligned} H_{\Phi_1}(\zeta) &\triangleq \Pi^{-1}(\zeta)H_\Phi(\zeta)V^{-1}(\zeta) = \Pi^{-1}(\zeta)\hat{\Pi}^{-1}(\zeta)\hat{C}_\Phi(\zeta)\hat{V}^{-1}(\zeta)V^{-1}(\zeta) \\ &= \Pi^{-1}(\zeta)\hat{\Pi}^{-1}(\zeta)\int_0^T \hat{\mathscr{D}}_1(\zeta,\nu)\hat{\mathscr{D}}_{M\Phi\underline{\Phi}'}(T,\zeta,\nu)\hat{a}_l(\nu)\hat{V}^{-1}(\zeta)V^{-1}(\zeta)\,\mathrm{d}\nu \end{aligned} \quad (8.166)$$

does not possess poles in $\mathscr{M}_\Phi$ or $\mathscr{M}_{\underline{\Phi}}$.

For the further proof , we consider separation (8.70)

$$H_\Phi(\zeta) = H_\Phi^+(\zeta) + H_\Phi^-(\zeta).$$

The left side of this equation, due to the above reasoning, does not possess poles in the set $\mathscr{M}_{\underline{\Phi}}$. Therefore, considering that the coloring filter $\Phi(p)$ is stable and the matrix $H_\Phi^+(\zeta)$ is also stable, we obtain that the matrix $H_\Phi^+(\zeta)$ does not possess poles in $\mathscr{M}_\Phi$ or in $\mathscr{M}_{\underline{\Phi}}$. Therefore, the right side of equation

$$H_\Phi^-(\zeta) = H_\Phi(\zeta) - H_\Phi^+(\zeta) \quad (8.167)$$

does not possess poles in $\mathscr{M}_{\underline{\Phi}}$. Since the matrix $H_\Phi^-(\zeta)$, by construction, does not possesses poles in $\mathscr{M}_\Phi$ too, so we conclude that the $H_\Phi^-(\zeta)$ does not possess poles in $\mathscr{M}_\Phi \cup \mathscr{M}_{\underline{\Phi}}$. Further, notice that we can derive from (8.166), (8.70) the relation

$$H_{\Phi 1}(\zeta) = \Pi^{-1}(\zeta)H_\Phi^+(\zeta)V^{-1}(\zeta) + \Pi^{-1}(\zeta)H_\Phi^-(\zeta)V^{-1}(\zeta) \tag{8.168}$$

The last expression, according to (8.71), can be written in the form

$$\theta_\Phi^0(\zeta) = \Pi^{-1}(\zeta)H_\Phi^+(\zeta)V^{-1}(\zeta) = H_{\Phi 1}(\zeta) - \Pi^{-1}(\zeta)H_\Phi^-(\zeta)\Gamma_\Phi^{-1}(\zeta)a_{\Phi 2}(\zeta). \tag{8.169}$$

On basis of the proven facts, both terms on the right side of this equation under the propositions of Theorem 8.4 do not possess poles in $\mathscr{M}_\Phi$. This statement means that in formula (8.71) all poles of the set $\mathscr{M}_\Phi$ are canceled. ■

## 8.6 General Properties of $\mathscr{H}_2$ Optimal Systems

(1)

**Theorem 8.5** *The matrices $a_{M\Phi}(\zeta)$ and $a_M(\zeta)$ in ILMFD (8.18), and also the matrix $a_l(\zeta)$ in the ILMFD*

$$C_2(I_\chi - \zeta e^{AT})^{-1} = a_l^{-1}(\zeta)b_l(\zeta) \tag{8.170}$$

*can be chosen in such a way that for the matrix $\Gamma_\Phi(\zeta)$, configured in factorization (8.74), the following holds:*

$$\deg \Gamma_\Phi(\zeta) \le \chi + \varkappa - 1. \tag{8.171}$$

*Proof* Using (8.39)–(8.41), we obtain an ILMFD of the form

$$\mathscr{D}_{M\Phi}(T, \zeta, t) = [k(\zeta)a_\Phi(\zeta)a_M(\zeta)]^{-1}b_{M\Phi}(\zeta, t), \tag{8.172}$$

where $k(\zeta)$ is any unimodular matrix. In particular, the matrix $k(\zeta)$ can be chosen such that

$$a_{M\Phi}(\zeta) \stackrel{\triangle}{=} k(\zeta)a_\Phi(\zeta)a_M(\zeta) \tag{8.173}$$

becomes row reduced. Then, due to

$$\deg \det[k(\zeta)a_\Phi(\zeta)a_M(\zeta)] = \alpha + \varkappa, \tag{8.174}$$

we obtain

$$\deg b_{M\Phi}(\zeta, t) \le \varkappa + \alpha - 1, \tag{8.175}$$

because the matrix $\mathscr{D}_{M\Phi}(T, \zeta, t)$ is strictly proper with respect to $\zeta$ for all $t$. Using (8.173), from (8.53), (8.54) we find

$$D_{M\Phi\underline{\Phi}'\underline{M}'}(T,\zeta,0) = a_M^{-1}(\zeta)[k(\zeta)a_\Phi(\zeta)]^{-1}K_{\Phi 1}(\zeta)[\widehat{k(\zeta)a_\Phi(\zeta)}]^{-1}\hat{a}_M^{-1}(\zeta), \quad (8.176)$$

where in relation to (8.55)

$$K_{\Phi 1}(\zeta) = \int_0^T b_{M\Phi}(\zeta,t)\hat{b}_{M\Phi}(\zeta,t)\,\mathrm{d}t. \quad (8.177)$$

Further notice that (8.176) and (8.56) yield

$$\begin{aligned} A_{M\Phi}(\zeta) &= a_l(\zeta)D_{M\Phi\underline{\Phi}'\underline{M}'}(T,\zeta,0)\hat{a}_l(\zeta) \\ &= a_l(\zeta)a_M^{-1}(\zeta)[k(\zeta)a_\Phi(\zeta)]^{-1}K_{\Phi 1}(\zeta)[\widehat{k(\zeta)a_\Phi(\zeta)}]^{-1}\hat{a}_M^{-1}(\zeta)\hat{a}_l(\zeta). \end{aligned} \quad (8.178)$$

Utilizing that

$$a_l(\zeta) = a_{M1}(\zeta)a_M(\zeta), \quad (8.179)$$

Equation (8.178) can be written in the form

$$A_{M\Phi}(\zeta) = a_{M1}(\zeta)[k(\zeta)a_\Phi(\zeta)]^{-1}K_{\Phi 1}(\zeta)[\widehat{k(\zeta)a_\Phi(\zeta)}]\hat{a}_{M1}(\zeta). \quad (8.180)$$

The matrix relation $a_{M1}(\zeta)[k(\zeta)a_\Phi(\zeta)]^{-1}$ is an IRMFD, because the polynomial matrices $a_{M1}(\zeta)$ and $k(\zeta)a_\Phi(\zeta)$ do not possess common eigenvalues. Therefore, in any ILMFD

$$a_{M1}(\zeta)[k(\zeta)a_\Phi(\zeta)]^{-1} = a_{\Phi 2}^{-1}(\zeta)a_{M2}(\zeta) \quad (8.181)$$

we have

$$a_{\Phi 2}(\zeta) \sim k(\zeta)a_\Phi(\zeta) \sim a_\Phi(\zeta), \quad a_{M2}(\zeta) \sim a_{M1}(\zeta). \quad (8.182)$$

Assume an arbitrary ILMFD (8.181) and any unimodular matrix $\ell(\zeta)$ of appropriate size. Then the matrix relation $[\ell(\zeta)a_{\Phi 2}(\zeta)]^{-1}[\ell(\zeta)a_{M2}(\zeta)]$ is also an ILMFD. Hereby, the matrix $\ell(\zeta)$ can be chosen such that the matrix $\ell(\zeta)a_{M2}(\zeta)$ is strictly proper. It follows from these considerations that the matrix $a_{M2}(\zeta)$ in (8.181) can be assumed row reduced. Moreover, due to $\deg\det a_{M2}(\zeta) = \chi - \alpha$, we also obtain

$$\deg a_{M2}(\zeta) \le \chi - \alpha. \quad (8.183)$$

From (8.170) and (8.171), we find

$$A_{M\Phi}(\zeta) = a_{\Phi 2}^{-1}(\zeta)K_\Phi(\zeta)\hat{a}_{\Phi 2}^{-1}(\zeta), \quad (8.184)$$

where

$$K_\Phi(\zeta) = \int_0^T [a_{M2}(\zeta)b_{M\Phi}(\zeta,t)][\widehat{a_{M2}(\zeta)b_{M\Phi}}(\zeta,t)]\,\mathrm{d}t \quad (8.185)$$

is a symmetric quasi-polynomial, which is nonnegative on the circle $|\zeta| = 1$. Moreover, it follows from estimates (8.175) and (8.183) that

$$\deg[a_{M2}(\zeta)b_{M\Phi}(\zeta, t)] \leq \chi + \varkappa - 1. \tag{8.186}$$

Hence, with the help of (8.185) we obtain the estimate

$$\operatorname{Deg} K_\Phi(\zeta) \leq \chi + \varkappa - 1. \tag{8.187}$$

Moreover, owing to the fact that the quasi-polynomial $K_\Phi(\zeta)$ is positive on the circle $|\zeta| = 1$, so from (8.187) emerges that for the factorization

$$K_\Phi(\zeta) = \Gamma_\Phi(\zeta)\hat{\Gamma}_\Phi(\zeta), \tag{8.188}$$

the estimate

$$\deg \Gamma_\Phi(\zeta) = \operatorname{Deg} K_\Phi(\zeta) \leq \chi + \varkappa - 1 \tag{8.189}$$

is valid. ■

(2)

**Theorem 8.6** *Let the conditions of Lemma 8.1 hold and the quasi-polynomial $K_\Phi(\zeta)$ (8.46) be positive on the circle $|\zeta| = 1$. Then*

$$\deg \det \Gamma_\Phi(\zeta) \leq \chi + \varkappa - n. \tag{8.190}$$

*Proof* From (8.178) and (8.184), we find

$$a_l(\zeta)D_{M\Phi\underline{\Phi'}\underline{M'}}(T, \zeta, 0)\hat{a}_l(\zeta) = a_{\Phi 2}^{-1}(\zeta)K_\Phi(\zeta)\hat{a}_{\Phi 2}^{-1}(\zeta), \tag{8.191}$$

from which we directly conclude

$$D_{M\Phi\underline{\Phi'}\underline{M'}}(T, \zeta, 0) = a_l^{-1}(\zeta)a_{\Phi 2}\mathscr{K}_\Phi(\zeta)\hat{a}_{\Phi 2}^{-1}(\zeta)a_l^{-1}(\zeta). \tag{8.192}$$

Calculating the determinant on both sides of the last equation leads to

$$\det D_{M\Phi\underline{\Phi'}\underline{M'}}(T, \zeta, 0) = \frac{\det K_\Phi(\zeta)}{\det a_l(\zeta)\det a_l(\zeta^{-1})\det a_{\Phi 2}(\zeta)\det a_{\Phi 2}(\zeta^{-1})}. \tag{8.193}$$

Hence,

$$\operatorname{exc}\det D_{M\Phi\underline{\Phi'}\underline{M'}}(T, \zeta, 0) = \operatorname{exc}\frac{\det K_\Phi(\zeta)}{\det a_l(\zeta)\det a_l(\zeta^{-1})\det a_{\Phi 2}(\zeta)\det a_{\Phi 2}(\zeta^{-1})}. \tag{8.194}$$

Since the product $M(p)\Phi(p)\Phi'(-p)M'(-p)$ is a strictly proper rational matrix, so owing to the above shown, the rational matrix $D_{M\Phi\underline{\Phi'}\underline{M'}}(T, \zeta, 0)$ is strictly proper, i.e.,

$$\text{exc}\, D_{M\Phi\underline{\Phi}'\underline{M}'}(T,\zeta,0) \geq 1. \tag{8.195}$$

Therefore, due to the fact that the matrix $M(p)\Phi(p)\Phi'(-p)M'(-p)$ has size $n \times n$, we conclude

$$\text{exc}\det D_{M\Phi\underline{\Phi}'\underline{M}'}(T,\zeta,0) \geq n. \tag{8.196}$$

Further, notice that the product in the denominator in formulae (8.193), (8.194) is a symmetric quasi-polynomial. Moreover, due to $\deg\det a_l(\zeta) = \chi$, $\deg\det a_{\Phi 2}(\zeta) = \varkappa$, and $\det a_l(0) \neq 0$, $\det a_{\Phi 2}(0) \neq 0$, we obtain

$$\text{Deg}[\det a_l(\zeta)\det a_l(\zeta^{-1})\det a_{\Phi 2}(\zeta)\det a_{\Phi 2}(\zeta^{-1})] = \chi + \varkappa. \tag{8.197}$$

From the preceding relations, we realize that $\det K_\Phi(\zeta)$ is a symmetric quasi-polynomial. Let $\text{Deg}\det K_\Phi(\zeta) = \gamma_\Phi$. Then, as in the conclusion of formula (7.142), we find

$$\text{exc}\,\frac{\det K_\Phi(\zeta)}{\det a_l(\zeta)\det a_l(\zeta^{-1})\det a_{\Phi 2}(\zeta)\det a_{\Phi 2}(\zeta^{-1})} = \chi + \varkappa - \gamma_\Phi. \tag{8.198}$$

From (8.194) with the help of (8.196) and (8.198) we arrive at the inequality

$$\chi + \varkappa - \gamma_\Phi \geq n,$$

or, equivalently

$$\text{Deg}\det K_\Phi(\zeta) = \gamma_\Phi \leq \chi + \varkappa - n. \tag{8.199}$$

In the case, when the quasi-polynomial $K_\Phi(\zeta)$ is positive on the circle $|\zeta| = 1$, formula (8.199) implies (8.190). ∎

(3)

**Theorem 8.7** *Under the conditions of Theorem 8.4, the characteristic polynomial of the $\mathcal{H}_2$ optimal system $\Delta_\Phi(\zeta)$ is, independently of the delay, a divisor of the polynomial*

$$\Delta^*_\Phi(\zeta) \triangleq \det\Pi(\zeta)\Delta_{M2}(\zeta)\det\Gamma_{\Phi_1}(\zeta), \tag{8.200}$$

*where the polynomial* $\det\Pi(\zeta)$ *is determined by formula (7.300)*

$$\det\Pi(\zeta) = \Delta_{L2}(\zeta)\rho_{L1}(\zeta), \tag{8.201}$$

*$\Delta_{M2}(\zeta)$ is stable polynomial (7.274), and the stable polynomial matrix $\Gamma_{\Phi_1}(\zeta)$ is the result of the factorization*

$$K_{\Phi 1}(\zeta) = \int_0^T b_{M\Phi}(\zeta,t)\hat{b}_\Phi(\zeta,t)\,\mathrm{d}t = \Gamma_{\Phi_1}(\zeta)\hat{\Gamma}_{\Phi_1}(\zeta). \tag{8.202}$$

*Proof* From (8.71) and (8.75), we find

$$\theta_\Phi^0(\zeta) = \Pi^{-1}(\zeta)H_\Phi^+(\zeta)\Gamma_\Phi^{-1}(\zeta)a_{\Phi 2}(\zeta). \tag{8.203}$$

Due to the propositions, the polynomial matrices $\Gamma_\Phi(\zeta)$ and $a_{\Phi 2}(\zeta)$ have disjunct sets of eigenvalues, so the matrix relation $\Gamma_\Phi^{-1}(\zeta)a_{\Phi 2}(\zeta)$ is an ILMFD. Hereby, for the IRMFD

$$\Gamma_\Phi^{-1}(\zeta)a_{\Phi 2}(\zeta) = a_{\Phi 4}(\zeta)\Gamma_{\Phi_2}^{-1}(\zeta) \tag{8.204}$$

the equivalence relations

$$a_{\Phi 4}(\zeta) \sim a_{\Phi 2}(\zeta), \quad \Gamma_{\Phi_2}(\zeta) \sim \Gamma_\Phi(\zeta) \tag{8.205}$$

are true. With the help of (8.204), formula (8.203) is transformed into

$$\theta_\Phi^0(\zeta) = \Pi^{-1}(\zeta)H_{\Phi 2}(\zeta)\Gamma_{\Phi_2}^{-1}(\zeta), \tag{8.206}$$

where

$$H_{\Phi 2}(\zeta) \stackrel{\triangle}{=} H_\Phi^+(\zeta)a_{\Phi 4}(\zeta). \tag{8.207}$$

Under the conditions of Theorem 8.4, the matrix $\theta_\Phi^0(\zeta)$ does not possess poles in the set $\mathscr{M}_\Phi$, and the same property, owing to

$$H_{\Phi 2}(\zeta) = \Pi(\zeta)\theta_\Phi^0(\zeta)\Gamma_{\Phi_2}(\zeta) \tag{8.208}$$

is true for the matrix $H_{\Phi 2}$.

On the other side, since the matrices $\hat{\Pi}^{-1}(\zeta)$ and $\hat{V}^{-1}(\zeta)$ do not possess stable poles, so with the help of Theorem 8.2, the poles of the stable matrix $H_\Phi^+$ are located in the set $\mathscr{M}_\Phi$. The same fact, on basis of (8.207), takes place for the matrix $H_{\Phi 2}(\zeta)$. The contradiction proves that the matrix $H_{\Phi 2}(\zeta)$ has no poles at all, i.e., it is a polynomial. Therefore, there exists the LMFD

$$H_{\Phi 2}(\zeta)\Gamma_{\Phi_2}^{-1}(\zeta) = \Gamma_{\Phi_3}^{-1}(\zeta)H_{\Phi 3}(\zeta), \tag{8.209}$$

where

$$\det \Gamma_{\Phi_3}(\zeta) \sim \det \Gamma_{\Phi_2}(\zeta) \sim \det \Gamma_\Phi(\zeta) \sim \Delta_{M2}(\zeta) \det \Gamma_{\Phi_1}(\zeta). \tag{8.210}$$

According to (8.206) and (8.209), we find the LMFD

$$\theta_\Phi^0(\zeta) = D_\Phi^{-1}(\zeta)M_\Phi(\zeta), \tag{8.211}$$

in which, by construction,

$$D_\Phi(\zeta) = \Gamma_{\Phi_3}(\zeta)\Pi(\zeta), \quad M_\Phi(\zeta) = H_{\Phi 3}(\zeta), \tag{8.212}$$

and

$$\det D_{\Phi}(\zeta) \sim \Delta^*_{\Phi}(\zeta). \tag{8.213}$$

Since at the same time we have ILMFD (8.79), so the relation

$$\frac{\det D_{\Phi}(\zeta)}{\det D_{\Phi 0}(\zeta)} = \frac{\Delta^*_{\Phi}(\zeta)}{\Delta_{\Phi 0}(\zeta)}$$

is a polynomial. ■

(4) In practical applications, the pair $(D_{\Phi}(\zeta), M_{\Phi}(\zeta))$ in expression (8.211) is irreducible, i.e., the matrix relation on the right side of (8.211) is an ILMFD. In this case

$$\det D_{\Phi 0}(\zeta) \sim \Delta_{\Phi 0}(\zeta) \sim \Delta^*_{\Phi}(\zeta). \tag{8.214}$$

When (8.214) is valid, it emerges from (8.200) that the set of poles (roots of the characteristic equation of the $\mathscr{H}_2$ optimal system) does not depend on the delay, and they can be classified into three groups:

Type I Roots of the polynomials $\Delta_{L2}(\zeta)$ and $\Delta_{M2}(\zeta)$ (fixed poles). From the formulae, determining the polynomials $\Delta_{L2}(\zeta)$ and $\Delta_{M2}(\zeta)$ we immediately find out that the fixed poles related to the polynomial $\Delta_{M2}(\zeta)$ do not depend on the form of the GDAC, and also do not depend on the coloring filter. Therefore, as before, for the existence of a stabilizing $\mathscr{H}_2$ optimal controller, it is not allowed that some of the fixed poles are located on the circle $|\zeta| = 1$.

Type II Roots of the polynomial $\det \Gamma_{\Phi_1}(\zeta)$. These poles do not depend on the form of the GDAC, but they depend on the form of the coloring filter.

Type III Roots of the polynomial $\rho_{L1}(\zeta)$. In contrast to the poles of Type II, these poles do not depend on the properties of the coloring filter, but depend on the form of the GDAC.

*Remark 8.1* Since the $H_\infty$ optimization problem is a marginal case of the $\mathscr{H}_2$ optimization problem under colored noise, so the fact on the independence of the set of fixed poles on the properties of the spectral density of the input signal is completely taken over, so that the set of fixed poles also exists for the $H_\infty$ optimal system. This means that the restrictions formulated for the location of the fixed poles are also required for optimal SD systems according to the $H_\infty$ criterion. For SISO systems, this fact was strictly proven in [4].

## References

1. E.N. Rosenwasser, *Linear Theory of Digital Control in Continuous Time* (Izd, LKI, St. Petersburg, 1989). (in Russian)
2. E.N. Rosenwasser, B.P. Lampe, *Computer Controlled Systems: Analysis and Design with Process-orientated Models* (Springer, London Berlin Heidelberg, 2000)
3. E.N. Rosenwasser, B.P. Lampe, *Multivariable Computer-controlled Systems–A Transfer Function Approach* (Springer, London, 2006)
4. K.Y. Polyakov, *Polynomial methods for direct design of optimal sampled-data control systems* (Saint Petersburg Maritime Technological University, Thesis for Doctor degree of Technological Sciences, 2006)

# Chapter 9
# $\mathscr{L}_2$ Tracking Problem for the Standard Model of SD Systems with Delay

**Abstract** In this chapter, we study the problem of the $\mathscr{L}_2$-optimization (SD tracking problem) of a stable standard SD system with delay. For this purpose, additional properties of the PTM $W_{zx}(\lambda, t)$ as a function of the argument $\lambda$ are studied. With the help of known relations (Rosenwasser and Lampe in Computer Controlled Systems: Analysis and Design with Process-orientated Models. Springer, London, 2000; Rosenwasser and Lampe in Multivariable Computer-controlled Systems—A Transfer Function Approach. Springer, London, 2006), connecting the PTM and the Laplace image of the output of the linear operator, an image of the output of the $\mathscr{S}_\tau$ system is constructed at zero initial energy. It allows us to formulate the $\mathscr{L}_2$-optimal tracking problem as a minimization problem for a quadratic functional on the set of stable system functions. In this case, unlike the case of $\mathscr{H}_2$-optimization problem, the constructed quadratic functional turns out to be singular, which leads to the impossibility of using the Wiener–Hopf method in the form described in the previous chapters. To overcome this difficulty, we use a special polynomial transformation of the functional to be minimized, which allows us to construct a set of $\mathscr{L}_2$-optimal control programs. In the conclusion of the chapter, the general properties of the $\mathscr{L}_2$-optimal system and the corresponding set of fixed poles are determined.

## 9.1 Operational Relations for the Standard Model of SD Systems with Delay

(1)

**Lemma 9.1** *Let*

$$W_{zx}(s,t) = \varphi_{L_\tau\mu}(T,s,t)\breve{R}_N(s)M(s)\mathrm{e}^{-s\tau_M} + K(s)\mathrm{e}^{-s\tau_K} \tag{9.1}$$

*be the PTM of the standard model (6.12) $\mathscr{S}_\tau$ from input $x(t)$ to output $z(t)$, determined by formula (6.67) for $\lambda = s$ and with the size $r \times \ell$. Then, there exists a real constant $\rho$ such that, for* Re$s > \rho$*, the matrix $W_{zx}(s,t)$ is analytic in $s$ for all $t$.*

*Proof* As follows from Theorem 6.4, the matrix $W_{zx}(s,t)$ is a meromorphic function of the argument $s$ for all $t$, and its poles are located in the set of roots of the equation



E. N. Rosenwasser et al., *Computer-Controlled Systems with Delay*,
https://doi.org/10.1007/978-3-030-15042-6_9

$$\det \breve{Q}(s, \breve{\alpha}, \breve{\beta}) = 0, \tag{9.2}$$

where the matrix $\breve{Q}(s, \breve{\alpha}, \breve{\beta})$ is determined by formula (6.95) for $\lambda = s$. For $e^{-sT} = \zeta$, with the help of (6.153) from (9.2), we obtain the algebraic equation

$$\det Q(\zeta, \alpha, \beta) = \det \begin{bmatrix} I_\chi - \zeta e^{AT} & 0_{\chi n} & -\zeta^{m_2+1} b(\zeta) \\ -C_2 & I_n & 0_{nq} \\ 0_{q\chi} & -\beta(\zeta) & \alpha(\zeta) \end{bmatrix} = 0. \tag{9.3}$$

From (9.3), we find

$$\det Q(\zeta, \alpha, \beta)|_{\zeta=0} = \alpha(0) \neq 0, \tag{9.4}$$

because we assumed the irreducibility and causality of the controller $(\alpha(\zeta), \beta(\zeta))$. It is seen from (9.4), that the roots $\zeta_i$ of Eq. (9.3) satisfy condition $|\zeta_i| > c_1$ where $c_1$ is a positive constant. Hereby, since the roots $p_i$ of Eq. (9.2) satisfy the relation $e^{-p_i T} = \zeta_i$, where $\zeta_i$ is one root of Eq. (9.3), so we obtain

$$e^{-\operatorname{Re} p_i T} > c_1,$$

or, equivalently,

$$\operatorname{Re} p_i < -\frac{1}{T} \ln c_1 \triangleq \rho. \tag{9.5}$$

Hence, for $\operatorname{Re} s > \rho$ the matrix $W_{zx}(s, t)$ does not possess poles, and therefore it is analytic. ■

**Corollary 9.1** *If the standard model (6.12) is stable, then $\rho < 0$. Indeed, the stability of the system $\mathscr{S}_\tau$ implies that in (9.5) we can choose $c_1 > 1$, thus $\rho < 0$.*

**Lemma 9.2** *For a stable model $\mathscr{S}_\tau$ and* Re$s \to \infty$, *the following estimate is valid:*

$$\|W(s, t)\| < \frac{c_2}{|s|}, \quad f = \text{const}. \tag{9.6}$$

*Here and below, $\|\cdot\|$ indicates some norm for number vectors, and its associated norm for number matrices.*

*Proof* At first we will show that for a stable model $\mathscr{S}_\tau$, the matrix $\breve{R}_N(s)$ emerging from (6.62) for $\lambda = s$, for sufficiently large $\operatorname{Re} s > 0$ is bounded. For this notice that from (6.195), for a stable model $\mathscr{S}_\tau$, we obtain

$$R_N(\zeta) = \breve{R}(s)\Big|_{e^{-sT}=\zeta} = \beta_{r0}(\zeta) a_l(\zeta) - a_r(\zeta)\theta(\zeta) a_l(\zeta), \tag{9.7}$$

where $\beta_{r0}(\zeta)$, $a_l(\zeta)$, and $a_r(\zeta)$ are polynomial matrices and $\theta(\zeta)$ is a stable rational matrix. Because all stable rational matrices of the argument $\zeta$ are causal, there exists the finite limit

$$\lim_{\zeta\to 0}\theta(\zeta) \triangleq \theta_0. \tag{9.8}$$

Therefore, for the matrix

$$\breve{R}_N(s) = R_N(\zeta)|_{\zeta=\mathrm{e}^{-sT}} = \beta_{r0}(\mathrm{e}^{-sT})a_l(\mathrm{e}^{-sT}) - a_r(\mathrm{e}^{-sT})\theta(\mathrm{e}^{-sT})a_l(\mathrm{e}^{-sT}) \tag{9.9}$$

there exists, for $\mathrm{Re}s \to \infty$, the finite limit

$$\lim_{\mathrm{Re}s\to\infty}\breve{R}_N(s) = \beta_{r0}(0)a_l(0) - a_r(0)\theta_0 a_l(0), \tag{9.10}$$

from which we conclude the boundedness of the matrix $\breve{R}_N(s)$ for sufficiently large $\mathrm{Re}s > 0$.

Further, we will show that the matrix

$$\varphi_{L_\tau\mu}(T,s,t) = \frac{1}{T}\sum_{k=-\infty}^{\infty} L_\tau(s+k\mathrm{j}\omega)\mu(s+k\mathrm{j}\omega)\mathrm{e}^{k\mathrm{j}\omega t} = \mathrm{e}^{-s\tau_L}\varphi_{L\mu}(T,s,t-\tau_L) \tag{9.11}$$

is bounded for sufficiently large $\mathrm{Re}s$.

For this notice that, from (6.18), (5.180), and (5.181) for $0 < t < T$, we obtain

$$\varphi_{L\mu}(T,s,t) = \mathrm{e}^{-st}\left[C_1(I_\chi - \mathrm{e}^{-sT}\mathrm{e}^{AT})^{-1}\breve{n}_{L\mu}(s,t) + D\breve{h}(s,t)\right], \tag{9.12}$$

where

$$\begin{aligned}\breve{n}_{L\mu}(s,t) &= \int_0^t \mathrm{e}^{A(t-\nu)}B_2\breve{h}(s,\nu)\,\mathrm{d}\nu + \mathrm{e}^{-sT}\mathrm{e}^{AT}\int_t^T \mathrm{e}^{A(t-\nu)}B_2\breve{h}(s,\nu)\,\mathrm{d}\nu,\\ \breve{h}(s,t) &= \sum_{i=0}^{P}\mathrm{e}^{-isT}h_i(t).\end{aligned} \tag{9.13}$$

From (9.12) and (9.13), we realize that the matrix $\varphi_{L\mu}(T,s,t)$ is uniformly bounded with respect to t in the interval $0 < t < T$ for sufficiently large $\mathrm{Re}s$. But then, using that the matrix $\varphi_{L_\mu}(T,s,t)$ is periodic in $t$, we conclude that this matrix is uniformly bounded with respect to t in the interval $-\infty < t < \infty$ for sufficiently large $\mathrm{Re}s$. Moreover, relation (9.11) says that analog properties hold for the matrix $\varphi_{L_\tau\mu}(T,s,t)$. Finally, the proven boundedness of the matrices $\breve{R}_N(s)$ and $\varphi_{L_\tau\mu}(T,s,t)$ verifies the claim of Lemma 9.2, because in (9.1), the matrices $M(s)$ and $K(s)$ are strictly proper. ■

(2)

**Theorem 9.1** *For the PTM (9.1) of the standard model $\mathscr{S}_\tau$, the following statements are true:*

*(i) The integral*

$$g(t, \nu) = \frac{1}{2\pi \mathrm{j}} \int_{c-\mathrm{j}\infty}^{c+\mathrm{j}\infty} W_{zx}(s, t) \mathrm{e}^{s(t-\nu)} \, \mathrm{d}s \tag{9.14}$$

*converges according to the primary value for all* $c > \rho$, *where* $c$ *is a real constant, and* $\rho$ *is the constant configured in Lemma 9.1, and it does not depend on the concrete choice of the value* $c$.

*(ii) The matrix* $g(t, \nu)$ *satisfies the causality condition*

$$g(t, \nu) = 0_{r\ell}, \quad t < \nu. \tag{9.15}$$

*(iii) The following estimate holds:*

$$\|g(t, \nu)\| < K\mathrm{e}^{\rho(t-\nu)}, \quad t > \nu, \tag{9.16}$$

*where* $K$ *is a constant.*

*(iv) The PTM* $W_{zx}(s, t)$ *is related to the matrix* $g(t, \nu)$ *by the formula*

$$W_{zx}(s, t) = \int_{-\infty}^{t} g(t, \nu) \mathrm{e}^{-s(t-\nu)} \, \mathrm{d}\nu, \quad \mathrm{Re}\, s > \rho, \tag{9.17}$$

*where the integral converges absolutely.*

*Proof* With the help of Lemmata 9.1 and 9.2, the proof runs in analogy to the proof in Chap. 14 in [1]. ■

**Definition 9.1** The matrix $g(t, \nu)$, defined by formula (9.14), is called impulse response of the standard model $\mathscr{S}_\tau$.

The impulse response $g(t, \nu)$ has an important property, which arise from the validity of the relation

$$g(t + T, \nu + T) = g(t, \nu), \tag{9.18}$$

which in turn emerges from the chain of equations

$$\begin{aligned} g(t + T, \nu + T) &= \frac{1}{2\pi \mathrm{j}} \int_{c-\mathrm{j}\infty}^{c+\mathrm{j}\infty} W_{zx}(s, t + T) \mathrm{e}^{-s(t-\nu)} \, \mathrm{d}\nu \\ &= \frac{1}{2\pi \mathrm{j}} \int_{c-\mathrm{j}\infty}^{c+\mathrm{j}\infty} W_{zx}(s, t) \mathrm{e}^{-s(t-\nu)} \, \mathrm{d}\nu. \end{aligned} \tag{9.19}$$

(3)

**Theorem 9.2** *Let the vector input signal* $x(t)$ *satisfy the conditions*

$$x(t) = 0_{\ell 1}, \quad t < t_1; \quad \|x(t)\| < \kappa \mathrm{e}^{\alpha t}, \quad t > t_1, \tag{9.20}$$

*where* $t_1, \kappa > 0$ *and* $\alpha$ *are real constants, and its elements be of bounded variation in each finite interval. Then for* $\mathrm{Re}\, s > \alpha$, *there exists the Laplace transform*

$$X(s) = \int_{-\infty}^{\infty} x(t)\mathrm{e}^{-st}\,\mathrm{d}t, \tag{9.21}$$

*and the following claims hold:*

*(i) The integral*

$$z(t) = \int_{-\infty}^{t} g(t,\nu)x(\nu)\,\mathrm{d}\nu \tag{9.22}$$

*converges absolutely for* $\operatorname{Re}s > \max\{\rho,\alpha\}$*, and it determines the solution of (6.12).*

*(ii) Solution (9.22) can also be represented as the integral in the complex plane*

$$z(t) = \frac{1}{2\pi\mathrm{j}}\int_{c-\mathrm{j}\infty}^{c+\mathrm{j}\infty} W(s,t)X(s)\mathrm{e}^{st}\,\mathrm{d}s, \quad \operatorname{Re}s > \max\{\rho,\alpha\}, \tag{9.23}$$

*where the integral converges absolutely.*

*Proof* With the help of Lemmata 9.1 and 9.2, the proof runs as in Chap. 14 in [1]. ∎

Further notice that formula (9.22) in a number of cases remains in force, even if restriction (9.20) does not hold, e.g., for $x(t) = I_\ell \mathrm{e}^{-st}$, $\operatorname{Re}s > \rho$, using (9.17), we obtain

$$z(t) = \int_{-\infty}^{t} g(t,\nu)\mathrm{e}^{s\nu}\,\mathrm{d}\nu = \mathrm{e}^{st}\int_{-\infty}^{t} g(t,\nu)\mathrm{e}^{-s(t-\nu)}\,\mathrm{d}\nu = \mathrm{e}^{st}W(s,t). \tag{9.24}$$

The integral on the right side of (9.24) converges absolutely, which follows from (9.16), and it defines a solution of Eq. (6.12).

(4) Relation (9.22) defines a linear operator

$$z(t) = \int_{-\infty}^{t} g(t,\nu)x(\nu)\,\mathrm{d}\nu = \int_{-\infty}^{\infty} g(t,\nu)x(\nu)\,\mathrm{d}\nu \stackrel{\triangle}{=} U[x(t)]. \tag{9.25}$$

The last relation is based on the causality condition (9.15).

**Lemma 9.3** *Operator (9.25) is T periodic in the sense of [1], i.e., it fulfills the relation*

$$z(t-T) = U[x(t-T)] \quad \forall t. \tag{9.26}$$

*Proof* From (9.25), we derive

$$U[x(t-T)] = \int_{-\infty}^{\infty} g(t,\nu)x(\nu-T)\,\mathrm{d}\nu = \int_{-\infty}^{\infty} g(t,\nu+T)x(\nu)\,\mathrm{d}\nu. \tag{9.27}$$

On the other side, substituting in (9.25) $t$ by $t-T$, we obtain

$$z(t-T)=\int_{-\infty}^{\infty} g(t-T,\nu)x(\nu)d\nu. \tag{9.28}$$

If we now take from (9.18)

$$g(t-T,\nu)=g(t,\nu+T), \tag{9.29}$$

then comparing formulae (9.27) and (9.28) leads to relation (9.26). ■

**Definition 9.2** Solution (9.25) of Eq. (6.12) is called solution with zero initial energy.

## 9.2 Laplace Transforms of Processes with Zero Initial Energy

(1)

**Theorem 9.3** *Under the conditions of Theorem 9.2, the following claims hold:*

*(i) Solution $z(t)$ with zero initial energy determined by formulae (9.22), (9.23), possesses the Laplace transform*

$$Z(s)=\int_{-\infty}^{\infty} z(t)\mathrm{e}^{-st}\,\mathrm{d}t, \tag{9.30}$$

*which converges absolutely in the half-plane* $\operatorname{Re} s > \max\{\rho,\alpha\}$ *and the inverse Laplace transform is*

$$z(t)=\frac{1}{2\pi\mathrm{j}}\int_{c-\mathrm{j}\infty}^{c+\mathrm{j}\infty} Z(s)\mathrm{e}^{st}\,\mathrm{d}s, \quad c>\max\{\rho,\alpha\}, \tag{9.31}$$

*where the value of the integral does not depend on the choice of the constant $c$, and the integral converges absolutely.*

*(ii) The image $Z(s)$ is related to the PTM $W_{zx}(s,t)$ and the image of the input $X(s)$ by the formula*

$$Z(s)=\int_0^T \left[\frac{1}{T}\sum_{k=-\infty}^{\infty} W_{zx}(s+k\mathrm{j}\omega,t)X(s+k\mathrm{j}\omega)\mathrm{e}^{k\mathrm{j}\omega t}\right]\mathrm{d}t. \tag{9.32}$$

*Proof* (i) This claim can be shown in analogy to the proof in [3].

(ii) We provide a formal proof of this claim. A strict mathematical proof is given in [1, 4].

From (9.26), we find for any integer $k$

$$z(t+kT)=U[x(t+kT)]. \tag{9.33}$$

Therefore, the DLT of the vector $z(t)$ due to (5.53) is defined by the relation

$$\breve{D}_z(T,s,t) = \sum_{k=-\infty}^{\infty} z(t+kT)\mathrm{e}^{-ksT} = \sum_{k=-\infty}^{\infty} U[x(t+kT)]\mathrm{e}^{-ksT}. \tag{9.34}$$

Suppose that the operators $\sum$ and $U$ are commutative, then

$$\breve{D}_z(T,s,t) = U\left[\sum_{k=-\infty}^{\infty} x(t+kT)\mathrm{e}^{-ksT}\right], \tag{9.35}$$

which can be written in the form

$$\breve{D}_z(T,s,t) = U\left[\breve{D}_x(T,s,t)\right], \tag{9.36}$$

where

$$\breve{D}_x(T,s,t) = \sum_{k=-\infty}^{\infty} x(t+kT)\mathrm{e}^{-ksT} \tag{9.37}$$

is the DLT of the vector $x(t)$. Let

$$\breve{D}_X(T,s,t) = \frac{1}{T}\sum_{k=-\infty}^{\infty} X(s+k\mathrm{j}\omega)\mathrm{e}^{(s+k\mathrm{j}\omega)t} \tag{9.38}$$

be the DLT of image $X(s)$ (9.21). Under the taken propositions, the vectors $\breve{D}_x(T,s,t)$ and $\breve{D}_X(T,s,t)$ are connected, as functions of the argument $t$, by the equivalence relation (5.68)

$$\breve{D}_x(T,s,t) \overset{t}{\sim} \breve{D}_X(T,s,t). \tag{9.39}$$

Thus, relation (9.36) can also be expressed by

$$\breve{D}_Z(T,s,t) = U\left[\breve{D}_X(T,s,t)\right], \tag{9.40}$$

where

$$\breve{D}_Z(T,s,t) = \frac{1}{T}\sum_{k=-\infty}^{\infty} Z(s+k\mathrm{j}\omega)\mathrm{e}^{(s+k\mathrm{j}\omega)t} \tag{9.41}$$

is the DLT of the image $Z(s)$. Using (9.38), we obtain

$$U\left[\breve{D}_X(T,s,t)\right] = U\left[\frac{1}{T}\sum_{k=-\infty}^{\infty} X(s+k\mathrm{j}\omega)\mathrm{e}^{(s+k\mathrm{j}\omega)t}\right]. \tag{9.42}$$

Assuming the commutativity of the operators $U$ and $\sum$, we achieve

$$U\left[\breve{D}_X(T,s,t)\right]=\frac{1}{T}\sum_{k=-\infty}^{\infty}U\left[\mathrm{e}^{(s+kj\omega)t}I_\ell\right]X(s+kj\omega). \tag{9.43}$$

Further notice that from (9.24), we obtain

$$U\left[\mathrm{e}^{st}I_\ell\right]=\int_{-\infty}^{t}g(t,\nu)\mathrm{e}^{s\nu}\,\mathrm{d}\nu=\mathrm{e}^{st}W_{zx}(s,t). \tag{9.44}$$

Hence,

$$U\left[\mathrm{e}^{(s+kj\omega)t}I_\ell\right]=\mathrm{e}^{(s+kj\omega)t}W_{zx}(s+kj\omega,t). \tag{9.45}$$

Therefore, with the help of (9.40) and (9.43), we find

$$D_Z(T,s,t)=\frac{1}{T}\sum_{k=-\infty}^{\infty}W_{zx}(s+kj\omega,t)X(s+kj\omega)\mathrm{e}^{(s+kj\omega)t}. \tag{9.46}$$

Since from (9.41), we easily derive

$$Z(s)=\int_0^T\breve{D}_Z(T,s,t)\mathrm{e}^{-st}\,\mathrm{d}t, \tag{9.47}$$

so from (9.46), we generate the relation which expresses the image $Z(s)$ by the PTM $W_{zx}(s,t)$ and image $X(s)$

$$Z(s)=\int_0^T\left[\frac{1}{T}\sum_{k=-\infty}^{\infty}W_{zx}(s+kj\omega,t)X(s+kj\omega)\mathrm{e}^{kj\omega t}\right]\mathrm{d}t. \tag{9.48}$$

∎

(2)

**Theorem 9.4** *Under the conditions of Theorem 9.3, the image of the process $Z(s)$ with zero initial energy is determined by the formula*

$$Z(s)=\mathrm{e}^{-s\tau_L}L(s)\mu(s)\breve{R}_N(s)\breve{D}_{MX}(T,s,-\tau_M)+K(s)X(s)\mathrm{e}^{-s\tau_K}, \tag{9.49}$$

*where $\breve{D}_{MX}(T,s,-\tau_M)$ is the sum of the series*

$$\breve{D}_{MX}(T,s,-\tau_M)=\frac{1}{T}\sum_{k=-\infty}^{\infty}M(s+kj\omega)X(s+kj\omega)\mathrm{e}^{-(s+kj\omega)\tau_M}. \tag{9.50}$$

*Proof* From (9.1), we obtain

$$W_{zx}(s+kj\omega,t)=\varphi_{L_\tau\mu}(T,s+kj\omega,t)\breve{R}_N(s+kj\omega)M(s+kj\omega)e^{-(s+kj\omega)\tau_M} + K(s+kj\omega)e^{-(s+kj\omega)\tau_K}. \tag{9.51}$$

For the further calculations, notice that formula (6.62) implies

$$\breve{R}_N(s+kj\omega)=\breve{R}_N(s). \tag{9.52}$$

Moreover, due to

$$\varphi_{L_\tau\mu}(T,s,t)=\frac{1}{T}\sum_{m=-\infty}^{\infty}L_\tau(s+mj\omega)\mu(s+mj\omega)e^{mj\omega t}, \tag{9.53}$$

we obtain

$$\varphi_{L_\tau\mu}(T,s+kj\omega,t)=\frac{1}{T}\sum_{m=-\infty}^{\infty}L_\tau(s+mj\omega+kj\omega)\mu(s+mj\omega+kj\omega)e^{mj\omega t}. \tag{9.54}$$

Substituting herein $m+k=n$, we find

$$\begin{aligned}\varphi_{L_\tau\mu}(T,s+kj\omega,t)&=\frac{1}{T}\sum_{n=-\infty}^{\infty}L_\tau(s+nj\omega)\mu(s+nj\omega)e^{(n-k)j\omega t}\\&=\varphi_{L_\tau\mu}(T,s,t)e^{-kj\omega t}.\end{aligned} \tag{9.55}$$

Inserting (9.52) and (9.55) into (9.51), we achieve

$$\begin{aligned}W_{zx}(s+kj\omega,t)&=\varphi_{l_\tau\mu}(T,s,t)\breve{R}_N(s)M(s+kj\omega)e^{-(s+kj\omega)\tau_M}e^{-kj\omega t}\\&\quad+K(s+kj\omega)e^{-(s+kj\omega)\tau_K}.\end{aligned} \tag{9.56}$$

Using this expression, we can write

$$\begin{aligned}&\frac{1}{T}\sum_{k=-\infty}^{\infty}W_{zx}(s+kj\omega,t)X(s+kj\omega)e^{kj\omega t}\\&\quad=\varphi_{L_\tau\mu}(T,s,t)\breve{R}_N(s)\frac{1}{T}\sum_{k=-\infty}^{\infty}M(s+kj\omega)X(s+kj\omega)e^{-(s+kj\omega)\tau_M}\\&\qquad+\frac{1}{T}\sum_{k=-\infty}^{\infty}K(s+kj\omega)X(s+kj\omega)e^{-(s+kj\omega)\tau_K}e^{kj\omega t}.\end{aligned} \tag{9.57}$$

With the help of (9.57), from (9.48), we find

$$Z(s) = \int_0^T \varphi_{L_\tau\mu}(T, s, t)\, dt\, \breve{R}_N(s)\breve{D}_{MX}(T, s, -\tau_M) + K(s)X(s)e^{-s\tau_K}. \qquad (9.58)$$

Observing (9.53), we obtain

$$\int_0^T \varphi_{L_\tau\mu}(T, s, t)\, dt = L_\tau(s)\mu(s), \qquad (9.59)$$

and with this result, from (9.58), we come to

$$Z(s) = L_\tau(s)\mu(s)\breve{R}_N(s)\breve{D}_{MX}(T, s, -\tau_M) + K(s)X(s)e^{-s\tau_K}, \qquad (9.60)$$

which is equivalent to (9.49). ■

## 9.3 Properties of the Image $Z(s)$

(1) In the further considerations, we propose that the input signal complies with the conditions of Theorem 9.2, and that the sum of series (9.38) is a rational function of the argument $e^{-sT}$ for all $t$. Then the vector

$$D_X(T, \zeta, t) = \breve{D}_X(T, s, t)\Big|_{e^{-sT}=\zeta}$$

is a rational function of the argument $\zeta$. In [2], those Laplace transforms are called pseudo-rational. It was also shown there that a wide class of input signals has those images, among them arbitrary signals of finite duration. For pseudo-rational images there exist, in $0 < t < T$, irreducible representations of the form

$$D_X(T, \zeta, t) \stackrel{\triangle}{=} \mathscr{D}_X(T, \zeta, t) = \frac{\sum_{i=0}^r \zeta^i f_i(t)}{d_X(\zeta)}, \qquad (9.61)$$

where $d_X(\zeta)$ is a polynomial, and $f_i(t)$ are known (vector) functions, where the elements of them are assumed to be functions of bounded variation. For $\zeta = e^{-sT}$, from (9.61), we find

$$\breve{D}_X(T, s, t) = \mathscr{D}_X(T, \zeta, t)|_{\zeta=e^{-sT}} = \frac{\sum_{i=0}^r e^{-isT} f_i(t)}{\breve{d}_X(s)}, \qquad (9.62)$$

where

$$\breve{d}_X(s) = d_X(\zeta)|_{\zeta=e^{-sT}}. \qquad (9.63)$$

Further on, we always propose that the polynomial $d_X(\zeta)$ does not possess roots on the circle $|\zeta| \le 1$. In this case, the function $\breve{d}_X(s)$ does not possess roots in the right half-plane Re$s \geq 0$.

(2) Below, the set of roots of the equation

$$\det \check{Q}(s, \check{\alpha}, \check{\beta}) = \det Q(\zeta, \alpha, \beta)|_{\zeta = \mathrm{e}^{-sT}} = 0, \tag{9.64}$$

where $Q(\zeta, \alpha, \beta)$ is the matrix (9.3), is denoted by $\mathscr{M}_Q$, and the set of roots of $\check{d}_X(s)$ by $\mathscr{M}_X$.

**Theorem 9.5** *The set of poles of the image $Z(p)$ (9.49) is contained in $\mathscr{M}_Q \cup \mathscr{M}_X$.*

*Proof* Denote

$$W_\ell(s) \triangleq \frac{1}{T}\int_0^T W_{zx}(s + \ell \mathrm{j}\omega, t)\mathrm{e}^{\ell \mathrm{j}\omega t}\,\mathrm{d}t, \tag{9.65}$$

where $\ell$ is any integer. Since the matrix $\check{Q}(s, \check{\alpha}, \check{\beta})$ does not change when $s$ is exchanged by $s + \ell \mathrm{j}\omega$, so from (6.98), we find that the set of poles $\mathscr{M}_\ell$ of the matrix $W_\ell(s)$ is located in the set $\mathscr{M}_Q$. Moreover, due to

$$X(s) = \int_0^T \check{D}_X(T, s, t)\mathrm{e}^{-st}\,\mathrm{d}t, \tag{9.66}$$

we realize from (9.62) and (9.63) that for any $\ell$, the poles of the vector

$$X(s + \ell \mathrm{j}\omega) = \int_0^T \check{D}_X(T, s, t)\mathrm{e}^{-(s+\ell \mathrm{j}\omega)t}\,\mathrm{d}t \tag{9.67}$$

are contained in the set $\mathscr{M}_X$. Now, formula (9.48) can be written in the form

$$Z(s) = \sum_{\ell=-\infty}^{\infty} W_\ell(s) X(s + \ell \mathrm{j}\omega). \tag{9.68}$$

Remember that the series on the right side converges absolutely for all $s$, except for the poles of its terms. Hence, the set of poles of the image $Z(s)$ is contained in the union of the sets $\mathscr{M}_Q$ and $\mathscr{M}_X$. ∎

(3) Formula (9.9) can be written in the form

$$\check{R}_N(s) = \check{\beta}_{r0}(s)\check{a}_l(s) - \check{a}_r(s)\check{\theta}(p)\check{a}_l(s), \tag{9.69}$$

where, as earlier, we denote

$$\begin{aligned} \check{\beta}_{r0}(s) &= \beta_{r0}(\mathrm{e}^{-sT}), \quad \check{\theta}(s) = \theta(\mathrm{e}^{-sT}), \\ \check{a}_l(s) &= a_l(\mathrm{e}^{-sT}), \quad \check{a}_r(s) = a_r(\mathrm{e}^{-sT}). \end{aligned} \tag{9.70}$$

Inserting (9.69) into (9.49), we obtain

$$Z(s) = \ell(s)\breve{\theta}(s)\breve{m}(s) + n(s), \tag{9.71}$$

where

$$\begin{aligned}
\ell(s) &\stackrel{\triangle}{=} -\mathrm{e}^{-s\tau_L} L(s)\mu(s)\breve{a}_r(s),\\
\breve{m}(s) &\stackrel{\triangle}{=} \breve{a}_l(s)\breve{D}_{MX}(T, s, -\tau_M),\\
n(s) &\stackrel{\triangle}{=} \mathrm{e}^{-s\tau_L} L(s)\mu(s)\breve{\beta}_{r0}(s)\breve{m}(s) + K(s)X(s)\mathrm{e}^{-s\tau_K}.
\end{aligned} \tag{9.72}$$

**Definition 9.3** Formula (9.71) is called representation of the image $Z(s)$ by the system function. The matrix $\ell(s)$ and the vectors $\breve{m}(s)$ and $n(s)$, defined by formulae (9.72), are the coefficients of this representation.

**Theorem 9.6** *The following claims hold.*

*(i) The matrix $\ell(s)$ is an integral function of the argument $s$.*
*(ii) The vectors $\breve{m}(s)$ and $n(s)$ are meromorphic functions of the argument $s$, and their sets of poles are contained in the set $\mathscr{M}_X$.*

*Proof* (i) Consider the series

$$\begin{aligned}
\breve{D}_\ell(T, s, t) &= \frac{1}{T}\sum_{k=-\infty}^{\infty} \ell(s + k\mathrm{j}\omega)\mathrm{e}^{(s+k\mathrm{j}\omega)t}\\
&= -\frac{1}{T}\sum_{k=-\infty}^{\infty} L(s + k\mathrm{j}\omega)\mu(s + k\mathrm{j}\omega)\mathrm{e}^{(s+k\mathrm{j}\omega)(t-\tau_L)}\breve{a}_r(s + k\mathrm{j}\omega).
\end{aligned} \tag{9.73}$$

Due to $\breve{a}_r(s + k\mathrm{j}\omega) = \breve{a}(s)$, from (9.73), we obtain

$$\begin{aligned}
\breve{D}_\ell(T, s, t) &= \left[-\frac{1}{T}\sum_{k=-\infty}^{\infty} L(s + k\mathrm{j}\omega)\mu(s + k\mathrm{j}\omega)\mathrm{e}^{(s+k\mathrm{j}\omega)(t-\tau_L)}\right]\breve{a}_r(s)\\
&= -D_{L\mu}(T, s, t - \tau_L)\breve{a}_r(s) = -\psi_1(s, t),
\end{aligned} \tag{9.74}$$

where $\psi_1(s, t)$ is the matrix (6.201) for $\lambda = s$, which, owing to Theorem 6.9 is an integral function of the argument $s$. Moreover, equation

$$\ell(s) = -\int_0^T \psi_1(s, t)\mathrm{e}^{-st}\,\mathrm{d}t \tag{9.75}$$

yields that $\ell(s)$ is an integral function of the argument $s$.

(ii) For the proof of the claim with respect to the vector $\breve{m}(s)$, notice that from (8.153) for $W_1(s) = M(s),\ W_2(s) = X(s)$ and $t = -\tau_M$, we achieve

$$\breve{D}_{MX}(T, s, -\tau_M) = \int_0^T \breve{D}_M(T, s, \nu)\breve{D}_X(T, s, -\tau_M - \nu)\,\mathrm{d}\nu. \tag{9.76}$$

Therefore, we can write

$$\begin{aligned}\breve{m}(s) &= \breve{a}_l(s)\breve{D}_{MX}(T, s, t - \tau_M) \\ &= \int_0^T \left[\breve{a}_l(s)\breve{D}_M(T, s, \nu)\right] \breve{D}_X(T, s, -\tau_M - \nu)\,\mathrm{d}\nu.\end{aligned} \tag{9.77}$$

From the proof of Theorem 6.9, we learned that the matrix in the square brackets does not possess poles. Moreover, from (9.62), it follows that the set of poles of the vector $\breve{D}_X(T, s, -\tau_M - \nu)$ for $0 < \nu < T$ is located in the set $\mathscr{M}_X$. Therefore, we obtain, from (9.77), that the set of poles of the vector $\breve{m}(s)$ also is contained in the set $\mathscr{M}_X$.

It remains to prove the claim with respect to $n(p)$. For this notice that from (9.71), we find

$$n(s) = Z(p)|_{\breve{\theta}(s)=0_{qn}}\,. \tag{9.78}$$

From (6.201) for $\lambda = s$, owing to Theorem 6.9, we obtain

$$W_{zx}(s, t)|_{\breve{\theta}(s)=0_{qn}} = \eta(s, t), \tag{9.79}$$

where $\eta(s, t)$, for each $t$, is an integral function of the argument $s$, i.e., it does not possess poles. Using (9.79), from (9.48), we derive

$$Z(s)|_{\breve{\theta}(s)=0_{qn}} = \int_0^T \left[\frac{1}{T}\sum_{k=-\infty}^{\infty} \eta(s + k\mathrm{j}\omega, t)X(s + k\mathrm{j}\omega)\mathrm{e}^{k\mathrm{j}\omega t}\right] \mathrm{d}t. \tag{9.80}$$

The series under the integral converges absolutely for all $s$, except for the poles of the vectors $X(s + k\mathrm{j}\omega)$. Further notice that from (9.62), we obtain

$$X(s) = \int_0^T \mathscr{D}_X(T, s, t)\mathrm{e}^{-st}\,\mathrm{d}t = \frac{\sum_{i=0}^{r} \mathrm{e}^{-isT} \int_0^T f_i(t)\mathrm{e}^{-st}\,\mathrm{d}t}{\breve{d}_X(s)}, \tag{9.81}$$

and, thus

$$X(s + k\mathrm{j}\omega) = \frac{\sum_{i=0}^{r} \mathrm{e}^{-isT} \int_0^T f_i(t)\mathrm{e}^{-(s+k\mathrm{j}\omega)t}\,\mathrm{d}t}{\breve{d}_X(s)}. \tag{9.82}$$

Therefore, the poles of the vector $X(p + k\mathrm{j}\omega)$ for all $k$ are located in the set $\mathscr{M}_X$. But then from (9.80), we directly conclude that the poles of the vector $n(p)$ are contained in the set $\mathscr{M}_X$. ∎

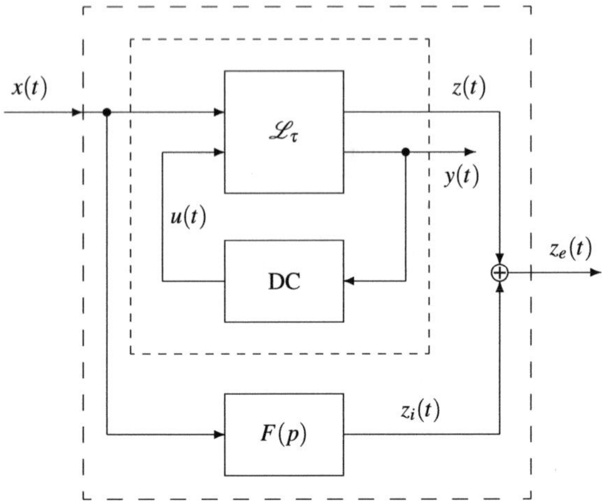


**Fig. 9.1** SD system with delay $\mathscr{S}_T$

## 9.4 Statement of the $\mathscr{L}_2$-Optimal Tracking Problem for the Standard Model of SD Systems with Delay

The subject under investigation in this section is the SD system with delay $\mathscr{S}_T$, shown in Fig. 9.1.

The dashed rectangle in Fig. 9.1 surrounds the standard model of the SD system with delay, which is shown in Fig. 6.1. Moreover, in Fig. 9.1, $F(s)$ is a transfer matrix describing the desired ideal behavior from the input vector $x(t)$ to the output signal $z_i(t)$. Then the error vector

$$z_e(t) = z(t) - z_i(t) \tag{9.83}$$

determines the deviation of the real output from the desired one. The amount of this deviation can be characterized by the integrated quadratic error

$$J = \int_0^\infty z_e'(t) z_e(t)\, \mathrm{d}t = \int_0^\infty [z(t) - z_i(t)]' [z(t) - z_i(t)]\, \mathrm{d}t, \tag{9.84}$$

where we suppose that the integral converges. Below, for the sake of simplicity, we always propose that the transfer matrix of the ideal behavior $F(s)$ is a stable and strictly proper rational $r \times \ell$ matrix, given by its minimal realization

$$F(s) = C_F(sI_\delta - A_F)^{-1} B_F, \tag{9.85}$$

although the described methods remain valid in more general situations.

To be concrete, we suppose, in addition to the above said, that the input signal $x(t)$ stimulates the system $\mathscr{S}_T$ only for $t > 0$, i.e., in (9.20) we have $t_1 = 0$. Moreover, suppose that in (9.20) $\alpha < 0$. In this case, for $t > 0$ we obtain, under zero initial energy,

$$\begin{aligned} z(t) &= \int_0^t g(t,\nu)x(\nu)\,\mathrm{d}\nu, \\ z_i(t) &= C_F \int_0^t \mathrm{e}^{A_F(t-\nu)} B_F x(\nu)\,\mathrm{d}\nu. \end{aligned} \tag{9.86}$$

**Definition 9.4** If (9.86) is true, then the number

$$\|\mathscr{S}_T\|_{\mathscr{L}_2} = +\sqrt{J} \tag{9.87}$$

is called the $\mathscr{L}_2$ norm of the system $\mathscr{S}_T$.

From the proposed stability of the transfer matrix $F(s)$, it follows that the system $\mathscr{S}_T$ is stable, if and only if the standard model $\mathscr{S}_\tau$ is stable. Hence, the set of stabilizing controllers for the system $\mathscr{S}_T$ coincides with the set of stabilizing controllers for the system $\mathscr{S}_\tau$. Considering the above said, we can formulate the topic of this section: the $\mathscr{L}_2$ optimization problem.

| |
|---|
| Let the matrix $\bar{W}_0(s)$ (6.23), the values of the delays $\tau_K$, $\tau_L$, $\tau_M$, and $\tau_N$, the equations of DC (6.10), and the transfer matrix of the ideal operator $F(s)$ be given. Find the transfer matrix of the discrete controller which stabilizes the standard model $\mathscr{S}_\tau$, and minimizes $\mathscr{L}_2$ norm (9.87). |

## 9.5 Construction of the Cost Functional

(1) Under the suppositions of Sect. 9.4, the error vector $z_e(t)$ has a Laplace transform under zero initial energy

$$Z_e(s) = Z(s) - Z_i(s), \tag{9.88}$$

in the half-plane Re$p > -\rho_1$, where $\rho_1$ is a positive constant. Here

$$\begin{aligned} Z(s) &= \mathrm{e}^{-s\tau_L} L(s)\mu(s)\breve{R}_N(s)\breve{D}_{MX}(T,s,-\tau_M) + K(s)X(s)\mathrm{e}^{-s\tau_K}, \\ Z_i(s) &= F(s)X(s). \end{aligned} \tag{9.89}$$

Applying the Parseval formula from the theory of Laplace transformation [5], integral (9.84) can be written as an integral in the frequency domain

$$J = \frac{1}{2\pi \mathrm{j}} \int_{-\mathrm{j}\infty}^{\mathrm{j}\infty} Z_e'(-s)Z_e(s)\,\mathrm{d}s = \frac{1}{2\pi \mathrm{j}} \int_{-\mathrm{j}\infty}^{\mathrm{j}\infty} [Z(s) - Z_i(s)]'\,[Z(s) - Z_i(s)]\,\mathrm{d}s. \tag{9.90}$$

From the above estimates, we realize that integral (9.90) converges absolutely. Therefore, integral (9.90) can be transformed by standard methods into an integral with finite limits

$$J = \frac{T}{2\pi \mathrm{j}} \int_{-\mathrm{j}\omega/2}^{\mathrm{j}\omega/2} \breve{D}_{\underline{Z}'_e Z_e}(T, s, 0)\, \mathrm{d}s, \tag{9.91}$$

where

$$\breve{D}_{\underline{Z}'_e Z_e}(T, s, 0) = \frac{1}{T} \sum_{k=-\infty}^{\infty} Z'_e(-s - k\mathrm{j}\omega) Z_e(s + k\mathrm{j}\omega). \tag{9.92}$$

(2) Using the representation of the image $Z(s)$ by the system function (9.71), we obtain, from (9.88),

$$Z_e(s) = \ell(s)\breve{\theta}(s)\breve{m}(s) + n_1(s), \tag{9.93}$$

where the matrix $\ell(s)$ and the vector $\breve{m}(s)$ are determined by formula (9.72), and the vector $n_1(s)$ is defined by the formula

$$n_1(s) = n(p) - F(s)X(s) = \mathrm{e}^{-s\tau_L} L(s)\mu(s)\breve{\beta}_{r0}(s)\breve{m}(s) + K_1(s)X(s), \tag{9.94}$$

in which

$$K_1(s) \stackrel{\triangle}{=} K(s)\mathrm{e}^{-s\tau_K} - F(s). \tag{9.95}$$

From (9.93), after transposition and renaming $s$ by $-s$, according to (9.72), (9.94), and (9.95), we find

$$Z'_e(-s) = \breve{m}'(-s)\breve{\theta}'(-s)\ell'(-s) + n'_1(-s), \tag{9.96}$$

where

$$\begin{aligned}
\ell'(-s) &= -\mathrm{e}^{s\tau_L}\breve{a}'_r(-s)\mu'(-s)L'(-s),\\
\breve{m}'(-s) &= \breve{D}'_{MX}(T, -s, -\tau_M)\breve{a}'_l(-s),\\
n'_1(-s) &= \mathrm{e}^{s\tau_L}\breve{m}'(-s)\breve{\beta}'_{r0}(-s)\mu'(-s)L'(-s) + X'(-s)K'_1(-s).
\end{aligned} \tag{9.97}$$

Multiplying expressions (9.93) and (9.96) results in

$$\begin{aligned}
Z'_e(-s)Z_e(s) = &\breve{m}'(-s)\breve{\theta}'(-s)\ell'(-s)\ell(s)\breve{\theta}(s)\breve{m}(s) + n'_1(-s)n_1(s)\\
&+ \breve{m}'(-s)\breve{\theta}'(-s)\ell'(-s)n_1(s) + n'_1(-s)\ell(s)\breve{\theta}(s)\breve{m}().
\end{aligned} \tag{9.98}$$

Hence, for any integer $k$ and when $s$ is substituted by $s + k\mathrm{j}\omega$, we obtain

$$\begin{aligned} Z_e'(-s-kj\omega)Z_e(s+kj\omega) = & \breve{m}'(-s)\breve{\theta}'(-s)\ell'(-s-kj\omega)\ell(s+kj\omega)\breve{\theta}(s)\breve{m}(s) \\ & + \breve{m}'(-s)\breve{\theta}'(-s)\ell'(-s-kj\omega)n_1(s+kj\omega) \\ & + n_1'(-s-kj\omega)\ell(s+kj\omega)\breve{\theta}(s)\breve{m}(s) \\ & + n_1'(-s-kj\omega)n_1(s+kj\omega), \end{aligned} \tag{9.99}$$

where we considered

$$\breve{m}(s+kj\omega) = \breve{m}(s), \quad \breve{\theta}(s+kj\omega) = \breve{\theta}(s). \tag{9.100}$$

Inserting (9.99) into (9.92), we achieve the following result:

$$\begin{aligned} \breve{D}_{\underline{Z_e'}Z_e}(T,s,0) = & \breve{m}'(-s)\breve{\theta}'(-s)\breve{D}_{\underline{\ell'}\ell}(T,s,0)\breve{\theta}(s)\breve{m}(s) \\ & + \breve{m}'(-s)\breve{\theta}'(-s)\breve{B}'(-s) + \breve{B}(s)\breve{\theta}(s)\breve{m}(s) + \breve{D}_{\underline{n_1'}n_1}(T,s,0). \end{aligned} \tag{9.101}$$

Herein

$$\begin{aligned} \breve{D}_{\underline{\ell'}\ell}(T,s,0) &= \frac{1}{T}\sum_{k=-\infty}^{\infty} \ell'(-s-kj\omega)\ell(s+kj\omega), \\ \breve{B}(s) \stackrel{\triangle}{=} \breve{D}_{\underline{n_1'}\ell}(T,s,0) &= \frac{1}{T}\sum_{k=-\infty}^{\infty} n_1'(-s-kj\omega)\ell(s+kj\omega), \\ \breve{B}'(-s) = \breve{D}'_{\underline{n_1'}\ell}(T,-s,0) = \breve{D}_{\underline{\ell'}n_1}(T,s,0) &= \frac{1}{T}\sum_{k=-\infty}^{\infty} \ell'(-s-kj\omega)n_1(s+kj\omega) \end{aligned} \tag{9.102}$$

and

$$\breve{D}_{\underline{n_1'}n_1}(T,s,0) = \frac{1}{T}\sum_{k=-\infty}^{\infty} n_1'(-s-kj\omega)n_1(s+kj\omega) \tag{9.103}$$

is a function which is not incorporated in the following calculations.

(3) We look for closed expressions for the matrix $\breve{D}_{\underline{\ell'}\ell}(T,s,0)$ and for the vectors $\breve{B}(s)$ and $\breve{B}'(-s)$. According to (9.72), we obtain

$$\ell'(-s)\ell(s) = \breve{a}_r'(-s)\mu'(-s)L'(-s)L(s)\mu(s)\breve{a}_r(s). \tag{9.104}$$

Using this expression with regard to $\breve{a}_r(s) = \breve{a}_r(s+kj\omega)$, from (9.102), we find immediately

$$\breve{D}_{\underline{\ell'}\ell}(T,s,0) = \breve{a}_r'(-s)\breve{D}_{\underline{\mu'L'}L\mu}(T,s,0)\breve{a}_r(s) = \breve{A}_L(s), \tag{9.105}$$

where $\breve{A}_L(s)$ is the matrix (7.64).

For the calculation of the vector $\breve{B}(s)$, notice that, with the help of (9.94),

$$n_1'(-s)\ell(s) = n'(-s)\ell(s) - X'(-s)F'(-s)\ell(s), \tag{9.106}$$

where

$$\begin{aligned} n'(-s)\ell(s) = &-\check{m}'(-s)\check{\beta}_{r0}'(-s)\mu'(-s)L'(-s)L(s)\mu(s)\check{a}_r(s) \\ &- X'(-s)K'(-s)L(s)\mu(s)\check{a}_r(s)e^{s(\tau_K-\tau_L)} \end{aligned} \tag{9.107}$$

$$X'(-s)F'(-s)\ell(s) = -e^{-s\tau_L}X'(-s)F'(-s)L(s)\mu(s)\check{a}_r(s). \tag{9.108}$$

With the help of Theorem 9.6, the matrix $\ell(s)$ is an integral function of the argument $s$, and the vector $n(s)$ is a meromorphic function with its poles located in the set $\mathscr{M}_X$. Therefore, the product $n'(-s)\ell(s)$ is a meromorph function with its poles located in the set $\mathscr{M}_{\underline{X}}$, the set of roots of the equation

$$\check{d}_X(-s) = 0. \tag{9.109}$$

Consider the sum of the series

$$\check{D}_{\underline{n}'\ell}(T,s,0) = \frac{1}{T}\sum_{k=-\infty}^{\infty} n'(-s-kj\omega)\ell(s+kj\omega). \tag{9.110}$$

Owing to (9.107) and (9.108), we obtain

$$\begin{aligned} &\check{D}_{\underline{n}'\ell}(T,s,0) = \\ &- m'(-s)\check{\beta}_{r0}'(-s)\frac{1}{T}\sum_{k=-\infty}^{\infty}\mu'(-s-kj\omega)L'(-s-kj\omega)L(s+kj\omega)\mu(s+kj\omega)\check{a}_r(s) \\ &\quad -\frac{1}{T}\sum_{k=-\infty}^{\infty}X'(-s-kj\omega)K'(-s-kj\omega)L(s+kj\omega)\mu(s+kj\omega)e^{s(\tau_K-\tau_L)}\check{a}_r(s) \\ &= -\check{m}'(-s)\check{\beta}_{r0}'(-s)\check{D}_{\underline{\mu}'\underline{L}'L\mu}(T,s,0)\check{a}_r(s) - \check{D}_{\underline{X}'\underline{K}'L\mu}(T,s,\tau_K-\tau_L)\check{a}_r(s). \end{aligned} \tag{9.111}$$

Observe from (9.111) that the sum of series (9.110) is a rational function of the argument $\zeta = e^{-sT}$. Moreover, it follows from the above considerations that the poles of the elements of absolutely convergent series (9.110) are located in the set $\mathscr{M}_{\underline{X}}$. Hence,

$$\check{D}_{\underline{n}'\ell}(T,s,0) = \frac{\check{V}_1(s)}{\check{d}_X(-s)}, \tag{9.112}$$

where $\check{V}_1(s)$ is a finite sum of the form

$$\check{V}_1(s) = \sum_{i=-\alpha}^{\beta} c_i e^{-isT}, \tag{9.113}$$

where $\alpha \geq 0, \beta \geq 0$ are integers and the $c_i$ are constant vectors. By analog reasoning, we find that the sum of the series

$$\begin{aligned}
&\breve{D}_{\underline{X}'\underline{F}'\ell}(T, s, 0) = \\
&= -\frac{1}{T}\sum_{k=-\infty}^{\infty} X'(-s-kj\omega)F'(-s-kj\omega)L(s+kj\omega)\mu(s+kj\omega)e^{-(s+kj\omega)\tau_L}\breve{a}_r(s) = \\
&= -\breve{D}_{\underline{X}'\underline{F}'L\mu}(T, s, -\tau_L)\breve{a}_r(s)
\end{aligned} \tag{9.114}$$

allows the representation

$$D_{\underline{X}'\underline{F}'\ell}(T, s, 0) = \frac{\breve{V}_2(s)}{\breve{d}_X(-s)\breve{d}_F(-s)}. \tag{9.115}$$

Herein

$$\breve{V}_2(s) = \sum_{i=-\gamma}^{\varepsilon} d_i e^{-isT}, \tag{9.116}$$

where $\gamma \geq 0$, $\varepsilon \geq 0$ are integers, the $d_i$ are constant vectors, and

$$\breve{d}_F(s) \stackrel{\triangle}{=} \det(I_\delta - e^{-sT}e^{A_F T}). \tag{9.117}$$

Regarding that (9.107) implies

$$\breve{D}_{\underline{n}_1'\ell}(T, s, 0) = \breve{D}_{\underline{n}'\ell}(T, s, 0) + \breve{D}_{\underline{X}'\underline{F}'L\mu}(T, s, -\tau_L)\breve{a}_r(s), \tag{9.118}$$

and using representations (9.112) and (9.115), we find expressions of the form

$$\breve{B}(s) = \breve{D}_{\underline{n}_1'\ell}(T, s, 0) = -\frac{\breve{V}(s)}{\breve{d}_X(-s)\breve{d}_F(-s)}. \tag{9.119}$$

Here $\breve{V}(s)$ is the finite sum

$$\breve{V}(s) = \sum_{i=-\eta}^{\theta} f_i e^{-isT}, \tag{9.120}$$

where $\eta \geq 0$, $\theta \geq 0$ are integers, and the $f_i$ are constant vectors. Substituting in (9.119) $s$ by $-s$, after transposition, we obtain

$$\breve{B}'(-s) = -\frac{\breve{V}'(-s)}{\breve{d}_X(s)\breve{d}_F(s)}. \tag{9.121}$$

(4) Inserting (9.101), (9.105), (9.119), and (9.121) in (9.91), we obtain

$$J = J_1 + J_2, \tag{9.122}$$

where

$$\begin{aligned} J_1 = \frac{T}{2\pi \mathrm{j}} \int_{-\mathrm{j}\omega/2}^{\mathrm{j}\omega/2} \Bigg[ & \breve{m}'(-s)\breve{\theta}'(-s)\breve{A}_L(s)\breve{\theta}(s)\breve{m}(s) - \\ & - \breve{m}'(-s)\breve{\theta}'(-s)\frac{\breve{V}'(-s)}{\breve{d}_X(s)\breve{d}_F(s)} - \frac{\breve{V}(s)}{\breve{d}_X(-s)\breve{d}_F(-s)}\breve{\theta}(s)\breve{m}(s) \Bigg] \mathrm{d}s \end{aligned} \tag{9.123}$$

and

$$J_2 = \frac{T}{2\pi \mathrm{j}} \int_{-\mathrm{j}\omega/2}^{\mathrm{j}\omega/2} \breve{D}_{\underline{n}_1' n_1}(T, s, 0)\, \mathrm{d}s. \tag{9.124}$$

The quantity $J_2$ does not depend on the choice of the discrete controller, because the vector $n_1(p)$ has this property. Therefore, the $\mathscr{L}_2$ optimization of system $\mathscr{S}_T$ leads to the minimization of the integral $J_1$.

After the substitution $\mathrm{e}^{-sT} = \zeta$, integral (9.123) can be written as the line integral

$$\begin{aligned} J_1 = \frac{1}{2\pi \mathrm{j}} \oint \Bigg[ & \hat{m}(\zeta)\hat{\theta}(\zeta)A_L(\zeta)\theta(\zeta)m(\zeta) \\ & - \hat{m}(\zeta)\hat{\theta}(\zeta)\frac{\hat{V}(\zeta)}{d_X(\zeta)d_F(\zeta)} - \frac{V(\zeta)}{d_X(\zeta^{-1})d_F(\zeta^{-1})}\theta(\zeta)m(\zeta) \Bigg] \frac{\mathrm{d}\zeta}{\zeta}, \end{aligned} \tag{9.125}$$

where, as usual,

$$A_L(\zeta) = \breve{A}_L(p)\Big|_{\mathrm{e}^{-sT}=\zeta}, \quad V(\zeta) = \breve{V}(p)\Big|_{\mathrm{e}^{-sT}=\zeta}, \quad d_F(\zeta) = \breve{d}_F(p)\Big|_{\mathrm{e}^{-sT}=\zeta}.$$

In formula (9.125), the integration is done over the circle $|\zeta| = 1$ in positive direction.

Moreover, in (9.125),

$$m(\zeta) = \breve{m}(s)|_{\mathrm{e}^{-sT}=\zeta} = \left[\breve{a}_l(s)\breve{D}_{MX}(T, s, -\tau_M)\right]_{\mathrm{e}^{-sT}=\zeta} = a_l(\zeta)D_{MX}(T, \zeta, -\tau_M). \tag{9.126}$$

For the further transformations, notice that formula (9.76) can be written in the form

$$\breve{D}_{MX}(T, s, -\tau_M) = \int_0^T \breve{\mathscr{D}}_M(T, s, \nu)\breve{D}_X(T, s, -\tau_M - \nu)\, \mathrm{d}\nu, \tag{9.127}$$

because the variable $\nu$ changes in the interval $0 < t < T$. With the help of this relation, from (9.126), we find

$$m(\zeta) = \int_0^T [a_l(\zeta)\mathscr{D}_M(T, \zeta, \nu)]\, D_X(T, \zeta, -\tau_M - \nu)\, \mathrm{d}\nu. \tag{9.128}$$

From (7.100) and (7.105), we obtain

$$a_l(\zeta)\mathscr{D}_M(T,\zeta,t) = a_{M1}(\zeta)b_M(\zeta,t), \tag{9.129}$$

where $b_M(\zeta,t)$ is a matrix, which depends polynomially on $\zeta$. Further, observe that for $qT < -\tau_M - \nu < (q+1)T$, the vector $D_X(T,\zeta,-\tau_M-\nu)$ permits the representation

$$\begin{aligned} D_X(T,\zeta,-\tau_M-\nu) &= \mathscr{D}_X(T,\zeta,-\tau_M-\nu-qT)\zeta^{-q} \\ &= \frac{\zeta^{-q}\sum_{i=0}^{r}\zeta^i f_i(-\tau_M-\nu-qT)}{d_X(\zeta)}, \end{aligned} \tag{9.130}$$

considering (9.61) and (5.55). Notice that $q \le 0$, so we obtain from the last expression

$$D_X(T,\zeta,-\tau_M-\nu) = \frac{N(\zeta,\nu)}{d_X(\zeta)}, \quad 0<\nu<T, \tag{9.131}$$

where $N(\zeta,\nu)$ is a vector which depends polynomially on $\zeta$. Inserting (9.129) and (9.131) into (9.128) yields

$$m(\zeta) = \frac{a_{M1}(\zeta)N_m(\zeta)}{d_X(\zeta)}, \quad \hat{m}(\zeta) = \frac{\hat{N}_m(\zeta)\hat{a}_{M1}(\zeta)}{d_X(\zeta^{-1})} \tag{9.132}$$

with a polynomial vector $N_m(\zeta)$. With the help of formula (9.132), functional (9.125) takes the form

$$\begin{aligned} J_1 = \frac{1}{2\pi \mathrm{j}}\oint \Bigg[ & \hat{N}_m(\zeta)\hat{a}_{M1}(\zeta)\hat{\theta}(\zeta)\frac{A_L(\zeta)}{d_X(\zeta)d_X(\zeta^{-1})}\theta(\zeta)a_{M1}(\zeta)N_m(\zeta) \\ & - \hat{N}_m(\zeta)\hat{a}_{M1}(\zeta)\hat{\theta}(\zeta)\frac{\hat{V}(\zeta)}{d_X(\zeta)d_X(\zeta^{-1})d_F(\zeta)} \\ & - \frac{V(\zeta)}{d_X(\zeta)d_X(\zeta^{-1})d_F(\zeta^{-1})}\theta(\zeta)a_{M1}(\zeta)N_m(\zeta)\Bigg]\frac{\mathrm{d}\zeta}{\zeta}, \end{aligned} \tag{9.133}$$

where $V(\zeta)$ is a known quasi-polynomial (or polynomial) vector.

## 9.6 Transformation and Minimization of the Cost Functional

(1) The direct application of the Wiener–Hopf method for the minimization of the functional (9.133) is difficult, due to the fact that this functional does not fulfill the conditions for Theorem 8.3 because it appears to be singular. In order to overcome these difficulties, the present section describes a method for transforming integral

(9.133) into a non-degenerated functional, for which the application of Theorem 8.3 is possible.

Let

$$a(\zeta) = \begin{bmatrix} a_1(\zeta) \\ \vdots \\ a_n(\zeta) \end{bmatrix} \tag{9.134}$$

be a polynomial column vector. According to the well-known Theorem of Hermite [6, 7], there exists an unimodular matrix $p(\zeta)$, such that

$$p(\zeta)a(\zeta) = \begin{bmatrix} d(\zeta) \\ 0 \\ \vdots \\ 0 \end{bmatrix} = d(\zeta)1_n, \tag{9.135}$$

where $d(\zeta)$ is a scalar polynomial and

$$1_n \triangleq \begin{bmatrix} 1 \\ 0 \\ \vdots \\ 0 \end{bmatrix} \tag{9.136}$$

is of dimension $n$.

**Lemma 9.4** *The polynomial $d(\zeta)$ in (9.135) is the greatest common divisor (GCD) of the elements of the column $a(\zeta)$.*

*Proof* Introduce

$$q(\zeta) \triangleq p^{-1}(\zeta) = \begin{bmatrix} q_{11}(\zeta) & \dots & q_{1n}(\zeta) \\ \vdots & \ddots & \vdots \\ q_{n1}(\zeta) & \dots & q_{nn}(\zeta) \end{bmatrix}. \tag{9.137}$$

Then, from (9.135), we obtain

$$a(\zeta) = d(\zeta)q(\zeta)1_n = d(\zeta) \begin{bmatrix} q_{11}(\zeta) \\ \vdots \\ q_{n1}(\zeta) \end{bmatrix} = \begin{bmatrix} a_1(\zeta) \\ \vdots \\ a_n(\zeta) \end{bmatrix}. \tag{9.138}$$

Here, we directly observe that the polynomial $d(\zeta)$ is a common divisor of all polynomials $a_1(\zeta), \dots, a_n(\zeta)$. Since the matrix $q(\zeta)$ is unimodular, so due to Theorem 1.15, the polynomials $q_{11}(\zeta), \dots, q_{n1}(\zeta)$ as a whole are coprime. This means that the polynomial $d(\zeta)$ is the greatest common divisor. ■

(2) Owing to Lemma 9.4, there exists an unimodular matrix $R(\zeta)$, such that

$$R(\zeta)a_{M1}(\zeta)N_m(\zeta) = d(\zeta)1_n, \tag{9.139}$$

where $d(\zeta)$ is the GCD of the elements of the vector $a_{M1}(\zeta)N_m(\zeta)$. Hence

$$\begin{aligned} a_{M1}(\zeta)N_m(\zeta) &= d(\zeta)Y(\zeta)1_n, \\ \hat{N}_m(\zeta)\hat{a}_{M1}(\zeta) &= d(\zeta^{-1})1_n'\hat{Y}(\zeta), \end{aligned} \tag{9.140}$$

where $Y(\zeta) \stackrel{\triangle}{=} R^{-1}(\zeta)$ is a known unimodular matrix. Substituting (9.140) in (9.133), we obtain

$$\begin{aligned} J_1 = \frac{1}{2\pi j}\oint \Bigg[ & 1_n'\hat{Y}(\zeta)\hat{\theta}(\zeta)\frac{A_L(\zeta)d(\zeta)d(\zeta^{-1})}{d_X(\zeta)d_X(\zeta^{-1})}\theta(\zeta)Y(\zeta)1_n \\ & - 1_n'\hat{Y}(\zeta)\hat{\theta}(\zeta)\frac{\hat{V}(\zeta)}{d_X(\zeta)d_X(\zeta^{-1})d_F(\zeta)}9 - \frac{V(\zeta)}{d_X(\zeta)d_X(\zeta^{-1})d_F(\zeta^{-1})}\theta(\zeta)Y(\zeta)1_n\Bigg]\frac{d\zeta}{\zeta}. \end{aligned} \tag{9.141}$$

Introduce the rational matrix

$$\theta_1(\zeta) \stackrel{\triangle}{=} \theta(\zeta)Y(\zeta), \tag{9.142}$$

then functional (9.141) takes the form

$$\begin{aligned} J_2 = \frac{1}{2\pi j}\oint \Bigg[ & 1_n'\hat{\theta}_1(\zeta)\frac{A_L(\zeta)d(\zeta)d(\zeta^{-1})}{d_X(\zeta)d_X(\zeta^{-1})}\theta_1(\zeta)1_n \\ & - \frac{1_n'\hat{\theta}_1(\zeta)\hat{V}(\zeta)}{d_X(\zeta)d_X(\zeta^{-1})d_F(\zeta)} - \frac{V(\zeta)\theta_1(\zeta)1_n}{d_X(\zeta)d_X(\zeta^{-1})d_F(\zeta^{-1})}\Bigg]\frac{d\zeta}{\zeta}. \end{aligned} \tag{9.143}$$

**Lemma 9.5** *The minimization problems for the functionals (9.143) and (9.133) are equivalent in the following sense. If the stable matrix $\theta^0(\zeta)$ minimizes functional (9.133), then the matrix $\theta_1^0(\zeta) = \theta^0(\zeta)Y(\zeta)$ is stable and it minimizes functional (9.143). If, however, the stable matrix $\theta_1^0(\zeta)$ minimizes functional (9.143), then the matrix $\theta^0(\zeta) = \theta_1^0(\zeta)Y^{-1}(\zeta) = \theta_1^0(\zeta)R(\zeta)$ is stable and minimizes functional (9.133).*

*Proof* Since in (9.142) the matrix $Y(\zeta)$ is unimodular, the matrices $\theta^0(\zeta)$ and $\theta_1^0(\zeta)$ are stable simultaneously. Denote functional (9.133) by $J_1(\theta)$, and functional (9.143) by $J_2(\theta)$. Moreover, introduce

$$J_{1\,\min} \stackrel{\triangle}{=} J_1(\theta^0(\zeta)), \quad J_{2\,\min} \stackrel{\triangle}{=} J_2(\theta_1^0(\zeta)).$$

Then, obviously $J_{2\,\min} \leq J_{1\,\min}$, because for $\theta_1^0(\zeta) \stackrel{\triangle}{=} Y(\zeta)\theta^0(\zeta)$, we obtain $J_2(\theta_1^0) = J_1(\theta^0)$. In the same way, we realize that $J_{1\,\min} \leq J_{2\,\min}$, and the proof is complete. ■

(3) Lemma 9.5 implies that the $\mathscr{L}_2$ optimization problem considered here leads to the minimization of functional (9.143). For the solution of this problem, represent the matrix $\theta_1(\zeta)$ in block form

$$\theta_1(\zeta) = \left[\,\Omega(\zeta)\ \Omega_1(\zeta)\,\right], \tag{9.144}$$

where $\Omega(\zeta)$ is the first column of the matrix $\theta_1(\zeta)$. Then, we can write

$$\theta_1(\zeta)1_n = \Omega(\zeta) \tag{9.145}$$

and the integral (9.143) can be written in such a form, that it only depends on the column $\Omega(\zeta)$

$$J_2 = \frac{1}{2\pi \mathrm{j}} \oint \left[ \hat{\Omega}(\zeta) \frac{A_L(\zeta)d(\zeta)d(\zeta^{-1})}{d_X(\zeta)d_X(\zeta^{-1})} \Omega(\zeta) \right. \\ \left. - \hat{\Omega}(\zeta) \frac{\hat{V}(\zeta)}{d_X(\zeta)d_X(\zeta^{-1})d_F(\zeta)} - \frac{V(\zeta)}{d_X(\zeta)d_X(\zeta^{-1})d_F(\zeta^{-1})} \Omega(\zeta) \right] \frac{\mathrm{d}\zeta}{\zeta}. \tag{9.146}$$

Let a stable rational $q \times 1$ matrix $\Omega^0(\zeta)$ exist, which minimizes functional (9.146), and $\Omega_1(\zeta)$ be any stable rational $q \times (n-1)$ matrix. Then the stable rational matrix

$$\theta_1^0(\zeta) = \left[\,\Omega^0(\zeta)\ \Omega_1(\zeta)\,\right] \tag{9.147}$$

minimizes functional (9.143), i.e., this functional actually depends only on the column $\Omega(\zeta)$. But then, owing to the above said, the stable matrix

$$\theta^0(\zeta) = \theta_1^0(\zeta)R(\zeta) = \left[\,\Omega^0(\zeta)\ \Omega_1(\zeta)\,\right]R(\zeta) \tag{9.148}$$

minimizes original functional (9.133). It is evident from (9.148) that the optimal system function $\theta^0(\zeta)$, and thus the optimal controller $(\alpha^0(\zeta), \beta^0(\zeta))$, depends on an arbitrary stable matrix $\Omega_1(\zeta)$ of size $q \times (n-1)$. This fact comes from the singularity of functional (9.133).

(4)

**Theorem 9.7** *Let the quasi-polynomial $A_L(\zeta)$ be positive on the circle $|\zeta| = 1$. Let also the polynomial $d(\zeta)$ in (9.139) be free of roots on the circle $|\zeta| = 1$. Then, there exists an optimal column $\Omega^0(\zeta)$, which can be constructed by the following algorithm:*

*(a) Perform the factorization*

$$A_L(\zeta) = \hat{\Pi}(\zeta)\Pi(\zeta), \tag{9.149}$$

*where $\Pi(\zeta)$ is a stable polynomial matrix, and*

$$d(\zeta)d(\zeta^{-1}) = c(\zeta)c(\zeta^{-1}), \tag{9.150}$$

*where $c(\zeta)$ is a stable polynomial. Then, build the factorization*

$$\frac{A_L(\zeta)d(\zeta)d(\zeta^{-1})}{d_X(\zeta)d_X(\zeta^{-1})} = \frac{A_L(\zeta)c(\zeta)c(\zeta^{-1})}{d_X(\zeta)d_X(\zeta^{-1})} = \hat{\Pi}_1(\zeta)\Pi_1(\zeta), \tag{9.151}$$

*where*

$$\Pi_1(\zeta) = \frac{\Pi(\zeta)c(\zeta)}{d_X(\zeta)} \tag{9.152}$$

*is a rational matrix, which is stable together with its inverse, because the polynomials $c(\zeta)$ and $d_X(\zeta)$ are stable.*

*(b) Calculate the rational vector*

$$H_F(\zeta) \triangleq \hat{\Pi}_1^{-1}(\zeta)\frac{\hat{V}(\zeta)}{d_X(\zeta)d_X(\zeta^{-1})d_F(\zeta)}. \tag{9.153}$$

*According to*

$$\hat{\Pi}_1(\zeta) = \frac{\hat{\Pi}(\zeta)c(\zeta^{-1})}{d_X(\zeta^{-1})}, \tag{9.154}$$

*from (9.153), find the rational vector*

$$H_F(\zeta) = \hat{\Pi}^{-1}(\zeta)\frac{\hat{V}(\zeta)}{d_X(\zeta)d_F(\zeta)c(\zeta^{-1})}. \tag{9.155}$$

*(c) Perform the separation*

$$H_F(\zeta) = H_F^+(\zeta) + H_F^-(\zeta), \tag{9.156}$$

*where $H_F^-$ is a strictly proper rational vector with poles located in the disc $|\zeta| < 1$, and the vector $H_F^+(\zeta)$ has the form*

$$H_F^+(\zeta) = \frac{N_F(\zeta)}{d_X(\zeta)d_F(\zeta)}, \tag{9.157}$$

*where $N_F(\zeta)$ is a polynomial vector.*

*(d) The optimal column $\Omega^0(\zeta)$ is determined by the formula*

$$\Omega^0(\zeta) = \frac{\Pi^{-1}(\zeta)N_F(\zeta)}{d_F(\zeta)c(\zeta)}. \tag{9.158}$$

*Proof* Applying factorization (9.151), we easily obtain the identity

$$\begin{aligned}\hat{\Omega}(\zeta)\frac{A_L(\zeta)c(\zeta)c(\zeta^{-1})}{d_X(\zeta)d_X(\zeta^{-1})}\Omega(\zeta) - \hat{\Omega}(\zeta)\frac{\hat{V}(\zeta)}{d_X(\zeta)d_X(\zeta^{-1})d_F(\zeta)}\\ - \frac{V(\zeta)}{d_X(\zeta)d_X(\zeta^{-1})d_F(\zeta^{-1})}\Omega(\zeta) = \quad (9.159)\\ = \left[\widehat{\Pi_1(\zeta)\Omega(\zeta) - H_F(\zeta)}\right][\Pi_1(\zeta)\Omega(\zeta) - H_F(\zeta)] - \hat{H}_F(\zeta)H_F(\zeta),\end{aligned}$$

which is analogous to (7.252). The further progress of the proof similar to the proof of Theorem 7.5. Hereby, relation (9.157) follows from (9.155), because the matrices $\hat{\Pi}_1^{-1}(\zeta)$ and $\hat{V}(\zeta)$ do not possess stable poles, and the polynomials $d_X(\zeta), d_F(\zeta)$ and $c(\zeta)$ are stable. Formula (9.158) is a consequence of the relation

$$\Omega^0(\zeta) = \Pi_1^{-1}(\zeta)H_F(\zeta) \tag{9.160}$$

and formula (9.152). ■

## 9.7 General Properties and Fixed Poles of $\mathscr{L}_2$ Optimal Systems

On basis of the above derived theoretical results, we are able to formulate a number of properties of $\mathscr{L}_2$ optimal systems, which are important for applications.

(1) The optimal system function $\theta^0(\zeta)$ of size $q \times n$, determined by the Wiener–Hopf method, depends on an arbitrary stable rational $q \times (n-1)$ matrix $\Omega_1(\zeta)$. This fact is a consequence of the singularity of the considered $\mathscr{L}_2$ optimization problem, and emerges from formula (9.148). Further on the matrix $\Omega_1(\zeta)$ is called system parameter.

(2)

**Theorem 9.8** *The Laplace image of the $\mathscr{L}_2$-optimal transient process does not depend on the choice of the parameter $\Omega_1(\zeta)$.*

*Proof* From (9.148) and (9.158), we find

$$\theta^0(\zeta) = \left[\,\Omega^0(\zeta)\;\; \Omega_1(\zeta)\,\right]R(\zeta) = \left[\,\frac{\Pi^{-1}(\zeta)N_F(\zeta)}{d_F(\zeta)c(\zeta)}\;\; \Omega_1(\zeta)\,\right]R(\zeta). \tag{9.161}$$

Substitute $\zeta$ by $e^{-sT}$, so

$$\breve{\theta}^0(s) = \left[\,\frac{\breve{\Pi}^{-1}(s)\breve{N}_F(s)}{\breve{d}_F(s)\breve{c}(s)}\;\; \breve{\Omega}_1(s)\,\right]\breve{R}(s). \tag{9.162}$$

Here, as earlier, we used the notation

$$\breve{f}(s) \stackrel{\triangle}{=} f(\zeta)\Big|_{\zeta=\mathrm{e}^{-sT}}. \tag{9.163}$$

Inserting (9.162) into (9.93), we obtain the image of the transient process

$$Z_e(s) = \ell(s)\left[\frac{\breve{\Pi}^{-1}(s)N_F(s)}{\breve{d}_F(s)\breve{c}(s)} \quad \breve{\Omega}_1(s)\right]\breve{R}(s)\breve{m}(s) + n_1(s). \tag{9.164}$$

From (9.132) and (9.139), we find

$$R(\zeta)m(\zeta) = R(\zeta)\frac{a_{M1}(\zeta)N_m(\zeta)}{d_X(\zeta)} = \frac{d(\zeta)}{d_X(\zeta)}1_n, \tag{9.165}$$

which after renaming $\zeta$ by $\mathrm{e}^{-sT}$ yields

$$\breve{R}(s)\breve{m}(s) = \frac{\breve{d}(s)}{\breve{d}_X(s)}1_n. \tag{9.166}$$

Inserting this expression into (9.164) leads to formula

$$Z_e^0(s) = \ell(s)\frac{\breve{\Pi}^{-1}(s)\breve{N}_F(s)\breve{d}(s)}{\breve{d}_F(s)\breve{d}_X(s)\breve{c}(s)} + n_1(s), \tag{9.167}$$

which does not depend on the choice of the system parameter $\Omega_1(\zeta)$. ■

(3) For a certain choice of the system parameter $\Omega_1(\zeta)$ assume the ILMFD

$$\theta^0(\zeta) = D_0^{-1}(\zeta)M_0(\zeta). \tag{9.168}$$

Then, as was stated earlier, the characteristic polynomial of the optimal system $\Delta^0(\zeta)$ satisfies the equivalence relation

$$\Delta^0(\zeta) \sim \det D_0(\zeta). \tag{9.169}$$

This relation, together with (9.148), allows the conclusion that the characteristic polynomial $\Delta^0(\zeta)$ does depend on the selection of the system parameter. On the other side, essential limitations take place, which are formulated in the next theorem.

**Theorem 9.9** *Assume for the optimal column $\Omega^0(\zeta)$ (9.158), the ILMFD*

$$\Omega^0(\zeta) = D_\Omega^{-1}(\zeta)M_\Omega(\zeta). \tag{9.170}$$

*Then, independent of the choice of the system parameter $\Omega_1(\zeta)$, the quantity*

$$\delta_T(\zeta) \stackrel{\triangle}{=} \frac{\det D_0(\zeta)}{\det D_\Omega(\zeta)} \tag{9.171}$$

*is a polynomial.*

*Proof* Under condition (9.168), the matrix

$$D_0(\zeta)\theta^0(\zeta) = D_0(\zeta)\theta_1^0(\zeta)R(\zeta) = \left[\, D_0(\zeta)\Omega^0(\zeta) \;\; D_0(\zeta)\Omega_1(\zeta) \,\right] R(\zeta) \quad (9.172)$$

is polynomial. Since the matrix $R(\zeta)$ is unimodular, so the matrix

$$D_0(\zeta)\theta_1^0(\zeta) = \left[\, D_0(\zeta)\Omega^0(\zeta) \;\; D_0(\zeta)\Omega_1(\zeta) \,\right] \quad (9.173)$$

is polynomial too, and, in particular, $D_0(\zeta)\Omega^0(\zeta)$ is a polynomial vector. Since the right side of (9.170) is an ILMFD, so

$$D_0(\zeta) = D_1(\zeta)D_\Omega(\zeta), \quad (9.174)$$

where $D_1(\zeta)$ is a polynomial matrix. Hence, independent of the choice of the system parameter,

$$\det D_0(\zeta) \sim \det D_1(\zeta) \det D_\Omega(\zeta). \quad (9.175)$$

Thus, relation (9.171) is a polynomial. ∎

(4) The achieved result allows us to formulate some important properties of the polynomial

$$\Delta_\Omega(\zeta) \triangleq \det D_\Omega(\zeta). \quad (9.176)$$

**Theorem 9.10** *The following claims hold:*

*(i) For the optimal vector $\Omega^0(\zeta)$ in (9.158) and (9.170), there exists an LMFD*

$$\Omega^0(\zeta) = \Pi_\Omega^{-1}(\zeta)N_\Omega(\zeta), \quad (9.177)$$

*where*

$$\det \Pi_\Omega(\zeta) \sim \det \Pi(\zeta) d_F(\zeta) \det c(\zeta) \quad (9.178)$$

*is a polynomial, which does not depend on the input signal $x(t)$.*

*(ii) Polynomial $\Delta_\Omega(\zeta)$ (9.176) is a divisor of the polynomial* $\det \Pi_\Omega(\zeta)$.

*Proof* Apply the Hermite transformation to the vector $N_F(\zeta)$. Then, in analogy to (9.138), we obtain

$$N_F(\zeta) = q_\Omega(\zeta) \begin{bmatrix} d_\Omega(\zeta) \\ 0 \\ \vdots \\ 0 \end{bmatrix}, \quad (9.179)$$

where $q_\Omega(\zeta)$ is an unimodular matrix, and $d_\Omega(\zeta)$ is the GCD of the elements of the vector $N_F(\zeta)$. Using this relation, from (9.158), we find

$$\Omega^0(\zeta) = \Pi^{-1}(\zeta) q_\Omega(\zeta) \begin{bmatrix} \dfrac{d_\Omega(\zeta)}{d_F(\zeta)c(\zeta)} \\ 0 \\ \vdots \\ 0 \end{bmatrix}. \tag{9.180}$$

Introduce the non-singular polynomial matrix

$$\Pi_\Omega(\zeta) \stackrel{\triangle}{=} \operatorname{diag}\{d_F(\zeta)c(\zeta), 1, \ldots, 1\} q_\Omega^{-1}(\zeta) \Pi(\zeta). \tag{9.181}$$

Obviously, the product

$$N_\Omega(\zeta) \stackrel{\triangle}{=} \Pi_\Omega(\zeta)\Omega^0(\zeta) = d_\Omega(\zeta) 1_n \tag{9.182}$$

is a polynomial vector. Hereby, since for matrix (9.181) condition (9.178) is fulfilled, so claim (i) is shown.

Claim (ii) can be seen by comparing LMFD (9.177) and ILMFD (9.170). ■

(5) In practical problems, the right side of formula (9.177) is normally an ILMFD. In that case, we obtain

$$\Delta_\Omega(\zeta) = \det D_\Omega(\zeta) \sim \det \Pi_\Omega(\zeta). \tag{9.183}$$

When (9.183) is true, the roots of the polynomial $\Delta_\Omega(\zeta)$ are called basic poles of the $\mathscr{L}_2$ optimal system. As follows from (9.178), the set of basic poles can be divided into three groups.

1. Roots of the polynomial $\det \Pi(\zeta)$. From the considerations in Sect. 7.7, we know that

$$\det \Pi(\zeta) \sim \det a_{L1}(\zeta)\rho_L(\zeta). \tag{9.184}$$

   Here

$$\det a_{L1}(\zeta) \sim \Delta_{L2}(\zeta), \tag{9.185}$$

   where $\Delta_{L2}(\zeta)$ is the stable polynomial (7.296), which only depends on the properties of the continuous elements of the standard model.
2. Roots of the polynomial

$$d_F(\zeta) \stackrel{\triangle}{=} \det(I_\rho - \zeta e^{A_F T}), \tag{9.186}$$

   which are determined only by the properties of the transfer matrix of ideal operator $F(p)$ (9.85).
3. Roots of the polynomial $c(\zeta)$ from factorization (9.150).
   As follows from (9.150), those poles are determined by the roots of the polynomial $d(\zeta)$ (9.139), which is the GCD of the elements of the vector $a_{M1}(\zeta)N_m(\zeta)$.

Assume

$$d(\zeta) = (\zeta - \zeta_1)^{\lambda_1} \cdots (\zeta - \zeta_\rho)^{\lambda_\rho} \tag{9.187}$$

where

$$|\zeta_i| > 1, \ (i = 1, \ldots, \eta); \quad |\zeta_i| < 1, \ (i = \eta + 1, \ldots, \rho), \tag{9.188}$$

then

$$c(\zeta) \sim (\zeta - \zeta_1)^{\lambda_1} \cdots (\zeta - \zeta_\eta)^{\lambda_\eta} (\zeta - \zeta_{\eta+1}^{-1})^{\lambda_{\eta+1}} \cdots (\zeta - \zeta_\rho^{-1})^{\lambda_\rho}. \tag{9.189}$$

Notice that, in applications, when the dimension $n$ of the output $y(t)$ in Eqs. (6.12) is greater than one, the polynomial $c(\zeta)$ normally converges to a constant. However, in the practically important case in which $n = 1$, the situation changes, because then in formula (9.132), the matrix $a_{M1}(\zeta)$ appears as a scalar polynomial, and

$$a_{M1}(\zeta) \sim \Delta_{M1}(\zeta), \tag{9.190}$$

where $\Delta_{M1}(\zeta)$ is the polynomial (7.99). Moreover, if we assume that in (9.140) the elements of the vector $N_m(\zeta)$ are coprime, what practically always takes place, then we obtain

$$d(\zeta) \sim \Delta_{M1}(\zeta), \tag{9.191}$$

and, consequently,

$$c(\zeta) \sim \Delta_{M2}(\zeta), \tag{9.192}$$

where $\Delta_{M2}(\zeta)$ is the stable polynomial (7.274). In this case, among the roots of the polynomial $\det \Pi_\Omega(\zeta)$, we find all fixed poles of the $\mathscr{H}_2$ optimal system.

## References

1. E.N. Rosenwasser, B.P. Lampe, *Computer Controlled Systems: Analysis and Design with Process-orientated Models* (Springer, London, 2000)
2. E.N. Rosenwasser, B.P. Lampe, *Multivariable Computer-controlled Systems—A Transfer Function Approach* (Springer, London, 2006)
3. B.P. Lampe, E.N. Rosenwasser, Singular $\mathscr{L}_2$-optimization problem and fixed poles for sampled-data systems with delay, in 7th *IFAC Symposium on Robust Control Design* (DK, Aalborg, 2012), pp. 172–177
4. E.N. Rosenwasser, B.P. Lampe, *Digitale Regelung in kontinuierlicher Zeit - Analyse und Entwurf im Frequenzbereich* (B.G. Teubner, Stuttgart, 1997)
5. B. van der Pol, H. Bremmer, *Operational Calculus Based on the Two-Sided Laplace Integral* (University Press, Cambridge, 1959)
6. F.R. Gantmacher, *The Theory of Matrices* (Chelsea, New York, 1959)
7. T. Kailath, *Linear Systems* (Prentice Hall, Englewood Cliffs, NJ, 1980)

# Chapter 10
# Mathematical Description and Optimization for the Generalized Standard Model of SD Systems with Delay

**Abstract** This chapter examines the generalized standard model (GSM) of a system with delay in which, unlike the standard model (SM), different components $z_i(t)$ of the output vector $\bar{z}(t)$ can have different time shifts $\tau_i$ to the input vector $x(t)$ and to the control vector $u(t)$. It is shown that the solution of the $\mathscr{H}_2$ optimization problem for the GSM can be reduced to solving an analog problem for the SM to which the results of Chaps. 7–9 could be applied.

## 10.1 Mathematical Description of the Generalized Standard Model

(1) When defining the standard model $\mathscr{S}_\tau$ of SD systems with delay, all components of the output vector $z(t)$ in (6.12) have the same time shift according to the state vector $v(t)$ and the control vector $u(t)$. However, for many real control systems, this requirement does not hold. The present chapter pays attention to the generalization of the optimization methods for those systems.

In the actual chapter, we suppose that the closed loop of the considered SD system, as in Chap. 6, again is described by the system of equations

$$
\begin{aligned}
\frac{\mathrm{d}v(t)}{\mathrm{d}t} &= Av(t) + B_1 x(t-\tau_1) + B_2 u(t-\tau_2), \\
y(t) &= C_2 v(t), \\
\xi_k &= y(kT), \quad (k = 0, \pm 1, \ldots), \\
\alpha(\zeta)\psi_k &= \beta(\zeta)\xi_k, \\
u(t) &= \sum_{i=0}^{P} h_i(t-kT)\psi_{k-i}, \quad kT < t < (k+1)T,
\end{aligned}
\tag{10.1}
$$

where all the notations and propositions of Sect. 6.1 are applied. For the output vector, we assume



E. N. Rosenwasser et al., *Computer-Controlled Systems with Delay*,
https://doi.org/10.1007/978-3-030-15042-6_10

$$\bar{z}(t) \stackrel{\triangle}{=} \begin{bmatrix} z_1(t) \\ \vdots \\ z_m(t) \end{bmatrix}, \tag{10.2}$$

where the vectors $z_i(t)$ of dimensions $r_i \times 1$ are determined by the equations

$$z_i(t) = C_{1i}v(t - \tau_{3i}) + D_i u(t - \tau_2 - \tau_{3i}), \quad (i = 1, \dots, m), \tag{10.3}$$

where the $C_{1i}$, $D_i$ are constant matrices of appropriate size, and $\tau_{3i}$ are real constants.

**Definition 10.1** The totality of Eqs. (10.1)–(10.3) is called the generalized standard model of SD system with delay. Further on, this system is denoted by $\mathscr{S}_g$.

(2) In the further investigations, for simplicity reasons, we assume in (10.2) $m = 2$, because the generalization to arbitrary $m$ is plain sailing. For $m = 2$ we obtain

$$\begin{aligned} z_1(t) &= C_{11}v(t - \tau_{31}) + D_1 u(t - \tau_2 - \tau_{31}), \\ z_2(t) &= C_{12}v(t - \tau_{32}) + D_2 u(t - \tau_2 - \tau_{32}). \end{aligned} \tag{10.4}$$

Applying (10.1) and (10.3), (10.4), with the help of the same transformations as in Sect. 6.2, we obtain a system of operator equations analogous to (6.16)

$$\begin{aligned} \bar{z}(t) &= \bar{K}_\tau(p)x(t) + \bar{L}_\tau(p)u(t), \\ y(t) &= M_\tau(p)x(t) + N_\tau(p)u(t), \\ \xi_k &= y(kT), \quad (k = 0, \pm 1, \dots), \\ \alpha(\zeta)\psi_k &= \beta(\zeta)\xi_k, \\ u(t) &= \sum_{i=0}^{P} h_i(t - kT)\psi_{k-i}, \quad kT < t < (k+1)T. \end{aligned} \tag{10.5}$$

Here we used the notations

$$\bar{K}_\tau(p) \stackrel{\triangle}{=} \begin{bmatrix} K_{1\tau}(p) \\ K_{2\tau}(p) \end{bmatrix}, \quad \bar{L}_\tau(p) \stackrel{\triangle}{=} \begin{bmatrix} L_{1\tau}(p) \\ L_{2\tau}(p) \end{bmatrix}, \tag{10.6}$$

where

$$\begin{aligned} K_{i\tau}(p) &\stackrel{\triangle}{=} K_i(p)\mathrm{e}^{-p\tau_{Ki}}, \quad (i = 1, 2), \\ L_{i\tau}(p) &\stackrel{\triangle}{=} L_i(p)\mathrm{e}^{-p\tau_{Li}}, \quad (i = 1, 2). \end{aligned} \tag{10.7}$$

Moreover,

$$\begin{aligned} K_i(p) &= C_{1i}(pI_\chi - A)^{-1}B_1, \quad (i = 1, 2), \\ L_i(p) &= C_{1i}(pI_\chi - A)^{-1}B_2 + D_i, \quad (i = 1, 2), \end{aligned} \tag{10.8}$$

where

$$\tau_{K1} = \tau_1 + \tau_{31},\ \tau_{K2} = \tau_1 + \tau_{32},$$
$$\tau_{L1} = \tau_2 + \tau_{31},\ \tau_{L2} = \tau_2 + \tau_{32}. \tag{10.9}$$

Finally, in (10.5) the quantities $M_\tau(p)$ and $N_\tau(p)$ are matrices independent on $i$, defined by formulae (6.17)–(6.19)

$$M_\tau(p) = M(p)\mathrm{e}^{-p\tau_M}, \quad N_\tau(p) = N(p)\mathrm{e}^{-p\tau_N}, \tag{10.10}$$

where

$$M(p) = C_2(pI_\chi - A)^{-1}B_1, \quad N(p) = C_2(pI_\chi - A)^{-1}B_2, \tag{10.11}$$

and

$$\tau_M = \tau_1, \quad \tau_N = \tau_2. \tag{10.12}$$

(3) Below, the system of operational Eq. (10.5) is denoted by $\mathscr{S}_g$. For the system $\mathscr{S}_g$, we can build the parametric transfer matrices (PTM) $W_{z_1x}(s,t)$ and $W_{z_2x}(s,t)$ from the input $x(t)$ to the outputs $z_1(t)$ and $z_2(t)$, respectively. These PTM are determined by formulae obtained in Chap. 6, as

$$W_{z_ix}(s,t) = \varphi_{L_{i\tau}\mu}(T,s,t)\breve{R}_N(s)M_\tau(s) + K_{i\tau}(s), \quad (i = 1,2), \tag{10.13}$$

where

$$\varphi_{L_{i\tau}\mu}(T,s,t) = \frac{1}{T}\sum_{k=-\infty}^{\infty} L_{i\tau}(s + k\mathrm{j}\omega)\mu(s + k\mathrm{j}\omega)\mathrm{e}^{k\mathrm{j}\omega t}, \quad (i = 1,2) \tag{10.14}$$

and

$$\breve{R}_N(s) = \breve{W}_d(s)\left[I_n - \breve{D}_{N\mu}(T,s,-\tau_N)\breve{W}_d(s)\right]^{-1} \tag{10.15}$$

is the matrix defined by formula (6.62) when $\lambda = s$. Obviously, for $m = 2$ the PTM $W_{\bar{z}x}(s,t)$ from the input $x(t)$ to output $\bar{z}(t)$ (10.2) takes the form

$$W_{\bar{z}x}(s,t) = \begin{bmatrix} W_{z_1x}(s,t) \\ W_{z_2x}(s,t) \end{bmatrix}. \tag{10.16}$$

Inserting here (10.13), we obtain

$$W_{\bar{z}x}(s,t) = \varphi_{\bar{L}_\tau\mu}(T,s,t)\breve{R}_N(s)M_\tau(s) + \bar{K}_\tau(s), \tag{10.17}$$

where

$$\varphi_{\bar{L}_\tau\mu}(T,s,t) \triangleq \begin{bmatrix} \varphi_{L_{1\tau}\mu}(T,s,t) \\ \varphi_{L_{2\tau}\mu}(T,s,t) \end{bmatrix} = \frac{1}{T}\sum_{k=-\infty}^{\infty} \bar{L}_\tau(s+k\mathrm{j}\omega)\mu(s+k\mathrm{j}\omega)\mathrm{e}^{k\mathrm{j}\omega t}. \tag{10.18}$$

(4) Since the stabilizability of the system $\mathscr{S}_g$ does not depend on the choice of the output vector $\bar{z}(t)$, so the stabilizability conditions and the set of stabilizing controllers, and also the system function $\theta(\zeta)$ for the generalized standard model $\mathscr{S}_g$, are the same which we obtained in Chap. 6 for the standard model $S_\tau$.

Owing to the fact that the matrices $M(s)$ and $N(s)$ for the standard model (6.12) and the generalized standard model (10.1)–(10.3) coincide, the matrix $\breve{R}_N(s)$ is achieved from (6.196) for $\lambda = s$

$$\breve{R}_N(s) = \breve{\beta}_{r0}(s)\breve{a}_l(s) - \breve{a}_r(s)\breve{\theta}(s)\breve{a}_l(s), \tag{10.19}$$

where the matrices $\breve{a}_l(s)$ and $\breve{a}_r(s)$ are determined by the relations in Sect. 6.5. After inserting (10.19) into (10.17), we obtain a representation of the PTM $W_{\bar{z}x}(s,t)$ by the system function in analogy to (6.199)

$$W_{\bar{z}x}(s,t) \triangleq \bar{\psi}(s,t)\breve{\theta}(s)\xi(s) + \bar{\eta}(s,t). \tag{10.20}$$

Here we used the notation

$$\bar{\psi}(s,t) \triangleq \begin{bmatrix} \psi_1(s,t) \\ \psi_2(s,t) \end{bmatrix}, \quad \bar{\eta}(s,t) \triangleq \begin{bmatrix} \eta_1(s,t) \\ \eta_2(s,t) \end{bmatrix}, \tag{10.21}$$

where in coincidence with (6.200)

$$\begin{aligned} \psi_i(s,t) &\triangleq -\varphi_{L_{i\tau}\mu}(T,s,t)\breve{a}_r(s), \\ \xi(s) &\triangleq \breve{a}_l(s)M(s)\mathrm{e}^{-s\tau_M}, \\ \eta_i(s,t) &\triangleq \varphi_{L_{i\tau}}(T,s,t)\breve{\beta}_{r0}(s)\xi(s) + K_i(s)\mathrm{e}^{-s\tau_{Ki}}. \end{aligned} \tag{10.22}$$

Applying (10.22), from (10.21) we obtain particular expressions for the matrices $\bar{\psi}(s,t)$ and $\bar{\eta}(s,t)$

$$\begin{aligned} \bar{\psi}(s,t) &= \begin{bmatrix} \varphi_{L_{1\tau}\mu}(T,s,t) \\ \varphi_{L_{2\tau}\mu}(T,s,t) \end{bmatrix} \breve{a}_r(s), \\ \bar{\eta}(s,t) &= \begin{bmatrix} \varphi_{L_{1\tau}\mu}(T,s,t) \\ \varphi_{L_{2\tau}\mu}(T,s,t) \end{bmatrix} \breve{\beta}_{r0}(s)\xi(s) + \begin{bmatrix} K_1(s)\mathrm{e}^{-s\tau_{K1}} \\ K_2(s)\mathrm{e}^{-s\tau_{K2}} \end{bmatrix}. \end{aligned} \tag{10.23}$$

**Theorem 10.1** *Matrices (10.22) do not possess poles, i.e., they are integral functions of the argument s.*

*Proof* The proof directly follows from Theorem 6.9, because owing to this theorem, the composite matrices (10.23) and also the matrix $\xi(s)$ are integral functions of the argument $s$. ∎

## 10.2 $\mathscr{H}_2$ Optimization of System $\mathscr{S}_g$

(1) In analogy to (7.20) introduce

$$A_{\bar{z}x}(s) \stackrel{\triangle}{=} \frac{1}{T}\int_0^T W'_{\bar{z}x}(-s,t)W_{\bar{z}x}(s,t)\,\mathrm{d}t. \tag{10.24}$$

Then, under excitation of the stable system $\mathscr{S}_g$ by vectorial white noise with uncorrelated components, the mean variance $\bar{d}_{\bar{z}}^0$ of the quasi-stationary process at the output $\bar{z}(t)$ is determined, analogously to (7.21), by the formula

$$\bar{d}_{\bar{z}}^0 = \frac{1}{2\pi\mathrm{j}}\int_{-\mathrm{j}\infty}^{\mathrm{j}\infty} \operatorname{trace} A_{\bar{z}x}(s)\,\mathrm{d}s. \tag{10.25}$$

**Definition 10.2** The number

$$\|\mathscr{S}_g\|_2 = +\sqrt{\bar{d}_{\bar{z}}^0} \tag{10.26}$$

is called the $\mathscr{H}_2$ norm for the generalized standard model of the SD system with delay.

(2) Using (10.17), from (10.24)–(10.26) we can achieve closed expressions for the quantity $\|\mathscr{S}_g\|_2^2$, in analogy to (7.38). Thereto, with some modifications, the calculations of Sect. 7.2 can be applied.

Since from (10.17) it follows

$$W'_{\bar{z}x}(-s,t) = M'_\tau(-s)\check{R}'_N(-s)\varphi'_{\bar{L}_\tau\mu}(T,-s,t) + \bar{K}'_\tau(-s), \tag{10.27}$$

so we obtain

$$\begin{aligned} W'_{\bar{z}x}(-s,t)W_{\bar{z}x}(s,t) &= \bar{K}'_\tau(-s)\bar{K}_\tau(s) \\ &+ M'_\tau(-s)\check{R}'_N(-s)\varphi'_{\bar{L}_\tau\mu}(T,-s,t)\varphi_{\bar{L}_\tau\mu}(T,s,t)\check{R}_N(s)M_\tau(s) \\ &+ M'_\tau(-s)\check{R}'_N(-s)\varphi'_{\bar{L}_\tau\mu}(T,-s,t)\bar{K}_\tau(s) \\ &+ \bar{K}'_\tau(-s)\varphi_{\bar{L}_\tau\mu}(T,s,t)\check{R}_N(s)M_\tau(s). \end{aligned} \tag{10.28}$$

After inserting this expression into (10.24), we find

$$
\begin{aligned}
A_{\bar{z}x}(s) =& M'_\tau(-s)\check{R}'_N(-s)\check{D}_{\bar{L}}(s)\check{R}_N(s)M_\tau(s) + \bar{K}'_\tau(-s)\bar{K}_\tau(s) \\
&+ M'_\tau(-s)\check{R}'_N(-s)Q'_{\bar{L}}(-s)\bar{K}_\tau(s) + \bar{K}'_\tau(-s)Q_{\bar{L}}(s)\check{R}_N(s)M_\tau(s),
\end{aligned}
\tag{10.29}
$$

where

$$
\begin{aligned}
\check{D}_{\bar{L}}(s) &\stackrel{\triangle}{=} \frac{1}{T}\int_0^T \varphi'_{\bar{L}_\tau\mu}(T,-s,t)\varphi_{\bar{L}_\tau\mu}(T,s,t)\,\mathrm{d}t, \\
Q_{\bar{L}}(s) &\stackrel{\triangle}{=} \frac{1}{T}\int_0^T \varphi_{\bar{L}_\tau\mu}(T,s,t)\,\mathrm{d}t = \frac{1}{T}\bar{L}_\tau(s)\mu(s), \\
Q'_{\bar{L}}(-s) &\stackrel{\triangle}{=} \frac{1}{T}\int_0^T \varphi'_{\bar{L}_\tau\mu}(T,-s,t)\,\mathrm{d}t = \frac{1}{T}\mu'(-s)\bar{L}'_\tau(-s).
\end{aligned}
\tag{10.30}
$$

From (10.29) and (10.25), we obtain explicit expressions for the quantity $\|\mathscr{S}_g\|_2^2$

$$
\begin{aligned}
\|\mathscr{S}_g\|_2^2 =& \frac{1}{2\pi\mathrm{j}}\int_{-\mathrm{j}\infty}^{\mathrm{j}\infty} \mathrm{trace}\Big[M'_\tau(-s)\check{R}'_N(-s)\check{D}_{\bar{L}}(s)\check{R}_N(s)M_\tau(s) + \bar{K}'_\tau(-s)\bar{K}_\tau(s) \\
&+ M'_\tau(-s)\check{R}'_N(-s)Q'_{\bar{L}}(-s)\bar{K}_\tau(s) + \bar{K}'_\tau(-s)Q_{\bar{L}}(s)\check{R}_N(s)M_\tau(s)\Big]\,\mathrm{d}s.
\end{aligned}
\tag{10.31}
$$

Using the relations derived at the beginning of this chapter, the formula achieved above can be represented in a detailed form. At first, due to (10.18), we obtain

$$
\varphi'_{\bar{L}_\tau\mu}(T,-s,t) = \frac{1}{T}\sum_{k=-\infty}^{\infty} \mu'(-s+k\mathrm{j}\omega)\bar{L}'_\tau(-s+k\mathrm{j}\omega)\mathrm{e}^{k\mathrm{j}\omega t},
\tag{10.32}
$$

so from (10.30), after integration by parts, we find

$$
\frac{1}{T}\int_0^T \varphi'_{\bar{L}_\tau\mu}(T,-s,t)\varphi_{L_\tau\mu}(T,s,t)\,\mathrm{d}t = \frac{1}{T}\check{D}_{\underline{\mu'\bar{L}'_\tau}\bar{L}_\tau\mu}(T,s,0),
\tag{10.33}
$$

where

$$
\check{D}_{\underline{\mu'\bar{L}'_\tau}\bar{L}_\tau\mu}(T,s,0) = \frac{1}{T}\sum_{k=-\infty}^{\infty} \mu'(-s-k\mathrm{j}\omega)\bar{L}'_\tau(-s-k\mathrm{j}\omega)\bar{L}_\tau(s+k\mathrm{j}\omega)\mu(s+k\mathrm{j}\omega).
\tag{10.34}
$$

For the further transformations, notice that, due to (10.6), (10.7),

$$
\bar{L}_\tau(s) = \begin{bmatrix} L_1(s)\mathrm{e}^{-s\tau_{L1}} \\ L_2(s)\mathrm{e}^{-s\tau_{L2}} \end{bmatrix}.
\tag{10.35}
$$

Hence,

$$
\bar{L}'_\tau(-s) = \begin{bmatrix} L'_1(-s)\mathrm{e}^{s\tau_{L1}} & L'_2(-s)\mathrm{e}^{s\tau_{L2}} \end{bmatrix},
\tag{10.36}
$$

and the multiplication of the last two expressions leads to a relation which is independent of the delays

$$\bar{L}'_\tau(-s)\bar{L}_\tau(s) = L'_1(-s)L_1(s) + L'_2(-s)L_2(s) = \bar{L}'(-s)\bar{L}(s). \tag{10.37}$$

Herein, we denoted

$$\bar{L}(s) \triangleq \begin{bmatrix} L_1(s) \\ L_2(s) \end{bmatrix}, \tag{10.38}$$

from which we directly derive

$$\bar{L}'(-s) = \begin{bmatrix} L'_1(-s) & L'_2(-s) \end{bmatrix}. \tag{10.39}$$

With the help of (10.34), (10.37) from (10.30), we find

$$\breve{D}_{\bar{L}}(T,s,0) = \frac{1}{T}\breve{D}_{\underline{\mu' \bar{L}'}\bar{L}\mu}(T,s,0) = \frac{1}{T}\breve{D}_{\underline{\mu' L_1'}L_1\mu}(T,s,0) + \frac{1}{T}\breve{D}_{\underline{\mu' L_2'}L_2\mu}(T,s,0) \tag{10.40}$$

In order to realize the further transformations, observe that from (10.30) and (10.18), we obtain

$$Q_{\bar{L}}(s) = \frac{1}{T}\int_0^T \varphi_{\bar{L}_\tau\mu}(T,s,t)\,\mathrm{d}t = \frac{1}{T}\bar{L}_\tau(s)\mu(s) = \begin{bmatrix} \frac{1}{T}L_1(s)\mu(s)\mathrm{e}^{-s\tau_{L1}} \\ \frac{1}{T}L_2(s)\mu(s)\mathrm{e}^{-s\tau_{L2}} \end{bmatrix} \tag{10.41}$$

and

$$Q'_{\bar{L}}(-s) = \begin{bmatrix} \frac{1}{T}\mu'(-s)L'_1(-s)\mathrm{e}^{s\tau_{L1}} & \frac{1}{T}\mu'(-s)L'_2(-s)\mathrm{e}^{s\tau_{L2}} \end{bmatrix}. \tag{10.42}$$

Moreover, from (10.6) and (10.7), we find

$$\bar{K}_\tau(s) = \begin{bmatrix} K_1(s)\mathrm{e}^{-s\tau_{K1}} \\ K_2(s)\mathrm{e}^{-s\tau_{K2}} \end{bmatrix}, \quad \bar{K}'_\tau(-s) = \begin{bmatrix} K'_1(-s)\mathrm{e}^{s\tau_{K1}} & K'_2(-s)\mathrm{e}^{s\tau_{K2}} \end{bmatrix}. \tag{10.43}$$

Therefore,

$$\bar{K}'_\tau(-s)Q_L(s) = \frac{1}{T}K'_1(-s)L_1(s)\mu(s)\mathrm{e}^{s(\tau_{K1}-\tau_{L1})} + \frac{1}{T}K'_2(-s)L_2(s)\mu(s)\mathrm{e}^{s(\tau_{K2}-\tau_{L2})}, \tag{10.44}$$

and thus

$$\begin{aligned} \bar{K}'_\tau(-s)Q_{\bar{L}}(s)\breve{R}_N(s)M_\tau(s) = {} & \frac{1}{T}K'_1(-s)L_1(s)\mu(s)\breve{R}_N(s)M(s)\mathrm{e}^{s(\tau_{K1}-\tau_{L1}-\tau_M)} \\ & + \frac{1}{T}K'_2(-s)L_2(s)\mu(s)\breve{R}_N(s)M(s)\mathrm{e}^{s(\tau_{K2}-\tau_{L2}-\tau_M)}. \end{aligned} \tag{10.45}$$

When applying the relation obtained from (10.9) and (10.12)

$$\tau_{Ki} - \tau_{Li} - \tau_M = -\tau_2 = -\tau_N, \quad (i = 1, 2), \tag{10.46}$$

then from (10.45), we find

$$\bar{K}'_\tau(-s)Q_{\bar{L}}(s)\breve{R}_N(s)M_\tau(s) = \\ \frac{1}{T}\left[K'_1(-s)L_1(s) + K'_2(-s)L_2(s)\right]\mu(s)\breve{R}_N(s)M(s)\mathrm{e}^{-s\tau_N}. \tag{10.47}$$

Introduce

$$\bar{K}(s) \stackrel{\triangle}{=} \begin{bmatrix} K_1(s) \\ K_2(s) \end{bmatrix}, \quad \bar{K}'(-s) = \begin{bmatrix} K'_1(-s) & K'_2(-s) \end{bmatrix}. \tag{10.48}$$

Then the formula (10.47) can be written in the form

$$\bar{K}'_\tau(-s)Q_{\bar{L}}(s)\mu(s)\breve{R}_N(s)M_\tau(s) = \frac{1}{T}\bar{K}'(-s)\bar{L}(s)\mu(s)\breve{R}_N(s)M(s)\mathrm{e}^{-s\tau_N}. \tag{10.49}$$

Hence,

$$M'_\tau(-s)\breve{R}'_N(-s)\mu'(-s)Q'_{\bar{L}}\bar{K}_\tau(s) = \frac{1}{T}M'(-s)\breve{R}'_N(-s)\mu'(-s)\bar{L}'(-s)\bar{K}(s)\mathrm{e}^{s\tau_N}, \tag{10.50}$$

where we took advantage from

$$\begin{aligned} K'_1(-s)L_1(s) + K'_2(-s)L_2(s) &= \bar{K}'(-s)\bar{L}(s), \\ L'_1(-s)K_1(s) + L'_2(-s)K_2(s) &= \bar{L}'(-s)\bar{K}(s). \end{aligned} \tag{10.51}$$

Finally, on basis of (10.40), (10.49), and (10.50), from (10.31) we find a relation analog to (7.38)

$$\begin{aligned} T\|\mathscr{S}_g\|_2^2 = \frac{1}{2\pi \mathrm{j}}\int_{-\mathrm{j}\infty}^{\mathrm{j}\infty} \operatorname{trace}\Big[&M'(-s)\breve{R}'_N(-s)\breve{D}_{\underline{\mu'\bar{L}'}\bar{L}\mu}(T, s, 0)\breve{R}_N(s)M(s) + \\ &+ T\bar{K}'(-s)\bar{K}(s) + M'(-s)\breve{R}'_N(-s)\mu'(-s)\bar{L}'(-s)\bar{K}(s)\mathrm{e}^{s\tau_N} \\ &+ \bar{K}'(-s)\bar{L}(s)\mu(s)\breve{R}_N(s)M(s)\mathrm{e}^{-s\tau_N}\Big]\,\mathrm{d}s. \end{aligned} \tag{10.52}$$

From this formula, with regard to (10.15), we come to the conclusion that the $\mathscr{H}_2$ norm of the generalized standard model $\mathscr{S}_g$ depends only on the delay in the closed loop $\tau_2 = \tau_N$, but does not depend on the values of the $\tau_{3i}$ or on $\tau_1 = \tau_M$. Moreover, all matrices configured in the expression under the integrand on the right side of (10.52), apart from the matrix $\breve{R}_N(s)$, do not depend on the choice of the discrete controller. At the same time, due to (10.15), under otherwise equal conditions, the functional (10.52) is uniquely determined by the transfer matrix of the discrete controller $\breve{W}_d(s)$. Therefore, the right side of formula (10.52), as earlier, can be understood as a functional defined over the set of transfer matrices of discrete controllers, which stabilize the closed loop. This allows us to formulate the $\mathscr{H}_2$ optimization problem for the generalized standard model $\mathscr{S}_g$.

> **$\mathscr{H}_2$ problem for $\mathscr{S}_g$**: Let matrices (10.38), (10.48), and (10.11), the value of the delay $\tau_2 = \tau_N$ and equations of GDAC (6.10) be given. Find the transfer matrix of a causal stabilizing discrete controller, which minimizes functional (10.52).

(3) Using (10.20) and (10.52), in analogy to Chap. 7, the norm $\|\mathscr{S}_g\|_2^2$ can be expressed by the system function. Hereby, notice that the right side of formula (10.52) can be generated from the right side of formula (7.38), when the matrix $L(p)$ is substituted by the matrix $\bar{L}(s)$ from (10.38), and the matrix $K(s)$ by the matrix $\bar{K}(s)$ from (10.48). Moreover, if we consider that the matrix $\breve{R}_N(s)$ for the models $\mathscr{S}_\tau$ and $\mathscr{S}_g$ is identical, then the representation of the cost functional by the system function can be obtained from formulae (7.79), (7.81) when renaming $L(s)$ by $\bar{L}(s)$, and $K(s)$ by $\bar{K}(s)$. Finally, we obtain the following expression for the cost functional:

$$TJ_1 = \frac{1}{2\pi \mathrm{j}} \oint \operatorname{trace}\left[\hat{\theta}(\zeta)A_{\bar{L}}(\zeta)\theta(\zeta)A_M(\zeta) - \hat{\theta}(\zeta)\hat{C}_g(\zeta) - C_g(\zeta)\theta(\zeta)\right]\frac{\mathrm{d}\zeta}{\zeta}, \tag{10.53}$$

where

$$\begin{aligned}
A_{\bar{L}}(\zeta) &\triangleq \hat{a}_r(\zeta)D_{\underline{\mu'\bar{L}'}\bar{L}\mu}(T,\zeta,0)a_r(\zeta),\\
A_M(\zeta) &\triangleq a_l(\zeta)D_{M\underline{M'}}(T,\zeta,0)\hat{a}_l(\zeta),\\
C_g(\zeta) &\triangleq A_M(\zeta)\beta_{r0}(\zeta)D_{\underline{\mu'\bar{L}'}\bar{L}\mu}(T,\zeta,0)a_r(\zeta) + a_l(\zeta)D_{M\underline{\bar{K}'}\bar{L}\mu}(T,\zeta,-\tau_N)a_r(\zeta).
\end{aligned} \tag{10.54}$$

Utilizing the results of Sects. 7.4 and 7.5, we can state some properties of the matrix (10.54), which are necessary for the further considerations.

(a) The matrix $A_M(\zeta)$ is a quasi-polynomial, which is nonnegative on the circle $|\zeta| = 1$, as it was proven in Theorem 7.1.
(b) The matrix $A_{\bar{L}}(\zeta)$ is a quasi-polynomial, which is nonnegative on the circle $|\zeta| = 1$. Indeed, from (10.38), (10.39) and (10.54) it follows that

$$A_{\bar{L}}(\zeta) = A_{L1}(\zeta) + A_{L2}(\zeta), \tag{10.55}$$

where

$$A_{Li}(\zeta) = \hat{a}_r(\zeta)D_{\underline{\mu'L_i'}L_i\mu}(T,\zeta,0)a_r(\zeta), \quad i = 1, 2. \tag{10.56}$$

Then, Theorem 7.2 says that each of the matrices $A_{Li}(\zeta)$ is a quasi-polynomial, and they are nonnegative on the circle $|\zeta| = 1$. Therefore, (10.55) yields that the matrix $A_{\bar{L}}(\zeta)$ is disposing of the same property.
(c) The matrix $C_g(\zeta)$ is a quasi-polynomial. This claim follows from the equation

$$C_g(\zeta) = C_{g1}(\zeta) + C_{g2}(\zeta), \tag{10.57}$$

where

$$
\begin{aligned}
C_{gi}(\zeta) &= A_M(\zeta)\beta_{r0}(\zeta)D_{\underline{\mu' L_i'} L_i\mu}(T,\zeta,0)a_r(\zeta) \\
&\quad + a_l(\zeta)D_{M\underline{K_i'} L_i\mu}(T,\zeta,-\tau_N)a_r(\zeta), \quad (i=1,2).
\end{aligned}
\tag{10.58}
$$

Since, due to Theorem 7.3, both matrices (10.58) are quasi-polynomials, so the matrix $C_g(\zeta)$ is a quasi-polynomial too.

(4) The next statement shows that the solution of the $\mathscr{H}_2$ optimization problem for the generalized standard model for SD systems with delay (10.1), (10.2), (10.4) leads to the solution of the $\mathscr{H}_2$ optimization problem, considered in Chap. 7, for a certain standard model.

**Theorem 10.2** *The $\mathscr{H}_2$ optimal system function $\theta^0(\zeta)$ of the generalized standard model (10.1), (10.2), (10.4), achieved by minimization of functional (10.53), coincides with the solution of the $\mathscr{H}_2$ optimization problem for the standard model for SD systems with delay in the closed loop*

$$
\begin{aligned}
\frac{\mathrm{d}v(t)}{\mathrm{d}t} &= Av(t) + B_1x(t) + B_2(t-\tau_2), \\
y(t) &= C_2v(t), \\
\bar{z}(t) &= \bar{C}_1v(t) + \bar{D}u(t-\tau_2), \\
\xi_k &= y(kT), \quad (k=0,\pm 1,\ldots), \\
\alpha(\zeta)\psi_k &= \beta(\zeta)\xi_k, \\
u(t) &= \sum_{i=0}^{P} h_i(t-kT)\psi_{k-i}, \quad kT<t<(k+1)T,
\end{aligned}
\tag{10.59}
$$

*where the matrices $\bar{C}_1$ and $\bar{D}$ of size $(r_1+r_2)\times\chi$ and $(r_1+r_2)\times m$, respectively, are determined by the relations*

$$
\bar{C}_1 \triangleq \begin{bmatrix} C_{11} \\ C_{12} \end{bmatrix}, \quad \bar{D} \triangleq \begin{bmatrix} D_1 \\ D_2 \end{bmatrix}.
\tag{10.60}
$$

*Proof* Transition from (10.59), (10.60) to the operational equations of form (6.16) yields

$$
\begin{aligned}
\bar{z}(t) &= \bar{K}(p)x(t) + \bar{L}(p)\mathrm{e}^{-p\tau_N}u(t), \\
y(t) &= M(p)x(t) + N(p)\mathrm{e}^{-p\tau_N}u(t), \\
\xi_k &= y(kT), \quad (k=0,\pm 1,\ldots), \\
\alpha(\zeta)\psi_k &= \beta(\zeta)\xi_k, \\
u(t) &= \sum_{i=1}^{P} h_i(t-kT)\psi_{k-i}, \quad kT<t<(k+1)T,
\end{aligned}
\tag{10.61}
$$

where $\bar{L}(p)$ and $\bar{K}(p)$ are the matrices (10.38) and (10.48); $M(p)$ and $N(p)$ are matrices (10.11), and $\tau_N = \tau_2$. We verify directly that functional (7.31) of the $\mathscr{H}_2$ optimal problem for standard model (10.59) coincides with the functional (10.31) for the generalized standard model (10.1), (10.2), (10.4). Since the closed loop in both models is one and the same, so the set of stabilizing controllers for them is identical. Therefore, the functional (7.79) for the standard model (10.59) coincides with the functional (10.53) for the generalized standard model (10.1), (10.2), (10.4). ■

(5) Theorem 10.2 suggests that the $\mathscr{H}_2$ optimization problem of the considered generalized standard model leads to the solution of the $\mathscr{H}_2$ optimization problem for the standard a]model (10.59). Therefore, the results of Chap. 7 can be applied for the solution of the considered problem. The relevant result is formulated next.

**Theorem 10.3** *Let the quasi-polynomials $A_{\bar{L}}(\zeta)$ and $A_M(\zeta)$ be positive on the circle $|\zeta| = 1$. Then, there exists an optimal system function $\theta_g^0(\zeta)$, which can be constructed with the help of the following Wiener–Hopf algorithm:*

*(a) Perform the factorizations*

$$\begin{aligned} A_{\bar{L}}(\zeta) &= \hat{a}_r(\zeta) D_{\underline{\mu' \bar{L}'} \bar{L}\mu}(T, \zeta, 0) \\ &= \hat{a}_r(\zeta) D_{\underline{\mu' L_1'} L_1\mu}(T, \zeta, 0) a_r(\zeta) + \hat{a}_r(\zeta) D_{\underline{\mu' L_2'} L_2\mu}(T, \zeta, 0) a_r(\zeta) \quad (10.62) \\ &= \hat{\Pi}_g(\zeta) \Pi_g(\zeta) \end{aligned}$$

*and*

$$A_M(\zeta) = \Gamma(\zeta) \hat{\Gamma}(\zeta), \quad (10.63)$$

*where $\Pi_g(\zeta)$ and $\Gamma(\zeta)$ are stable polynomial matrices.*

*(b) For the rational matrix*

$$H_g(\zeta) \triangleq \hat{\Pi}_g^{-1}(\zeta) \hat{C}_g(\zeta) \hat{\Gamma}(\zeta) \quad (10.64)$$

*build the separation*

$$H_g(\zeta) = H_{g+}(\zeta) + H_{g-}(\zeta), \quad (10.65)$$

*where $H_{g+}(\zeta)$ is a polynomial matrix, and $H_{g-}(\zeta)$ is a strictly proper rational matrix with poles located inside the disc $|\zeta| < 1$.*

*(c) The optimal system function is determined by the formula*

$$\theta_g^0(\zeta) = \Pi_g^{-1}(\zeta) H_{g+}(\zeta) \Gamma^{-1}(\zeta). \quad (10.66)$$

*Proof* The proof of Theorem 10.3 completely repeats the proof of Theorem 7.5, when the matrix $L(p)$ is substituted by $\bar{L}(p)$, and the matrix $K(p)$ by $\bar{K}(p)$. ■

(6) When the optimal system function $\theta_g^0(\zeta)$ is constructed, then with the help of the methods described in Theorem 7.4, the transfer matrix of the optimal discrete

controller $W_d(\zeta)$ can be calculated. Moreover, as already was stated in Theorem 7.4, if we have the ILMFD

$$\theta_g^0(\zeta) = D_g^{-1}(\zeta)M_g(\zeta), \tag{10.67}$$

then the characteristic polynomial of the $\mathscr{H}_2$ optimal generalized standard model $\Delta_g^0(\zeta)$ satisfies the condition

$$\Delta_g^0(\zeta) \sim \det D_g(\zeta). \tag{10.68}$$

**Theorem 10.4** *The following claims hold.*

*(i) The characteristic polynomial $\Delta_g^0(\zeta)$ is a divisor of the polynomial*

$$\Delta_g(\zeta) \stackrel{\Delta}{=} \det \Pi_g(\zeta) \det \Gamma(\zeta), \tag{10.69}$$

*which does not depend on the delays of the system.*

*(ii) If in (10.66) the pairs $(\Pi_g(\zeta), H_{g+}(\zeta)$ and $[\Gamma(\zeta), H_{g+}(\zeta)]$ are irreducible, then*

$$\Delta_g^0(\zeta) \sim \Delta_g(\zeta). \tag{10.70}$$

*Proof* (i) Owing to Lemma 2.6, there exists the LMFD

$$H_{g+}(\zeta)\Gamma^{-1}(\zeta) = \Gamma_1^{-1}(\zeta)H_1(\zeta), \tag{10.71}$$

in which $\det \Gamma_1(\zeta) \sim \det \Gamma(\zeta)$. Hereby, with the help of (10.66), we obtain the LMFD

$$\Pi_g^{-1}(\zeta)H_{g+}(\zeta)\Gamma^{-1}(\zeta) = \left[\Gamma_1(\zeta)\Pi_g(\zeta)\right]^{-1}H_1(\zeta). \tag{10.72}$$

Comparing LMFD (10.72) with ILMFD (10.67), we find out that the relation

$$\frac{\det \Gamma_1(\zeta) \det \Pi_g(\zeta)}{\det D_g(\zeta)} \sim \frac{\det \Gamma(\zeta) \det \Pi_g(\zeta)}{\det \Delta_g^0(\zeta)}$$

generates a polynomial.

(ii) We will show that in the most common case, when the pairs $(\Pi_g(\zeta), H_+(\zeta))$ and $[\Gamma(\zeta), H_+(\zeta)]$ are irreducible, then the right side of Eq. (10.72) is an ILMFD. Indeed, in this case the right side of formula (10.71) is an ILMFD, and the pair $(\Gamma_1(\zeta), H_1(\zeta))$ is irreducible. Therefore, there exist polynomial matrices $a(\zeta)$ and $b(\zeta)$ of size $q \times q$ and $n \times q$, respectively, such that

$$\Gamma_1(\zeta)a(\zeta) + H_1(\zeta)b(\zeta) = I_q. \tag{10.73}$$

Moreover, the irreducibility of the pair $(\Pi_g(\zeta), H_+(\zeta))$ implies the existence of polynomial matrices $c(\zeta)$ and $d(\zeta)$, for which

$$\Pi_g(\zeta)c(\zeta) + H_{g+}(\zeta)d(\zeta) = I_q. \tag{10.74}$$

Substituting (10.74) in (10.73), with the help of (10.71), we obtain the chain of equations

$$\begin{aligned}\Gamma_1(\zeta)a(\zeta)+H_1(\zeta)b(\zeta) &= \Gamma_1(\zeta)I_q c(\zeta) + H_1(\zeta)b(\zeta)\\ &= \Gamma_1(\zeta)\left[\Pi_g(\zeta)c(\zeta) + H_{g+}(\zeta)d(\zeta)\right]a(\zeta) + H_1(\zeta)b(\zeta)\\ &= \Gamma_1(\zeta)\Pi_g(\zeta)\left[c(\zeta)a(\zeta)\right] + H_1(\zeta)\left[\Gamma(\zeta)d(\zeta)a(\zeta) + b(\zeta)\right] = I_q,\end{aligned} \tag{10.75}$$

from which we find out that the pair $(\Gamma_1(\zeta)\Pi_g(\zeta), H_1(\zeta))$ is irreducible, i.e., the right side of formula (10.72) is an ILMFD. Since the right side of (10.67) is an ILMFD too, so relation (10.70) is true. ∎

(7) Let us consider the properties of the polynomial $\Delta_g(\zeta)$ in more detail. At first, we have

$$\det \Gamma(\zeta) \sim \Delta_{M2}(\zeta)\rho_{M1}(\zeta), \tag{10.76}$$

where $\Delta_{M2}(\zeta)$ is the stable polynomial from (7.274), and $\rho_{M1}(\zeta)$ is the stable polynomial from factorization (7.266). For studying the properties of the polynomial $\det \Pi_g(\zeta)$, owing to the relation

$$A_{\bar{L}}(\zeta) = \hat{a}_r(\zeta)D_{\underline{\mu'\bar{L}'}\bar{L}\mu}(T,\zeta,0)a_r(\zeta), \tag{10.77}$$

we can use the considerations of Sect. 7.5, when substituting $L(p)$ by $\bar{L}(p)$. In that case, utilizing (10.38) and (10.8), we obtain

$$\bar{L}(p) = \bar{C}_1(pI_\chi - A)^{-1}B_2 + \bar{D}, \tag{10.78}$$

where, as above,

$$\bar{C}_1 = \begin{bmatrix} C_{11} \\ C_{12} \end{bmatrix}, \quad \bar{D} = \begin{bmatrix} D_1 \\ D_2 \end{bmatrix}. \tag{10.79}$$

In general, the pair $[A, \bar{C}_1]$ is not observable. Then, besides (10.78), there exists a minimal realization analogous to (7.154)

$$\bar{L}(p) = \bar{C}_{\bar{L}}(pI_\eta - A_{\bar{L}})^{-1}B_{\bar{L}} + \bar{D}, \tag{10.80}$$

where $\eta < \chi$. Hereby, due to Lemma 7.6, the relation

$$\delta_{\bar{L}}(p) \triangleq \frac{\det(pI_\chi - A)}{\det(pI_\eta - A_{\bar{L}})} \tag{10.81}$$

defines a polynomial. If now we assume the extensions

$$\begin{aligned} \det(pI_\chi - A) &= (p-p_1)^{\nu_1}\cdots(p-p_\rho)^{\nu_\rho}, \quad \nu_1+\cdots+\nu_\rho=\chi, \\ \det(pI_\eta - A_{\bar L}) &= (p-p_1)^{\mu_1}\cdots(p-p_\rho)^{\mu_\rho}, \quad \mu_1+\cdots+\mu_\rho=\eta, \end{aligned} \tag{10.82}$$

then

$$\delta_{\bar L}(p) = (p-p_1)^{\nu_1-\mu_1}\cdots(p-p_\rho)^{\nu_\rho-\mu_\rho}. \tag{10.83}$$

(8) For $0 < t < T$ and applying the GDAC

$$u(t) = \sum_{i=0}^{P} h_i(t-kT)\psi_{k-i}, \quad kT < t < (k+1)T \tag{10.84}$$

from (7.172), (7.173) by means of renaming the matrix $L(p)$ by the matrix $\bar L(p)$, we can generate the two equivalent formulae

$$D_{\bar L\mu}(T,\zeta,t) = \mathscr{D}_{\bar L\mu}(T,\zeta,t) = \begin{cases} \bar C_1 e^{At}(I_\chi - \zeta e^{AT})^{-1} m_l(A) + r_{\bar L}(\zeta,t), \\ C_{\bar L} e^{A_{\bar L}t}(I_\eta - \zeta e^{A_{\bar L}T})^{-1} m_{\bar L}(A_{\bar L}) + r_{\bar L}(\zeta,t), \end{cases} \tag{10.85}$$

where

$$\begin{aligned} m_l(A) &= \sum_{i=0}^{P} e^{-iAT}\int_0^T e^{-A\nu}B_2 h_i(\nu)\,d\nu, \\ m_{\bar L}(A_{\bar L}) &= \sum_{i=0}^{P} e^{-iA_{\bar L}T}\int_0^T e^{-A_{\bar L}\nu}B_{\bar L} h_i(\nu)\,d\nu \end{aligned} \tag{10.86}$$

and $r_{\bar L}(\zeta,t)$ is a matrix, which depends polynomially on $\zeta$.

Further on, we suppose that the period $T$ is non-pathological, and that the pairs $(e^{-AT}, m_l(A))$ and $(e^{-A_{\bar L}T}, m_{\bar L}(A_{\bar L}))$ are minimal, and a condition analogous to (7.155) is fulfilled. Then, Lemma 7.7 yields that, from the validity of the IRMFD

$$\begin{aligned} (I_\chi - \zeta e^{AT})^{-1} m_l(A) &= b_r(\zeta) a_r^{-1}(\zeta), \\ (I_\eta - \zeta e^{A_{\bar L}T})^{-1} m_{\bar L}(A_{\bar L}) &= b_{\bar L}(\zeta) a_{\bar L}^{-1}(\zeta), \end{aligned} \tag{10.87}$$

it follows

$$a_r(\zeta) = a_{\bar L}(\zeta) a_{\bar L1}(\zeta), \tag{10.88}$$

where $a_{\bar L1}(\zeta)$ is a polynomial matrix satisfying

$$\det a_{\bar L1}(\zeta) \sim \Delta_{\bar L}(\zeta) \stackrel{\triangle}{=} \frac{\det(I_\chi - \zeta e^{AT})}{\det(I_\eta - \zeta e^{A_{\bar L}T})}. \tag{10.89}$$

Apart from the above said, from the content of Sect. 7.5, we have

$$A_{\bar L}(\zeta) = \hat a_{\bar L1}(\zeta) A_{\bar L1}(\zeta) a_{\bar L1}(\zeta), \tag{10.90}$$

where under the conditions of Theorem 10.3, the matrix $A_{\bar{L}1}(\zeta)$ is a quasi-polynomial positive on the circle $|\zeta| = 1$, and the matrix $a_{\bar{L}1}(\zeta)$ is free of eigenvalues located on this circle.

Then by analogy to (7.295), the factorization

$$\det a_{\bar{L}1}(\zeta) \det a_{\bar{L}1}(\zeta^{-1}) = \Delta_{\bar{L}2}(\zeta)\Delta_{\bar{L}2}(\zeta^{-1}) \tag{10.91}$$

is possible, where $\Delta_{\bar{L}2}(\zeta)$ is a stable polynomial. Then, if in (10.83)

$$\mathrm{Re}\, p_i < 0, \quad (i = 1, \ldots, \phi); \quad \mathrm{Re}\, p_i > 0, \quad (i = \phi + 1, \ldots, \rho),$$

then

$$\begin{aligned}\Delta_{\bar{L}2}(\zeta) \sim (1 - \zeta \mathrm{e}^{p_1 T})^{\nu_1 - \mu_1} \cdots (1 - \zeta \mathrm{e}^{p_\phi T})^{\nu_\phi - \mu_\phi} \times \\ (1 - \zeta \mathrm{e}^{-p_{\phi+1} T})^{\nu_{\phi+1} - \mu_{\phi+1}} \cdots (1 - \zeta \mathrm{e}^{-p_\rho T})^{\nu_\rho - \mu_\rho}.\end{aligned} \tag{10.92}$$

Since under the taken propositions the factorization

$$A_{\bar{L}1} = \hat{\Pi}_{\bar{L}1}(\zeta)\Pi_{\bar{L}1}(\zeta) \tag{10.93}$$

is possible with a stable matrix $\Pi_{\bar{L}1}(\zeta)$, so from (10.90), we achieve factorization (10.62)

$$A_{\bar{L}}(\zeta) = \hat{\Pi}_g(\zeta)\Pi_g(\zeta), \tag{10.94}$$

where

$$\det \Pi_g(\zeta) = \Delta_{\bar{L}2}(\zeta) \det \Pi_{\bar{L}1}(\zeta). \tag{10.95}$$

From (10.76) and (10.95), we come to the conclusion that the set of fixed poles for the $\mathscr{H}_2$ optimization problem is the union of the sets of roots of the polynomials $\Delta_{M1}(\zeta)$ and $\Delta_{\bar{L}2}(\zeta)$.

## 10.3 $\mathscr{H}_2$ Optimization of the Generalized Standard Model for Colored Noise

(1) In this section, we investigate the $\mathscr{H}_2$ optimization problem for generalized standard model (10.1)–(10.3) for $m = 2$, in the case when the system $\mathscr{S}_g$ is excited by colored noise $x(t)$, which can be regarded as the response of a filter with transfer matrix $\Phi(s)$ (8.1) to vectorial white noise $\rho(t)$. In this case, according to the input $\rho(t)$ and output $\bar{z}(t)$, the investigated system in analogy to (8.2) is configured to the parametric transfer matrix

$$W_{\bar{z}\rho}(s, t) = W_{\bar{z}x}(s, t)\Phi(s), \tag{10.96}$$

where $W_{\bar{z}x}(s,t)$ is PTM (10.17). From (10.96) and (10.17), we obtain

$$W_{\bar{z}\rho}(s,t) = \varphi_{\bar{L}_\tau\mu}(T,s,t)\breve{R}_N(s)M_\tau(s)\Phi(s) + \bar{K}_\tau(s)\Phi(s). \tag{10.97}$$

This formula means that the PTM $W_{\bar{z}\rho}(s,t)$ is generated from the PTM $W_{\bar{z}x}(s,t)$ (10.17) when substituting the matrix $M_\tau(s)$ by the matrix $M_\tau(s)\Phi(s)$, and the matrix $\bar{K}_\tau(s)$ by the matrix $\bar{K}_\tau(s)\Phi(s)$. In particular, we obtain

$$M_\tau(s) = M(s)\Phi(s)\mathrm{e}^{-s\tau_M}, \quad \bar{K}_\tau(s)\Phi(s) = \begin{bmatrix} K_1(s)\Phi(s)\mathrm{e}^{-s\tau_{K1}} \\ K_2(s)\Phi(s)\mathrm{e}^{-s\tau_{K2}} \end{bmatrix}. \tag{10.98}$$

Let the generalized standard system $\mathscr{S}_g$ be stable. Then the quasi-stationary response to white noise $\rho(t)$ represents itself as a periodic nonstationary stochastic process with the variance

$$d_{\bar{z}}(t) = \frac{1}{2\pi\mathrm{j}}\int_{-\mathrm{j}\infty}^{\mathrm{j}\infty} \operatorname{trace}\left[W'_{\bar{z}\rho}(-s,t)W_{\bar{z}\rho}(s,t)\right] \mathrm{d}s, \tag{10.99}$$

or, more detailed

$$d_{\bar{z}} = \frac{1}{2\pi\mathrm{j}}\int_{-\mathrm{j}\infty}^{\mathrm{j}\infty} \operatorname{trace}\left[\Phi'(-s)W'_{\bar{z}x}(-s,t)W_{\bar{z}x}(s,t)\Phi(s)\right] \mathrm{d}s. \tag{10.100}$$

Then, the mean variance $\bar{d}_{\bar{z}}$ is determined by

$$\bar{d}_{\bar{z}} = \int_{-\mathrm{j}\infty}^{\mathrm{j}\infty} \operatorname{trace}\left[\Phi'(-s)A_{\bar{z}\rho}(s)\Phi(s)\right] \mathrm{d}s, \tag{10.101}$$

where

$$A_{\bar{z}x}(s) = \frac{1}{T}\int_0^T W'_{\bar{z}x}(-s,t)W_{\bar{z}x}(s,t)\,\mathrm{d}t. \tag{10.102}$$

Hereby, in analogy with (8.8), the number

$$\|\mathscr{S}_c\|_2 \triangleq +\sqrt{\bar{d}_{\bar{z}}} \tag{10.103}$$

is called the $\mathscr{H}_2$ norm of the generalized standard model $\mathscr{S}_c$ with colored input signal $x(t)$. On basis of the above reasonings, we are able to formulate the $\mathscr{H}_2$ optimization problem for the generalized standard model of SD system with delay and colored noise input.

> **$\mathscr{H}_2$ problem for generalized standard model under colored noise:** Let the generalized standard model (10.1), (10.2), (10.4) for a SD system with delay, and in addition the transfer matrix for the input filter $\Phi(s)$ be given. Find the transfer function $W_d(\zeta)$ of a causal discrete controller which stabilizes the closed loop and yields a minimum of the norm (10.103).

(2) For the construction of the functional, the minimization of which yields the solution of the formulated problem, the relations in Chap. 8 can be applied, when the matrix $M_\tau(s)$ is substituted by $M_\tau(s)\Phi(s)$ and the matrix $K_\tau(s)$ by the matrix $\bar{K}_\tau(s)\Phi(s)$. Hereby, we have to assume that the poles of the matrices $M_\tau(s)\Phi(s)$ and $\bar{K}_\tau(s)\Phi(s)$ satisfy all propositions of Chap. 8.

Following this idea, the representation of PTM $W_{\bar{z}r}(s,t)$ from the input $g(t)$ to the output $\bar{z}(t)$ by the system function can be written in the form

$$W_{\bar{z}\rho}(s,t) = \bar{\psi}(s,t)\breve{\theta}(s)\xi_\Phi(s) + \bar{\eta}_\Phi(s,t), \tag{10.104}$$

where the matrix $\bar{\psi}(s,t)$ is determined by the first formula in (10.23), and moreover

$$\begin{aligned}\xi_\Phi(s) &\overset{\triangle}{=} \xi(s)\Phi(s),\\ \bar{\eta}_\Phi(s,t) &\overset{\triangle}{=} \bar{\eta}(s,t)\Phi(s).\end{aligned} \tag{10.105}$$

Herein, $\xi(s)$ is the matrix determined by (6.202), and the matrix $\bar{\eta}(s,t)$ is given by the second formula in (10.23). More detailed, we can write

$$\begin{aligned}\xi_\Phi(s) &= \breve{a}_l(s)M(p)\Phi(s)\mathrm{e}^{-s\tau_M},\\ \bar{\eta}_\Phi(s,t) &= \begin{bmatrix}\varphi_{L_{1\tau}\mu}(T,s,t)\\ \varphi_{L_{2\tau}\mu}(T,s,t)\end{bmatrix}\breve{\beta}_{r0}(s)\breve{a}_l(s)M(s)\Phi(s)\mathrm{e}^{-s\tau_M} + \begin{bmatrix}K_1(s)\Phi(s)\mathrm{e}^{-s\tau_{K1}}\\ K_2(s)\Phi(s)\mathrm{e}^{-s\tau_{K2}}\end{bmatrix}.\end{aligned} \tag{10.106}$$

Further transformations, with analogy in Sect. 8.2, lead to the minimization of the functional

$$J_{g\Phi} \overset{\triangle}{=} \frac{1}{2\pi \mathrm{j}}\oint trace\left[\hat{\theta}(\zeta)A_{\bar{L}}(\zeta)\theta(\zeta)A_{M\Phi}(\zeta) - \hat{\theta}(\zeta)\hat{C}_{g\Phi}(\zeta) - C_{g\Phi}(\zeta)\theta(\zeta)\right]\frac{\mathrm{d}\zeta}{\zeta}, \tag{10.107}$$

where the matrices $A_{\bar{L}}(\zeta)$ and $A_{M\Phi}(\zeta)$ are determined by (10.54) and (8.14), respectively. Moreover, the matrix $C_{g\Phi}(\zeta)$ is obtained from (8.14), when $L(s)$ is substituted by $\bar{L}(s)$ and $K(s)$ by $\bar{K}(s)$. As a result, we find that in (10.107)

$$\begin{aligned}A_{\bar{L}}(\zeta) &= \hat{a}_r(\zeta)D_{\underline{\mu'\bar{L}'}\bar{L}\mu}(T,\zeta,0)\hat{a}_r(\zeta),\\ A_{M\Phi}(\zeta) &= a_l(\zeta)D_{M\Phi\underline{\Phi'M'}}(T,\zeta,0)\hat{a}_l(\zeta),\\ C_{g\Phi}(\zeta) &= a_l(\zeta)D_{M\Phi\underline{\Phi'M'}}(T,\zeta,0)\hat{a}_l(\zeta)\hat{\beta}_{r0}(\zeta)D_{\underline{\mu'\bar{L}'}\bar{L}\mu}(T,\zeta,0)a_r(\zeta)\\ &\quad + a_l(\zeta)D_{M\Phi\underline{\Phi'\bar{K}'}\bar{L}\mu}(T,\zeta,-\tau_N)a_r(\zeta).\end{aligned} \tag{10.108}$$

(3) Analyzing formulae (10.107) and (10.108), we find out that the functional (10.107) coincides with the functional (8.13) for the model (10.59) for colored input. Utilizing that the closed loop in models (10.1) and (10.59) are the same, we can conclude that, in the case of the $\mathscr{H}_2$ problem, for the minimization of functional (10.107) we can apply directly the result of Chap. 8, when the matrix $L(s)$ is substituted by $\bar{L}(s)$ and the matrix $K(s)$ by $\bar{K}(s)$. Especially, from Theorem 8.3, we obtain the following theorem.

**Theorem 10.5** *Let standard model (10.59) satisfy the conditions of Theorem 8.3. Then the following claims hold.*

*(i) There exists the factorization (10.62)*

$$A_{\bar{L}}(\zeta) = \hat{\Pi}_g(\zeta)\Pi_g(\zeta) \tag{10.109}$$

*with a stable polynomial matrix $\Pi_g(\zeta)$. Moreover, there exists the factorization (8.68)*

$$A_{M\Phi}(\zeta) = V(\zeta)\hat{V}(\zeta), \tag{10.110}$$

*where the rational matrix $V(\zeta)$ is stable together with its inverse.*

*(ii) The optimal system function $\theta^0(\zeta)$ can be constructed with the help of the following algorithm:*

*(a) For the rational matrix*

$$H_{g\Phi}(\zeta) \triangleq \hat{\Pi}_g^{-1}(\zeta)\hat{C}_{g\Phi}(\zeta)\hat{V}(\zeta) \tag{10.111}$$

*build the separation*

$$H_{g\Phi}(\zeta) = H_{g\Phi}^{+}(\zeta) + H_{g\Phi}^{-}(\zeta), \tag{10.112}$$

*where the rational matrix $H_{g\Phi}^{-}(\zeta)$ is strictly proper and its poles are all poles of the matrix $H_{g\Phi}(\zeta)$, which are in the disc $|\zeta| < 1$.*

*(b) The optimal system function $\theta_g^0(\zeta)$ is determined by the formula*

$$\theta_g^0(\zeta) = \Pi^{-1}(\zeta)H_g^{+}(\zeta)V^{-1}(\zeta). \tag{10.113}$$

■

(4) Using the detected connection between the $\mathscr{H}_2$ optimization problem for the standard model and the generalized standard model, on basis of the results in Chap. 8, we are able to state a number of general properties of the $\mathscr{H}_2$ optimal generalized standard models $\mathscr{S}_g$ under colored input noise.

(a) If the standard model (10.59) fulfills the conditions of Theorem 8.4, then the set of poles of matrix (10.113) does not contain external poles.

(b) The set of fixed poles of the $\mathscr{H}_2$ optimal system, under the conditions of Theorem 8.4, is determined by the relations of Sect. 8.6 applied to the standard model (10.59).

## 10.4 $\mathscr{L}_2$ Optimization of the Generalized Standard Model

(1) This section deals with a special case of $\mathscr{L}_2$ tracking problem for the generalized standard model, i.e., search for a quadratic optimal transient process.

**Definition 10.3** As $\mathscr{L}_2$ norm of the generalized standard model (10.1), (10.2), (10.4), we understand the number

$$\|\mathscr{S}_g\|_{\mathscr{L}_2} = +\sqrt{\int_0^\infty \bar{z}'(t)\bar{z}(t)\,\mathrm{d}t}, \tag{10.114}$$

where $\bar{z}(t)$ is the output vector under zero initial energy.

According to the input vector $x(t)$, we propose below the conditions of Chap. 9. In particular, we assume that the vector $x(t) = 0_{l1}$ for $t < 0$, and the image $X(s)$ is pseudo-rational, such that formulae (9.61), (6.92) are true, and the roots of function $\breve{d}_\chi(s)$ (9.63) are located in the open left half-plane.

In the special case of the standard model (6.12), the $\mathscr{L}_2$ norm is determined by formula (9.84) for $z_i(t) = 0_{r1}$, i.e.,

$$F(s) = 0_{rl}, \tag{10.115}$$

where $r = r_1 + r_2$.

(2) With the help of the Parseval formula, from (10.114), we realize that

$$\|S_c\|^2_{\mathscr{L}_2} = \frac{1}{2\pi\mathrm{j}}\int_{-\mathrm{j}\infty}^{\mathrm{j}\infty} \bar{Z}'(-s)Z(s)\,\mathrm{d}s, \tag{10.116}$$

where

$$\bar{Z}(s) = \int_0^\infty \bar{z}(t)\mathrm{e}^{-st}\,\mathrm{d}t \tag{10.117}$$

is the Laplace image of the output $\bar{z}(t)$. The right side of formula (10.116) can be expressed by the parameter of the original generalized standard model. For this, we first find closed expressions for the image $\bar{Z}(s)$.

Let, in harmony with (10.13),

$$W_{z_i x}(s,t) = \varphi_{L_{i\tau}\mu}(T,s,t)\breve{R}_N(s)M_\tau(s) + K_{i\tau}(s), \quad (i = 1, 2) \tag{10.118}$$

be the PTM of the generalized standard model from the input $x(t)$ to the output $z_i(t)$. Then, due to the results in Chap. 9, the image of the output $z_i(t)$ for zero initial energy is determined by the formula following from (9.49):

$$Z_i(s) = L_{i\tau}(s)\mu(s)\breve{R}_N(s)\breve{D}_{MX}(T, s, -\tau_M) + K_{i\tau}(s)X(s). \tag{10.119}$$

Hence, the image $\bar{Z}(s)$ of the output $\bar{z}(t)$ is determined by

$$\bar{Z}(s) = \begin{bmatrix} Z_1(s) \\ Z_2(s) \end{bmatrix}. \tag{10.120}$$

Inserting herein (10.119), we obtain

$$\bar{Z}(s) = \begin{bmatrix} L_1(s)e^{-s\tau_{L1}} \\ L_2(s)e^{-s\tau_{L2}} \end{bmatrix} \mu(s)\breve{R}_N(s)\breve{D}_{MX}(T, s, -\tau_M) + \begin{bmatrix} K_1(s)e^{-s\tau_{K1}} \\ K_2(s)e^{-s\tau_{K2}} \end{bmatrix} X(s), \tag{10.121}$$

which, when using notations (10.6), (10.7), can be written in the form

$$\bar{Z}(s) = \bar{L}_\tau(s)\mu(s)\breve{R}_N(s)\breve{D}_{MX}(T, s, -\tau_M) + \bar{K}_\tau(p)X(s). \tag{10.122}$$

(3) Applying (10.19) and (9.71), (9.72) each image $Z_i(s)$ can be expressed by the system function

$$Z_i(s) = \ell_i(s)\breve{\theta}(s)\breve{m}(s) + n_i(s), \quad (i = 1, 2), \tag{10.123}$$

where

$$\begin{aligned} \ell_i(s) &= -e^{-s\tau_{Li}} L_i(s)\mu(s)\breve{a}_r(s), \\ \breve{m}(s) &= \breve{a}_l(s)\breve{D}_{MX}(T, s, -\tau_M), \\ n_i(s) &= e^{-s\tau_{Li}} L_i(s)\mu(s)\breve{\beta}_{r0}(s)\breve{m}(s) + K_i(s)X(s)e^{-s\tau_{Ki}}. \end{aligned} \tag{10.124}$$

It follows from Theorem 9.6 that the matrices $\ell_i(s)$ are integral functions of the argument $s$, and the vectors $\breve{m}(s)$ and $n_i(s)$ are meromorphic functions with the set of their poles contained in the set of roots of the equation

$$\breve{d}_X(s) = 0, \tag{10.125}$$

which is denoted, as in Chap. 9, by $\mathscr{M}_X$.

From (10.120) and (10.123), we obtain the representation of the image $\bar{Z}(s)$ by the system function

$$\bar{Z}(s) = \bar{\ell}(s)\breve{\theta}(s)\breve{m}(s) + \bar{n}(s), \tag{10.126}$$

where

$$\begin{aligned}\bar{\ell}(s) &\triangleq \begin{bmatrix}\ell_1(s)\\ \ell_2(s)\end{bmatrix} = -\bar{L}_\tau(s)\mu(s)\breve{a}_r(s),\\ \bar{n}(s) &\triangleq \begin{bmatrix}n_1(s)\\ n_2(s)\end{bmatrix} = \bar{L}_\tau(s)\mu(s)\breve{\beta}_{r0}(s)\breve{m}(s) + \bar{K}_\tau(s)X(s).\end{aligned} \tag{10.127}$$

The above said directly implies that the matrix $\bar{\ell}(s)$ is an integral function of the argument $s$, and the vector $\bar{n}(s)$ is a meromorphic function with its poles located in the set $\mathscr{M}_X$ of roots of Eq. (10.125).

(4) From (10.126) after substituting $s$ by $-s$ and transposition, we find

$$\bar{Z}'(-s) = \breve{m}'(-s)\breve{\theta}'(-s)\bar{\ell}'(-s) + \bar{n}'(-s), \tag{10.128}$$

where

$$\begin{aligned}\bar{\ell}'(-s) &= -\breve{a}'_r(-s)\mu'(-s)\bar{L}'_\tau(-s),\\ \breve{m}'(-s) &= \breve{D}'_{MX}(T,-s,-\tau_N)\breve{a}'_l(-s),\\ \bar{n}'(-s) &= \breve{m}'(-s)\breve{\beta}'_{r0}(-s)\mu'(-s)\bar{L}'_\tau(-s) + X'(-s)\bar{K}'_\tau(-s).\end{aligned} \tag{10.129}$$

After multiplying formulae (10.126) and (10.129), we obtain

$$\begin{aligned}\bar{Z}'(-s)\bar{Z}(s) =& \breve{m}'(-s)\breve{\theta}'(-s)\bar{\ell}'(-s)\bar{\ell}(s)\breve{\theta}(s)\breve{m}(s) + \bar{n}'(-s)\bar{n}(s)\\ &+ \breve{m}'(-s)\breve{\theta}'(-s)\bar{\ell}'(-s)\bar{n}(s) + \bar{n}'(-s)\bar{\ell}(s)\breve{\theta}(s)\breve{m}(s).\end{aligned} \tag{10.130}$$

For the further transformations, notice that according to (10.127), (10.129)

$$\bar{\ell}'(-s)\bar{\ell}(s) = \breve{a}'_r(-s)\mu'(-s)\bar{L}'_\tau(-s)\bar{L}_\tau(s)\mu(s)\breve{a}_r(s). \tag{10.131}$$

Since due to (10.37)

$$\bar{L}'_\tau(-s)\bar{L}_\tau(s) = \bar{L}'(-s)\bar{L}(s), \tag{10.132}$$

where $\bar{L}(s)$ is the matrix (10.38), so the expression (10.131) can be represented in a form independent of the delays

$$\bar{\ell}'(-s)\bar{\ell}(s) = \breve{a}'_r(-s)\mu'(-s)\bar{L}'(-s)\bar{L}(s)\mu(s)\breve{a}_r(s). \tag{10.133}$$

Further calculation with regard to (10.127) (10.129) yields

$$\begin{aligned}\bar{n}'(-s)\bar{\ell}(s) =& -\breve{m}'(-s)\breve{\beta}'_{r0}(-s)\mu'(-s)\bar{L}'_\tau(-s)\bar{L}_\tau(s)\mu(s)\breve{a}_r(s)\\ &- X'(-s)\bar{K}'_\tau(-s)\bar{L}_\tau(s)\mu(s)\breve{a}_r(s).\end{aligned} \tag{10.134}$$

Applying (10.6), (10.7), we find

$$\begin{aligned} \bar{K}'_\tau(-s) &= \left[\, K'_1(-s)e^{s\tau_{K1}} \;\; K'_2(-s)e^{s\tau_{K2}} \,\right], \\ \bar{L}_\tau(s) &= \begin{bmatrix} L_1(s)e^{-s\tau_{L1}} \\ L_2(s)e^{-s\tau_{L2}} \end{bmatrix}. \end{aligned} \tag{10.135}$$

Hence,

$$\bar{K}'_\tau(-s)\bar{L}_\tau(s) = K'_1(-s)L_1(s)e^{s(\tau_{K1}-\tau_{L1})} + K'_2(-s)L_2(s)e^{s(\tau_{K2}-\tau_{L2})}. \tag{10.136}$$

Utilizing from (10.9), (10.12)

$$\tau_{K1} - \tau_{L1} = \tau_{K2} - \tau_{L2} = \tau_1 - \tau_2 = \tau_M - \tau_N, \tag{10.137}$$

then from (10.136) we achieve the expression

$$\bar{K}'_\tau(-s)\bar{L}_\tau(s) = \bar{K}'(-s)\bar{L}(s)e^{s(\tau_1-\tau_2)}, \tag{10.138}$$

where $\bar{K}(s)$ is the matrix (10.48). After inserting (10.132) and (10.138) into (10.134), we arrive at

$$\begin{aligned} \bar{n}'(-s)\bar{\ell}(s) = &- \check{m}'(-s)\check{\beta}'_{r0}(-s)\mu'(-s)\bar{L}'(-s)\bar{L}(s)\mu(s)\check{a}_r(s) \\ &- X'(-s)\bar{K}'(-s)\bar{L}(s)\mu(s)\check{a}_r(s)e^{s(\tau_1-\tau_2)}, \end{aligned} \tag{10.139}$$

from which we come to

$$\begin{aligned} \bar{\ell}'(-s)\bar{n}(s) = &- \check{a}'_r(-s)\mu'(-s)\bar{L}'(-s)L(s)\mu(s)\check{\beta}_{r0}(s)\check{m}(s) \\ &- \check{a}'_r(-s)\mu'(-s)\bar{L}'(-s)\bar{K}(s)X(s)e^{-s(\tau_1-\tau_2)}. \end{aligned} \tag{10.140}$$

Formulae (10.139) and (10.140) can be achieved from (9.107), when the matrix $L(s)$ is exchanged by the matrix $\bar{L}(s)$, and the matrix $K(s)$ by $\bar{K}(s)$. With the help of analog calculations we can derive closed expressions for the product $\bar{n}'(-s)\bar{n}(s)$. On basis of (10.127) and (10.129), we find

$$\begin{aligned} \bar{n}'(-s)\bar{n}(s) = &\left[\check{m}'(-s)\check{\beta}'_{r0}(-s)\mu'(-s)\bar{L}'_\tau(-s) + X'(-s)\bar{K}'_\tau(-s)\right] \\ &\times \left[\bar{L}_\tau(s)\mu(s)\check{\beta}_{r0}(s)\check{m}(s) + \bar{K}_\tau(s)X(s)\right] \\ = &\,\check{m}'(-s)\check{\beta}'_{r0}(-s)\mu'(-s)\bar{L}'_\tau(-s)\bar{L}_\tau(s)\mu(s)\check{\beta}_{r0}(s)\check{m}(s) \\ &+ \check{m}'(-s)\check{\beta}'_{r0}(-s)\mu'(-s)\bar{L}'_\tau(-s)\bar{K}_\tau(s)X(s) \\ &+ X'(-s)\bar{K}'_\tau(-s)\bar{L}_\tau(s)\mu(s)\check{\beta}_{r0}(s)\check{m}(s) + X'(-s)\bar{K}'_\tau(-s)\bar{K}_\tau(s)X(s). \end{aligned} \tag{10.141}$$

When applying

$$\bar{K}'_\tau(-s)\bar{K}_\tau(s) = \bar{K}'(-s)\bar{K}(s), \tag{10.142}$$

which is held from (10.6), (10.7), and in addition relations (10.132), (10.138), then expression (10.141) takes the form

$$\begin{aligned}\bar{n}'(-s)\bar{n}(s) &= \breve{m}'(-s)\breve{\beta}'_{r0}(-s)\mu'(-s)\bar{L}'(-s)\bar{L}(s)\mu(s)\breve{m}(s)\\ &+ \breve{m}'(-s)\breve{\beta}'_{r0}(-s)\mu'(-s)\bar{L}(-s)\bar{K}(s)X(s)\mathrm{e}^{-s(\tau_1-\tau_2)}\\ &+ X'(-s)\bar{K}'(-s)\bar{L}(s)\mu(s)\breve{\beta}_{r0}(s)\breve{m}(s)\mathrm{e}^{s(\tau_1-\tau_2)} + X'(-s)\bar{K}'(-s)K(s).\end{aligned} \tag{10.143}$$

(5) Summarizing the obtained relations, we can formulate the next result.

**Theorem 10.6** *The $\mathscr{L}_2$ optimal system function and the $\mathscr{L}_2$ optimal transfer matrix of the discrete controller for the generalized standard model (10.1), (10.2), (10.4) coincide with the $\mathscr{L}_2$ optimal system function and transfer matrix of the $\mathscr{L}_2$ optimal discrete controller for the standard model of the SD system with delay of form (6.12)*

$$\begin{aligned}\frac{\mathrm{d}v}{\mathrm{d}t} &= Av(t) + B_1x(t-\tau_1) + B_2u(t-\tau_2),\\ y(t) &= C_2v(t),\\ \bar{z}_1(t) &= \bar{C}_1v(t) + \bar{D}u(t-\tau_2),\\ \xi_k &= y(kT), \quad (k = 0, \pm 1, \ldots),\\ \alpha(\zeta)\psi_k &= \beta(\zeta)\xi_k,\\ u(t) &= \sum_{i=0}^{P} h_i(t-kT)\psi_{k-i}, \quad kT < t < (k+1)T,\end{aligned} \tag{10.144}$$

*where $\bar{C}_1$ and $\bar{D}$ are matrices (10.60).*

*Proof* Passing from Eq. (10.144) to operator equations of the form (6.16), using (6.17)–(6.19), we obtain

$$\begin{aligned}\bar{z}_1(t) &= K^*_\tau(p)x(t) + L^*_\tau(p)u(t),\\ y(t) &= M_\tau(p)x(t) + N_\tau(p)u(t),\\ \xi_k &= y(kT), \quad (k = 0, \pm 1, \ldots),\\ \alpha(\zeta)\psi_k &= \beta(\zeta)\xi_k,\\ u(t) &= \sum_{i=0}^{P} h_i(t-kT)\psi_{k-i}, \quad kT < t < (k+1)T.\end{aligned} \tag{10.145}$$

Herein

$$K^*_\tau(p) = K^*(p)\mathrm{e}^{-p\tau_{K^*}}, \quad L^*_\tau(p) = L^*(p)\mathrm{e}^{-p\tau_{L^*}}, \tag{10.146}$$

where

$$K^*(p) = \bar{C}_1(pI_\chi - A)^{-1}B_1, \quad L^*(p) = \bar{C}_1(pI_\chi - A)^{-1}B_2 + \bar{D} \tag{10.147}$$

and

$$\tau_{K^*} = \tau_1, \quad \tau_{L^*} = \tau_2. \tag{10.148}$$

Hereby, owing to (10.146), (10.38), and (10.48)

$$K^*(p) = \bar{K}(p), \quad L^*(p) = \bar{L}(p). \tag{10.149}$$

Moreover, in (10.145) the matrices $M_\tau(p)$ and $N_\tau(p)$ are determined by formulae (10.10)–(10.12), i.e.,

$$M_\tau(p) = M(p)\mathrm{e}^{-p\tau_M}, \quad N_\tau(p) = N(p)\mathrm{e}^{-p\tau_N}, \tag{10.150}$$

where

$$M(p) = C_2(pI_\chi - A)^{-1}B_1, \quad N(p) = C_2(pI_\chi - A)^{-1}B_2, \tag{10.151}$$

with

$$\tau_M = \tau_1, \quad \tau_N = \tau_2. \tag{10.152}$$

The solution of the $\mathscr{L}_2$ optimization for the system (10.145) can be obtained as special case of the relations in Chap. 9 when the condition (10.115) is fulfilled. Hereby, from (9.98) under conditions (10.146)–(10.152) and (10.126), (10.128), we find

$$\begin{aligned} \bar{Z}_1'(-s)\bar{Z}_1(s) =& \breve{m}'(-s)\breve{\theta}'(-s)\ell'(-s)\ell(s)\breve{\theta}(s) + \bar{n}'(-s)\bar{n}(s) \\ &+ \breve{m}'(-s)\breve{\theta}(s)\ell'(-s)\bar{n}(s) + \bar{n}'(-s)\ell(s)\breve{\theta}(s)\breve{m}(s), \end{aligned} \tag{10.153}$$

where the matrix $\ell(s)$ and the vectors $\breve{m}(s)$ and $\bar{n}(s)$ are defined in harmony with (9.72). The right side of formula (9.153) coincides structurally with the right side of formula (10.130). Therefore, it is easy to verify that the products $\bar{\ell}'(-s)\bar{\ell}(s)$, $\bar{n}'(-s)\bar{n}(s)$ and $\bar{\ell}'(-s)\bar{n}(s)$, configured in (10.130), can be obtained from the products $\ell'(-s)\ell(s)$, $n'(-s)n(s)$ and $\ell'(-s)n(s)$ in formula (10.153), when substituting the matrix $L(s)$ by $\bar{L}(s)$ and the matrix $K(s)$ by $\bar{K}(s)$. Therefore, the right side of formula (10.130) is generated from the right side of (10.153) with the help of analog substitutions, and the same is transferred, due to (10.116), to the expressions for the $\mathscr{L}_2$ norm. This fact proves the claim of Theorem 10.6. ∎

(6) It follows from Theorem 10.6 that for the solution of the $\mathscr{L}_2$ optimization problem for the generalized standard model of SD systems with delay (10.1), (10.2), (10.4), we can apply all relations from Chap. 9, when conditions (10.148), (10.152) are valid, and we rename $L(s)$ by $\bar{L}(s)$ and $K(s)$ by $\bar{K}(s)$. Moreover, in the considered case, all statements in Chap. 9 still hold, in particular, the construction of solution of the $\mathscr{L}_2$

optimization problem. Of course, the transferability involves the statements about the singularity of the $\mathscr{L}_2$ optimization problem and the properties of the fixed poles.

## 10.5 Mathematical Description of a Single-Loop Multivariable SD Systems with Delay as Generalized Standard Model

(1) The present section considers the $\mathscr{H}_2$ optimization problem for the single-loop SD system $\mathscr{S}_0$, the structure of which is shown in Fig. 10.1

In Fig. 10.1, the notation $W_i(p)$, $(i = 1, 2, 3)$ stands for rational matrices of dimensions $\ell \times m$, $\kappa \times \ell$ and $n \times \kappa$, respectively, and $\varkappa$ is a real constant. The properties of the matrices $W_i(p)$ will be determined below. Moreover, $\nu_i$, $(i = 1, 2, 3)$ are nonnegative constants, and as earlier, DC is the digital controller, described by the equations

$$\begin{aligned} \xi_k &= y(kT), \quad (k = 0, \pm 1, \ldots), \\ \alpha(\zeta)\psi_k &= \beta(\zeta)\xi_k, \\ u(t) &= \sum_{i=0}^{P} h(t - kT)\psi_{k-i}, \quad kT < t < (k+1)T. \end{aligned} \tag{10.154}$$

As output of the system in Fig. 10.1, we consider the vector

$$\bar{z}(t) \triangleq \begin{bmatrix} z_1(t) \\ z_2(t) \end{bmatrix} \begin{matrix} \ell \\ \kappa \end{matrix} = \begin{bmatrix} \varkappa z(t) \\ z_2(t) \end{bmatrix}. \tag{10.155}$$

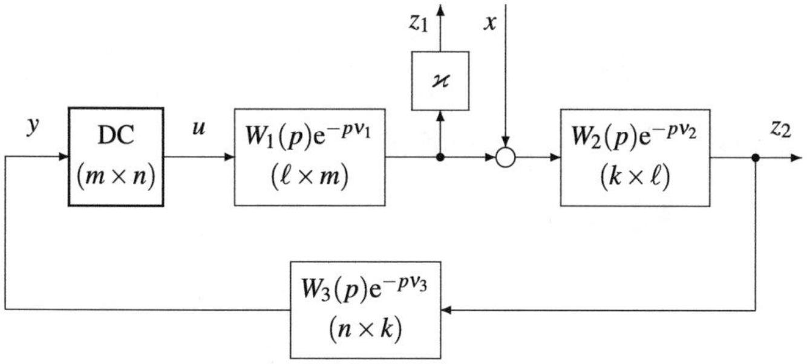


**Fig. 10.1** Block-diagram of SD System

Further on, we assume that the considered system satisfies all conditions for model standardization according to the outputs $z_1(t)$ and $z_2(t)$, advised in Examples 6.8 and 6.9.

(2) The PTM $W_{\bar{z}x}(s,t)$ from the input $x(t)$ to the output $\bar{z}(t)$ is determined by the relation

$$W_{\bar{z}x}(s,t) = \overset{\ell}{\begin{bmatrix} W_{z_1x}(s,t) \\ W_{z_2x}(s,t) \end{bmatrix}} \begin{matrix} \ell \\ \kappa \end{matrix}, \tag{10.156}$$

where $W_{z_ix}(s,t)$ is the PTM from the input $x(t)$ to the output $z_i(t)$. From the results of Example 6.2, we know that

$$\begin{aligned} W_{z_2x}(s,t) =& \mathrm{e}^{-\lambda(\nu_2+\nu_1)}\varphi_{W_2W_1\mu}(T,s,t-\nu_2-\nu_1)\check{R}_N(s)W_3(s)W_2(s)\mathrm{e}^{-s(\nu_3+\nu_2)} \\ &+ W_2(s)\mathrm{e}^{-s\nu_2}, \end{aligned} \tag{10.157}$$

which can be written in the standard form (6.67)

$$W_{z_2x}(s,t) = \varphi_{L_{2\tau}\mu}(T,s,t)\check{R}_N(s)M_\tau(s) + K_{2\tau}(s), \tag{10.158}$$

where

$$\begin{aligned} K_{2\tau}(s) &= K_2(s)\mathrm{e}^{-s\tau_{K2}}, \;\; L_{2\tau}(s) = L_2(s)\mathrm{e}^{-s\tau_{L2}}, \\ M_\tau(s) &= M(s)\mathrm{e}^{-s\tau_M}, \quad N_\tau(s) = N(s)\mathrm{e}^{-s\tau_N}. \end{aligned} \tag{10.159}$$

Here

$$\begin{aligned} K_2(s) &= W_2(s), & L_2(s) &= W_2(s)W_1(s), \\ M(s) &= W_3(s)W_2(s), & N(s) &= W_3(s)W_2(s)W_1(s), \end{aligned} \tag{10.160}$$

are rational matrices of dimensions $\kappa\times\ell$, $\kappa\times m$, $n\times\ell$, and $n\times m$, respectively, and moreover

$$\begin{aligned} \tau_{K2} &= \nu_2, & \tau_{L2} &= \nu_2+\nu_1, \\ \tau_M &= \nu_3+\nu_2, & \tau_N &= \nu_3+\nu_2+\nu_1. \end{aligned} \tag{10.161}$$

In analogy to the results of Example 6.3, we obtain

$$W_{z_1x}(s,t) = \varkappa\varphi_{L_{1\tau}}\mu(T,s,t)\check{R}_N(s)M_\tau(p), \tag{10.162}$$

where

$$L_{1\tau}(s) = L_1(s)\mathrm{e}^{-s\tau_{L1}}, \quad M_\tau(s) = M(s)\mathrm{e}^{-s\tau_M}, \quad N_\tau(s) = N(s)\mathrm{e}^{-s\tau_N}. \tag{10.163}$$

Here

$$L_1(s) = W_1(s), \quad M(s) = W_3(s)W_2(s), \quad N(s) = W_3(s)W_2(s)W_1(s), \tag{10.164}$$

and according to (6.244)

$$\tau_{K1} = 0, \quad \tau_{L1} = \nu_1, \quad \tau_M = \nu_3 + \nu_2, \quad \tau_N = \nu_3 + \nu_2 + \nu_1. \tag{10.165}$$

Inserting (10.158) and (10.162) into (10.156), an expression for the PTM of the investigated system $\mathscr{S}_0$ is achieved, which has the form

$$W_{\bar{z}x}(s,t) = \varphi_{\bar{L}_\tau \mu}(T,s,t)\breve{R}_N(s)M_\tau(s) + \bar{K}_\tau(s), \tag{10.166}$$

where

$$\begin{aligned} \bar{K}_\tau(s) &= \begin{bmatrix} O_{\ell\ell} \\ W_2(s)\mathrm{e}^{-s\nu_2} \end{bmatrix}, & \bar{L}_\tau(s) &= \begin{bmatrix} \varkappa W_1(s)\mathrm{e}^{-s\nu_1} \\ W_2(s)W_1(s)\mathrm{e}^{-s(\nu_2+\nu_1)} \end{bmatrix}, \\ M_\tau(s) &= W_3(s)W_2(s)\mathrm{e}^{-s(\nu_3+\nu_2)}, & N_\tau(s) &= W_3(s)W_2(s)W_1(s)\mathrm{e}^{-s(\nu_3+\nu_2+\nu_1)}. \end{aligned} \tag{10.167}$$

Considering matrix (10.166) as the PTM of the generalized standard model (10.1), (10.2), (10.4), we introduce, with respect to (10.38) and (10.48), the matrices

$$\begin{aligned} \bar{K}(s) &= \overset{\ell}{\begin{bmatrix} O_{\ell\ell} \\ W_2(s) \end{bmatrix}} \begin{matrix} \ell \\ \kappa \end{matrix}, & \bar{L}(s) &= \overset{m}{\begin{bmatrix} \varkappa W_1(s) \\ W_2(s)W_1(s) \end{bmatrix}} \begin{matrix} \ell \\ \kappa \end{matrix}, \\ M(s) &= W_3(s)W_2(s), & N(s) &= W_3(s)W_2(s)W_1(s), \end{aligned} \tag{10.168}$$

and also the extended matrix

$$\bar{W}_0(s) = \begin{bmatrix} \bar{K}(s) & \bar{L}(s) \\ M(s) & N(s) \end{bmatrix} = \begin{array}{c} \begin{matrix} \quad\ell & \qquad\qquad m \end{matrix} \\ \begin{bmatrix} O_{\ell\ell} & \varkappa W_1(s) \\ W_2(s) & W_2(s)W_1(s) \\ W_3(s)W_2(s) & W_3(s)W_2(s)W_1(s) \end{bmatrix} \end{array} \begin{matrix} \ell \\ \kappa \\ n \end{matrix}. \tag{10.169}$$

**Lemma 10.1** *Let the matrices $W_i(s)$ be independent, i.e.,*

$$\text{Mind}\, N(s) = \text{Mind}\, W_3(s) + \text{Mind}\, W_2(s) + \text{Mind}\, W_1(s). \tag{10.170}$$

*Then*

$$\text{Mind}\, \bar{W}_0(s) = \text{Mind}\, N(s) = \text{Mind}\, W_1(s) + \text{Mind}\, W_2(s) + \text{Mind}\, W_3(s), \tag{10.171}$$

*i.e., the matrix $N(s)$ dominates in the matrix $\bar{W}_0(s)$.*

*Proof* From (10.169), we easily obtain

$$\bar{W}_0(s) = W_{03}(s)W_{02}(s)W_{01}(s), \tag{10.172}$$

where

$$W_{03}(s) \triangleq \begin{array}{c} \begin{matrix} \ell & \kappa & \kappa \end{matrix} \\ \begin{bmatrix} I_\ell & 0 & 0 \\ 0 & I_\kappa & 0 \\ 0 & 0 & W_3(s) \end{bmatrix} \end{array} \begin{matrix} \ell \\ \kappa \\ n \end{matrix} = \operatorname{diag}\{I_\ell, I_\kappa, W_3(s)\}$$

$$W_{02}(s) \triangleq \begin{array}{c} \begin{matrix} \ell & \ell \end{matrix} \\ \begin{bmatrix} 0 & \varkappa I_\ell \\ W_2(s) & W_2(s) \\ W_2(s) & W_2(s) \end{bmatrix} \end{array} \begin{matrix} \ell \\ \kappa \\ \kappa \end{matrix} \tag{10.173}$$

$$W_{01}(s) \triangleq \begin{array}{c} \begin{matrix} \ell & m \end{matrix} \\ \begin{bmatrix} I_\ell & 0 \\ 0 & W_1(s) \end{bmatrix} \end{array} \begin{matrix} \ell \\ \ell \end{matrix} = \operatorname{diag}\{I_\ell, W_1(s)\}.$$

Introduce

$$\gamma_i \triangleq \operatorname{Mind} W_i(s), \quad (i = 1, 2, 3). \tag{10.174}$$

Then we will show that for matrices (10.173)

$$\operatorname{Mind} W_{0i}(s) = \gamma_i, \quad (i = 1, 2, 3). \tag{10.175}$$

Assume the ILMFD

$$W_i(s) = a_i^{-1}(s) b_i(s), \quad (i = 1, 2, 3). \tag{10.176}$$

Then by definition of ILMFD

$$\deg \det a_i(s) = \gamma_i. \tag{10.177}$$

For the matrix $W_{03}(s)$, we obtain the LMFD

$$W_{03}(s) = a_{03}^{-1}(s) b_{03}(s), \tag{10.178}$$

where

$$a_{03}(s) = \operatorname{diag}\{I_\ell, I_\kappa, a_3(s)\}. \tag{10.179}$$

From (10.179) it follows that $\deg \det a_{03}(s) = \deg \det a_3(s) = \gamma_3$ and, consequently due to the properties of LMFD,

$$\operatorname{Mind} W_{03}(s) \leq \deg \det a_{03}(s) = \gamma_3. \tag{10.180}$$

On the other side, from (10.173) and Corollary 2.6, we obtain

$$\operatorname{Mind} W_{03}(s) \geq \operatorname{Mind} W_3(s) = \gamma_3. \tag{10.181}$$

Comparing the last two inequalities proves that Mind $W_{01}(s) = \gamma_3$. Analogously, we can show that Mind $W_{03}(s) = \gamma_1$.

For the remaining proof of the relations (10.175), notice that the matrix $W_{02}(s)$ can be written in the form

$$W_{02}(s) = \begin{bmatrix} I_\ell & 0_{\ell\kappa} & 0_{\ell\kappa} \\ 0_{\kappa\ell} & I_\kappa & I_\kappa \\ 0_{\kappa\ell} & 0_{\kappa\kappa} & I_\kappa \end{bmatrix} W_{04}(s), \tag{10.182}$$

where

$$W_{04}(s) = \begin{bmatrix} 0_{\ell\ell} & \varkappa I_\ell \\ 0_{\kappa l} & 0_{\kappa\ell} \\ W_2(s) & W_2(s) \end{bmatrix}. \tag{10.183}$$

Since the first factor on the right side of (10.182) is a non-singular matrix, we obtain

$$\text{Mind } W_{02}(s) = \text{Mind } W_{04}(s).$$

Moreover, from

$$W_{04}(s) = \left[\text{diag}\{I_\ell, I_\kappa, a_2(s)\}\right]^{-1} \begin{bmatrix} I_\ell & \varkappa I_\ell \\ 0_{\kappa\ell} & 0_{\kappa\ell} \\ b_2(s) & b_2(s) \end{bmatrix},$$

we find

$$\text{Mind } W_{04}(s) \leq \gamma_2.$$

At the same time, from (10.183), it directly follows that

$$\text{Mind } W_{04}(s) \geq \gamma_2.$$

The comparison of the last two inequalities yields

$$\text{Mind } W_{02}(s) = \text{Mind } W_{04}(s) = \gamma_2,$$

which completes the proof of relations (10.175).

For the further proof notice that (10.172) and (10.175) imply

$$\text{Mind } \bar{W}_0(s) \leq \gamma_1 + \gamma_2 + \gamma_3. \tag{10.184}$$

At the same time, when the condition of independence (10.170) is fulfilled, from (10.169) we obtain

$$\text{Mind } \bar{W}_0(s) \geq \text{Mind } N(s) = \gamma_1 + \gamma_2 + \gamma_3. \tag{10.185}$$

The last two inequalities verify Lemma 10.1. ■

(3) From the conditions derived in Sect. 6.8 for model standardizability of the system $\mathscr{S}_0$ simultaneously for the outputs $z_1(t)$ and $z_2(t)$, we realize that

$$\begin{aligned} &\operatorname{exc} W_2(s) \geq 1, \qquad \operatorname{exc}[W_3(s)W_2(s)] \geq 1, \\ &\operatorname{exc}[W_3(s)W_2(s)] \geq 1, \operatorname{exc}[W_3(s)W_2(s)W_1(s)] \geq 1, \\ &\operatorname{exc} W_1(s) \geq 0. \end{aligned} \tag{10.186}$$

Therefore, from (10.168), (10.169) we find

$$\begin{aligned} \operatorname{exc}\bar{K}(s) \geq 1,\ \operatorname{exc}\bar{L}(s) \geq 0, \\ \operatorname{exc}M(s) \geq 1,\ \operatorname{exc}N(s) \geq 1. \end{aligned} \tag{10.187}$$

Hereby, if the matrix $\bar{L}(s)$ in (10.168) is written in the form

$$\bar{L}(s) = \overset{m}{\begin{bmatrix} L_1(s) \\ L_2(s) \end{bmatrix}} \begin{matrix} \ell \\ \kappa \end{matrix}, \tag{10.188}$$

then we also obtain

$$\operatorname{exc}L_1(s) \geq 0, \quad \operatorname{exc}L_2(s) \geq 1. \tag{10.189}$$

Assume the minimal realization

$$N(s) = C_2(sI_\chi - A)^{-1}B_2, \tag{10.190}$$

where $\chi = \gamma_1 + \gamma_2 + \gamma_3$. Then, it follows from Lemma 10.1 and the statements of Chap. 2 that the matrix $\bar{W}_0(s)$ (10.169) can be represented in the form

$$\bar{W}_0(s) = \begin{bmatrix} \overset{\ell}{\bar{C}_1(sI_\chi - A)^{-1}B_1} & \overset{m}{\bar{C}_1(sI_\chi - A)^{-1}B_2 + \bar{D}} \\ C_2(sI_\chi - A)^{-1}B_1 & C_2(sI_\chi - A)^{-1}B_2 \end{bmatrix} \begin{matrix} \ell+\kappa \\ n \end{matrix}. \tag{10.191}$$

Comparing relations (10.169) and (10.191), we find

$$\begin{aligned} \bar{K}(s) = \bar{C}_1(sI_\chi - A)^{-1}B_1,\quad \bar{L}(s) = \bar{C}_1(sI_\chi - A)^{-1}B_2 + \bar{D}, \\ M(s) = C_2(sI_\chi - A)^{-1}B_1,\quad N(s) = C_2(sI_\chi - A)^{-1}B_2. \end{aligned} \tag{10.192}$$

Assume

$$\bar{C}_1 = \overset{\chi}{\begin{bmatrix} \bar{C}_{11} \\ \bar{C}_{12} \end{bmatrix}} \begin{matrix} \ell \\ \kappa \end{matrix} \quad \bar{D} = \overset{m}{\begin{bmatrix} \bar{D}_1 \\ \bar{D}_2 \end{bmatrix}} \begin{matrix} \ell \\ \kappa \end{matrix} \tag{10.193}$$

then, with respect to (10.169) in addition to (10.192), we achieve

$$\bar{K}(s) = \overset{\ell}{\begin{bmatrix} K_1(s) \\ K_2(s) \end{bmatrix}} \begin{matrix} \ell \\ \kappa \end{matrix} = \begin{bmatrix} \bar{C}_{11}(sI_\chi - A)^{-1}B_1 \\ \bar{C}_{12}(sI_\chi - A)^{-1}B_1 \end{bmatrix} = \overset{\ell}{\begin{bmatrix} 0_{\ell\ell} \\ W_2(s) \end{bmatrix}} \begin{matrix} \ell \\ \kappa \end{matrix} \tag{10.194}$$

and also

$$\bar{L}(s) = \begin{bmatrix} L_1(s) \\ L_2(s) \end{bmatrix} = \begin{bmatrix} \bar{C}_{11}(sI_\chi - A)^{-1}B_2 + \bar{D}_1 \\ \bar{C}_{12}(sI_\chi - A)^{-1}B_2 \end{bmatrix} \begin{matrix} \ell \\ \kappa \end{matrix} = \overset{m}{\begin{bmatrix} \varkappa W_1(s) \\ W_2(s)W_1(s) \end{bmatrix}} \begin{matrix} \ell \\ \kappa \end{matrix} \tag{10.195}$$

where we used that, due to (10.190),

$$\bar{D}_2 = 0_{m\kappa}. \tag{10.196}$$

From (10.193)–(10.195) and (10.169) we derive

$$\begin{aligned} \bar{C}_{11}(sI_\chi - A)^{-1}B_1 &= 0_{\ell\ell}, & \bar{C}_{12}(sI_\chi - A)^{-1}B_1 &= W_2(s), \\ \bar{C}_{11}(sI_\chi - A)^{-1}B_2 + \bar{D}_1 &= \varkappa W_1(s), & \bar{C}_{12}(sI_\chi - A)^{-1}B_2 &= W_2(s)W_1(s). \end{aligned} \tag{10.197}$$

(4) The next theorem summarizes the above considerations.

**Theorem 10.7** *When relations (10.197) and also the independence condition (10.170) are fulfilled, PTM (10.166) coincides with the PTM of the generalized standard model for the SD system with delay described by equations of form (10.1), (10.2), (10.4)*

$$\begin{aligned} \frac{dv(t)}{dt} &= Av(t) + B_1x(t-\tau_1) + B_2u(t-\tau_2), \\ y(t) &= C_2v(t), \\ z_1^0(t) &= \bar{C}_{11}v(t-\tau_{31}) + \bar{D}_1u(t-\tau_2-\tau_{31}), \\ z_2^0(t) &= \bar{C}_{12}v(t-\tau_{32}), \\ \xi_k &= y(kT), \quad (k = 0, \pm 1, \ldots), \\ \alpha(\zeta)\psi_k &= \beta(\zeta)\xi_k, \\ u(t) &= \sum_{i=0}^{P} h_i(t-kT)\psi_{k-i}, \quad kT < t < (k+1)T, \end{aligned} \tag{10.198}$$

*where*

$$\begin{aligned} \tau_1 = \tau_M = \nu_3 + \nu_2, \quad \tau_2 = \tau_N = \nu_3 + \nu_2 + \nu_1, \\ \tau_{31} = -\nu_3 - \nu_2, \quad \tau_{32} = -\nu_3. \end{aligned} \tag{10.199}$$

*Proof* Using formulae (10.13), build the PTM $W_{z_1^0x}(s,t)$ and $W_{z_2^0x}(s,t)$ of the generalized standard model for SD systems with delay (10.198). Then, with respect to formulae (10.197) and (10.199), we easily achieve

$$W_{z_1^0x}(s,t) = W_{z_1x}(s,t), \quad W_{z_2^0x}(s,t) = W_{z_2x}(s,t),$$

where $W_{z_1x}(s,t)$ and $W_{z_2x}(s,t)$ are the PTM (10.162) and (10.158), respectively. Therefore, we obtain

$$W_{\tilde{z}x}(s,t) \triangleq \begin{bmatrix} W_{\zeta_1x}(s,t) \\ W_{z_2x}(s,t) \end{bmatrix} = \begin{bmatrix} W_{z_1^0x}(s,t) \\ W_{z_2^0x}(s,t) \end{bmatrix} = W_{z^0x}(s,t),$$

where $W_{z^0x}(s,t)$ is the PTM of system (10.198) from input $x(t)$ to output $z^0(t) = \left[z_1^0(t)\ z_2^0(t)\right]'$. ∎

## 10.6 $\mathcal{H}_2$ Optimization of Single-Loop SD Systems with Delay

(1) Extending Definition 6.8 to the generalized standard model for $m = 2$, the single-loop system $\mathcal{S}_0$ is called model standardizable, if there exists a generalized standard model $\mathcal{S}_g$ (10.1), (10.2), (10.4), the PTM of which coincides with the PTM (10.166). Hereby, the original system $\mathcal{S}_0$ and the corresponding generalized model are considered to be identical.

From Theorem 10.7 and the content of Sect. 10.5, we know that the system $\mathcal{S}_0$ is model standardizable if the following conditions are fulfilled.

(a) The matrices

$$\begin{aligned} M(s) &= W_3(s)W_2(s), & N(s) &= W_3(s)W_2(s)W_1(s), \\ K_2(s) &= W_2(s), & L_2(s) &= W_2(s)W_1(s) \end{aligned} \tag{10.200}$$

are strictly proper.

(b) The matrix

$$L_1(s) = \varkappa W_1(s) \tag{10.201}$$

is at least proper.

(c) The independence condition (10.170) is fulfilled, such that the matrix $N(s) = W_3(s)W_2(s)W_1(s)$ dominates in matrix (10.169)

$$\bar{W}_0(s) = \begin{bmatrix} O_{\ell\ell} & \varkappa W_1(s) \\ W_2(s) & W_2(s)W_1(s) \\ W_3(s)W_2(s) & W_3(s)W_2(s)W_1(s) \end{bmatrix}. \tag{10.202}$$

The condition of type (d) in Theorem 6.10 is always fulfilled for system $\mathcal{S}_0$.

(2) In the following considerations, we assume that conditions (a)–(c) are satisfied. Moreover, if not explicitly stated different, we always assume that the sampling period

$T$ is non-pathological according to the union of the sets of poles of the matrices $W_i(s)$, $(i = 1, 2, 3)$.

When these propositions hold, the $\mathcal{H}_2$ optimization problem for the system $\mathcal{S}_0$ is equivalent to the $\mathcal{H}_2$ optimization problem for the generalized standard model (10.198), (10.199), which on its part, owing to Theorem 10.2, coincides with the solution of the $\mathcal{H}_2$ optimization problem for the standard model with delay in the closed loop, which is obtained from (10.198) and (10.199) for $\tau_1 = \tau_{31} = \tau_{32} = 0$ and takes the form

$$\begin{aligned}
\frac{\mathrm{d}v(t)}{\mathrm{d}t} &= Av(t) + B_1x(t) + B_2u(t - \nu_1 - \nu_2 - \nu_3),\\
y(t) &= C_2v(t),\\
\bar{z}_1(t) &= \bar{C}_{11}v(t) + \bar{D}_1u(t - \nu_1 - \nu_2 - \nu_3),\\
z_2(t) &= \bar{C}_{12}v(t),\\
\xi_k &= y(kT), \quad (k = 0, \pm 1, \ldots),\\
\alpha(\zeta)\psi_k &= \beta(\zeta)\xi_k,\\
u(t) &= \sum_{i=0}^{P} h_i(t - kT)\psi_{k-i}, \quad kT < t < (k+1)T,
\end{aligned} \tag{10.203}$$

where the matrices $\bar{C}_{11}$, $\bar{C}_{12}$ and $\bar{D}_1$ are determined in (10.193).

In harmony to the content of Chap. 7, the $\mathcal{H}_2$ optimal system function $\theta^0(\zeta)$ for the standard model (10.203) can be determined as a result of the minimization of the functional of form (7.79)

$$J_1 = \frac{1}{2\pi \mathrm{j}} \oint \operatorname{trace}\left[\hat{\theta}(\zeta)A_{\bar{L}}(\zeta)\theta(\zeta)A_M(\zeta) - \hat{\theta}(\zeta)\hat{C}(\zeta) - C(\zeta)\theta(\zeta)\right]\frac{\mathrm{d}\zeta}{\zeta}, \tag{10.204}$$

where, as in (7.81),

$$\begin{aligned}
A_{\bar{L}}(\zeta) &= \hat{a}_r(\zeta)D_{\underline{\mu'\bar{L}'}\bar{L}\mu}(T, \zeta, 0)a_r(\zeta),\\
A_M(\zeta) &= a_l(\zeta)D_{M\underline{M'}}(T, \zeta, 0)\hat{a}_l(\zeta),\\
C(\zeta) &= A_M(\zeta)\hat{\beta}_{r0}(\zeta)D_{\underline{\mu'\bar{L}'}\bar{L}\mu}(T, \zeta, 0)a_r(\zeta) + a_l(\zeta)D_{M\underline{\bar{K}'}\bar{L}\mu}(T, \zeta, -\tau_N)a_r(\zeta).
\end{aligned} \tag{10.205}$$

Further, notice that from (10.168), (10.169), we obtain the formulae

$$\begin{aligned}
N(s) &= W_3(s)W_2(s)W_1(s), & N'(-s) &= W_1'(-s)W_2'(-s)W_3'(-s),\\
M(s) &= W_3(s)W_2(s), & M'(-s) &= W_2'(-s)W_3'(-s),\\
\bar{L}(s) &= \begin{bmatrix} \varkappa W_1(s) \\ W_2(s)W_1(s) \end{bmatrix}, & \bar{L}'(-s) &= \begin{bmatrix} \varkappa W_1'(-s) & W_1'(-s)W_2'(-s) \end{bmatrix},\\
\bar{K}(s) &= \begin{bmatrix} 0_{\ell\ell} \\ W_2(s) \end{bmatrix}, & \bar{K}'(-s) &= \begin{bmatrix} 0_{\ell\ell} & W_2'(-s) \end{bmatrix},
\end{aligned} \tag{10.206}$$

from which we find

$$\begin{aligned}\bar{L}'(-s)L(s) &= \varkappa^2 W_1'(-s)W_1(s) + W_1'(-s)W_2'(-s)W_2(s)W_1'(-s),\\ M(s)M'(-s) &= W_3(s)W_2(s)W_2'(-s)W_3'(-s),\\ \bar{K}'(-s)\bar{L}(s) &= W_2'(-s)W_2(s)W_1(s),\\ M(s)\bar{K}'(-s)\bar{L}(s) &= W_3(s)W_2(s)W_2'(-s)W_2(s)W_1(s).\end{aligned} \tag{10.207}$$

Thus, for the single-loop system $\mathscr{S}_0$, formulae (10.205) take the form

$$\begin{aligned}A_{\bar{L}}(\zeta) &= \varkappa^2 \hat{a}_r(\zeta)D_{\mu' \underline{W_1'} W_1 \mu}(T,\zeta,0)a_r(\zeta) + \hat{a}_r(\zeta)D_{\mu' \underline{W_1'}\, \underline{W_2'} W_2 W_1 \mu}(T,\zeta,0)a_r(\zeta),\\ A_M(\zeta) &= a_l(\zeta)D_{W_3 W_2 \underline{W_2'}\, \underline{W_3'}}(T,\zeta,0)\hat{a}_l(\zeta),\\ C(\zeta) &= A_M(\zeta)\hat{\beta}_{r0}(\zeta)\left[\varkappa^2 D_{\mu' \underline{W_1'} W_1 \mu}(T,\zeta,0) + D_{\mu' \underline{W_1'}\, \underline{W_2'} W_2 W_1 \mu}(T,\zeta,0)\right]a_r(\zeta)\\ &\quad + a_l(\zeta)\left[D_{W_3 W_2 \underline{W_2'} W_2 W_1 \mu}(T,\zeta,-\tau_N)a_r(\zeta)\right].\end{aligned} \tag{10.208}$$

The matrices $a_l(\zeta)$, $a_r(\zeta)$, $\beta_{r0}(\zeta)$ configured in (10.208), according to Sect. 6.5, can be determined in the following way. The matrix $a_l(\zeta)$ is achieved from the ILMFD (6.172)

$$W_N(\zeta) = a_l^{-1}(\zeta)b_l(\zeta,-\tau_N).$$

Let $(\alpha_{l0}(\zeta), \beta_{l0}(\zeta))$ be a basic controller for the process $(a_l(\zeta), b_l(\zeta,-\tau_n))$, such that the matrix

$$Q_l(\zeta,\alpha_{l0},\beta_{l0}) = \begin{bmatrix} a_l(\zeta) & -b_l(\zeta,-\tau_N\\ \beta_{l0}(\zeta) & \alpha_{l0}(\zeta)\end{bmatrix} \tag{10.209}$$

becomes unimodular. Then the matrices $a_r(\zeta)$ and $\beta_{r0}(\zeta)$ can be found from the equality

$$Q_l^{-1}(\zeta,\alpha_{l0},\beta_{l0}) = \begin{bmatrix} \alpha_{r0}(\zeta) & -b_r(\zeta,-\tau_N\\ \beta_{r0}(\zeta) & a_r(\zeta)\end{bmatrix}. \tag{10.210}$$

According to the derived relations, the construction of the $\mathscr{H}_2$ optimal function $\theta^0(\zeta)$ for the system $\mathscr{S}_0$ leads to the minimization of the functional (10.204), in which relation (10.208) holds, and the matrices $a_l(\zeta)$ and $a_r(\zeta)$ are determined from (10.209) and (10.210). Hereby, the application of the Wiener–Hopf method yields the following statement.

**Theorem 10.8** *Let the quasi-polynomials $A_{\bar{L}}(\zeta)$ and $A_M(\zeta)$ in (10.208) be positive on the circle $|\zeta| = 1$. Then the $\mathscr{H}_2$ optimal system function $\theta^0(\zeta)$ can be constructed with the help of the algorithm (10.62)–(10.66). Hereby, the characteristic polynomial of the $\mathscr{H}_2$ optimal system is determined according to formulae (10.70), and the transfer matrix of the optimal controller $W_d^0(\zeta)$ can be constructed by one of the methods described in Theorem 7.4.* ■

(3) Consider the problem of constructing the set of fixed poles for the $\mathscr{H}_2$ optimal system $\mathscr{S}_0$, when we select outputs of form (10.155). According to what was shown above this problem leads to the construction of the set of fixed poles for the standard model (10.203). For this purpose, we use the result of Sects. 7.4 and 7.5 together with relations (10.206). Since the matrices $M(s)$ and $N(s)$ for the standard model (10.203) are the same as in Sect. 7.8, so the set of fixed poles, connected with the matrix $M(s)$, is determined in the same manner as in Examples 7.1–7.3.

Assume the minimal realizations

$$\begin{aligned} N(s) &= C_2(sI_\chi - A)^{-1}B_2, \quad \chi = \gamma_3 + \gamma_2 + \gamma_1, \\ M(s) &= W_3(s)W_2(s) = C_M(sI_{\gamma_3+\gamma_2} - A_M)^{-1}B_M, \end{aligned} \tag{10.211}$$

where it was used that the matrices $W_3(s)$ and $W_2(s)$ are independent, and due to (10.175) we had $\operatorname{Mind} W_3(s) = \gamma_3$ and $\operatorname{Mind} W_2(s) = \gamma_2$. Hereby, the roots of the polynomial

$$\delta_M(s) = \frac{\det(sI_\chi - A)}{\det(sI_{\gamma_3+\gamma_2} - A_M)} \tag{10.212}$$

define the subset of fixed poles depending on the matrix $M(s)$. Furthermore, the results of Sect. 7.8 allow

$$\delta_M(s) = \psi_1(s), \tag{10.213}$$

where $\psi_1(s)$ is the McMillan denominator of the matrix $W_1(s)$. Moreover, when we assume the expansion

$$\psi_1(s) = (s - p_1)^{q_1} \cdots (s - p_\rho)^{q_\rho}, \tag{10.214}$$

where all numbers $p_i$ are distinct, then from the facts, shown in Chap. 7, we come to the following conclusions.

If among the numbers $p_i$ there are some on the imaginary axis, then the $\mathscr{H}_2$ optimization problem does not possess a solution on the set of causal stabilizing controllers. When, however,

$$\operatorname{Re} p_i < 0, \ (i = 1, \dots, r), \quad \operatorname{Re} p_i > 0, \ (i = r + 1, \dots, \rho), \tag{10.215}$$

takes place, then the subset of fixed poles $\zeta_{Mi}$, connected with the matrix $M(s)$, coincides with the set of roots of the polynomial

$$\psi_{1T}(\zeta) \stackrel{\triangle}{=} (1 - \zeta e^{p_1 T})^{q_1} \cdots (1 - \zeta e^{p_r T})^{q_r} (1 - \zeta e^{-p_{r+1} T})^{q_{r+1}} \cdots (1 - \zeta e^{-p_\rho T})^{q_\rho}. \tag{10.216}$$

(4) Assume, in addition to the minimal realization (10.211), the minimal realization

$$\bar{L}(s) = \begin{bmatrix} \varkappa W_1(s) \\ W_2(s)W_1(s) \end{bmatrix} = \bar{C}_L(sI_\delta - \bar{A}_L)^{-1}\bar{B}_L + \bar{D}, \tag{10.217}$$

where

$$\delta \overset{\triangle}{=} Mind\bar{L}(s).$$

**Lemma 10.2** *The following claims are valid.*

*(i) The following equality holds:*

$$\delta = \gamma_2 + \gamma_1. \tag{10.218}$$

*(ii) The IRMFD of the matrix $\bar{L}(s)$ can be represented in the form*

$$\bar{L}(s) = b_{\bar{L}}(s)\,[d_1(s)d_3(s)]^{-1}. \tag{10.219}$$

*Herein*

$$\det d_1(s) \sim \psi_1(s), \qquad \det d_3(s) \sim \psi_2(s), \tag{10.220}$$

*where $\psi_i(s)$ are the McMillan denominators of the matrices $W_i(ss)$, $(i = 1, 2)$.*

*Proof* (i) Represent the matrix (10.217) in the form

$$\bar{L}(s) = \bar{L}_0(s)W_1(s), \tag{10.221}$$

where

$$\bar{L}_0(s) \overset{\triangle}{=} \begin{bmatrix} \varkappa I_\ell \\ W_2(s) \end{bmatrix}. \tag{10.222}$$

With the help of the earlier calculated examples, we can easily show that

$$\text{Mind}\,\bar{L}_0(s) = \text{Mind}\,W_2(s) = \gamma_2.$$

Therefore, from (10.221) we achieve

$$\text{Mind}\,\bar{L}(s) \le \text{Mind}\,\bar{L}_0(s) + \text{Mind}\,W_1(s) = \gamma_2 + \gamma_1.$$

On the other side, from (10.217), we obtain

$$\text{Mind}\,\bar{L}(s) \ge \text{Mind}[W_2(s)W_1(s)] = \gamma_2 + \gamma_1.$$

Comparing both last inequalities proves formula (10.218).

(ii) Assume the IRMFD

$$W_i(s) = c_i(s)d_i^{-1}(s), \quad (i = 1, 2), \tag{10.223}$$

where owing to the properties of IRMFD, we obtain

$$\det d_i(s) = \psi_i(s), \quad (i = 1, 2),$$

where $\psi_i(s)$ is the associated McMillan denominator, and thus

$$\deg\det d_i(s) = \gamma_i. \tag{10.224}$$

Using (10.223), the matrix $\bar{L}(s)$ can be represented in the form of the RMFD

$$\bar{L}(s) = \begin{bmatrix} \varkappa c_1(s)d_3(s) \\ c_2(s)c_3(s) \end{bmatrix} [d_1(s)d_3(s)]^{-1}, \tag{10.225}$$

where the polynomial matrices $c_3(s)$ and $d_3(s)$ are defined from the IRMFD

$$d_2^{-1}(s)c_2(s) = c_3(s)d_3^{-1}(s). \tag{10.226}$$

Hence (10.218) together with (10.224) allow the conclusion that the right side of (10.225) is an IRMFD, satisfying condition (10.220). ■

Assume for the minimal realizations of the matrix $N(s)$ and $\bar{L}(s)$ (10.211) and (10.217) the controllable pairs $(\mathrm{e}^{-AT}, m(A))$ and $(\mathrm{e}^{-\bar{A}_L T}, m_{\bar{L}}(\bar{A}_L))$, where

$$\begin{aligned} m(A) &= \sum_{i=0}^{P} \mathrm{e}^{-iAT} \int_0^T \mathrm{e}^{-A\nu} B_2 h_i(\nu)\,\mathrm{d}\nu, \\ m_{\bar{L}}(\bar{A}_L) &= \sum_{i=0}^{P} \mathrm{e}^{-i\bar{A}_L T} \int_0^T \mathrm{e}^{-\bar{A}_L \nu} \bar{B}_L h_i(\nu)\,\mathrm{d}\nu. \end{aligned} \tag{10.227}$$

Then, according to (10.218) and the results of Sect. 7.5, the subset of fixed poles of the $\mathscr{H}_2$ optimal standard model (10.203), related to the matrix $\bar{L}(s)$, coincides with the set of roots of the polynomial

$$\delta_{\bar{L}}(s) = \frac{\det(sI_\chi - A)}{\det(sI_{\gamma_2+\gamma_1} - \bar{A}_L)}. \tag{10.228}$$

This relation is equivalent to the formula

$$\delta_{\bar{L}}(s) = \frac{\psi_N(s)}{\psi_{\bar{L}}(s)}, \tag{10.229}$$

where $\psi_N(s)$ and $\psi_{\bar{L}}(s)$ are the McMillan denominators of the matrices $N(s)$ and $\bar{L}(s)$, respectively. In Sect. 7.8, it was shown that under the conditions of independence (10.170)

$$\psi_N(s) = \psi_3(s)\psi_2(s)\psi_1(s). \tag{10.230}$$

Moreover, from statement (ii) in Lemma 10.2, we find

$$\psi_{\bar{L}}(s) = \psi_2(s)\psi_1(s), \tag{10.231}$$

and formula (10.229) allows the conclusion

$$\delta_{\bar{L}}(s) = \psi_3(s). \tag{10.232}$$

Hence, assuming

$$\psi_3(s) = (s - q_1)^{\lambda_1} \cdots (s - q_\eta)^{\lambda_\eta},$$

where all numbers $q_i$ are distinct, the set of fixed poles $\zeta_{\bar{L}i}$ related to the matrix $\bar{L}(s)$ establishes itself as the set of roots of the equation

$$\psi_{3T}(\zeta) \stackrel{\triangle}{=} (1 - \zeta e^{q_1 T})^{\lambda_1} \cdots (1 - \zeta e^{q_\ell T})^{\lambda_\ell} (1 - \zeta e^{-q_{\ell+1} T})^{\lambda_{\ell+1}} \cdots (1 - \zeta e^{-q_\eta T})^{\lambda_\eta}, \tag{10.233}$$

where we assumed

$$\text{Re} q_i < 0,\ (i = 1, \ldots, \ell), \quad \text{Re} q_i > 0,\ (i = \ell + 1, \ldots, \eta). \tag{10.234}$$

Combining formulae (10.212) and (10.228), we can state that, when we have the minimal realizations

$$\begin{aligned} W_1(s) &= C_1^*(sI_{\gamma_1} - A_1)^{-1} B_1^* + D_1, \\ W_3(s) &= C_3^*(sI_{\gamma_3} - A_3)^{-1} B_3^* + D_3(\zeta), \end{aligned}$$

where $D_1$ is a constant matrix and $D_3(\zeta)$ is a polynomial matrix, then under conditions which practically always take place in applications, the set of fixed poles for the $\mathscr{H}_2$ optimal system $\mathscr{S}_0$ is the roots of the equation

$$\det(sI_{\gamma_1} - A_1) \det(sI_{\gamma_3} - A_3) = 0. \tag{10.235}$$

## 10.7 $\mathscr{H}_2$ Optimization of a Single-Loop SD System with Delay Under Colored Noise

(1) If the vector input $x(t)$ in Fig. 10.1 is the response of a coloring filter with transfer matrix $\Phi(s)$ (8.1) to vectorial white noise $\rho(t)$, then the PTM of the system $\mathscr{S}_0$ from the input $\rho(t)$ to output $\bar{z}(t)$ (10.155) is equal to

$$W_{\bar{z}\rho}(s, t) = W_{\bar{z}x}(s, t)\Phi(s), \tag{10.236}$$

where $W_{\bar{z}\rho}(s, t)$ is the PTM from the input $\rho(t)$ to the output $\bar{z}(t)$, which owing to (10.166), (10.167), can be represented in the form

$$W_{\bar{z}\rho}(s, t) = \varphi_{\bar{L}_\tau \mu}(T, s, t)\check{R}_N(s) M_\tau(s)\Phi(s) + \bar{K}_\tau(s)\Phi(s). \tag{10.237}$$

We directly realize that the PTM (10.237) formally coincides with PTM (8.3) of standard SD system (10.203) under excitation by the colored input $x(t)$. Therefore, the solution of the $\mathscr{H}_2$ optimization problem for the system $\mathscr{S}_0$ under colored input can be obtained completely with the help of the relations in Chap. 8, when the matrix $L_\tau(s)$ is substituted by the matrix

$$\bar{L}_\tau(s) = \begin{bmatrix} \varkappa W_1(s)\mathrm{e}^{-s\nu_1} \\ W_2(s)W_1(s)\mathrm{e}^{-s(\nu_2+\nu_1)} \end{bmatrix},$$

the matrices $M(s)$ and $N(s)$ by the matrices

$$M(s) = W_3(s)W_2(s), \quad N(s) = W_3(s)W_2(s)W_1(s), \tag{10.238}$$

and the matrix $K_\tau(s)$ by the matrix

$$\bar{K}_\tau(s) = \begin{bmatrix} 0_{\ell\ell} \\ W_2(s)\mathrm{e}^{-s\nu_2}(s) \end{bmatrix}. \tag{10.239}$$

(2) Since the set of stabilizing controllers does not depend on the form of the input signal, so as a result of the taken substitutions, we arrive at the minimization problem of the functional, which is achieved from (8.13) by performing the mentioned substitutions

$$TJ_{\bar{\Phi}} = \frac{1}{2\pi \mathrm{j}} \oint \operatorname{trace}\left[\hat{\theta}(\zeta)A_{\bar{L}}(\zeta)\theta(\zeta)A_{M\bar{\Phi}}(\zeta) - \hat{\theta}(\zeta)\hat{C}_{\bar{\Phi}}(\zeta) - C_{\bar{\Phi}}(\zeta)\theta(\zeta)\right]\frac{\mathrm{d}\zeta}{\zeta}, \tag{10.240}$$

where the matrix $A_{\bar{L}}(\zeta)$ is determined by (10.205).

Moreover, the matrix $A_{M\bar{\Phi}}(\zeta)$ can be calculated by formulae obtained from (8.56) and (10.238)

$$A_{M\bar{\Phi}}(\zeta) = a_l(\zeta)D_{M\Phi\underline{\Phi}'M'}(T,\zeta,0)\hat{a}_l(\zeta) = a_l(\zeta)D_{W_3W_2\Phi\underline{\Phi}'\underline{W_2}'\underline{W_3}'}(T,\zeta,0)a_l(\zeta), \tag{10.241}$$

and the matrix $C_{\bar{\Phi}}(\zeta)$ is derived from (10.208) in the form

$$\begin{aligned} C_{\bar{\Phi}}(\zeta) &= A_{M\Phi}(\zeta)\hat{\beta}_{r0}(\zeta)\left[\varkappa^2 D_{\underline{\mu}'\underline{W_1}'W_1\mu}(T,\zeta,0) + D_{\underline{\mu}'\underline{W_1}'\underline{W_2}'W_2W_1\mu}(T,\zeta,0)\right]a_r(\zeta) \\ &\quad + a_l(\zeta)D_{W_3W_2\Phi\underline{\Phi}'\underline{W_2}'W_2W_1\mu}(T,\zeta,-\nu_1-\nu_2-\nu_3)a_r(\zeta). \end{aligned} \tag{10.242}$$

Moreover, if the matrices $A_{\bar{L}}(\zeta)$ and $A_{M\Phi}(\zeta)$ satisfy analog conditions as in Theorem 8.3, then the following claims hold.

(i) There exist the factorizations

$$A_{\bar{L}}(\zeta) = \hat{\Pi}_1(\zeta)\Pi_1(\zeta), \quad A_{M\Phi}(\zeta) = V(\zeta)\hat{V}(\zeta), \tag{10.243}$$

where $\Pi_1(\zeta)$ is a stable polynomial matrix and $V(\zeta)$ is a rational matrix which is stable together with its inverse.

(ii) The optimal system function $\theta^0(\zeta)$ is determined by formulae analog to (8.69)–(8.71).

(iii) Under conditions that are normally fulfilled in applications, the set of fixed poles of the $\mathscr{H}_2$ optimal system $\mathscr{S}_0$ under colored noise does not depend on the concrete form of this signal and it coincides with the set of roots of Eq. (10.235).

## 10.8 $\mathscr{L}_2$ Optimization of a Single-Loop SD System with Delay

(1) Applying expression (10.166) for the PTM of the system $\mathscr{S}_0$ for output (10.155)

$$W_{\bar{z}x}(s,t) = \varphi_{\bar{L}_\tau \mu}(T,s,t)\breve{R}_N(s)M_\tau(s) + \bar{K}_\tau(s), \tag{10.244}$$

and in addition formula (9.32), we obtain an expression for the image of the process $\bar{z}(t)$ for zero initial energy in a form analog to (9.49)

$$\begin{aligned}\bar{Z}(s) = &\begin{bmatrix} \varkappa W_1(s)e^{-s\nu_1} \\ W_2(s)W_1(s)e^{-s(\nu_2+\nu_1)} \end{bmatrix} \mu(p)\breve{R}_N(s)\breve{D}_{W_3W_2X}(T,s,-\nu_3-\nu_2) \\ &+ \begin{bmatrix} O_{\ell\ell} \\ W_2(s)X(s)e^{-s\nu_2} \end{bmatrix}.\end{aligned} \tag{10.245}$$

Using formula (10.19), from (10.245), we can build an image $\bar{Z}(s)$ similar to (10.126) by the system function

$$\bar{Z}(s) = \bar{\ell}_1(s)\breve{\theta}(s)\breve{m}(s) + \bar{n}_1(s), \tag{10.246}$$

in which

$$\begin{aligned}\bar{\ell}_1(s) &= \begin{bmatrix} \ell_{11}(s) \\ \ell_{12}(s) \end{bmatrix} = -\begin{bmatrix} \varkappa W_1(s)e^{-s\nu_1} \\ W_2(s)W_1(s)e^{-s(\nu_2+\nu_1)} \end{bmatrix} \mu(s)\breve{a}_r(s), \\ \breve{m}(s) &= \breve{a}_l(s)\breve{D}_{W_3W_2X}(T,s,-\nu_3-\nu_2), \\ \bar{n}_1(s) &= \begin{bmatrix} \varkappa W_1(s)e^{-s\nu_1} \\ W_2(s)W_1(s)e^{-s(\nu_2+\nu_1)} \end{bmatrix} \nu(s)\breve{\beta}_{r0}(s)\breve{m}_1(s) + \begin{bmatrix} O_{\ell\ell} \\ W_2(s)e^{-s\nu_2} \end{bmatrix} X(s).\end{aligned} \tag{10.247}$$

We can directly verify that the image (10.246) formally coincides with the image $Z(s)$ (10.126), (10.127) for

$$\bar{L}(s) = \begin{bmatrix} \varkappa W_1(s)e^{-s\nu_1} \\ W_2(s)W_1(s)e^{-s(\nu_1+\nu_2)} \end{bmatrix}, \qquad \bar{K}(s) = \begin{bmatrix} 0_{\ell\ell} \\ W_2(s)e^{-s\nu_2} \end{bmatrix},$$
$$M(s) = W_3(s)W_2(s), \qquad N(s) = W_3(s)W_2(s)W_1(s). \tag{10.248}$$

Using realizations (10.192) for matrices (10.248), we easily conclude that the images (10.246), (10.247) coincide with images of the form (10.126) for the standard system

$$\begin{aligned}
\frac{dv(t)}{dt} &= Av(t) + B_1 x(t - \nu_3 - \nu_2) + B_2 u(t - \nu_3 - \nu_2 - \nu_1), \\
y(t) &= C_2 v(t), \\
\bar{z}(t) &= \bar{C}_1 v(t) + \bar{D} u(t - \nu_3 - \nu_2 - \nu_1), \\
\xi_k &= y(kT), \quad (k = 0, \pm 1, \ldots), \\
\alpha(\zeta)\psi_k &= \beta(\zeta)\xi_k, \\
u(t) &= \sum_{i=0}^{P} h_i(t - kT)\psi_{k-i}, \quad kT < t < (k+1)T.
\end{aligned} \tag{10.249}$$

Since the $\mathscr{L}_2$ norm is uniquely determined by the form of the process image at zero initial energy, the $\mathscr{L}_2$ optimal system function for the system $\mathscr{S}_0$ coincides with the $\mathscr{L}_2$ optimal system function for the standard system (10.249). Therefore, the solution of the $\mathscr{L}_2$ optimization problem for the system $\mathscr{S}_0$ can be obtained with the help of the results from Sect. 10.4. Hereby, all specifics of the solution of the $\mathscr{L}_2$ optimization problem are preserved, which are stated in Sects. 10.4 and 9.7.

# Appendix A
# Choice of the Sampling Period for Modal Control of Sampled-Data Systems with Generalized Hold and Control Delay

## A.1 System Description and Problem

(1) Consider the single-loop sampled-data system $\mathscr{S}_\tau$ shown in Fig. A.1.

In Fig. A.1, the quantity $W(s)$ is the transfer function of the continuous process

$$W(s) = \frac{m(s)}{d(s)}, \tag{A.1}$$

where

$$\begin{aligned} m(s) &= m_1 s^{n-1} + m_2 s^{n-2} + \cdots + m_n, \\ d(s) &= s^n + d_1 s^{n-1} + \cdots + d_n \end{aligned} \tag{A.2}$$

are coprime real polynomials, and $\tau$ is a real constant satisfying

$$\tau = rT + \theta = (r+1)T - \gamma, \tag{A.3}$$

where $0 \leq \theta < T, \quad 0 < \gamma = T - \theta \leq T$, and $T$ is the sampling period. Hereby, $r \geq 0$ is an integer. The transfer function of the process $W(s)$ can be presented with the help of the minimal realization

$$W(s) = C(sI_n - A)^{-1} B, \tag{A.4}$$

where $A$ is a real $n \times n$ matrix, and $B,\ C$ are real vectors of size $n \times 1$ and $1 \times n$, respectively. Hereby, the pair $(A, B)$ is controllable and the pair $(A, C)$ is observable. Moreover, $I_n$ is the $n \times n$ identity matrix. Representation (A.4) corresponds to the state-space equations



E. N. Rosenwasser et al., *Computer-Controlled Systems with Delay*,
https://doi.org/10.1007/978-3-030-15042-6

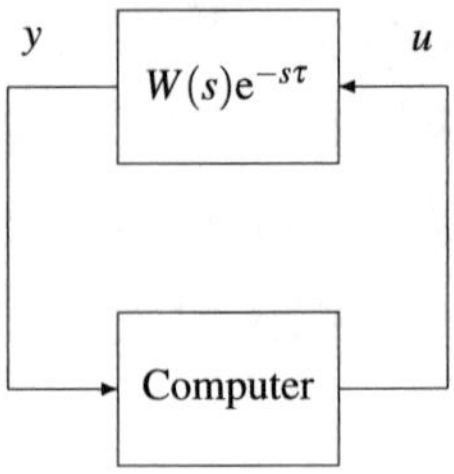


**Fig. A.1** Structure of SD system $\mathscr{S}_\tau$

$$\begin{aligned}\frac{dv(t)}{dt} &= Av(t) + Bu(t-\tau),\\ y(t) &= Cv(t),\end{aligned} \tag{A.5}$$

which are used in the following.

(2) Moreover, in Fig. A.1 the computer is described by the equations

$$\xi_k = y(kT), \quad (k = 0, \pm 1, \ldots), \tag{A.6}$$

$$\alpha_0\psi_k + \alpha_1\psi_{k-1} + \cdots + \alpha_\ell\psi_{k-\ell} = \beta_0\xi_k + \beta_1\xi_{k-1} + \cdots + \beta_\ell\xi_{k-\ell}, \tag{A.7}$$

$$u(t) = \sum_{i=0}^{R} h_i(t-kT)\psi_{k-i}, \quad kT < t < (k+1)T. \tag{A.8}$$

Equation (A.6) describes the analog-to-digital converter (ADC), and (A.7) the control program. Below it is always assumed that

$$\alpha_0 \neq 0, \tag{A.9}$$

which is called causality condition of the program. If we introduce the shift operator $g$ for one step back, according to

$$g\xi_k = \xi_{k-1}, \quad g\psi_k = \psi_{k-1}, \tag{A.10}$$

then the control program can be written in polynomial form

$$\alpha(g)\psi_k = \beta(g)\xi_k, \tag{A.11}$$

where

$$\begin{aligned}\alpha(g) &= \alpha_0 + g\alpha_1 + \cdots + g^\ell\alpha_\ell,\\ \beta(g) &= \beta_0 + g\beta_1 + \cdots + g^\ell\beta_\ell\end{aligned}$$

are real polynomials. In the following considerations, the operator $g$ is identified with the complex variable $\zeta$. Hereby, the polynomial pair $(\alpha(\zeta), \beta(\zeta))$ is called discrete controller.

Equation (A.8) describes a generalized digital-to-analog converter (GDAC) of order $R$, or shortly an R-order hold (ROH). Herein, the $h_i(t)$ are real functions of bounded variation on the interval $0 \leq t \leq T$.

Under assumption (A.3), from (A.8) on basis of the results in Chap. 4, we find

$$u(t-\tau) = \begin{cases} \sum_{i=0}^{R} h_i(t-kT+\gamma)\psi_{k-r-i-1}, & kT < t < kT+\theta, \\ \sum_{i=0}^{R} h_i(t-kT-\theta)\psi_{k-r-i}, & kT+\theta < t < (k+1)T. \end{cases} \tag{A.12}$$

Overall, the introduced relations establish the system of differential–difference equations

$$\begin{aligned} \frac{\mathrm{d}v(t)}{\mathrm{d}t} &= Av(t) + Bu(t-\tau), \\ y(t) &= Cv(t), \\ \xi_k &= y(kT), \quad (k = 0, \pm 1, \ldots), \\ \alpha(\zeta)\psi_k &= \beta(\zeta)\xi_k, \\ u(t-\tau) &= \begin{cases} \sum_{i=0}^{R} h_i(t-kT+\gamma)\psi_{k-r-i-1}, & kT < t < kT+\theta, \\ \sum_{i=0}^{R} h_i(t-kT-\theta)\psi_{k-r-i}, & kT+\theta < t < (k+1)T. \end{cases} \end{aligned} \tag{A.13}$$

(3) Denote

$$v(kT) = v_k, \tag{A.14}$$

then from the first equation in (A.13), we find

$$v(t) = \mathrm{e}^{A(t-kT)}v_k + \int_{kT}^{t} \mathrm{e}^{A(t-\nu)}u(\nu-\tau)\,\mathrm{d}\nu B. \tag{A.15}$$

For $t = (k+1)T$, we obtain

$$v_{k+1} = \mathrm{e}^{AT}v_k + \int_{kT}^{(k+1)T} \mathrm{e}^{A(kT+T-\nu)}u(\nu-\tau)\,\mathrm{d}\nu B. \tag{A.16}$$

Then, applying (A.12) yields

$$\begin{aligned} v_{k+1} = \mathrm{e}^{AT}v_k &+ \int_{kT}^{kT+\theta} \mathrm{e}^{A(kT+T-\nu)} \sum_{i=0}^{R} h_i(\nu-kT+\gamma)\psi_{k-r-i-1}B \\ &+ \int_{kT+\theta}^{(k+1)T} \mathrm{e}^{A(kT+T-\nu)} \sum_{i=0}^{R} h_i(\nu-kT-\theta)\psi_{k-r-i}B. \end{aligned} \tag{A.17}$$

Substitute

$$\nu - kT + \gamma = \mu. \tag{A.18}$$

Then

$$\int_{kT}^{kT+\theta} \mathrm{e}^{A(kT+T-\nu)} h_i(\nu - kT + \gamma)\,\mathrm{d}\nu = \mathrm{e}^{A(T+\gamma)} \int_{\gamma}^{T} \mathrm{e}^{-A\mu} h_i(\mu)\,\mathrm{d}\mu. \tag{A.19}$$

In analogy,

$$\nu - kT - \theta = \mu, \tag{A.20}$$

leads to

$$\int_{m+\theta}^{(k+1)T} \mathrm{e}^{A(kT+T-\nu)} h_i(\nu - kT - \theta)\,\mathrm{d}\nu = \mathrm{e}^{A\gamma} \int_{0}^{\gamma} \mathrm{e}^{-A\mu} h_i(\mu)\,\mathrm{d}\mu. \tag{A.21}$$

With the help of (A.19) and (A.21), from (A.17), after substituting $k$ by $k-1$, we find

$$\begin{aligned} \upsilon_k = \mathrm{e}^{AT}\upsilon_{k-1} + \mathrm{e}^{A(T+\gamma)} \int_{\gamma}^{T} \mathrm{e}^{-A\mu} \sum_{i=0}^{R} h_i(\mu)\,\mathrm{d}\mu\psi_{n-r-i-2}B \\ + \mathrm{e}^{A\gamma} \int_{0}^{\gamma} \mathrm{e}^{-A\mu} \sum_{i=0}^{R} h_i(\mu)\psi_{n-r-i-1}B, \end{aligned} \tag{A.22}$$

which, when using operator $\zeta$, can be written in the form

$$a(\zeta)\upsilon_k - (b_1(\zeta) + b_2(\zeta))B\psi_k = 0_{n,1}, \tag{A.23}$$

where $a(\zeta)$, $b_1(\zeta)$, $b_2(\zeta)$ are polynomial matrices of the form

$$\begin{aligned} a(\zeta) &= I_n - \zeta\mathrm{e}^{AT}, \\ b_1(\zeta) &= \mathrm{e}^{A(T+\gamma)}\zeta^{r+2} \sum_{i=0}^{R} \zeta^i \int_{\gamma}^{T} \mathrm{e}^{-A\mu} h_i(\mu)\,\mathrm{d}\mu, \\ b_2(\zeta) &= \mathrm{e}^{A\gamma}\zeta^{r+1} \sum_{i=0}^{R} \zeta^i \int_{0}^{\gamma} \mathrm{e}^{-A\mu} h_i(\mu)\,\mathrm{d}\mu. \end{aligned} \tag{A.24}$$

In (A.23) and further on, $0_{p,q}$ is the $p \times q$ zero matrix. Adding to (A.23) the remaining Eq. (A.13), we achieve the system of difference equations

$$\begin{aligned} a(\zeta)v_k - (b_1(\zeta) + b_2(\zeta))B\psi_k &= 0_{n,1}, \\ -Cv_k + \xi_k &= 0, \\ -\beta(\zeta)\xi_k + \alpha(\zeta)\psi_k &= 0. \end{aligned} \tag{A.25}$$

(4) As follows from the results of Chap. 4, when causality (A.9) is present and external disturbances are absent, then the properties of the system $\mathscr{S}_\tau$ are determined by its poles, which are the eigenvalues of the polynomial matrix

$$Q(\zeta, \alpha, \beta) = \begin{bmatrix} a(\zeta) & 0_{n,1} & -(b_1(\zeta) + b_2(\zeta))B \\ -C & 1 & 0 \\ 0 & -\beta(\zeta) & \alpha(\zeta) \end{bmatrix}. \tag{A.26}$$

By construction, the poles of system $\mathscr{S}_\tau$ are the roots of the characteristic equation

$$\Delta_\tau(\zeta) = \det Q(\zeta, \alpha, \beta) = 0. \tag{A.27}$$

Hereby, in harmony with Chap. 4, the modal control problem leads to the solution of the DPE

$$\Delta_\tau(\zeta) = \det Q(\zeta, \alpha, \beta) = \det \begin{bmatrix} a(\zeta) & 0_{n,1} & -(b_1(\zeta) + b_2(\zeta))B \\ -C & 1 & 0 \\ 0 & -\beta(\zeta) & \alpha(\zeta) \end{bmatrix} \sim \Delta(\zeta) \tag{A.28}$$

depending on the polynomials $\alpha(\zeta)$ and $\beta(\zeta)$ for a given polynomial $\Delta(\zeta)$.

(5) For a given matrix $A$ and vectors $B, C$ in representation (A.4), and given functions $h_i(t)$ and known value of the delay $\tau$, the properties of DPE (A.28) depend on the choice of the sampling period $T$. Below, values of the sampling period for which Eq. (A.28) has a solution for any polynomial $\Delta(\zeta)$, i.e., the system $\mathscr{S}_\tau$ becomes modal controllable, are called non-pathological. Values of the sampling period, for which the system $\mathscr{S}_\tau$ does not become modal controllable, following [1, 2], are called pathological.

The present appendix provides necessary and sufficient conditions for the sampling period to become non-pathological. Moreover, a method for constructing the set of pathological sampling periods is also presented, together with a number of general properties of the system $\mathscr{S}_\tau$ for the case when the sampling period is pathological.

## A.2 Irreducibility Criterion for a Pair of Polynomial Matrices

(1) In the following considerations, such as in the main part of the book, denote by $\mathbb{R}^{nm}$ and $\mathbb{C}^{nm}$ the set of constant $n \times m$ matrices with real and complex elements, respectively. Further on, $\mathbb{R}^{nm}[\lambda]$ describes the set of $n \times m$ real polynomial matrices.

The matrix $A(\lambda) \in \mathbb{R}^{nn}[\lambda]$, as earlier, is called nonsingular, if $\det A(\lambda) \neq 0$. Let the matrix $A(\lambda) \in \mathbb{R}^{nn}[\lambda]$ be nonsingular, and $\det A(\lambda)$ not be constant. Moreover, let the different roots of the characteristic equation

$$\Delta_A(\lambda) = \det A(\lambda) = 0 \tag{A.29}$$

be the numbers

$$\lambda_1, \ldots, \lambda_\rho \tag{A.30}$$

with multiplicity $\nu_1, \ldots, \nu_\rho$, respectively. Numbers (A.30) are called eigenvalues of the matrix $A(\lambda)$. By definition of the eigenvalues, we obtain

$$\operatorname{rank} A(\lambda_i) = r_i < n, \quad (i = 1, \ldots, \rho). \tag{A.31}$$

Equation (A.31) implies the existence of vectors $\ell_i \in C^{1,n}$, $(i = 1, \ldots, \rho)$ satisfying

$$\ell_i A(\lambda_i) = 0_{1,n}, \quad (i = 1, \ldots, \rho). \tag{A.32}$$

The vectors $\ell_i$ arising in (A.32), are called (left) eigenvectors of the matrix $A(\lambda)$, according to the eigenvalue $\lambda_i$. Further on, the set of those vectors is denoted by $L_i$, $(i = 1, \ldots, \rho)$. As is known [3], the set $L_i$ is a linear space, and its dimension $\dim L_i$ is determined by

$$\dim L_i = n - \operatorname{rank} A(\lambda_i), \quad (i = 1, \ldots, \rho). \tag{A.33}$$

(2) Let the matrices $A(\lambda) \in \mathbb{R}^{nn}[\lambda]$ and $B(\lambda) \in \mathbb{R}^{nm}[\lambda]$ build the horizontal pair $(A(\lambda), B(\lambda))$, and the matrices $A(\lambda) \in \mathbb{R}^{nn}[\lambda]$, $C(\lambda) \in \mathbb{R}^{mn}[\lambda]$ build the vertical pair $(A(\lambda), C(\lambda)$. Obviously, if the pair $(A(\lambda), C(\lambda))$ is a vertical pair, the the transposed matrices $A'(\lambda)$, $C'(\lambda)$ build a horizontal pair. Therefore, below we consider only horizontal pairs, because analogue properties for vertical pairs can be stated by transposition.

Below, following Chap. 1, the pair $(A(\lambda), B(\lambda))$ is called nonsingular if the matrix $A(\lambda) \in \mathbb{R}^{nn}[\lambda]$ is nonsingular and the set of its eigenvalues is not empty. The pair $(A(\lambda), B(\lambda))$ can be configured to the extended matrix $N_{AB}(\lambda) \in \mathbb{R}^{n,n+m}[\lambda]$ according to

$$N_{AB}(\lambda) = \begin{bmatrix} A(\lambda) & B(\lambda) \end{bmatrix}. \tag{A.34}$$

As earlier, the pair $(A(\lambda), B(\lambda))$ is called irreducible from left, and the matrix $N_{AB}(\lambda)$ a latent, if for all $\lambda$

$$\operatorname{rank} N_{AB}(\lambda) = n. \tag{A.35}$$

However, if there exists a number $\lambda^*$ with

$$\operatorname{rank} N_{AB}(\lambda^*) < n, \tag{A.36}$$

then, as already mentioned in Chap. 1, the matrix $N_{AB}(\lambda)$ is called latent, and the number $\lambda^*$ is its latent number. The pair $(A(\lambda), B(\lambda))$ is called reducible from left, if the corresponding matrix $N_{AB}(\lambda)$ is latent. Hereby, the latent numbers of the matrix $N_{AB}(\lambda)$ are also called latent numbers of the pair $(A(\lambda), B(\lambda))$.

(3) The following theorem provides a criterion for the irreducibility of the pair $(A(\lambda), B(\lambda))$, which is not contained among the irreducibility criterions presented in Chap. 1.

**Theorem A.1** *For the irreducibility of the pair $(A(\lambda), B(\lambda))$ it is necessary and sufficient that for any eigenvalue $\lambda_i$ of the matrix $A(\lambda)$ from the series (A.30) and any eigenvector $\ell_i \in L_i$, $(i = 1, \ldots, \rho)$ the following condition holds*

$$\ell_i B(\lambda_i) \neq 0_{1,m}. \tag{A.37}$$

*Proof* The proof is provided in some steps.

(a) We show that any latent number $\lambda^*$ of the pair $(A(\lambda), B(\lambda))$ is an eigenvalue of the matrix $A(\lambda)$. Indeed, by definition of latent numbers, we obtain

$$\operatorname{rank} N_{AB}(\lambda^*) = \operatorname{rank} \left[ A(\lambda^*) \; B(\lambda^*) \right] < n. \tag{A.38}$$

Hence

$$\operatorname{rank} A(\lambda^*) < n, \tag{A.39}$$

because otherwise (A.38) does not hold. Formula (A.39) shows that $\lambda^* = \lambda_k$, where $\lambda_k$ is one of the numbers in series (A.30).

(b) From (A.38) for $\lambda^* = \lambda_k$, we find

$$\operatorname{rank} N_{AB}(\lambda_k) = \operatorname{rank} \left[ A(\lambda_k) \; B(\lambda_k) \right] < n. \tag{A.40}$$

Formulae (A.39), (A.40) imply the existence of a constant vector $\ell^*$, where

$$\ell^* N_{AB}(\lambda^*) = \ell^* N_{AB}(\lambda_k) = \ell^* \left[ A(\lambda_k) \; B(\lambda_k) \right] = 0_{1,n+m}. \tag{A.41}$$

Hence $\ell^*$ is an eigenvector of the matrix $A(\lambda)$ according to the eigenvalue $\lambda_k$. Indeed, Eq. (A.41) can be written in the form

$$\left[ \ell^* A(\lambda_k) \; \ell^* B(\lambda_k) \right] = 0_{1,n+m}. \tag{A.42}$$

Thus

$$\ell^* A(\lambda_k) = 0_{1,n}, \tag{A.43}$$

and $\ell^* \in L_k$.

(c) Considering (a) and (b), we are able to prove Theorem A.1. At first we show that under condition (A.37) the pair $(A(\lambda), B(\lambda))$ is irreducible. Indeed, if the number $\lambda$ is different from an eigenvalue of $A(\lambda)$, then $\operatorname{rank} A(\lambda) = n$ and

$$\operatorname{rank} N_{AB}(\lambda) = \operatorname{rank}\left[\, A(\lambda)\ B(\lambda) \,\right] = n. \tag{A.44}$$

However, if $\lambda = \lambda_k \in L_k$, then under condition (A.37) there does not exist a constant vector $\ell^*$, such that (A.42) is fulfilled. This means, that for all $k = 1, \ldots, \rho$

$$\operatorname{rank} N_{AB}(\lambda_k) = n. \tag{A.45}$$

Together, relations (A.44), (A.45) show that for all $\lambda$ Eq. (A.35) holds, and thus the pair $(A(\lambda), B(\lambda))$ is irreducible. The sufficiency of the conditions in Theorem A.1 is shown.

For the proof of the necessity of the conditions in Theorem A.1, it is enough to show that

$$\ell_i B(\lambda_i) = 0_{1,m} \tag{A.46}$$

for one value of $i$ from the series $i = 1, 2, \ldots, \rho$, implies the reducibility of the pair $(A(\lambda), B(\lambda))$. This claim follows from the fact that under condition (A.46)

$$\ell_i \left[\, A(\lambda_i)\ B(\lambda_i) \,\right] = \left[\, \ell_i A(\lambda_i)\ \ell_i B(\lambda_i) \,\right] = \left[\, 0_{nn}\ 0_{nm} \,\right]. \tag{A.47}$$

Hence, the pair $(A(\lambda), B(\lambda))$ is reducible.

(4) From the content of Chap. 1, it follows that, for a reducible pair $(A(\lambda), B(\lambda))$, there exists a set of non-singular matrices $D(\lambda) \in \mathbb{R}^{nn}[\lambda]$, such that

$$A(\lambda) = D(\lambda) A_1(\lambda), \quad B(\lambda) = D(\lambda) B_1(\lambda), \tag{A.48}$$

where $A_1(\lambda) \in \mathbb{R}^{nn}[\lambda],\ B_1(\lambda) \in \mathbb{R}^{nm}[\lambda]$. The matrix $D(\lambda)$, arising in (A.48) is called a common divisor (CD) of the matrices $A(\lambda)$, $B(\lambda)$. Under (A.48), we obtain

$$N_{AB}(\lambda) = D(\lambda) \left[\, A_1(\lambda)\ B_1(\lambda) \,\right]. \tag{A.49}$$

A CD $D(\lambda)$ is called greatest common divisor (GCD) of the pair $(A(\lambda), B(\lambda))$, if the matrix

$$N_{A_1 B_1}(\lambda) = \left[\, A_1(\lambda)\ B_1(\lambda) \,\right] \tag{A.50}$$

is a latent. Below a GCD is denoted by $D^0(\lambda)$.

**Theorem A.2** *The set of latent numbers of the matrix $N_{AB}(\lambda)$ coincides with the set of eigenvalues of the GCD $D^0(\lambda)$.*

*Proof* By definition of the GCD, we obtain

$$N_{AB}(\lambda) = D^0(\lambda)\begin{bmatrix} A_1(\lambda) & B_1(\lambda) \end{bmatrix}, \tag{A.51}$$

where the matrix $\begin{bmatrix} A_1(\lambda) & B_1(\lambda) \end{bmatrix}$ is alatent. Hence, for all $\lambda$

$$\operatorname{rank}\begin{bmatrix} A_1(\lambda) & B_1(\lambda) \end{bmatrix} = n. \tag{A.52}$$

Let $\lambda^*$ be a latent number of the matrix $N_{AB}(\lambda)$, then

$$\operatorname{rank} N_{AB}(\lambda^*) < n. \tag{A.53}$$

Due to (A.51), (A.52) the last relation can only take place if

$$\operatorname{rank} D(\lambda^*) < n, \tag{A.54}$$

i.e., the number $\lambda^*$ is an eigenvalue of the GCD $D^0(\lambda)$. Conversely, if $\lambda^*$ is an eigenvalue of the GCD $D^0(\lambda)$, then (A.54) takes place, and from (A.51) we conclude that $\operatorname{rank} N_{AB}(\lambda^*) < n$, i.e., $\lambda^*$ is a latent number.

## A.3 Auxiliary Relations

(1) Assume the series of different eigenvalues of the matrix $A(\lambda) \in \mathbb{R}^{nn}[\lambda]$ in the form (A.30), where

$$\operatorname{rank} A(\lambda_i) = n - 1, \quad (i = 1, \ldots, \rho). \tag{A.55}$$

In this case the matrix $A(\lambda)$ is called simple. From (A.55) and (A.33), it follows that for a simple matrix $A(\lambda)$ the dimension of the set of eigenvectors $L_i$ satisfies

$$\dim L_i = 1, \quad (i = 1, \ldots, \rho). \tag{A.56}$$

From this equation, it follows that if $\ell_i^0$ is an eigenvector of the matrix $A(\lambda)$ according to the eigenvalue $\lambda_i$, then the complete set $L_i$ consists of vectors of the form

$$\ell_i = d\ell_i^0, \quad (i = 1, \ldots, \rho), \quad d = \text{const.} \neq 0. \tag{A.57}$$

Notice, that if the eigenvalue series of the simple matrix $A(\lambda)$ has the form (A.30) with multiplicity $\nu_1, \ldots, \nu_\rho$, then the series of elementary divisors of the matrix $A(\lambda)$ are

$$(\lambda - \lambda_1)^{\nu_1}, \ldots, (\lambda - \lambda_\rho)^{\nu_\rho}. \tag{A.58}$$

Let $U \in \mathbb{R}^{nn}$ be a constant matrix and

$$U_\lambda = \lambda I_n - U \tag{A.59}$$

its characteristic matrix. Then as eigenvalues of the polynomial matrix $U_\lambda$ we understand the eigenvalues of the matrix $U$, and the eigenvectors of the matrix $U_\lambda$ are the eigenvectors of the matrix $U$. Hereby, the matrix $U$ is called cyclic, if the matrix $U_\lambda$ is simple. This means that for a cyclic matrix $U$ the series of eigenvalues satisfies relations (A.30), (A.58), and we obtain

$$\operatorname{rank} U_\lambda = \operatorname{rank}(\lambda_i I_n - U) = n - 1, \quad (i = 1, \dots, \rho). \tag{A.60}$$

Notice, that for an eigenvector $\ell_i \in L_i$ of the matrix $U_\lambda$, from relation

$$\ell_i U_\lambda = 0_{1,n} \tag{A.61}$$

it always follows that

$$\ell_i U = \lambda_i \ell_i. \tag{A.62}$$

(2)

**Theorem A.3** *The following claims hold.*

*(i) Matrix A in representation (A.4) is cyclic.*
*(ii) For any eigenvector $\ell_k$ of the matrix A we have*

$$\ell_k B \neq 0, \quad (k = 1, \dots, \rho), \tag{A.63}$$

*where the vector B is configured in representation (A.4), and $\rho$ is the number of different eigenvalues of matrix A.*
*(iii) The equation*

$$\det(sI_n - A) = d(s) \tag{A.64}$$

*holds, where $d(s)$ is the numerator of the transfer function $W(s)$, configured in (A.2).*

*Proof* (i) By definition, the pair $A$, $B$ in representation (A.4) is controllable. Due to the PBH test [3], it follows immediately that the pair $(\lambda I_n - A, B)$ is irreducible, i.e., for all $\lambda$

$$\operatorname{rank}\left[\, \lambda I_n - A \ \ B \,\right] = n. \tag{A.65}$$

In particular, if the series of various eigenvalues of the matrix $A$ has form (A.30), then we obtain

$$\operatorname{rank}\left[\, \lambda_i I_n - A \ \ B \,\right] = n, \quad (i = 1, \dots, \rho). \tag{A.66}$$

If the matrix $A$ would be not cyclic, then there would exist a value $i_0$, $(1 \le i_0 \le \rho)$, such that

$$\operatorname{rank}(\lambda_{i0} I_n - A) < n - 1 \tag{A.67}$$

However, this inequality implies

$$\operatorname{rank}\left[\,\lambda_{i0}I_n - A\ B\,\right] < n, \tag{A.68}$$

because $\operatorname{rank} B = 1$. Thus, for all $1 \le i \le \rho$, we obtain $\operatorname{rank}(\lambda_i I_n - A) = n - 1$.

(ii) Let $\lambda_k$ be an eigenvalue of the matrix $A$, which due to i) is cyclic, and $\ell_k$ should be an eigenvector for $\lambda_k$. Then

$$\ell_k\left[\,\lambda_k I_n - A\ B\,\right] = \left[\,0_{nn}\ \ell_k B\,\right]. \tag{A.69}$$

If we assume that $\ell_k B = 0$, then

$$\ell_k\left[\,\lambda_k I_n - A\ B\,\right] = 0_{1,n+1}$$

and the number $\lambda_k$ turns out to be a latent number of the matrix $\left[\,\lambda I_n - A\ B\,\right]$. Hence, the polynomial pair $(\lambda I_n - A, B)$ is reducible, and on owing to the theorem of Hautus, the pair $(A, B)$ is not controllable, which contradicts the primary assumption. Thus (A.63) is true.

(iii) Formula (A.3) can be written in the form

$$W(s) = \frac{CA_{\Pi}(s)B}{\det(sI_n - A)}, \tag{A.70}$$

where $A_{\Pi}(s)$ is the corresponding adjoint matrix. Since the matrix $A$ is cyclic, the pair $A, B$ is controllable and the pair $A, C$ observable, so according to results in [4], fraction (A.70) is irreducible. Hence

$$\det(sI_n - A) = Kd(s), \tag{A.71}$$

where $K \ne 0$ is a constant. Using that the coefficient of $s^n$ in $d(s)$ and in $\det(sI_n - A)$ coincide, we obtain formula (A.64). ■

(3) Let $A$ be a cyclic matrix, and $T$ a real constant. Then, if the series of eigenvalues of the matrix $A$ has form (A.30), so the series of eigenvalues of the matrix $\mathrm{e}^{AT}$ has the form [3, 5]

$$\mathrm{e}^{\lambda_1 T}, \ldots, \mathrm{e}^{\lambda_\rho T}. \tag{A.72}$$

**Theorem A.4** *Let all different eigenvalues of the matrix A written in series (A.30) satisfy the relations*

$$\mathrm{e}^{\lambda_i T} \ne \mathrm{e}^{\lambda_k T}, \quad (i \ne k;\quad i, k = 1, \ldots, \rho). \tag{A.73}$$

*Then,*

*(i) The matrix* $e^{AT}$ *is cyclic.*
*(ii) The set of eigenvectors of matrix* $e^{AT}$*, according to eigenvalues (A.72) coincides with the set* $L_i$ *of eigenvectors of the matrix A according to eigenvalues (A.30). Moreover, when* $\ell_i$ *is an eigenvector of the matrix A according to the eigenvalue* $\lambda_i$*, then*

$$\ell_i e^{AT} = e^{\lambda_i T} \ell_i. \tag{A.74}$$

*(iii) Polynomial matrix*

$$a(\zeta) = I - \zeta e^{AT} \in \mathbb{R}^{nn}[\zeta] \tag{A.75}$$

*is simple.*

*Proof* (i) As was shown in [6], when condition (A.73) holds, then the pair $e^{AT}$, $B$ is controllable. Therefore, repeating the arguments for the proof of claim (i) in Theorem A.3, we find out that $e^{AT}$ is cyclic.

(ii) As was shown in [2], any eigenvector $\ell_i$ of the matrix $A$ for the eigenvalue $\lambda_i$, is an eigenvector of the matrix $e^{AT}$ for the eigenvalue $e^{\lambda_i T}$. This follows from the succession of equations

$$\begin{aligned}\ell_i e^{AT} &= \ell_i \left(I_n + \frac{AT}{1!} + \frac{A^2T^2}{2!} + \cdots\right) = \ell_i + \frac{\ell_i \lambda_i T}{1!} + \frac{\ell_i \lambda_i^2 T^2}{2!} + \cdots \\ &= \ell_i \left(1 + \frac{\lambda_i T}{1!} + \frac{\lambda_i^2 T^2}{2!} + \cdots\right) = e^{\lambda_i T} \ell_i. \end{aligned} \tag{A.76}$$

If condition (A.73) holds, then due to (i) the matrix $e^{AT}$ is cyclic and the set of eigenvectors of this matrix, according to eigenvalues (A.72), has the dimension one. Therefore, the sets of eigenvectors of matrix $e^{AT}$, according to the eigenvalues (A.72), coincide with the sets $L_i$.

(iii) Because under condition (A.73) matrix $e^{AT}$ is cyclic, so

$$\operatorname{rank}(e^{\lambda_i T} I_n - e^{AT}) = n - 1, \quad (i = 1, \dots, \rho), \tag{A.77}$$

and, thus

$$\operatorname{rank}(I_n - e^{-\lambda_i T} e^{AT}) = n - 1, \quad (i = 1, \dots, \rho). \tag{A.78}$$

Since the number $\zeta_i = e^{-\lambda_i T}$ is an eigenvalue of matrix $a(\zeta)$ (A.75), so from (A.78) we conclude that the matrix $a(\zeta)$ is simple. ■

## A.4 Choosing Sampling Period for Modal Controllability and Stabilizability of System $\mathcal{S}_\tau$

(1) The next theorem provides necessary and sufficient condition for the modal controllability of system $\mathcal{S}_\tau$, or equivalently, necessary and sufficient conditions for non-pathological sampling periods.

**Theorem A.5** *For the sampling period $T$ to be non-pathological for the system $\mathcal{S}_\tau$, the following conditions are necessary and sufficient:*

*(i) Condition (A.73) holds.*
*(ii) Eigenvalues (A.30) satisfy*

$$H(T, \lambda_i) \neq 0, \quad (i = 1, \ldots, \rho), \tag{A.79}$$

*where $H(T, \lambda)$ is the equivalent transfer function of higher order hold (A.8), defined by*

$$H(T, \lambda) = \sum_{i=0}^{R} \mathrm{e}^{-i\lambda T} \int_0^T \mathrm{e}^{-\lambda T} h_i(t)\, \mathrm{d}t. \tag{A.80}$$

*Proof* *Sufficiency*. As follows from Chap. 3, for the solvability of DPE (A.28) for an arbitrary polynomial $\Delta(\zeta)$ according to the polynomials $\alpha(\zeta)$, $\beta(\zeta)$, it is necessary and sufficient, that the polynomial pairs $(a(\zeta), (b_1(\zeta) + b_2(\zeta))B)$, and $[a(\zeta), C]$ are irreducible. Therefore, for the necessity of conditions (i) and (ii) we have to show, that under conditions (A.73) and (A.79) the relevant polynomial pairs are irreducible. For that reason, at first notice, that by construction the pair $A, C$ is observable, i.e.,

$$\operatorname{rank}\left[\mathrm{e}^{\lambda_i T} I_n - \mathrm{e}^{A' T}, C'\right] = n, \quad (i = 1, 2, \ldots, \rho). \tag{A.81}$$

Hence

$$\operatorname{rank}\left[\, I_n - \mathrm{e}^{-\lambda_i T} \mathrm{e}^{A' T}\; C' \,\right] = \operatorname{rank}\left[\, a'(\mathrm{e}^{-\lambda_i T}),\, C' \,\right] = n, \quad (i = 1, 2, \ldots, \rho), \tag{A.82}$$

which implies that the pair $(a(\zeta), C)$ is irreducible. Now consider the pair $(a(\zeta), b_1(\zeta) + b_2(\zeta)B)$, where the matrices $a(\zeta), b_1(\zeta)$ and $b_2(\zeta)$ are given by formulae (A.24). Notice first, that the set of eigenvectors of the matrix $a(\zeta) = I_n - \zeta \mathrm{e}^{AT}$ coincides with the set eigenvectors of the matrices $A$ and $\mathrm{e}^{AT}$. Indeed, owing to the above shown, the set of eigenvectors of the matrices $A$ and $\mathrm{e}^{AT}$ coincide, when relation (A.73) is true. Moreover, when $\ell_i$ is an eigenvector of the matrix $A$, then

$$\ell_i a(\mathrm{e}^{-\lambda_i T}) = \ell_i (I - \mathrm{e}^{-\lambda_i T} \mathrm{e}^{AT}) = \ell_i - \mathrm{e}^{-\lambda_i T} \ell_i \mathrm{e}^{AT} = 0_{1,n}. \tag{A.83}$$

Therefore, $\ell_i$ is an eigenvector of the matrix $a(\zeta)$, according to the eigenvalue $\zeta_i = \mathrm{e}^{-\lambda_i T}$. Now, using that the matrix $a(\zeta)$ is simple, then the sets of eigenvectors of the matrices $A$, $\mathrm{e}^{AT}$ and $a(\zeta) = I_n - \zeta \mathrm{e}^{AT}$ coincide. Taking (A.24), consider the extended matrix

$$N_{a,b_1+b_2}(\zeta) = \left[\, I_n - \zeta e^{AT} \;\; [b_1(\zeta) + b_2(\zeta)]B \,\right]. \tag{A.84}$$

Inserting herein the expression

$$\zeta = \zeta_k = \mathrm{e}^{-\lambda_k T}, \quad (k = 1, 2, \ldots, \rho), \tag{A.85}$$

where the number $\zeta_k$ is an eigenvalue of the matrix $a(\zeta)$, we find

$$N_{a,b_1+b_2}(\zeta_k) = \left[\, I_n - \mathrm{e}^{-\lambda_k T}\mathrm{e}^{AT} \; \left[b_1(\mathrm{e}^{-\lambda_k T}) + b_2(\mathrm{e}^{-\lambda_k T})\right] B \,\right]. \tag{A.86}$$

From (A.24) after inserting (A.85), we obtain

$$\begin{aligned} b_1(\mathrm{e}^{-\lambda_k T}) &= \mathrm{e}^{A\gamma}\mathrm{e}^{AT}\mathrm{e}^{-(r+2)\lambda_k T}\sum_{i=0}^{r}\mathrm{e}^{-i\lambda_k T}\int_{\gamma}^{T}\mathrm{e}^{-A\mu}h_i(\mu)\,\mathrm{d}\mu, \\ b_2(\mathrm{e}^{-\lambda_k T}) &= \mathrm{e}^{A\gamma}\mathrm{e}^{-(r+1)\lambda_k T}\sum_{i=0}^{r}\mathrm{e}^{-i\lambda_k T}\int_{0}^{\gamma}\mathrm{e}^{A\mu}h_i(\mu)\,\mathrm{d}\mu. \end{aligned} \tag{A.87}$$

Multiplying expression (A.87) by the eigenvector $\ell_k$, and noticing

$$\ell_k\mathrm{e}^{A\gamma} = \mathrm{e}^{\lambda_k\gamma}\ell_k, \quad \ell_k\mathrm{e}^{AT} = \mathrm{e}^{\lambda_k T}\ell_k, \quad \ell_k\mathrm{e}^{-A\mu} = \mathrm{e}^{-\lambda_k\mu}\ell_k, \tag{A.88}$$

we obtain

$$\begin{aligned} \ell_k b_1(\mathrm{e}^{-\lambda_k T}) &= \mathrm{e}^{\lambda_k\gamma}\mathrm{e}^{-(r+1)\lambda_k T}\sum_{i=0}^{r}\mathrm{e}^{-i\lambda_k T}\int_{\gamma}^{T}\mathrm{e}^{-\lambda_k\mu}h_i(\mu)\,\mathrm{d}\mu\;\ell_k B, \\ \ell_k b_2(\mathrm{e}^{-\lambda_k T}) &= \mathrm{e}^{\lambda_k\gamma}\mathrm{e}^{-(r+1)\lambda_k T}\sum_{i=0}^{r}\mathrm{e}^{-i\lambda_k T}\int_{0}^{\gamma}\mathrm{e}^{-\lambda_k\mu}h_i(\mu)\,\mathrm{d}\mu\;\ell_k B. \end{aligned} \tag{A.89}$$

From (A.89), we immediately derive

$$\ell_k(b_1(\zeta_k) + b_2(\zeta_k)) = \mathrm{e}^{\lambda_k\gamma}\mathrm{e}^{-(r+1)\lambda_k T}H(T, \lambda_k), \tag{A.90}$$

where $H(T, \lambda)$ is function (A.80). Multiplying (A.84) by the eigenvector $\ell_k$, we obtain with the help of (A.90)

$$\ell_k N_{a,b_1+b_2}(\zeta_k) = \left[\, 0_{1,n} \;\; \mathrm{e}^{\lambda_k\gamma}\mathrm{e}^{-(r+1)\lambda_k T}H(T, \lambda_k)\ell_k B \,\right] \tag{A.91}$$

where

$$H(T, \lambda_k) = \sum_{i=0}^{r} \mathrm{e}^{-i\lambda_k T} \int_0^T \mathrm{e}^{-\lambda_k \mu} h_i \mu \, \mathrm{d}\mu \; \ell_k B. \tag{A.92}$$

On basis of earlier shown

$$\ell_k B \neq 0, \quad (k = 1, 2, \ldots, \rho). \tag{A.93}$$

Moreover, by proposition

$$H(T, \lambda_k) \neq 0, \quad (k = 1, 2, \ldots, \rho). \tag{A.94}$$

Therefore, Theorem A.1 yields that the pair $(a(\zeta), [b_1(\zeta) + b_2(\zeta)]B)$ is irreducible. This fact completes the proof of sufficiency.

*Necessity*. For the proof of necessity, we show that when one of the conditions (i) or (ii) is violated, the pair $(I - \zeta \mathrm{e}^{AT}, [b_1(\zeta) + b_2(\zeta)]B)$ is reducible. Assume at first that condition (i) does not hold and for two conjugate complex numbers $\lambda_\phi$ and $\lambda_\psi$ of multiplicity $\nu$ we have

$$\lambda_\phi = u + \mathrm{j}\omega_0, \quad \lambda_\psi = u - \mathrm{j}\omega_0, \tag{A.95}$$

and the condition

$$\mathrm{e}^{\lambda_\phi T} = \mathrm{e}^{\lambda_\psi T}, \tag{A.96}$$

holds, or equivalently

$$T(\lambda_\phi - \lambda_\psi) = 2d\pi \mathrm{j}, \quad \mathrm{j} = \sqrt{-1}, \tag{A.97}$$

where $d \neq 0$ is an integer. Consequently,

$$T = T_d = \frac{d\pi}{\omega_0} \tag{A.98}$$

and

$$\mathrm{e}^{\lambda_\phi T} = \mathrm{e}^{\lambda_\psi T} = (-1)^d \mathrm{e}^{\frac{du\pi}{\omega_0}} = z_0. \tag{A.99}$$

Since the matrix $A$ is cyclic, so eigenvalues (A.95) according to the elementary divisors are

$$(\lambda - \lambda_\phi)^\nu, \quad (\lambda - \lambda_\psi)^\nu. \tag{A.100}$$

Hereby, as follows from [5], the matrix $zI_n - \mathrm{e}^{AT}$ has the real eigenvalue $z_0 \neq 0$ of multiplicity $2\nu$, which corresponds to two elementary divisors

$$(z - z_0)^\nu, (z - z_0)^\nu. \tag{A.101}$$

As follows from Chap. 1, when (A.101) is fulfilled, then the reciprocal matrix $a(\zeta) = I_n - \zeta \mathrm{e}^{AT}$ has the real eigenvalue

$$\zeta_0 = z_0^{-1} = (-1)^d \mathrm{e}^{-\frac{d\nu\pi}{\omega_0}}. \tag{A.102}$$

with multiplicity $2\nu$, for which two elementary divisors are configured

$$(\zeta - \zeta_0)^\nu, \quad (\zeta - \zeta_0)^\nu. \tag{A.103}$$

From (A.103), we conclude

$$\operatorname{rank} a(\zeta_0) = \operatorname{rank}(I_n - \zeta_0 \mathrm{e}^{AT}) < n - 1, \tag{A.104}$$

and this fact means

$$\operatorname{rank}\left[a(\zeta_0)\ [b_1(\zeta_0) + b_2(\zeta_0)]B\right] < n, \tag{A.105}$$

because $\operatorname{rank} B = 1$. Relation (A.105) means, that under condition (A.96) the pair $(R(\zeta), [b_1(\zeta) + b_2(\zeta)]B)$ is reducible, i.e., the necessity of condition (A.73) is shown.

Assume now, that condition (A.73) is true, but for the eigenvalue $\lambda_k$ of the matrix $A$ we obtain $H(T, \lambda_k) = 0$. Then from (A.91), we find

$$\ell_k N_{a,b_1+b_2}(\zeta_k) = \left[0_{1,n}\ 0_{1,m}\right], \tag{A.106}$$

what contradicts the irreducibility of the pair $(a(\zeta), [b_1(\zeta) + b_2(\zeta)]B)$. ■

(2) The just considered eigenvalues of matrix $A$, for which pathological periods exist and none of the conditions (i) or (ii) of Theorem A.5 hold, are called critical.

Poles of the system $\mathscr{S}_T$, corresponding to critical eigenvalues of matrix $A$ and pathological sampling periods, are called pathological. From the proof of Theorem A.5, it follows that all critical eigenvalues can be partitioned into two types. To the first type, we can count all complex eigenvalues of matrix $A$. Each pair of such eigenvalues (A.95) generates a set of pathological sampling periods, defined by formula (A.98). These periods do not depend on the vectors $B$ and $C$ in representation (A.4). Pathological poles of system $\mathscr{S}_\tau$ corresponding to eigenvalues of the first type are always real. Critical eigenvalues of the second type are the roots of equation $H(T, \lambda) = 0$. In general, they can be real as well as complex.

(3) The following theorem determines general properties of system $\mathscr{S}_\tau$ in the case, when the sampling period is pathological.

**Theorem A.6** *Let the pathological sampling period $T^*$ exist as a set of critical eigenvalues $\lambda_1^*, \ldots, \lambda_\eta^*$ of the matrix A, building the set $\mathscr{M}^*$. Then, independent on the selection of the discrete controller $(\alpha(\zeta), \beta(\zeta))$, the characteristic polynomial $\Delta_\tau(\zeta)$ of system $\mathscr{S}_\tau$ satisfies condition*

$$\Delta_\tau(\zeta) = \det \begin{bmatrix} a(\zeta) & 0_{n,1} & -[b_1(\zeta) + b_2(\zeta)]B \\ -C & 1 & 0 \\ 0 & -\beta(\zeta) & \alpha(\zeta) \end{bmatrix} = \Delta_1(\zeta)\Delta_2(\zeta), \qquad \text{(A.107)}$$

*where $\Delta_1(\zeta)$ is a polynomial, which roots are all numbers $\zeta_i^*$, determined by*

$$\zeta_1^* = e^{-\lambda_i^* T}, \quad (i = 1, \ldots, \eta). \qquad \text{(A.108)}$$

*Proof* From the proof of Theorem A.5, it follows that numbers (A.108) are contained in the set of latent numbers of matrix$N_{a,b_1+b_2}(\zeta)$. Therefore, due to Theorem A.2

$$N_{a,b_1+b_2}(\zeta) = f_1^*(\zeta)\left[a^*(\zeta)\ -b^*(\zeta)\right], \qquad \text{(A.109)}$$

where $f_1^*(\zeta) \in \mathbb{R}^{nn}[\zeta]$ is a matrix whose eigenvalues are numbers (A.108), and matrix

$$\left[a^*(\zeta)\ b^*(\zeta)\right] \in R^{n,n+1}[\zeta] \qquad \text{(A.110)}$$

is a latent. From (A.109), we directly derive

$$a(\zeta) = f_1^*(\zeta)a^*(\zeta), \quad -[b_1(\zeta) + b_2(\zeta)]B = -f_1^*(\zeta)b^*(\zeta). \qquad \text{(A.111)}$$

With the help of (A.111), from (A.26) we obtain

$$Q(\zeta, \alpha, \beta) = \operatorname{diag}\{f_1^*(\zeta), 1, 1\} \begin{bmatrix} a^*(\zeta) & 0_{n1} & -b^*(\zeta) \\ -C & 1 & 0 \\ 0 & -\beta(\zeta) & \alpha(\zeta) \end{bmatrix}. \qquad \text{(A.112)}$$

Since latent numbers of the matrix $(I_n - \zeta e^{A'T} - C')$ can be only eigenvalues satisfying condition (A.102), so analogue properties are true for the pair $[a^*(\zeta), -C]$. Hence

$$\begin{bmatrix} a^*(\zeta) \\ -C \end{bmatrix} = \begin{bmatrix} a_1^*(\zeta) \\ -c^*(\zeta) \end{bmatrix} f_2^*(\zeta), \qquad \text{(A.113)}$$

where $f_2^*(\zeta) \in \mathbb{R}^{nn}[\zeta]$ is a matrix, which eigenvalues are located in the number set (A.108). Using (A.113), from (A.112) we find

$$Q(\zeta, \alpha, \beta) = \operatorname{diag}\{fA_1^*(\zeta), 1, 1\} \begin{bmatrix} a_1^*(\zeta) & 0_{n,1} & -b^*(\zeta) \\ -c^*(\zeta) & 1 & 0 \\ 0 & -\beta(\zeta) & \alpha(\zeta) \end{bmatrix} \operatorname{diag}\{f_2^*(\zeta), 1, 1\}. \qquad \text{(A.114)}$$

Calculating the determinant on the left and right side of (A.114) yields formula (A.107). ■

(4) It follows from Theorem A.6, that in case of a pathological sampling period, the pathological poles, independently on the form of the discrete controller, arrive in

the set of roots of the characteristic equation of system $\mathscr{S}_\tau$. Further on, eigenvalues (A.30), for which $\text{Re}\lambda_i < 0$, ($\text{Re}\lambda_i \geq 0$) are called stable (unstable). Poles of the system $\mathscr{S}_\tau$ are called stable, if

$$|\zeta_i| > 1, \tag{A.115}$$

and unstable, if $|\zeta_i| \leq 1$. In Chap. 4 it was shown, that for the asymptotic stability of system $\mathscr{S}_\tau$ when (A.9) holds, it is necessary and sufficient, that all poles (roots of characteristic Eq. (A.27)) are stable. Therefore, Theorem A.6 implies, that for a given pathological sampling period, system $\mathscr{S}_\tau$ can be made asymptotically stable in the case and only in the case, when among the critical eigenvalues of matrix $A$, or what is equivalent, under the pathological poles of system $\mathscr{S}_\tau$, there are no unstable ones. This reason leads us to the following general statements.

(a) If the sampling period is non-pathological, then system $\mathscr{S}_\tau$ is modal controllable, and therefore stabilizable.
(b) If the sampling period is pathological, then system $\mathscr{S}_\tau$ is not modal controllable. Then for the stabilizability of system $\mathscr{S}_\tau$ it is necessary and sufficient, that no unstable eigenvalues are among the critical eigenvalues of matrix $A$.
(c) Any pair of conjugate complex eigenvalues of matrix $A$ (A.95) generates a series of pathological sampling periods (A.98). Moreover, if the set of eigenvalues of form (A.95) contains unstable ones, for chosen sampling period (A.98), system $\mathscr{S}_\tau$ is unstabilizable.
(d) Let for parameter $T$ equation

$$H(T, \lambda) = 0 \tag{A.116}$$

possess roots $\lambda_k^0(T)$, building the set $\mathscr{M}^0$. Then period $T^0$ becomes pathological, if for at least one value of $k$, we have $\lambda_k \in \mathscr{M}^0$. Moreover, if eigenvalue $\lambda_k$ is unstable, then for $T = T^0$, system $\mathscr{S}_\tau$ is non-stabilizable.
(e) The construction of the set of pathological sampling periods does not depend on the value of the delay $\tau$. This fact is a consequence of the circumstance that conditions (i) and (ii) in Theorem A.5 do not depend on the delay.

## A.5 Examples

*Example A.1* Consider system $\mathscr{S}_\tau$ with zero-order hold

$$u(t) = \psi_k, \quad kT < t < (k+1)T, \tag{A.117}$$

that corresponds to the case, when in (A.7) we apply $r = 0$ and $h_0(t) = 1$. Hence, function $H(T, \lambda)$ becomes

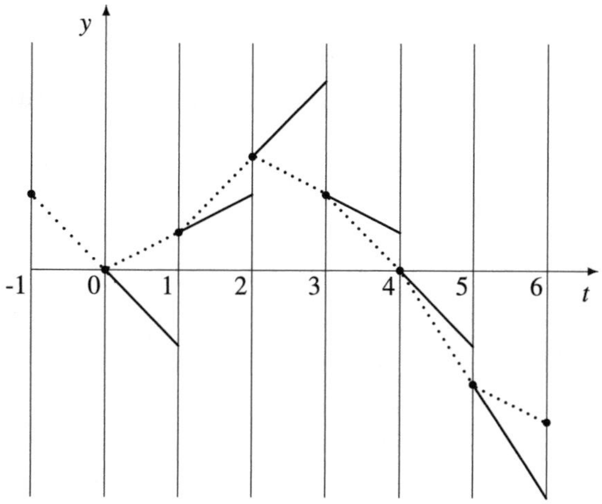


**Fig. A.2** Mode of operation for extrapolating first-order hold

$$H(T,\lambda)=\frac{1-\mathrm{e}^{-\lambda T}}{\lambda}. \tag{A.118}$$

As roots of function (A.118), we find the numbers

$$\lambda_k^0(T)=\frac{2k\pi j}{T}, \quad (k=\pm 1,\pm 2,\ldots). \tag{A.119}$$

If some of roots (A.119) are located in set of eigenvalues (A.30), then period $T$ is pathological, and for those values of the sampling period the system $\mathscr{S}_\tau$ is not stabilizable.

*Example A.2* Consider system $\mathscr{S}_\tau$ with GDAC of first order, where its function principle is shown in Fig. A.2. The equations of such GDAC has the form

$$u(t)=\psi_k+\frac{\psi_k-\psi_{k-1}}{T}(t-kT), \quad kT<t<(k+1). \tag{A.120}$$

This equation can be represented in form (A.8) with $r=1$,

$$h_0(t)=1+\frac{t}{T}, \quad h_1(t)=-\frac{t}{T} \tag{A.121}$$

and function $H(T,\lambda)$ (A.80) takes the form

$$H(T,\lambda)=\int_0^T \mathrm{e}^{-\lambda t}(1+\frac{t}{T})\,\mathrm{d}t-\frac{\mathrm{e}^{-\lambda T}}{T}\int_0^T \mathrm{e}^{-\lambda t}t\,\mathrm{d}t. \tag{A.122}$$

From this relation, after some elementary transformations, we find

$$H(T, \lambda) = \frac{(1 - e^{-\lambda T})^2 (T\lambda + 1)}{T\lambda^2}. \tag{A.123}$$

As roots of this function, we obtain numbers (A.119), and in addition the number

$$\lambda_0^0 = -\frac{1}{T}. \tag{A.124}$$

Hence, for system $\mathscr{S}_\tau$ with GDAC (A.120), in addition to the pathological periods according to the eigenvalues of matrix $A$, there exists an additional pathological period issued from the negative real eigenvalues of this matrix (if such kind exist). If $\bar{\lambda}_i$, $(i = 1, \ldots, \eta)$ are these eigenvalues, then the corresponding pathological sampling periods are determined by

$$\bar{T}_i = -\frac{1}{\bar{\lambda}_i}, \quad (i = 1, 2, \ldots, \eta). \tag{A.125}$$

The pathological periods $\bar{T}_i$ are completed independently on $i$ by the unique stable pathological pole of system $\mathscr{S}_\tau$

$$\zeta_0^0 = \mathrm{e}.$$

## References

1. K.J. Åström, B. Wittenmark, *Computer Controlled Systems: Theory and Design*, 3rd edn. (Prentice-Hall, Englewood Cliffs, 1997)
2. T. Chen, B.A. Francis, *Optimal Sampled-Data Control Systems* (Springer, Berlin, 1995)
3. T. Kailath, *Linear Systems* (Prentice Hall, Englewood Cliffs, NJ, 1980)
4. E.N. Rosenwasser, B.P. Lampe, *Algebraische Methoden zur Theorie der Mehrgrößen-Abtastsysteme* (Universitätsverlag, Rostock, 2000). ISBN 3-86009-195-6
5. F.R. Gantmacher, *The Theory of Matrices* (Chelsea, New York, 1959)
6. R.E. Kalman, Y.C. Ho, K. Narendra, Controllabiltiy of linear dynamical systems. Contrib. Theory Differ. Equ. **1**, 189–213 (1963)

# Appendix B
# *DIRSD*–A Toolbox for Direct Design of SD Systems

## B.1 Introduction

The *DIRSD* toolbox was first introduced in 1999 [1] as a MATLAB package to solve analysis and synthesis problems for sampled-data systems using the parametric transfer function approach. The toolbox has evolved through the years, with the most important updates being the development of the MIMO version DIRECTSDM [2] and the definitive version described in [3].

The latest version [3] was a mature project: albeit it focused on SISO systems, it included options to use either polynomial methods or the lifting technique [4] to solve design problems and a comprehensive set of examples and demos integrated with the MATLAB help explorer.

This appendix describes the current version of the toolbox, presenting some usage examples. The software can be downloaded from the site www.rt.uni-rostock.de/ccsd.

## B.2 Structure of the Toolbox

### *B.2.1 System Representation*

To have a common structure for the analysis and synthesis problems for sampled-data systems, the *DIRSD* toolbox uses the standard representation of a multi-input, multi-output sampled-data system shown in Fig. B.1.

Here $\mathbf{K}(s)$, $\mathbf{L}(s)$, $\mathbf{M}(s)$, and $\mathbf{N}(s)$ represent transfer matrices of continuous elements, $\mathbf{C}(\zeta)$ is a discrete controller and $\mathbf{H}(s)$ represents a hold device. The current version of the toolbox supports only SISO controllers, therefore $C(\zeta)$ and $H(s)$ must be scalar transfer functions, and $y(t)$, $\varepsilon_k$, $\nu_k$, and $u(t)$ scalar signals. The dimensions of the continuous elements should be compatible with these.



E. N. Rosenwasser et al., *Computer-Controlled Systems with Delay*,
https://doi.org/10.1007/978-3-030-15042-6

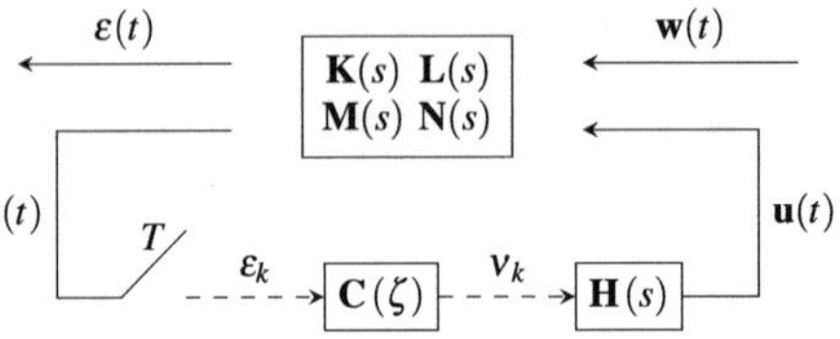


**Fig. B.1** Standard sampled-data system

## *B.2.2 Classes*

The *DIRSD* toolbox defines classes for the manipulation of polynomials and rational matrices. Polynomials are represented as objects of the class `poln`, whereas rational matrices are represented by means of objects derived from standard classes included in the control systems toolbox.

### B.2.2.1 Polynomial Objects: the `poln` class

This class is used to define polynomial objects. Each object of this class is represented by a coefficient vector and a variable used to indicate whether it is a polynomial in continuous time (variables `s` or `p`) or discrete time (variables `d`, `z` or `q`). Instead of the coefficient vector, the polynomial can also be created by providing its roots and gain, i.e., the coefficient of the term with highest degree.

The basic arithmetic operations have been implemented for objects of this class. Other useful methods provided by the toolbox are the following:

| | |
|---|---|
| `poln` | Constructor method for objects of the class `poln`. |
| `norm` | Norm of the coefficient vector of a polynomial. |
| `deg` | Highest and lowest degree of a polynomial. |
| `polyval` | Evaluates a polynomial. |
| `derive` | Finds the derivative of a polynomial. |
| `polyder` | Evaluates the derivative of a polynomial. |
| `comden` | Common denominator of two polynomials. |
| `coprime` | Removes common factors of two polynomial, making them coprime. |
| `mtimesn` | Product of a polynomial and a scalar. |
| `sfactor` | Stable factor of a Hermitian polynomial. |

### B.2.2.2 Rational Matrices: classes `sdzpk` and `stdf`

Previous versions of the *DIRSD* toolbox used the standard classes `zpk` and `tf`, included in the control systems toolbox of MATLAB, to represent rational matrices. In order to add extra functionality, some of the standard functions were overloaded. The change in the OOP model introduced with MATLAB version 7.3 (Release 2008a) made this approach obsolete. To recover the functionality whilst avoiding the need to rewrite much code, *DIRSD* 3.5 defines two subclasses, `sdzpk` and `stf`, which inherit all the methods of the original classes and for which the particular methods needed are redefined.

## *B.2.3 Functions*

*DIRSD* 3.5 includes a vast amount of functions used for the direct polynomial design of sampled-data systems. Some of the most commonly used are listed below, classified by its purpose.

**Polynomial Equations**

| | |
|---|---|
| `dioph` | Simple Diophantine polynomial equation. |
| `dioph2` | Solves a special polynomial equations of the form $XA + \tilde{X}B + YC = 0$. |
| `diophsys` | Solves a system of two Diophantine polynomial equations. |

**Discrete Transformations**

| | |
|---|---|
| `ztrm` | Modified Z-transform for a transfer matrix. |
| `dtfm` | Discrete Laplace transform with a hold. |
| `dtfm2` | Special symmetric discrete Laplace transform with a hold. |

**Analysis of Sampled-Data Systems**

| | |
|---|---|
| `charpol` | Characteristic polynomial of sampled-data system. |
| `sdmargin` | Stability margin for sampled-data system. |
| `sdl2err` | Integral quadratic error for sampled-data systems. |
| `sdh2norm` | H2-norm of sampled-data systems. |
| `sdhinorm` | Hinf-norm for sampled-data systems |

**Polynomial Design**

| | |
|---|---|
| `ch2` | H2-optimal continuous-time controller. |
| `polquad` | Polynomial minimization of a quadratic functional. |
| `whquad` | Wiener–Hopf minimization of frequency-domain quadratic functionals. |
| `sdh2` | H2-optimal controller for sampled-data systems. |
| `sdl2` | L2-optimal controller for sampled-data systems. |
| `polopth2` | Solution to polynomial H2-minimization problem. |

**Miscellaneous Functions**

| | |
|---|---|
| `bilintr` | Bilinear transformation for SISO system. |
| `improper` | Separate improper (polynomial) part of a rational matrix. |
| `sector` | Find degree of stability and oscillation for closed-loop poles. |

| | |
|---|---|
| `sumzpk` | Reliable summation of zpk models with common poles. |
| `sfactor` | Spectral factorization for polynomials and transfer functions. |
| `separss` | Proper separation (state-space technique). |
| `separtf` | Proper separation (polynomial equations technique). |

## B.3 Examples

In this section, we present two examples that can give the reader a feeling of the usage and capabilities of the toolbox. The examples described in the appendix of the monograph [5] and in the papers [1] and [6] can also be reproduced to obtain a wider picture of the functionality included in *DIRSD*.

### *B.3.1 $H_2$ Optimization of a System with Delay Using Zero- and First-Order Hold*

This case study consists on solving the $H_2$ optimization problem for a sampled-data system affected by a time delay, using two different hold devices. The objective of the $H_2$ optimization is to find a transfer function $C(\zeta)$ of the LTI digital controller described in the complex variable $\zeta = z^{-1} = e^{-sT}$, such that the minimum value of the mean variance of the output $\varepsilon(t)$ is obtained. This mean variance is defined as

$$\bar{d}_\varepsilon = \int_0^T d_\varepsilon(t)dt \tag{B.1}$$

where $d_\varepsilon(t) = d_\varepsilon(t+T)$ is the instantaneous variance of the signal $\varepsilon(t)$ [5].

We would like to compare the results (the order of the controller and the optimal cost) using each hold device.

The system under study is shown in Fig. B.2. It consists of a continuous plant $P(s)$, a digital controller $C(\zeta)$, a hold device $H(s)$ and a pure delay representing an time needed for computation, communication, or transport. The plant is taken as $P(s) = \frac{1}{s(s+1)}$, and a sampling period $T = 1$ s is used. A value of $\tau = 0.1$ s is set for the delay.

The two options considered for the hold device are a zero-order hold (ZOH) for which

$$u(t) = v_k \quad \text{for } kT \le t < (k+1)T \tag{B.2}$$

and a first-order hold (FOH) for which

$$u(t) = v_k + (t - kT)\,\frac{v_k - v_{k-1}}{T} \tag{B.3}$$

**Fig. B.2** Structure of the system

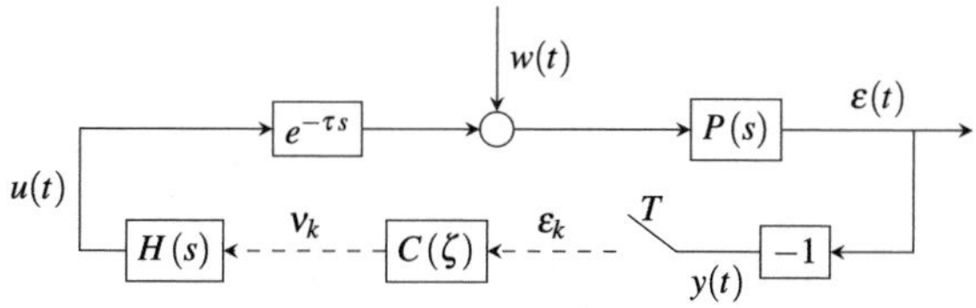


for $kT \leq t < (k+1)T$.

In order to use the DirecSD toolbox, the system must be converted to the standard form of MIMO sampled-data systems shown in Fig. B.1. Since we are dealing with a SISO system, all matrices and signals are scalar quantities. The scalar blocks in this case are configured as

$$\begin{array}{ll} K(s) = P(s) & L(s) = P(s)\,e^{-\tau s} \\ M(s) = -P(s) & N(s) = -P(s)\,e^{-\tau s} \end{array}$$

The commands needed to run the example are as follows. At first, we set up the system:

```
>>T = 1; %Sampling Period
>>H0 = tf(1,[1 0]); %Zero-order hold
>>H1(:,:,1,1)=ss(1);%First-order hold
>>H1(:,:,2,1)=ss(1);%First-order hold
>>P = zpk([],[0 -1],1); %Plant
>>tau = 0.1;
>>Fdelay = tf(1);
>>Fdelay.iodelay = tau;
>>Pdelay = P*Fdelay;
>>sys = [P Pdelay; -P -Pdelay];
```

The last line creates a rational matrix, which represents a system in the standard form. The function sdh2 is used to synthesize the controllers and to find the optimal cost.

The commands issued are

```
>>[Kzohd, err0d] = sdh2(sys,T,[],H0);
>>[Kfohd, err1d] = sdh2(sys,T,[],H1);
```

For the ZOH, the toolbox reports the controller

$$C_0(\zeta) = \frac{-193.39(\zeta - 3.964)}{(\zeta + 171.1)(\zeta + 1.143)} \tag{B.4}$$

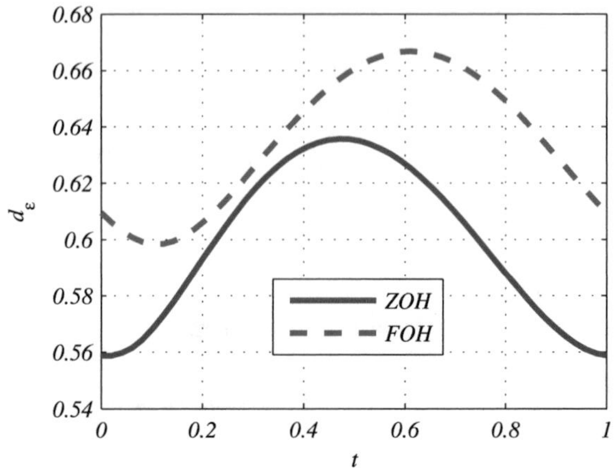


**Fig. B.3** Comparison of the inter-sample variance for the cases with ZOH and FOH

with a minimum cost of $\bar{d}_{\varepsilon 0} = 0.6003$. On the other hand, using the FOH, we obtain the controller

$$C_1(\zeta) = \frac{200(\zeta - 3.819)}{(\zeta + 118.4)(\zeta - 2.705)(\zeta + 0.9415)} \tag{B.5}$$

with an optimal cost $\bar{d}_{\varepsilon 1} = 0.6538$.

These results show that for this particular case the introduction of a FOH does not represent any advantage over the use of a ZOH. Besides the higher order of the reported controller, the performance index worsens when the FOH is used. While a first-order interpolation would give smoother results, the necessary causality of the real hold element makes the FOH an extrapolator, which explains its worse performance when compared to the ZOH. To have a more decisive comparison, Fig. B.3 presents the inter-sample variance $\bar{d}_{\varepsilon}(t)$ for both cases.

To obtain this plot, we used the function `sdh2norm` which finds either the mean variance or the variance at a certain point within the sampling interval. For this particular case we issued the following commands to *DIRSD* 3.5:

```
>>t = linspace(0,T,50);
>>errs0d = sdh2norm(sys,Kzohd,t,H0);
>>errs1d = sdh2norm(sys,Kfohd,t,H1);
```

in order to obtain the variance at 50 equidistant points within the sampling interval. The results were then plotted to obtain the result observed in Fig. B.3.

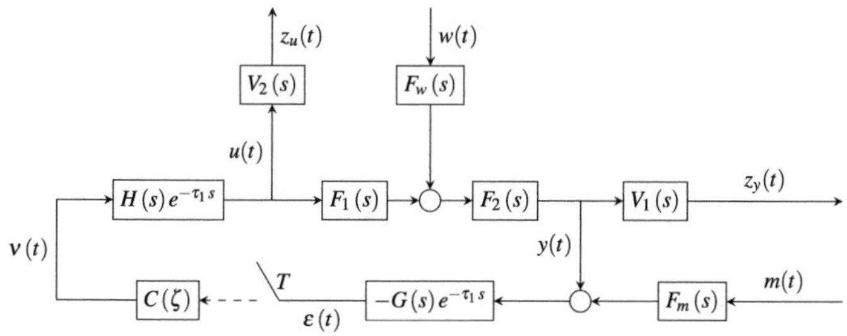


**Fig. B.4** Structure of the system

## B.3.2 $H_2$ Optimization of a Generic System with Two Delays

This example considers the $H_2$ optimization of a closed-loop sampled-data system, as shown in Fig. B.4. This is the most general case for a single-loop system with a SISO controller.

For optimization purposes, the output of the system is taken as

$$\mathbf{z}(t) = \begin{bmatrix} z_y(t) \\ z_u(t) \end{bmatrix} \tag{B.6}$$

and the cost function to be minimized is

$$J = \bar{d}_z = \bar{d}_{zy} + \bar{d}_{zu}. \tag{B.7}$$

In order to use the *DIRSD* toolbox, the system must be transformed into the standard form of Fig. B.1. Given that the input of the system is

$$\mathbf{x}(t) = \begin{bmatrix} w(t) \\ m(t) \end{bmatrix}, \tag{B.8}$$

we obtain

$$\mathbf{K}(s) = \begin{bmatrix} V_1(s)\,F_2(s)\,F_w(s) & 0 \\ 0 & 0 \end{bmatrix} \tag{B.9}$$

$$\mathbf{L}(s) = \begin{bmatrix} V_1(s)\,F_2(s)\,F_1(s)\,H(s)\,e^{-\tau_1 s} \\ V_2(s)\,H(s)\,e^{-\tau_1 s} \end{bmatrix} \tag{B.10}$$

$$\mathbf{M}(s) = -\begin{bmatrix} G(s)F_2(s)F_w(s)e^{-\tau_2 s} & G(s)F_m(s)e^{-\tau_2 s} \end{bmatrix} \tag{B.11}$$

$$\mathbf{N}(s) = -G(s)F_2(s)F_1(s)H(s)e^{-(\tau_1+\tau_2)s}. \tag{B.12}$$

Consider a simple case in which the disturbance and measurement noise are both withe noise. This means that the forming filters $F_w(s)$ and $F_m(s)$ are both equal to 1. The weights of the output signals are taken as $V_1 = 1$ and $V_2 = 2$. The other components take the following values:

$$F_1(s) = \frac{1}{s+1}, \quad G(s) = 1,$$
$$F_2(s) = \frac{2}{s+2}, \quad H(s) = \frac{0.25}{s+0.25}.$$

The two time delays are $\tau_1 = 0.05$ and $\tau_2 = 0.1$, whereas the sampling period is taken as $T = 0.5$.

This example is run by entering the following commands:

```
>>F1 = tf(1,[1 1]);
>>F2 = tf(2,[1 2]);
>>Fm = 1;
>>Fw = 1;
>>tau1 = 0.05;
>>tau2 = 0.1;
>>G = -tf(1,'iodelay',tau2);
>>H = tf(0.25,[1 0.25],'iodelay',tau1);
>>V1 = 1;
>>V2 = 2;
>>T = 0.5;
>>H0 = tf(1,[1 0]);
>>K = [V1*F2*Fw 0; 0 0];
>>L = [V1*F2*F1*H; V2*H];
>>M = [G*F2*Fw G*Fm];
>>N = G*F2*F1*H;
>>sys = [K L; M N];
>>[C, err] = sdh2(sys,1,[],H0)
```

Considering these values, the optimal controller is found as

$$C_{opt}(\zeta) = \frac{0.88964(\zeta - 1.284)(\zeta - 2.718)}{(\zeta + 3.721)(\zeta - 3.134)(\zeta - 13.81)} \tag{B.13}$$

with an optimal cost $J_{opt} = 0.99995$.

## References


1. K.Y. Polyakov, E.N. Rosenwasser, B.P. Lampe, DirectSD – a toolbox for direct design of sampled-data systems, in *Proceedings of IEEE International Symposium of CACSD'99*, Kohala Coast, Island of Hawai'i, Hawai'i, USA (1999), pp. 357–362
2. K.Y. Polyakov, E.N. Rosenwasser, B.P. Lampe, *DirectSDM* - a toolbox for polynomial design of multivariable sampled-data systems, in *Proceedings of IEEE International Symposium of Computer Aided Control Systems Design*, Taipei, Taiwan (2004), pp. 95–100

3. K.Y. Polyakov, E.N. Rosenwasser, B.P. Lampe, *DIRECTSD 3.0* toolbox for MATLAB, in *Proceedings of IEEE Conference on Computer Aided Control Systems Design*, Munich, Germany (2006), pp. 1946–1951
4. T. Chen, B.A. Francis, *Optimal Sampled-Data Control Systems* (Springer, Berlin, 1995)
5. E.N. Rosenwasser, B.P. Lampe, *Computer Controlled Systems: Analysis and Design with Process-orientated Models* (Springer, Berlin, 2000)
6. K.Y. Polyakov, *Polynomial Methods for Direct Design of Optimal Sampled-data Control Systems* (Saint Petersburg Maritime Technological University, Thesis for Doctor degree of Technological Sciences, 2006)

# Appendix C
# GarSD–A MATLAB Toolbox for Analysis and Design of Sampled-Data System with Guaranteed Performance

## C.1 Introduction

For the construction of control systems under practical conditions, one of the most important problems consists in guaranteeing the required performance under real stochastic excitations. Study of the actual literature shows, that at present for the solution of those problems traditional approach dominates in engineering practice. This means that the control strategy is found on basis of design procedures for different concrete excitations. According to sampled-data systems, procedures for optimal design constitute important developments for the realization of that approach. Several textbooks deal with that topic, e.g., [1, 2, 3, 4], and of course the monograph at hand.

However, at the solution of practical problems in connection with stochastic optimization of control systems, we are confronted with parametric or structural uncertainties according to the excitation [5]. Therefore, applying optimal design procedures for the solution of practical problems, when structural uncertainties appear, requires the selection of characteristic excitations with the worst effect on the system. In case of parametric uncertainties, we will investigate an excitation with characteristics that correspond to a certain set of excitations, which can happen during operation. Those excitations, with the type or average characteristics, can essentially deviate from the real excitations, that really affect the system under certain operation regimes. Therefore, the functioning of the system under excitations with type, or average characteristics under real conditions can lead to unpredictable dissatisfactions.

In connection with the above explained facts, the fundamental design problem arises in control theory, how the required system performance can be guaranteed for a given scope of stochastic excitations, derived from its real operation conditions. Those systems are called systems with guaranteed performance. One of the first works in this field is the monograph [6], where the general idea of guaranteed performance according to structural or parametric uncertainties in the stochastic excitation has been formulated.



E. N. Rosenwasser et al., *Computer-Controlled Systems with Delay*,
https://doi.org/10.1007/978-3-030-15042-6

As a subclass of robust systems, systems with guaranteed performance ensure the function under a certain class of stochastic excitations with time-varying characteristics. In practice, this means the preservation of necessary acceptable qualities of the systems functioning at a certain change of operation conditions, or at the discrepancy of these conditions with settlement. Therefore, while solving practical problems in connection with the design of control systems, the application of the method with guaranteed performance seems to be convenient.

Previously, methods for guaranteed performance are considered in literature mainly for applications to stationary continuous processes with analogue controllers. Various special points when realizing this approach, have been elaborated in the monographs [5, 7]. Moreover, the solutions are compared with optimal systems,

The considerations in the present monograph are based on the parametric transfer function (PTF) concept, that allows to develop exact methods for the direct investigation of sampled-data systems in continuous time, and it permits to extend the guaranteed performance methods to sampled-data systems.

Here, the investigations show that SD systems of guaranteed performance prove to be free of a number of restrictions, which occur, when the traditional approach on basis of optimal design methods is used. Especially, for SD systems of guaranteed performance do not possess fixed poles [8–10], which can restrict the dynamics of the transfer processes in the control loop, and that cannot be avoided in optimal systems. Moreover, papers [11, 12] verify, that often the order of the system with guaranteed performance can be reduced in relation to the order of the optimal system.

The present appendix provides an approach for analysis and design of SD systems by joining the methods in [5, 7] for guaranteed performance with the PTF concept. This approach permits to describe the dynamics of the SD system completely in continuous time. Moreover, the approach does not need approximations according to the characteristics of the stochastic excitations. The design does not utilize concrete excitations or approximate characteristics, but only takes the set of excitations of the given class. Papers [13–15] present basic considerations and some specifics of this approach.

Methods for analysis and design of SISO SD systems with guaranteed performance, on basis of the described approach, have been realized in form of the MATLAB toolbox GarSD. Earlier versions of the toolbox are described in [16–18]. Application examples for the solution of practical problems, including digital control of continuous processes are considered in [19]. The appendix at hand contains a short description of the guaranteed performance problem for SD systems and an overview of the capabilities of the GarSD toolbox and algorithms for its application. Moreover, the appendix provides some examples for the solution of the design problem for SD systems with guaranteed performance. The toolbox can be downloaded from the site www.rt.uni-rostock.de/ccsd.

## C.2 Guaranteed Performance Problem for SD Systems

### *C.2.1 General Problem Statement for SD Systems with Guaranteed Performance*

Consider the linear SD system $\mathscr{W}$ with one continuous input $g(t)$ and vector output $y(t) \in \mathbb{R}^n$, according to the structure shown in Fig. C.1. The overall system $\mathscr{W}$ is periodic even if the process within the system is time-invariant.

The dynamics of system $\mathscr{W}$ are described by the PTF $W(s, t, Z)$ with the components $W_i(s, t, Z)$, $(i = 1, \ldots, n)$ from input $g(t)$ to outputs $y_i(t)$, respectively. Moreover, $Z$ is a vector of constructive parameters of the system. Let a set $\mathscr{Z}$ of admissible values for the elements of this vector be given. In practice the set $\mathscr{Z}$ consists of such vectors $Z$, that guarantee the asymptotic stability of the system and meet certain other requirements.

Assume that for the considered system, a certain vector $Z \in \mathscr{Z}$ has been chosen. Moreover, assume excitation $g(t)$ as stationary stochastic signal given by the spectral density $S_g(s)$. Then the output signals $y_i(t)$ in quasi-stationary mode are periodically nonstationary random signals. A subsumed characteristic of those signals are the mean variances $\bar{d}_i(Z)$ [3], depending on the vector $Z$

$$\bar{d}_i(Z) = \frac{1}{\pi} \int_0^\infty \bar{A}_i(\nu, Z) \tilde{S}_g(\nu)\, \mathrm{d}\nu, \tag{C.1}$$

where $\nu$ is the frequency and $\tilde{S}_g(\nu) = S_g(\mathrm{j}\nu)$—the even function, determining the spectral density. Moreover

$$\bar{A}_i(\nu, Z) = \frac{1}{T} \int_0^T A_i^2(\nu, t, Z)\, \mathrm{d}t \tag{C.2}$$

satisfies $\bar{A}_i(\nu, Z) > 0$ for all $\nu$ and $Z$, where $A_i(\nu, t, Z)$ is the parametric frequency response (PFR) defined by substituting $s = \mathrm{j}\nu$ in the corresponding PTF $W_i(s, t, Z)$. The PFR of sampled-data systems can also be determined by experiments [20].

As a stochastic measure for the system performance in the considered case, we can use the scalar criterion

$$J(Z) = \sum_{i=1}^{n} \rho_i \bar{d}_i(Z), \tag{C.3}$$

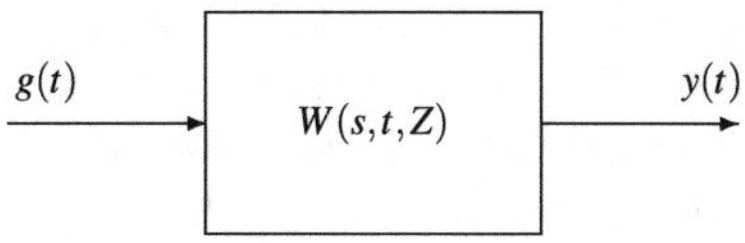


**Fig. C.1** Linear SIMO periodic system $\mathscr{W}$

where $\rho_i > 0$ are real weighting coefficients. Obviously, $J(Z) \geq 0$ and the performance of the system in quasi-stationary mode is better, if the value of $J(Z)$ is smaller. The technical function of the system is guaranteed for $J(Z) \leq J_{\max}$, so that this restriction is hard.

Assume parametrical or structural uncertainties according to the excitation $g(t)$. Instead of its exact spectral density $S_g(s)$, we only know that it belongs to a class of stationary stochastic excitations $\mathscr{M}$. In case of a fixed vector $Z \in \mathscr{Z}$ and generalized characteristics by the class $\mathscr{M}$, let us find numbers

$$\bar{D}_i(Z) \geq \bar{d}_i(Z) \quad \forall g(t) \in \mathscr{M}. \tag{C.4}$$

The number $\bar{D}_i(Z)$ is called estimate of the mean variance of the system output over the class $\mathscr{M}$. Then for the performance estimation of the system over the class $\mathscr{M}$, we can formulate the scalar criterion

$$E(Z) = \sum_{i=1}^{n} \rho_i \bar{D}_i(Z). \tag{C.5}$$

Assume that we have found a vector $Z_{gar} \in \mathscr{Z}$, for which criterion (C.5) yields a minimal value. In that case, the system with parameter $Z_{gar}$ is called system with guaranteed performance according to the class $\mathscr{M}$, and the related controller is called controller with guaranteed performance. The problem of finding a vector $Z_{gar}$, that minimizes (C.5) over the class $\mathscr{M}$ is called guaranteed performance problem.

If it happens that for vector $Z_{gar}$

$$E(Z_{gar}) \leq J_{\max}, \tag{C.6}$$

then the system with parameter $Z_{gar}$ for any excitation of class $\mathscr{M}$ works successful with guarantee.

## *C.2.2 Description of SD Systems for the Solution of the Guaranteed Performance Problem*

The `GarSD` toolbox solves guaranteed performance problems for single-loop models of standardizable SD systems with a structure according to Fig. C.2. The structure meets a great number of systems arising from practical problems. The process with transfer function $F(s)$ should work under external stationary stochastic excitation $g(t)$. The value of the output $y_1(t)$ is determined by a sensor with transfer function $L(s)$. The measured value $z(t)$ is sent to the digital controller, consisting of analog-to-digital converter (ADC), control program $C(\zeta)$, and generalized digital-to-analogue converter (GDAC). The ADC generates the sequence $\xi_k$. Hereby, we suppose that

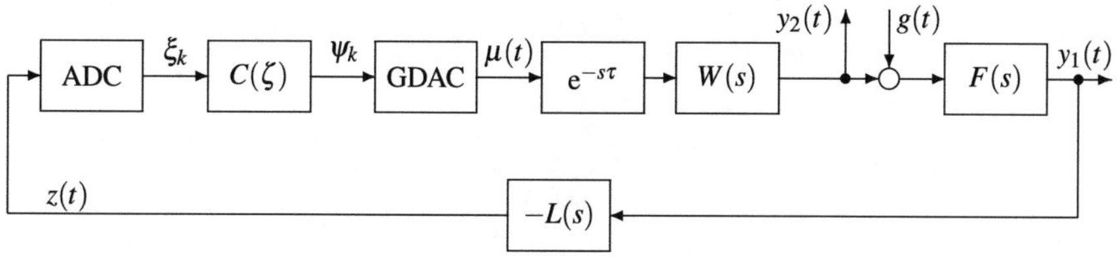


**Fig. C.2** Structure of sampled-data systems covered by toolbox `GarSD`

the sampling period $T$ of the ADC is non-pathological. The control sequence $\psi_k$ is generated by the control algorithm

$$C(\zeta) = \frac{\psi(\zeta)}{\xi(\zeta)} = \frac{\sum_{r=0}^{R} \beta_r \zeta^r}{\sum_{r=0}^{R} \alpha_r \zeta^r} = \frac{\beta(\zeta)}{\alpha(\zeta)}, \tag{C.7}$$

where $\beta_r$ and $\alpha_r$ are real coefficients with $\alpha_0 \neq 0$, $R \geq 0$ is an integer, namely the order of the controller, $\zeta = \mathrm{e}^{-sT}$ is the shift operator for one step backward, and $\beta(\zeta), \alpha(\zeta)$ are polynomials. The GDAC transforms the digital control signal $\psi_k$ into the analogue signal $\mu(t)$ according to

$$\mu(t) = \sum_{\ell=0}^{H} \psi_{k-\ell} h_\ell(t - kT), \qquad kT < t \leq (k+1)T, \tag{C.8}$$

where $H \geq 0$ is an integer, namely the order of the GDAC, and $h_\ell(t)$ are shape-giving functions. The toolbox `GarSD` considers a special cases of GDAC the well-known interpolator and extrapolator [21].

The delay of the control loop is aggregated in one block, and we split it into

$$\tau = \tau_0 + \eta T, \tag{C.9}$$

where $\eta$ is a nonnegative integer and $0 \leq \tau_0 < T$.

The delayed analogue control signal $\mu(t - \tau)$ arrives at the actuator with transfer function $W(s)$. The actuator output $y_2(t)$ governs the process with transfer function $F(s)$ and output $y_1(t)$.

Suppose that all continuous elements in the system are strictly causal, and the TF $F(s)$ is strictly proper irreducible rational functions and $W(s)$ and $L(s)$ are strictly proper irreducible rational functions or constant. Moreover, we assume, that the product $F(s)W(s)L(s)$ is irreducible and does not possess multiple poles.

The dynamics of the system in Fig. C.2 is described by the PTF $W_1(s, t)$ and $W_2(s, t)$ from input $g(t)$ to the outputs $y_1(t)$ and $y_2(t)$, respectively. These PTF have the form

$$W_1(s,t) = F(s)\left[1 - \frac{L(s)\tilde{C}(s)\tilde{\mathscr{D}}_{G_H F W}(T,s,t-\tau)\mathrm{e}^{-st}}{\tilde{\lambda}(s)}\right],$$

$$W_2(s,t) = \frac{F(s)L(s)\tilde{C}(s)\tilde{\mathscr{D}}_{G_H W}(T,s,t-\tau)\mathrm{e}^{-st}}{\tilde{\lambda}(s)}, \tag{C.10}$$

where

$$\tilde{\lambda}(s) = 1 + \tilde{C}(s)\tilde{\mathscr{D}}_{G_H F W L}(T,s,-\tau) \tag{C.11}$$

is the characteristic function, assigning the stability of the system, and $\tilde{C}(s) = C(\zeta)|_{\zeta=\mathrm{e}^{-sT}}$. Moreover, in (C.10), (C.11) the quantities $\mathscr{D}_{G_H W}(T,s,t)$, $\mathscr{D}_{G_H F W}(T,s,t)$, $\mathscr{D}_{G_H F W L}(T,s,t)$ are the discrete Laplace transforms (DLT) of the functions $G_H(s)W(s)$, $G_H(s)F(s)W(s)$, and $G_H(s)F(s)W(s)L(s)$, respectively, where $G_H(s)$ is the forming element of the DAC. In [22, 23] it was shown that for any strictly proper rational function $X(s)$, the DLP can be represented in the form

$$\tilde{\mathscr{D}}_X(T,s,t) = \frac{1}{T}\sum_{k=-\infty}^{\infty} X(s+k\mathrm{j}\omega)\mathrm{e}^{(s+k\mathrm{j}\omega)t}, \qquad \omega = \frac{2\pi}{T}. \tag{C.12}$$

However, in this treatment we employ closed expressions for the DLP.

Assume that the excitation $g(t)$ belongs to the class $\mathscr{M}$ of stationary stochastic disturbances given by their generalized characteristics. Then the indices of guaranteed function of the system in Fig. C.2 are the estimates $\bar{D}_1(Z)$ and $\bar{D}_2(Z)$ for the outputs $y_1(t)$ and $y_2(t)$, respectively. Hence, optimization criterion (C.5) becomes

$$E(Z) = \bar{D}_1(Z) + \rho D_2(Z). \tag{C.13}$$

The toolbox GarSD provides methods for determining the transfer function of digital controller (C.7) for systems according to Fig. C.2, such that criterion (C.13) is minimal over class $\mathscr{M}$. Moreover, several restrictions can be respected, for instance, limits for the degree of the controller, degree of stability for the closed loop, or the value of the constant part in the control signal $y_1(t)$ in the steady-state regime.

## *C.2.3 Constructing Models for Classes of Stationary Stochastic Excitations*

For the construction of models for the class $\mathscr{M}$ of stationary stochastic excitations in order to solve the guaranteed performance problem, we assume that each excitation of class $\mathscr{M}$ roughly fulfills the following requirements: The excitations should be continuous, in a weaker sense stationary signals, possessing $N$ stationary moments and should be $N$ times differentiable.

Let us have a structural uncertain excitation, so that we only know, that $g(t)$ belongs to the class $\mathscr{M}$, given by generalized characteristics, but the general form of the spectral density $S_g(s)$ is unknown. In this situation, the class $\mathscr{M}$ is defined by the collectivity of $N+1$ limiting integrated generalized moments $d_l$, $l = 0, \ldots, N$ of the spectral density according to the basis functions $u_l(\nu)$

$$d_l = \frac{1}{\pi}\int_0^{\infty} u_l(\nu)\tilde{S}_g(\nu)\,\mathrm{d}\nu, \qquad l = 0, \ldots, N, \tag{C.14}$$

where the basis functions $u_l(\nu)$ and the real constants $0 < d_l < \infty$ are given .

In [6, 7] it was shown, that the form of the basis functions $u_l(\nu)$ should be chosen according to the best approximation of the square of the frequency response of the investigated system of linear combination of this basis function. There also was shown that this condition is satisfied by basis functions of the form

$$u_l(\nu) = \nu^{2l}. \tag{C.15}$$

In that case generalized moments (C.14) of the spectral density receive the meaning of the variance of the excitation and $N$ of its derivatives, while commonly according to class $\mathscr{M}$ we assume that these derivatives exist and the values of their variances are finite. The collection of numbers $\{d_l\}$ defines the class $\mathscr{M}$ of nonparametric stochastic continuous in a weak sense stationary excitations. Estimates for the numbers $d_l$ for a concrete class of excitations can be determined by experiments [5, 7, 10], and they are really convenient for the solution of practical problems.

In practical applications, we often can assume that the spectral density of the excitations of the considered class $\mathscr{M}$ are distributed on a limited region, for instance, there exists a real number $0 < \beta_S < \infty$ satisfying

$$\tilde{S}_g(\nu) \approx 0, \quad \text{for } \nu > \beta_S, \quad \forall S_g(\mathrm{j}\nu) \in \mathscr{M}. \tag{C.16}$$

Therefore, while solving practical problems, the numbers $d_l$ can be calculated by

$$d_l \approx \frac{1}{\pi}\int_0^{\beta_S} \nu^{2l}\tilde{S}_g(\nu)\,\mathrm{d}\nu, \tag{C.17}$$

and a model of class $\mathscr{M}$ is given by $N+1$ numbers $\{d_l\}$ (C.17) and number $\beta_S$ (C.16).

Now consider the case of parametric uncertainties, when the concrete spectral density $S_g(s)$ ($\tilde{S}_g(\nu)$) of the excitation is unknown, but we know that it belongs to an admissible (in general infinite) class $\mathscr{M}$. Assume that for any value of the frequency $\nu$

$$\tilde{S}_{0g}(\nu) = \max_{\mathscr{M}}(\tilde{S}_g(\nu)), \quad \forall \nu, \tag{C.18}$$

i.e., the values of the function $\tilde{S}_{0g}(\nu)$, for any value of the frequency will certainly not take lower values than any of the spectral densities of the class $\mathcal{M}$. Then the function $\tilde{S}_{0g}(\nu)$ is called envelope of the spectral densities. In case of parametric uncertainties, the envelope of the spectral densities defines the class of stochastic excitations. The toolbox `GarSD` provides both possibilities to define the class $\mathcal{M}$.

## *C.2.4 Calculating the Functional of Guaranteed Performance*

In the toolbox `GarSD` the calculation of the criterion for guaranteed performance of sampled-data systems is based on an approximation method, that is considered in [5, 7]. This method is shortly described below.

Assume the situation of structural uncertainties in the excitation, and the class $\mathcal{M}$ is given by the $N+1$ numbers $\{d_l\}$, $(l = 0, \ldots, N)$ and the limit $\beta_S$. Moreover, let us have chosen for some reasons a vector $Z \in \mathcal{Z}$, and we have constructed the PTF $W_i(s, t, Z)$ and function $\bar{A}_i(\nu, Z)$ (C.2). Suppose, that on the interval $[0, \beta_S]$ we have approximations $\bar{C}_k(\nu, Z)$ for the functions $\bar{A}_i(\nu, Z)$, such that

$$\bar{C}_i(\nu, Z) \approx \bar{A}_i(\nu, Z), \quad \bar{A}_i(\nu, Z) \leq \bar{C}_i(\nu, Z) \quad \forall \nu \in [0, \beta_S]. \tag{C.19}$$

Moreover, $\bar{C}_i(\nu, Z)$ should have the form

$$\bar{C}_i(\nu, Z) = \sum_{l=0}^{N} c_{li}(Z)\nu^{2l}, \tag{C.20}$$

where $c_{li}(Z)$ are a real coefficients depending on $Z$.

Then it can be shown that mean variance estimate $\bar{D}_k(Z)$ is determined by

$$\bar{D}_i(Z) = \sum_{l=0}^{N} c_{li}(Z)d_l, \tag{C.21}$$

and the guaranteed performance criterion can be calculated by formula (C.5).

Now assume the situation, that parametric uncertainties of the excitation take place, when the class $\mathcal{M}$ is defined by the envelope spectral density $\tilde{S}_{0g}(\nu)$. Then the realization of an approximate method can be modified with the goal to receive performance estimations. Let for a given vector $Z \in \mathcal{Z}$ the poles related to the $i$th output are restricted to the interval $[0, \beta_i]$. Consider the collection of numbers

$$d'_{li} = \frac{1}{\pi} \int_0^{\beta_i} \tilde{S}_{0g}(\nu)\nu^{2n}\, d\nu. \tag{C.22}$$

The numbers $d'_{li}(Z)$ are called derived variances. The totality of $N+1$ derived variances $d'_{li}(Z)$, $l=0,\dots,N$, defines a class $\mathscr{M}'\subset\mathscr{M}$ of excitations, which has influence on the guaranteed functioning of the system according to the $i$th outputs for a given parameter vector $Z$.

Then for the calculation of an estimate $\bar{D}_i(Z)$, it is sufficient to approximate $\bar{A}_i(\nu,Z)$ by the function $\bar{C}'_i(\nu,Z)$ of form (C.20), which has to fulfill condition (C.19) only on the interval $[0,\beta_i]$. Hereby, estimate $\bar{D}_i(Z)$ can be determined by

$$\bar{D}_i(Z)=\sum_{l=0}^{N} c'_{li}(Z)d'_i. \tag{C.23}$$

Various realization aspects of the approximation method for calculating estimates of guaranteed performance for sampled-data systems are considered more detailed in [13–15].

### *C.2.5 Construction of the set $\mathscr{Z}$ and Minimization of the Guaranteed Performance Criterion for SD Systems*

In the framework of the optimization problem according to the criterion for guaranteed performance of SD systems, we search for a causal stabilizing digital controller with a transfer function of form (C.7), that minimizes criterion (C.5). In order to guaranty that each vector $Z\in\mathscr{Z}$ uniquely defines a transfer function (C.7) of an admissible controller, we build set $\mathscr{Z}$ in the following way.

In [24], it was shown that substituting $\zeta=\mathrm{e}^{-sT}$, characteristic function (C.11) $\tilde{\lambda}(s)$ permits the representation

$$\tilde{\lambda}(s)\Big|_{\mathrm{e}^{-sT}=\zeta}=\lambda(\zeta)=1+\frac{\beta(\zeta)}{\alpha(\zeta)}\frac{b(\zeta)}{a(\zeta)}, \tag{C.24}$$

where $b(\zeta)$ and $a(\zeta)$ are coprime polynomials [24] achieved from DLT $\tilde{\mathscr{D}}_{G_H WFL}(T,s,t)$ for $t=-\tau$ and substitution $\zeta=\mathrm{e}^{-sT}$

$$\tilde{\mathscr{D}}_{G_H WFL}(T,s,-\tau)\Big|_{\zeta=\mathrm{e}^{-sT}}=\mathscr{D}_{G_H WFL}(T,\zeta,-\tau)=\frac{b(\zeta)}{a(\zeta)}. \tag{C.25}$$

Also in [24], it was shown that for system with structure of Fig. C.2, for $\tau_0>0$ the degree of polynomial $a(\zeta)$ is the added degree of the continuous elements of the system, and

$$\deg b(\zeta)=\deg a(\zeta)+(\eta+1)+H, \tag{C.26}$$

so that for processes with delay, we always have

$$\deg b(\zeta) > \deg a(\zeta). \tag{C.27}$$

When (C.24) is true, then the set of polynomial pairs $(\alpha(\zeta), \beta(\zeta))$, building the set of transfer functions for stabilizing controllers of form (C.7), is defined by the set of solutions of the Diophantine equation

$$\beta(\zeta)b(\zeta) + \alpha(\zeta)a(\zeta) = \Delta(\zeta), \tag{C.28}$$

where $\Delta(\zeta)$ is a stable polynomial, i.e., its roots are located outside the unit circle.

As mentioned above, the designer is interested in finding a controller of least possible order. For that, in the `GarSD` toolbox the degree of $\Delta(\zeta)$ can be chosen as

$$\deg \Delta(\zeta) \leq \deg b(\zeta) + \deg a(\zeta) - 1. \tag{C.29}$$

In this case Eq. (C.28) is proper [25]. The set of solutions of such a polynomial equation is always not empty, and there exists a unique minimal solution $\{\alpha_0(\zeta), \beta_0(\zeta)\}$, while with refer to (C.27)

$$\deg \beta_0(\zeta) \leq \deg \alpha_0(\zeta). \tag{C.30}$$

This solution corresponds to unique causal controller (C.7), that with regard to (C.30) has the order

$$R_{\min} = \deg \alpha_0(\zeta) \leq R_0 = \deg b(\zeta) - 1. \tag{C.31}$$

The integer $R_0$ is called the basic order of the controller. Thus, each polynomial $\Delta(\zeta)$ (C.29) uniquely corresponds to a causal stabilizing controller of form (C.7) with basic order $R_0$.

In the case, when the order of the searched controller is restricted by the number $R_0$, the optimization criterion of guaranteed performance is investigated over the set of stabilizing controllers of basic order, and the set $\mathscr{Z}$ of the optimizing parameters is build by the vectors

$$Z = (\zeta_1, \ldots, \zeta_\delta)', \tag{C.32}$$

where $\delta = \deg \Delta(\zeta)$. The numbers $\zeta_i$, $(i = 1, \ldots, \delta)$ are located outside the unit circle and are the roots of the polynomial $\Delta(\zeta)$.

If the demanded order of the controller with guaranteed performance is $R_{\max} > R_0$, then the optimization is performed over the set of controllers disposing of the transfer function with polynomials

$$\beta_{\max}(\zeta) = \beta_0(\zeta) + \xi(\zeta)a(\zeta), \qquad \alpha_{\max}(\zeta) = \alpha_0(\zeta) - \xi(\zeta)b(\zeta), \tag{C.33}$$

where $\xi(\zeta)$ is any polynomial satisfying $\deg \xi(\zeta) \leq R_{\max} - \deg a$. In this case, the parameter set $\mathscr{Z}$ is the set of vectors

$$Z = (\zeta_1, \ldots, \zeta_\delta, \xi_0, \ldots, \xi_p)', \quad \delta = \deg \Delta(\zeta), \; p = \deg \xi(\zeta), \tag{C.34}$$

where, as earlier $\zeta_i$ are the roots of the stable polynomial $\Delta(\zeta)$, but $\xi_i$ are the coefficients of the arbitrary polynomial $\xi(\zeta)$.

When the controller should have an order lower than $R_0$, the optimization has to performed over a special subset of controllers of basic order. A procedure for the selection of such subsets as it was realized in the GarSD toolbox, is described for instance in [11].

In a number of cases, an integral part of the controller is required to ensure zero steady-state error in the presence of steady-state parts of the excitation $g(t)$ or when uncertainties occur in the process models. Then, the controller needs a pole at $\zeta = 1$. The set of controllers $C_{int}(\zeta)$ including an integrator is parametrized by the polynomial pairs $(\alpha_{int}(\zeta), \beta_{int}(\zeta))$, which are solutions of the Diophantine equation [26, 27]

$$\beta_{int}(\zeta)b(\zeta) + (\zeta - 1)\alpha_{int}(\zeta)a(\zeta) = \Delta(\zeta), \tag{C.35}$$

and the transfer function of the controller appears as

$$C_{int}(\zeta) = \frac{\beta_{int}(\zeta)}{(\zeta - 1)\alpha_{int}(\zeta)}. \tag{C.36}$$

The GarSD toolbox solves the optimization over the set $\mathscr{X}$ numerically, by minimizing criterion (C.5). Here genetic algorithms are applied. If required, the optimization process allows to respect constraints of pole location of the closed loop and also of the admissible steady-state error $\varepsilon$ of the output in case of non-centralized excitation $g(t)$.

## C.3 Optimization of SD System According to Guaranteed Performance Using the GarSD Toolbox

### *C.3.1 General Description and Basic Possibilities of the Toolbox GarSD*

The described approach to guaranteed performance design of SD systems is realized in form of the MATLAB toolbox GarSD. Various modifications of the GarSD toolbox have been considered in [16–19]. Essential innovations of the actual toolbox version are the inclusion of computational delay and the usage of generalized digital-to-analog converters.

All the program modules which were entered to the package GarSD are grouped on functional purpose in the separate groups, intended for forming and the analysis of the generalized characteristics of classes of stochastic disturbances and for the mathematical description, analysis and optimization of SD systems of the guaranteed accuracy.

In addition, the GarSD toolbox contains a group of program modules for numeric optimization algorithms. The last modules can be used separately for the solution of optimization problems with no relation to SD systems. Moreover, the toolbox provides a group of auxiliary program modules, in order to adapt the toolbox to the solution of analysis and design problems for SD control systems for maritime dynamical objects.

The object-oriented possibilities of MATLAB have been employed. The construction of new classes in the toolbox bases on the standard object `tf` of the *Control System Toolbox*.

For administration and processing of data of sampled-data systems, the toolbox established the object `system`. It contains information about the continuous elements of the system (in form of the standard object `tf`), about the sampler ADC and the type of GDAC, and also about the controller. The object is generated by means of the construct `sys`, as

```
system = sys (Cont,Reg,tau,ord,dac).
```

The argument `Cont` is a $1 \times 3$ vector, containing standard LTI objects `tf`, dedicated to the continuous elements $F(s)$, $W(s)$, $L(s)$, respectively. The argument `Reg` is a discrete LTI object `tf`, when the controller is given, but a quantity of type `double`, holding the sampling period $T$ of the sampler, when the transfer function of the controller is unknown. The argument `tau` of type `double` communicates the value of the delay $\tau$. The argument `ord` of type `double` means the order of the GDAC. Finally, the argument `dac` of type `double` states the type of GDAC (interpolator or extrapolator).

An object `disturb` for data managing in the class of excitations can be generated by the construct `spt`. As arguments of this construct the envelope of the spectral density $S_{og}(s)$ ($\tilde{S}_{og}(\nu) = S_{og}(s)|_{s=\mathrm{j}\nu}$) arise, or alternatively the numbers (C.14), (C.17). The format in this construction is as follows.

For given $\tilde{S}_{og}(\nu)$ at $N$ points supply

```
>> disturb = spt (info);
```

where `info` contains $2 \times N$ `double` variables, where the first row is the abscissa, and the second row the ordinate of the spectral density $\tilde{S}_{og}(\nu)$. If $\tilde{S}_{og}(\nu)$ is given as strictly proper rational transfer function, then the entry is

```
>> disturb = spt (num,den);
```

where `num` and `den` are vectors containing the polynomial coefficients of the numerator and denominator of $\tilde{S}_{og}(\nu)$. When the excitations are non-regular ocean waves, then the envelope of the spectral densities can be defined in exponential form as

$$\tilde{S}_{og}(\nu) = A\nu^{-m}\mathrm{e}^{-B\nu^{n}},$$

where $A$, $B$, $m$, $n$ are real constants. For those problems, the command `spt` takes the form

```
>> disturb = spt (A,B,m,n);
```

where `A, B, m, n` are `double` objects.

When $S_{og}(s)$ is given by the forming filter $F_{ff}(s)$, then the input command is

```
>> disturb = spt (F),
```

where `F` is a `tf` object, according to the transfer function $F_{ff}(s)$.

In cases, where the class $\mathscr{M}$ is determined by the collection of $N+1$ numbers (C.16) $d_i$, $(i = 0, \ldots, N)$, and when informations about the width of the spectrum $\beta_S$ is known, command `spt` takes the form

```
>> disturb = spt (D);
```

or

```
>> disturb = spt (D,beta);
```

respectively, where `D` is a `double` vector of size $1 \times N+1$ for the numbers $d_i$, and `beta` is a `double` object containing $\beta_S$.

The generated objects `disturb` and `system` are utilized in various program modules of the toolbox for analysis and design of systems with guaranteed performance.

For analysis of systems with guaranteed performance, toolbox `GarSD` applies approximation methods. Algorithms for approximate methods are collected in a group of modules, where the special function `aprsys` controls its application. The input command sounds

```
>> D = aprsys(k,system,disturb);
```

where `k` is the number of system output, and `D` is the estimate of the mean variance for the $k$th output, and both are `double` objects. The selection of concrete algorithms in the approximation methods in the command `aprsys` comes automatically in dependence on the properties of the objects `system` and `disturb`.

Stability is an important feature of SD control systems. In `GarSD` the stability is investigated by the function

```
>> U = isstsys(system)
```

or

```
>> P = isstsys(system,arg).
```

In the first case, the output `U` is a logic variable holding the value `1` in case of asymptotic stability of the system `system`, and `0` else. In the second case, the output `P` is a `double` vector containing the poles of the system `system`. The argument `arg` of type `string` adjusts the kind of representing the poles in the complex planes $s$, $\zeta$, or $z = e^{sT}$.

A mean value of a stepped signal, e.g., the output $y_1(t)$ in a quasi-steady-state mode, when as GDAC a zero-order hold is used, can be calculated by the function `errorstatsys` in form of

```
>> error=errorstatsys(system);
```

where `error` is of type `double` and holds the mean value of the signal $y_1(t)$ of system `system`.

For the solution of design problems for SD systems with guaranteed performance, toolbox `GarSD` provides the function `regelgarsys`. The function is called by the command

```
>> [sysGar,D1,D2,E,P] = regelgarsys(type,deg,T, ...
                                system,disturb,rho,num);
```

The input arguments `system` and `disturb` define the parameters of the system to be optimized, and the class of input disturbances, respectively. The remaining input arguments are defined as follows. The `string` variable `type` defines the class $\mathscr{X}$ over which the optimization procedure is running and takes the value `"sta"`, in the case when the optimization is fulfilled over the set of causal stabilizing controllers without additional constraints. If the optimization needs to include a discrete integrator into the TF of the controller, then argument `type` takes the value `"int"`. If all controllers have to hold $y_1(t)$ inside a certain mean stationary deviation for step input, but without inclusion of an integral part, argument `type` takes the value `"lim"`.

Argument `deg` is a parameter of the set $\mathscr{X}$. If the optimization is taken over the set of causal stabilizing controllers of order $R > R_0$, then `deg` is of type `double` and holds the number $R$. If the optimization is taken over the set of causal stabilizing controllers of basic order $R_0$, then `deg` is a `string` with the value `"min"`. If, however, the optimization should run over the set of causal stabilizing controllers of reduced order $R < R_0$, then `deg` takes the numbers `"min1"`, `"min2"` , etc., according to the order reduction by `1,2, ...` in relation to the basic order. When the optimization is taken over the set controllers with basic order, that ensure the compensation of constant deviations of the excitation, then the argument `deg` of type `double` contains a value, either the stability degree of the system (when the controller includes a discrete integrator), or the greatest admissible value of the mean deviation. The remaining arguments of the function `regelgarsys` are the sampling period `T` of type `double`, the argument `rho` of type `double`, representing the weighting factor $\rho$ in criterion (C.5), and the argument `num` of type `double`, containing the number of iterations for numerical optimization procedures.

The output arguments of function `regelgarsys` are scheduled as follows. In analogy to the object `system` the object `sysGar` is related to the system with guaranteed performance optimized over the class `disturb`. The arguments `D1` and `D2` are of type `double` and hold estimates for guaranteed performance $\bar{D}_1$, $\bar{D}_2$ of the mean variances of the signals $y_1(t)$ and $y_2(t)$, respectively, for the system `sysGar` over the class `disturb`. The argument `E` is a scalar object of type `double` keeping the value of criterion (C.5). Finally, argument `P` is a `double` vector, storing the poles of system `sysGar`.

Moreover, the `GarSD` toolbox contains functions that allow to draw for fixed $t$ the PFR according to a system `system`, and also to build for this system the DLT $\tilde{\mathscr{D}}_{G_H WFL}(T, s, t)$.

An additional possibility of the `GarSD` toolbox consists in construction and transformation of the spectral density of the excitation, raised by irregular ocean waves on moved objects.

## *C.3.2 Examples for the Optimal SD Control System Design According to Guaranteed Performance*

*Example C.1* In the first example, we design a course stabilizing SD system with guaranteed performance for the vessel Montebella, that moves under various irregular waves [28]. The system can be approximately described by the structure in Fig. C.2.

In the structure of Fig. C.2, the vessel under control corresponds to process $F(s)$. In [28], it was shown that for the actual digital controller design, a simplified Nomoto model [29] from the disturbance to the course can be used. It has the form

$$F(s) = \frac{k_\beta}{s(1 + T_\beta s)}, \tag{C.37}$$

where for speed 20 kn, we obtain $k_\beta = 0.069\,\mathrm{s}^{-1}$ and $T_\beta = 18.219\,\mathrm{s}$.

The element $W(s)$ corresponds to the rudder engine, and the element $L(s)$ is the gyro. These elements are assumed to be

$$W(s) = \frac{1}{s+2}, \qquad L = 1. \tag{C.38}$$

Moreover, the computational delay is assumed as $\tau = 0.1875\,\mathrm{s}$, so that according to (C.9), we have $\eta = 0$. In the actual case, the output signals are the deviation from the required course $y_1(t)$ and the rudder angle $y_2(t)$. The parameters of the excitation on the ship by ocean waves $g(t)$ depend on their intensity and concrete values cannot be defined in advance. Therefore, parametric uncertainty of the excitation is assumed. Let class $\mathscr{M}$ be centralized noise given by the numbers

$$d_0 = 0.315, \quad d_1 = 1.507 \tag{C.39}$$

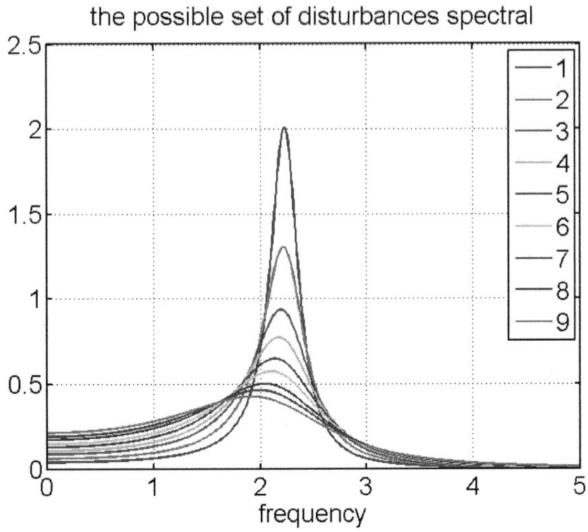


**Fig. C.3** Spectral densities for excitations by ocean waves of various intensities belonging to class $\mathscr{M}$

**Table C.1** Coefficients of spectral densities belonging to class $\mathscr{M}$

| Nr. | 1 | 2 | 3 | 4 | 5 | 6 | 7 | 8 | 9 |
|---|---|---|---|---|---|---|---|---|---|
| $\alpha_1$ | 1.00 | 1.55 | 2.20 | 2.71 | 3.30 | 3.81 | 4.50 | 5.00 | 5.66 |
| $\alpha_2$ | 9.90 | 9.80 | 9.60 | 9.40 | 9.10 | 8.80 | 8.30 | 7.90 | 7.30 |
| $\alpha_3$ | 25.00 | 25.20 | 25.40 | 25.60 | 25.80 | 26.00 | 26.20 | 26.40 | 26.60 |

which are the variance of the excitation and its first derivative. Figure C.3 presents some possible spectral densities of the differentiated excitation to the considered vessel according to different intensities of the waves. All spectra in the figure have the form

$$S(\nu) = \frac{\alpha_1}{\nu^4 + \alpha_2 \nu^2 + \alpha_3} \tag{C.40}$$

and they are approximate models of the ocean waves investigated in [30]. Values $\alpha_1$, $\alpha_2$, and $\alpha_3$ according to the spectral densities in Fig. C.3 are listed in Table C.1. For the considered process, we want to design a controller with guaranteed performance of basic order without additional restrictions. Moreover, as GDAC we will apply a first-order hold described by

$$\mu(t) = \xi_k + \frac{\xi_k - \xi_{k-1}}{T} \cdot (t - kT), \quad kT < t < (k+1)T. \tag{C.41}$$

For this hold, we obtain in (C.8) $H = 1$, $h_0(t) = 1 + \frac{t}{T}$, $h_1(t) = -\frac{t}{T}$. For the sampling period we choose $T = 1,5\,\mathrm{s}$.

Then for DLT $\mathcal{D}_{G_H WFL}(T, \zeta, -\tau)$, according to (C.25), we find

$$\mathcal{D}_{G_H WFL}(T, \zeta, -\tau) = \zeta \frac{\left(-4.096\zeta^4 - 5.659\zeta^3 + 41.239\zeta^2 + 11.924\zeta\right) \cdot 10^{-4} + 1.676 \times 10^{-8}}{-0.435\zeta^3 + 1.828\zeta^2 - 2.393\zeta + 1}, \tag{C.42}$$

that means in the given case, we have $\deg \beta(\zeta) = 3$, which coincides with the added orders of the continuous elements of the system. Moreover, in harmony with (C.26) $\deg \alpha(\zeta) = 5$, and due to (C.31) the basic order of the controller is $R_0 = 4$. In the actual case, for vector $Z$ of optimization parameters according to (C.32), and due to (C.29), we obtain $\deg \Delta = 7$. Therefore, $Z = \left[\, \zeta_1, \ldots, \zeta_7 \,\right]'$, with the constraints $|\zeta_i| > 1\, \forall i$. Every admissible $Z$ corresponds uniquely to a controller $C_{gar}(\zeta)$ of basic order $R_0$ and form (C.7), when $A(\zeta)$ and $B(\zeta)$ are the minimal solution of proper polynomial Eq. (C.28), which in the actual case takes the form

$$\begin{aligned}\Big[\Big(-4.096\zeta^4 - 5.659\zeta^3 + 41.239\zeta^2 + 11.924\zeta\Big) \cdot 10^{-4} + 1.676 \cdot 10^{-8}\Big] A(\zeta) \\ - \Big(0.435\zeta^3 - 1.828\zeta^2 + 2.393\zeta - 1\Big) B(\zeta) = \sum_{i=1}^{7} (\zeta - \zeta_i).\end{aligned} \tag{C.43}$$

The numerical optimization procedure searches for the vector $Z_0$ of the set $\mathcal{Z}$, that the corresponding controller $C_0(\zeta)$ yields the lowest value of criterion (C.5). In the actual case, it yields

$$Z_{gar} = \left[\, e^{11.8200T} \; e^{0.7449T} \; e^{0.5520T} \; e^{0.0293T} \; e^{0.0306T} \; e^{0.1250T} \; e^{0.1409T} \,\right]' \tag{C.44}$$

and applying (C.43), we find the causal stabilizing controller with guaranteed performance of order $R_0 = 4$

$$C_{gar}(\zeta) = \frac{0.7658\zeta^2 - 1.6485\zeta + 0.8881}{-0.0007\zeta^4 - 0.2039\zeta^3 + 1.1458\zeta^2 - 1.9211\zeta + 1.0000} \tag{C.45}$$

Using this controller, for the class of excitations $\mathcal{M}$ determined by numbers (C.39), we can estimate $\bar{D}_1(Z_{gar}) = 0.0248\,\text{rad}^2$, $\bar{D}_2(Z_{gar}) = 0.0022\,\text{rad}^2$, and for criterion (C.5) with $\rho = 2000$, the value $\bar{E}(Z_{gar}) = 4.507$ is guaranteed. This means that for any excitation of class $\mathcal{M}$, the mean standard deviation of the course $\bar{\sigma}_1 = \sqrt{\bar{D}_1(Z_{gar})} = 9.03^\circ$ is guaranteed. Hereby, the value of the mean standard deviation of the rudder angle with guarantee is not higher than $\bar{\sigma}_2 = \sqrt{\bar{D}_2(Z_{gar})} = 2.71^\circ$. Figures C.4 and C.5 illustrate the simulated motion of the considered vessel under various excitations of class $\mathcal{M}$, i.e., 1, 4, and 7 in Fig. C.3. Hereby, Fig. C.4 shows the deviation from the desired course, and Fig. C.5 the applied rudder angle. The dotted

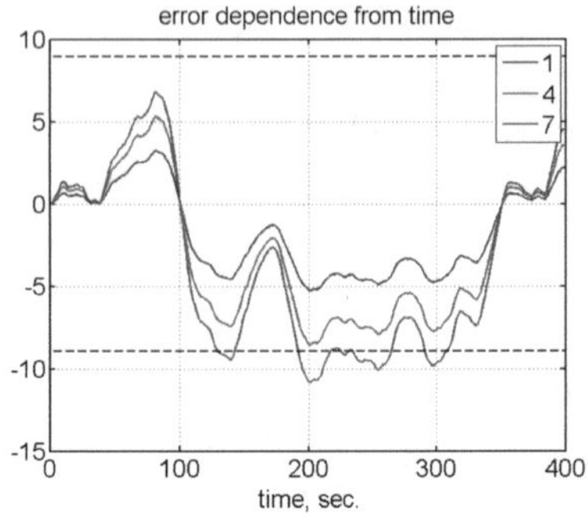


**Fig. C.4** Course deviation under control with guaranteed performance

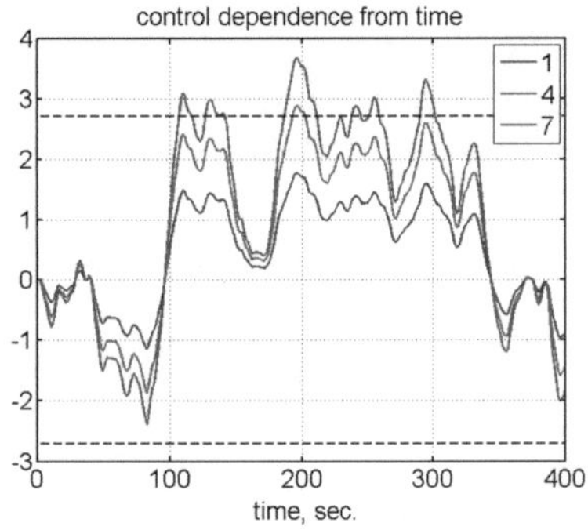


**Fig. C.5** Rudder angle under control with guaranteed performance

lines indicate the limits for guaranteed MSD $\bar{\sigma}_1$ and $\bar{\sigma}_2$ over class $\mathscr{M}$. The plots verify for the selected disturbances, that the motions and the effort are in compliance with the required limits.

*Example C.2* For the vessel Montebella, we want to design a digital controller with guaranteed performance, that completely compensates or limits the effect of constant parts in the excitation. Let the input $g(t)$, be the effect of irregular ocean waves belonging to class $\mathscr{M}$, defined by numbers (C.39), but not centralized. Moreover, assume the characteristics of the continuous elements of the system: vessel $F(s)$, rudder engine $W(s)$ and measuring system $L(s)$, as well as sampling period $T$ and delay $\tau$ as in the preceding example, but for the sake of simplicity the GDAC is only a zero-order hold, i.e., $H = 0$.

Then function $\mathscr{D}_{G_H WFL}(T, \zeta, -\tau)$ can be derived from DLT (C.25) as

$$\mathscr{D}(T, \zeta, -\tau) = \zeta \, \frac{9.344 \cdot 10^{-7}\zeta^3 + 8.448 \cdot 10^{-4}\zeta^2 + 2.892 \cdot 10^{-3}\zeta + 6.026 \cdot 10^{-4}}{-0.435\zeta^3 + 1.828\zeta^2 - 2.393\zeta + 1}, \tag{C.46}$$

Then in the actual case, in the same way as earlier, we obtain $\deg \beta(\zeta) = 3, \deg \alpha = 4$, and basic order $R_0 = 3$.

Let us start with the design of a controller with guaranteed performance that completely compensates the effect of constant parts in the excitation. Hereby, a stability degree of $\theta = 0.01$ is required. Vector $Z$ of optimization parameters, as earlier, now consists of numbers (C.32), where with regard to (C.29) $\deg \Delta = 6$, i.e., $Z = \left[\, \zeta_1 \, \ldots \, \zeta_6 \,\right]'$. Each vector $Z$ is connected with one and only one controller $C_{int}(\zeta)$ of order $R_0 + 1 = 4$ with transfer function of form (C.36), where the polynomials $\alpha_{int}(\zeta)$ and $\beta_{int}(\zeta)$ are determined by Diophantine Eq. (C.33), which in the actual case becomes

$$\zeta \left(9.344 \cdot 10^{-7}\zeta^3 + 8.448 \cdot 10^{-4}\zeta^2 + 2.892 \cdot 10^{-3}\zeta + 6.026 \cdot 10^{-4}\right) \beta_{int}(\zeta) - \left(0.435\zeta^3 - 1.828\zeta^2 + 2.393\zeta - 1\right)(\zeta - 1)\alpha_{int}(\zeta) = \sum_{i=1}^{6} (\zeta - \zeta_i). \tag{C.47}$$

Set $\mathscr{Z}$ of admissible vectors is defined by condition $|\zeta_i| > 1 \; \forall i$, and also by the degree of stability $\theta$. The optimization procedure searches for such numbers $\zeta_i$ that the corresponding controller $C_{int}(\zeta)$ yields a minimal value of criterion (C.5) over class $\mathscr{M}$ defined by numbers (C.39). In the actual case for $\rho = 2000$ criterion, (C.5) takes it minimal value on set $\mathscr{Z}$ at

$$Z_{int} = \left[\, e^{14.240T} \; e^{0.900T} \; e^{0.770T} \; e^{0.120T} \; e^{0.010} \; e^{0.060} \,\right]'. \tag{C.48}$$

Applying here (C.47), we find the the causal stabilizing controller with guaranteed performance

$$C_{int}(\zeta) = \frac{-33.0043\zeta^3 + 137.6127\zeta^2 - 181.9220\zeta + 77.5444}{-0.0001\zeta^4 - 0.0641\zeta^3 - 0.0982\zeta^2 - 0.8376\zeta + 1.0000}. \tag{C.49}$$

Notice that the TF of controller (C.47) possesses a pole at $\zeta = 1$, so that the closed loop removes steady-state deviations.

For some practical problems, a complete compensation of of constant deviations is not required, but the output deviation in steady state is bounded by a certain number $\varepsilon$. Under the condition of the actual example, we illustrate the design of a digital controller with guaranteed performance of basic order that ensures $\varepsilon < 1$.

Now function $\mathscr{D}_{G_H FWL}(T, \zeta, -\tau)$ has form (C.46), because, as earlier, basic order becomes $R_0 = 3$, and vector of optimization parameters $Z$ is determined by (C.32), where in harmony with (C.29) $\deg \Delta = 6$. As before, $Z = \left[\, \zeta_1 \, \ldots \, \zeta_6 \,\right]'$. Each vector $Z$ uniquely corresponds to a controller $C_{lim}(\zeta)$ of order $R_0$ and form (C.7), where the polynomials $\alpha_{lim}(\zeta)$ and $\beta_{lim}(\zeta)$ are defined as minimal solution of Diophantine Eq. (C.35), which in the actual case appears as

$$\zeta\left(9.344\cdot 10^{-7}\zeta^3+8.448\cdot 10^{-4}\zeta^2+2.892\cdot 10^{-3}\zeta+6.026\cdot 10^{-4}\right)\beta_{lim}(\zeta)$$
$$-\left(0.435\zeta^3-1.828\zeta^2+2.393\zeta-1\right)\alpha_{lim}(\zeta)=\sum_{i=1}^{6}(\zeta-\zeta_i). \tag{C.50}$$

Moreover, set $\mathscr{Z}$ of admissible vectors $Z$ satisfy $|\zeta_i|>1\ \forall i$, and is additional restricted by quantity $\varepsilon$: in set $\mathscr{Z}$ during numerical optimization are only such values allowed that guarantee that the steady deviation of signal $y_1(t)$ is bounded by $\varepsilon$. The numerical optimization procedure for $\rho=2000$ provides

$$Z_{lim}=\left[\mathrm{e}^{10.178T}\ \mathrm{e}^{8.039T}\ \mathrm{e}^{1.550T}\ \mathrm{e}^{1.450T}\ \mathrm{e}^{0.3500T}\ \mathrm{e}^{0.200T}\right]'. \tag{C.51}$$

Applying (C.50) we find the causal stabilizing controller with guaranteed performance, that ensures in the steady-state mode that the constant deviation of signal $y_1(t)$ is not greater than 0.1. This controller has the transfer function

$$C_{lim}(\zeta)=\frac{86.4687\zeta^2-283.5114\zeta+216.5451}{0.0002\zeta^3+0.1681\zeta^2+0.7191\zeta+1}. \tag{C.52}$$

The dynamics of the systems with controllers (C.49) and (C.52) can be observed in Figs. C.6 and C.7, where the responses of the closed loops $y_1(t)$ to $y_2(t)$ to a unit step for $g(t)$ are plotted. We realize that system with controller (C.49) completely removes the constant deviation, but in system with controller (C.52), the constant deviation does not exceed bound 0.1. Hereby, the order of controller (C.52) is at one lower than of controller (C.49).

The solution of considered example by means of toolbox GarSD contains the following essential steps.

Data for given elements of the system are built by

```
>> k_beta = 0.069; T_beta = 18.219;
>> F = tf(k_beta,[T_beta 1 0]); W = tf(2,[1 2]);
```

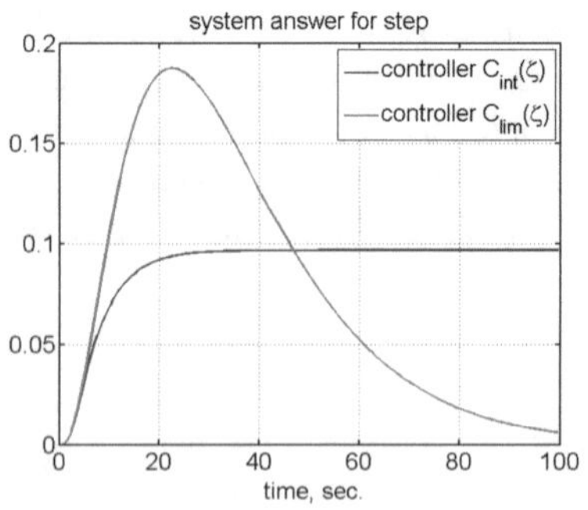


**Fig. C.6** Step response of output signal $y_1(t)$

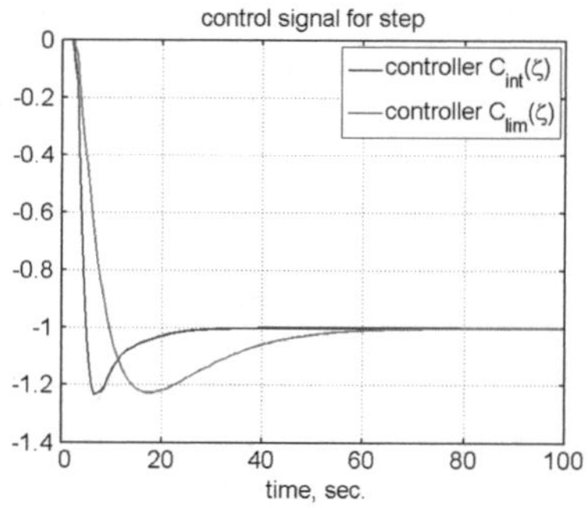


**Fig. C.7** Step response of control signal $y_2(t)$

```
    L = tf(1); T = 1.5;
>> system = sys ([F W L],T,tau,1,0);
```

The class $\mathscr{M}$ of stochastic excitations, given by numbers (C.39), is generated by the command

```
>> d0 = 0.315 ; d1 = 1.507 ; disturb = spt([d0 d1]);
```

Assign the number of iterations in the numerical optimization procedure to 50, an design a controller with guaranteed performance of basic order according to criterion (C.5) with weight $\rho = 2000$. This task is solved by the command

```
>> [sysGar,D1,D2,E,P] = regelgarsys('sta','min',T,...
   system,disturb,2000,50);
```

The operation generates object sysGar, which is structurally analogue to object system, but is a system with guaranteed performance. The object sysGar contains an element

```
>> sysGar.C
Transfer function: 0.8881 z^4 - 1.648 z^3 + 0.7658 z^2
-------------------------------------------------
z^4 - 1.921 z^3 + 1.146 z^2 - 0.2039 z - 0.0007211
Sampling time: 1.5
```

that corresponds to controller (C.45). Moreover, we obtain

```
>> D1
ans =
0.0248
>> D2
ans =
0.0022
```

as values of the estimates (in rad$^2$), $\bar{D}_1(Z_{gar})$ and $\bar{D}_2(Z_{gar})$ guaranteed by controller (C.45), with respect to $y_1(t)$ and $y_2(t)$. The elements of vector $Z_{gar}$, that minimizes criterion (C.5), are stored in vector `P`

```
>> P'
ans =
-11.8200 -0.7449 -0.5520 -0.0293 -0.0306 -0.1250 -0.1409.
```

These numbers are the poles (in complex $s$ plane) of calculated system with guaranteed performance `sysGar` and refer to the elements of vector (C.44).

For the design of controllers with guaranteed performance, compensating the influence of constant parts in the excitations and ensuring zero statistical mean in steady state, we call

```
>> system = sys ([F W L],T,tau,0,0);
>> [sysGarInt,D1Int,D2Int,EInt,PInt] = ...
    regelgarsys('int',0.01,T,system,disturb,2000,50);
```

The command provides object `sysGarInt`, that is structurally analogue to the original object `system`, containing the element

```
>> sysGarInt.C
Transfer function:
77.54 z^4 - 181.9 z^3 + 137.6 z^2 - 33 z
-----------------------------------------------------
z^4 - 0.8376 z^3 - 0.09823 z^2 - 0.06409 z - 7.089e-005
Sampling time: 1.5
```

that corresponds to controller (C.49).

Under the condition of the investigated example, toolbox `GarSD` allows the design of a controller with guaranteed performance, which ensures that in the steady-state the deviation of signal $y_1(t)$ does not exceed the value $\varepsilon = 0.1$. This is achieved by the command

```
>> [sysGarLim,D1Lim,D2Lim,ELim,PLim] = ...
    regelgarsys('lim',0.1,T,system,disturb,2000,50);
```

The command generates the object `sysGarLim` with an analogue structure as the original object `system`, that contains the element

```
>> sysGarLim.C
Transfer function:
216.5 z^3 - 283.5 z^2 + 86.47 z
----------------------------------------
z^3 + 0.7191 z^2 + 0.1681 z + 0.0001857
Sampling time: 1.5
```

that corresponds to controller (C.52).

## References


1. K.J. Åström, B. Wittenmark, *Computer Controlled Systems: Theory and Design*, 3rd edn. (Prentice-Hall, Englewood Cliffs, 1997)
2. T. Chen, B.A. Francis, *Optimal Sampled-Data Control Systems* (Springer, Berlin, 1995)
3. E.N. Rosenwasser, B.P. Lampe, *Computer Controlled Systems: Analysis and Design with Process-orientated Models* (Springer, Berlin, 2000)
4. E.N. Rosenwasser, B.P. Lampe, *Multivariable Computer-controlled Systems—A Transfer Function Approach* (Springer, London, 2006)
5. A.V. Nebylov, *Ensuring Control Accuracy* (Springer, Berlin, 2004)
6. V.A. Besekerskii, A.V. Nebylov, *Robust Systems in Automatic Control* (Nauka, Moscow, 1983). (in Russian)
7. A.V. Nebylov, *Warranting of Accuracy of Control* (Nauka, Moscow, 1998). (in Russian)
8. B.P. Lampe, E.N. Rosenwasser, $\mathscr{H}_2$-optimization and fixed poles for sampled-data systems with generalized hold. Autom. Remote Control **73**(1), 31–55 (2012)
9. B.P. Lampe, E.N. Rosenwasser, $\mathscr{L}_2$-optimization and fixed poles for sampled-data systems with generalized digital to analog converters. Autom. Remote Control **75**(1), 34–56 (2014)
10. B.P. Lampe, E.N. Rosenwasser, $\mathscr{L}_2$-optimization and fixed poles of multivariable sampled-data systems with delay. Int. J. Control **88**(4), 815–831 (2015). (Published online Dec 2014)
11. V.O. Rybinskii, B.P. Lampe, E.N. Rosenwasser, Design of digital controllers of reduced order for periodic sampled-data systems with delay according to guaranteed H2 norm, in *Proceedings of 8th European Nonlinear Dynamics Conference ENOC*, 6–11 July 2014 (Austria, Vienna, 2014), p. 2014
12. V.O. Rybinskii, Optimizing a sampled data submarine motion control system over a set of reduced order controllers. Gyroscopy Navig. **7**, 197–203 (2016)
13. V.O. Rybinskii, B.P. Lampe, K.Y. Polyakov, E.N. Rosenwasser, Control with guaranteed performance for two-rate sampled-data systems under stochastic disturbances, in *Proceedings of 17th IFAC Worl Congress*, Seoul, KR (2007), pp. 15291–15296
14. V.O. Rybinskii, B.P. Lampe, K.Y. Polyakov, E.N. Rosenwasser, A.A. Shylovskii, Control with guaranteed precision in time-delayed sampled-data systems, in *Proceedings of 3rd MSC: IEEE Multi-conference on Systems and Control*, St. Petersburg Russia (2009)
15. V.O. Rybinskii, B.P. Lampe, E.N. Rosenwasser, Advanced statistical analysis with guaranteed estimation of current performance for sampled-data systems with delay and generalized higher order hold, in *Proceedings of 9th IFAC Conference on Control Applications in Marine Systems*, Osaka, Japan (2013)
16. V.O. Rybinskii, E.N. Rosenwasser, The Matlab toolbox for synthesis of guaranteed accuracy sampled-data systems, in *Proceedings of Design of Scientific and Engineering Applications in Matlab*, Moscow, Russia (2002), pp. 471–482. (in Russian)
17. V.O. Rybinskii, B.P. Lampe, E.N. Rosenwasser, Design of digital ship motion control with guaranteed performance, in *Proceedings of 49 International Wissensch. Kolloquium*, vol. 1, Ilmenau, Germany (2004), pp. 381–386
18. V.O. Rybinskii, B.P. Lampe, E.N. Rosenwasser, Extension of the GarSD-toolbox: polynomial method for guaranteed performance design of sampled-data control systems with non-zero static error and pure delays, in *Proceedings of IEEE International Symposium on Computer-Aided Control System Design (CACSD)*, Denver, CO, USA (2011)
19. V.O. Rybinskii, The Matlab toolbox for synthesis of guaranteed accuracy digital control systems for marine dynamic objects. Morskie intellekualnye tehnologii **1**(19), 39–44 (2013). (in Russian)
20. V.B. Sommer, B.P. Lampe, E.N. Rosenwasser, Experimental investigations of analog-digital control systems by frequency methods. Autom. Remote Control **55**(Part 2), 912–920 (1994)
21. W.B. Cleveland, First-order-hold interpolation digital-to-analog converter with application to aircraft simulation. NASA Technical Note, NASA TN D-8331, Washington D.C. (1982)

22. E.N. Rosenwasser, *Linear Theory of Digital Control in Continuous Time* (Nauka, Moscow, 1994). (in Russian)
23. E.N. Rosenwasser, B.P. Lampe, *Digitale Regelung in kontinuierlicher Zeit - Analyse und Entwurf im Frequenzbereich* (B.G. Teubner, Stuttgart, 1997)
24. V.O. Rybinskii, Selecting a Sampling Period for a Sampled-Data Submarine Motion Control System. Gyroscopy Navig. **1**, 57–63 (2013)
25. L.N. Volgin, *Optimal Discrete Control of Dynamic Systems* (Nauka, Moscow, 1986). (in Russian)
26. V.O. Rybinskii, B.P. Lampe, J. Ladisch, E.N. Rosenwasser, A.V. Smolnikov, Assurance of digital control accuracy at non-centralized stochastic actions, in *Proceedings of 8th International Conference of Fast Sea transportation*, SPb, Russia (2005)
27. V.O. Rybinskii, B.P. Lampe, E.N. Rosenwasser, Digital Control with guaranteed performance under non-centred stochastic disturbances, in *Proceedings of 7th IFAC Conference on Manoeuvring and Control of Marine Craft*, Lisbon, Portugal (2006)
28. J. Ladisch, *Anwendung moderner Regelungskonzepte fur Kurs- und Bahnfuhrungssysteme in der Seeschifffahrt*. Ph.D. thesis, University of Rostock (2005)
29. K. Nomoto, T. Taguchi, K. Honda, S. Hirano, On the steering qualities of ships. Int. Shipbuild. Prog. **4**, 354–370 (1957)
30. T.I. Fossen, *Marine Control Systems. Marine Cybernetics*. (Trondheim, 2002)

# Bibliography

## Chapter References


1. K.J. Åström, B. Wittenmark, *Computer Controlled Systems: Theory and Design*, 3rd edn.(Prentice-Hall, Englewood Cliffs, NJ, 1997)
2. A.E. Barabanov, *Synthesis of Minimax Controllers* (State University Press, St. Peterburg, 1996). (in Russian)
3. V.A. Besekerskii, A.V. Nebylov, *Robust Systems in Automatic Control.* (Nauka, Moscow, 1983). (in Russian)
4. T. Chen, B.A. Francis, *Optimal Sampled-data Control Systems* (Springer, Berlin, Heidelberg, New York, 1995)
5. W.B. Cleveland, First-order-hold interpolation digital-to-analog converter with application to aircraft simulation. *NASA technical note, NASA TN D-8331*, Washington, D. C., 1982.
6. Th.I. Fossen. *Marine Control Systems*. (Marine Cybernetics. Trondheim, 2002)
7. V.N. Fomin, *Control Methods for Discrete Multidimensional Processes* (University Press, Leningrad, 1985). (in Russian)
8. F.R. Gantmacher, *The Theory of Matrices* (Chelsea, New York, 1959)
9. E.G. Gilbert, Controllability and observability in multivariable control systems. *SIAM J. Control*, A(1), 128–151 (1963)
10. T. Hagiwara, M. Araki, FR-operator approach to the $\mathscr{H}_2$-analysis and synthesis of sampled-data systems. IEEE Trans. Autom. Control, **AC-40**(8), 1411–1421 (1995)
11. H. Kwakernaak, R. Sivan, *Linear Optimal Control Systems* (Wiley-Interscience, New York, 1972)
12. T. Kailath, *Linear Systems* (Prentice Hall, Englewood Cliffs, NJ, 1980)
13. R.E. Kalman, Y.C. Ho, K. Narendra, Controllabiltiy of linear dynamical systems. Contributions to the Theory of Differential Equations **1**, 189–213 (1963)
14. J. Ladisch, *Anwendung moderner Regelungskonzepte fur Kurs- und Bahnfuehrungssysteme in der Seeschifffahrt*. PhD thesis, University of Rostock, 2005
15. B.P. Lampe, E.N. Rosenwasser, Controllability and observability of discrete-time models for sampled continuous-time processes with higher order holds and delay. Autom. Remote Control **68**(4), 593–609 (Apr 2007)
16. B.P. Lampe, M.A. Obraztsov, E.N. Rosenwasser, $\mathscr{H}_2$-norm computation for stable linear continuous-time periodic systems. Archives of Control Sciences **14**(2), 147–160 (2004)
17. B.P. Lampe, E.N. Rosenwasser, Polynomial modal control for sampled-data systems with delay. Autom. Remote Control **69**(6), 953–967 (Jun 2008)




E. N. Rosenwasser et al., *Computer-Controlled Systems with Delay*,
https://doi.org/10.1007/978-3-030-15042-6

18. B.P. Lampe, E.N. Rosenwasser, $\mathscr{H}_2$-optimization and fixed poles for sampled-data systems with generalized hold. Autom. Remote Control **73**(1), 31–55 (2012)
19. B.P. Lampe, E.N. Rosenwasser, $\mathscr{L}_2$-optimization and fixed poles for sampled-data systems with generalized digital to analog converters. Autom. Remote Control **75**(1), 34–56 (2014)
20. B.P. Lampe, E.N. Rosenwasser, $\mathscr{L}_2$-optimization and fixed poles of multivariable sampled-data systems with delay. Int. J. Control, **88**(4), 815–831 (2015). (Published online Dec 2014).
21. B.P. Lampe, E.N. Rosenwasser, Singular $\mathscr{L}_2$-optimization problem and fixed poles for sampleddata systemswith delay, in 7th *IFAC Symposium on Robust ControlDesign* (DK, Aalborg, 2012), pp. 172–177
22. B.P. Lampe, M.A. Obraztsov, E.N. Rosenwasser, Statistical analysis of stable FDLCP systems by parametric transfer matrices. Int. J. Control **78**(10), 747–761 (Jul 2005)
23. B.P. Lampe, E.N. Rosenwasser, $\mathscr{H}_2$-optimisation of MIMO sampled-data systems with pure time-delays. Int. J. Control **82**(10), 1899–1916 (2009)
24. B.P. Lampe, E.N. Rosenwasser, $\mathscr{H}_2$-optimization of time-delayed sampled-data systems on basis of the parametric transfer matrix method. Autom. Remote Control **71**(1), 49–69 (2010)
25. A.V. Nebylov, *Warranting of Accuracy of Control*. (Nauka, Moscow, 1998). (in Russian)
26. A.V. Nebylov, *Ensuring Control Accuracy*. (Springer, Berlin, Heidelberg, 2004)
27. K. Nomoto, T. Taguchi, K. Honda, S. Hirano, On the steering qualities of ships. Int. Ship building Prog. **4**, 354–370 (1957)
28. K.Y. Polyakov, *Polynomial Methods for Direct Design of Optimal Sampled-data Control Systems* (Saint Petersburg Maritime Technological University, Thesis for Doctor degree of Technological Sciences, 2006)
29. K.Y. Polyakov, E.N. Rosenwasser, B.P. Lampe, DirectSD – a toolbox for direct design of sampled-data systems. *Proceedings of the IEEE International Symposium CACSD'99*, (Kohala Coast, Island of Hawai'i, Hawai'i, USA, 1999), pp. 357–362
30. K.Y. Polyakov, E.N. Rosenwasser, B.P. Lampe, `DIRECTSD 3.0` toolbox for MATLAB. In *Proceedings of the IEEE Conference on Computer Aided Control Systems Design*, (Munich, Germany, Oct 2006), pp. 1946–1951
31. V.M. Popov, *Hyperstability of Control Systems* (Springer, Berlin, 1973)
32. E.N. Rosenwasser, *Linear Theory of Digital Control in Continuous Time* (Izd. LKI, St. Petersburg, 1989). (in Russian)
33. E.N. Rosenwasser *Linear Theory of Digital Control in Continuous Time*. (Nauka, Moscow, 1994). (in Russian)
34. E.N. Rosenwasser, B.P. Lampe, *Digitale Regelung in kontinuierlicher Zeit - Analyse und Entwurf im Frequenzbereich*. (B.G. Teubner, Stuttgart, 1997)
35. E.N. Rosenwasser, B.P. Lampe, *Computer Controlled Systems: Analysis and Design with Process-orientated Models*. (Springer, London Berlin Heidelberg, 2000)
36. E.N. Rosenwasser, B.P. Lampe, *Multivariable Computer-controlled Systems—A Transfer Function Approach*. (Springer, London, 2006)
37. V.O. Rybinskii, E.N. Rosenwasser, The Matlab toolbox for synthesis of guaranteed accuracy sampled-data systems. In *Proceedings of the Design of scientific and engineering applications in Matlab*, (Moscow, Russia, May 2002), pp. 471–482. (in Russian)
38. V.O. Rybinskii, B.P. Lampe, E.N. Rosenwasser, Design of digital ship motion control with guaranteed performance. In *Proceedings of the 49th International Wissensch Kolloquium*, vol 1, (Ilmenau, Germany, 2004), pp. 381–386
39. V.O. Rybinskii, B.P. Lampe, J. Ladisch, E.N. Rosenwasser, A.V. Smolnikov, Assurance of digital control accuracy at non-centralized stochastic actions. In *Proceedings of the 8th International Conference of Fast sea transportation*, (SPb, Russia, June 2005)
40. V.O. Rybinskii, B.P. Lampe, E.N. Rosenwasser, Digital Control with guaranteed performance under non-centred stochastic disturbances. In *Proceedings of the 7th IFAC Conference on Manoeuvring and Control of Marine Craft*, (Lisbon, Portugal, Sept. 2006)

41. V.O. Rybinskii, B.P. Lampe, K.Y. Polyakov, E.N. Rosenwasser, Control with Guaranteed Performance for Two-rate Sampled-data Systems under Stochastc Disturbances. In *Proceedings of the 17th IFAC Worl Congress*, (Seoul, KR, July 2007), pp. 15291–15296
42. V.O. Rybinskii, B.P. Lampe, K.Y. Polyakov, E.N. Rosenwasser, A.A. Shylovskii, Control with Guaranteed Precision in Time-delayed Sampled-data Systems. In *Proceedings of the 3rd MSC: IEEE Multi-conference on Systems and Control*, (St. Petersburg Russia, July 2009)
43. V.O. Rybinskii, B.P. Lampe, E.N. Rosenwasser, Extension of the GarSD-toolbox: Polynomial method for guaranteed performance design of sampled-data control systems with non-zero static error and pure delays. In *Proceedings of the IEEE International Symposium on Computer-Aided Control System Design (CACSD)*, (Denver, CO, USA, Sept 2011), pp. 450–455
44. V.O. Rybinskii, The Matlab toolbox for synthesis of guaranteed accuracy digital control systems for marine dinamic objects. *Morskie intellekualnye tehnologii*, **1**(19), 39–44 (2013). (in Russian)
45. V.O. Rybinskii, Selecting a Sampling Period for a Sampled-Data Submarine Motion Control System. Gyroscopy and Navigation, **1**, 57–63 (2013)
46. V.O. Rybinskii, B.P. Lampe, E.N. Rosenwasser, Advanced statistical analysis with guaranteed estimation of current performance for sampled-data systems with delay and generalized higher order hold. In *Proceedings of the 9th IFAC Conference on Control Applications in Marine Systems*, (Osaka, Japan, Sept. 2013)
47. V.O. Rybinskii, Optimizing a Sampled Data Submarine Motion Control System over a Set of Reduced Order Controllers. Gyroscopy and Navigation, **7**, 197–203 (2016)
48. V.B. Sommer, B.P. Lampe, E.N. Rosenwasser, Experimental investigations of analog-digital control systems by frequency methods. Autom. Remote Control **55**(Part 2), 912–920 (1994)
49. E.C. Titchmarsh, *The Theory of Functions*, 2nd edn. (University Press, Oxford, 1997) Reprint
50. J.S. Tsypkin, *Sampling Systems Theory* (Pergamon Press, New York, 1964)
51. B. van der Pol, H. Bremmer, *Operational Calculus Based on the Two-sided Laplace Integral* (University Press, Cambridge, 1959)
52. L.N. Volgin, *Optimal Discrete Control of Dynamic Systems*. (Nauka, Moscow, 1986). (in Russian)
53. L.A. Zadeh. Frequency analysis of variable networks. Proc. IRE, 39(March):291–299 (1950)

# Further Readings

1. M. Araki, T. Hagiwara, Y. Ito. Frequency response of sampled-data systems II. Closed-loop considerations. In *Proceeding of the 12th IFAC Triennial World Congress*, vol 7, (Sydney, 1993), pp. 293–296
2. M. Araki, Y. Ito, Frequency response of sampled-data systems I. Open-loop considerations. In *Proceeding of the 12th IFAC Triennial World Congress*, vol 7, (Sydney, 1993), pp. 289–292
3. F.A. Aliev, V.B. Larin, K.I. Naumenko, V.I. Suntsev, *Optimization of Linear Time-invariant Control Systems* (Naukova Dumka, Kiev, 1978). (in Russian)
4. B.D.O. Anderson, Controller design: Moving from theory to practice. IEEE Control Syst. **13**(4), 16–24 (1993)
5. B.D.O. Anderson, J.B. Moore, *Optimal Filtering* (Prentice-Hall, Englewood Cliffs, NJ, 1979)
6. K.J. Åström, *Introduction to Stochastic Control Theory* (Academic Press, NY, 1970)
7. B.A. Bamieh, J.B. Pearson, A general framework for linear periodic systems with applications to $\mathscr{H}_\infty$ sampled-data control. IEEE Trans. Autom. Control, **AC-37**(4), 418–435 (1992)
8. S. Bochner, *Lectures on Fourier Integrals* (University Press, Princeton, NJ, 1959)
9. I. Boroday, V. Mohrenschildt, et al. *Behavior of Ships in Ocean Waves*. (Sudostroyenie, Leningrad, 1969)

10. I. K. Boroday, V.V. Nezetaev, *Application Problems of Dynamics for Ships on Waves.* (Sudostroyenie, Leningrad, 1989)
11. B.W. Bulgakov, *Vibrations* (GITTL, Moscow, 1954). (in Russian)
12. F.M. Callier, C.A. Desoer, *Linear System Theory* (Springer, New York, 1991)
13. S.S.L. Chang, *Synthesis of Optimum Control Systems* (McGraw Hill, New York, Toronto, London, 1961)
14. T. Chen, B.A. Francis, *Optimal Sampled-data Control Systems.* (Springer-Verlag, Berlin, Heidelberg, New York, 1995)
15. R.E. Crochiere, L.R. Rabiner, *Multirate Digital Signal Processing* (Prentice-Hall, Englewood Cliffs, NJ, 1983)
16. C.E. de Souza, G.C. Goodwin, Intersample variance in discrete minimum variance control. IEEE Trans. Autom. Control, **AC-29**, 759–761 (1984)
17. L. Dai, *Singular Control Systems* (Lecture notes in Control and Information Sciences. Springer-Verlag, New York, 1989)
18. J.A. Daletskii, M.G. Krein, *Stability of Solutions of Differential Equations in Banach-space* (Nauka, Moscow, 1970). (in Russian)
19. J.C. Doyle, Guaranteed margins for LQG regulators. IEEE Trans. Autom. Control, **AC-23**(8), 756–757 (1978)
20. G. Doetsch, *Anleitung zum praktischen Gebrauch der Laplace Transfor mation und z-Transformation* (Oldenbourg, München, Wien, 1967)
21. G.F. Franklin, J.D. Powell, H.L. Workman, *Digital Control of Dynamic Systems* (Addison Wesley, New York, 1990)
22. D.K. Faddeev, V.N. Faddeeva. *Numerische Methoden der Linearen Algebra.* (Oldenbourg, München, 1979). (L. Bittner)
23. A. Feuer, G.C. Goodwin, Generalised sample and hold functions—frequency domain analysis of robustness, sensitivity and intersampling difficulties. *IEEE Trans. Autom. Contr*, **AC-39**(5), 1042–1047 (1994)
24. G.C. Goodwin, M.E. Salgado, Frequency domain sensitivity functions for continuous-time systems under sampled-data control. Automatica **30**(8), 1263–1270 (1994)
25. M.J. Grimble, *Robust Industrial Control: Optimal Design Approach for Polynomial Systems* (International Series in Systems and Control Engineering, Prentice Hall International (UK) Ltd, Hemel Hempstead, Hertfordshire, 1994)
26. M.J. Grimble, V. Kučera (eds.), *Polynomial Methods for Control Systems Design* (Springer, London, 1996)
27. M. Günther, *Kontinuierliche und zeitdiskrete Regelungen* (B.G. Teubner, Stuttgart, 1997)
28. M.E. Halpern, Preview tracking for discrete-time SISO systems. IEEE Trans. Autom. Control, **AC-39**(3), 589–592 (1994)
29. U. K. Herne. *Methoden zur rechnergestützten Analyse und Synthese von Mehrgrößenregelsystemen in Polynommatrizendarstellung*. PhD thesis, University of Bochum, 1988
30. M.A. Jevgrafov, *Analytic Functions* (Nauka, Moscow, 1965). (in Russian)
31. R. Kalman, J.E. Bertram, A unified approach to the theory of sampling systems. J. Franklin Inst. **267**, 405–436 (1959)
32. P.T. Kabamba, S. Hara, Worst-case analysis and design of sampled-data control systems. IEEE Trans. Autom. Control, **AC-38**(9), 1337–1358 (1993)
33. V.J. Katkovnik, R.A. Polucektov, *Discrete multidimensional control* (Nauka, Moscow, 1966). (in Russian)
34. S. Karlin, *A first course in Stochastic Processes* (Academic Press, New York, 1966)
35. R.E. Kalman, Mathematical description of linear dynamical systems. *SIAM J. Control*, A(1):152–192 (1963)
36. U. Keuchel. *Methoden zur rechnergestützten Analyse und Synthese von Mehrgrößensystemen in Polynommatrizendarstellung*. PhD thesis, University of Bochum, (1988)

37. P.P. Khargonekar, N. Sivarshankar, $\mathscr{H}_2$-optimal control for sampled-data systems. Systems & Control Letters **18**, 627–631 (1992)
38. B.C. Kuo, D.W. Peterson, Optimal discretization of continuous-data control systems. Automatica **9**(1), 125–129 (1973)
39. V. Kučera, *Discrete Linear Control* (The Polynomial Approach. Academia, Prague, 1979)
40. V. Kučera, *Analysis and Design of Discrete Linear Control Systems* (Prentice Hall, New York, 1991)
41. B.P. Lampe, U. Richter, Digital controller design by parametric transfer functions - comparison with other methods. In *Proceedings of the 3rd International Symposium Methods Models Autom Robotics*, vol 1, (Miedzyzdroje, Poland, 1996), pp. 325–328
42. B.P. Lampe, E.N. Rosenwasser, Design of hybrid analog-digital systems by parametric transfer functions. In *Proceedings of the 32nd CDC*, (San Antonio, TX, 1993), pp. 3897–3898
43. B.P. Lampe, E.N. Rosenwasser, Best digital approximation of continuous controllers and filters in $\mathscr{H}_2$. In *Proceedings of the 41st KoREMA*, vol 2, (Opatija, Croatia, 1996), pp. 65–69
44. B.P. Lampe, E.N. Rosenwasser, Best digital approximation of continuous controllers and filters in $\mathscr{H}_2$. AUTOMATIKA **38**(3–4), 123–127 (1997)
45. B.P. Lampe, E.N. Rosenwasser, Closed formulae for the $\mathscr{L}_2$-norm of linear continuous-time periodic systems. In *Proceedings of the IFAC Workshop on Periodic Control Systems*, (Yokohama, Japan, Sep 2004), pp. 231–236
46. B.P. Lampe, E.N. Rosenwasser, $\mathscr{H}_2$-optimization of sampled-data systems with linear periodic process. Part I: Parametric transfer matrix and its properties. Autom. Remote Control **77**(8), 1334–1350 (2016)
47. B.P. Lampe, E.N. Rosenwasser, $\mathscr{H}_2$-optimization of sampled-data systems with linear periodic process. Part II: $\mathscr{H}_2$-optimization of system $\mathscr{S}_T$ on basis of the Wiener-Hopf method. Autom. Remote Control **77**(9), 1524–1543 (2016)
48. B.P. Lampe, E.N. Rosenwasser, $\mathscr{H}_2$-optimization of sampled-data systems with linear periodic process. Part I: Parametric transfer matrix and its properties. Autom. Remote Control, **77**(8), 1334–1350 (2016)
49. B.P. Lampe, E.N. Rosenwasser, $\mathscr{H}_2$-optimization of sampled-data systems with linear periodic process. Part II: $\mathscr{H}_2$-optimization of system $\mathscr{S}_T$ on basis of the Wiener-Hopf method. Autom. Remote Control **77**(9), 1524–1543 (2016)
50. B.P. Lampe, Strukturelle Instabilität in linearen Systemen - Frequenzgangsmethoden auf dem Prüfstand der Mathematik, *Mitteilungen der Mathematischen Gesellschaft in Hamburg*, volume XVIII (Hamburg, Germany, 1999), pp. 9–26
51. B.P. Lampe, E.N. Rosenwasser, Forward and backward models for anomalous linear discrete-time systems. In *Proceedings 9th IEEE International Conference Methods Models Autom Robotics*, (Miedzyzdroje, Poland, Aug 2003), pages 369–373
52. B.P. Lampe, E.N. Rosenwasser, $\mathscr{H}_2$ optimization and fixed poles of sampled-data systems under colored noise. Eur. J. Control **19**(3), 222–234 (2013)
53. B.P. Lampe, E.N. Rosenwasser, Polynomial solution to stabilisation problem for multivariable sampled-data systems with time delay. Autom. Remote Control **67**(1), 105–114 (2006)
54. B.P. Lampe, E.N. Rosenwasser, Stabilization of multidimensional digital control systems with delay. Dokl. Akad. Nauk. **405**(2), 180–183 (2005). (in Russian)
55. B.P. Lampe, E.N. Rosenwasser, Modal control for sampled-data systems with generalized higher-order hold and time-delay. In *Proceedings of IFAC Workshop on Time-Delay Systems*, (L'Aquila, Italy, July 2006), pp. WeA2_c(1–6)
56. B.P. Lampe, E.N. Rosenwasser, Causal stabilisation of forward models of discrete LTI processes under constraints on the set of elementary divisors of the characteristic matrix. Int. J. Control **81**(6), 1002–1012 (2008)
57. B.P. Lampe, E.N. Rosenwasser, Modal control in sampled-data systems with high-order generalized holds. Dokl. Mathematics **75**(3), 479–482 (Jun 2007)

58. B.P. Lampe, E.N. Rosenwasser, Unterordnung und Dominanz rationaler Matrizen. Automatisierungstechnik **53**(9), 434–444 (2005)
59. B.P. Lampe, U. Richter, Experimental investigation of parametric frequency response. In *Proceeding of the 4th International Symposium Methods Models Autom Robotics*, (Miedzyzdroje, Poland, 1997), pp. 341–344
60. B.P. Lampe, E.N. Rosenwasser, Application of parametric frequency response to identification of sampled-data systems. In *Proceeding of the 2nd International Symposium Methods Models Autom Robotics*, vol 1, (Miedzyzdroje, Poland, 1995), pages 295–298
61. B.P. Lampe, E.N. Rosenwasser, Parametric transfer functions for sampled-data systems with time-delayed controllers. In *Proceedings of the 36th IEEE Conference Decision Control*, (San Diego, CA, 1997), pp. 1609–1614
62. B.P. Lampe, E.N. Rosenwasser, Polynomial stabilization of sampled-data systems with higher-order hold. In *Proceedings of the 16th CIE*, (Cayo Santamaria Beach, Cuba, Jun 2011), , pp. 1–6
63. B.P. Lampe, E.N. Rosenwasser, Sampled-data systems: The $L_2$−induced operator norm. In *Proceedings of the 4th International Symposium Methods Models Autom Robotics*, (Miedzyzdroje, Poland, 1997), pp. 205–207
64. B.P. Lampe, E.N. Rosenwasser, Laplace transforms for MIMO SD systems with delay. Arch. Control Sci. **16(LII)**(4), 363–381 (2006). (published in 2007)
65. F.H. Lange, *Signale und Systeme*, volume 1–3. (Verlag Technik, Berlin, 1971).
66. V.B. Larin, K.I. Naumenko, V.N. Suntsov, *Spectral Methods for Design of Linear Systems with Feedback*. (Naukova Dumka, Kiev, 1971). (in Russian)
67. V.B. Larin, K.I. Naumenko, V.N. Suntsov, *Spectral Methods for Design of Linear Systems with Feedback* (Naukova Dumka, Kiev, 1971). (in Russian)
68. B. Lennartson, Sampled-data control for time-delayed plants. Int. J. Control **48**, 1601–1614 (1989)
69. B. Lennartson, T. Söderström, Investigation of the intersample variance in sampled-data control. Int. J. Control **50**, 1587–1602 (1989)
70. B. Lennartson, T. Söderström, and Sun Zeng-Qi. Intersample behavior as measured by continuous-time quadratic criteria. Int. J. Control, **49**(3), 2077–2083 (1989)
71. B. Lennartson, Sampled-data control for time-delayed plants. Int. J. Control, **48**, 1601–1614 (1989)
72. B. Lennartson, T. Söderström, Investigation of the intersample variance in sampled-data control. Int. J. Control, **50**, 1587–1602 (1989)
73. B. Lennartson, T. Söderström, Sun Zeng-Qi, Intersample behavior as measured by continuous-time quadratic criteria. Int. J. Control, **49**(3), 2077–2083 (1989)
74. O. Lingärde, B. Lennartson, Frequency analysis for continuous-time systems under multirate sampled-data control. In *Proceedings off the 13th IFAC Triennial World Congress*, vol 2a–10, 5, (San Francisco, USA, 1996), pp. 349–354
75. O. Lingärde and B. Lennartson. Frequency analysis for continuous-time systems under multirate sampled-data control. In *Proceedings of the 13th IFAC Triennial World Congress*, vol 2a–10, 5, (San Francisco, USA, 1996), pp. 349–354
76. N.N. Lusin, Matrix theory for studying differential equations. Avtomatika i Telemechanika **5**, 4–66 (1940). (in Russian)
77. N.N. Lusin, Matrizentheorie zum Studium von Differentialgleichungen. Avtomatika i Telemechanika **5**, 4–66 (1940)
78. J. Lunze, *Regelungstechnik 2—Mehrgrößensysteme* (Digitale Regelung. Springer, Berlin, Heidelberg, 1997)
79. D.G. Luenberger, Dynamic equations in descriptor form. IEEE Trans. Autom. Control, **AC-22**(3), 312–321 (1977)
80. J. Lunze, *Robust Multivariable Feedback control*. (Akademie-Verlag, Berlin, 1988)

81. J. Lunze, *Regelungstechnik 2 - Mehrgrößensysteme, Digitale Regelung*. (Springer-Verlag, Berlin, Heidelberg, 1997)
82. N.N. Lusin, Matrix theory for studying differential equations. Avtomatika i Telemechanika, **5**, 4–66 (1940). (in Russian)
83. N.N. Lusin, Matrizentheorie zum Studium von Differentialgleichungen. Avtomatika i Telemechanika, **5**, 4–66 (1940)
84. A.G. Madievski, B.D.O. Anderson, A lifting technique for sampled-data controller reduction for closed-loop transfer function consideration. In *Proceedings of the 32nd IEEE Conference Decision Control*, (San Antonio, TX, 1993), pp. 2929–2930
85. J.M. Maciejowski, *Multivariable Feedback Design*. (Addison-Wesley, Wokingham, England a.o., 1989)
86. J.M. Maciejowski, *Predictive Control - with Constraints*. (Pearson Education Lim., Harlow, England, 2002)
87. A.G. Madievski, B.D.O. Anderson, A lifting technique for sampled-data controller reduction for closed-loop transfer function consideration. In *Proceedings of the 32nd IEEE Conference Decision Contrrol*, (San Antonio, TX, 1993), pp. 2929–2930
88. S.G. Michlin, *Vorlesungen über lineare Integralgleichungen*. (Dt. Verlag d. Wissenschaften, Berlin, 1962)
89. L. Mirkin, Z.J. Palmer, D. Shneiderman, Dead-time compensation for systems with multiple I/O delay: A loop shifting approach. IEEE Trans. Autom. Control, **56**(11), 2542–2554 (2011)
90. L. Mirkin, T. Shima, G. Tadmor, Analog loop shifting in $H_2$ optimization of input-delay sampled-data systems. *52nd IEEE Conference on Decision and Control*, (Firenze, 2013), pp. 4725–4729
91. L. Mirkin, T. Shima, G. Tadmor, Sampled-Data $H_2$ Optimization of Systems With I/O Delays via Analog Loop Shifting. IEEE Trans. Autom. Control, **59**(3), 787–791 (2014)
92. L. Mirkin, T. Shima, G. Tadmor, Analog loop shifting in $H_2$ optimization of input-delay sampled-data systems. *52nd IEEE Conference on Decision and Control*, (Firenze, 2013), pp. 4725–4729
93. L. Mirkin, T. Shima, G. Tadmor, Sampled-Data $H_2$ Optimization of Systems With I/O Delays via Analog Loop Shifting. IEEE Trans. Autom. Control **59**(3), 787–791 (2014)
94. L. Mirkin, Z.J. Palmer, D. Shneiderman, Dead-time compensation for systems with multiple I/O delay: A loop shifting approach. IEEE Trans. Autom. Contr **56**(11), 2542–2554 (2011)
95. A.V. Nebylov, *Differentiation of a Signals with Limited Derivatives Dispersions in Noise*. Avtomatika i Telemechanika, **3**, 164–167 (1992)
96. A.V. Nebylov, V.N. Kalinichenko, Analysis of the source data for designing the linear filter of a signal with constrained variances of derivatives. Autom. remote control **7**, 1133–1142 (2000)
97. A.V. Nebylov, *Measuring Parameters of a Plane near the Sea Surface*. (Saint Petersburg State University Academic Press, St. Petersburg, 2000). (in Russian)
98. A.V. Nebylov, V.I. Kulakova, Guaranteed estimation of signals with bounded variances of derivatives. Autom. remote control **1**, 76–88 (2008)
99. V.G. Pak, V.N. Fomin, *Linear Quadratic Optimal Control Problem under known Disturbance I. Abstract Linear Quadratic Problem under known Disturbance*. Preprint VINITI, N2063-B97, St. Petersburg, 1997. (in Russian)
100. K. Parks, J.J. Bongiorno. Modern Wiener-Hopf design of optimal controllers – Part II: The multivariable case. IEEE Trans. Autom. Control, **AC-34**(6), 619–626 (1989)
101. U. Petersohn, H. Unger, W. Wardenga, Beschreibung von Multirate-Systemen mittels Matrix kalkül. *AEÜ*, **48**(1), 34–41 (1994)
102. I.I. Priwalow, *Einführung in die Funktionentheorie*, 3rd edn. (B.G. Teubner, Leipzig, 1967)
103. K.Y. Polyakov, E.N. Rosenwasser, B.P. Lampe, Quasipolynomial low-order digital controller design using genetic algorithms. In *Proceedings of the 9th IEEE Mediterranian Conference on Control and Automation*, (Dubrovnik, Croatia, June 2001), pp. WM1–B5

104. J. Raisch, *Mehrgrößenregelung im Frequenzbereich* (R. Oldenbourg Verlag, München, 1994)
105. J.R. Ragazzini, G.F. Franklin, *Sampled-data Control Systems* (McGraw-Hill, New York, 1958)
106. J.R. Ragazzini, L.A. Zadeh, The analysis of sampled-data systems. AIEE Trans. **71**, 225–234 (1952)
107. K.S. Rattan, Digitalization of existing control systems. *IEEE Trans. Autom. Contr.* **AC-29**, 282–285 (1984)
108. K.S. Rattan, Compensating for computational delay in digital equivalent of continuous control systems. *IEEE Trans. Autom. Contr.* **AC-34**, 895–899 (1989)
109. G. Roppenecker. Fortschr.-Ber. VDI-Z. In *Vollständige modale Synthese linearer Systeme und ihre Anwendung zum Entwurf strukturbeschränkter Zustandsrückführungen*, number 59 in 8. VDI-Verlag, Düsseldorf, 1983
110. E.N. Rosenwasser, *Vibrations of Nonlinear Systems* (The method of integral equations. Army Foreign Science and Technology Center, Charlottesville, Virginia, USA, 1971)
111. E.N. Rosenwasser, *Lyapunov Indices in Linear Control Theory* (Nauka, Moscow, 1977). (in Russian)
112. E.N. Rosenwasser, *Polynomial and rational matrix*, Modal control of discrete objects. (SMTU Press, St. Petersburg, 2013). (in Russian)
113. E.N. Rosenwasser, *Vibrations of Nonlinear Systems* (The method of integral equations. Army Foreign Science and Technology Center, Charlottesville, Virginia, USA, 1971)
114. E.N. Rosenwasser, System characteristics of linear periodic puls operators. Soviet. Phys. Dokl. **36**(9), 614–616 (1991). (in Russian)
115. E.N. Rosenwasser, *Polynomial and rational matrix* (Modal control of discrete objects. SMTU Press, St. Petersburg, 2013). (in Russian)
116. E.N. Rosenwasser, P.G. Fedorov, B.P. Lampe. Construction of MFD-representation of real rational transfer matrices on basis of normalisation procedure. In *International Conference on Computer Methods for Control Systems*, (Szczecin, Poland, 1997), pp. 39–42
117. E.N. Rosenwasser, P.G. Fedorov, B.P. Lampe, Construction of state-space model with minimal dimension for multivariable system on basis of transfer matrix normalization procedure. In *Proceedings of 5th International Symposium Methods Models Autom Robotics*, vol 1, (Miedzyzdroje, Poland, 1998), pp. 235–238
118. E.N. Rosenwasser, K.Y. Polyakov, B.P. Lampe, Frequency domain method for $\mathscr{H}_2$−optimi zation of time-delayed sampled-data systems. Automatica **33**(7), 1387–1392 (1997)
119. E.N. Rosenwasser, K.Y. Polyakov, B.P. Lampe, Optimal discrete filtering for time-delayed systems with respect to mean-square continuous-time error criterion. Int. J. Adapt. Control Signal Process. **12**, 389–406 (1998)
120. E.N. Rosenwasser, K.Y. Polyakov, B.P. Lampe, Comments on "A technique for optimal digital redesign of analog controllers". IEEE Trans. Control Syst. Technol. **7**(5), 633–635 (September 1999)
121. E.N. Rosenwasser, K.Y. Polyakov, B.P. Lampe, Application of Laplace transformation for digital redesign of continuous control systems. IEEE Trans. Autom. Control **4**(4), 883–886 (April 1999)
122. E.N. Rosenwasser, B.P. Lampe, W. Drewelow, T. Jeinsch, Standardizability and $H_2$-optimi zation of sampled-data systems with mulitple delays. Autom. Remote Control **80**(3), 413–428 (2019)
123. V.O. Rybinskii, B.P. Lampe, Accuracy estimation for digital control systems at incomplete information about stochastic input disturbances. In B.P. Lampe, (ed), *Maritime Systeme und Prozesse*. (Universitätsdruckerei, Rostock, 2001), pp. 43–52
124. V.O. Rybinskii, B.P. Lampe, E.N. Rosenwasser, Design of digital controllers of reduced order for periodic sampled-data systems with delay according to guaranteed H2 norm. In *Proceedings of the 8th European Nonlinear Dynamics Conference ENOC 2014*, (Vienna, Austria, 2014), July 6-11

125. M.F. Sågfors, *Optimal Sampled-Data and Multirate Control*. PhD thesis, Faculty of Chemical Engineering, Åbo Akademi University, Finland, 1998
126. H. Schwarz, *Optimale Regelung und Filterung - Zeitdiskrete Regelungssysteme* (Akademie-Verlag, Berlin, 1981)
127. L.S. Shieh, B.B. Decrocq, J.L. Zhang, Optimal digital redesign of cascaded analogue controllers. Optimal Control Appl. Methods **12**, 205–219 (1991)
128. L.S. Shieh, J.L. Zhang, J.W. Sunkel, A new approach to the digital redesign of continuous-time controllers. Control Theory Adv. Techn. **8**, 37–57 (1992)
129. R.F. Stengel, *Stochastic Optimal Control* (Theory and application. J. Wiley & Sons Inc, New York, 1986)
130. I.Z. Shtokalo, Generalisation of symbolic method principal formula onto linear differential equations with variable coefficients. Dokl. Akad. Nauk SSR **42**, 9–10 (1945). (in Russian)
131. V.B. Sommer, B.P. Lampe, E.N. Rosenwasser, Experimental investigations of analog-digital control systems by frequency methods. *Autom. Remote Control*, **55**(Part 2), 912–920 (1994)
132. H. Unger, U. Petersohn, S. Lindow. Zur Beschreibung hybrider Multiraten-Systeme mittels Matrixkalküls. *FREQUENZ*, 1997. (einger.)
133. J. Tou, *Digital and Sampled-Data Control Systems* (McGraw-Hill, New York, 1959)
134. H.L. Trentelmann, A.A. Stoorvogel, Sampled-data and discrete-time $\mathscr{H}_2$−optimal control. In *Proceedings of the 32nd Conference Decision and Control*, (San Antonio, TX, 1993), pp. 331–336
135. L.N. Volgin, *Optimal Discrete Control of Dynamic Systems* (Nauka, Moscow, 1986). (in Russian)
136. E.T. Whittaker, G.N. Watson, *A Course of Modern Analysis*, 4th edn. (University Press, Cambridge, 1927)
137. D.V. Yakubovich, Algorithm for supplementing a rectangular polynomial matrix to a quadratic matrix with given determinant. Kybernetika i Vychisl. **23**, 85–89 (1984)
138. R.A. Yackel, B.C. Kuo, G. Singh, Digital redesign of continuous systems by matching of states at multiple sampling periods. Automatica **10**, 105–111 (1974)
139. Y. Yamamoto. A function space approach to sampled-data systems and tracking problems. IEEE Trans. Autom. Control, **AC-39**(4), 703–713 (1994)
140. Y. Yamamoto and P. Khargonekar. Frequency response of sampled-data systems. IEEE Trans. Autom. Control, **AC-41**(2), 161–176 (1996)
141. D.C. Youla, H.A. Jabr, J.J. Bongiorno (Jr.). Modern Wiener-Hopf design of optimal controllers. Part II: The multivariable case. IEEE Trans. Autom. Control, **AC-21**(3), 319–338 (1976)
142. L.A. Zadeh, Circuit analysis of linear varying-parameter networks. J. Appl. Phys. **21**(6), 1171–1177 (1950)
143. P. Zhang, S. X. Ding, G. Z. Wang, and D. H. Zhou. Fault detection in multirate sampled-data systems with time-delays. In *Proceedings of 15th IFAC Triennial World Congress*, volume Fault detection, supervision and safety of technical processes, page REG2179, Barcelona, (2002)
144. K. Zhou, J.C. Doyle, K. Glover, *Robust and Optimal Control* (Prentice-Hall, Englewood Cliffs, NJ, 1996)

# Index

**A**
Addition, 29
Algebra-difference equation, 159

**B**
Backward model, 139
  causal, 230
  discrete, 139
Backward transfer matrix, 231
Basic controller, 87, 91
  dual, 91
  left, 87
  right, 90
Basic division, 62
Basic representation, 98
  controller, 98
Bézout's identity, 91
Block
  LTI, 232
Block matrix, 67
  strictly proper, 73
  triangular, 184

**C**
Characteristic equation, 176
  right, 90
Characteristic matrix, 6, 86, 176
  forward
    left, 118
  left, 85
  right, 86
Control algorithm, 231
Controllability, 187
  modal, 260
    complete, 177
  pair, 223
Controller, 85
  backward (B-controller), 158
  digital, 108, 161
  discrete, 158
  forward
    left, 118
  left, 85
  right, 86
  stabilizing, 153
Control problem, 176
  B-modal, 109
  causal, 176
  F-modal, 119
  $\mathscr{H}_2$ optimal, 347
  modal, 85
    causal, 110
    causal backward, 108
    right, 86
Coprime, 4

**D**
Defect
  normal, 10
Degree, 5
  greatest possible, 4
  minimal, 61
  minimal polynomial, 196
  polynomial matrix, 5
  quasi-polynomial matrix, 309
Denominator, 30
  left, 41
  right, 41
Descriptor system, 159
Determinant, 5
Determinant divisor, 10


E. N. Rosenwasser et al., *Computer-Controlled Systems with Delay*,
https://doi.org/10.1007/978-3-030-15042-6

greatest, 10
Difference equation, 106
matrix
system, 253
Differential-difference equation, 147
Dimension, 6
realization, 58
minimal, 60
Discretization, 167
Disturbance, 165
Division
basic, 62
Divisor, 4
common, 4, 398
greatest, 4, 398
elementary, 11, 12
left, 18, 73
greatest, 18
right, 73
greatest, 18

**E**
Eigenvalue, 6
Element, 36
continuous, 230
forming, 220
transfer function, 220
transfer matrix, 224
hold, 157
inverse, 29
LTI, 232
zero, 29
Entire part, 4
Equation
backward, 158
forward, 159
latent, 13
operator, 176
Equivalence, 7
left, 17
pseudo-, 7
right, 17
strict, 21
Excitation, 411
external, 259
input, 247
noise
colored, 347
Exponential-periodic (EP), 226

**F**
Field, 29
complex number, 11
Forward model, 181
discrete, 153
system, 182
Fraction, 29
improper, 31
irreducible, 33, 35
strictly proper, 32
proper, 31
at least, 31
strictly, 31
rational, 30
reducible, 35
scalar, 35
Function
continuous
piecewise, 166
exponential-periodic (EP), 226, 242
fractional-rational, 29
index, 201
integral, 256
meromorph, 427
periodic, 204
rational, 29
at least proper, 129
stable, 129
rational-periodic (RP), 209
scalar, 197
system, 264
optimal, 326
stable, 340
transfer, 102

**H**
Hermitian matrix, 310

**I**
Image, 203
Laplace, 402
pseudo-rational, 425
Index, 226
Initial energy, 425
Initial values, 159
Input
exponential-periodic (EP), 228
Integrator
double, 130
Invariant polynomial, 22
Irreducibility, 45

**J**
Jordan
block, 25
canonical form, 22
chain, 183

**L**
Laplace transform, 235
discrete (DLT), 203
inverse, 382
two-sided, 291
Linear time invariant (LTI), 226

**M**
Matrix, 4
adjoint, 195
alatent, 13
anomal, 5
arbitrary, 112
B-characteristic, 107
B-transfer, 159
block, 67
strictly proper, 73
triangular, 184
block diagonal, 11
causal, 83
strictly, 80
characteristic, 6, 21, 86, 176
forward, 118
left, 85
right, 86
column reduced, 16
compound, 18
constant, 4
controllability, 23
covariance, 291
diagonal, 7
equivalent, 6, 43
left, 42
right, 42
exponential-periodic, 244
F-characteristic
left, 118
F-transfer, 159
generating, 196
Hermitian, 310
horizontal, 65
identity, 19
inverse, 92
invertible, 113, 115, 159
latent, 13
non-degenerated, 14
non-singular, 5
alatent, 13
number, 6
polynomial, 63
horizontal, 22
left equivalent, 17
non-singular, 42, 61
reciprocal, 78
regular, 6
right equivalent, 17
stable, 111, 114
unstable, 111
proper, 34
at least, 34
strictly, 34
pseudo-equivalent, 7, 43
quadratic, 7
quasi-polynomial, 308, 309
symmetric, 318
rational, 29
block, 66
fractional, 34
McMillan form, 38
periodic, 209
shortly, 34
stable, 111
standard form, 35
strictly proper, 63, 206
unstable, 111
rational periodic, 209
reciprocal, 24
rectangular, 12
regular, 5
row reduced, 16
RP - rational periodic, 294
singular, 5
stabilizing, 115
stable, 112
polynomial, 111
rational, 111
$T$-periodic, 218
transfer, 59, 102, 117
backward, 231
parametric, 241
unimodular, 7
unstable
polynomial, 111
rational, 111
zero, 5
Matrix function, 195
McMillan
denominator, 38, 40

index, 40
multiplicity
maximal, 39
numerator, 38
McMillan form, 38
MFD=matrix fraction description, 41
irreducible(IMFD), 41
left(LMFD), 41
irreducible(ILMFD), 41
right(RMFD), 41
irreducible(IRMFD), 41
Minimal polynomial, 22, 195
Minor, 10
non-singular, 6
Modal control, 131
Model, 101
backward
causal, 230
discrete, 106
backward, 139
forward, 130
parametric, 209, 221
forward, 181
linear, 157, 270
matrix
discrete, 253
process
discrete, 190
SD
standard, 337
stable, 378
standard, 237
generalized, 410
stable, 292
Model standardizability, 280
Model standardizable, 279
Multiplication, 29
matrix, 91
Multiplicity, 3
maximal, 39
pole, 40
root, 3

**N**
Non-pathological, 306
Number, 3
complex, 3
conjugate, 3
latent, 39
real, 12
Numerator, 30
left, 41
right, 41

**O**
Observability, 187, 237
Operator, 106
differential, 226
ideal, 391
linear, 162
pure delay, 162
shift, 106
backward, 158
forward, 117
$T$-periodic, 162
transposition, 17
Order, 6
left
polynomial matrix, 15
matrix, 5
right
polynomial matrix, 16
Original, 25
Output
exponential-periodic (EP), 230

**P**
Pair, 4
controllability, 223
controllable, 22
completely, 58
degenerated, 20
horizontal, 18
non-singular, 60
irreducible, 23
minimal, 317
noncontrollable, 304
non-degenerated, 18
observable, 23
completely, 58
polynomial, 85
reducible, 53
uncontrollability, 223
uncontrollable, 223
vertical, 18
non-singular, 60
Parametric discrete model, 221
Parametric transfer matrix, 241, 245, 292, 346, 409
Partial fraction decomposition, 192
Partial fraction expansion, 36
Pencil, 20
equivalent, 21
strictly, 21

regular, 21
Period, 202
non-pathological, 202
sampling, 128
Pole, 39
basic, 405
fixed, 337
stable, 374
Polynomial, 3
admissible, 150
stable, 152
causal, 109
characteristic, 6
division
basic, 62
invariant, 22
basic, 24
matrix, 4
minimal, 22
monic, 3
stable, 125
quasi-, 307
reciprocal, 23
monic, 121
scalar, 23
stabilizing, 152
stable, 111
unstable, 116
Polynomial matrix, 4
column reduced, 16
horizontal, 22
left equivalent, 17
non-singular, 42
reciprocal, 78
regular, 6
right equivalent, 17
row reduced, 16
stable, 111, 114
Polynomial matrix description, 130
irreducible, 137
strictly causal homogeneous, 262
minimal, 137
non-singular, 132
real, 58
strictly causal, 138
Process, 130
backward, 106
discrete, 106
causal, 107
strictly, 107
continuous, 165
deadbeat, 123
discrete, 117
finite dimensional, 158
dual, 91
strictly causal, 112
dual right, 264
irreducible, 115
forward, 140
discrete, 117
strictly causal, 152
forward (F-process), 117
irreducible, 87
strictly causal, 115
left, 85
irreducible, 89
linear
one dimensional, 140
LTI, 235
modulated, 215
non-singular, 107
quasi-stationary, 292
reducible, 97
right, 86
irreducible, 89
stochastic, 292
periodic nonstationary, 292
stationary, 291
transient, 403
quadratic optimal, 425
transitional, 292
Pulse frequency response
displaced, 204

**Q**

Quasi-polynomial, 307
nonnegative, 307
positive, 307
symmetric, 307

**R**

Rank, 6
normal, 6
maximal, 40
Realization, 58, 184
irreducible, 188
minimal, 58, 189
related, 59
system, 184
Regime
exponential-periodic (EP)
stationary, 249
steady, 247
steady state, 230

Remainder, 4
  left, 62
  right, 62
Representation, 14
  basic, 98
  general, 87
  irreducible, 30
  minimal, 71
  standard, 35
Response
  exponential-periodic (EP), 229
    steady, 226
Roots
  outer, 358
Row, 15
  block, 66

**S**
Sampled-data system, 235
  closed, 237
  continuous–discrete, 279
  multivariable
    single-loop, 431
  optimal, 375
  single-loop, 270
  stable, 293
  structural standardizable, 279
Sampling period, 128, 292
  non-pathological, 306
Separation, 33
  minimal, 34
  rational matrix, 317
Signal
  continuous, 164
  exponential, 271
  exponential-periodic (EP), 232
  input, 162
    colored, 422
    discrete-time, 157
    exponential-periodic (EP), 226
  output, 390
    exponential-periodic (EP), 227
  stochastic
    stationary, 293
  vector, 157
Smith canonical form, 7
Solution
  exponential-periodic (EP), 228, 229
Spectral density, 291, 375
Spectrum, 6
Stability, 114
Standard form, 30
  rational matrix, 35
Standard model, 291
  generalized, 410
  stable, 292
Standard sampled-data system, 294, 445
Stroboscopic property, 230, 247
Subordination, 70
  right, 71
Sylvester condition, 310
System, 11
  $H_\infty$ optimal, 375
  $\mathscr{H}_2$ optimal, 340
  $\mathscr{L}_2$ optimal, 405
  closed, 85
  control, 190
  discrete
    closed, 131
  general, 184
  linear, 255
  LTI, 226
  minimal, 184
  modal controllable
    completely, 177
  model-standardizable, 279
  optimal, 324
  sampled-data, 184
    closed, 237
    continuous–discrete, 279
    single-loop, 270
  single-loop, 438
  stabilizable, 180
  stable, 179
  standard, 292
    generalized, 422
  structural standardizable, 286
System function, 265
  $\mathscr{H}_2$ optimal, 335
  $\mathscr{L}_2$ optimal, 429
  optimal, 326
  stable
    optimal, 340
System parameter, 402

**T**
Trace of a matrix, 292
Transfer function, 102
Transfer matrix, 59, 102, 117, 159
  backward, 231
  $\mathscr{L}_2$ optimal, 429
  parametric, 241

**V**
Value, 8
  latent, 13
  primary, 380
Variance, 292
  mean, 294
Vector, 165
  control, 165
  error, 390
  input, 235
  number
    finite dimensional, 179
  optimal, 404
  output, 165
  periodic, 228
  polynomial, 397
  polynomial column, 398
  process, 165
  quasi-polynomial, 397
  rational, 401
    strictly proper, 401
  state, 173
    extended, 182

**W**
White noise, 422
  vector, 345
  vectorial, 411
    uncorrelated, 293
Wiener–Hopf algorithm, 417
Wiener–Hopf method, 323

**Z**
Zero element, 29
Zero matrix, 5

Printed in the United States
By Bookmasters